Light Scattering by Particles in Water

Theoretical and Experimental Foundations

Light Scattering by Particles in Water

Theoretical and Experimental Foundations

Miroslaw Jonasz
MJC Optical Technology
Beaconsfield
Quebec
Canada

And

Georges R. Fournier
DRDC Valcartier
Québec QC
Canada

Amsterdam • Boston • Heidelberg • London
New York • Oxford • Paris • San Diego
San Francisco • Singapore • Sydney • Tokyo

Academic Press is an imprint of Elsevier

Academic Press is an imprint of Elseiver
84 Theobald's Road, London WCIX 8RR, UK
Radarweg 29, PO Box 211, 1000 AE Amsterdam, The Netherlands
The Boulevard, Langford Lane, Kidlington, Oxford OX5 1GB, UK
30 Corporate Drive, Suite 400, Burlington, MA 01803, USA
525 B Street, Suite 1900, San Diego, CA 92101-4495, USA

First edition 2007

ISBN-13: 978-0-12-388751-1
ISBN-10: 0-12-388751-8

For information on all Academic Press publications
visit our website at books.elsevier.com

Acknowledgement

In the rush of the final copy editing, a thankful acknowledgement to Prof. Emmanuel Boss, who has kindly and thoroughly reviewed an early version of the manuscript, was regretfully omitted.

Table of Contents

Preface

Optical modeling of the interaction of light with small particles has applications in virtually every branch of environmental sciences. This is a consequence of the importance of this interaction in many natural processes occurring in natural environments. For example, particles significantly contribute to the transfer of sunlight through the atmosphere and the ocean, with vital implications for the climate of our planet.

Models of the interaction of light with small particles, light scattering models for short, are frequently needed by the analytical sciences, because such models are the basis of rapid, non-contact, and non-destructive particle characterization methods. These methods proved successful in many branches of science and technology (e.g., *Jonasz* 1991a). However, the development of an optical model of light scattering by particles poses significant problems because of the complex characteristics which these particles may exhibit. Just to hint at this complexity, we point to the extremely wide ranges of properties of naturally occurring particles, such as those dispersed in seawater, as compared with many other populations of particles. For example, the sizes of particles important for the interaction of light with seawater span 5 decades (e.g., *Stramski* and Kiefer 1991). The particles may have complex shapes and structures, ranging from structured needles to irregular complexes of organic substances with imbedded mineral grains.

A successful light scattering model correctly predicts light scattering properties of particles when using realistic assumptions about the relevant characteristics of the particles (size, shape, structure, refractive index, . . .). In an ideal situation, the success of such a model would be complete if the model, through an inversion algorithm, could retrieve accurate physical and chemical characteristics of the particles from light scattering and/or absorption data. In real situations, this inverse problem is ill posed mathematically because many particle ensembles can give rise to very similar light scattering properties. This severely limits the development of and makes it difficult to verify such models. Consequently, matching a limited set of experimental data with calculated results is not a guarantee of general applicability of a model of light scattering. The development and verification of a successful model may require consideration of several sets of theoretical and experimental constraints. Unfortunately, relevant data and knowledge are widely dispersed throughout literature of many unrelated branches of science, a testimony to the breadth of interest in the roles of particles in environmental and other processes. For example, the title count of periodicals used in preparation of this work exceeds 80. Such a wide literature breadth is not easy to follow, resulting in needless repetition of efforts and ignorance of relevant information, even in the age of the web search engines.

We feel that an essential part of the development of a light scattering model and of its verification can be much simplified if such constraints and, in a more general outlook, foundations of such physically acceptable models are comprehensively discussed and critically assessed in one work, affording the researcher a unified view. It is through the work leading to the precursor of this book (*Jonasz* 1992) that we ourselves gained a new perspective on the light scattering models of marine particles and on the characteristics of these particles (e.g., *Jonasz* and Fournier 1996).

No work similar to this one in its purpose has yet been published to our knowledge. Of other related works, some are specifically devoted to the theory of light scattering by small particles in general (e.g., *Bohren* and Huffman 1983), modeling of light scattering itself (*Barber* and Hill 1990), or are parts of larger reviews, devoted to mainly to marine optics (*Jerlov* 1976, 1968), modeling of the light field in the sea (*Mobley* 1994), marine physics (e.g., *Dera* 1992), or optical aspects of marine biology (for example, *Kirk* 1983a).

In this work, we attempt to focus on the theoretical and experimental foundations of the study and on the modeling of light scattering by particles in water and critically evaluate the key constraints of light scattering models applicable to such particles. We begin with a brief review of the theoretical fundamentals of the interaction of light with condensed matter. We then present the basic optical properties of pure water and the physical principles that explain them as well as discuss specific features of pure seawater and the most common components of natural waters. In order to clarify and put in focus some of the basic physical principles of scattering by large ensembles of particles, we employ a simple model ourselves. The purpose of this model is to allow us to explain the physical theory basis of some of the most important features found in the experimental data.

Finally we discuss implications of these fundamental issues on the modeling of light scattering by marine particles. The reader can interpret these implications according to his/her point of view. For example, if the reader is interested in the experimental constraints, he/she may use this discussion to formulate an efficient measurement program. If, on the other hand, the reader is interested in modeling alone, he/she may use this discussion to specify a set of constraints that are essential for the success of the model development.

In reviewing the experimental constraints, we begin with a detailed discussion of the measurement techniques and experimental data on light scattering by particles in natural waters. The great majority of these data—and thus the focus of our discussion—regard marine particles. We put a particular stress on the discussion of available experimental data because a light scattering model must be able to faithfully reproduce measurement data. We conclude the overview of experimental constraints by discussing the independent variables of light scattering models: the experimental data on the size distribution of these particles, their optical properties, such as the refractive index (composition) and its structure, as well as the particle shape. Again, we deliberately focus on the discussion of methods of obtaining the various experimental data and their limitations, as these topics tend to be overlooked in an understandable but usually troublesome desire to support one's approach to modeling of physical processes.

Given the vast territory that we felt needs to be covered, we tried to keep a precarious balance between limiting the discussion of many topics to a minimum that could be incomprehensible to some readers and opening floodgates to a multi-volume treatise that would cover all aspects in their due detail. This dilemma could perhaps be most succinctly illustrated by the following story. One of us once encountered a paper in an electronics magazine where an electronic engineer recalled his experiences as a young radio enthusiast. Wanting to build a radio receiver, that author procured a book with a promising title of *How to build a radio* or the like. The first chapter (on the vacuum diode) was very easy to understand! The second, on the triode, was not too bad either—a logical progression from the first, and so on, up to the penthode. Unfortunately, a rapid buildup of his understanding of the topic was abruptly halted at a following chapter entitled, say, *A superheterodyne receiver on five penthodes*. It took that engineer several years of university studies to realize that it was not his fault in not being able to jump across the abyss that the author of that radio amateur book created for his readers. We hope that most readers of this book will appreciate the balance we tried to achieve, and a finite number (who will not) will hopefully be stimulated to study the references we listed and get the satisfaction they missed in reading this book.

There is another problem posed by the wide range of topics covered in this book, namely that of confusing nomenclature. Notations in many of the subtopic fields tended to evolve surprisingly independently, creating historically enshrined conventions for the names of the various quantities. Given a limited span of the Latin and Greek alphabets, this led to the usage of the same notations for different quantities and various notations for the same or similar quantities. We tried to wade through this "notation swamp" by adhering to traditional notations when discussing topics in their "native" fields, but in many cases this was not possible. We include a list of major symbols as a help in solving the notation puzzles and hope that readers will appreciate our predicament and will not treat our solution to it as a shortcoming of this work.

Incidentally, the nomenclature problems are not limited to notation only. Names of quantities have also fallen prey to this independent development of (confusing) nomenclatures. One example that comes immediately to mind is the intensity of light. In physics, which is the reference frame for discussing light scattering models, this term is traditionally reserved for the power of radiation per unit area.

In radiometry, which is involved in measuring light scattering, the term intensity refers to the power of radiation per unit solid angle.

We hope that readers of this book will take notice of the problems that are created by a "disintegrated" approach to naming physical variables and attempt, in their own work, to identify the variables in sufficient detail to avoid creation of confusion and misunderstandings in the minds of readers of their publications.

Lastly, we thank many researchers for supplying unpublished data and for valuable discussions. We especially thank Dr Dariusz Stramski for his comments on the early version of the manuscript. We also acknowledge the support for this research by DRDC Valcartier and MJC Optical Technology.

Chapter 1

Basic principles of the interaction of light with matter

1.1. Introduction

The physical basis for all the phenomena we will be studying in this book is the fundamental theory of the interaction of light with matter. This theory has arguably the most distinguished history and protagonists in all of physics. The latest version of the theory is known as quantum electrodynamics or QED and was presented in its current form by Richard Feynman (1918–1988) in 1949 (*Feynman* 1949). He was the latest in a long list of physicists whose work span three centuries. This list includes Christiaan Huyghens (1629–1695), Isaac Newton (1642–1727), Joseph von Fraunhofer (1787–1826), Jean Augustin Fresnel (1788–1827), James Maxwell (1831–1879), Max Planck (1858–1947), Albert Einstein (1879–1955), Niels Bohr (1885–1962), and Paul Dirac (1902–1984).

Throughout its long history, the basic physical picture of light alternated between a particle model and a wave-based model, referred to as the duality problem. We now realize that this historical alternation of models is not a question of esthetics or fashion. It is due to the fact that a significant set of experiments are most naturally explained by treating light as discrete particles (photons), while another equally significant set of experiments finds its most natural explanation by treating light as a wave. The most significant aspect of the particle-like behavior of light is the photoelectric effect (for example, *Hecht* 1987). The wave-like behavior is best manifested by the production of fringes and oscillations when different sources of light interact (for example, *Crawford* 1968).

1.2. The quantum field model

In the QED model, light consists of photon particles which travel and interact with matter in a highly localized manner. The theory allows us only to predict the probabilities of finding these photons at any given point in space–time. These probability distributions, their interactions, and dynamics follow a wave description.

The duality problem is probably brought into sharpest focus by the simple double slit experiment. In that classical experiment, light from a single monochromatic source illuminates an opaque screen in which two closely spaced narrow slits have been cut. The pattern of the light transmitted through the screen is typically recorded on a photographic film positioned a short distance away from that screen. If the light source is intense, the pattern one sees on the film is an alternation of light and dark bands that are spaced and have an intensity distribution that matches precisely the interference pattern one would expect of a wave being transmitted through both apertures. If we now sufficiently reduce the intensity of the light source, we will reach a point where after a short exposure we see single well-separated points on the film. The interaction of light with the film is always well localized in space and is a clear manifestation of its particle-like nature: individual photons are absorbed by individual crystals in the film. If we now perform a series of experiments in which the exposure time is progressively increased, the points representing the impact of individual photons will start to cluster in specific areas on the film. Some zones remain dark with no impacts, while others have more than their share. The clustering pattern follows precisely the same wave-like interference pattern mentioned above. In the limit, a very long exposure time picture of a very low-intensity source will be identical with a short exposure of a high-intensity source. This effect is correctly captured by quantum field theory which states that light is emitted as photons that interact with matter in a highly localized manner but that the probability distributions of these interactions follow a wave-like behavior.

1.3. Basic quantum electrodynamics

To quantify the order of magnitude of the phenomena we are talking about here, let us first note that the energy of a photon, E_p, is given by (for example, *Feynman* 1962)

$$E_\mathrm{p}(\nu) = h\nu = h\frac{c}{\lambda} \tag{1.1}$$

and its momentum is given by

$$\vec{\mathbf{p}} = \frac{E_\mathrm{p}}{c}\mathbf{i} = \frac{h\nu}{c}\mathbf{i} = \frac{h}{\lambda}\mathbf{i} = \frac{h}{2\pi}\mathbf{k} \tag{1.2}$$

where $h = 6.6260693 \times 10^{-34}$ J sec is the Planck's constant, ν [Hz] the frequency of light, $c = 3 \times 10^8\,\mathrm{m/sec}$ the velocity of light in vacuum, λ the wavelength of light, and $\mathbf{i}$ is a unit vector in the direction of propagation of the photon, vector $\mathbf{k}$, with a magnitude of $2\pi/\lambda$, specifies the direction of propagation of the wave and is fixed by the momentum of the photon. The power of 1 W at a wavelength of 500 nm (green light) corresponds therefore to a photon flux of about 2.5×10^{18} photons/s. As we shall see later, a photon is always associated with a state of the electromagnetic field that can be represented by a plane vector wave (*Feynman* 1962)

$$\mathbf{A}(\mathbf{k}, \omega, \hat{\mathbf{j}}) = \hat{\mathbf{j}}\sqrt{\frac{hc^2}{\omega}}\, e^{-i\mathbf{k}\bullet\mathbf{r}+i\omega t} \tag{1.3}$$

In the above equation, angular frequency, $\omega = 2\pi\nu$, is the frequency of the photon (i.e., the frequency of light) and is fixed by its energy, E_p, $\hat{\mathbf{j}}$ represents one of two possible spin or polarization states. The amplitude of photon wave in (1.3) has been normalized to a unit probability of finding one photon per unit volume.

The wave nature of the photon implies that only a finite number of states can exist in a finite volume of space. The number of states per unit volume can be computed by considering the number of plane waves that can satisfy periodic boundary conditions in a cube of unit volume. These boundary conditions require that the wavelength of the photon be such that the field repeats itself at opposite faces of the cube. For traveling waves, this is equivalent to requesting their continuity across space. After some simple algebra (*Feynman* 1962), the following expression can be obtained for the density of photon states in a frequency interval $d\nu$.

$$s(\nu, \Omega)d\nu = 4\pi\frac{\nu^2}{c^3}\,d\nu \tag{1.4}$$

The state density is obviously isotropic: the number of states per unit solid angle, Ω, is the same in every direction. This number is simply obtained by dividing (1.4) by 4π. The number of states per unit volume contained within a given solid angle $d\Omega$, another quantity that is also frequently of interest, is thus obtained by multiplying (1.4) by $d\Omega/4\pi$:

$$s(\nu, \Omega)\,d\nu\,d\Omega = \frac{\nu^2}{c^3}\,d\nu\,d\Omega \tag{1.5}$$

While it is admittedly complete, QED is also complex to use in all but the simplest of situations. For this reason and as a matter of convenience, physicists use routinely many different simplified models to discuss and analyze the interaction of light with matter, secure in the knowledge that if some serious ambiguity arises it can be resolved (at least in principle) by resorting to the full QED. This approach can be confusing to someone not familiar with the field. It is not uncommon to find in the literature a paper that discusses some aspect of its results from the point of view of light photons and then uses the wave picture of light to compute some other properties of the results.

In order to familiarize the reader with this approach, we will discuss in the following sections the elementary interactions of light with matter from various points of view. We will also try to outline the fundamental concepts of the most frequently used simplified models and outline their respective domains of validity.

In its simplest expression, the formalism of QED assumes that in the absence of interaction with matter, a photon can be fully described by a plane vector wave with a given frequency, direction of propagation, and an integer spin of $+1$ or -1. The complete ensemble of all such plane waves forms what is called the set of base states of the photon. As long as there are no interactions with particles of other types, such as electrons, the state of the whole radiation field is fully described for all times by simply assigning the appropriate number of photons to every plane wave or base state. This number of photons can only vary if they interact with free electrons or electrons bound to the nuclei of atoms or molecules. These time-invariant states of the pure light field are also sometimes known as the eigenstates of the field.

What consistent picture can one use for a plane vector wave with spin of ±1? One can imagine a vector oriented perpendicular to the direction of propagation whose tip is spinning around the axis of propagation and whose amplitude is modulated at the wave frequency. By convention (e.g., *Bohren* and Huffman 1983, pp. 44–45), the rotation is counterclockwise when looking along the direction of propagation for a particle with a spin of $+1$ and clockwise for spin -1. In classical optics, these base states correspond to right and left circularly polarized electromagnetic waves respectively. They are the free space solution of the electromagnetic vector and scalar potentials.

As it was mentioned, this is the simplest representation. In QED, we can in principle choose any set of basis states which are most convenient for the solution of a given problem as long as this set is complete in the sense that it can reproduce any combination of the above plane wave unity spin basis states. The number of photons in each of these new basis states will be different than the ones in the standard set. This can be viewed as analogous to performing a coordinate system change in classical physics. As an example of such a base set change, consider two linearly polarized waves as base states. These perpendicular polarization states are each formed by a linear combination of two circularly polarized waves with an appropriate phase difference.

Introduction of matter into the picture requires that a representation of matter be given in the QED formalism. This begins with the representation of the base states of the electron. These states are once again given by plane waves but with a half integer spin of $+1/2$ or $-1/2$ this time. They are the solution of the Dirac equation with positive energy. This modification gives rise to waves with four components. The Dirac equation also has a set of solutions with negative energy and four components that represents the positron, the anti-matter equivalent of the electron. These terms must be included to properly account for phenomena such as spontaneous emission and, at higher energies, particle creation and annihilation. As in the case of photons, the number of electron and positrons in a given state does not vary with time if they do not interact with another particle.

The final and most difficult part of QED is the description and computation of the interaction between the photons and the free or bound electrons and positrons. This interaction gives rise to solutions in which the number of photons and electrons/positrons varies with time. The various particles are either exchanged between states or created and destroyed. A consistent approach to these complex phenomena was only developed in the last 50 years, with Richard Feynman first presenting its simplest computational version in 1949 (*Feynman* 1949). It would be well beyond the scope of this book to delve into this theory in further detail. For our purposes, it will suffice to describe the most frequent types of interactions and some of their key characteristics.

1.3.1. Emission and absorption

In order to help compute and keep track of these fundamental interactions, Feynman came up with a graphical representation now known as Feynman diagrams. In these diagrams, the time axis is vertical and the space portion of the interaction is represented schematically in one dimension along the horizontal axis. Figure 1.1 shows a set of two simplified Feynman diagrams of the emission and absorption processes. The wavy lines represent the photons and the straight lines represent the free or bound electrons. The graph on the left-hand side of the figure is a schematic of both the spontaneous and stimulated emission processes. N_i is the number of photons present before the interaction. In the emission process, a new photon is generated and the electron loses a corresponding amount of energy in the form of either kinetic energy if we are dealing with a free electron or potential energy if the electron is bound to an atom or a molecule. The change in potential energy generally occurs in discrete increments since bound electrons exist in a discrete spectrum of eigenstates of the atom or molecule. The required angular

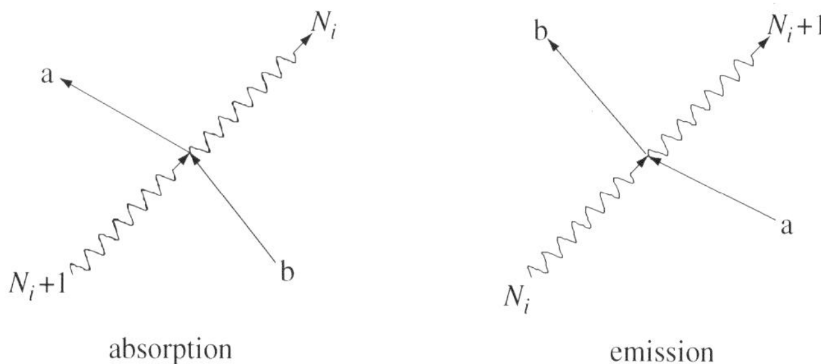

Figure 1.1. Simplified Feynman diagrams for the absorption (left) and emission (right) of a photon by a bound or free electron. The electron changes from state a to state b for emission and from state b to state a for absorption. Time runs along the vertical axis and space along the horizontal axis. The incoming and outgoing photons travel in the same direction (state). As is clear from these diagrams, absorption is the time reverse of emission.

momentum to generate the photon spin is also given away or taken up by the electron. The initial state and final states of the electron are denoted symbolically as a and b.

In the limit where N_i is 0, a photon can still be emitted in a process known as spontaneous emission. As there are no photons to start with, the electron is pictured as interacting instead with a randomly fluctuating electromagnetic field that permeates all of space. In the QED model, this vacuum fluctuation field arises because of the continuous creation and almost simultaneous destruction of virtual particles: electron/positron pairs. The energy required for this process is pictured as being furnished by the unavoidable uncertainty in energy of even the vacuum over a sufficiently short time ($\Delta E \, \Delta t < h$). These virtual particles are not themselves measurable since they exist for such a short time, but their secondary effects are definitely calculable and measurable. Arguably, the most significant of these effects is the spontaneous production of radiation by excited atoms and molecules. This interaction is the source of all the naturally occurring radiation and also, by far, of most of the man-made light.

When $N_i > 0$, there are already photons present in a given state of the field. These photons also interact with the electrons through a process known as stimulated emission. This process is the source of laser radiation. It is used in many modern optical measurement and imaging devices as it can produce radiation with narrower spectral band, better coherence, better collimation, shorter pulses, and higher intensity than spontaneous emission.

The stimulated emission process is tied to a fundamental property of the photon, its integer spin. Particles with integer spin are known collectively as bosons. The photon is the best-known boson. Stimulated emission turns out to be a fundamental property of any boson. Assume first that a process involving the creation or scattering of a boson in an empty final state of the field occurs with a probability, p, per unit time. There is then always a corresponding probability $N_i p$ of creating or scattering a boson in a final state of the field that already contains N_i bosons.

Since there is an enhanced probability for a photon of being emitted in an already occupied state, one can set up situations that strongly favor the build-up of the number of photons in a particular state. For a photon, as we mentioned before, a state is defined as a given direction of propagation with a given energy and spin. In a laser, this increase in the number of photons is generally achieved by allowing the radiation to build up in a closed cavity, containing a gain medium, with mirrors at both ends. After several reflections, only the photons propagating very near the direction of the axis normal to the mirrors are still being used to stimulate emission of other photons by the gain medium. Other photons simply leave the cavity. It should be noted that another important condition must be satisfied for stimulated emission to increase the number of photons: the rate of stimulated emission should be larger than the rate of absorption. In the context of the present work, stimulated emission is only of interest as the mechanism required to produce sources of photons in optical measurement instrumentation, and we will only discuss it further when we analyze such optical instruments and their characteristics and limitations.

The Feynman diagram on the left-hand side of Figure 1.1 shows the absorption process. $N_i + 1$ photons in one initial state of the field interact with an electron. One of the photons is absorbed. This leaves N_i photons in the final state of the field, which in this case is identical to the initial state. Note that the absorption diagram is precisely the time-reversed diagram of the emission process. It can be shown that the equations of QED, like their classical electrodynamics counterpart, are symmetrical in time. The results under time reversal must therefore be identical. Given our previous analysis of the stimulated emission probability, the corresponding probability of absorption of a photon from a state of the field that contains $N_i + 1$ photons is therefore $(N_i + 1)p$. After absorbing the photon, the electron will gain either kinetic energy if it is already moving freely in space or potential energy if it is bound to a nucleus. Note that the probability of absorption of a photon per initial photon in the field is constant and equal to p. Thus, given a medium with a uniform density of bound absorbing electrons which is much larger than the number of photons, the number of photons will decay exponentially since the number of electrons in the medium will be to a first approximation unchanged. This is often referred to as the Beer–Lambert law.

1.3.2. Scattering

In the case of absorption and emission, the final state of the field is the same as the initial state of the field. Only the number of photons in that field and the state of the electron vary after the interaction. The other possible interactions that include changes in the field and electron states are shown in Figure 1.2. In these

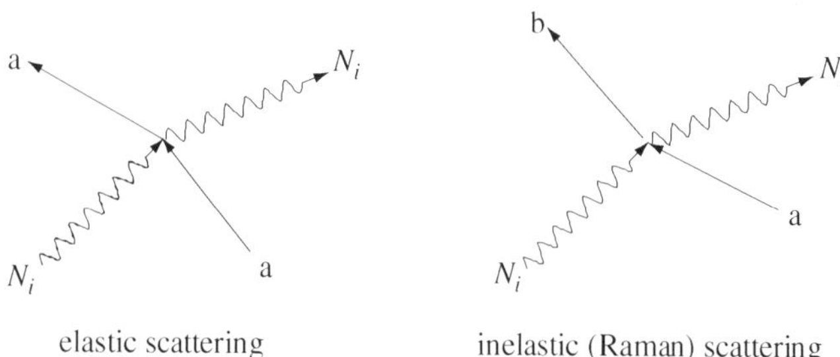

Figure 1.2. Simplified Feynman diagrams for the elastic scattering (left) and inelastic (Raman) scattering (right) of a photon by a bound or free electron. Following elastic scattering, the state of the incoming photon changes. So does the direction of motion of the electron to account for the momentum exchange. However, the bound state of the electron does not change. For inelastic (Raman) scattering, there is an additional change of the bound state of the electron from state a to state b.

interactions, the photon is absorbed and immediately re-emitted, generally in a different direction.

The left-hand side of Figure 1.2 shows a simplified diagram of this type of interaction in which the outgoing and incoming photons have the same energy. It is generally referred to as elastic scattering. The electron once again picks up the momentum required by the law of momentum conservation. This elastic scattering is by far the most frequent photon–electron interaction in nature and is the source, directly or indirectly, of almost all the scattering of light in the natural waters. Most of this book will be concerned with its experimental and theoretical study.

Two features of the QED solution for this interaction are of particular interest and will be used later. At low energies, where one can neglect relativistic effects, the angular pattern of scattered photons is identical to the well-known dipole scattering pattern first derived on the basis of classical electrodynamics by Rayleigh and Thompson (*van de Hulst* 1957, pp. 63–84). Furthermore, each scattered photon is delayed in time, and this delay shows up itself as a phase difference between the incoming and scattered light waves.

The right-hand side of Figure 1.2 is a simplified diagram of a scattering interaction in which the outgoing and incoming photons have different energies. This type of interaction is called inelastic or Raman scattering. It also occurs in natural waters, and its importance was only recently recognized (e.g., *Stavn* and Weidemann 1992). It acts to modify the spectrum of light in the deep ocean by generating a significant excess of yellow and red radiation over what would be expected, given the absorption spectrum of water.

There is a significant confusion in the literature from the various fields of optics that is caused by different interpretations of the names of the various types of light scattering. If in doubt, please refer to an illuminating discussion by Young of this subject (*Young* 1981).

We will not delve into QED any further. The interested reader can find a reasonably simple and at the same time complete account in the published lecture notes of R. P. *Feynman* (1962). The results of QED are simple and elegant when one considers a single interaction. However, this theory does not easily lend itself to dealing with large ensembles of interactions. In such a case, a hybrid approach is generally adopted. In that approach, the properties of the medium are obtained directly from QED or in simple cases from standard quantum theory, and the radiation field is treated by using Maxwell's equations and classical electrodynamics. At energies low enough to neglect relativistic effects, the accuracy of these models largely exceeds the precision of present instrumentation.

1.4. Incoherent scattering

We will now sketch a way in which such a passage to the classical limit can be viewed. Consider first an ensemble of randomly positioned identical atoms or molecules separated on average from one another by a distance much greater than the wavelength of the incident light. In that case, the amount of interference between the photons generated or scattered by each interaction will be minimized. Their mutual interactions can be neglected simply because they are far apart from each other. The overall effect will be well approximated by a sum of the individual interactions. In that case, light intensities can then be summed and mutual interference neglected.

Let us consider the case of a beam of photons propagating along an axis z and absorbed by such an ensemble of identical atoms or molecules. From our previous discussion, recall that the probability of an individual absorption interaction is proportional to the number of photons present in the initial state of the field. If we assume that $W_{a-b}(\nu, \hat{\mathbf{j}})$ is the probability of spontaneous emission per unit time and per unit frequency interval in a transition from state a to state b, neglect mutual interference effects, sum over intensities, and properly account for the density of photon states, we obtain the following simple results:

$$I = \pi(\nu)\, c\, h\nu \tag{1.6}$$

$$\begin{aligned} \frac{dI}{dz} &= \frac{1}{c}\frac{dI}{dt} \\ &= -I\, n_a \frac{W_{a-b}(\nu, \hat{\mathbf{j}})}{c\omega(\nu, \hat{\mathbf{j}})} \\ &= -I\, n_a\, \sigma_{abs}(\nu) \\ &= -Ia(\nu) \end{aligned} \tag{1.7}$$

where I is the intensity of the beam, $\pi(\nu)$ the number density of photons per unit volume, c the speed of light, $h\nu$ the individual photon energy, n_a the number density of absorbers per unit volume, σ_{abs} the absorption cross-section, and $a(\nu)$ the total absorption coefficient at a given frequency of light, ν, in units of inverse distance. The above equation can be integrated to result in what is called the Beer–Lambert law:

$$I = I_0\, e^{-a(\nu)\, z} \tag{1.8}$$

A note of caution must be sounded here about the names of the various quantities related to the propagation of electromagnetic waves. In physics, the symbol, I, and name *intensity* has traditionally been used for the power flux of the electromagnetic wave per unit area of surface perpendicular to the wave vector. In oceanography (*Anonymous* 1985, *Morel* and Smith 1982), atmospheric sciences (*Raschke* 1978), and applied optics, this quantity is described by the term *irradiance* and denoted by E (in the older literature it may be denoted by H). The term *intensity, I,* in that second context, denotes the power of the electromagnetic wave per unit solid angle. As the symbol E is in the present context used for the energy and the electric field, we retain here the traditional physical terminology and will return to the applied optics terminology in the later chapters.

A similar approach can be taken to evaluate the effect of elastic scattering on a parallel beam of photons by an incoherent ensemble of atoms or molecules. Once again, the beam propagates along the z-axis. In elastic scattering, the final state of the photon after the interaction has a finite probability of having any given propagation direction. In order to properly evaluate the total loss of photons from a beam, we must therefore sum over all possible directions in space (final photon states). For an ensemble of scatterers whose individual scattering patterns are axially symmetrical, such as spheres, or for an ensemble of randomly oriented scattering particles, we obtain:

$$\begin{aligned} \frac{dI}{dz} &= -I\,2\pi \int_0^{\pi} \beta(\nu, \theta) \sin\theta \, d\theta \\ &= -I\,b(\nu) \end{aligned} \tag{1.9}$$

$$I = I_0\, e^{-b(\nu)\, z} \tag{1.10}$$

where $b(\nu)$ is the volume scattering function at a given frequency in units of inverse distance. Function $\beta(\nu, \theta)$ represents the angular distribution of scattered light in units of inverse distance times inverse solid angle. We stress here that, I is the intensity of the photon beam corresponding to the initial state of the field, i.e., energy, direction of propagation, and spin (polarization). Although in elastic scattering the photon energy does not change, the last two parameters do. Note that in equation (1.9), an average has been performed over the azimuth angle, measured from an arbitrary plane (say, the scattering plane, containing the incident and scattered directions) about the incident direction. This average is obviously meaningful only if either the pattern of each individual scattering is axially symmetric or the particles of the ensemble, over which the average is carried out, are randomly oriented. When the angular distribution, $\beta(\nu, \theta)$, is normalized by dividing with the scattering coefficient, the result is called the phase function, $p(\nu, \theta)$.

The volume scattering function, $\beta(\nu, \theta)$, is arguably the most important data required when one needs to compute the light field in scattering media. A discussion of its computation and methods of measurement will form an important portion of this book. Finding this function is the first required step in computing the evolution of any light field in turbid medium.

Finally, in cases where both absorption and scattering are present, the incoming light beam is attenuated as the sum of both coefficients. This sum is called the attenuation (extinction) coefficient $c(\nu)$:

$$\begin{aligned} I &= I_0 \, e^{-[a(\nu)+b(\nu)]z} \\ &= I_0 \, e^{-c(\nu)z} \end{aligned} \tag{1.11}$$

$$c(\nu) = a(\nu) + b(\nu) \tag{1.12}$$

Symbol c in the above equation should not be confused with that used to denote the velocity of light that we used in (1.6) and (1.7) in a time-honored tradition in physics.

1.5. Coherent scattering

The discussion up to this point has assumed sufficient distances between the scattering particles so that the re-scattering by a particle of light scattered by a different particle is negligible. We also assumed that the particles are randomly distributed in space so that the interference terms between the scattered waves can be neglected. This requires the mean distance between scattering centers, at a minimum, be greater than the wavelength of the incoming light. For visible light, 500 nm in the green region of the spectrum, this condition is far from satisfied by the molecules of any gas at atmospheric pressure. The situation is even worse for a liquid or a solid. The mean intermolecular distance in a gas at atmospheric pressure and density is about 3.0 nm. Even in the UV region of the spectrum at 300 nm, there are 100 particles per wavelength. In water, the mean intermolecular spacing is approximately 0.3 nm. We are therefore looking at a minimum of 1000 particles per wavelength across the entire UV to visible spectrum. In all solids, liquids, or gases at atmospheric pressure, scattering from atomic or molecular interaction is thus highly coherent. The mutual interference terms in fact dominate and must be accounted for.

Detailed computation of the coherence effects is beyond the scope of the present work. The key effects can however be explained on the basis of some simple physical arguments. *Fabelinskii* (1968, pp.1–17) gives a particularly clear and simple presentation of the problem, and we will use a similar approach. As shown in Figure 1.3, consider the plane wave associated with a photon as it gets scattered by closely spaced molecules. Along or very near the wavefront A–A′, we can always find two neighboring molecules that are spaced apart such that their scattered waves in a direction θ are $\lambda/2$ out of phase and cancel each other by destructive interference. The spacing, d, required for this condition to hold is:

$$d = \frac{\lambda}{2 \sin \theta} \tag{1.13}$$

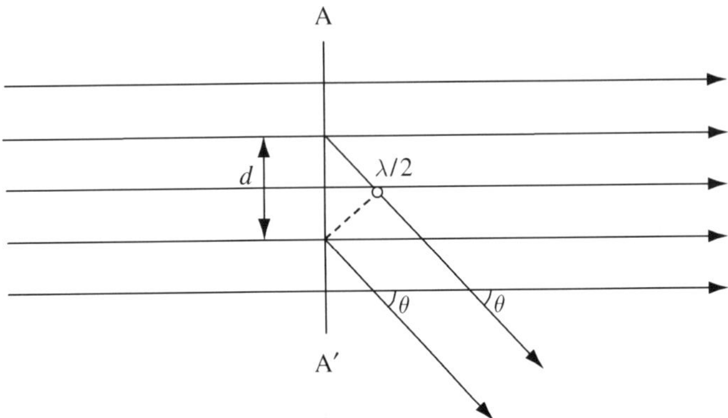

Figure 1.3. Coherent scattering of a light beam by a dense homogeneous medium. For a given scattering angle θ different from 0 and π, one can always find two corresponding scattering regions with a spacing such that their scattered light has a phase difference of half a wavelength. This results in destructive interference for light scattered in all but the forward ($\theta = 0$) and backward ($\theta = \pi$) direction.

As follows from (1.13), for any angle different from 0 or π, we can always find at a finite distance d molecules that radiate in opposite phases. There is however no finite distance for the wave propagating at either an angle 0 or an angle π. Thus, the scattered waves do not cancel in those two special directions, and from the previous argument, in a homogeneous medium of infinite lateral extent, light can only be propagated either in the forward ($\theta = 0$) or in the backward ($\theta = \pi$) direction.

1.5.1. Molecular optics and the concept of refractive index

Deep in the bulk of the homogeneous medium, the light propagating in the backward direction is cancelled by subsequent scattering by the molecules lying in that direction. After several stages of backscattering, re-backscattering and re-re-backscattering, only the forward-propagated wave remains (*James* and Griffiths 1992). As we stated before, each scattering event introduces a delay between the incoming wave and the scattered wave. Because of the repeated scattering events, the speed of the wave propagating this way in the medium is therefore slower than it would have been in vacuum. The ratio of the speed of light in vacuum to the speed of light in the medium is called the (absolute) real refractive index and is usually denoted by n. The vector potential associated with the propagating wave inside the homogeneous medium can be represented as:

$$\mathbf{A}(\mathbf{k}, \omega, \hat{\mathbf{j}}) = \hat{\mathbf{j}}\sqrt{\frac{hc^2}{\omega}}\, e^{-in\mathbf{k}\bullet\mathbf{r}+i\omega t} \tag{1.14}$$

where the only difference between this equation and (1.4) (vector potential, **A**, of a wave in vacuum) is in the presence of the refractive index, n, in the exponent. This vector potential wave can be related to a classical electromagnetic wave with an electric vector $\mathbf{E}(\mathbf{k}, \omega)$ by the standard Coulomb gauge transformation (*Feynman* 1962)

$$\mathbf{E} = -\frac{1}{c}\frac{\partial \mathbf{A}}{\partial t} \tag{1.15}$$

where c is here the velocity of light.

A beam of photons propagating inside a homogeneous medium that does not absorb light and where the molecules are close enough to enforce the coherence of light scattering can thus be represented by a classical electromagnetic wave with a real refractive index, n.

$$\mathbf{E}(\mathbf{k}, \omega, \hat{\mathbf{j}}) = \hat{\mathbf{j}}E_0, e^{-in\mathbf{k}\bullet\mathbf{r}+i\omega t} \tag{1.16}$$

In this picture, the photon energy density (the number density of photons times the photon energy) becomes equal to the energy density of the wave.

$$\pi(\nu)h\nu = \frac{1}{8\pi}E_0^2 \tag{1.17}$$

Following our previous discussion, the excess delay over the vacuum case, which can be expressed as $(n-1)$, is to first order proportional to the number density of scattering particles. This is true only in the limit where one accounts for the first order of interference and specifically neglects all the back-reaction (multiple backscattering of backscattering) terms. A simple derivation of this limit can be found in *van de Hulst* (1957, pp. 32–33). In simple parametric form, the refractive index is thus given by:

$$(n-1)\frac{W}{\rho} = A_{\mathrm{m}} \tag{1.18}$$

where W is the molecular weight of the substance, ρ the density of the medium in units of mass per unit volume, and A_{m} the molar refractivity, a constant for a given wavelength and temperature. This formulation is only valid in the limit of very small values of $(n-1)$ such as those for gases. For air at 15°C and atmospheric pressure, the refractive index at a wavelength of 500 nm is 1.0002781 and (1.18) can be used. In solids or liquids, the density of scattering molecules is approximately 1000 times greater. For pure water at 500 nm, the refractive index $n = 1.33$. In that situation, one must include all the back-reaction terms that lead to the Lorentz–Lorenz formula (*Born* and Wolf 1980, pp. 98–108).

$$\frac{n^2-1}{n^2+2}\frac{W}{\rho} = A_{\mathrm{m}} \tag{1.19}$$

where the symbols have each the same definition as in (1.18). This form is quite accurate for a great variety of substances, and we will use it later in this book.

The effect of absorption can also be simply included in the case of closely spaced molecules. Since the absorption of a photon does not produce any immediate radiation, there is no interference term between the absorbers and no coherence effects no matter how closely spaced the absorbers become. Absorption simply reduces the amplitude of the propagating waves. This effect can be accounted for by allowing the refractive index to become a complex number, n, with real part, n', representing the change of the wave velocity and the imaginary part, n'' representing the damping of the wave by absorption:

$$n = n' - in'' \tag{1.20}$$

If, $n'' = 0$, then the magnitude of n is equal to n'. In such a case, we shall use n interchangeably with n'. It follows that the electric field of the wave in material represented by the complex refractive index (in respect of the scalar magnitude of the field, E) can be expressed as follows:

$$\begin{aligned} E &= E_0 e^{-inkz+i\omega t} \\ &= E_0 e^{-in'kz+i\omega t} e^{-n''kz} \end{aligned} \tag{1.21}$$

where $e^{-n''kz}$ is the damping factor. The imaginary part, n'', of the complex index is directly related to the absorption coefficient used in (1.8). Indeed, the intensity, I, is defined as $\langle EE^* \rangle$, where brackets $\langle\rangle$ denote the time average over an interval much greater than the wave period and the asterisk denotes the conjugate of a complex variable. Thus, we have from (1.21):

$$\begin{aligned} I &= \langle EE^* \rangle \\ &= E_0^2 \langle (e^{-in'kz+i\omega t} e^{-n''kz})(e^{-in'kz+i\omega t} e^{-n''kz})^* \rangle \\ &= I_0 e^{-2n''kz} \end{aligned} \tag{1.22}$$

and, by comparing with (1.8), we have

$$\begin{aligned} a &= 2n''k \\ &= \frac{4\pi}{\lambda} n'' \end{aligned} \tag{1.23}$$

All the substances we will be concerned with here have indices of refraction with a very small imaginary part, n'', almost never exceeding 10^{-2}. Despite its low magnitude, this value indicates a significant absorption of light in relatively thin layers of material. At this value of the imaginary part of the refractive index, a layer of material of thickness as thin as 0.1 mm would absorb 92% of the light and look pitch black.

1.5.2. Classical electromagnetic wave theory

Coherent scattering discussed above is also responsible for another important phenomenon, the partial reflection of photons at the boundary between two different scattering media. Consider first the interface between vacuum and a homogeneous scattering medium of index n. The backward propagating wave is canceled only deep in the bulk of the medium, but not close to the interface: there are simply no medium molecules prior to the interface to effect such cancellation. The fraction of light backscattered by the first layer of molecules at the interface is obviously not cancelled. Subsequent fractions backscattered by deeper layers can also only be diminished in amplitude by the limited number of shallower layers of scatterers lying between the deeper layer and the surface of the medium. The net effect of this partial cancellation of backward coherent scattering is that each interface reflects a certain fraction of the light that strikes it (*James* and Griffiths 1992). This fraction, represented by the reflection coefficient, depends on the refractive index. This is not surprising since as we have seen, the refractive index is directly related to the coherent scattering. This reflection phenomenon also occurs at the interface between two different media each with its own refractive index. Here again, the reflection occurs because of an incomplete cancellation of the backward propagating wave. In that case, the reflected fraction of the light depends on the ratio of the indices across the interface.

The results for the dependence of this reflection coefficient on the ratio of the indices of refraction are identical to those obtained by simply matching boundary conditions that require continuity of the electromagnetic wave across an interface between the two media (*van de Hulst* 1957, p. 204). These formulae are known as the Fresnel reflection coefficients in honor of their discoverer, Jean Augustin Fresnel.

In fact, as expected, all the classical results of electromagnetic theory are identical with the coherent scattering model. This includes the law of reflection, Snell's law of refraction, and the Fresnel coefficients. The particular virtue of this model is to explain in a consistent fashion how they arise as a result of coherent molecular scattering. The study of coherent multiple scattering to explain the optical properties of materials is called molecular optics and is still the subject of research papers (*Reali* 1982, *Lalor* and Wolf 1972) and expository papers (*Ballenegger* and Weber 1999, *Fearn* et al. 1996, *James* and Griffiths 1992, *Reali* 1992) which the interested reader should consult for a more rigorous presentation. Born and Wolf also present some of the basic results in their classic textbook (*Born* and Wolf 1980, pp. 98–108).

Figure 1.4 shows the geometry of the reflection and transmission from an interface whose normal is at an angle θ_i with respect to the incoming light beam. We have:

$$\theta_i = \theta_r \tag{1.24}$$

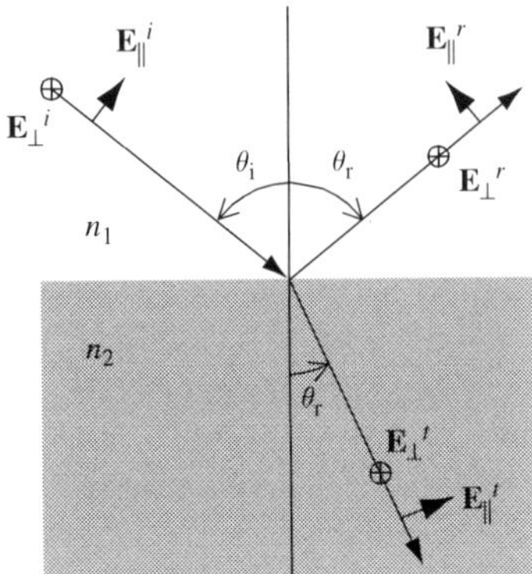

Figure 1.4. Geometry of the reflection from and transmission through an interface between two media with different indices of refraction n_1 and n_2. $\mathbf{E}_{\|}$ denotes the polarization component with the electric field parallel in the propagation plane (paper plane). $\mathbf{E}_{\perp}$ is the polarization component perpendicular to the propagation plane. That plane is defined by the incidence, reflection, and refraction directions.

$$\begin{aligned} \sin\theta_i &= \frac{n_2'}{n_1'}\sin\theta_t \\ &= n'\sin\theta_t \end{aligned} \tag{1.25}$$

The second equation describes what is called the Snell law of refraction (e.g., *Hecht* 1987). The symbol n' in the last line of (1.25) from now on will denote the real relative index which is defined as the ratio of the refractive index of the medium into which the light is transmitted to the refractive index of the medium in which the incident light propagates. All the solutions of the wave equations of electrodynamics (for example, *Kerker* 1969) can be cast in terms of this relative refractive index. This concept is particularly important when one is considering the scattering of light by particles in water. The absolute refractive index of the scattering particle may be substantially different from unity, but since water has the absolute real part of the refractive index of roughly $n' = 1.33$ in the visible, the relative refractive index, n', of the particles rarely exceeds 1.1 to 1.2. We shall see that the closeness of the relative index to unity simplifies many results.

The reflection coefficients, which shall interest us most (see the complete set of reflection and transmission coefficients, e.g., in *Hecht* 1987), are most easily expressed in terms of linearly polarized electromagnetic waves. A complete set of two such polarization components is defined by the propagation plane that contains the directions of the incident, reflected waves, and refracted waves (Figure 1.4). Incidentally, this plane is unambiguously defined only for the oblique incidence at the interface. If the wave incidence direction is perpendicular to the interface, so are the directions of the reflected and refracted (transmitted) waves, assuming that the medium following the interface is not birefringent. Thus, the propagation plane can be any plane that contains the wave direction. Consequently, we shall expect that the polarization of the incident wave does not matter in the normal incidence case, simply by considering the symmetry of the incidence geometry. In this case, we have (e.g., *Hecht* 1987):

$$r = \frac{1 - \dfrac{n_2}{n_1}}{1 + \dfrac{n_2}{n_1}} = \frac{1 - n}{1 + n} \tag{1.26}$$

$$R = r \cdot r^* = \left| \frac{1 - n}{1 + n} \right|^2 = \frac{(n' - 1)^2 + n''^2}{(n' + 1)^2 + n''^2} \tag{1.27}$$

where r is the reflection coefficient for the wave amplitude, r^* its complex conjugate, and R the corresponding reflection coefficient for light intensity.

The situation is markedly different for oblique incidences (Figure 1.4). The wave component with polarization perpendicular to the propagation plane is expected to retain its polarization direction after reflection and refraction, as its electric vector is parallel to the interface, which locally can be treated as a plane. However, the polarization direction of the perpendicular component changes both on reflection and refraction. Thus, for all oblique angles, the reflection coefficient is different for each polarization (e.g., *Hecht* 1987):

$$r_{||} = \frac{\cos \theta_i - n cos\ \theta_t}{\cos \theta_i + n cos\ \theta_t} \tag{1.28}$$

$$r_{\perp} = \frac{n cos \theta_i - cos\ \theta_t}{n cos \theta_i + cos\ \theta_t} \tag{1.29}$$

$$\begin{aligned} R_{||} &= r_{||} \cdot r_{||}^* \\ R_{\perp} &= r_{\perp} \cdot r_{\perp}^* \end{aligned} \tag{1.30}$$

Consider now for simplicity, media that do not absorb light, i.e., the ones with a real refractive index, n. In the case where the electric vector is perpendicular to the propagation plane, and if $n > 1$, the reflection coefficient rises monotonically from the value given by (1.26) and (1.27) at normal incidence to unity at grazing incidence. If, $n < 1$, as in the case of an air–water interface at the wall of a bubble in water, the reflection coefficient reaches the value of 1 for an oblique incidence angle, $\theta_i = \theta_C$ (critical angle) when $\theta_t = \pi/2$. Hence, from the Snell law (1.25), we have:

$$\sin \theta_C = n' = \frac{n'_2}{n'_1} \tag{1.31}$$

For example, for water and air ($n' \cong 1.33$), the critical angle $\theta_C \cong 48.7°$.

When the electric vector lies in the propagation plane, the reflection coefficient decreases until, if there is no absorption, it reaches 0 at what is known as the Brewster's angle, θ_B. This is the angle defined by the following condition:

$$\theta_t = \frac{\pi}{2} - \theta_i \tag{1.32}$$

Thereafter, the reflection coefficient also rises monotonically to unity at grazing incidence. For a real refractive index, $n = n' + i0$, the Brewster's angle is given by:

$$\tan \theta_B = n' = \frac{n'_2}{n'_1} \tag{1.33}$$

this follows from (1.28) with $r_{||}$ set to 0. For example, for water ($n' \cong 1.33$), the Brewster angle $\theta_B \cong 53°$. If $r_{||} = 0$, then the reflected wave vanishes for the parallel polarization, i.e., the reflected wave completely polarized at the Brewster angle. Note that at the Brewster angle, the direction of the electric vector of the refracted wave is aligned with the direction of the reflected wave. This nicely fits with the explanation of refraction and reflection of light in terms of coherent scattering of light by the medium following the interface, although the usual superficial explanation of this process is flawed, as pointed out by *Doyle* (1985, and references therein) who also gives the correct treatment.

For homogeneous media, (1.24) to (1.33) form the basis of geometrical optics approximation. In that approximation, one still refers to the concepts of wavefronts and intensities but one assumes that the wavelength is so small that interference effects can be neglected. Light beams are viewed as bundles of rays propagating in straight lines between interfaces. Reflection and refraction of these ray bundles only occur at those same interfaces and are not subject to any interference phenomenon. We will use many of the results of geometric optics throughout this book to obtain useful approximations to several scattering problems.

1.5.3. Scattering by fluctuations of the refractive index

The analysis of coherent scattering leads to another very important conclusion: there should not be any significant scatter of ultraviolet, visible, and infrared radiation in pure liquids such as water. As we have seen, in dense matter with many scattering centers in a distance of one wavelength, the first-order effects can be simply described by the use of a complex refractive index and classical electrodynamics. In that approach, a wave incident at the interface of two media is partially reflected back into the first medium and partially transmitted with a different phase velocity through the second medium. If that medium absorbs light, the wave is also attenuated, i.e., its amplitude decreases with distance traveled in the medium.

This picture is only valid to the extent that we neglect local fluctuations in the density of scattering particles in the medium. Although on average that density is constant, the laws of statistical mechanics imply the existence of fluctuations in particle densities that occur at all fluctuation size scales (*Fabelinskii* 1968). As the refractive index is a function of material density, the density change leads to a change in the refractive index of the substance. These index fluctuations are random, and the fluctuations occurring in neighboring volumes are statistically independent of one another. The scattering from these fluctuations is therefore incoherent.

The probability of occurrence of a fluctuation of a given amplitude and size is given by a standard Boltzmann distribution over the total excess energy required to produce this change from the mean values (*Kerker* 1969). This mechanism is the dominant source of light scattering in dense media. The equations describing this effect were first obtained by *Einstein* (1910) and *Smoluchowski* (1908). *Morel* (1974) has given a thorough account of this effect for pure water and pure seawater. More recently, *Buiteveld* et al. (1994) revisited these results for pure water using recent data on the refractive index and its derivative with respect to pressure. The scattering formula they obtained closely matches the experimental data given by Morel. We will present and review in detail this and more recent work in the field in the next chapter.

1.5.4. Scattering by aerosols and hydrosols

The other important source of light scattering in both atmospheric air and natural waters are the suspended particles called aerosols in air and hydrosols in water. Each of these particles is composed of a large number of molecules in either solid crystal or liquid form. Thus each aerosol or hydrosol is assumed to contain enough molecules that it can be treated as a macroscopic object with its own refractive index. Under these conditions, light scattering from an individual particle can be accurately treated within the framework of standard electromagnetic theory. In most naturally occurring circumstances, the number density of these particles is low enough that the mutual coherence of the light they scatter can be neglected.

Most naturally occurring aerosols consist of water droplets condensed around a solid core of sand or a particle of salt that might have subsequently dissolved. The amount of water accreted by each aerosol core is directly related to the relative humidity. Because the water droplet is held together by surface tension, these aerosols are almost perfect spheres. Their refractive index is close to that of either pure or seawater. The maximum size of aerosols is limited by the balance of the force of gravity against the aerodynamic drag due to the vertical component of the local airflow.

The nature of the particles suspended in the water column is much more complex. In the open ocean, these particles are mainly biological in origin and consist of everything from viruses to chlorophyll-containing phytoplankton and organic detritus from marine animals. Nearer to shore, the water column also contains substantial amounts of sand and silt particles washed away from nearby land or shallow coastal zones by wave and tide action or by rivers. The shapes of these particles are not constrained by surface tension and can vary enormously. Particles with shapes approximated by cylinders, platelets, oblate, and prolate spheroids are all present in significant numbers. A substantial portion of this book is devoted to the study of the optical properties, shape, and size distributions of these particles. Because of the effect of buoyancy and density, the maximum size of particles which can be suspended in water for any significant length of time is also much larger than in air, and the velocity of these particles relative to the water flow is much slower. Compared to aerosols, hydrosols have a much lower relative refractive index since the refractive index of water is already close to 1.33 across most of the UV, visible, and near IR. Fortunately, this property of closeness of the relative index to unity substantially simplifies some of the scattering results. As in the case of aerosols, rigorous solutions to the scattering of light by some types of particles can be obtained by standard electromagnetic theory. The number density (concentration) of hydrosols is also generally low enough that the rules of incoherent scattering apply.

1.6. Basic scattering formalism

We have so far described scattering in a qualitative manner. We will now present the basic formalism of the scattering of a wave by a particle of finite size and otherwise arbitrary shape and composition. In-depth treatment of the scattering of light by small particles can be found in monographs by *van de Hulst* (1957), *Kerker* (1969), and more recently, by *Bohren* and Huffman (1983). The first results we will discuss only depend on the fact that we are dealing with a wave and can be obtained without reference to the exact nature of the wave. It could be a sound wave, a light wave, or an electron matter wave.

1.6.1. Scalar waves

Following *van de Hulst* (1957), consider a plane scalar wave propagating along the positive z direction and interacting with a particle of finite size. The coordinate system is chosen such that its origin is inside the particle. The incoming wave is given by:

$$u_0 = e^{-ikz+i\omega t} \tag{1.34}$$

In the distant field, the scattered wave is a spherical outgoing wave whose amplitude is inversely proportional to the distance. The scattered wave amplitude may then be written as:

$$u = S(\theta, \varphi)\frac{e^{-ikr+i\omega t}}{ikr} \tag{1.35}$$

By using the dimensionless product kr, the scattering amplitude function of the particle remains a pure number. This practice of using the dimensionless product of the characteristic size of the scattering particle and the wave number k of the incoming wave (with the wavelength λ):

$$k = \frac{2\pi}{\lambda} \tag{1.36}$$

as a unit is used throughout all of scattering theory. It allows one to express results in a manner that is independent of the absolute size. The scattering angle θ is measured from the direction of the incoming wave to the direction of observation of the scattered wave and is defined in the plane containing both the incoming wave and the scattered wave directions. The azimuth angle φ is defined as the orientation of the scattering plane about the incoming wave direction. Substituting (1.34) into (1.35), we can relate the amplitude of the scattered wave to the amplitude of the incoming wave as follows:

$$u = S(\theta, \varphi)\frac{e^{-ikr+ikz}}{ikr}u_0 \tag{1.37}$$

The scattering amplitude function is complex and can also be written as:

$$S(\theta, \varphi) = se^{i\eta} \tag{1.38}$$

In this expression, s is positive and σ is real. Both are functions of the same angles as the scattering amplitude. Parameter η is the phase change. A negative value of $(\eta - \pi/2)$ indicates a phase lag that can be interpreted as the delay between the incident and scattered wave we talked about in the previous sections. The scattered wave intensity is proportional to the square of the modulus of the amplitude, and the scattered fraction of the incoming beam power is therefore

$$\begin{aligned}\frac{I_{\text{scat}}}{I_0} &= \frac{|S(\theta, \varphi)|^2}{k^2r^2} \\ &= \frac{s^2(\theta, \varphi)}{k^2}\frac{1}{r^2} \\ &= \frac{\sigma_{\text{scat}}(\theta, \varphi)}{r^2}\end{aligned} \tag{1.39}$$

where

$$\sigma_{\text{scat}}(\theta, \varphi) = \frac{s^2(\theta, \varphi)}{k^2} \tag{1.40}$$

is the differential scattering cross-section of a single particle. If the scattering pattern s has an axial symmetry, we shall also define an axially symmetrical differential cross-section:

$$\sigma_{\text{scat}}(\theta) = 2\pi \frac{s^2(\theta)}{k^2} \tag{1.41}$$

with the factor 2π accounting for the integration of the two-dimensional differential angular scattering cross-section $\sigma_{\text{scat}}(\theta, \ \varphi)$ over φ that varies in a range of 0 to 2π.

By integrating this expression over all directions, one obtains the total scattering cross-section C_{scat} for a single particle. Another often used quantity is the scattering efficiency, Q_{scat}. It is defined as the ratio of the actual scattering cross-section to the geometric cross-section of the particle. For an arbitrarily shaped particle, the geometric cross-section area is defined as the projection or shadow of the particle onto a plane perpendicular to the direction of propagation of the incoming beam of light.

$$\begin{aligned} C_{\text{scat}} &= \int_0^{2\pi} \int_0^{\pi} \frac{s^2(\theta, \varphi)}{k^2 r^2} r^2 \sin\theta \, d\theta \, d\varphi \\ &= \frac{1}{k^2} \int_0^{2\pi} \int_0^{\pi} s^2(\theta, \varphi), \sin\theta \, d\theta \, d\varphi \end{aligned} \tag{1.42}$$

$$Q_{\text{scat}} = \frac{C_{\text{scat}}}{A_{\text{shadow}}} \tag{1.43}$$

If the particle does not absorb light, as is the case when the refractive index is real, there is another important way in which the total cross-section can be evaluated without having to carry out the integrals in (1.42). Consider the total intensity seen by a telescope looking back toward the scattering particle from a great distance away. The amplitude observed through the telescope will be given by the sum of the amplitude of the original plane wave and of the spherical wave scattered by the particle in the forward direction. Squaring the modulus of this sum will give the intensity of the light remaining in the initial beam direction. The difference between the total initial beam intensity intercepted by the telescope and the intensity remaining when scattering occurs is the total amount of power lost by the beam through both scattering and absorption. This quantity is called the attenuation (extinction) cross-section, C_{attn}. After evaluating the intensity captured by the telescope in both cases (*van de Hulst* 1957), a truly simple yet remarkable formula results.

$$C_{\text{attn}} = \frac{4\pi}{k^2} \text{Re}\{S(0)\} \tag{1.44}$$

This important result is called the optical theorem and was first discovered by van de Hulst. It was later found to be a basic property of the process of scattering and to be applicable even to the scattering of elementary particles in high-energy accelerators. If the refractive index of the particle is real, no absorption can occur in the particle and $C_{attn} = C_{scat}$. In that case, we thus can write that

$$\begin{aligned} C_{scat} &= \frac{4\pi}{k^2}\mathrm{Re}\{S(0)\} \\ &= \frac{1}{k^2}\int_0^{2\pi}\int_0^{\pi} s^2(\theta,\varphi),\sin\theta\, d\theta\, d\varphi \end{aligned} \tag{1.45}$$

We will often be concerned with evaluating the total scattering coefficient and (1.44) will many times allow us to do this more simply than by attempting to evaluate the integrals in (1.42).

1.6.2. Polarization effects

It should now be obvious that the scattering amplitude function contains all the information required for the complete solution of the scattering problem for a scalar wave. The formalism described in (1.1) to (1.38) can be generalized to polarized transverse waves. In the most general case, a matrix of four different scattering amplitudes will relate linearly the complex amplitudes of the two possible incoming polarization components to the complex amplitude of the two outgoing polarization states. For the intensities, this will give rise to a scattering matrix of 16 elements, each a function of the scattering angle. As we shall see in later chapters, different symmetry properties of the scattering particles will reduce the number of independent elements of the matrix.

In the important case of homogeneous spheres, the scattering amplitude matrix is diagonal. This symmetry leads to the simplest possible result for a polarized incident beam. There are only two amplitude functions, and these functions only depend on the scattering angle θ. For the component of the electric field perpendicular to the plane of scattering (defined by the incidence and observation directions), we can write

$$E_{\perp} = S_1(\theta)\frac{e^{-ikr+ikz}}{ikr}E_{\perp 0} \tag{1.46}$$

while for the component of the electric field parallel to the scattering plane we have

$$E_{||} = S_2(\theta)\frac{e^{-ikr+ikz}}{ikr}E_{||0} \tag{1.47}$$

The intensity of the scattered light for the case of polarization perpendicular to the scattering plane is

$$\frac{I_\perp}{I_0} = \frac{|S_1(\theta)|^2}{k^2 r^2} \tag{1.48}$$

For polarization parallel to the scattering plane, we have

$$\frac{I_\parallel}{I_0} = \frac{|S_2(\theta)|^2}{k^2 r^2} \tag{1.49}$$

Incident natural (unpolarized) light is a mixture of equal amounts of each polarization and therefore

$$\frac{I}{I_0} = \frac{1}{2}\frac{|S_1(\theta)|^2 + |S_2(\theta)|^2}{k^2 r^2} \tag{1.50}$$

i.e., as it follows from comparing (1.39) and the above equation, we have:

$$|S(\theta)|^2 = \frac{1}{2}(|S_1(\theta)|^2 + |S_2(\theta)|^2) \tag{1.51}$$

1.6.3. Dipole and Rayleigh scattering from small particles

If the particle is much smaller than the wavelength of light, the exciting field due to the incoming wave is uniform across the volume of the particle. Thus, the particle responds to the field as a whole. This allows a considerable simplification. Solutions for particles of many different shapes have been obtained with relative ease (*van de Hulst* 1957, *Kerker* 1969). This type of analysis for particles of characteristic dimensions much smaller than the wavelength was first carried out by Rayleigh and the field bears his name. Scattering from small particles is often called Rayleigh scattering. For a sphere, the solution is the same as that of an oscillating dipole.

$$S_1(\theta) = i k^3 a' V \tag{1.52}$$

$$S_2(\theta) = i k^3 a' V \cos(\theta) \tag{1.53}$$

where V is the volume of the particle and a' is called the average volume polarizability.

For a uniform particle material, this polarizability is related to the refractive index by the Lorentz–Lorenz equation (*van de Hulst* 1957) and represents the coordinated response of the electrons in the medium to the electric field of the exciting wave. In this case, the particle shape does not matter.

In some cases such as liquids or crystals, the polarizabilty may be a vector or tensor rather than a scalar. In other words, the medium may partially respond to an excitation in a different direction than that of the exciting wave's electric vector. In the scalar case, for a particle of radius a, the amplitudes of the scattered waves for the two polarization states are given by

$$S_1(\theta) = i\frac{n^2 - 1}{n^2 + 2}x^3 \tag{1.54}$$

$$S_2(\theta) = i\frac{n^2 - 1}{n^2 + 2}x^3 \cos\theta \tag{1.55}$$

where

$$\begin{aligned} x &= k\,a \\ &= \frac{2\pi a}{\lambda} \end{aligned} \tag{1.56}$$

where λ is the wavelength of light in the medium surrounding the particle.

When dealing with spheres, the dimensionless product of the wave number, k, and the particle radius, a, is the most natural unit and is frequently assigned the symbol x. We will follow this convention throughout this book.

For natural light, the intensity of the scattered wave can be expressed as follows:

$$\frac{I(x,\theta)}{I_0} = \left|\frac{n^2 - 1}{n^2 + 2}\right|^2 x^6 \frac{1 + \cos^2\theta}{2k^2r^2} \tag{1.57}$$

Hence, in the small-particle approximation, the one-dimensional differential scattering cross-section, $\sigma_{scat}(x, \theta)$, of a particle can be expressed as:

$$\sigma_{scat}(x,\theta) = \frac{2\pi}{k^2}\left|\frac{n^2 - 1}{n^2 + 2}\right|^2 x^6(1 + \cos^2\theta) \tag{1.58}$$

according to (1.41).

We will see later that, to a good approximation, the difference between the angular responses of the two states of polarization applies also to larger particles with relative index close to unity. The scattering amplitude for parallel polarization will, to a first approximation, be the same as for perpendicular polarization except that it will be multiplied by a the cosine factor.

For a real refractive index, the total scattering cross-section and scattering efficiencies are therefore

$$
\begin{aligned}
C_{\text{scat}} &= \frac{8}{3}\left|\frac{n^2-1}{n^2+2}\right|^2 \frac{\pi}{k^2} x^6 \\
&= \frac{8}{3}\left|\frac{n^2-1}{n^2+2}\right|^2 x^4 \pi a^2
\end{aligned}
\tag{1.59}
$$

$$
\begin{aligned}
Q_{\text{scat}} &= \frac{8}{3}\left|\frac{n^2-1}{n^2+2}\right|^2 x^4 \\
&= \frac{8}{3}\left|\frac{n^2-1}{n^2+2}\right|^2 (2\pi)^4 \frac{a^4}{\lambda^4}
\end{aligned}
\tag{1.60}
$$

Note that the scattering cross-section and efficiency for small particles scale as the inverse fourth power of the wavelength. This is the well-known Rayleigh relationship. Thus, at a given wavelength, the scattering cross-section increases for very small particles as the sixth power of the particle radius. The scattering efficiency therefore increases as the fourth power of the particle size. We will see later in this book that as the size parameter $x = kr$ becomes larger than 1, the rate of increase of the scattering efficiency will diminish: the scattering efficiency becomes proportional to the second power of the particle size. As the particle becomes even larger and the size parameter further increases, the scattering efficiency first oscillates about and eventually settles to a value close to 2. We will later see that the value of this asymptote can be derived by using diffraction theory and the Babinet principle.

The scattering of sunlight by air in the atmosphere follows such an inverse power law in respect of the wavelength of light. It also follows the polarization-related angular dependence. The power law explains, among other effects, the blue color of the sky (in conjunction with the spectral response of the human eye) and the polarization distribution of skylight. As we discussed before, Rayleigh originally believed that the presence of small aerosols accounted for these effects. Rayleigh was queried by Maxwell himself in 1871 on the possibility of measuring the size of molecules by using his formulae for scattering (*Kerker* 1969, pp. 30–31). Having studied the problem, he became aware of the coherent nature of the scattering from molecules and of the fact that they have a minimal contribution to atmospheric scattering if no account is taken of the variation of index induced by spontaneous density fluctuations. A completely satisfactory formulation would have to wait for *Smoluchowski*'s (1908) and Einstein's (1910) contributions.

It should be noted that for this scattering from fluctuations in air and most gases at atmospheric density, the angular dependence as a function of polarization remains the same dipole type as that given in (1.49) to (1.54). This is due to the overwhelming probability of occurrence of the smaller fluctuations and to the fact that the observed dipole angular dependence arises mainly from the uniformity of the electric field of the exciting wave across the largest dimension of the randomly oriented scattering volumes.

In the case of water, a small residual component of the parallel scattering component can be seen at 90° to the incoming beam. This results from a small anisotropy of the polarizability caused by the interaction of groups of water molecules. The scattering correction due to this term is called the Cabannes factor and takes the following form (*Kerker* 1969, pp. 574–594):

$$(1+\cos^2\theta) \Rightarrow \frac{6+6\delta}{6-7\delta}\left(1+\frac{1-\delta}{1+\delta}\cos^2\theta\right) \tag{1.61}$$

where δ is the anisotropy factor. This parameter is particularly difficult to measure because of its smallness (~0.039 according to *Farinato* and Roswell 1975) and the amount of stray light present in most experimental setups.

As was stated above, the observed dipole-type angular dependence of light scattering arises mainly from the uniformity of the electric field of the exciting wave across the largest dimension of the scattering particles. Indeed, this angular dependence can also be seen in the low-energy limit of the QED solution to the scattering cross-section of a free electron (*Feynman* 1962). The same scattering pattern holds for the scattering of photons by bound electrons. For interaction with a bound electron, the amplitude of the cross-section will also depend on the binding energy and the spectrum of states of the molecular or atomic potential well in which it resides. This dipole-type angular pattern is also precisely the one that is used in the molecular optics approach to derive the refractive index and its relationship to transmitted and refracted waves.

1.7. The diffraction approximation

There are some conditions under which light can be approximated as a combination of two independent, uncoupled scalar waves, one for each polarization state. The scattering solutions are then identical with the proviso that the results of the scalar wave we associate with the polarization component parallel to the scattering plane must be multiplied by the cosine term. Particles with a relative index close to unity have exactly this behavior. The vast majority of the particles found in water are in this category.

One can initially neglect polarization effects in this approximation because the two polarization waves are coupled only at the surface of the particle. Consider a particle that is large compared to the wavelength and for which we can use the geometric optics approximation. In that case, the coupling problem is due to the different reflectivity as a function of polarization at each point of the surface of the scattering particle. In general, a ray will have its polarization rotated after it exits from the particle because of the different amplitudes reflected and transmitted at the surface. This result follows directly from (1.28) and (1.29) and some elementary geometry. As shown in (1.26), if the relative refractive index is close to unity, the total amount reflected at the interface will be very small. In fact, the coupling between polarization states due to scattering vanishes when n approaches 1.

From these arguments, we would expect scalar wave theory to be a useful approximation to the more complex polarized transverse wave theory. This is often the case. A further simplification occurs if we realize that in scattering theory we are usually not interested in an accurate prediction of the details of the field near the particle, but only in a calculation of its asymptotic behavior far away from the particle.

The study of the propagation of scalar waves through or around obstacles at distances far removed from the obstacle is called diffraction theory. Quite a few of its results will be extremely useful in the analysis and interpretation of the behavior of light scattering by marine particles.

Fresnel proposed the first ad hoc version of diffraction theory in 1820. Kirkchoff (1824–1887) derived a more rigorous and exact version of the theory 40 years later (e.g., *Hecht* 1987). The simple form Fresnel gave the theory assumes that, starting from a given wavefront, the behavior of the propagating wave can be computed by treating each element of the wavefront surface as the source of a spherical wave that interferes with all the other spherical waves coming from that surface. When no obstacles to the wavefront are present, this leads to standard geometrical optics results. When an obstacle is present, part of the wave is cancelled, and the remaining portion propagates (diffracts) in the obstacle's shadow zone.

1.7.1. Scattering by an aperture

Consider an infinite plane wave propagating around and through an obstacle such as a particle. Let us first assume that the obstacle is absorbing and that no light propagates through it. It is a simple matter to compute the amplitude at every point of the wavefront immediately after the obstacle. The points in the geometric shadow immediately after the obstacle have zero amplitude. All the other points of the wavefront are undisturbed. Starting from these conditions, it is now possible to compute the radiation field at infinity as a function of angle. For the case of a circular shadow of radius a, the resulting angular scattering pattern at infinity is given by (*van de Hulst* 1957):

$$S(\theta) = k^2 a^2 \frac{J_1\left(k\,a\,2\sin\frac{\theta}{2}\right)}{k\,a\,2\sin\frac{\theta}{2}} \tag{1.62}$$

$$\frac{I(\theta)}{I_0} = \frac{S^2(\theta)}{k^2 r^2} \tag{1.63}$$

where J_1 is the first-order Bessel function. The last two equations give the behavior of light diffracted around a particle and therefore do not depend on refractive index. Using the optical theorem, (1.44), and (1.62) we immediately find that, for a particle much larger than the wavelength of light, the attenuation cross-section C_{attn} is equal to twice the geometric area and the attenuation efficiency is therefore 2. To compute the scattering cross-section, C_{scat}, we can integrate the scattered intensity (1.63) over all directions in the standard way [see (1.42)]. For a particle much larger than the wavelength, we find that the scattering cross-section is precisely equal to the geometric area of the shadow. The light removed from the geometric shadow (i.e., through complete absorption) contributes a factor of 1 to the attenuation efficiency and the light that is deflected around the shadow contributes the remaining factor of 1. This puzzling equality of the amount of light removed by the geometric shadow and the amount deflected by diffraction around the obstacle creating the shadow is a fundamental property of wave propagation. It applies to diffraction around an object with a geometric shadow of any shape whatsoever. We will shortly see that it is a fundamental consequence of the principle of superposition for the solution of linear wave equations.

If the particle we are studying is partially transparent, then some of the light is refracted, i.e., it passes through the particle. Because of the phase delay due to the refractive index, even a fully transparent particle will produce a distortion of the wavefront which will show up as a distribution of phase difference in the "shadow" zone behind the particle. If there were some simple way to compute the distribution in amplitude and phase difference with respect to the incoming wave due to a partially transparent particle, we could replace the shadow zone with an amplitude and phase distribution and then solve again for the scattering pattern at infinity.

We could also solve the problem as the sum of a diffraction term and a refraction term. The contribution to the amplitude at infinity of the light that was diffracted around the particle is first computed by assuming the particle is fully absorbing. In that case, the light field amplitude is set to zero at every point in the shadow zone and is undisturbed from its initial value everywhere else in the plane immediately behind the particle. The amplitude in the far field due to this situation is then calculated. Then, for a partially transparent particle, the contribution due to the pure refracted term can also be estimated by computing with some approximate method the amplitude and phase difference distribution in the shadow zone and by setting the amplitude and phase to zero everywhere in the rest of the plane. The far field solution is then computed again. Because this term is dependent on the phase difference accumulated by passing through the particle, it will explicitly involve the refractive index. The full solution of the scattered light field amplitude at infinity due to the diffraction and refraction is then obtained by simply summing the result of both cases. We are allowed to do this because of the linearity of diffraction theory that allows the superposition of solutions. This superposition principle also applies to full electromagnetic theory. In fact, we used it already in discussing incoherent scattering.

1.7.2. Babinet's principle and large particle scattering

The principle of superposition leads to another very important result in scattering theory. The total amount of light absorbed and scattered by a particle much larger than the wavelength is precisely twice the amount that one would calculate from the geometric area of the shadow. This means that in the limit of very large particles which do not absorb light (the refractive index is purely real), the scattering efficiency is 2. This result is a direct consequence of what is called Babinet's principle (*van de Hulst* 1957, p. 105).

This principle is based on the following argument involving a simple superposition experiment. In the first case, replace the particle with an opaque disk with the same size and shape as the geometric shadow of the particle and call the resulting scattered amplitude function Ψ_1. In the second case, compute the amplitude function Ψ_2 resulting from an infinite opaque screen with a hole in the same location and with the same shape and size as the disk of the previous case. By the principle of superposition, the sum of the amplitudes Ψ_1 and Ψ_2 must be equal to the initial undisturbed wave amplitude Ψ_0. We therefore have

$$\Psi_1 = \Psi_0 - \Psi_2 \tag{1.64}$$

As we have seen previously in the derivation of the optical theorem (1.44), the total field after a scattering event is always given by the sum of the incident wave amplitude, in this case Ψ_0, and the scattered wave amplitude. From (1.64), we can see that the amplitude of the scattered wave for the opaque disk is equal to minus the amplitude Ψ_2 of the wave that passed through a hole of the same shape in an opaque screen. The total amount of energy going through this hole is obviously equal to its surface area times the intensity of the incident wave $|\Psi_0|^2$. The implication of this result is that the amount of energy removed from the initial beam by the light diffracted around a particle is the same as the amount of energy that passes through an aperture of the same area.

If a particle is sufficiently large, we can safely assume that it will remove from the incident beam all the light that fall on it by either absorption or scattering. The contribution of this term to the total attenuation cross-section will be equal to the area of the shadow of the particle. Since, as we have just demonstrated, the amount of light diffracted around the particle is the same as the amount that passes through an aperture of the area of the shadow of the particle, the diffraction contribution to the total attenuation cross-section will also be equal to the geometric area of the shadow. From the previous arguments, the total attenuation cross-section for large particles of arbitrary shape will tend to be twice the area of the geometric shadow. The attenuation efficiency of a large particle will therefore approach 2.

$$\begin{aligned} C_{\text{attn}}(\infty) &\rightarrow 2A_{\text{shadow}} \\ Q_{\text{attn}}(\infty) &\rightarrow 2 \end{aligned} \tag{1.65}$$

As pointed out by *Bohren* and Huffman (1983), the diffracted light contribution to attenuation is concentrated in a narrow range of angles near the forward direction. This implies that the acceptance angle of equipment designed to measure the attenuation from large particles must be much smaller than the width of the forward diffraction peak [see (1.63)].

The limiting behavior described in (1.65) is extremely general and applies to particles of all shapes and compositions. Many particles in the natural waters are very large when compared to the wavelength, and their contribution to the scattering coefficient of these waters is often dominant. In those cases, Babinet's principle leads to several very significant general results.

As we will see later, for particles with a relative refractive index close to unity, the diffraction term, the refraction term, and terms arising from their mutual interference are by far the dominant contributions to the scattering at angles less than 90°. For the large particles, i.e., those with characteristic dimensions of the order of a few wavelengths of light and more, the diffraction term is dominant in the forward direction and in the first few degrees around it. The first-and higher-order refractive terms then take over, and they account for most of the scattering function over the balance of the forward scattering hemisphere. One further effect must be accounted for to model the scattering of the large particles in and around the backward direction. One must explicitly take into account the contribution to the scattering pattern of the reflection from both the front and back surfaces of the particle. As we will show later, this can be done to a reasonable level of accuracy by using geometrical optics and avoiding singularities by accounting for simple diffraction effects.

Diffraction theory is extremely useful in developing simple models that allow one to understand the origin of some of the features of scattering functions of seawater. It is also useful in allowing one to compute approximate results for complex shapes or for particles of very large sizes where there is no exact solution or where numerical solutions are extremely difficult or outright impossible to obtain with present-day computing power.

1.8. Conclusion

This concludes our very succinct survey of the basic theory of the interaction of light with matter. We have merely sketched the fundamental concepts involved, because we considered the clarity and simplicity of the explanations to be more important than completeness. We quoted only the most important results, and all of those are required to understand and interpret the data and experiments that we will present in the following chapters. The interested reader will find in the references a simple but fairly complete presentation of this fascinating subject that, by definition, attempts to account for all we see around us.

1.9. Problems

Derive the expression for the number density of photons states per unit volume [equations (1.4) and (1.5)] by using the wave continuity condition: the slope and amplitude of the incoming wave into the unit volume must match the slope and amplitude of the outgoing wave from the unit volume, i.e., periodic boundary conditions with an arbitrary phase: $\exp[ik_x\ (x+L) = \exp[ik_x\ x]$ applied to three dimensions (x, y, and z), where L is the edge length of the unit volume cube.

Chapter 2

Optical properties of pure water, seawater, and natural waters

2.1. Introduction

Many constituents are involved in the interaction of light with naturally occurring bodies of water. First and foremost, there is water itself, which even in its purest form exhibits a complex absorption spectrum and a significant amount of scattering from refractive index fluctuations. The addition of the various salts, which are present in seawater, gives rise to an extra absorption in the far ultraviolet (UV) and an increase of the amount of scattering due to the occurrence of small variations in salt concentration that result in additional refractive index fluctuations.

The second most significant optical component of natural waters is a mixture of various dissolved organic matters (DOM) known collectively as yellow substance, also referred to as *Gelbstoff* that is the German name for yellow substance. Its yellow color in sunlight comes from the strong absorption in the UV and blue regions of the spectrum of the various complex organic compounds that make up this residue of biological activity. This yellow substance originates from the metabolism and breakdown of various living organisms. Some of it is leached from land and carried to the ocean by rivers and surface waters. Some of it, produced by the breakdown of viruses, bacteria, and plankton, is the direct byproduct of biological activity in the open ocean. Concentration of this substance varies from the almost negligible in the purest oceanic waters to significantly high near the coasts, changing the color of the water column from its characteristic mid-ocean deep blue to green. Its presence is the primary factor controlling the spectrum of visible light as a function of depth in the ocean.

Large quantities of suspended particles are also present in all natural water bodies. They consist of both biological components such as plankton and mineralogical matter from crushed rocks such as quartz, silica, sand, and silt. These particles are in a sense the main subject matter of this book since they are almost always the dominant source of the scattering found in water. Their contribution to the total scattering occurring in natural waters is much larger than that due to the only other term Einstein–Smoluchowski scattering due to the refractive index fluctuations. These suspended particles are large enough for the concept of refractive index to apply to their material. Thus, their scattering properties can be studied by using standard electromagnetic theory concepts.

As mentioned previously, the shape of many of these particles can be approximated by those of cylinders, platelets, oblate, and prolate spheroids. All such shapes are present in significant numbers. Almost all these particles have a low relative refractive index that allows for some very significant further simplification of the scattering solutions. The number density per unit volume of water of these particles is also low enough that the rules of incoherent scattering apply to them.

In this chapter, we will study the scattering and absorption from both pure and salt water. We will also briefly look at the effect of DOM on the absorption spectrum. An accurate picture of these properties is important for any further work since the effects of these components must be removed from the scattering and absorption data for a dispersion of particles before it can be analyzed and compared to a light scattering theory.

Furthermore, as we have seen in Chapter 1, the theoretical solutions for light scattering by a particle are based on the ratio of the complex refractive index of the particle to the complex refractive index of the surrounding medium. Hence, accurate values of the refractive index for the medium are a prerequisite for any work.

We will first give a simple picture of the molecular structure of water. We will retain only the elements which will help in understanding the theoretical basis of the behavior of the optical properties. As mentioned by many authors, if water was not such a common substance, it would be the subject of endless fascination for its bizarre physical and optical properties. It perhaps should not be overly surprising that the substance which sustains life on this planet should show such complexity. For example, it is one of the only substances which is less dense as a solid than as a liquid. This among other things prevents the water bodies in cold climates from freezing from the bottom up. It is also a highly efficient polar solvent, which presents extreme difficulties to experimentalist trying to obtain samples pure enough for reliable measurements of absorption and scattering. On the positive side, this same efficiency as a solvent allows enough DOM to be present in even the purest natural waters to protect marine organisms from excessive irradiation by UV light.

2.2. Physical properties and the intermolecular potential

The water molecule consists of an oxygen atom core with two hydrogen atoms attached at an angle of 102.5° at a distance of 0.0957 nm (*Kjaergaard* et al. 1994) when at rest with respect to each other. The molecule is polar because there is a slight excess of positive charge at the tip of the hydrogen bonds and a compensating excess of negative charge on the back of the oxygen core. What is the best model for the potential acting between water molecules is still a subject of current research in chemical physics (*Errington* and Panagiotopoulos 1998, *Wallqvist* and Berne 1993). When used in conjunction with simple many-particle molecular dynamics simulations, such model potentials now give an adequate account of most of the physical properties of water. In all models, the charge distribution of the water molecule is represented by simple coulomb or inverse radial potentials centered at points slightly offset from the atomic centers. On top of this set of coulomb potentials, one adds a central potential with a repulsive core and an attractive outer portion. This potential accounts for the van der Walls forces that produce the adhesion of molecules in a liquid. The repulsive core prevents the molecules from collapsing on top of one another. A schematic of the repulsive core of water is shown in Figure 2.1.

The simplest mathematical expression of this potential is called a Lennard–Jones 6–12 potential since the core is modeled by an inverse twelfth power repulsion and the outer portion an inverse sixth power attraction.

$$V_{\mathrm{d}} = 4\varepsilon\left[\left(\frac{\chi}{r}\right)^{12} - \left(\frac{\chi}{r}\right)^{6}\right] \tag{2.1}$$

For water $\chi = 0.31655\,\mathrm{nm}$ and $\varepsilon = 543.5\,\mathrm{cm}^{-1}$. This potential explains satisfactorily the compressibility, surface tension, viscosity, and other physical and thermodynamic properties of water. We will also see later that the energy change

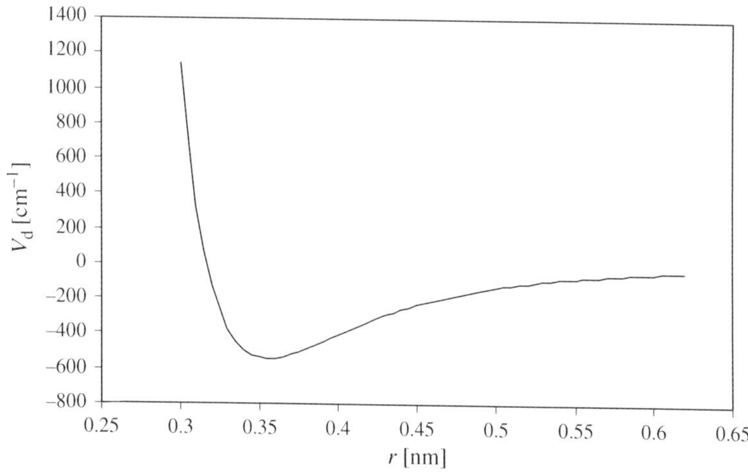

Figure 2.1. Lennard–Jones 6–12 potential joining a pair of water molecules. The energy is expressed in inverse wavelength units. The equilibrium point (the minimum of $-543.4\,\mathrm{cm}^{-1}$) is at 0.355 nm and the binding energy is $543.5\,\mathrm{cm}^{-1}$. This type of dimer potential explains most of the physical properties of water and some of the radiative properties.

during collisions between water molecules due to this intermolecular potential explains the far wings of the electronic absorption spectrum of water. The spectral wing due to this effect is the source of the dominant term of the ultraviolet absorption of water from 200 to 380 nm.

The geometry of the charge distribution accounts for the tetrahedral structure of ice crystals. The hydrogen atom of one molecule is attracted by the oxygen atom of the other molecule and simultaneously repulsed by its hydrogen atoms. This leads to a structure where the second molecule attaches itself in a plane at 56° to the plane of the first molecule (*Dera* 1992, pp.59–64). This type of bond between water molecules is called a translinear hydrogen bond. It forms the most stable pair of water molecules. Such pairs are called a water dimer. The angle between the molecular planes of a single dimer is close to 60°. Thus groups of water dimers form a tetrahedral structure. Such structures are the building blocks of ice crystals. Surprisingly, traces of this tetrahedral grouping subsist in the liquid state even at temperatures far removed from the melting point. This leads to the existence of large open clusters of water molecules in the liquid state. As the temperature of liquid water increases, these clusters break up into smaller units that can be more closely packed. This clustering and breakup phenomenon is the explanation for the increasing density of water from 0 to 4°C. As the clusters break up due to increased thermal motion, the water molecules pack themselves closer, leading to a higher density of the liquid. In the limit, one could consider that water contains two types of entities, clusters on the one side and single molecules on the other. As temperature increases, the number of clusters diminish and the number of single water molecules increases. This mechanism has recently been invoked to explain some of the opposite temperature dependence of several features in the red and near infrared absorption spectrum of water (*Pegau* et al. 1997).

2.3. Radiative properties and the intramolecular potential

The infrared and visible portion of the absorption spectrum are explained by the vibrations and rotations of the hydrogen nuclei within the water molecules, while the far UV spectrum is due to both a transition between the electronic energy levels of a single molecule and its broadening by a collision interaction with another water molecule. Figure 2.2 shows the three modes of vibration of the water molecule. When both hydrogen atoms oscillate simultaneously toward and away

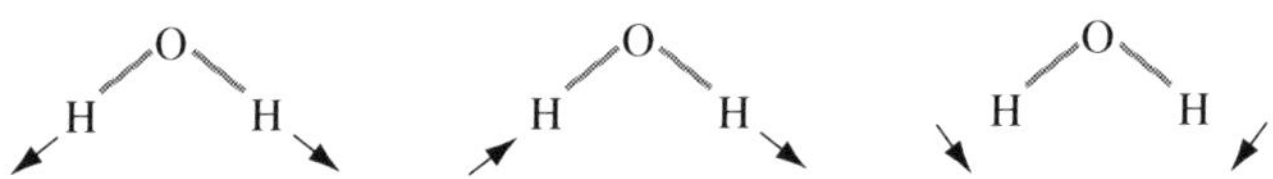

Figure 2.2. Normal modes of vibration of the nuclei of the water molecule. From the left: the symmetric stretch mode, the asymmetric stretch mode, and the transverse or scissors mode.

from the oxygen atom, we have the symmetric stretch mode. When one hydrogen atom moves toward the oxygen core while the other moves away from it, we have the asymmetric stretch mode. When both hydrogen atoms move toward and away from each other while maintaining their distance from the oxygen core, we have the transverse or scissors mode. The transverse mode involves the expenditure of much less energy than either of the stretch modes. These vibrational motions of the atomic nuclei within the molecule occur because of the form of the potential joining each of the atomic nuclei to the other nuclei of the individual molecules.

We will now sketch the basic elements of quantum theory required to explain the significant features of the absorption spectrum of water. Our aim is to obtain simple formulas whose functional forms are constrained by appropriate theoretical considerations. We will attempt to retain all the significant features of the phenomena while simplifying the model as much as possible.

Going back to the fundamentals outlined in Chapter 1, let us first consider the problem of an electron bound to an atomic nucleus. The full solution involves the Dirac equation with its four components. Two of the components are only significant at high energies and represent the relativistic correction to the equations of motion of the electron. At low energies, the remaining two components are given by the Pauli equation (*Feynman* 1962, pp. 6–10). These two components can be further simplified to the solution of a single equation, the Schroedinger equation, multiplied by a set of 2 by 2 matrices. The matrices handle the symmetry and anti-symmetry relations required to properly model the half integer spin of the electron. The solution of the Schroedinger equation gives the probability amplitude $\psi(\mathbf{r}, \mathbf{t})$, generally a complex function, of finding the electron at any point in space–time. The product of this probability amplitude and its complex conjugate $\psi\psi^*$ gives the actual probability of finding the electron, a positive definite quantity. This probability is called the wave function.

When solving for the motion of an electron in the potential around a nucleus, one finds that the electron can only exist for a significant time in a set of discrete energy levels $\psi_i(\mathbf{r})$. These are the stationary states of the atom. If there were no perturbation, an electron would remain indefinitely in one of these states of motion. The electromagnetic field of the photon can induce transitions between these energy levels. The probability of a transition from state n to state m due to the perturbation of a photon can be computed to first order by evaluating the dipole moment $\mathbf{M}$ induced by the transition between the states (*Herzberg* 1950, pp. 18–22).

$$\mathbf{M} = \sum_{\mathrm{i}} e\mathbf{r}_{\mathrm{i}} \tag{2.2}$$

Then a transition matrix $\mathbf{R}^{nm}$ can be constructed as follows:

$$\mathbf{R}^{nm} = \int \psi_n \, \mathbf{M} \, \psi_m \, d\tau \tag{2.3}$$

Finally, the absorption coefficient, a, can be expressed with the following equation:

$$a = N_m \frac{8\pi^3 \nu_{nm}}{3h\ c} |\mathbf{R}^{nm}|^2 \tag{2.4}$$

where e is the electron charge, N_m the number density of atoms in state m, and ν_{nm} the frequency corresponding to the energy difference between state n and state m. The integrals are carried out over all space for a single electron. In the case where several electrons orbit around a nucleus, the wave functions must be solved by accounting for the mutual influence of the electrons on each other.

The exact, multidimensional wave function for a multielectron atom gives the probability of jointly finding electron 1 at position 1, electron 2 at position 2, etc. The integrals in (2.3) are then carried out over all the $3j$ dimensional configuration space for j electrons. This complexity has led to a plethora of approximate methods of solution. Most of these methods find a systematic way of representing the influence of the inner electrons of the atom on the outer electron as an effective potential in which this electron orbits. The trick is in making sure that this potential is self-consistent with the motion of all the other electrons. For atoms, the best known of these approximations is called the Hartree–Fock method (e.g., *Hurley* 1976).

The situation is obviously substantially more complicated for molecules where at least two nuclei are involved and one or several electrons can contribute to binding them together. In the case of an atom, the problem could simply be solved in the center of mass frame of reference. Because of the large mass difference between the nucleus and the electrons, the effect of nuclear motion could be neglected to a high order of accuracy. This is obviously not a viable option for a molecule where the nuclei can move with respect to one another.

Fortunately, Born and Oppenheimer found that the nuclei of a molecule move very slowly in comparison with the electrons; thus, one could de-couple the motion of nuclei and electrons. The nuclei, to a substantial accuracy, could be considered to move in a potential well created by an instantaneous readjustment of the binding electrons. A reasonable method of solution consists in fixing the nuclei in a given spatial arrangement with respect to each other and computing the binding energy due to the electrons. After a small displacement of the nuclei, the binding energy is again computed. In this fashion, an n-dimensional map of the binding potential is built up. This approximation implies that the wave function can be represented as the product of an electronic term that gives the probability amplitude of finding the binding electrons in a given set of positions for each position of the nuclei with respect to each other and a term that gives the probability amplitude of finding the nuclei at that precise relative position:

$$\psi_{m,v,j}^{tot} = \psi_m^{el}(r, \mathbf{r}_i)\psi_{m,v,j}^{vib-rot}(r) \tag{2.5}$$

For a molecule with two nuclei such as O_2, the electronic potential energy binding the two nuclei together or repelling the two nuclei can be completely represented on a one-dimensional graph with the internuclear spacing as the ordinate. For a molecule of water H_2O, a two-dimensional map of the potential is required. The relative motion of the nuclei in this potential well is then computed according to the rules of quantum mechanics. The equation for $\psi_{m,v,j}^{vib-rot}(r)$ is the same as the standard Schroedinger equation. However, in this case, the potential energy term is pre-computed from a set of solutions of the electronic part of the wave function, for all possible relative positions of the nuclei. This wave function is used to describe in detail the vibrational and rotational states of motion of the molecule.

2.3.1. Basics of electronic molecular transitions

Another very important consequence of the instantaneous rearrangement of the electrons relative to nuclear motion is that, during the transition of an electron from one electronic energy level to another, the nuclei do not move with respect to one another. This is called the Franck–Condon principle, and it has important consequences for the spectral shape of molecular electronic transitions.

Figure 2.3 is a simplified one-dimensional schematic energy diagram of the internuclear potential of the ground X^1A_1 and the $^1B_1$ first excited electronic states of water (*Quickenden* and Irvin 1980) as a function of the oxygen–hydrogen internuclear separation. In reality for water, we should be talking about energy surfaces. This would however unduly complicate the arguments without changing the important conclusions. For the sake of simplicity, we will in future discuss molecular absorption phenomena with reference to diatomic bond between pairs of atoms. Any significant problems arising from this simplification will be noted.

In the case of liquid water, this is in fact a very good approximation, and it corresponds to what is called the local mode model (*Kjaergaard* et al. 1994). The local mode model treats water as a jumble of loosely coupled O–H bonds. The coupling between the O–H bonds is handled by a one-dimensional representation equivalent to the bending or scissors mode of water. This model works well because, as we have seen, there is substantial coupling between many water molecules, and the net effect is to make the ensemble behave to a first approximation like a soup of diatomic bonds.

The $X^1A_1 \rightarrow {}^1B_1$ transition shown in Figure 2.3 gives rise to an absorption continuum centered at 176 nm in water vapor and at 147 nm, i.e., shifted to the UV, in liquid water. Note that the nuclei in the lower electronic level are attracted to each other, while the nuclei in the upper electronic level are always repelled from one another. This simply means that a molecule in that upper state dissociates. According to the Franck–Condon principle, electronic transitions are represented by a vertical line since the nuclei do not move during the transition.

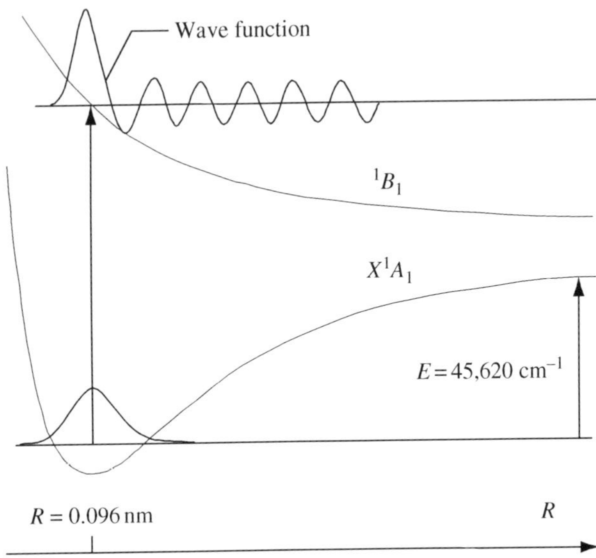

Figure 2.3. Schematic diagram of the energy levels involved in the first electronic transition in water. The stable potential for the X^1A_1 lower state is approximated by the Morse potential of the O–H stretch mode of the water molecule in the local mode model. The energy is expressed in inverse wavelength units. The equilibrium point is at the internuclear distance $R = 0.096\,\text{nm}$ and the binding energy is $45\,620\,\text{cm}^{-1}$. The upper $^1B_1$ state is unstable and leads to immediate disassociation. The wave function of the bound lowest vibrational level is represented schematically as a Gaussian and the wave function of the free upper state as a plane wave, with a peak amplitude at the classical turning point of the potential. Following the Franck–Condon principle, the electronic transition is represented as a vertical line at $R = 0.096\,\text{nm}$.

As a consequence, the transition matrix element between diatomic electronic states is given by:

$$\mathbf{R}^{nm} = \int \psi^{vib-rot}_{n,vn,jn}(r)\ \mathbf{M}_{\text{nm}}(r)\ \psi^{vib-rot}_{m,vm,jm}(r)\ dr \tag{2.6}$$

where $\mathbf{M}_{nm}(r)$ is computed for each internuclear spacing by integrating over the electron configuration at that spacing.

$$\mathbf{M}_{nm}(r) = \int \psi^{el}_{n}(r, \mathbf{r}_{\text{i}})\left[\sum e\mathbf{r}_{\text{i}}\right]\psi^{el}_{m}(r, \mathbf{r}_{\text{i}})\ d\mathbf{r}_{\text{i}} \tag{2.7}$$

Internuclear motion can be decomposed into vibration and rotation components and the wave functions in (2.6) are the wave functions describing this vibration–rotation motion of the nuclei relative to one another. The Franck–Condon principle is expressed by the use of the same internuclear radius value in both the lower and upper electronic states.

As in the case of the electronic states, there are stationary bound states of nuclear vibration and rotation (*Herzberg* 1950, pp. 66–145). The vibration–rotation wave functions are obtained by the solution of a Schroedinger-type equation for the motion of the nuclei in the potential well of the binding electrons. We can model the potential of a bound state near its minimum as a simple harmonic oscillator. The solution for the lowest energy state of this bound oscillator is given by:

$$\psi_0^2 = \sqrt{\frac{\alpha}{\pi}} \exp[-\alpha(r - r_0)^2] \tag{2.8}$$

$$\alpha = \frac{4\pi^2 \mu \nu_{osc}}{h} = \frac{2\pi\sqrt{\mu n''}}{h} \tag{2.9}$$

$$V = \frac{n''(r - r_0)^2}{2} \tag{2.10}$$

$$\mu = \frac{m_1\, m_2}{m_1 + m_2} \tag{2.11}$$

where V is the assumed internuclear potential, μ the reduced mass of the molecule whose atoms have masses m_1 and m_2, and ν_{osc} the frequency of emission of the fundamental vibrational transition. This solution is generally a good description of the lowest vibrational state. The internuclear potential is in fact asymmetric, and the wave functions for the higher vibrational states will progressively depart from the harmonic model. A more general model for which there are analytic solutions is given by the Morse potential. As we will later see, this form of the potential accounts very nicely for the effects of asymmetry and the finite well depth of the potential. However, for the lowest energy state of deep wells, its solution also tends to a Gaussian.

For a free state, such as first electronic upper state of water, the wave function tends at large nuclear separation to the plane wave solution representing the free motion of a particle. At small nuclear separations, the relative potential energy of the free molecular components rises. Near the point where this potential energy becomes equal to the relative kinetic energy, the relative motion of the free components of the molecule slows down and ultimately reverses direction. This means that the components spend a considerable time at that location, and the wave function has a correspondingly large peak (large probability) near the intercept internuclear distance (Figure 2.3). In fact, this is the only area from which there is a significant contribution to the integral in (2.6). Everywhere else, the contribution to the integral of the positive half cycles of the plane wave is almost perfectly cancelled by their negative counterparts. This implies that in practice, we can approximate to first order the wave function of the upper state by a Dirac delta function centered at the intercept point r. This leads to the following result for the absorption coefficient a:

$$a = N_m \frac{8\pi^3 \nu_{nm}}{3h\ c} \left| \psi^{\text{vib-rot}}_{m,vm,jm}(r) \mathbf{M}_{nm}(r) \right|^2 \tag{2.12}$$

The potential of the upper state decreases with increasing internuclear separation. We can associate with each internuclear radius a transition frequency proportional to the energy difference between the bound ground state and the potential energy of the dissociative upper state at that separation. Assuming that this upper state potential can be approximated by an inverse square function of internuclear distance in the range of separations where there is a significant probability of finding the molecule in the bound lower state leads to the following approximation for the absorption spectrum:

$$a(\nu) = N_m \frac{8\pi^3 \nu}{3h\ c} \sqrt{\frac{\alpha}{\pi}} \exp[-\alpha(r - r_0)^2] \, |\mathbf{M}(r)|^2 \tag{2.13}$$

with

$$\nu = \frac{K_n}{r^2} \tag{2.14}$$

$$r = \sqrt{\frac{K_n}{\nu}} \tag{2.15}$$

$$r_0 = \sqrt{\frac{K_n}{\nu_0}} \tag{2.16}$$

If we further assume that the electronic dipole moment does not vary significantly over the zone where the integral is significant we obtain:

$$a(\nu) = N_m \frac{8\pi^3 \nu}{3h\ c} |\mathbf{M}(r_0)|^2 \sqrt{\frac{\alpha}{\pi}} \exp[-\alpha\, K_n (\frac{1}{\sqrt{\nu}} - \frac{1}{\sqrt{\nu_0}})^2] \tag{2.17}$$

Equation (2.17) is the basis for the following rule of thumb: far away from its peak, the absorption coefficient of a continuous molecular electronic transition decreases approximately as an exponential of the frequency. This behavior is called Urbach's rule (*Quickenden* and Irvin 1980). We will use it later in this chapter to explain certain features of the UV absorption spectrum of water.

2.3.2. The vibrational component of molecular transitions

As we have seen above, there are stationary bound states of nuclear vibration and rotation (*Herzberg* 1950, pp. 66–145). Figure 2.4 is a simplified schematic of the distribution of vibrational levels and their rotational sublevels.

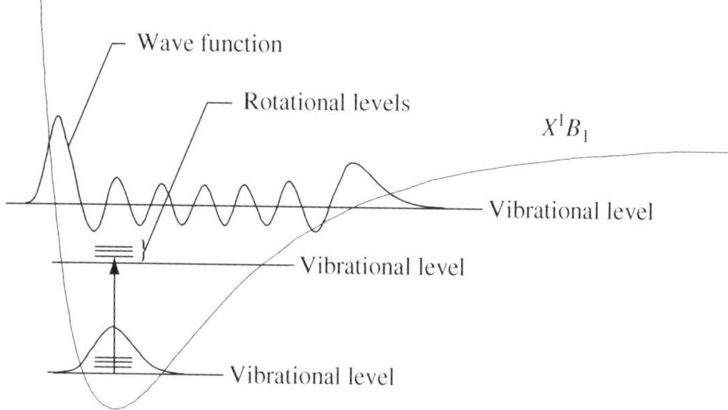

Figure 2.4. Schematics of the vibrational-rotational energy level structure of the lower X^1A_1 electronic state of water. Once again, the Morse potential of the O–H stretch mode of the water molecule in the local mode model is used. The wave function of the lowest vibrational level is represented schematically as a Gaussian and the wave function of one of the higher vibrational level is shown schematically as an oscillating function with asymmetric peaks in amplitude at the classical turning points of the potential.

The vibration–rotation wave functions are obtained by the solution of a Schroedinger-type equation for the motion of the nuclei in the potential well of the binding electrons. Dipole transitions between these states can occur. However, for a molecule, the dipole moment is now a function of internuclear separation and is evaluated by the formula in (2.7) except that the same electronic function is used for both lower and upper states

$$\mathbf{M}_m(r) = \int \psi_m^{el}(r, \mathbf{r}_i) \left[\sum e\mathbf{r}_i \right] \psi_m^{el}(r, \mathbf{r}_i)\, d\mathbf{r}_i \tag{2.18}$$

In a molecule that dissociates into neutral components, the dipole moment will rise from zero at small internuclear separation to a maximum and then fall to zero as the separation increases. For the purposes of calculation, the dipole moment is generally assumed to vary approximately linearly over the region where the vibrational wave functions have significant amplitudes (*Herzberg* 1950). The probability of a transition between the vibrational states of the same electronic state is thus given by:

$$\left| \mathbf{R}^{\text{vib-rot}} \right|^2 = \left| \int \psi_{v,j}^{\text{vib-rot}}(r)\, \mathbf{M}_m(r)\, \psi_{v',j'}^{\text{vib-rot}}(r)\, dr \right|^2 \tag{2.19}$$

By assuming that the dipole moment varies linearly with intermolecular radius and substituting the wave function solutions of the harmonic oscillator into (2.19), it can be shown that transitions can only occur between vibrational energy levels whose quantum number differs by plus or minus one unit. The transitions from the lowest $v = 0$ state to the $v = 1$ state are called the fundamental mode of vibration.

An asymmetric function is in fact a better fit to the actual internuclear potential. In that case, the solution of (2.19) shows that there is a small probability of a transition occurring between any two vibrational levels. The transition from the lowest vibrational level $v = 0$ to $v = j$ is called the jth-1 overtone of the mode. The probability of a transition drops precipitously as the difference in the vibrational quantum number increases.

As we mentioned before, a reasonable approximation of the actual internuclear potential is given by the following Morse potential:

$$V = U_0[1 - e^{-A(r-r_0)}]^2 \tag{2.20}$$

for which there exist analytic solutions of the vibrational wave functions. Parameter U_0 denotes the dissociation energy. Distances r and r_0 are defined by (2.15) and (2.16). The notation we will adopt follows that of *Nieto* and Simmons (1979). The Morse potential has the following vibrational energy level distribution:

$$G(v) = \omega_e(v + 1/2) - \omega_e x_e(v + 1/2)^2 \tag{2.21}$$

The parameters ω_e and $\omega_e x_e$ are usually measured by molecular spectroscopy. The radial asymmetry of the potential is what leads to the non-linear second term in the energy level distribution. The observed vibrational overtone spectrum is therefore given by:

$$\Delta\nu_{n,0} = j\,[(\omega_e - \omega_e x_e) - \omega_e x_e\, j] \tag{2.22}$$

Patel and Tam (1979) use a formula of this type to explain spectral location of the shoulders and peaks observed in the red and near infrared part of the absorption spectrum of liquid water. Using the local mode model, the same analysis was carried out by Kjaergaard and colleagues for the infrared overtones of water vapor (*Kjaergaard* et al. 1994).

The observed frequency intervals are related to the parameters of the potential as follows:

$$U_0 = \frac{\omega_e^2}{4\omega_e x_e} = \Lambda^2 \varepsilon_0 \tag{2.23}$$

where

$$\Lambda = \frac{\omega_e}{2\omega_e x_e} \tag{2.24}$$

We also have

$$\varepsilon_0 = \omega_e x_e = \frac{h A^2}{8\pi^2 c\mu} \tag{2.25}$$

Hence

$$A = \sqrt{\frac{8\pi^2 c\mu}{h}}\,(\omega_e x_e)^{1/2} \tag{2.26}$$

Finally

$$\mu = \frac{m_1\, m_2}{m_1 + m_2} \tag{2.27}$$

where μ is once again the reduced mass of the molecule. Two parameters of the potential, U_0 and a, can be fixed from the vibrational spectrum data. The third parameter, r_0, (2.20), can only be fixed by an analysis of the rotational component of the spectrum.

By inserting the analytic solutions for the Morse wave functions in (2.19) and carrying out the integral by numerical or analytic means, we can obtain an expression for the ratio of the intensities of the jth harmonic to the jth + 1 harmonic. This ratio form is convenient, since it does not require us to know the absolute value of the rate of change of the molecular dipole moment of the electronic state.

$$O_j^{j+1} = \frac{\Delta\nu_{j+1,0}}{\Delta\nu_{j,0}}\left|\frac{(2\lambda - 2j - 3)\,(2\lambda - j - 1)\,j^2}{(2\lambda - 2j - 1)\,(2\lambda - j - 2)^2(j+1)}\right| \tag{2.28}$$

In Table 2.1, the overtone intensity ratio is evaluated using the parameters of the equivalent O–H bond in the local mode model for water (the *Theory* column) and compared with experimental data for water vapor (*Kjaegaard* et al. 1994) and for liquid water (*Kou* et al. 1993, *Curcio* and Petty 1951). The general tendencies are well modeled, and it is therefore a reasonable assumption that the features of the absorption spectrum of liquid water in the red and infrared are due to the vibrational overtone modes. In fact, we will later see that such vibration–rotation transitions are the dominant source of the absorption spectrum of water from 450 nm to the infrared.

2.3.3. The rotational component of molecular transitions

We have so far purposely neglected the rotational effects so as to not confuse the issues raised by the physics of vibration. In formal terms, the effect of rotation enters into the vibrational–rotational wave equations as a centrifugal energy term added to the internuclear potential. As the perturbation due to this term is small, we can to a good first approximation de-couple the vibrational solutions from the rotational solutions and approximate the resulting wave function as a product of a pure vibration wave function with a pure rotation term (*Herzberg* 1950, pp. 109–110). More sophisticated approximations can be obtained by using these

Table 2.1. Comparison between theory and experiment for the ratio of the intensities of the overtone transitions of water in both vapor and liquid phases.

Experiment					**Theory**
Water vapor		**Liquid water**			
Overtone wavelength ratio (nm/nm)	***Kjaergaard* et al. (1994)**	**Overtone wavelength ratio (nm/nm)**	***Kou* et al. (1993), *Pope* and Fry (1997)**	***Curcio* and Petty (1951)**	
942/1379	0.031	976/1453	0.015	0.018	0.044
723/942	0.060	755/976	0.059	0.059	0.067
592/723	0.093	605/755	0.082	–	0.094
–	–	514/605	0.154	–	0.120
–	–	449/514	0.250	–	0.154

The first and third columns give the wavelength ratios of the overtone. Note a small wavelength shift between the liquid and vapor states. The intensity ratios (in the *Theory* column) for the vapor phase were computed from the experimental data on oscillator strengths (*Kjaergaard* et al. 1994). The ratios for the liquid phases are the ratios of the intensities measured at the stated wavelengths in the respective references. These wavelengths correspond to either peaks or shoulders in the absorption spectrum. The theoretical results are given by (2.28) with a value of the parameter of 23.6 computed from the Morse potential parameters of the O–H stretch mode of the water molecule in the local mode model.

wave functions in a perturbation expansion of the full equation. The de-coupled wave function solution will be perfectly adequate for the purposes of discussing the absorption spectrum features for liquid water. The many other approximations implicit in the treatment of water as a loosely coupled ensemble of O–H bonds result in much larger errors and uncertainties than any potential contribution of the vibration–rotation coupling terms. The simplifications that follow from this de-coupling make it very worthwhile. Note that the Morse potential can only be solved analytically for the case of no rotation. However, this is clearly not a problem in the present approach.

A body whose angular motion can be approximated by a single dominant moment of inertia is called a simple rotator. This is a very good approximation for diatomic molecules. The situation is of course considerably more complex for a bent tri-atomic molecule such as water. However, it turns out that the results of the simple rotator can be generalized to this case by applying them to each of the moments of inertia axes separately and correcting for second-order coupling. Details can vary, but the results of the pure rotator form the fundamental building block of the vibration–rotation spectrum. The overall spectral behavior of this rotator indicates what are the appropriate functional forms to use in the analysis and modeling of the spectrum. The energy level distribution for this pure rotator is given by:

$$F(J) = BJ(J+1) \tag{2.29}$$

$$B = \frac{h}{8\pi^2 cI} \tag{2.30}$$

$$I = r_0 \frac{m_1\, m_2}{m_1 + m_2} \tag{2.31}$$

where F(J) is the energy of rotational level J, B is the rotational energy constant in units of cm^{-1}, I is the moment of inertia, and r_0 is defined by (2.16). An analysis of the transition moment integrals for the rotational levels shows that the only allowed transitions are those that change the rotational quantum number J by plus or minus one unit.

The actual vibrational–rotational energy levels are given by the simple sum of the vibrational energy in (2.22) and the rotational energy in formula (2.29), with the appropriate vibrational and rotational quantum numbers v and J inserted in their respective equations. The frequency difference between any two vibrational–rotational levels is thus given by:

$$\nu = \nu_0 + B[J'(J'+1) - J''(J''+1)] \tag{2.32}$$

The transitions with $\Delta J = +1$ form a band called the R-branch that lies at higher frequencies than the pure vibrational transition (at $J = 0$).

$$\nu_r = \nu_0 + 2\mathrm{B}\,(J''+1) \tag{2.33}$$

The transitions with $\Delta J = -1$ also form a band called the P-branch that lies at lower frequencies than the pure vibrational transition (at $J = 0$).

$$\nu_p = \nu_0 - 2B\,J'' \tag{2.34}$$

The simple rotational spectrum thus shows up as a series of lines lying on either side of the pure vibrational transition (at $J = 0$). The same rotational R and P branch structure will be found with each vibrational overtone. It is possible to evaluate the relative distribution of intensities in the separate rotational lines by simple statistical mechanics (*Herzberg* 1950, pp.124–128).

$$I_{\mathrm{rot}} \propto \frac{hcB}{KT}\,\nu(J' + J'' + 1)\exp\left[-\mathrm{B}J''(J''+1)\frac{hc}{KT}\right] \tag{2.35}$$

where K is the Boltzmann constant and T is the absolute temperature. The expression above is correct for low-pressure gases and gives two combs of narrow lines above and below the $J = 0$ limit. The P branch has a missing line at $J = 0$. The P and R branch are asymmetric. As the temperature increases, the two bands extend further out from the pure vibrational transition frequency, their peak amplitudes are reduced, and they become more symmetrical. The distance between the peaks is given by a simple formula:

$$\Delta\nu^{\mathrm{p-r}} = \sqrt{\frac{8BKT}{hc}} \tag{2.36}$$

At high pressure or in the liquid state, each rotational line is broadened by the direct effect of collisions, and the discrete substructure is lost. It is interesting to note that even while this substructure is lost, the distance between the peaks can still be used to extract information about the rotational constant. It is possible to approximate the effect of this broadening by first replacing the discrete energy spectrum by its continuous counterpart.

$$I_{\mathrm{rot}} = N_{m,v=0}\frac{8\pi^3\nu}{3h\ c}\left|\mathbf{R}^{\mathrm{vib-rot}}\right|^2 Z_{\mathrm{r}}\left(\left|\frac{\Delta\nu}{2B}\right|\right)\exp\left[\frac{\Delta\nu}{2B}\left(1-\frac{\Delta\nu}{2B}\right)\frac{hcB}{KT}\right] \tag{2.37}$$

The equation above is valid for both P and R branch. $N_{m,v=0}$ is the number of molecules per unit volume in the lowest vibrational state of the electronic level m. Z_{r} is a normalization factor that should be computed such that the integral over all frequencies of the P and R branches is unity. In order to be consistent with the collision broadening model, the gap between the branches must be filled in to some extent. Since the intensity in the near wings of collision broadened lines is inversely proportional to $(\Delta\nu)^2$, and since the spacing between the modes scales as in (2.36), the amount of filling in of the gap must be inversely proportional to the product of rotational constant and temperature. These considerations immediately lead to the following expression:

$$\begin{aligned} I_{\mathrm{rot}} &= N_{m,v=0}\frac{8\pi^3\nu}{3h\ c}\left|\mathbf{R}^{\mathrm{vib-rot}}\right|^2 Z_{\mathrm{r}}\left(\left|\frac{\Delta\nu}{2B}\right|+\frac{\Delta_0}{BT}\right) \\ &\times\exp\left[\frac{\Delta\nu}{2B}\left(1-\frac{\Delta\nu}{2B}\right)\frac{hcB}{KT}\right] \end{aligned} \tag{2.38}$$

$$\begin{aligned} Z_{\mathrm{r}} &= \frac{1}{2B}\frac{hcB}{KT} \\ &\times\left\{1+\sqrt{\frac{\pi\, hc\, B}{4KT}}\exp\left(\frac{hc\, B}{4KT}\right)\left[\frac{2\Delta_0}{BT}+\mathrm{Erf}\left(\sqrt{\frac{hc\, B}{4KT}}\right)\right]\right\}^{-1} \end{aligned} \tag{2.39}$$

The gap filling proportionality factor is Δ_0 and is treated as an adjustable parameter. Figure 2.5 shows how this rotational structure varies with temperature

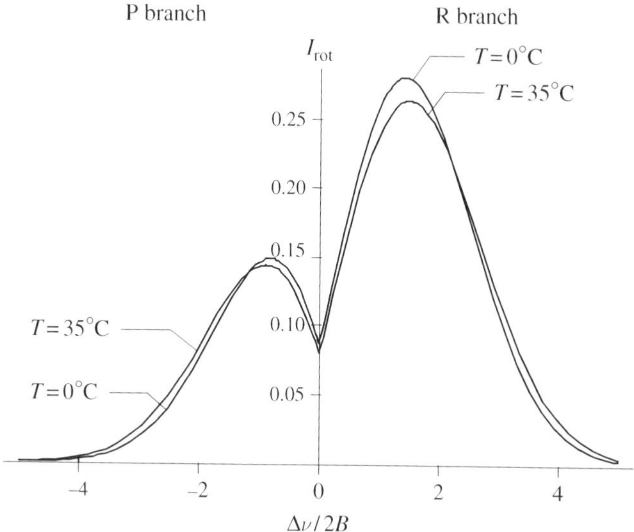

Figure 2.5. The distribution of intensity of rotational P and R branch transitions of liquid water according to (2.38). Each line representative of gaseous state (not shown) is broadened in the liquid state so that individual transitions meld into a continuous band. Note the asymmetry between the branches. Note also the broadening and increased spacing between the peaks as the temperature increases from 0 to 35°C. This behavior of the rotational structure explains some of the temperature dependence of the absorption coefficient and the success in modeling the effect by pairs of Gaussian functions (*Pegau* et al. 1997).

from 0 to 35°C. The parameters of the O–H bond are those used in the figure, and the fill factor was set at 0.5. The asymmetry between the P and R branch is still evident. It is obvious from the figure that one could easily approximate the form by a set of two Gaussians if the fill factor is small. In the limit of higher fill factors, a single Gaussian could be used to model adequately the whole band. We will later see that this is the approach that Pegau and colleagues (*Pegau* et al. 1997) took in their effort to model the red and near infrared portion of the absorption spectrum of water and its temperature dependence. The above considerations show why this simple Gaussian model was effective.

With this, we have in hand the basic tools that will be required in the further analysis of the visible and infrared portion of the absorption spectrum of pure water.

2.4. The intrinsic scattering of pure water

We will now turn our attention to the formulation and evaluation of the intrinsic scattering of water and seawater. We need to address this problem now before we discuss absorption any further since two of the experimental techniques used to evaluate absorption by pure water actually measure the total attenuation of water. Absorption is obtained from the attenuation data by subtracting out the intrinsic scattering contribution.

We will structure our discussion along the lines of Morel's excellent review of the field (*Morel* 1974) and incorporate when necessary some other significant results not discussed by him. Some of the details of the calculation will follow a very clear discussion given by *Kerker* (1969). As mentioned previously, the intrinsic scattering is a consequence of random variation of the refractive index due to microscopic density fluctuations for pure liquids and to both density and composition fluctuations in solutions such as seawater. For a system at constant volume and temperature, the probability distribution of these fluctuations is given by the standard Boltzmann distribution of the excess energy required to bring about the change. This excess energy is called the Helmholtz free energy, E_{H}. The probability of occurrence of a fluctuation with ΔE_{H} energy is given by:

$$p(\Delta E_{\mathrm{H}}) = \exp\left(-\frac{\Delta E_{\mathrm{H}}}{KT}\right) \tag{2.40}$$

For small deviations about the average, the change in free energy can be expressed as a Taylor series in terms of density, ρ:

$$\Delta E_{\mathrm{H}} = \left(\frac{\partial \Delta E_{\mathrm{H}}}{\partial \rho}\right)_{\mathrm{T,V}} \Delta\rho + \frac{1}{2}\left(\frac{\partial^2 \Delta E_{\mathrm{H}}}{\partial \rho^2}\right)_{\mathrm{T,V}} (\Delta\rho)^2 + \cdots \tag{2.41}$$

The first term of the expansion is strictly zero since the free energy is by definition always a minimum at the equilibrium condition. If we substitute the second term in (2.40), we obtain the probability distribution of the density fluctuation squared. We can immediately evaluate the mean square value:

$$\begin{aligned}\overline{(\Delta\rho)^2} &= \frac{\int_0^{\infty} (\Delta\rho)^2 \exp\left[-\left(\frac{\partial^2 \Delta E_{\mathrm{H}}}{\partial \rho^2}\right)_{\mathrm{T,V}} \frac{(\Delta\rho)^2}{2KT}\right] d(\Delta\rho)}{\int_0^{\infty} \exp\left[-\left(\frac{\partial^2 \Delta E_{\mathrm{H}}}{\partial \rho^2}\right)_{\mathrm{T,V}} \frac{(\Delta\rho)^2}{2KT}\right] d(\Delta\rho)} \\ &= \frac{KT}{\left(\frac{\partial^2 \Delta E_{\mathrm{H}}}{\partial \rho^2}\right)_{\mathrm{T,V}}}\end{aligned} \tag{2.42}$$

Using certain standard thermodynamic relations for liquids (e.g., *Kerker* 1969) we get

$$\frac{\overline{(\Delta\rho)^2}}{\Delta V} = KT\,\beta_T\,\rho^2 \tag{2.43}$$

Equation (2.43) relates the mean square fluctuation for a volume element ΔV to the product of the density squared and the isothermal compressibility, β_T. If we know an experimental or theoretical relationship between density and refractive index, (2.43) allows us to explicitly evaluate the mean square excess in refractive index for a volume element. The volume over which the index fluctuations occur is considered large when compared with molecular dimensions but small enough when compared to the wavelength so that these fluctuations can be substituted in the appropriate expressions for scattering by small particles (*Kerker* 1969).

For small-particle scattering, the volume scattering function, $\beta(\theta)$, and the scattering coefficient b can be expressed as follows (*Morel* 1974):

$$\beta(\theta) = \beta_{\text{iso}}(\pi/2)\frac{6+6\delta}{6-7\delta}\left(1+\frac{1-\delta}{1+\delta}\cos^2\theta\right) \tag{2.44}$$

where δ is the ratio of the intensities of the two linearly polarized components at a scattering angle of $\pi/2$. If the medium's response is isotropic, this factor is zero. Anisotropic response gives rise to a small but finite value of this depolarization factor. $\beta_{\text{iso}}(\pi/2)$ is the isotropic portion of the 90° scattering.

The total scattering coefficient b, i.e., the integral over all solid angles of the volume scattering function, is expressed as:

$$b = \frac{8\pi}{3}\beta_{\text{iso}}(\pi/2)\frac{6+6\delta}{6-7\delta}\,\frac{2+\delta}{1+\delta} \tag{2.45}$$

The value of the volume scattering function at 90° is sometimes also called the Rayleigh ratio, R, and is given for random small (real) refractive index fluctuations over a volume ΔV by:

$$R = \beta_{\text{iso}}(\pi/2) = \frac{\pi^2}{2\lambda^4}\Delta V\ \overline{\langle\Delta n'^2\rangle^2} \tag{2.46}$$

If we assume that the fluctuations in the refractive index are due to density fluctuations, even without knowing an explicit form for their relationship, we can immediately write down that:

$$\overline{\langle\Delta n^2\rangle^2} = \left(\frac{dn'^2}{d\rho}\right)^2\overline{\langle\Delta\rho\rangle^2} \tag{2.47}$$

Substituting (2.44) and (2.47) in (2.46) we obtain:

$$\beta_{\text{iso}}(\pi/2) = \frac{\pi^2}{2\lambda^4}KT\,\beta_T\left(\rho\frac{dn'^2}{d\rho}\right)^2 \tag{2.48}$$

We can evaluate the expression above by making direct use of the experimental measurements of refractive index, n', and of its rate of change with pressure at constant temperature.

$$\beta_{\text{iso}}(\pi/2) = \frac{2\pi^2}{\lambda^4} K T n'^2 \frac{1}{\beta_{\text{T}}} \left(\frac{\partial n'}{\partial p} \right)_T^2 \tag{2.49}$$

Equations (2.44), (2.45), and (2.49) along with basic experimental data on compressibility, refractive index, and its rate of change with pressure are sufficient to evaluate the intrinsic scattering term for pure water.

Buiteveld et al. (1994) recently proposed a basic set of formulas for the various terms required to evaluate (2.49) as a function of wavelength and temperature at atmospheric pressure. This expression is adequate for the correction of experimental attenuation results in order to obtain the absorption coefficient for pure water. For the general case of data obtained in seawater at great depth, comprehensive formulas are needed and can be obtained by a more sophisticated approach.

Buiteveld et al. suggest the following expressions as the most reliable to use in the evaluation of the various terms of (2.49). The dependence of the isothermal compressibility, in Pa^{-1}, on temperature, T_c[°C], where the subscript c simply indicates the centigrade temperature scale, is given as a quadratic fit to the data of *Lepple* and Millero (1971).

$$\beta_{\text{T}} = (5.062271 - 0.03179 T_c + 0.000407 T_c^2) \times 10^{-10} \tag{2.50}$$

For the refractive index, they use the formula given by *McNeil* (1977), with the values of the coefficients listed in Table 2.2.

$$\begin{aligned} n'(\lambda, T_c, S) = {} & n_0 + n_1 \lambda^{-2} + n_2 \lambda^{-4} \\ & + n_3 T^2 + n_4 S (n_5 + n_6 T) \end{aligned} \tag{2.51}$$

where S is the salinity in parts per thousand. We give the full formula of McNeil in section 6.2.

McNeil (1977) also gives terms that model the pressure dependence of refractive index, n', but these are not accurate enough to use in evaluating the derivative of

Table 2.2. Coefficients of equation (2.51) for the refractive index of seawater.

Coefficient	**Value**	**Coefficient**	**Value**
n_0	1.3247E + 00	n_4	4.0E − 05
n_1	−3.3E + 03	n_5	5.0E + 00
n_2	−3.2E + 07	n_6	2.0E − 02
n_3	−2.5E − 06	–	–

n' as a function of pressure. Instead, Buiteveld et al. use the directly measured values of this derivative from two sources. First, the temperature dependence of the refractive index derivative, in Pa^{-1}, is modeled on the data of *O'Connor* and Schlupf (1967) taken from 5 to 35°C with a helium–neon laser at 633 nm.

$$\frac{\partial n'}{\partial p}(633, T_c) = (1.61857 - 0.005785\, T_c) \times 10^{-10} \tag{2.52}$$

A linear approximation of $\partial n'/\partial p$, in Pa^{-1}, as a function of wavelength is then used to fit the 20°C data of *Evtyushenkov* and Kiyachenko (1982).

$$\frac{\partial n'}{\partial p}(\lambda, 20°\mathrm{C}) = (1.5989 - 0.000156\, \lambda) \times 10^{-10} \tag{2.53}$$

The data from (2.52) and (2.53) are combined in the following geometric average type formula.

$$\frac{\partial n'}{\partial p}(\lambda, T_c) = \frac{\dfrac{\partial n}{\partial p}(\lambda, 20) \times \dfrac{\partial n}{\partial p}(633\,\mathrm{nm}, T_c)}{\dfrac{\partial n}{\partial p}(633\,\mathrm{nm}, 20°\mathrm{C})} \tag{2.54}$$

By substituting (2.50), (2.51), and (2.54) into (2.49), we can evaluate the isotropic component of the Rayleigh ratio. To complete the process and evaluate the full expressions for the volume scattering function and the total scattering coefficient, we need to fix the depolarization ratio δ. Morel, given the data available to him, determined the average of the depolarization ratio found in several experiments as 0.09. *Buiteveld* et al. (1994) chose instead the value of 0.051 found under argon–ion laser illumination at 514.5 nm and broadband analysis by *Farinato* and Roswell (1975) in their careful experiments. Farinato and Roswell obtained a value close to that of previous experimenters ($\delta = 0.104$) when they used the argon laser with all its different wavelenghts operating at the same time. When they used a single wavelenght of 514.5 nm with the same broadband type of detectors, they obtained a value of 0.051 for the depolarization ratio. This surprising result prompted them to go through a very careful analysis of the sources of error, and they concluded that both stray light and contributions from angles other than 90° due to a too large a detector acceptance angle were at the root of the problems encountered by other experimenters. They confirmed their results further by using a medium band filter (22.5 nm) and a narrow band filter (0.46 nm). With a medium band filter, they obtained a value of 0.045 and with the narrow band filter a value of 0.039. Since the narrow band filtering eliminates more of the stray light, the value obtained with the narrow band filter must be considered as the most accurate. Table 2.3 shows a comparison of the theory with

Table 2.3. Comparison between theory and experiment for the value of the scattering function $\beta(\pi/2)$.

	Experiment	**Theory**			
	***Morel* (1974), *Pike* et al. (1975)**	***Morel* (1974)**	***Buiteveld* et al. (1994) (refractive index formula from: *Quan* and Fry 1995)**		
λ (nm)		$\delta = 0.09$	$\delta = 0.09$	$\delta = 0.051$	$\delta = 0.039$
366.0	4.53	5.32	5.06	4.64	4.52
405.0	2.90	3.42	3.32	3.05	2.97
436.0	2.12	2.49	2.45	2.25	2.19
546.0	0.83	0.94	0.97	0.89	0.86
578.0	0.66	0.73	0.76	0.70	0.68
633.0	0.49[a]	–	0.52	0.48	0.47

[a] *Pike* et al. (1975)

The first column gives the experimental values found by *Morel* (1974) and by *Pike* et al. (1975) at 22°C. The second column contains the theoretical estimate by *Morel* (1974) with a value of 0.09 for the depolarization ratio, δ. The third column gives the value computed with the *Buiteveld* et al. (1994) formula using the approximation of *Quan* and Fry (1995) for the refractive index and a depolarization ratio of 0.09. The last two columns use the same formulas as that column, but with the depolarization ratio measured by *Farinato* and Roswell (1975) for broad band illumination ($\delta = 0.051$) and for a narrow band illumination ($\delta = 0.039$). This last value leads to the best fit of the theory with experiment for reasons discussed in the text.

values of $\delta = 0.09$, 0.051, and 0.039 with the experimentally measured volume scattering function at 90° which is expressed as follows

$$\beta(\pi/2) = \beta_{\text{iso}}(\pi/2)\frac{6+6\delta}{6-7\delta} \tag{2.55}$$

The experimental values of *Morel* (1974) and *Pike* et al. (1975) are shown in the first column of Table 2.3, the second column is the theoretical value determined by Morel with the ratio set at 0.09. The third column is the theoretical values using the formula given by Buiteveld and colleagues (*Buiteveld* et al. 1994) also with $\delta = 0.09$. The fourth and fifth columns use the same formula but with a value of the depolarization ratio of 0.051 and 0.039 respectively. The best agreement between experiment and theory is found with this last value.

The expressions we have so far are adequate to compute the density fluctuation scattering for pure water at atmospheric pressure. In the case of an electrolytic solution such as seawater, an additional term must be taken into account. The thermodynamic variation of the concentration of the electrolyte in a volume element leads to an additional refractive index fluctuation. Details of the derivation of this term are given by *Morel* (1974). It is the only significant additional term in an electrolytic solution, and its contribution should be isotropic for small ions such as Cl^- and Na^+.

$$\beta_{\mathrm{iso}}^{\mathrm{cf}}(\pi/2) = \frac{2\pi^2}{\lambda^4}\frac{W\,S}{\eta_i N_a} n'^2_0 \left(\frac{\partial n'}{\partial S}\right)^2_{\mathrm{P,T}} \tag{2.56}$$

In (2.56), W is the molecular weight of the electrolyte, 58.4 for pure NaCl, N_a is Avogadro's number, η_i is the number of ions, S is the concentration of electrolyte in grams per gram, n'_0 is the refractive index of the pure solvent, and n' is the refractive index of the solution. In the case of seawater, S would be almost equal to the salinity. Experimental results outlined by *Morel* (1974) show that pure NaCl solutions exhibit an excess amount of scattering that grows linearly with salt concentration. A similar increase in the attenuation of NaCl solutions in the visible portion has been found by *Ravisankar* et al. (1988). The percent rate of relative increase as a function of concentration is 20% for $S = 0.035$ or

$$\Delta\beta(\pi/2) = \beta_{\mathrm{s}}(\pi/2)\frac{0.2}{0.035}S \tag{2.57}$$

Real and artificial seawater have a larger rate of increase than equivalent NaCl solutions. This can be attributed to the fact that seawater is a mixture of different salts. Many cations other than Na^+ are present and also several anions other than Cl^-. The average molecular weight is larger than that of pure NaCl, and the difference in molecular weight accounts for almost all the excess scattering without having to use the full theory for a multi-component system (*Morel* 1974).

$$\Delta\beta(\pi/2) = \beta_{\mathrm{s}}(\pi/2)\frac{0.3}{0.035}S \tag{2.58}$$

A direct theoretical evaluation of this term is possible with a minor extension of the work of *Buiteveld* et al. (1994). Recently *Quan* and Fry (1995) carried out a new analysis of empirical formulae for refractive index of pure water and seawater and showed some serious systematic errors in the forms of *McNeil* (1977) and *Matthäus* (1974). Quan and Fry suggest an alternate form valid for pure water and seawater at atmospheric pressure (see section 6.2). The results obtained by using this form for the refractive index of pure water in the scattering equations are identical within a fraction of a percent to those obtained with the McNeil's equation for $P = 0$ (2.51). The derivative as a function of salinity is however different, and the form suggested by *Quan* and Fry (1995) incorporates a wavelength-dependent part not found in McNeil expression.

$$\begin{aligned}\left(\frac{\partial n'}{\partial S}\right)_{\mathrm{P,T}} &= 1.779\times10^{-4} - 1.05\times10^{-6}T_{\mathrm{c}} + 1.6\times10^{-8}T_{\mathrm{c}}^2 \\ &\quad + \frac{0.01155}{\lambda}\end{aligned} \tag{2.59}$$

Table 2.4. Basic composition in dissolved mineral salts of seawater (e.g., *Chamberlin* 1899) (for a more modern account, see p. 252 in *Millero* 2001).

Dissolved mineral salts	**Chemical formula**	**Fractional abundance f_i**	**Number of ions n_i**	**Molecular weight W_i**	$\frac{f_i W_i}{n_i}$
Sodium chloride	NaCl	0.7776	2	55.4	22.706
Magnesium chloride	$MgCl^2$	0.1089	3	95.1	3.452
Magnesium sulfate	$MgSO^4$	0.0473	2	120.3	2.845
Calcium sulfate	$CaSO^4$	0.0360	2	136.0	1.673
Potassium sulfate	K_2SO^4	0.0246	3	174.0	1.427
Calcium carbonate	$CaCO^3$	0.0034	2	100.0	0.170
Magnesium bromide	$MgBr^2$	0.0022	3	184.3	0.135

The total abundance is 0.035 g of salt per kg of water. The sum of the last column, 32.41, is the number that must be entered into equation (2.56) for the electrolyte concentration fluctuation.

Table 2.4 gives the basic composition in dissolved mineral salts of seawater based on the data of *Chamberlin* (1899). The first column lists the salts in order of abundance. The second column gives the chemical formula. The third column is the fractional abundance. The fourth is the number of ions into which the salt dissolves. The fifth column is the molecular weight of the salt and the sixth column is the product of the fractional abundance and the molecular weight divided by the number of ions. The sum of this column is the result that should enter directly into equation (2.56) as W. For seawater of the given composition, this term is equal to 32.41. Combining this result with the derivative as a function of salinity from equation (2.59) into equation (2.56), one can compute directly the contribution of the electrolyte concentration fluctuation to the scattering. We find that for seawater with a salinity $S = 0.035$ at a temperature of 25°C, this computed excess scattering contribution is 31% at 366.0 nm and 29% at 546.0 nm. This matches the results of *Morel* (1974) to within experimental error. We have managed to achieve this result without having to resort to the full complexities of concentration fluctuations in multi-component solutions. Our treatment of seawater as a two-component solution with an equivalent electrolyte accounts for the experimental results.

We are now in the rather satisfactory position of being able to accurately evaluate the scattering of both pure water and pure seawater at atmospheric pressure from the near UV to the near infrared portions of the spectrum over the temperature range from 0 to 35°C. We are therefore able to subtract the inherent scattering contribution from the results of any experiment that measures the optical attenuation of pure water or seawater and to thereby obtain an accurate estimate of the absorption.

We are however still not able to evaluate the intrinsic scattering contribution at substantial pressures. This capability is required when we have to subtract the pure water scattering contribution from the phase function measured at depth. This subtraction is a delicate task, and it is easy to find a contribution from pure water scattering that exceeds the experimental measurements of total scattering if one is not very careful and does not properly account for the effect of increased pressure as a function of water depth (*Kullenberg* 1984). In order to obtain the depth dependence of the pure water or seawater scattering, we will need to generalize our approach.

So far, we have used experimental data for the refractive index and separate experimental data for its derivative as a function of pressure or density. This is because the fit to the refractive index data as a function of pressure and density is not sufficiently precise to allow an accurate computation of its derivatives. A sufficiently accurate expression for the refractive index as a function of density or pressure would remedy this situation. We could then evaluate all the derivatives directly from the index formula. In the case of pure water, Henryk Eisenberg (*Eisenberg* 1965) obtained the most accurate expression for the refractive index as a function of wavelength, density, and temperature. This expression is indeed accurate enough to allow for direct evaluation of the scattering by using the derivative as a function of density.

The expression found by Eisenberg is based on the observation of two experimental facts about the Lorentz–Lorenz formula (*Eisenberg* 1965). First define the left-hand side of the Lorentz–Lorenz formula as:

$$f(n') = \frac{n'^2 - 1}{n'^2 + 2} \tag{2.60}$$

This obviously implies that:

$$n' = \left[\frac{1 + 2f(n')}{1 - f(n')}\right]^{1/2} \tag{2.61}$$

For water and many other liquids, the derivative of the logarithm of $f(n')$ as a function of pressure divided by the isothermal compressibility is a constant on the order of unity.

$$\frac{1}{\beta_T}\left(\frac{\partial \ln f(n')}{\partial p}\right)_T = B \tag{2.62}$$

For water specifically, the derivative of the logarithm of $f(n)$ as a function of temperature is a linear function of the volume expansion coefficient:

$$-\left(\frac{\partial \ln f(n')}{\partial T}\right)_p = B\left(\frac{\partial \ln \rho}{\partial T}\right)_p + C \tag{2.63}$$

Integrating formally these two relations leads to the following form for the refractive index of water:

$$f(n', \lambda, T) = A_t(\lambda) \left(\frac{\rho}{\rho_{T=4°C}} \right)^{B(\lambda)} \exp[-C(\lambda)\,T_c] \qquad (2.64)$$

This relationship is particularly attractive because it can be extrapolated to the standard definition of polarizability in terms of refractive index, and the three adjustable parameters can be related directly to the thermodynamic properties of the liquid. According to simple theory, for an ideal substance $B = 1$ and $C = 0$. For most liquids, it is indeed true that $C = 0$. Water is a notable anomaly in that respect. However, it should be noted that for all liquids, B is significantly different from unity. B and C express in a compact manner the effect of the complex internal structure of water on the index. In (2.64), the various parameters have been evaluated at a temperature of 4 °C. By using the data of *Tilton* and Taylor (1938) and *Waxler* et al. (1964), transforming to absolute refractive index with respect to vacuum, *Eisenberg* (1965) obtains values for the three constants at 13 different wavelengths ranging from 400 to 700 nm. He claims an accuracy of better than one part in a million for each separate wavelength. The expression also fits the data of Waxler and colleagues up to a pressure of 1000 bars.

Note that the first term of the right side of (2.64) can also be evaluated by substituting the formula given by *Quan* and Fry (1995) into (2.60) after having corrected it to the absolute index with respect to vacuum. Comparing this expression with the values obtained by *Eisenberg* (1965) yields a difference of 50 parts in a million at 4 °C. This result is actually consistent with the absolute level of accuracy claimed by Quan and Fry. We fitted polynomials in wavelength to *Eisenberg*'s data for the three parameters at a temperature of 4 °C as expressed below (the coefficients are listed in Table 2.5):

$$A_t(\lambda) = a_0 + a_1\lambda^{-1} + a_2\lambda^{-2} + a_3\lambda^{-3} \qquad (2.65)$$

$$B(\lambda) = b_0 + b_1\lambda^{-1} + b_2\lambda^{-2} \qquad (2.66)$$

$$C(\lambda) = c_0 + c_1\lambda + c_2\lambda^2 + c_3\lambda^3 \qquad (2.67)$$

Table 2.5. Coefficients of polynomials in λ [nm] relevant to equations (2.65), (2.66), and (2.67) for the refractive index of pure water based on the data of *Eisenberg* (1965).

Coefficient	A_t	**Coefficient**	B	**Coefficient**	C
a_0	1.95279E − 01	b_0	8.99098E − 01	c_0	5.24968E − 05
a_1	8.81846E + 00	b_1	0.0E + 00	c_1	8.27948E − 09
a_2	−2.34716E + 03	b_2	−7.11567E + 03	c_2	−1.5343E − 10
a_3	6.10753E + 05	b_3	–	c_3	6.85878E − 14

The accuracy of the results is better than one part in 10^6 for A and one part in 10^4 for both B and C. The index and its derivative with respect to density can now both be computed. Using (2.48) and (2.64), we find that:

$$\beta_{\text{iso}}(\pi/2) = \frac{\pi^2}{2\lambda^4} KT\, \beta_{\text{T}} \left(3\text{B}(\lambda) \frac{f(n', \lambda)}{[1 - f(n', \lambda)]} \right)^2 \tag{2.68}$$

To evaluate this expression for pure water, we only need to know the density as a function of pressure and temperature. A general expression proposed by *Mamaev* (1975) can be used for this purpose:

$$\begin{aligned} \rho(S_w, T_c, 0) = {} & \rho(0, 4°, 0) \\ & + 0.028152 - 7.35 \times 10^{-5} T_c - 4.69 \times 10^{-6} T_c^2 \\ & + (0.802 - 0.002\, T_c)(S_w - 0.035) \end{aligned} \tag{2.69}$$

where T_c is the temperature in degrees Celsius.

This expression can also be used as an approximation for seawater. In the general case of larger pressures, up to 1000 bars, the isothermal compressibility of water and seawater is to first order constant as a function of pressure. We therefore have to a first approximation:

$$\frac{\rho(S_w, T_c, p)}{\rho(S_w, T_c, 0)} = \exp(\beta_{\text{T}} p) \tag{2.70}$$

where the isothermal compressibility is given by the fit to the data of *Lepple* and Millero (1971), equation (2.54). More accurate results can be obtained by using the full expressions of the international equation of state for seawater (*Dera* 1992, pp. 137–139). The dominant effect of salinity on index is in the change of density computed with Mamaev's expression (*Mamaev* 1975). The dominant contribution to the derivative of index with salinity can therefore be evaluated by using a straightforward combination of equations (2.61), (2.64), and (2.68). From this and our previous results for the composition of seawater, we can evaluate the excess scattering due to concentration fluctuation.

With this last algorithm, we can obtain a satisfactory approximation to the intrinsic scattering of pure water and seawater under all practical conditions of temperature salinity and pressure. Its effect can thus be subtracted from attenuation measurements to obtain an estimate of absorption. The intrinsic scattering phase function can also be subtracted from the experimental phase function to isolate the scattering phase function due to particles suspended in the water column.

2.5. Measurements of the absorption of pure water

The measurement of an accurate absorption spectrum for pure water has proven over the years to be a much more arduous task than anyone who attempted it could reasonably foresee. The principal difficulty has been in obtaining and storing sufficiently pure samples of water. Being highly polar, water is almost an ideal solvent. Moreover, many of the organic compounds present in a water sample have boiling points very close to that of water. Many of these organic substances and some ions are entrained during the evaporation process by water molecules due to their polar nature. This makes it difficult to obtain pure enough samples by simple repeated distillation. Significant efforts must be made to chemically eliminate both ions and organic materials before and during distillation. Long-term storage is also a significant problem. Pure water will attack even amorphous quartz and Pyrex glassware to recuperate some of its missing ions. Some of the best recent measurements have used specialized apparatus with reverse osmosis systems for producing on demand reagent grade water. *Pope* and Fry (1997) note that pure type I reagent grade water stored in Pyrex shows a measurable increase in absorption of $0.0006\,\mathrm{m}^{-1}$ per day.

The second significant problem in obtaining accurate measurements of the absorption of water is the extraordinary transparency of pure water in the blue region of the spectrum. In that region, molecular scattering is of the same order of magnitude as absorption and an effective means of removing the effect of this intrinsic scattering, either by computation or by the experimental method of measurement itself, must be used. This small value of the absorption coefficient is what severely constrains the purity required from the water samples.

Reliable and consistent sets of measurements have only recently been obtained for the visible and near infrared region of the spectrum. Some discrepancies are still obvious in the UV. We will see that most of these problems of consistency between measurements in that spectral region can be resolved by carefully accounting for the effect of dissolved oxygen in a water sample.

There are several excellent review and analysis articles that discuss in detail the measurement techniques and results for the absorption of pure water (*Morel* 1974, *Jerlov* 1968, *Smith* and Baker 1981). We will not try to duplicate these or attempt here a complete review of the new measurements. Instead, we will concentrate on analyzing a subset of significant new experiments and some older results that are still valid even in the light of this new information.

We will also use the theory developed in the first part of this chapter to explain and discuss these results in a consistent framework. We will use some elements of this theory to obtain plausible analytical forms to fit several parts of the absorption spectrum. When possible, we will show the directions in which the results can be generalized. We will also analyze what effects could profitably be measured and what pitfalls are to be expected in carrying out these measurements, particularly in the still-under-debate UV region of the spectrum.

2.5.1. Some significant absorption measurements

The most recent and complete set of measurements in the near infrared are those of *Kou* et al. (1993) taken at 22 °C. Their results were obtained by the differential beam attenuation technique. They cover the range from 2.5 micron to 667 nm. These measurements overlap the data of *Pope* and Fry (1997) that extends from 725 to 380 nm. The measurements of Pope and Fry were obtained with samples also at 22 °C. These measurement sets agree well within their quoted error bars over their entire range of overlap. The data of Pope and Fry were obtained with a completely different technique, the integrating cavity meter, which directly measures the absorption coefficient and does not need to be corrected for the effect of intrinsic scattering. Taken together, these data sets form the most reliable estimate of the absorption spectrum of pure water at 22 °C from 2.5 micron to 380 nm.

Pegau et al. (1997) as well as *Trabjerg* and Højerslev (1996) measured a small but significant temperature dependence of the absorption coefficient. The effect is linear in temperature, and the magnitude of the slope is a function of wavelength with maximum values occurring at or near the previously mentioned shoulders of the spectrum. The peak temperature gradient they measured occurred around 740 nm along with a smaller but definite effect at 600 nm. The magnitude of the gradient in absorption as a function of temperature at these two wavelengths is in fact proportional to the magnitude of the absorption coefficient itself at the same wavelengths. For these measurements, *Pegau* et al. (1997) used a reflecting-tube absorption meter. They found also that the absorption coefficient showed no significant dependence on salinity over the visible portion of the spectrum. Interestingly, they also managed to fit some portions of the absorption spectrum and its derivative as a function of temperature with a set of Gaussian functions. As we saw, this type of fitting procedure can be justified by theory.

Few reliable measurement sets exist in the UV. In the region between 195 and 320 nm, the definitive set of data is that of *Quickenden* and Irvin (1980) taken at 25 °C. They used the differential attenuation technique. They took stringent precautions during the preparation of the samples, including an oxidation step to remove all organic constituents. Most importantly, they ensured that no oxygen was present in the water by saturating the samples with helium. Helium itself has no absorption in the spectral zone under study (*Herzberg* 1950).

The strong effects of oxygen on the absorption of liquid water in the UV were thoroughly studied in a remarkable experiment by Tait and co-workers (*Heidt* and Johnson 1957). They determined that dissolved oxygen increased the absorption in the 190 to 220 nm region far more than would be expected on the basis of pure molecular oxygen absorption.

They also inferred that dissolved oxygen most likely forms two types of molecular bridges with water and that these compound absorbers have different absorption spectra and different concentrations as a function of temperature. They noted that the effect persists, albeit much weaker, at least to 250 nm, which was the limit of their measurements. In fact, there is no reason to suspect that the effect of dissolved oxygen does not extend down into the visible region of the spectrum. The far wing of this effect, like the far wing of the pure UV spectrum of water, is due to broadening by collisions between molecules that form unstable dimers as shown in Figure 2.6. This mechanism was thoroughly analyzed by *Szudy* and Bayliss (1975) who give a simple functional form for this far wing. This form could be used to extend the range of results beyond that of the experimental measurements.

The data from *Quickenden* and Irvin (1980) are sufficient to obtain a good fit of the Szudy and Bayliss form of the far wing of the pure water with no oxygen. As pointed out by *Kopelevich* (1976), an equally reliable set of values for air saturated water at 19°C was obtained by *Grundinkina* (1956). The results extend from 210 to 350 nm. Once again, extraordinary precautions were taken in this experiment. The water was triply distilled in silica subsequent to a first distillation from a solution of $KMnO_4$ and $Ba(OH)_2$. The $KMnO_4$ serves to oxidize all the organic compounds in the solution, while the $Ba(OH)_2$ precipitates the salt and sulfate

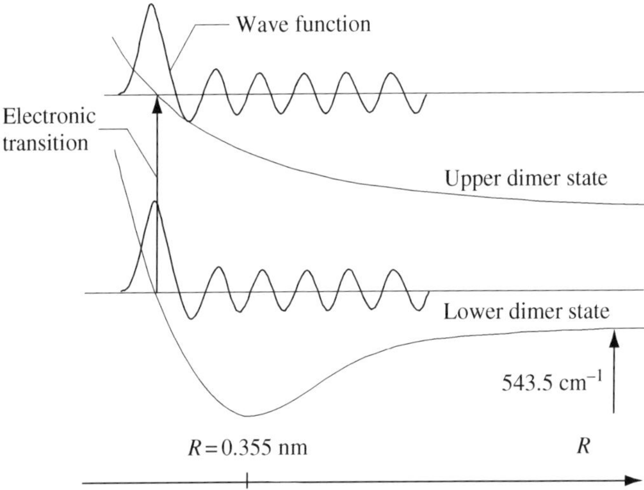

Figure 2.6. Schematic diagram of the electronic transition of a dimmer collision pair (water–water: for which the lower state parameters are given, or water–oxygen). The transition is vertical (i.e., it follows the Franck–Condon principle). If the rate of increase in energy of the upper state as a function of the molecular separation, R, is smaller than the rate for the lower state at the same separation, R, the transition energy will be smaller than the transition energy of isolated molecules. This energy shift gives rise to an absorption wing on the long wavelength side of an electronic molecular transition. The value of R at which the transition will take place obviously depends on the relative kinetic energy of the dimer (*Szudy* and Bayliss 1975).

ions. Using the Tait formula, the effect of oxygen can be added to the data of *Quickenden* and Irvin (1980) and compared directly with the data of *Grundinkina* (1956) over the interval from 210 to 220 nm. In this comparison, both data sets fall within their respective quoted errors.

This confirms that the data of Quickenden and Irvin for oxygen-free water and the data of Grundinkina for air saturated water are consistent with one another and with the results of Tait for the effect of oxygen on the absorption spectrum of pure water. The fit to the data of Quickenden and Irvin by a formula of the type derived by *Szudy* and Bayliss (1975) can be used to estimate the results of pure oxygen-free water at the same temperature as that at which the results of Grundinkina were taken. The difference between the results of Grundinkina and this fit can now be used to extrapolate the formulas given by Tait for the oxygen enhancement effect to 320 nm and beyond.

Other more recent results by *Boivin* et al. (1986) also confirm the accuracy of Grundinkina's early work. These workers, using various laser sources, obtained attenuation values at 254, 313, and 366 nm. Their measurements also agree with those of Grudinkina within the quoted experimental errors. Their measurement technique is a direct attenuation method using 50 cm long cell and an empty reference cell for window transmission correction.

2.5.2. Optical measurement instruments

In this section, we will discuss in more detail the various experimental techniques used to gather the data mentioned in the previous section. The aim of these short analyses is to allow us to evaluate the relative accuracy of the different experiments and the magnitude and reliability of the corrections which must be applied to the data. We will see that some techniques are more applicable to cases where the absorption is medium to high, while some others are suitable even for very low absorption values.

The first technique we will study is the differential attenuation method. There are many variations of this method. A simplified diagram of a typical experimental apparatus is given in Figure 2.7. This type of arrangement is common to many experiments. Light from a source, often a mercury or quartz–iodine lamp, is passed through a monochromator and collimated by a lens to obtain nearly monochromatic, parallel beam. The beam is split into two beams which pass through a reference cell and a sample cell respectively. These beams are recombined at a single detector after they pass through the respective cells. A light chopper is used to sequentially illuminate either the sample or reference cells. This permits one to (1) use a single detector to measure the power transmitted by the reference and the sample cells and to avoid problems with differences between responsivities of two detectors, and (2) apply synchronous (phase-locked) detection techniques to reject the effect of ambient light.

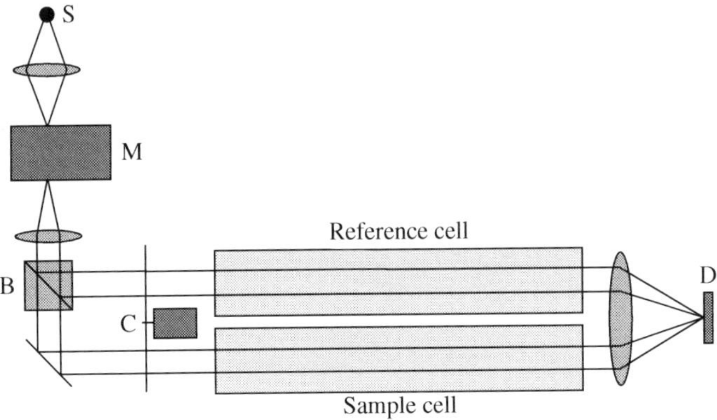

Figure 2.7. Schematic diagram of a typical double-path (differential) attenuation meter. Light from a source S, typically a mercury or quartz–iodine lamp, is passed through a monochromator, M, and collimated by a lens to obtain a nearly monochromatic parallel beam. The beam is split into a reference and a sample beam with a beamsplitter, B. Following the cells, these beams are recombined (here with a lens) at a single detector. A light chopper, C, alternates illumination between the reference and sample cells. This permits one to (1) use a single detector to measure light power transmitted by each cell and thus avoid problems with unequal responsivities of two detectors, and (2) apply a synchronous detection (phase-locked) technique to suppress the effect of the ambient light.

In the experiments of *Quickenden* and Irvin (1980), three identical cells with path lengths of 10.00, 5.00, and 1.00 cm were used in pairs to provide path differences of 9.00 and 5.00 cm. The procedure involved a first wavelength scan of a pair of empty cells followed by a scan of a pair of water-filled cells. The longer cell was always in the sample beam and the shorter cell in the reference beam. They carefully aligned the cells normal to the beams. By combining the results of both scans and computing the effect of cell length on the quantity of multiply reflected light collected by the instrument, they established a procedure that allowed them to correct accurately for the slight differences in the absorbances and reflectivities of the cell windows. *Grundinkina* (1956) used two cells with a path difference of 7.97 cm and did not check cell matching like Quickenden and Irvin.

Boivin et al. (1986) used yet another variant of the technique with lasers as the light sources. A single beam was used with data taken both with the cell empty and the cell filled with water. Their cell was 12.5 cm in diameter and 50 cm in length, considerably larger than those used by *Grundinkina* (1956) as well as by *Quickenden* and Irvin (1980). The laser beams were 2.5 cm in diameter. The windows were tilted at 5° with respect to the laser beam axis. This avoided the multiple reflection problems encountered by Quickenden and Irvin. *Boivin* et al. (1986) explicitly corrected for the effect of the different water–air interface by using the Fresnel reflection coefficients along with data for the index of water and fused silica.

Trabjerg and Højerslev (1996) used the same basic technique but replaced the monochromator with a broadband source and used an optical multichannel detector with a spectral range of 400 to 760 nm split into 670 channels. The cells they used were each 1 m long, and an accurate temperature difference could be maintained between the reference and sample cells. This apparatus allowed them to measure the effect of temperature on the spectrum of water.

The infrared data of *Kou* et al. (1993) were obtained by using a Fourier-transform spectrometer with a spectral resolution of 16 cm^{-1} and an absolute wavelength calibration. They inserted different cell lengths in the spectrometer and computed the attenuation coefficient by using the logarithm of the ratio of the transmittances. This technique again cancels out most of the effects of reflection and absorption losses from the windows. The optical path lengths in water varied from 100 μm to 20 cm. They were selected such that the attenuation coefficient could be determined from transmittance values that ranged from 20 to 60%, a region in which the transmittance error is minimized. In the spectral region from 670 nm to 2.5 μm, the intrinsic scattering is negligible with respect to absorption. In that spectral range, one can with negligible error assume that attenuation is entirely due to absorption. Their results are higher by approximately 10% in their zone of overlap from 670 to 800 nm than those estimated by *Smith* and Baker (1981). They are however in much closer agreement with the results of *Pope* and Fry (1997).

Except in cases were the absorption completely dominates the scattering, such as in the far red and near infrared, all the experiments mentioned above require a correction to account for the effect of intrinsic scattering. The results are particularly sensitive to this correction in the blue and green regions of the spectrum where the absorption of water is at a minimum. Fry and co-workers (*Fry* et al. 1992a) found an elegant technique to go around this problem and measure the absorption coefficient directly with an integrating cavity absorption meter. Later, *Pope* and Fry (1997) used this apparatus to measure the spectrum of pure water and obtain some startling new results that are now accepted as the standard. They found that water transmits much more than was previously thought in the blue and green regions of the spectrum.

Figure 2.8 is a schematic diagram of an integrating cavity absorption meter. The meter consists actually of three concentric and closed cavities created by an arrangement of thick enclosures of a highly optically diffusive material such as Spectralon (Labsphere, Inc.). The inner cavity (I) contains the sample whose absorption one wants to measure. The middle cavity (II) is composed of the thick translucent diffusing material itself. The outer cavity (III) is an air-filled space between the outer and inner diffusing enclosures. Light is brought into cavity III by optical fiber cables going through the wall of the external enclosure. Diffuse reflection of light from the outer and inner wall of cavity III ensures that the

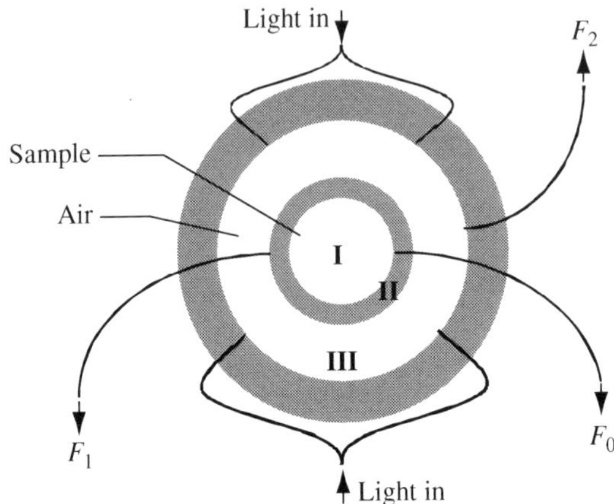

Figure 2.8. Schematic diagram of an integrating cavity absorption meter. This meter, developed by *Fry* et al. (1992a), consists of three concentric cavities created by an arrangement of thick-wall enclosures (shaded) made of light diffusing material (Spectralon, Labsphere, Inc.). The inner cavity (I) contains the sample whose absorption one wants to measure and is created by enclosure cavity (II). The middle cavity (II) is the inner Spectralon enclosure itself. The third cavity (III) is an air space between the outer and inner Spectralon enclosures. Light is brought into cavity III by optical fiber cables going through the wall of the external enclosure and terminated at its inner surface. The measured light is obtained through a set of three optical fiber cables, each sampling scalar irradiance at an outer surface of a cavity: F_0 for cavity I, F_1 for cavity II, and F_2 for cavity III.

illumination which penetrates cavities II and I is uniform. The measured light is sampled through three sets of optical fiber cables, each terminated at an outer surface of a cavity. The first set samples irradiance F_0 at the outer surface of cavity I. The second set samples irradiance F_1 at the outer surface of cavity II, and finally the third set samples the outgoing scalar irradiance F_2 from enclosure III.

Using the law of conservation of energy within each cavity, relationships between these irradiances and the absorption coefficient can be derived. In order to do this, *Fry* et al. (1992a) first assume that the irradiance in the inner cavity is isotropic and homogeneous.

They then assume that the radiance at any point in cavities II and III is proportional to the radiance at other points in the respective cavities. In fact, the radiance in these cavities is also nearly uniform and isotopic, but it is not necessary to assume so. From energy conservation in cavity I we obtain:

$$C_1 F_1 A_0(1 - r_0) = F_0 A_0(1 - r_0) + F_0 A_0 + 4a_0 F_0 V_0 \tag{2.71}$$

where C_1 is the proportionality constant for cavity I, A_0 is the area of the fiber optic collector set in the wall of cavity I, r_0 is the reflectivity of the wall, V_0 is the volume of cavity I, and a_0 is the absorption coefficient of the sample in cavity I. The term on the left-hand side is the power supplied to cavity I through its wall. The right-hand side describes the various power losses. The first term is the power leaving the cavity through the wall. The second term is the power absorbed by the detector and the third term is the power absorbed by the sample. This last term is the correct form for a convex cavity of otherwise arbitrary shape. By re-arranging (2.71), we obtain:

$$a_0 = \frac{1}{4F_0V_0}\left[C_1A_0(1-r_0)\frac{F_1}{F_0} - A_0(1-r_0) - A_0\right] \tag{2.72}$$

Since the signals from the detectors S_0 and S_1 are also directly proportional to F_0 and F_1, we can therefore write that:

$$a_0 = K_1\frac{S_1}{S_0} + K_2 \tag{2.73}$$

where K_1 and K_2 are instrumental constants that can be measured at each wavelength by using a minimum of two samples with known calibrated absorption. The instrument therefore allows for an absolute calibration against a standard set of solutions. *Fry* et al. (1992a) show that a similar relation can be derived from the energy conservation expression for the second cavity. This leads to a second linear equation with its own set of instrument calibration factors. In practice, both equations can be used as a check of instrumental consistency.

In the course of their experiments, *Pope* and Fry (1997) used a master solution of 1.0 mg l^{-1} of Irgalan Black and Alcian Blue to generate by dilution a set of 19 reference samples with absorption ranging from 0.01 to 8.0 m^{-1}. This set of solutions served as the instrument calibration standard. The instrumental constants were then determined for each wavelength by linear regression analysis of the data obtained with this standard solution set. The integrating cavity absorption meter has been used to obtain the most reliable results for the absorption coefficient of water from 380 to 725 nm. As it obviates the need for subtracting the amount of molecular scattering, the approach is particularly attractive to use in spectral zones where the absorption coefficient is low. The accuracy of the method is directly traceable to the accuracy of the absorption standard and to the achievable water purity. No other assumptions need to be invoked to ensure the precision of the results. Fry and colleagues claim that the instrument is capable of an accuracy of better than ± 0.001 m^{-1}.

The last experimental technique we will look at is the reflecting tube absorption meter. This type of meter was used by Pegau and colleagues (*Pegau* et al. 1997) to obtain their results on the temperature and salinity dependence of pure water and seawater. This kind of instrument measures the absorption relative to a reference sample.

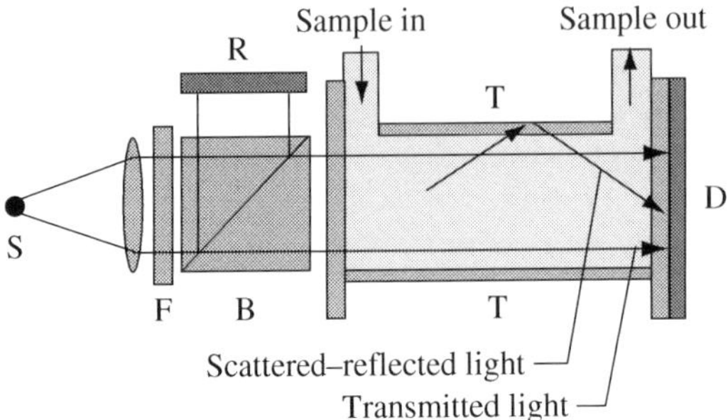

Figure 2.9. Schematic of a reflecting tube absorption meter (ac-9, Wet Labs, Inc.). A lens collimates light from a source S into a parallel beam. The beam which passes through a narrow band filter F is divided by beam splitter B into a reference and sample parts. The reference part is collected by a reference detector R. The other part the beam passes through a sample cell surrounded by a reflecting tube T. Water flows in from orifices located near the entrance window and flows out by similar orifices located near the detector D. The detector consists of a photodiode placed against the back of a translucent diffuser. This detector measures the beam power transmitted through the sample cell and also power scattered by the sample and reflected by the reflective tube. Most reflection takes place at the outer air–silica surface of the tube. Light scattered at angles of up to 41° from the beam axis and hitting the wall of the tube undergoes total internal reflection at that interface. After one or more reflections, this light is be collected by the detector. A typical length of a reflecting tube is 25 cm.

Figure 2.9 is a schematic of a basic implementation of that technique. Light from a source is collimated by a lens and passes through a narrow band filter onto a beam splitter. Part of the beam power is collected by a reference detector. The other part passes through sample cell that is surrounded by a reflecting tube. Water flows in from orifices located near the entrance window and flows out by similar orifices located near the detector. The detector consists of a photodiode placed against the back of a translucent diffuser located at the far end of the cell. This detector measures light power transmitted by the sample and also light power scattered by the sample and reflected at the reflective tube made of silica. Most reflection takes place at the air–silica interface outside the cell. Light scattered at angles of up to 41° from the beam axis and hitting the air–silica interface will undergo total internal reflection. After one or more reflections, this light will be collected by the detector. The typical length of a reflecting wall cavity is 25 cm.

Most of light scattering in natural seawater is due to medium-size and large particles. This scattering is therefore concentrated in the forward direction, and a well-designed reflecting tube will collect almost all this signal back on the detector. This type of absorption meter is thus particularly well suited for *in situ* oceanic measurement of absorption. *Kirk* (1983b) used a Monte Carlo model to evaluate the performance of this type of absorption meter. He found that the positive measurement errors (the measured absorption coefficient is always greater than the true value) increase with the ratio of scattering to absorption coefficients at a rate which depends on the form of the scattering phase function. Performance is improved by increasing the diameter of the reflective tube.

Typically, a pure water sample, passed through a 0.2-μm poresize filter to remove bubbles and residue particulate matter, is used as a calibration blank. The subsequent measurements are taken with the measurements from the blank sample subtracted. The remaining error in the relative absorption measurement is due to variations in the portion of the total scatter that was not reflected back to the detector.

Zaneveld et al. (1994) have investigated several calibration and correction procedures for this type of light absorption meter. The most reliable calibration procedure involves a simultaneous measurement of the attenuation of light by the sample to obtain an estimate of the scattering coefficient. In the case of measurements of pure water or seawater that concern us here, intrinsic scattering dominates, and its variation as a function of temperature or salinity is already negligible in the spectral areas of interest. To put their measurements on an absolute scale, *Pegau* et al. (1997) used the absorption values estimated by *Smith* and Baker (1981). This leads to a 10% overestimate when compared with the data of *Kou* et al. (1993) and *Pope* and Fry (1997).

2.5.3. The preparation of pure water samples

Experimental techniques such as the differential attenuation method among others have been used for many years by careful experimentalists. The major part of the large discrepancies found in the past between the result of many investigators cannot be ascribed to limitations in the accuracy of the instruments. Rather, as we mentioned before, most of the problems encountered in practice seem to be directly related to the considerable difficulties in obtaining truly pure water. Some difficulties and questions of procedure still remain today, particularly in the methods of preparing samples for analysis in the UV.

Partly as a caveat against believing the results of casual experimentation, we will outline some of the preparation techniques that have led to what appears to be the most reliable absorption measurements to date.

It should first be noted that for measurements in the infrared and near infrared above 700 nm where the absorption coefficient exceeds $0.5\,m^{-1}$, standard triply distilled water is of sufficient quality to give data that agree within 10% for the various experiments reported here. The problems and discrepancies show up most starkly in the UV and blue green regions of the spectrum where the absorption coefficient can be as small as $0.005\,m^{-1}$.

Quickenden and Irvin (1980) give the most painstakingly complete description of their purification method. We will briefly summarize it here. As stated by them, their aim was to eliminate all organic impurities because of their high absorption coefficients in the UV region. The first stage was a supply of standard laboratory grade well-filtered deionized water. The second stage consisted of distillation from a Pyrex glass still with a silica encased immersion heater. The third stage was a further distillation in a two-step glass still with a solution of 4×10^{-3} mol dm^{-3} of $KMnO_4$ and 5×10^{-2} mol dm^{-3} of KOH, both of analytical reagent grade, as the starting material for the distillation. This solution was prepared with the water produced in the second stage. As mentioned before, the $KMnO_4$ serves to oxidize the organic compounds in the solution, while the KOH precipitates the remaining salt and sulfate ions. During this last stage, medical grade oxygen, pre-filtered through sintered glass with a 1.3-μm pore size and passed through pure water, was bubbled through both stages of distillation to assist in oxidation, remove the gaseous impurities, and prevent contamination by airborne particles. The still was sealed from air. The distillate was stored in sealed 1-l Pyrex flasks, with the first 50 cc of distillate discarded. Measurements or further processing were then carried out within 24 h.

In a fourth stage, the distillate from stage 3 was further distilled in a single-stage amorphous silica cell with the same bubbling of medical oxygen. The resulting water vapor flowed through a 39-cm long high-temperature zone (600°C) to further oxidize any remaining contaminants. The same sealing procedures were followed as in step 3.

In the final process, nitrogen was bubbled through the final distillate by the same method as described above. The nitrogen filter water was from the same distillate as the sample being prepared to avoid possible contamination by droplet entrainment. This final water samples were then kept sealed in a nitrogen atmosphere throughout the measurement process. The purpose of this last step is to remove the dissolved oxygen from the water. Dissolved oxygen was shown by *Heidt* and Johnson (1957) to strongly absorb in the UV region of the spectrum. *Quickenden* and Irvin (1980) also describe in great detail the procedures required to clean and keep free from contaminants all the containers and materials that enter in contact with the treated water at all stages in the experiment. The interested reader should consult the original article for a complete description.

Quickenden and Irvin measured the sample attenuation after each stage in the process by the differential method described previously. Over the spectral range of 200 to 320 nm and down to a value of attenuation of 0.01 m^{-1}, they conclude that water prepared by the three-stage process seems as optically pure as that prepared by the four-stage process as long as all samples are properly deoxygenated.

Grundinkina (1956) followed a very similar procedure starting with deionized water and distilling it from a solution of $KMnO_4$ and $Ba(OH)_2$. Subsequent to this first distillation, the water was triply distilled in silica still. No special precautions were taken to eliminate dissolved oxygen. Like Quickenden and Irvin, she used a differential attenuation meter. The attenuation was measured after each distillation step. The improvement in the quality of the water between the first and second distillation is quite substantial. The results for the first two steps are the average of nine measurements taken from three separate samples. For the final step, the improvements are smaller but still significant. For this step, the results of an average of 32 measurements from eight different samples are presented. Except for the absence of dissolved oxygen-removal process, the water quality obtained by Grundinkina should normally correspond to the same quality as that obtained by Quickenden and Irvin after step 3 of their purification. The difference between the results of *Grundinkina* (1956) and *Quickenden* and Irvin (1980) is a direct measure of the effect of dissolved oxygen on the UV attenuation of water from 210 to 320 nm. We will later see how these results can be used to extend the wavelength range of the work of *Heidt* and Johnson (1957) on the effect of oxygen (200 nm to 215 nm) .

The other sets of UV measurements were carried out by *Boivin* et al. (1986). Their first-stage water was obtained from a Nanopure II deionizer (Sybron Barnstead). In the second purification stage, they used a OrganicPure (D3600) Sybron Barnstead system to remove the organic contaminants. The third stage consisted of a fused silica double-distillation system from Quartz et Silice. They also did not deoxygenate their water. Their results at 254 and 313 agree within the quoted experimental error with those of Grundinkina. Both data sets are therefore consistent with one another.

Pope and Fry (1997) as well as *Sogandares* and Fry (1997) used very high-quality commercial water purification systems from both Culligan and Millipore to produce reagent grade type I water. This is the highest purity available from the best standard laboratory water purification systems. As an indication of its purity, the resistivity of this type of water peaks at 18 MΩ cm. Water at this level of purity will over days leach significant amounts of ions from Pyrex. This imposes strict limits on the time one can store water before conducting experiments. In all cases, the samples were produced and measured in the shortest time possible. Considerable effort was also expended in thoroughly cleaning all the glassware and containers used in the experiment. The combination of this type I reagent grade water and of the accuracy inherent in the integrating cavity absorption meter led to what are probably the definitive measurements to this date of the absorption of water in the visible and near UV. These measurements show an extraordinary small minimum absorption of $0.0044\,m^{-1}$ at 420 nm. It should also be noted that the water used in these experiments was not deoxygenated in any way. This leaves open the possibility that the true minimum absorption of pure water could lie even deeper at even shorter wavelength.

2.6. Analysis of the infrared and visible absorption spectrum

As mentioned before, the infrared and the visible part of the absorption spectrum of pure water above 450 nm is completely dominated by the vibrational–rotational transitions of the water molecule. As shown in Figure 2.2, water has three normal modes of vibration: the symmetric stretch, the asymmetric stretch, and the bending or scissors mode. Each of these modes can absorb radiation in the first excited vibrational level and also in higher excited states giving rise to a strong fundamental band and a series of progressively weaker overtone bands. Absorption can also occur to mixed or combination modes where both the scissors and one or both of the other normal modes are simultaneously excited in the either in the fundamental or overtone modes.

If all these modes had similar absorption strengths, it would lead to an undifferentiated mess from which little information could be garnered. Fortunately, as shown by *Kjaergaard* et al. (1994), it turns out that the pure asymmetric stretch mode and its overtones dominate. The symmetric stretch mode is generally a factor of 10 smaller for the same overtone. The frequencies of both modes are very close to one another. Their rotational components overlap. The first combination mode which involves excitation of the fundamental of the scissors mode along with the fundamental and subsequent overtones of either of the other two modes is also approximately an order of magnitude weaker than the corresponding pure mode. Since, however, the various absorption peaks of this first combination mode fall between the peaks of the pure stretch modes, they are easily recognizable in the absorption spectrum of either gaseous or liquid water.

Table 2.6 shows the different vibrational states predicted and observed for water in both the gaseous and liquid forms. The data for the gaseous from are taken from *Kjaergaard* et al. (1994) and references quoted therein. The optical transition strengths are given in terms of what are called oscillator strengths. These oscillator strengths are directly proportional to the absorption coefficients. In the first column, the first number (vs) denotes the overtone of the symmetric or anti-symmetric stretch mode involved in the transition and the second number (vb) denotes whether the scissors (bending) mode is also involved in the transition. The data for liquid water are displayed in the last two columns and come from the work of *Kou* et al. (1993) and of *Pope* and Fry (1997). Note the shift toward longer wavelengths for the same transition when going from the vapor to the liquid state.

The absorption data for liquid water from 2.5 μm to 380 nm (*Kou* et al. 1993, *Pope* and Fry 1997) are plotted in Figure 2.10 in terms of photon energy in units of cm^{-1}. This scale was chosen because it shows up the regularity in the absorption pattern of water. In the case of liquid water where the rotational bands are broadened and the individuals levels are therefore indistinguishable, the symmetric and anti-symmetric modes appear as one structure. The second apparent structure is the scissors stretch combination mode. The various levels of both overtone spectra are indicated by arrows in Figure 2.10. Given this simplification of the structure, *Patel* and Tam (1979) came up with a simple anharmonic formula to account for the spectral location, in cm^{-1}, of the peaks and shoulders of absorption in water:

$$\Delta\nu = \nu(3620 - 63\nu) \tag{2.74}$$

Table 2.6. Overtone modes of water in the vapor and liquid phases (*Kjaegaard* et al. 1994).

Overtone level (vs, vb)	Water vapor				Liquid water	
	Symmetric stretch mode	λ_1 [nm]	Anti-symmetric stretch	λ_2 [nm]	a[cm^{-1}]	λ_w [nm]
(1,0)	5.6×10^{-7}	2734	8.1×10^{-6}	2662	12026.	2811
(1,1)	4.2×10^{-8}	1910	9.1×10^{-7}	1876	136.7	1930
(2,0)	6.9×10^{-8}	1389	6.4×10^{-7}	1379	32.8	1453
(2,1)	1.6×10^{-9}	1141	4.0×10^{-8}	1135	1.28	1196
(3,0)	2.4×10^{-9}	943	1.9×10^{-8}	942	0.49	976
(3,1)	5.5×10^{-11}	824	1.4×10^{-9}	823	0.042	850
(4,0)	9.6×10^{-11}	723	1.2×10^{-9}	723	0.029	755
(4,1)	1.6×10^{-11}	652	1.0×10^{-10}	652	0.0046	662
(5,0)	2.3×10^{-11}	592	9.9×10^{-11}	592	0.0026	605
(5,1)	–	–	–	–	5.7×10^{-4}	550
(6,0)	–	–	–	–	4.0×10^{-4}	514
(6,1)	–	–	–	–	1.1×10^{-4}	474
(7,0)	–	–	–	–	9.2×10^{-5}	449
(7,1)	–	–	–	–	4.4×10^{-5}	418
(8,0)	–	–	–	–	6.6×10^{-5}	401

In the first column, abbreviation vs denotes the stretch mode vibrational number and vb denotes the bending mode vibrational number. The second and fourth columns contain experimental values of the oscillator strength (a quantity proportional to the absorption coefficient) for the vibrational transitions in the vapor phase. The third and fourth columns give the wavelength of the symmetric and anti-symmetric transitions in the vapor phase. The absorption coefficient in the liquid phase (*Kou* et al. 1993, *Pope* and Fry 1997) and the respective wavelength are given in the sixth and seventh column. Note that the ratios of the oscillator strengths of the higher overtones in the vapor phase are similar to the ratios of the absorption coefficients of the corresponding overtones in the liquid phase.

In order to obtain the corresponding spectrum of peaks and shoulders for the combination mode, we only need to add the frequency of the fundamental of the scissors mode, 1645.0 cm^{-1}. Although much smaller, these peaks and shoulders are easily detectable in the high-quality spectrum obtained by *Kou* et al. (1993) and by *Pope* and Fry (1997).

From the infrared to the visible, the amplitudes of the absorption associated with each mode decrease. As was shown in Table 2.4, the ratio of the magnitude from overtone to overtone follows closely the sequence one computes from a simple anharmonic oscillator model. This regular progression is broken as a minimum is reached around 400 nm (25 000 cm^{-1}), and the absorption starts increasing again toward the UV. This increased absorption is due to line broadening in the far wing

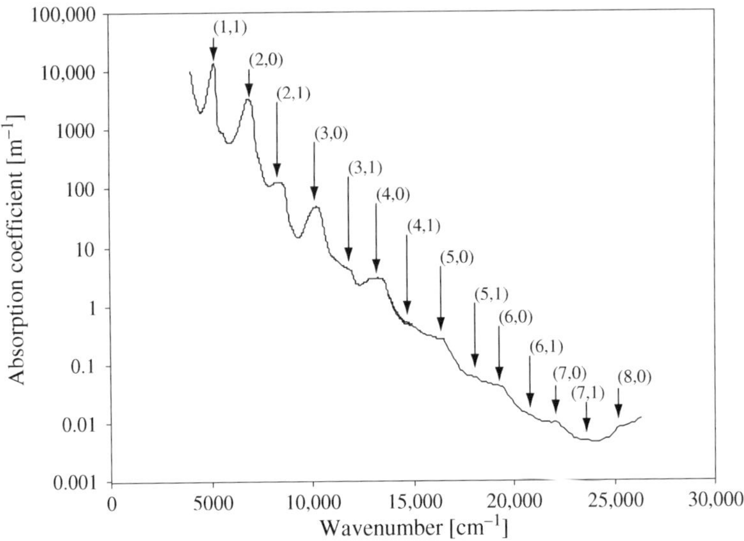

Figure 2.10. Absorption data for liquid water at 22°C from 2.5 μm to 380 nm plotted in terms of photon energy (wave number) in units of cm^{-1}. This scale was chosen because it accentuates a regularity in the absorption pattern of water. The symmetric and anti-symmetric stretch modes are those identified by $(n, 0)$. The scissors stretch combination modes are indicated by $(n,1)$. The various levels of both overtone spectra are indicated by the appropriate arrows.

of the first electronic absorption level of water. We will study this in further detail in the next section.

The spectrum shown in Figure 2.10 was taken at a temperature of 22°C. The overall structure is relatively insensitive to both temperature and salinity. However, small variations can be seen as a function of temperature in certain zones of the spectra. Some temperature dependence is to be expected even from the simplest vibrational–rotational model. Most of the effect comes from the redistribution of the available population of absorbers between the various energy levels of the first few lowest energy states. Since each level has a different absorption spectrum, a change in their relative populations will lead to a change in the spectrum.

The temperature and salinity dependence of the absorption spectrum was most recently studied in detail by *Trabjerg* and Højerslev (1996) and *Pegau* et al. (1997). Both found that the variation at any given wavelength of the absorption as a function of temperature was linear to the accuracy of the experimental results. The value of the slope was found to be a strong function of wavelength. The most significant effect occurs around 740 nm with a measured slope of 0.01 $m^{-1}\,°C^{-1}$. A second smaller peak is shown by Trabjerg and Højerslev to occur around 604 nm. Its magnitude is a factor of 10 smaller. Interestingly, the ratio of the peak values of the slopes of the temperature dependence scales as the ratio of the absorption coefficient itself. This behavior is consistent with what would be expected from a simple vibrational–rotational model.

In an effort to better represent the behavior as a function of wavelength of the temperature dependence, Pegau and colleagues decided to model the absorption spectrum as a set of Gaussian functions. Once this was done, they could simply assign a temperature dependence to the amplitudes of each Gaussian. They found that they needed up to four Gaussians to explain the structures around 740 nm and two Gaussians to model the peak around 660 nm. Given the complexity of the rotational spectrum, this is not surprising. The remarkable success of this approach is due to the fact that, as we have seen, simple individual rotational bands can be well approximated either by a single Gaussian or by a pair of Gaussians.

Pegau et al. (1997) used a reflecting tube absorption meter of the type described previously. These instruments measure absorption values relative to pure water. Absolute values can only be determined by reference to a standard absorption spectrum. The standard they used was the spectrum of *Smith* and Baker (1981). The values of Smith and Baker are approximately 5 to 15% lower than those reported by both *Kou* et al. (1993) and *Pope* and Fry (1997). This discrepancy prompted us to re-compute their fit and re-scale the temperature dependence appropriately. We have also extended the results to encompass the whole of the visible spectrum from 800 down to 380 nm.

The parameters of the Gaussians are given in Table 2.7 and Figure 2.11 is a graph as a function of wavelength. The solid line is the fit and the dots are experimental data (*Pope* and Fry 1997: 380 to 727.5 nm, *Kou* et al. 1993: 728 to 800 nm). The initial locations of the Gaussians were chosen by using Patel and Tam's formula (*Patel* and Tam 1979).

As expected, only minor wavelength adjustments were required afterward. The Gaussians we used were normalized. We used the following formulas:

$$a_w(\lambda, T) = a_w(\lambda, T_r) + \Psi_T(\lambda)(T - T_r) \tag{2.75}$$

$$a_w(\lambda, T_r) = \sum \frac{M}{\sigma} \exp\left[-\frac{(\lambda - \lambda_0)^2}{2\sigma^2}\right] \tag{2.76}$$

$$\Psi_T(\lambda) = \sum M_T \frac{M}{\sigma} \exp\left[-\frac{(\lambda - \lambda_0)^2}{2\sigma^2}\right] \tag{2.77}$$

The total absorption is the sum of the absorption at a reference temperature T_r and of a pure linear function of the temperature difference. The slope of the linear variation of the temperature is computed as a fraction M_T of the absorption at the reference temperature. For the pure stretch mode overtones, the fraction is constant at 0.0045. For the stretch and bend combination mode overtones, the fraction is also constant with a value of 0.002. In addition to the pure stretch and combination modes, four additional Gaussians were required to fit the zone from

Table 2.7. Gaussian parameters for the fit of the overtone modes of water [equations (2.75) through (2.77)].

Overtone level (vs, vb)	Amplitude, M [nm/m]	λ_0 [nm]	σ [nm]	Temperature slope fraction, M_T
(*, *)	47.48	795	29.87	−0.0010
(*, *)	23.33	775	24.79	−0.0010
(4,0)	35.07	744	20.28	0.0062
(*, *)	1.794	740	5.48	0.0045
(*, *)	9.216	697	28.22	−0.0010
(4,1)	4.955	669	24.78	0.0020
(*, *)	2.341	638	20.08	−0.0040
(5,0)	3.574	610	18.40	0.0045
(5,1)	1.310	558	22.84	0.0020
(6,0)	0.3359	517	13.52	0.0045
(6,1)	0.2010	485	19.27	0.0020
(7,0)	0.1161	449	18.86	0.0045
(7,1)	0.0138	415	15.79	0.0020
(8,0)	0.03839	396	20.88	0.0045
(8,1)	0.2219	370	21.09	0.0020

In the first column vs denotes the stretch mode vibrational number and vb the bending mode vibrational number. The extra levels used to complete the fit (see text) are denoted by (*, *). The third column gives the center wavelength of the Gaussian. Differences from the *Patel* and Tam (1979) form are the results of the fitting process. The corresponding standard deviation, σ, is given in the fourth column. The fraction by which the Gaussians must be multiplied to obtain the slope of the temperature dependence M_T is shown in the last column.

700 to 800 nm and one other Gaussian was required at 638 nm. These additional Gaussians are required simply because of the existence of a minimum of two rotational branches for each vibrational transition, each of which can require a separate Gaussian for an adequate model.

This simple approach yields an impressive fit to the visible and near infrared spectrum. The worst apparent discrepancies occur around the absorption shoulders and are particularly noticeable around 600 and 450 nm. These discrepancies are probably once again due to attempting to model the rotational substructure by simple Gaussians. The absorption data of Pope and Fry are the first that are sufficiently accurate to show clearly the appearance of the combination modes.

The variation of the spectrum of water with salinity was also found to be linear by *Pegau* et al (1997). This dependence was found to be much smaller than for temperature, however a clear effect could be demonstrated around 750 and 412 nm. The peak coefficients in units of m^{-1} are 0.012 S at 412 nm, −0.027 S at 715 nm, and 0.064 S at 750 nm, where S is the salinity in units of g l^{-1}. The data in all instances are too sparse to allow a Gaussian fitting approach.

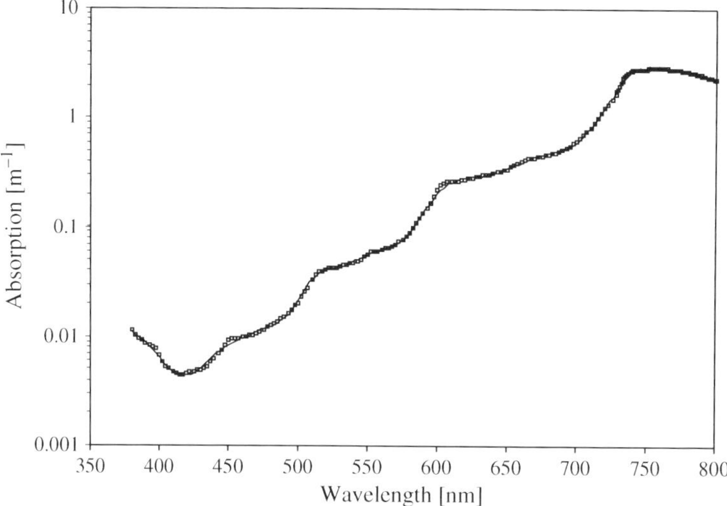

Figure 2.11. The absorption coefficient of pure water at 22°C as a function of wavelength from 380 to 800 nm. The solid line is the fit using a set of Gaussians whose parameters are given in Table 2.7 and the dots are the experimental data (*Pope* and Fry 1997: 380 to 727.5 nm, *Kou* et al. 1993: 728 to 800 nm). The initial locations of the Gaussians for the fit were chosen by using Patel and Tam's formula (*Patel* and Tam 1979).

As we have seen, the visible and infrared spectrum of water can be explained and modeled in some satisfying detail. Further work will be required to obtain a better fit to the fine structure of the spectrum. The simple Gaussians used will first need to be replaced by more realistic functions. These functions exhibit an explicit temperature dependence. Their predicted temperature dependence will need to be matched to the experimental observations. If computation from theory of the ratio of the overtones of the pure and combination bands is improved, it is conceivable that the relative amplitudes of all the modes could also be fixed by fitting them to parameters directly related to the internuclear potential itself. Given the high quality of the data that has recently become available, achieving a good model for the absorption spectrum of water in the visible and infrared seems within reach.

2.7. Analysis of the UV absorption spectrum

In order to extend the results further into the blue and UV, the effects of the first electronic transition at 150 nm and its extension by line broadening to the near UV and blue region of the spectrum absorption must be analyzed in detail. We will see that the situation in the UV is not yet in a satisfactory state as for the visible and infrared, both from the theory and from the experimental point of view. It should be noted that in all cases where attenuation was measured, we used the intrinsic scattering formula of Buiteveld and colleagues (*Buiteveld* et al. 1994) with the *Quan* and Fry (1995) refractive index to evaluate the total intrinsic scattering and subtract it from the attenuation to obtain the net absorption coefficient.

We will start our analysis from the short wavelength side of the spectrum and gradually progress back to the blue region. The best results obtained so far in the UV are undoubtedly those of *Quickenden* and Irvin (1980). Their data extend from 196 to 320 nm. The long wavelength limit of the sharp rise that marks the core of the continuous first electronic transition $X^1A_1 - {}^1B_1$ of water, starts around 205 nm. *Ghormley* and Hochanadel (1971) also obtained oxygen-free absorption data from 180 to 215 nm. Their results agree within experimental error with those of Quickenden and Irvin over their zone of overlap. However, results of Ghormley and Hochanadel extend much further on the short wavelength side and therefore allow a more reliable fit to be obtained for the falling edge of the core of the transition.

Ghormley and Hochanadel also give several absorption curves taken at different temperatures which clearly show the occurrence of a strong wavelength shift of the edge toward longer wavelength as the sample is heated. This wavelength shift toward longer wavelength is explained by the redistribution of the population between the low-lying vibrational states of the ground electronic state (*Herzberg* 1950). The absorption spectra produced by the higher vibrational states are both broader due to the increased spacing between the radial turning points of the vibration and shifted to longer wavelength due to the reduction in the total energy of the transition by the energy difference between the vibrational states. The net effect is a broadening and shift of the absorption spectrum. The following function was found to fit their data and is compatible with the theoretical form resulting from the previous arguments and (2.17)

$$a = 7.067 \times 10^{-40} \exp\left(119.0 \frac{\lambda_r}{\lambda - \Delta\lambda}\right) \tag{2.78}$$

$$\Delta\lambda = 0.0465(T_c - 25.0) \tag{2.79}$$

where λ_r is a reference or scale wavelength arbitrarily chosen as 150 nm and T_c is the temperature in °C. Both sets of data show a sharp change of slope occurring around 210 nm. This second slope is much less pronounced and stretches out right to the limit of the measurements. No structure is evident in this wavelength interval. *Quickenden* and Irvin (1980) computed the amount of intrinsic scattering in an effort to explain these results. The pure molecular scattering term turned out to be insufficient to explain the far wing in the attenuation spectrum. They assumed the existence of excess scattering from very small particle. Given the care with which they carried out the preparation of their samples, excess small-particle scattering is an unlikely source.

In dense media, molecules are subjected to frequent collisions. During these collisions they behave as a dimer pair with an intermolecular potential given by an equation like (2.1). During the course of the collisions, particles which have some kinetic energy with respect to one another will undergo transitions that are shifted in energy from their unperturbed state by the energy difference between the dimer potentials at the turning point of the lower state potential. This turning point is defined as the point where the relative kinetic energy of the pair of water molecules is equal to the potential energy of the dimer in its ground state. If the repulsive portion of the dimer potential of the upper state increases more slowly as a function of intermolecular separation than the repulsive portion of the lower state, the transition shifts will lead to the appearance of a featureless wing on the long wavelength side of the absorption line or continuum. *Szudy* and Bayliss (1975) managed to simplify the evaluation of the probability of transition during collisions and obtained a simple formula to account for this effect. For dimer potentials with an inverse 12th power repulsive portion, Szudy and Bayliss show that the appropriate form of the collision broadened far wing is given by:

$$a(\nu) = \frac{K_1}{\nu^{5/4}} e^{-K_2 Z_c} \tag{2.80}$$

where

$$Z_c = \left[\frac{\nu^{11/12}}{\sqrt{T}} \right]^{2/3} \tag{2.81}$$

with T being the absolute temperature in K. We have used this form to fit the far wing UV data of *Quickenden* and Irvin (1980) at the stated temperature. With the frequency expressed in inverse centimeters, the resulting values for the constants are:

$$a(\nu) = \frac{5.0 \times 10^6}{\nu^{5/4}} \exp[-0.076 Z_c] \tag{2.82}$$

The lower curve in Figure 2.12 shows both the data sets of *Ghormley* and Hochanadel (1971) and *Quickenden* and Irvin (1980) at 25°C and their fit with the expressions given above. The simple fits are excellent. The great advantage of basing them on theory is the ability to handle the temperature dependence explicitly. Equations (2.79), (2.80), and (2.82) are a complete fit to the absorption spectrum of oxygen-free pure water.

The effect of oxygen on pure water absorption between 200 and 215 nm was carefully studied in a remarkably thorough paper by *Heidt* and Johnson (1957). They first noted that the absorption by oxygen dissolved in water was much greater than the absorption of an equivalent amount of gaseous oxygen. They attributed this increased absorption in solution to the occurrence of weak bonds between the oxygen and water molecules.

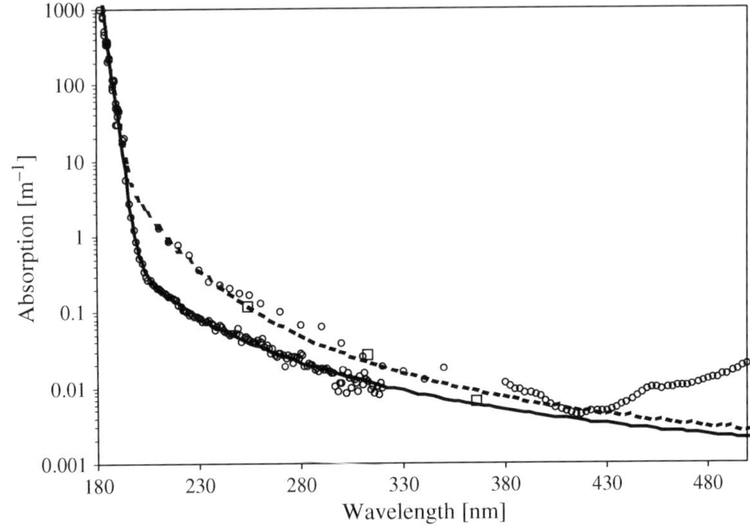

Figure 2.12. The lower group of data and thick curve show both the data sets of *Ghormley* and Hochanadel (1971) and *Quickenden* and Irvin (1980) at 25°C for water free of dissolved oxygen as well as the fit using the simple expressions given in the text. The upper curve (dotted line) is the result of the addition of this fit with the fit for the effect of dissolved oxygen. The upper group of data show absorption of air saturated water at 19°C (*Grundinkina* 1956). Data points from the work of *Boivin* et al. (1986) are shown as squares about the dotted curve. All appropriate corrections were made, and the fit to data for air saturated water was computed for 19°C. The fit was then corrected to 22°C and compared with the data of *Pope* and Fry (1997, far right group of data points). This fit is indistinguishable from the one at 19°C.

By studying the temperature dependence, they deduced the existence of two different binding states O_2' and O_2'' and established their absorption spectrum and the variation of their relative concentration as a function of temperature and amount of dissolved oxygen. Denoting by α the fraction of dissolved O_2 present as the first state O_2' and by ε' the specific absorption spectrum associated with that state, the total specific absorption is given by:

$$a_{\text{air}} = 0.1917 a_{o2} \tag{2.83}$$

$$a_{o2} = S_{o2}\varepsilon \tag{2.84}$$

$$\varepsilon = \varepsilon'\alpha + (1-\alpha)\varepsilon'' \tag{2.85}$$

$$\alpha = \frac{K_0}{1+K_0} \tag{2.86}$$

$$K_0 = 15850.0 \, \exp\left(-\frac{2993.4}{T}\right) \tag{2.87}$$

where a_{air} is the excess UV water absorption in m^{-1} due to the amount of oxygen dissolved in water when air is present. a_{o2} is the excess absorption due to dissolved O_2. S_{o2} is the solubility in moles per liter of pure oxygen in water. The solubility of pure O_2 is given by:

$$S_{o2} = 0.0127692 - \frac{9.06725}{T} + \frac{1677.41}{T^2} \tag{2.88}$$

where T is the temperature in Kelvin. Over the wavelength range studied by *Heidt* and Johnson (1957), the spectrum of the excess molar absorption due to dissolved oxygen can be represented by:

$$\varepsilon' = 4.68 \times 10^{-6} \exp\left(28.0\frac{\lambda_r}{\lambda}\right) \tag{2.89}$$

$$\varepsilon'' = 2.04 \times 10^{-10} \exp\left(40.6\frac{\lambda_r}{\lambda}\right) \tag{2.90}$$

where λ_r is once again a reference or scale wavelength arbitrarily chosen as 150 nm. When all the above expressions are used to evaluate at 19 °C the excess absorption due to dissolved oxygen and added to the fit obtained previously for oxygen-free water, the results agree with the UV absorption data of *Grundinkina* (1956) over the 200 to 215 nm range. Unfortunately, the data of Heidt and Johnson do not extend to longer wavelengths and simply extrapolating from expressions (2.89) and (2.90) has little chance being correct since the same sharp slope change due to line broadening should also be present in the spectra of both dissolved oxygen states.

The most promising approach to resolving this problem is to simply use the data of *Grundinkina* (1956) and *Boivin* et al. (1986) to extend the range where the difference between oxygen-free and air saturated water absorption can be evaluated. This difference can also be fitted with an expression of the type suggested by *Szudy* and Bayliss (1975). The final result is given by:

$$\varepsilon' = \frac{1.36 \times 10^{13}}{\nu^{5/4}} \exp(-0.144 Z_c) \tag{2.91}$$

$$\varepsilon'' = \frac{3.79 \times 10^{12}}{\nu^{5/4}} \exp(-0.144 Z_c) \tag{2.92}$$

It was not necessary to add another exponential term of the form found in equations (2.89) and (2.90) to fit the results to first order. To obtain equations (2.91) and (2.92), the long wavelength fit to the oxygen excess absorption was apportioned for each state according to the ratio of the measured amplitudes at 215 nm. All appropriate corrections were made, and the fit to air saturated water was computed for the same temperature as that of Grundinkina's experiment, 19°C.

The upper dotted curve in Figure 2.12 is the result of the addition of this fit to the fit for oxygen-free water. This curve represents the fitted UV absorption corrected to 22°C. The data points of *Boivin* et al. (1986) are the small squares close to the dotted curve. The group of points on the far right of the graph are the data of Pope and Fry also taken at 22°C. The extrapolation of the results of Grudinkina (1956) and *Boivin* et al. (1986: squares in Figure 2.12) exceeds very slightly the minimum absorption measured by Pope and Fry at 420 nm. This error is well within the variation expected given the experimental errors of the data of Grundinkina and of Boivin and colleagues.

If we keep in mind that the data shown in Figure 2.12 were collated from several experiments that took place over a time period of 40 years, the agreement is quite satisfactory. It gives confidence that the explanations for the spectral features of the absorption spectrum are on a sound footing and raises hopes that we should soon be able to accurately represent the complete spectrum by a compact set of simple formulas.

To complete this program, new detailed measurements need be taken to cover the blue and UV spectral regions with oxygen removed and oxygen present at various temperatures and concentrations. As mentioned by *Quickenden* and Irvin (1980), particular attention should be given in all cases to ensuring as complete a removal of organic compounds by at least one oxidative step. As can be seen from the data, the current minimum absorption is $0.0044\,m^{-1}$ at 420 nm for air saturated water. The potential minimum for oxygen-free water could be considerably smaller, and the level of sensitivity and accuracy required to carry out an accurate absorption experiment on this type of water is daunting still today. The most promising approach so far appears to be the integrating cavity absorption meter in combination with improved water purification systems.

2.8. Organic substances dissolved in the water column: *Gelbstoff*

The organic residue from the biological processes occurring in natural waters and from organic matter entrained from land to the ocean by rivers absorbs predominantly in the UV and the blue and therefore appears yellow, thus the name *Gelbstoff* or yellow substance given to it by the first investigators. The same substance is also referred to as CDOM, an acronym that stands for chromophoric or colored dissolved organic matter.

The amount of CDOM present in water is often the dominant factor in determining the apparent color of ocean water. Since Gelbstoff absorbs more in the UV and blue regions of the spectrum, as its concentration increases, the apparent color of the water column will slowly change from blue-violet in very clear open ocean waters to green in ocean waters nearer to shore. The measurements of CDOM are carried out by first carefully filtering the seawater with 0.2 μm to a maximum of 0.4 μm pore size filters. The absorption of what has passed through the filters is then measured and the absorption of pure water subtracted to determine the intrinsic absorption spectrum of CDOM.

As demonstrated by *Carder* et al. (1989), the two main components of this mixture are humic and fulvic acids. The absorption coefficients for both acids decrease exponentially as a function of wavelength.

$$a_{\mathrm{f}}(\lambda) = C_{\mathrm{f}}\ 35.959\ \exp(-0.01105\lambda) \tag{2.93}$$

$$a_{\mathrm{h}}(\lambda) = C_{\mathrm{h}}\ 18.828\ \exp(-0.0189\lambda) \tag{2.94}$$

$$\frac{C_{\mathrm{h}}}{C_{\mathrm{h}} + C_{\mathrm{f}}} \approx 0.1 \tag{2.95}$$

where λ is the wavelength in nanometers and the concentration of fulvic C_{f} and humic acids C_{h} in milligrams per cubic meter. The absorption coefficients themselves are in inverse meters. These formulas along with the relationship between the concentration of humic and fulvic acid are though to apply to open ocean water (*Haltrin* 1999).

The real situation is considerably more complex than described above, and some significant variability has been observed, particularly when one works closer to shore and in the UV (*Højerslev* and Aas 2001, 1998). The small-particle fraction remaining in the filtrate can also affect the results significantly (*Aas* 2000).

2.9. An important special case: chlorophyll

In opposition to *Gelbstoff*, chlorophyll is only present in photosynthetic cells and quickly decomposes when freed into water. It is thus an exception in the context of our discussion of absorption. However, even if it is contained within the body of scattering particles, we will mention it here because the resulting absorption, even though due to particles, is one of the dominant factors influencing the overall absorption measured in the ocean.

According to simple models (*Haltrin* 1999), the total absorption coefficient of seawater can be written as:

$$a(\lambda) = a_{\mathrm{w}}(\lambda) + a_{\mathrm{f}}(\lambda) + a_{\mathrm{h}}(\lambda) + a_{\mathrm{c}}(\lambda) \tag{2.96}$$

where a_{c} is the absorption due to chlorophyll. This absorption is generally expressed (*Haltrin* 1999) most conveniently in the following form:

$$a_{\mathrm{c}}(\lambda) = a_{\mathrm{c}}^{0}(\lambda) \left(\frac{C_{\mathrm{c}}}{C_{\mathrm{c}}^{0}} \right)^{0.602} \tag{2.97}$$

where $a_c^0(\lambda)$ is the specific absorption coefficient of chlorophyll at a reference concentration $C_c^0 = 1\,\mathrm{mg\ m^{-3}}$.

The non-linear dependence on concentration is due to the fact that the chlorophyll is contained in cells. The total concentration is proportional to the total volume occupied by the cells, while the absorption is proportional to the total projected area of the same cells. This is strictly true in the case where the individual cells absorb almost all the light that is incident on them, i.e., their absorption efficiency Q_{abs} is close to 1. In that case, the absorption coefficient should be proportional to the 2/3 power of the concentration. This is very close to the situation described by (2.97).

Figure 2.13 shows the absorption coefficient of phytoplankton from 400 to 800 nm at 1 mg Chl*a* m^{-3} according to an expression developed by Bricaud and colleagues by using the world ocean data (*Bricaud* et al. 1995). It shows clearly the two characteristic absorption peaks at 440 and 715 nm. This peculiar structure is what gives chlorophyll-bearing plants their characteristic green color.

We have briefly presented all the significant elements required to compute the absorption coefficient of seawater. In an interesting approach, *Haltrin* (1999) has recently attempted to correlate all the concentrations together so that a single parameter, chlorophyll concentration, could be used to fix the inherent optical

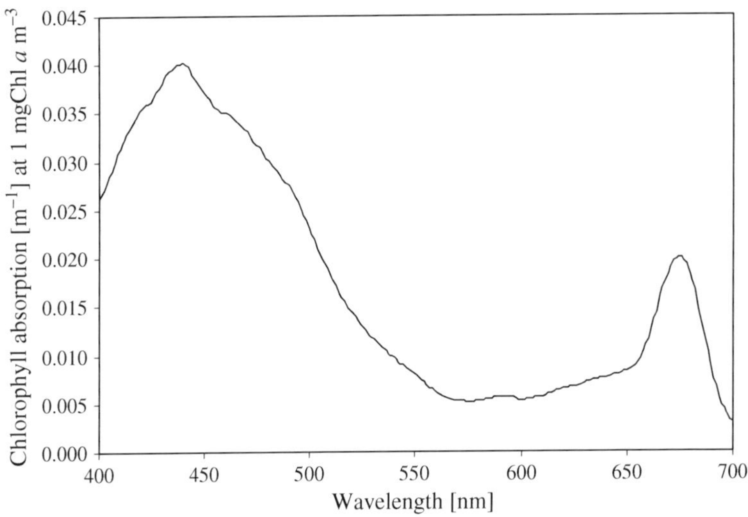

Figure 2.13. The absorption coefficient of phytoplankton at a concentration of chlorophyll *a* of 1 mg m^{-3}. The curve was calculated according to an expression developed by *Bricaud* et al. (1995) by using data obtained in various regions of the world ocean. Their data represent a chlorophyll *a* concentration range of 0.02 to 25 mg m^{-3}. The curve shows clearly the two characteristic absorption peaks at 440 and 690 nm. This peculiar structure is what gives chlorophyll-bearing plants their characteristic green color.

properties of seawater. He found that the following simple relations apply in ocean waters where the chlorophyll concentration is less than 12 mg m^{-3}.

$$C_{\mathrm{f}} = 1.74098\, C_{\mathrm{c}}\ \exp\left(0.12327\frac{C_{\mathrm{c}}}{C_{\mathrm{c}}^{0}}\right) \tag{2.98}$$

$$C_{\mathrm{h}} = 0.19334\, C_{\mathrm{c}}\ \exp\left(0.12343\frac{C_{\mathrm{c}}}{C_{\mathrm{c}}^{0}}\right) \tag{2.99}$$

He also found that his model is consistent with a relationship proposed by *Gordon* and Morel (1983), relating the ocean surface concentration of chlorophyll to diffuse reflectance. This relationship can in fact be used to estimate the chlorophyll concentration in the surface layer of the ocean by simply measuring the ratio of the diffuse reflectance, R, in the green at 550 nm to that in the blue at 440 nm.

$$C_{\mathrm{c}} = 1.92\left[\frac{R(550)}{R(440)}\right]^{1.8} \tag{2.100}$$

These relationships allow the computation of approximations to the absorption properties of seawater. Such simple models are very useful in many remote sensing applications and in establishing consistent sets of optical properties for modeling the underwater light field. A consistent estimate of the potential operating range of an optical instrument is also obviously invaluable to both the instrument designer and the experimenter trying to optimize the experiment either by choosing appropriate instruments or by appropriate operating settings for these instruments.

2.10. Problems

The material of which some scattering particles are made is assumed to have an absorption coefficient that is constant as a function of wavelength. For an ensemble of these particles whose size distribution follows a power law with a negative exponent [$f(D) = kD^{-m}$, where k and m are constants], derive the wavelength dependence of the absorption coefficient due to these particles and the wavelength dependence of the extinction (attenuation) coefficient.

Chapter 3

General features of scattering of light by particles in water

3.1. Introduction

We have so far seen how to compute the intrinsic scattering and the scattering from particles much smaller than the wavelength. Scattering in seawater and natural waters is due to ensembles of particles of many sizes, shapes, and structures. We therefore need to carefully study scattering from larger particles. There are several good books on the subject of scattering by particles on the order of and much larger than the wavelength (*Bohren* and Huffman 1983, *Kerker* 1969, *Deirmendjan* 1969, *van de Hulst* 1957). An interested reader may also want to consult reviews of the field which are published from time to time (e.g., *Jones* 1999). We will attempt not to treat the subject in as much detail as they do but instead to construct a base which will permit us to reach certain general conclusions applicable to modeling of light scattering by particles in water.

As we mentioned before, the scattering of light by natural waters is the result of interactions of a large number of different particles with the incident light. In seawater, particles larger than the wavelength of light are in general well separated from one another. This, along with their random distribution in space, causes results of their interactions with light to be incoherent. Thus, one can simply sum the scattering properties of a group of particles to obtain the overall effect of these particles acting together. Our aim is to obtain as much insight as possible into the behavior of this total scattering. The fact that we are dealing with integrals over large groups of scattering particles actually simplifies that behavior. Many detailed complex features of single-particle scattering virtually disappear from the ensemble scattering, and we are left with a simpler situation. This leaves open the possibility of using some powerful simplifying assumptions that retain only the dominant features of single-particle scattering to obtain results which match closely the experiments and more detailed theories.

In order to do this, we will try to gain as much insight into the basic physics of scattering. We will then carefully study various approximations based on these basic physical processes. Obviously, the efficiency and accuracy of approximations can only be gauged against exact results. With this purpose clearly in mind, we will acquaint ourselves with the basic methods of solution for obtaining exact solutions to the problem of scattering from a single particle. Using some of these exact solutions, we will study precisely what features are captured and which are neglected by the various approximations. This will allow us evaluate the suitability of these approximations for computing the integrated scattering from typical particle distributions in seawater.

3.2. An inventory of solutions

Many important features of electromagnetic scattering are difficult to model by other means than an exact treatment. In this category, one can include the details of the polarization effects, particularly in the case of non-spherical particles, and the structure of the resonances occurring both in the body and on the surface of the particle. The effect of these resonances is particularly noticeable in the backscatter direction.

The edge effect is another problem that is difficult to account for except with an exact theory. This effect produces an excess amount of scattering over what would be expected from the sum of diffraction and refraction terms. This phenomenon occurs because of the requirement that the electric and magnetic field components tangential to the surface of the scattering particle be continuous across the surface. This matching condition imposes an additional distortion on the incident field near the edge of the particle.

All these difficulties imply that there is still a very real need for exact solutions. In the following sections, we will briefly describe some of the more common cases for which exact solutions or methods of solution have been established.

3.2.1. Exact analytic solutions: spheres

The general analytic method of obtaining exact solutions for the scattering of an electromagnetic wave proceeds as follows. It can be shown (*Stratton* 1941) that in the case of wave equations with a sinusoidal time dependence, the electric and magnetic field vectors can be replaced by two other vectors that can themselves be derived from a scalar function. This scalar function also satisfies the homogeneous wave equation. The new vectors are called the vector spherical harmonics and are defined as follows (for example, *Bohren* and Huffman 1983).

$$\mathbf{M} = \nabla \times (\mathbf{r}\psi) \tag{3.1}$$

$$\mathbf{N} = \frac{\nabla \times \mathbf{M}}{k} \tag{3.2}$$

If in the case of spheres we choose the radius vector of the sphere to be the vector $\mathbf{r}$, $\mathbf{M}$ and $\mathbf{N}$ are the vector spherical harmonics. $\mathbf{M}$ is everywhere tangential to all the spherical surfaces defined by $|\mathbf{r}| = \text{constant}$. In the case of a cylinder, we would choose the axis of the cylinder, as the vector $\mathbf{r}$. The resulting $\mathbf{M}$ and $\mathbf{N}$ would then become the cylindrical vector harmonics. In all cases, the function ψ must satisfy the homogeneous scalar wave equation.

It is well known that the three-dimensional homogeneous wave equation is solvable in a variety of geometries by the method of separation of variables. In those geometries, the solution can be expressed as a product of three one-dimensional functions. In the case of the wave equation, the set of geometries is in fact completely specified. It includes the case of cylindrical, spherical, spheroidal, and ellipsoidal coordinates.

Once the form of the solution has been expressed as a product in the appropriate coordinate system, we need to determine the values of a group of arbitrary coefficients. These values are obtained by solving a set of field-matching conditions that must be satisfied along the surface of the particle. The electric and magnetic field components tangential to the surface of the scattering particle must be continuous across the surface. These conditions involve relationships between vector potentials representing the incident plane wave, the scattered wave outside the particle, and the field inside the particle. These conditions are in fact sufficient to fix the value of all the arbitrary coefficients of the formal solution.

Since the field-matching conditions must be applied along and across the boundary of the particle, by using the method outlined above we can in principle only solve for cylinders, spheres, spheroids, and ellipsoids. With some substantial additional work, we can also solve for layered objects with any of these shapes. An arbitrary number of layers of various materials can be used (*Gurwich* et al. 2000, *Bhandari* 1985, *Kerker* 1969). Solutions for spheres, cylinders, and spheroids have been extensively studied.

The first exact solution for the scattering of electromagnetic radiation from dielectric particles of arbitrary size was obtained for spheres by several workers at the turn of the century. One of the most complete presentation was given by Gustav Mie in 1908 (*Mie* 1908), and ever since the scattering from dielectric particles has been called rightly or wrongly Mie scattering. The term *Mie scattering* is sometimes also used for light scattering by non-spherical particles. For those interested in further studying the complex history of this field, *Kerker* (1969) gives an illuminating resume.

For reference, the independent solutions of the spherical vector potential are structured as follows.

$$\psi_{\mathrm{e}\ pq} = \cos(q\phi) P_p^q(\cos\theta) z_p(kr) \tag{3.3}$$

$$\psi_{\mathrm{o}\ pq} = \sin(q\phi) P_p^q(\cos\theta) z_p(kr) \tag{3.4}$$

$$\psi_p(x) = x\, j_p(x) \tag{3.5}$$

$$\chi_p(x) = x\ y_p(x) \tag{3.6}$$

$$\zeta_p(x) = \psi_p(x) + i\chi_p(x) \tag{3.7}$$

The $\psi_{\mathrm{e}\ pq}$ and $\psi_{\mathrm{o}\ pq}$ are the even and odd solutions for the various values of the indices, p and q, both assuming integer values, arising from the process of separation of variables. The spherical Bessel functions (e.g., *Abramowitz* and Stegun 1964) are denoted by j_p and y_p. Symbol z_p denotes any of the spherical Bessel functions j_p, y_p. The solutions as functions of the azimuth angle, ϕ, in a plane perpendicular to the axis of symmetry imposed by the direction of propagation of the incident wave are given by simple sine and cosine combinations. In the radial direction, the solutions are expressed in terms of the Ricatti–Bessel functions ψ_p and χ_p. In the scattering plane, which contains the incident and scattered directions, the angular distribution is given by the associated Legendre polynomials $P_p^q(\cos\theta)$ (e.g., *Abramowitz* and Stegun 1964).

The final solution of the scattering problem is obtained by equating the coefficients of the expansions in vector spherical harmonics of the plane incident wave, the scattered wave, and the field inside the particle (*Kerker* 1969, *van de Hulst* 1957). The expansion of the incident plane wave contains only vector spherical harmonics with $q = 1$. All the other terms are identically zero (e.g., *Bohren* and Huffman 1983). The resulting equations therefore only involve terms with $q = 1$ and can be easily solved. This process results in simple expressions for the various vector spherical harmonic expansion coefficients. In those expressions, n is the complex refractive index, and the particle size is expressed by a dimensionless size parameter

$$x = kr = \frac{2\pi r}{\lambda} \tag{3.8}$$

where r is, for now, the particle radius and λ is the wavelength of light in the medium surrounding the sphere. This procedure results in two sets of equations for two coefficients each. The two coefficients of the first set are the only ones involving the amplitude of the scattered wave at infinity, and in fact their knowledge is sufficient to completely determine all the scattering parameters (*Bohren* and Hufmann 1983). The coefficients a_p are the amplitudes of the vector spherical harmonics with even symmetry and the b_p are the corresponding amplitudes for the odd symmetry terms.

$$a_p = \frac{\psi_p'(nx)\psi_p(x) - n\psi_p(nx)\psi_p'(x)}{\psi_p'(nx)\zeta_p(x) - n\psi_p(nx)\zeta_p'(x)} \tag{3.9}$$

$$b_p = \frac{n\psi_p'(nx)\psi_p(x) - \psi_p(nx)\psi_p'(x)}{n\psi_p'(nx)\zeta_p(x) - \psi_p(nx)\zeta_p'(x)} \tag{3.10}$$

The second pair of coefficients is related to the amplitude of the field inside the particle (*van de Hulst* 1957):

$$c_p = \frac{i}{\psi_p'(nx)\zeta_p(x) - n\psi_p(nx)\zeta_p'(x)} \tag{3.11}$$

$$d_p = \frac{i}{n\psi_p'(nx)\zeta_p(x) - \psi_p(nx)\zeta_p'(x)} \tag{3.12}$$

These coefficients are not used here and are mentioned only for the sake of completeness. The scattering amplitude functions are given by (*Bohren* and Huffmann 1983, *van de Hulst* 1957):

$$S_1(\theta) = \sum_{p=1}^{\infty} \frac{2p+1}{p(p+1)} \left\{ a_p \frac{P_p^1(\cos\theta)}{\sin\theta} + b_p \frac{d}{d\theta} P_p^1(\cos\theta) \right\} \tag{3.13}$$

$$S_2(\theta) = \sum_{p=1}^{\infty} \frac{2p+1}{p(p+1)} \left\{ b_p \frac{P_p^1(\cos\theta)}{\sin\theta} + a_p \frac{d}{d\theta} P_p^1(\cos\theta) \right\} \tag{3.14}$$

Each scattering amplitude function refers to one of the two orthogonal states of linear polarization of the scattered wave. The normalized scattered intensities of these polarizations are:

$$i_1(\theta) = |S_1(\theta)|^2 \tag{3.15}$$

$$i_2(\theta) = |S_2(\theta)|^2 \tag{3.16}$$

The total intensity of light scattered in an arbitrary direction with respect to the direction of an incident polarized wave of unity irradiance is:

$$F(\theta, \phi) = i_2(\theta)\cos^2\phi + i_1(\theta)\sin^2\phi \tag{3.17}$$

Note that here the intensity ($W\,sr^{-1}$) and irradiance ($W\,m^{-2}$) are radiometric quantities. As we noted it previously, radiometric irradiance corresponds to intensity in the traditional physical nomenclature. We stress that the physical intensity ($W\,m^{-2}$) is not the radiometric intensity.

By integrating over ϕ, from (1.43) and (1.50), we have the differential scattering cross-section of a homogeneous sphere:

$$\begin{aligned} \sigma_{scat}(\theta) &= \frac{2\pi}{k^2} \frac{i_1(\theta) + i_2(\theta)}{2} \\ &= \frac{\pi}{k^2} [i_1(\theta) + i_2(\theta)] \end{aligned} \tag{3.18}$$

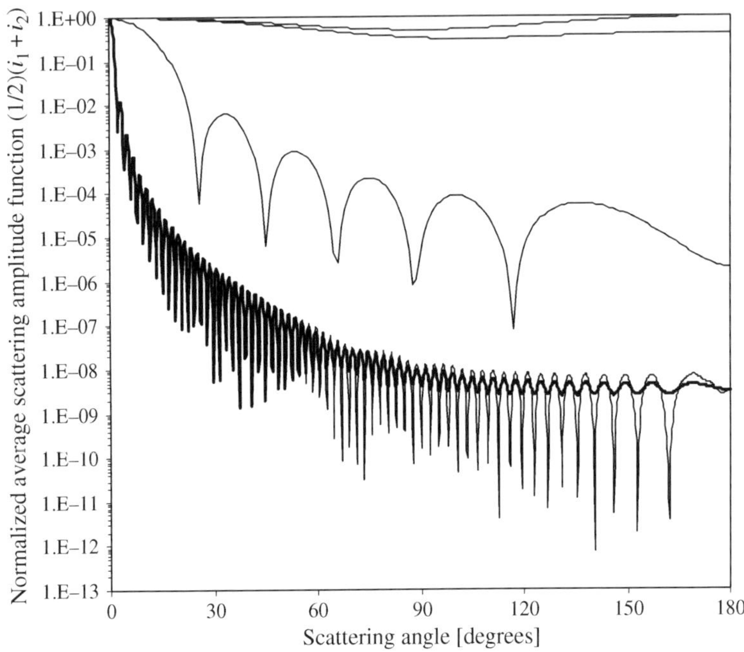

Figure 3.1. Sample (normalized) average scattering amplitude function $(1/2)(i_1+i_2)=M_{11}$ for homogeneous spheres with the size parameters $x=0.1$ (top curve) 1, 10, and 100 (bottom curve) and the refractive index $n=n'-in''=1.01-i0$ (except for the thick curve which refers to $n=1.01-i0.005$). The calculations were performed in double-precision arithmetic with programs developed by MJC Optical Technology. We stress that these are samples only and that one should refrain from generalizations apart from those discussed in the text. The plotting angle increment is 1° for $x=0.1$, 1, and 10 and 0.2° for $x=100$. The depths of the sharp resonant minima for the $x=10$ and $x=100$ curves are somewhat distorted by the coarse angle increment ($x=100$ for $n''=0.005$). The curves for $n''=0.005$ for the smaller spheres are not shown because they essentially overlap with those for non-absorbing spheres ($n'=0$). The values of un-normalized M_{11} at the scattering angle of 0° for the thin curves (top to bottom) are 4.43×10^{11}, 4.46×10^{-5}, 4.51×10^{1}, and 3.56×10^{7}. That value for the thick curve is 2.29×10^{7}.

Sample exact scattering cross-sections, as $(1/2)(i_1+i_2)$, of homogeneous spheres for sizes and refractive indices that span ranges relevant to particles in water are shown in Figure 3.1 and Figure 3.2. The quantity of $(1/2)(i_1+i_2)$ is the element M_{11} of the scattering (or Mueller) matrix that we referred to earlier. That matrix completely specifies scattering of light of arbitrary polarization by a particle. We will discuss in Chapter 4 in more detail how the scattering matrix is defined, used, and measured.

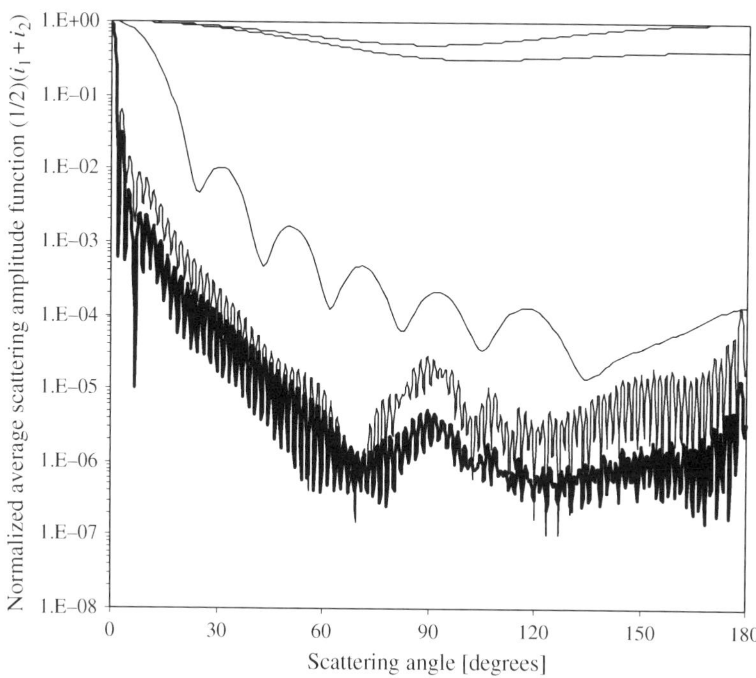

Figure 3.2. Sample (normalized) average scattering amplitude function $(1/2)(i_1+i_2)=M_{11}$ for homogeneous spheres with the size parameters $x=0.1$ (top curve), 1, 10, and 100 (bottom curve) and the refractive index, $n=n'-in''=1.1-i0$ (except for the thick curve which refers to $n=1.1-i0.005$). The calculations were performed in double-precision arithmetic with programs developed by MJC Optical Technology. We stress that these curves are samples only and that one should refrain from generalizations apart from those discussed in the text near Figure 3.1. Note that the vertical scale has changed as compared with Figure 3.1. The plotting angle increment is 1° for $x=0.1$, 1, and 10 and 0.2° for $x=100$. The depths of the sharp resonant minima for the $x=10$ and $x=100$ curves are somewhat distorted by the coarse angle increment ($x=100$ for $n''=0.005$). The values of un-normalized M_{11} at the scattering angle of 0° for the thin curves (top to bottom) are 4.28×10^{-9}, 4.60×10^{-3}, 4.10×10^{3}, and 2.21×10^{7}. The value for the thick curve is 2.56×10^{7}.

These figures should be regarded strictly as samples and one should refrain from generalizations of the scattering pattern behavior apart from perhaps three features: (1) the angular patterns of light scattering become increasingly forward-peaked as the sphere size increases, (2) the frequency of oscillations with the scattering angle increases with the sphere size, and (3) these oscillations may be significantly damped for absorbing particles.

The oscillations in the scattering patterns, especially for the large spheres, are very sensitive to numerical interrelations between the sphere size parameter, x, and the complex refractive index, m. The relationship between these oscillations and the sphere size was used for sizing single spheres (*Steiner* et al. 1999, *Crouse* and Latimer 1990). In particular, Steiner et al., who analyzed the periodicity of the angular oscillations of the scattering pattern, established the following linear relationship between the size parameter, x, and the number of oscillations per degree, v:

$$\nu = 0.00483x \tag{3.19}$$

for $50 \leq x \leq 500$ and $1.3 \leq n' \leq 1.75$ (relative to the surrounding medium). In the refractive index range that we are concerned with, the oscillation frequency is lower than that and decreases with the refractive index. For example, at $n' = 1.01$, we have approximately five oscillations per 120° at $x = 10$, i.e., $dv/dx \sim 0.0042$, while at $n' = 1.1$, by similar account, we have $dv/dx \sim 0.0044$. This frequency is important in determining the step size in a *brute force* integration of the differential scattering cross-section over the particle size.

The total scattering cross-section is by definition given by the integral of the intensity over all angles:

$$C_{\text{scat}} = \frac{\pi}{k^2} \int_0^{\pi} \{i_1(\theta) + i_2(\theta)\} \sin\theta \, d\theta \tag{3.20}$$

Note that (3.20) shows the result in which integration over the azimuth angle, ϕ, has been already performed. Substituting equations (3.13) and (3.14) into (3.15) and (3.16) respectively, and using the orthogonality properties of the Legendre polynomials, one obtains:

$$C_{\text{scat}} = \frac{2\pi}{k^2} \sum_0^{\infty} (2p+1) \left(|a_p|^2 + |b_p|^2 \right) \tag{3.21}$$

We note in passing that such integration between any two angles has been performed analytically (*Wiscombe* and Chýlek 1977 – for any interval of θ, *Pendleton* 1982 – into a conical solid angle about any θ), resulting in summation of a series similar to that of equation (3.21).

From the optical theorem (Chapter 1), the total attenuation cross-section is directly related to the scattering amplitude function in the forward direction. Since for the sphere the amplitude functions are equal to each other at $\theta = 0$, and

$$\left(\frac{P_p^1(\cos\theta)}{\sin\theta} \right)_{\theta \to 0} = \left(\frac{d}{d\theta} P_p^1(\cos\theta) \right)_{\theta \to 0} = \frac{1}{2} p(p+1) \tag{3.22}$$

the total amplitude function in the forward direction ($\theta = 0$) becomes

$$S(0) = \frac{1}{2} \sum_1^{\infty} (2p+1)(a_p + b_p) \tag{3.23}$$

Thus, by using the optical theorem (1.44), the attenuation cross-section for spheres can be written down in the following form:

$$\begin{aligned} C_{\text{attn}} &= \frac{4\pi}{k^2}\text{Re}\{S(0)\} \\ &= \frac{2\pi}{k^2}\sum_{1}^{\infty}(2p+1)\text{Re}(a_p+b_q) \end{aligned} \tag{3.24}$$

The absorption cross-section can be obtained as the difference between the attenuation and scattering cross-sections.

It very often more illuminating to write the above results in terms of the corresponding efficiencies which are defined as the ratio of the real cross-section to the geometric cross-section. The various efficiencies are thus given by:

$$\begin{aligned} Q_{\text{attn}} &= \frac{4}{x^2}\text{Re}\{S(0)\} \\ &= \frac{2}{x^2}\sum_{1}^{\infty}(2p+1)\text{Re}(a_p+b_q) \end{aligned} \tag{3.25}$$

where we make again the use of the optical theorem, and by

$$\begin{aligned} Q_{\text{scat}} &= \frac{1}{x^2}\int_{0}^{\pi}\{i_1(\theta)+i_2(\theta)\}\sin\theta\, d\theta \\ &= \frac{2}{x^2}\sum_{1}^{\infty}(2p+1)\left\{|a_p|^2+|b_p|^2\right\} \end{aligned} \tag{3.26}$$

$$\begin{aligned} Q_{\text{abs}} &= Q_{\text{attn}} - Q_{\text{scat}} \\ &= \frac{2}{x^2}\sum_{1}^{\infty}(2p+1)\left[\text{Re}\left(a_p+b_p\right)-\left(|a_p|^2+|b_p|^2\right)\right] \end{aligned} \tag{3.27}$$

This completes our basic sketch of the exact scattering solution for spherical particles of arbitrary size and refractive index.

The solution to light scattering by spheres is the simplest exact form. However, even in this basic case, it is difficult to obtain physical insight from the mathematical form of the solution itself. The only asymptote that is reasonably straightforward to treat is the small particle case. This involves studying only the first few coefficients in equations (3.13) and (3.14). As expected, in this regime, the exact solution converges to the Rayleigh scattering results, and we recuperate all the features of the simple dipole approximation.

Ever since the publication of the basic solution around the turn of the century, most of the effort devoted to this problem by researchers has been concentrated on developing efficient and reliable methods of evaluating the functions involved in equations (3.9) through (3.14). This effort to obtain efficient means of computing these functions was started by Rayleigh himself and pursued by several other distinguished physicists and mathematicians such as Debye and Watson (*Watson* 1952). This effort has led to the production of a significant body of mathematical work on Bessel functions and on procedures for their efficient numerical approximations (*Abramowitz* and Stegun 1964). With the advent of computers, these techniques now allow fast computations of the Mie solution to be carried out to extremely large sizes and indices of refraction. Several good codes have been made available in the public domain (e.g., *Bohren* and Huffman 1983). The availability of an exact result allows for exploration of the basic physics involved in the scattering of light in order to develop simplified approaches based on physical insights. We will pursue this approach after having discussed other cases where exact solutions can be obtained.

3.2.2. Exact analytic solutions: cylinders and spheroids

In order of simplicity, the next solution is that for the infinite cylinder. By an infinite cylinder, one actually means a çylinder long enough that the end effects can be neglected. In this case, the results for the various cross-sections and efficiencies are of course normalized per unit length of cylinder.

The solution for the case of normal incidence on a cylinder is very similar to the solution for the sphere except that the coefficients now involve the standard Bessel functions and the cylindrical vector harmonics instead of the Riccati–Bessel functions used for spheres (*Bohren* and Huffman 1983). The full solution for an arbitrary angle of incidence is however considerably more complicated. In that case, the full scattering amplitude function is represented by a matrix of four functions. At angles away from normal incidence and for moderate indices of refraction, all four of these functions are of the same order of magnitude. The cross terms represent the significant couplings that exist between the modes of polarization of the incident and scattered waves.

These polarization effects comprise some of the more notable differences from the results obtained for spheres. For spheres, the scattering amplitude and the attenuation cross-sections are the same for both polarization states. For cylinders, the polarization states of the scattered light are much more strongly coupled, even in the forward direction for the case of normal incidence. The scattering and attenuation cross-sections for cylinders depend on the initial state of polarization. However, we should note that these differences between light scattering by spheres and cylinders depend on the magnitude of the refractive index. For a small value of the relative index, i.e., one close to unity, these polarization effects, while still present, are much less significant. For relative indices of refraction on the order of 1.1, such as is the case for most particles in water, the differences are much smaller and less noticeable.

Even though it requires considerably more computing power, results can also be obtained for ensembles of randomly oriented cylinders. The process of integration is not straightforward, and considerable care must be exercised not to badly skew the results. These difficulties are due to the fact that the scattering from infinite cylinders occurs on infinitely thin cones whose axis of rotation is coincident with the axis of the cylinder and whose apex angles are equal to the angle of incidence of the radiation. These scattering cones are infinitely thin because the cylinders are infinitely long. Correctly handling and weighing these singularities requires a robust integration method (*Haracz* et al. 1985).

Not all procedures reported in the literature are correct, and some care must be exercised when using results for randomly oriented ensembles of cylinders. Furthermore, different normalization schemes are used to compute the scattering and absorption efficiencies. A similar care should be exercised for all non-spherical particles.

Sample normalized angular scattering patterns for a representative selection of monodisperse randomly oriented cylinders are shown in Figure 3.3 and Figure 3.4. Note that averaging over random orientations did not wipe out deep resonances. As for the spheres, these scattering patterns should be regarded as samples only and one should refrain from generalizations except perhaps for those issues mentioned in a brief cautionary note near Figure 3.1.

The only other particle shape for which a practical solution can be obtained by the method of separation of variables is that of a spheroid. The solution is similar to that for spheres. However, the separation is first carried out in spheroidal coordinates and only then are the fields expanded in vector spherical harmonics as in the case of spheres (*Bohren* and Huffman 1983).

Asano and Yamamoto (1975) worked out the details of this method of solution. Subsequently, *Asano* (1979) and *Asano* and Sato (1980) obtained many numerical results for spheroids of various shapes, orientations, and refractive indices. This method is in fact able to handle larger particles with more extreme eccentricity than other approaches. *Voshchinnikov* and Farafonov (1985) also obtained an exact solution. They used a different basis for the expansion which permitted to significantly simplify the derivation.

Even though the solution for spheroids is formally exact, many terms of the series of coefficients must be computed for spheroids of any reasonable size. The spheroidal functions are extremely difficult to compute accurately, and they suffer from several convergence problems (*Abramowitz* and Stegun 1964). The problems are tractable, but a great deal of care and effort is required. The computations are both lengthy and complicated. Recently, *Voshchinnikov* and Farafonov (1985) claimed to have developed an approach which is 10 times more efficient than that

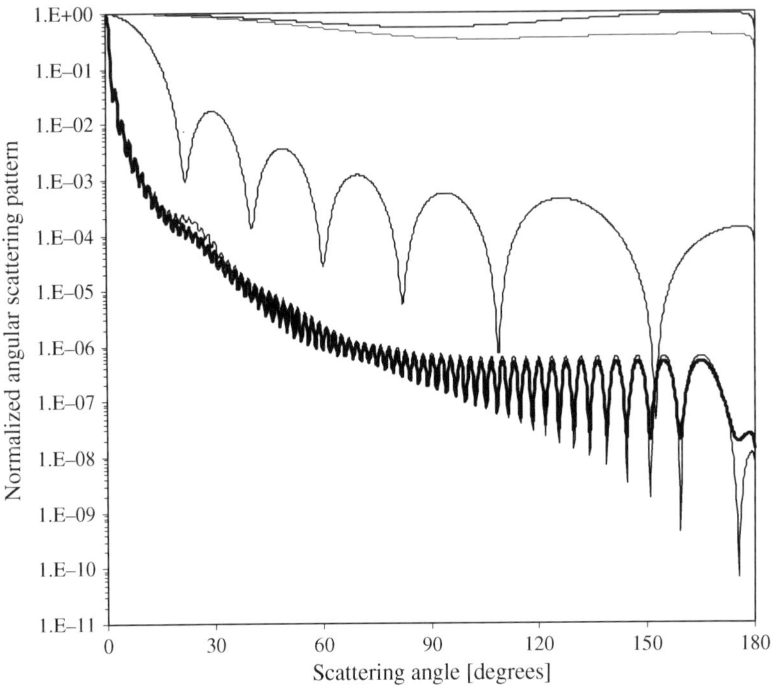

Figure 3.3. Sample normalized angular scattering patterns for monodisperse homogeneous randomly oriented infinite cylinders with size parameters $x = 0.1$ (top curve), 1, 10, and 100 (bottom curve), and the refractive index $n = n' - in'' = 1.01 - i0$ (except of the thick curve which refers to $n = 1.01 - i0.001$). These curves are samples only and one should refrain from generalizations apart from those discussed in the text near Figure 3.1. Note that the vertical scale has changed as compared with Figure 3.1. The plotting angle increment is 0.2°. The depths of the sharp resonant minima for the $x = 10$ and $x = 100$ curves are somewhat distorted by the coarse angle increment. The values of un-normalized phase function at a scattering angle of 0° for the thin curves (top to bottom) are: 6.62×10^{-2}, 8.36×10^{-2}, 6.07×10^{-1}, and 5.66×10^{0}. The value for the thick curve is 5.87×10^{0}. At the selected value of $n'' = 0.001$, the effect of light absorption is relatively minor and can only be appreciated for the cylinders with the largest radius ($x = 100$).

of *Asano* and Yamamoto (1975) for small values of the spheroid's axial ratio and 100 times at large values.

The computational effort is compounded by the fact that one needs to perform averages over orientation and also often integrate over size and shape distributions. If the problem was only one of raw computing power, the method would still be attractive given the ready availability of powerful computers. However, the complexity of the solution and the unavailability of good computer codes is probably what has prevented its ready acceptance.

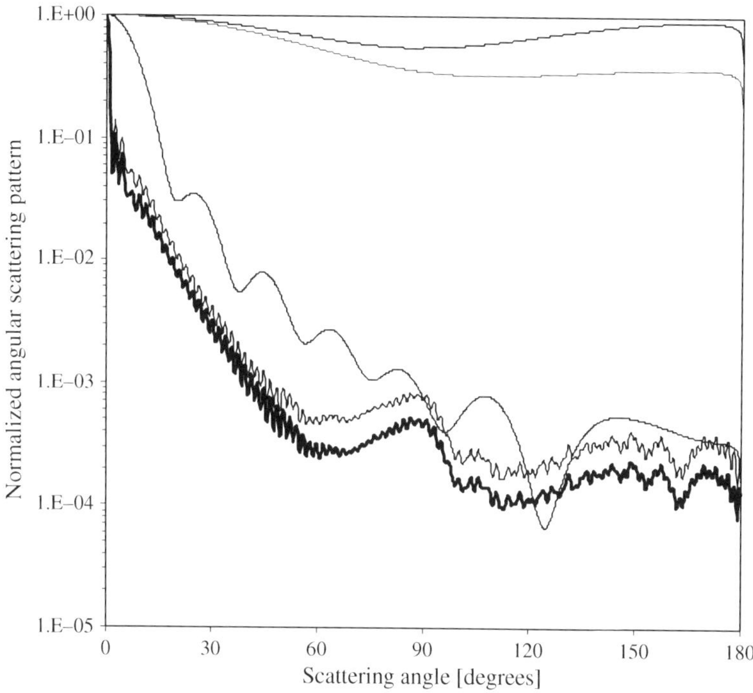

Figure 3.4. Sample normalized phase function for monodisperse homogeneous randomly oriented infinite cylinders with size parameters $x = 0.1$ (top curve), 1, 10, and 100 (bottom curve), and the refractive index $n = n' - in'' = 1.1 - i0$ (except of the thick curve which refers to $n = 1.1 - i0.001$). These curves are samples only and one should refrain from generalizations apart from those discussed in the text near Figure 3.1. Note that the vertical scale has changed as compared with Figure 3.3. The plotting angle increment is 0.2°. The values of un-normalized phase function at a scattering angle of 0° for the thin curves (top to bottom) are: 6.59×10^{-2}, 8.34×10^{-2}, 6.22×10^{-1}, and 3.32×10^{0}. The value for the thick curve is 4.01. At the selected value of $n'' = 0.001$, the effect of absorption is relatively minor and can only be appreciated for the cylinders with the largest radius ($x = 100$).

3.2.3. General solutions for arbitrarily shaped particles

3.2.3.1. T-matrix

The other reason that the exact solution methods for non-spherical particles such as that of *Asano* and Yamamoto (1975) has been neglected is the concurrent development of a powerful method to solve the scattering problem for arbitrary particle shapes. It is particularly well adapted to particles with cylindrical symmetry such as spheroids and finite cylinders. It was originally proposed by *Waterman* (1971) who called it the T-matrix method.

The method is also based on an expansion in vector spherical harmonics. The coefficients relating the incident wave to the scattered wave form as in the case of cylinders, a matrix: the transition matrix between the expansion coefficients in spherical harmonics of the incident wave and the expansion coefficients of the scattered waves. The equations relating the coefficients are derived by evaluating integral relations over the arbitrary surface of the particle.

In this approach, particles with an axis of symmetry lead to two-dimensional transition matrices. The order of the matrix required to achieve a given precision depends on the shape and size of the particle. The solution for the sphere is exact in this formalism, and in that case the matrix is purely diagonal. *Barber* and Yeh (1975) gave an alternative derivation to that of Waterman and also extensively investigated homogeneous spheroids and finite cylinders with spherical end caps. *Barber* and Hill (1990) have written a very nice monograph that both clearly describes the method and also presents several good computer programs to solve for scattering from oriented and randomly oriented spheroids and finite cylinders. The method is relatively efficient, but it can still be overwhelmingly time consuming for randomly oriented large particles with significant eccentricity. For example, gathering a modest database of attenuation, scattering, and absorption efficiencies with size parameter up to 30, spheroid aspect ratio of 2, and real part of the refractive index up to 1.8 took at least 10^{13} floating point operations. This is still a significant amount of work even by today's computing standards.

3.2.3.2. Finite-difference time domain (FDTD)

A more general method for solving scattering from inhomogeneous objects of any shape was developed by *Yee* (1966, see also a more contemporary review of this method by *Taflove* and Umashankar 1989). Yee found a stable numerical way of directly solving Maxwell's equations over a bounded domain by using two alternate three-dimensional rectangular grids. Appropriate boundary conditions have also been developed to prevent reflection of the scattered wave from the boundary. The numerical solution is first obtained in the near field at the boundary surface. These fields are then propagated into the far field by the free space Green's function.

The method is both time and memory consuming which explains its original neglect. Its straightforwardness and the recent advances in computing power have made it more popular. Its great virtue is that it allows one to obtain the scattering solution for very complex entities such as cells with multiple organelles that act as internal absorbing and scattering centers from within the particle (cell) itself (*Dunn* and Richards-Kortum 1996). A fairly detailed survey of the extensive literature on this method can be found in *Schlager* and Schneider (1995).

3.2.3.3. Discrete dipole approximation

The last method we will mention is the discrete dipole approximation developed by *Purcell* and Pennypacker (1973). It is a conceptually simple technique where an arbitrary body is decomposed into sub-regions, each being small enough to respond to the incident field as a dipole. The excited dipoles interact with one another and the resulting field, including retardation effects due to the finite speed of propagation of the field between the various dipoles, can be computed in a straightforward fashion.

Draine and Flatau (1994) and Draine (2000) have given fairly complete reviews of the technique (see also section 6.4.1). Unfortunately, this technique is limited to absolute values of the complex refractive index less than 4 and to maximum sizes of the particle of less than $\sim 5\lambda$, where λ is the wavelength in the medium surrounding the particle. The speed and memory requirements of the technique scale roughly as the volume of the scattering particle and as the absolute value of the refractive index m. The particle shape can obviously be arbitrary, and the technique has been applied for complex forms (Draine 2000). It should be noted that even though the size range accessible to this technique is limited, this technique can give valuable insight into polarization effects from complex shapes and into field distortions due to edge effects.

As was noted some time ago by *Bohren* and Huffman (1983), all exact methods become almost prohibitively time consuming when averages over large ensembles of particles are required. This is in fact an irony, since the situation should become simpler when averages are studied. All traces of the fine details of scattering from individual particles are erased, and only the basic elements of the scattering phenomenon remain. When dealing with large ensembles, simple structures always seem to emerge. This hints at the possibility of understanding the light scattering behavior of ensembles of particles by considering only a well-chosen subset of basic physical processes.

In the next section, armed with exact solutions, we will attempt to analyze what are the basic features of scattering and which of these survive averaging over the types of particle distributions found in natural waters.

It is worth to note that the scattering of light by the particles made of subunits has been solved rigorously for aggregates of spheres, such as a sphere doublet (e.g., *Fuller* 1991). An analytical solution has been developed also for clusters of spheres in the Rayleigh domain (*Mackowski* 1995, 1991) and of arbitrary size (e.g., *Xu* 1995, *Fuller* 1991). *Botet* et al. (1997) proposed an approximation in which each of the spheres is experiencing a mean field, the same for all spheres, but the wave scattered by the aggregate is obtained by the summation of the waves scattered by all spheres, as in the exact theory. Monte Carlo modeling of light scattering by a fractal aggregate of spheres has also been tried (e.g., *Deng* et al. 2004). This latter work illustrates that the cooperative scattering in a fractal aggregate combines to extinguish oscillations of the phase function that in a solid sphere are due to interference effects and also flatten the phase function of the aggregate as compared with that of a solid particle of the same size.

3.3. Basic structures in scattering

As we saw in the last section, the scattering of light by homogeneous particles depends on the particle shape and size as well as on the value of its complex refractive index relative to the surrounding medium. If we wish to understand and simplify the scattering model to extract the basic structures, it is necessary to clearly delimit the range of both the real and imaginary parts of the indices of refraction we expect for most of the particles found in natural waters. This step is a prerequisite to selecting the appropriate methods of approximation.

To refresh our memory, let us write once again the equation for propagation of a plane wave in a medium that can also absorb the wave energy.

$$\begin{aligned} E &= E_0 e^{-in\frac{2\pi l}{\lambda}} \\ &= E_0 e^{-in'\frac{2\pi l}{\lambda}} e^{-n''\frac{2\pi l}{\lambda}} \\ &= E_0 e^{-in'\zeta} e^{-n''\zeta} \end{aligned} \tag{3.28}$$

where

$$\zeta = \frac{2\pi l}{\lambda} \tag{3.29}$$

where λ is the wavelength of light in the surrounding medium and

$$n = n' - in'' \tag{3.30}$$

Parameter l is the distance the field has penetrated into the medium. The complex relative refractive index is n. The ratio of the speed of propagation in the medium relative to the speed of propagation outside of the medium is the real part, n', of the relative refractive index. The distance traveled in the medium is often conveniently expressed in terms of a dimensionless parameter, ζ, (3.29), given by 2π times the number of wavelengths traveled (l/λ).

The irradiance of the light wave, which is given by the absolute square of the field, E, is thus attenuated by twice the imaginary part of the index times the dimensionless distance traveled in the medium:

$$I = EE^* = E_0^2\ e^{-2n''\zeta} = I_0\ e^{-\frac{4\pi n''}{\lambda}l} = I_0\ e^{-al} \tag{3.31}$$

The last term in (3.31) expresses the Beer–Lambert law of attenuation of light by a medium with an absorption coefficient, a. Thus, equation (3.31) relates the absorption coefficient of the medium to the imaginary part of the refractive index as follows:

$$a = \frac{4\pi n''}{\lambda} \tag{3.32}$$

The real part of the refractive index, n', of the majority of the species of phytoplankton assumes values between 1.05 and 1.1 relative to water. The imaginary part of the refractive index, n'', which is directly related to absorption of light by phytoplankton, varies from roughly 0.001 at 550 nm to 0.01 at 435 nm, the wavelength where chlorophyll absorbs the most. For minerals such as those composing sand and some aerosols, the real part of the index in the visible portion of the spectrum is around 1.5, which translates to a relative index in water of 1.12. The imaginary part is on the order of $n'' = 0.0001$. For soil and Saharan dust, the imaginary part can range from 0.003 to 0.01. We discuss experimental data supporting these statements and relevant measurement methods in Chapter 6.

Given these values, we can immediately conclude that in virtually all cases we will be concerned with, the real part of the refractive index is close to 1 and with a small imaginary part. This immensely simplifies the approaches one can take to evaluate scattering of light by water-borne particles. By far, the most important consequence of the closeness to 1 of the real part of the index is that the direction of propagation of radiation is almost unchanged after traversing an interface at an arbitrary angle. Using Snell's law of refraction (1.25), it can be shown that the net angle of deflection of a light ray through an interface is given by:

$$\Delta\theta \approx (n' - 1)\theta \tag{3.33}$$

Once we have assumed that light paths are not deflected, it is relatively straightforward to compute a good approximation of the field around and inside the particle. However, we must also neglect any field distortion that would occur at the edge of the particle. Such distortion is due to the boundary conditions that require matching of tangential components of the fields at the surface of the particle. We assume that the portion of the incident plane wave that does not directly impinge on the particle continues to propagate as a plane wave with no phase difference. The parts of the wave that impinge on the particle acquire a phase difference proportional to the distance they travel inside the particle. The distance traveled to any point inside the particle is given by the length of a straight line drawn parallel to the direction of the incident wave and extending from the given point back to the point of entry. The effect of a spherical particle on the phase of the plane wave is shown in Figure 3.5.

The phase difference is given by:

$$\Delta\eta(x, y, z) = (n' - 1)z(x, y) \tag{3.34}$$

where z is the distance traveled through the particle and (x, y) are the coordinates of the point of entry. The total phase difference after passing through the particle is obviously a function of the coordinates of the point of entry and also

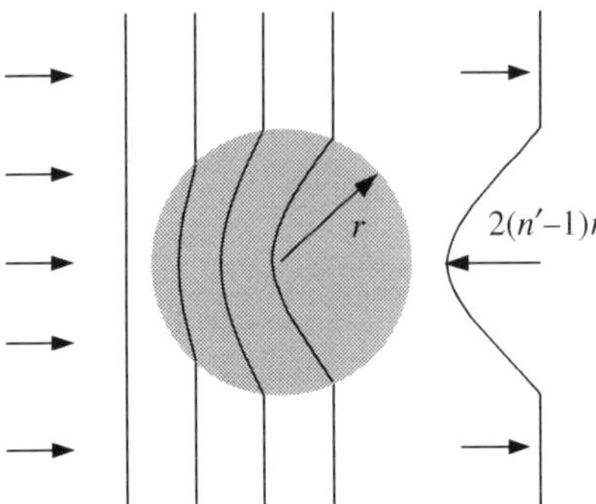

Figure 3.5. Phase fronts of a light wave traveling through a sphere of radius r. The wave slows down while traveling through the particle. The accumulated phase difference is proportional to the total distance traveled through the particle and is a function of the point of entry. The phase difference between the light passing through the center of the sphere and the light passing outside the sphere is $2(n' - 1)r$.

of the particle orientation in the case of non-spherical particles. If the particle absorbs light, then the amplitude of the light wave is reduced in addition to the modification of the wave phase. The field at any point in the particle can be approximated by:

$$\begin{aligned} E(x, y, z) &= E_0 \exp[-n''z(x, y)]\exp[i(n' - 1)z(x, y)] \\ &= E_0 \exp[-i\Delta\eta(x, y, z)] \\ &= E_0 \exp[-i(n - 1)z(x, y)] \end{aligned} \tag{3.35}$$

The field just beyond the particle is obviously given by equations (3.34) and (3.35) with $z(x, y)$ now the complete distance traveled through the particle by a ray entering at coordinates (x, y). The part of the incident field that did not penetrate the particle is assumed undisturbed.

Using the approximation of 'straight-through' propagation, we have been able to approximate with reasonable accuracy both the field inside the particle and in a plane just beyond the particle. From this point, one can attempt to compute the scattered field by two distinct approaches.

One can use the field computed in the plane just beyond the particle and propagate it to infinity using the Huyghens principle from diffraction theory (*van de Hulst* 1957). This approximation is known as the anomalous diffraction approximation. In this approach, the scattering function is given by carrying out the following integral over the projected area of the particle, i.e., the area where the phase and possibly the amplitude of the incident wave has been changed.

$$S(\alpha, \beta) = \frac{1}{2\pi} \iint \left(1 - e^{-i\Delta\eta(x,y,z)}\right) e^{-ix2\sin(\alpha/2)}\ e^{-iy2\sin(\beta/2)}\ dx\ dy \tag{3.36}$$

Note that (3.36) is written in Cartesian coordinates. This is done to keep the results in a form that is sufficiently general to apply to any particle shape. Variables x, y, and z are also all in the same dimensionless form shown in equation (3.29). Angles α and β define the direction of the scattered radiation in the (x, z) and (y, z) planes respectively. It is a straightforward matter to transform the coordinate system to take advantage of any symmetry that the particle may possess.

The second approach is to use the integral formulation of scattering (*Klett* and Sutherland 1992). The scattered wave far away from the particle again takes the form of a scattering amplitude vector multiplying a spherical wave function. The scattering amplitude vector is everywhere normal to the radius vector of the spherical wave. This vector can be computed by directly integrating the internal field over the volume of the particle. *Klett* and Sutherland give a very simple outline of the method. If the internal field is approximated by straight-line propagation through the particle with the same phase delay and amplitude decay as in the case of anomalous diffraction, the integral method is known as the Wentzel–Kramers–Brillouin (WKB) solution in honor of its proponents in quantum mechanics. It is a particular form of a more general approximation technique known as the eikonal approximation (EA). By using more general approximation than the straight-line propagation, the EA can be generalized to handle larger indices of refraction. The interested reader should consult a series of articles by *Chen* (1993, 1989, 1984) and by *Chen* and Smith (1992). For water-borne particles, we will always be dealing with a refractive index close to unity, i.e. $|n-1| << 1$, and the results from the anomalous diffraction approximation (ADA) do apply very well to this case. In fact, as pointed out by *Sharma* (1992), the usual restriction of $|n-1| << 1$ can be relaxed to include a larger class of particles with the refractive index that fulfill a condition of $|n-1|^2 << |n+1|^2$. We will therefore not pursue the integral over internal fields approaches further except to mention a truly remarkable result achieved by *Chen* (1993) for spheres, which allows a simple formulation for the scattering amplitude function for spheres with moderate refractive index.

3.3.1. Anomalous diffraction results

To compare the results from the approximation with the exact case, we will first study the scattering by spheres. This will allow us to use the simple Mie theory outlined in section 3.2.1. Because of the symmetry, the scattering amplitude will only be a function of the polar angle θ. Transforming to polar coordinates, we obtain the following expressions for the scattering amplitude:

$$S(\theta) = \frac{k^2}{2\pi} 2\pi\, a^2 \int_0^{2\pi} \int_0^{\pi/2} (1 - e^{-i\rho\cos\tau}) e^{-iz\sin\tau\cos\varphi} \sin\tau\cos\tau d\tau d$$

$$= \frac{x^2}{2\pi} \int_0^{2\pi} \int_0^{\pi/2} (1 - e^{-i\rho\cos\tau}) e^{-iz\sin\tau\cos\varphi} \sin\tau\cos\tau \; d\tau \; d\varphi \qquad (3.37)$$

$$= x^2 \int_0^{\pi/2} (1 - e^{-i\rho\cos\tau}) J_0(z\sin\tau) \sin\tau\cos\tau \; d\tau$$

where:

$$\rho = 2(n-1)x = \rho' - i2n''x \qquad (3.38)$$

$$\rho' = 2(n'-1)x \qquad (3.39)$$

$$z = 2x\sin\frac{\theta}{2} \qquad (3.40)$$

As shown in Figure 3.6, x is the size parameter of a sphere with radius r [see equation (3.8)]. Parameter ρ' is the complex phase difference of the central

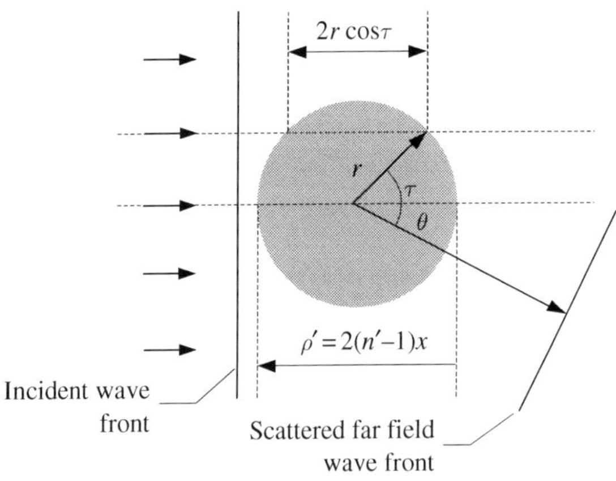

Figure 3.6. Diagram of the parameters required to compute the anomalous diffraction formulas for scattering by a sphere or radius r in a direction θ. The distance traveled through the particle is $2r\cos\tau$. The maximum phase difference is $\rho' = 2(n'-1)x$ where x is the size parameter $x = 2\pi r/\lambda$.

ray through the sphere and ρ is its real part. Parameter n'' is the imaginary part of the refractive index which represents the damping of the incident light wave amplitude by absorption in the particle. Parameter z is the product of the size parameter times the polar angle of the direction in which we want to compute the scattering amplitude. Parameter τ is the angle between the direction of the incoming wave and the radius vector from the center of the sphere to the exit point of a ray offset from the center by a distance $r \sin \tau$. Thus, $\rho \cos\tau$ is the corresponding total phase difference picked up by that ray after it has traversed the sphere.

Unfortunately, the last integral in (3.37) that is required to evaluate the scattering amplitude at any angle must either be approximated or carried out numerically. We will postpone this analysis and concentrate for the moment on the evaluation of the attenuation cross-section. According to the optical theorem (1.44), we only need to evaluate the scattering amplitude in the forward direction to find the attenuation cross-section. This particular integral is straightforward to obtain.

$$\begin{aligned} S(0) &= x^2 \int_0^{\pi/2} (1 - e^{-i\rho\cos\tau}) \sin\tau \cos\tau \, d\tau \\ &= x^2 \left[\frac{1}{2} + \frac{\exp(-\omega)}{\omega} + \frac{\exp(-\omega) - 1}{\omega^2} \right] \end{aligned} \tag{3.41}$$

where

$$\begin{aligned} \omega &= i\rho \\ &= i2(n - 1)x \\ &= i2(n' - 1)x + 2n''x \end{aligned} \tag{3.42}$$

Hence, the attenuation cross-section, C_{attn}, can be calculated as follows by using the optical theorem:

$$\begin{aligned} C_{\text{attn}} &= \frac{4\pi}{k^2} \text{Re}\{S(0)\} \\ &= \pi r^2 \text{Re}\{S(0)\} = \pi r^2 \, Q_{\text{attn}} \end{aligned} \tag{3.43}$$

The attenuation efficiency, Q_{attn}, is thus expressed as follows:

$$Q_{\text{attn}} = 4\text{Re}\left\{ \frac{1}{2} + \frac{\exp(-\omega)}{\omega} + \frac{\exp(-\omega) - 1}{\omega^2} \right\} \tag{3.44}$$

The total energy absorbed inside the particle is also computed easily. The integral of the decrease in intensity of the incident wave as it crosses the particle is given by

$$C_{\text{abs}} = 2\pi\, a^2 \int_0^{\pi/2} (1 - e^{-4n''x\cos\tau}) \sin\tau \cos\tau \; d\tau \tag{3.45}$$

$$Q_{\text{abs}} = 2\left[\frac{1}{2} + \frac{\exp(-4n''x)}{4n''x} + \frac{\exp(-4n''x) - 1}{(4n''x)^2}\right] \tag{3.46}$$

The factor of $4n''x$ in (3.45) arises because the irradiance of a light wave is given by the absolute square of its amplitude. In the ADA, the scattering cross-section is thus most easily expressed as the difference between the attenuation cross-section and the absorption cross-section.

Figure 3.7 shows the attenuation efficiency Q_{attn} for non-absorbing homogeneous spheres as a function of ρ (the real part of the maximum phase difference) as compared with results obtained with Mie theory for $n' = 1.05$, 1.1, and 1.15. The attenuation efficiency from (3.44) is the lowest of the curves. There is good agreement between the exact theory and the approximation. The slight increase of

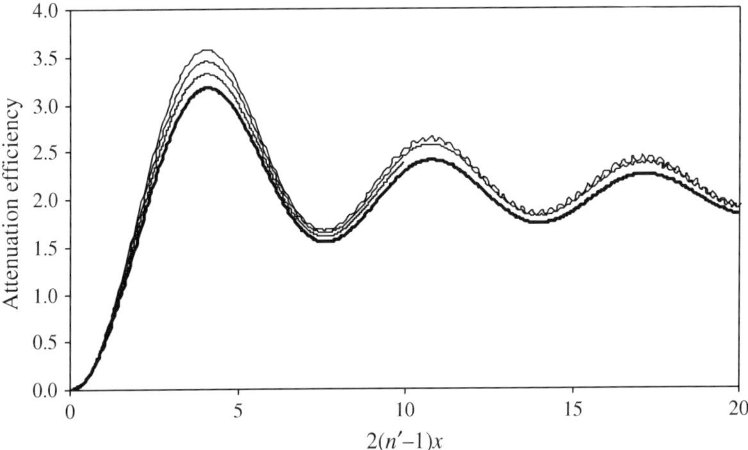

Figure 3.7. Attenuation efficiency Q_{ext} for non-absorbing spheres as a function of $\rho' = 2(n' - 1)x$, the real part of the phase difference through the center of the sphere with non-dimensional size x. The results from the exact Mie code are shown for real refractive index, n', values of 1.15, 1.1, and 1.05 (beginning with the top curve), a range typical of particles found in seawater. The attenuation efficiency from formula (3.44) is given for comparison (the lowest, thick curve). The discrepancy between the exact code and the results of the approximation is caused by neglecting the edge effects in the derivation of the latter formula.

the value of the attenuation efficiency computed by the exact code over the results of the approximation is due to the neglect of the edge effects here. The structure of alternating maxima and minima is created by the interference between the portion of the light wave which is diffracted around the particle and the portion of the wave which is refracted through the particle. The initial rise of the attenuation efficiency to a value around 2 followed by a series of damped maxima and minima is a universal feature that can be seen in the attenuation graphs for particles of all shapes.

Another universal feature is the scaling of these features as a function of the characteristic phase difference through the particle. To briefly illustrate this point, Figure 3.8 shows the corresponding plot of the exact solution for the attenuation efficiency of a set of non-absorbing randomly oriented infinite cylinders with the same indices as in the case of the spheres. The result of the ADA for randomly oriented cylinders, which we derive in section 3.3.3, is also shown for comparison. Once again, the agreement is satisfactory. By considering the interference effects, we would expect the maxima and minima to be found at integer numbers of half cycles of the real part of the average phase difference through the particle (*Bohren* and Hufmann 1983). The spacing between the successive peaks and valleys does follow that pattern.

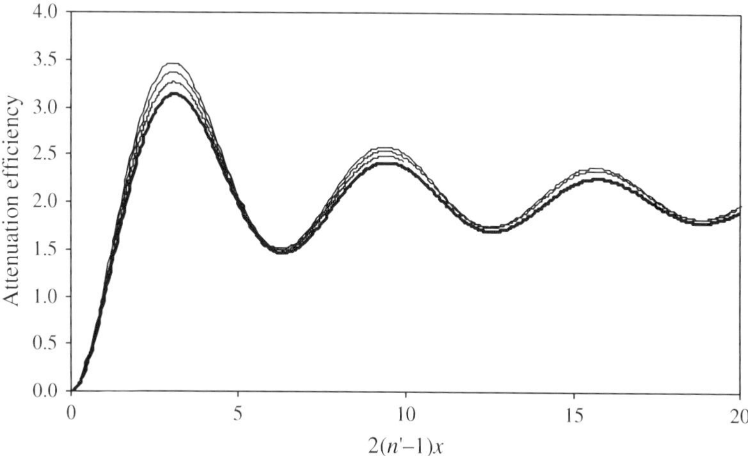

Figure 3.8. Attenuation efficiency, Q_{attn}, for an ensemble of non-absorbing randomly oriented cylinders as a function of $\rho' = 2(n' - 1)x$, the real part of the phase difference through the center of a cylinder (at incidence normal to the cylinder axis) with non-dimensional radius x and the axis normal to the incident light direction. The results from calculation based on an exact analytical solution are shown for refractive index values of 1.15, 1.1, and 1.05 (beginning with the top curve), a range typical of particles found in seawater. The anomalous diffraction result is represented by the lowest (thick) curve.

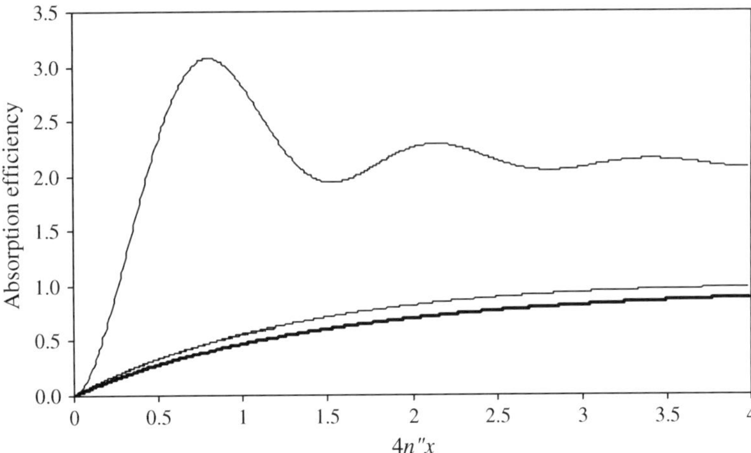

Figure 3.9. Absorption efficiency Q_{abs} for spheres as a function of $4n''x$, the damping factor for the irradiance of the wave that passed through the center of the sphere. The results from an exact Mie code are shown for the imaginary part, n'', of the refractive index values of 0.001, 0.003, and 0.01, a range typical of aquatic particles. The anomalous diffraction result [equation (3.46)] is slightly lower than the Mie code results. This is due to the neglect of the deflection of the light rays inside the particle. Such deflection increases the mean path length through the particle and thus the absorption efficiency. For reference, the attenuation efficiency for a complex index of $n = n' - in'' = 1.1 - i0.01$ is also shown (the top oscillating curve).

Figure 3.9 shows the absorption efficiency as a function of $4n''x$, the damping factor for the intensity of the wave that passed through the center of the sphere. The graph shows curves resulting from the exact Mie code for n'' values of 0.001, 0.003, and 0.01, representing a typical range for water-borne particles. The curve for the ADA runs slightly lower than those obtained with the Mie code. The absorption efficiency simply rises initially linearly with particle size and then saturates. Once again, the minor differences between the exact results and the approximation can be attributed to the neglect of the edge effect.

The higher running oscillating curve is the attenuation efficiency for the $n = n' - in'' = 1.1 - i0.01$ case. Note the damping of maximum and minimum amplitude of the oscillations when compared to the case with no absorption. This damping is due to a reduction of the interference term between diffraction and refraction by the absorption of the light refracted through the particle. Note also that the asymptote for the absorption efficiency does not approach unity as is predicted by the simple anomalous diffraction formula (3.46). The reason for this discrepancy is the neglect of the Fresnel reflection of light from the surface of the particle. This reflected light is not absorbed by the particle. The reflection term

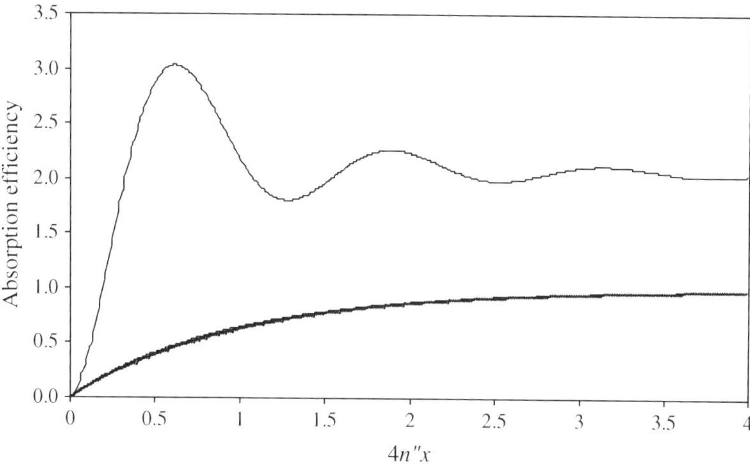

Figure 3.10. Absorption efficiency Q_{abs} for monodisperse randomly oriented cylinders as a function of $4n''x$, the damping factor for the irradiance of the wave that passed through the center of a cylinder with non-dimensional size x and axis normal to the direction of incident radiation. The results from an exact code are shown for the real part, n'', index values of 0.001, 0.003, and 0.01 and the anomalous diffraction result all shown by the lower 'curve', a group of overlapping, almost indistinguishable curves. For reference, the attenuation efficiency is shown for a complex index of $n = n' - in'' = 1.1 - i0.01$ (the top curve).

combines with the diffracted and refracted terms. For large particles, we will see later that this reflected term actually dominates the backscattering.

Figure 3.10 is the corresponding graph of the absorption efficiency as a function of $4n''x$ for randomly oriented cylinders. Once again, the n'' values range from 0.001 to 0.01. The upper curve is the attenuation efficiency for the $n = n' - in'' = 1.1 - i0.01$ case. Except for a slight change in the shape of the absorption curve, all the features are the same as in the case of the sphere. The agreement between the approximation and theory is excellent. The curves can barely be distinguished from one another.

The amount of light reflected from the particle will be the same as in the case of the sphere. This equality of the reflected component for spheres and randomly oriented cylinders is based on a fundamental geometrical property. We will see later that this equality extends to all randomly oriented convex bodies. This fact has profound implications for the evaluation of backscatter from particles with a small refractive index.

All the results we have obtained so far were based on the evaluation of the scattering amplitude in the forward direction. As we mentioned earlier, the integral required to evaluate the scattering amplitude at any angle, as derived in equation (3.37), must be either approximated or carried out numerically. If it could be carried out, this integral would give the full scattering phase function at any angle due to the diffracted and refracted light. A very fine analysis of the problem was carried out by *van de Hulst* (1957). After considerable effort, he obtained the following expression.

$$S(\theta) = x^2\left[\frac{J_1(z)}{z} + \frac{i\rho'\zeta_1(y)}{y^2} + \frac{J_0(z)}{\rho^2}\right] \tag{3.47}$$

$$\zeta_1(y) = \left[\frac{\sin y}{y} - \cos y\right] + i\left[\frac{\cos y}{y} - \sin y\right] \tag{3.48}$$

$$y = \sqrt{z^2 + \rho^2} \tag{3.49}$$

where z is defined by (3.40) (also note a new definition of y).

Out of the three possibilities for the angular variable suggested as equivalent by van de Hulst, we have used both here and in equation (3.37) the form $2\sin(\theta/2)$. As pointed out by *Chen and Smith* (1992), this form accounts best for internal phase differences between the incident wave propagating straight through the particle and the scattered wave in any given direction. It also has the advantage of being usable at all scattering angles from 0 to π. In the limit of $\theta = 0$, equation (3.47) is identical with that for scattering in the forward direction [see (3.41)].

The first term in (3.47) is the diffraction term. It is given as the first-order Bessel function of the first kind divided by the value of its argument z. It has the same form as the classical diffraction from a circular aperture. Note that it is independent of the value of the refractive index. This is exactly as expected, since diffracted light does not penetrate the particle and therefore cannot be influenced by the value of its refractive index.

The second term, in the limit of large-particle sizes, is equal to the refracted light term that is calculated by the classical geometric optics approach. The extra components of that term that come into play for smaller particle sizes account for the first-order effects of internal diffraction of the refracted light. As expected, the term is explicitly dependent on the refractive index since in this case we are evaluating the effect of the particle on the light that has gone through it.

The last term is the first of a series of terms that handle progressively higher order interactions between the refracted and diffracted components.

In the limit of the large-particle size, only two terms remain significant in the scattering function.

$$S(\theta) = x^2\left[\frac{J_1(z)}{z} - \frac{i\rho'\exp(-i\,y)}{y^2}\right] \tag{3.50}$$

In that limit, for a particle with a real refractive index, the scattering cross-section becomes:

$$C_{\text{scat}} = \frac{2\pi}{k^2} \int_0^{\pi} |S(\theta)|^2 \sin\theta \, d\theta \tag{3.51}$$

$$\begin{aligned} \sigma_{\text{scat}}(\theta) &= \frac{2\pi}{k^2} |S(\theta)|^2 \\ &= \pi r^2 \left[2\frac{J_1^2(x\Theta)}{\Theta^2} + 2\frac{4\mu^2}{(\Theta^2 + 4\mu^2)^2} \right] \end{aligned} \tag{3.52}$$

where $\mu = n' - 1$ and,

$$\Theta = 2 \sin\frac{\theta}{2} \tag{3.53}$$

hence, $z = x\Theta$.

The first term in the right side of the second line of (3.52) accounts for diffraction around the sphere and the second term is the refraction through the sphere. In the limit of the large x and small μ, each of the two terms integrates approximately to unity:

$$\begin{aligned} \int_0^{\pi} 2\frac{J_1^2(x\Theta)}{\Theta^2} \sin\theta \, d\theta &= \int_0^{\pi} 2\frac{J_1^2\left(x 2\sin\frac{\theta}{2}\right)}{\left(2\sin\frac{\theta}{2}\right)^2} \sin\theta \, d\theta \\ &\approx \int_0^{\infty} 2\frac{J_1^2(x\theta)}{(x\theta)^2} (x\theta) \, d(x\theta) = 1 \end{aligned} \tag{3.54}$$

$$2\int_0^{\pi} \frac{4\mu^2}{\left[\left(2\sin\frac{\theta}{2}\right)^2 + 4\mu^2\right]^2} \sin\theta \, d\theta \approx \int_0^{\infty} \frac{2}{\left[\left(\frac{\theta}{2}\mu\right)^2 + 1\right]^2} \left(\frac{\theta}{2}\mu\right) d\left(\frac{\theta}{2}\mu\right) = 1 \tag{3.55}$$

The total scattering cross-section is thus equal to twice the geometric cross-section of the sphere, and the scattering efficiency, Q_{scat}, equals 2. This is precisely the result one would have obtained from a combination of geometric optics to evaluate refraction and diffraction theory to compute the light propagating around the particle.

The general form of the refraction term can be explained very simply. First assume that the sphere acts as a lens. The well-known lens maker formula then predicts that the equivalent focal length of the sphere with a radius, r, is:

$$f = \frac{r}{2(n' - 1)} \tag{3.56}$$

The angle θ_0 at which the rays hitting the outer perimeter of the sphere arrive at the focus is therefore:

$$\theta_0 = \frac{r}{f} = 2(n' - 1) = 2\mu \Rightarrow \theta_0^2 = 4\mu^2 \tag{3.57}$$

This is precisely the angle at which the refractive term formula predicts that the irradiance will begin to sharply decrease. The refractive term is thus consistent with the physical picture of the spherical particle acting to the first order of approximation as a classical thin lens.

By using geometric optics, one can also modify the refraction term to take into account approximately the effect of finite absorption (*van de Hulst* 1957). If the real part of the refractive index is small, one obtains for the refractive term alone:

$$\sigma_{\text{scat}}^{\text{refr}}(\theta) = \pi\, r^2 2 \frac{4\mu^2 \exp\left(-\frac{2}{3}4n''x\right)}{\left(\Theta^2 + 4\mu^2\right)^2} \tag{3.58}$$

$$\begin{aligned} C_{\text{scat}}^{\text{refr}} &= \int_0^{\pi} \sigma_{\text{scat}}^{\text{refr}}(\theta) \sin\theta \, d\theta \\ &\approx \pi\, r^2 \exp\left(-\frac{2}{3} 4n''x\right) \end{aligned} \tag{3.59}$$

We used the weighed mean of the distance traveled by light inside the sphere times the absorption coefficient of the sphere's material, $a = 4\pi n''/\lambda$, as the magnitude of the exponential damping of the intensity of the refracted light. In this first-order approximation, we neglected the slight modification of the angular form of the refractive term by absorption. Note that in the limit of the large values of $n''x$, the entire refraction term vanishes. Once again, this is precisely what one would expect. In that limit, the sphere will look like an opaque disk, and the only contribution to its scattering will be from the diffraction term. Note also that the amount of absorption required to produce the damping of the refractive term is in complete agreement with formulas (3.45) and (3.46) for small values of $4n''x$.

3.3.2. The reflection terms

As mentioned before, the amount of light reflected by the particle is not taken into account in the previous formulation for scattering. However, the division into pure diffractive and refractive terms allows a simple means of incorporating the effect of reflection. In this picture, the refractive term is reduced by the total amount of reflection ω. The diffractive term is untouched. The total scattering cross-section must thus at a minimum incorporate the following terms:

$$\sigma_{\text{scat}}(\theta) = \sigma_{\text{scat}}^{\text{diff}}(\theta) + (1-\omega)\sigma_{\text{scat}}^{\text{refr}}(\theta) + \omega\sigma_{\text{scat}}^{\text{refl}}(\theta) \tag{3.60}$$

where ω is the fraction of light reflected by the surfaces.

For large particles, the diffractive term dominates in a narrow angular zone near the forward direction. In the case of small real refractive index, the angular spread of the refractive term is also roughly limited to a narrow cone of vertex angle 2μ as shown in equation (3.57). On the other hand, if the sphere were purely reflecting, its scattering pattern would be isotropic. A partially reflecting sphere does not have an isotropic scattering pattern, but nevertheless this pattern is much broader than a scattering pattern due to either diffraction or refraction. These considerations imply that the reflective term becomes dominant in the backward scattering hemisphere.

The reflective term is relatively easy to evaluate. Before we proceed, we must note two important theorems presented by *van de Hulst* (1957). The first theorem states that: *the scattering pattern caused by reflection on large convex particles with random orientation is identical with the scattering pattern of a large sphere made of the same material and with the same surface condition.*

The second theorem, attributed to Cauchy (e.g., *Vickers* and Brown 2001), allows an easy evaluation of the average geometrical cross-section of randomly oriented particles. It reads: *the average geometrical cross-section of a convex particle is 1/4 of its surface area.*

A convex body can be of any shape as long as each element of the inner part of its surface can "see" every other element of the inner part of the same surface. The surface condition can vary from completely smooth, which leads to specular (i.e., mirror-like) reflection, or rough, which leads to uniform (i.e., Lambertian) reflection. The proof follows from the simple fact that the normals to elements of a random convex surface have the same distribution as the normals of a sphere.

The first theorem means that for randomly oriented particles, the reflection term from the first surface encountered by the incoming light wave (front surface) contains no information about the shape of a particle. In general, this is only true for the reflection from the first surface of the particle. If the particle is transparent, a reflection of almost equal amplitude will be produced by the second surface of the particle encountered by the wave. It has in fact been pointed out that to first order, the backscattering coefficient of a particle can be computed by assuming an interference between the amplitudes of the first and second surface reflections (*Klett* and Sutherland 1992, *Kerker* 1969). In the case of a small real refractive index, the direction of propagation of light is assumed not to be affected by the particle, whether a light ray starts outside or inside the particle. Thus, rays reflected from the second surface will exit the particle along their respective reflection directions. This implies that to first order, for randomly oriented convex particles with small real refractive index, the reflection from the second surface of the particle will also be independent of particle shape.

Thus, for particles typically found in natural waters, both the first and second surface reflection terms contain no information about the shape of the particle. As we will see, the angular variation and the magnitude of light scattering in the back hemisphere can be directly related to the value of the refractive index and to the condition of the surface of the particle (rough or smooth). Since the reflection term dominates the scattering in the back hemisphere, we come to a very important conclusion that the phase function in the back hemisphere is almost independent of the particle shape.

One notable exception to the above statement occurs in the case of perfect spheres where multiple internal reflections can lead to a very sharp increase in the amount of light scattered at and around 180° (*Kerker* 1969). However, it can easily be shown that almost any surface imperfections will substantially destroy these "anomalous" reflections. Increases in the backscattering direction have been observed, but they are nowhere near the amounts predicted by Mie codes. In fact, most of these increases can be accounted for by the roughness of the surface of the particle.

For a sphere, the scattered light reflected from the first surface is given by (*van de Hulst* 1957, *Kerker* 1969):

$$\begin{aligned}\sigma_{\text{scat}}^{\text{refl},0}(\theta) &= \pi r^2 \left[\frac{1}{4}\left(|r_{||}|^2 + |r_{\perp}|^2\right)\right] \\ &= \left(\frac{\hat{s}}{4}\right)\left[\frac{1}{4}\left(|r_{||}|^2 + |r_{\perp}|^2\right)\right]\end{aligned} \tag{3.61}$$

$$\begin{aligned}C_{\text{scat}}^{\text{refl},0} &= \left(\frac{\hat{s}}{4}\right)\int_0^{\pi}\left[\frac{1}{4}\left(|r_{||}|^2 + |r_{\perp}|^2\right)\right]\sin(\theta)d\theta \\ &= \left(\frac{\hat{s}}{4}\right)\frac{\omega_{||0} + \omega_{\perp 0}}{2}\end{aligned} \tag{3.62}$$

where the notation refl,0 is used to show that we are dealing with the terms involving direct reflections from the first surface of the particle, and $r_{||}$ and $r_{\perp}$ are reflectances of the surfaces for the polarization respectively parallel and perpendicular to the scattering plane. We will discuss these parameters shortly in more detail.

We have also used the second theorem and replaced the geometric cross-section of the sphere by one-fourth the mean surface area per particle, $\hat{s}$, of an ensemble of randomly oriented convex particles of otherwise arbitrary shape. According to the first theorem on reflection from randomly oriented convex particles, this replacement is all that is needed for equations (3.61) and (3.62) to apply to any set of randomly oriented convex particles.

In the above equations, $r_{||}$ is the amplitude reflection coefficient for polarization parallel to the plane defined by the incident and scattered radiation and $r_{\perp}$ is the corresponding amplitude reflection coefficient for polarization perpendicular to that plane. These parameters are defined in the following equations, with θ_i being an angle measured between the incident ray and the normal to the surface of the particle and θ_t measured between the transmitted ray and the surface normal:

$$r_\perp = \frac{\cos\theta_i - n cos\theta_t}{\cos\theta_i + n cos\theta_t} \tag{3.63}$$

$$r_{||} = \frac{n\cos\theta_i - cos\theta_t}{n\cos\theta_i + cos\theta_t} \tag{3.64}$$

Snell's law of refraction through a plane surface relates both angles to one another:

$$\sin\theta_i = n\sin\theta_t \tag{3.65}$$

With $|n-1| << 1$, the scattering angle, i.e., the total angular change, θ, of direction is given at refraction by:

$$\theta = |n-1|\,\theta_i \tag{3.66}$$

For reflection, we have

$$\theta = \pi - 2\theta_i \tag{3.67}$$

For particles with no absorption and the real part of the refractive index close to unity, the scattering cross-sections for the light reflected from the second surface of the particles and afterward transmitted out of the particle is given approximately by the following expressions:

$$\sigma_{\text{scat}}^{\text{refl},2}(\theta) = \frac{\hat{s}}{4}\left\{\frac{1}{4}\left[|r_{||}|^2\left(1-|r_{||}|^2\right)^2 + |r_\perp|^2\left(1-|r_\perp|^2\right)^2\right]\right\} \tag{3.68}$$

$$\begin{aligned} C_{\text{scat}}^{\text{refl},2} &= \frac{\hat{s}}{4}\int_0^\pi \left\{\frac{1}{4}\left[|r_{||}|^2\left(1-|r_{||}|^2\right)^2 + |r_\perp|^2\left(1-|r_\perp|^2\right)^2\right]\right\}\sin(\theta)d\theta \\ &= \frac{\hat{s}}{4}\,\frac{\omega_{||2}+\omega_{\perp 2}}{2} \end{aligned} \tag{3.69}$$

The notation refl,2 means that we are dealing here with terms involving reflection from the second surface of the particle of the light transmitted through the first surface. This reflection from the second surface is followed by transmission of the reflected light straight through the outer surface of the particle. For a real refractive index (no absorption occurring inside the particle), the integrals in (3.61), (3.62), (3.68), and (3.69) can be carried out to yield rational functions (*van de Hulst* 1957):

$$\omega_{\perp,0} = \frac{1}{3}\,\frac{(3n+1)(n-1)}{(n+1)^2} \tag{3.70}$$

$$\omega_{\parallel 0} = \frac{1}{(n^2+1)^3(n^2-1)^2} \times \Big\{ (n^4-1)(n^6-4n^5-7n^4+4n^3-n^2-1) + 2n^2\Big[(n^2-1)^4 \ln\Big(\frac{n-1}{n+1}\Big) + 8n^2(n^4+1)\ln(n)\Big]\Big\} \tag{3.71}$$

$$\omega_{\perp,2} = \frac{8}{35}\frac{\left[35n^3+21n^2+7n+1\right](n-1)}{(n+1)^6} \tag{3.72}$$

$$\begin{aligned}\omega_{\parallel 2} = -\frac{16}{15(n^2-1)^6(n^2+1)^7}\Big\{ & n^4(n^2+1)(n-1)(79n^{17}-131n^{16}+270n^{15}-430n^{14} \\ & +3094n^{13}-2170n^{12}+2626n^{11}-3010n^{10}+9784n^9-3460n^8+3010n^7 \\ & -2626n^6+4674n^5-590n^4+430n^3-270n^2+225n+15) \\ & +120n^6(3n^{16}+44n^{12}+98n^8+44n^4+3)\ln\left(\frac{\sqrt{n^2-1}}{n-1}\right) \\ & +15n^4(n^4+1)(n^{16}+60n^{12}+262n^8+60n^4+1)\ln\left(\frac{n-1}{n^2(n+1)}\right)\Big\}\end{aligned} \tag{3.73}$$

For the sake of simplicity, we chose to approximate the second surface reflection coefficient, $\omega_{\perp 2}$, by assuming it would be in the same ratio to the front surface reflection term, $\omega_{\perp 0}$, as the corresponding coefficients for the other polarization state. This procedure considerably simplifies the expressions, and it results in a maximum relative error of less than 1% for all cases.

We have so far derived our results under the assumption that there was no significant absorption occurring inside the particle. The effect of absorption can be approximated by the following simple prescription. First, compute an effective index n_{eff} that accounts to first order for the increase in reflectivity due to the contribution of the imaginary part of the refractive index of the absorbing material.

$$n_{\text{eff}} = \sqrt{n'^2 + n''^2} \tag{3.74}$$

Second, account for the damping of the reflection from the second surface by using an attenuation along the average pathway through the particle, i.e., by a factor of $\exp(-16n''\hat{x}/3)$:

$$\sigma_{\text{scat}}^{\text{refl}}(\theta) = \sigma_{\text{scat}}^{\text{refl},0}(\theta) + \sigma_{\text{scat}}^{\text{refl},2}(\theta)\exp\left(-\frac{16}{3}n''\hat{x}\right) \tag{3.75}$$

$$\begin{aligned} C_{\text{scat}}^{\text{refl}} &= C_{\text{scat}}^{\text{refl},0} + C_{\text{scat}}^{\text{refl},2}\exp\left(-\frac{16}{3}n''x\right) \\ &= \omega_t \frac{\hat{s}}{4} \end{aligned} \tag{3.76}$$

$$\omega_t = \frac{\omega_{\parallel 0}+\omega_{\perp 0}}{2} + \frac{\omega_{\parallel 2}+\omega_{\perp 2}}{2}\exp\left(-\frac{16}{3}n''\hat{x}\right) \tag{3.77}$$

The approximations used in equations (3.75) to (3.77) are phenomenological. For the damping term, we have used twice the amount of damping used to account for the reduction of the refractive term due to absorption inside a sphere [see equations (3.58) and (3.59)]. This increase represents doubling of the reflected light path as compared with that of refracted light. The refracted light leaves the particle after just one traverse (i.e., through the second surface), while light reflected at the second surface must traverse the particle again before it leaves the particle through the first surface. For lack of an exact term, the size parameter, x, used in equations (3.75) to (3.77), is the equivalent relative size of a sphere with a volume equal to the mean volume per particle of the ensemble of randomly oriented convex bodies. This approach assumes that the rays transmitted to and reflected from the second surface will have a length distribution equivalent to that of a sphere. This isotropy assumption is the only assumption consistent with the general theorems stated at the beginning of this section.

In fact, the real absorption term would not only damp the integrated amplitude but would also modify the pattern as a function of angle of the scattering cross-section. We are not taking into account this second-order effect in the present model. We should note that the model correctly describes the extreme cases of no absorption and total absorption of radiation inside the particle. In both these cases, equations (3.61), (3.62), (3.68), and (3.69) are correct because of the theorem on reflection from randomly oriented convex bodies. Since equations (3.75) to (3.77) are constrained to the correct solution in both the no-absorption and total absorption limits, and because the approximation is also consistent with the assumptions made in the derivation of the fundamental theorems, we expect its behavior to follow closely the experimental results in all cases.

We have so far managed to derive a form for the scattering from reflections by both the front and back surfaces of randomly oriented large convex particles. This term is independent of the particle shape distribution. The results obtained up to this point depend only on three assumptions. The particles are so large that we can neglect the effects of internal diffraction and approximate the scattering pattern by geometric optics, i.e., ray tracing. The real part of the refractive index is so close to unity that the ADA assumption of straight-line propagation through the particle surfaces applies. Finally, we assume that the surfaces of the convex particles are optically smooth, and their reflections are therefore specular.

Most large particles found in water, of either biological or mineral origin, are not expected to have perfectly smooth surfaces. Fortunately, we can modify our analysis to handle the case of particle with rough surfaces that reflect diffusely (*van de Hulst* 1957). This reflection pattern is sometimes called Lambertian diffusion. The rough surface required to produce a diffuse reflection pattern is obviously not convex on the small scale. For the rest of our analysis to hold, the surface roughness scale must be small compared to the size of the particle so that the approximation of convexity still applies to the bulk of the particle. The resulting form for diffuse scattering from the front and back surfaces of randomly oriented convex particles is:

$$\sigma_{\text{scat}}^{\text{rough}}(\theta) = \omega_t \frac{\hat{s}}{4} \frac{4}{3\pi} (\sin\theta - \theta\cos\theta) \tag{3.78}$$

$$C_{\text{scat}}^{\text{rough}} = \omega_t \frac{\hat{s}}{4} \tag{3.79}$$

The scattering pattern from diffuse surfaces goes to zero in the forward direction and peaks at 180° in the backscatter direction. The ratio of its value at 180° to its value at 90° is approximately 3. This type of behavior is often seen in the measured oceanic scattering functions.

In practice, we expect most particles to have a reflection behavior with a mix of both specular and diffuse components. This mixing can be modeled by linearly interpolating between the specular and diffuse cases. In order to do this, we will introduce a roughness parameter R that will vary from a value of 0.0 for perfectly smooth particles to 1.0 for particles with perfectly diffusing surfaces.

$$\sigma_{\text{scat}}^{\text{mixed}}(\theta) = (1-R)\sigma_{\text{scat}}^{\text{refl}}(\theta) + R\sigma_{\text{scat}}^{\text{rough}}(\theta) \tag{3.80}$$

$$C_{\text{scat}}^{\text{mixed}} = \omega_t \frac{\hat{s}}{4} \tag{3.81}$$

Note that value of the roughness parameter does not change the magnitude of the total scattering cross-section due to reflections. It only changes the scattering pattern as a function of angle. Figure 3.11 shows the scattering cross-section of particles due to reflection divided by the average geometric projected area for a randomly polarized beam. The descending thin curve is for a real refractive index of 1.1 and smooth particles with a roughness parameter of 0. Note that in this case, the reflected radiation pattern comes down almost monotonically to the value of the Fresnel coefficient in the backscatter direction at 180°. The curve is almost flat as a function of the angle from 100° to 180°. The thick curve is obtained for a particle with the same index and a roughness parameter of 0.25. Note the initial decrease to a minimum at 100° followed by a smooth rise to 180°. Finally, the ascending thin curve that rises monotonically from 0° to 180° represents particles with perfectly diffusing surfaces and a roughness parameter of 1. The form of the scattering curves in the zone extending from 90° to 180° is almost unchanged in the case where all the radiation is absorbed in the particle. The amplitude is simply reduced by half.

It has often been said that radiation scattered in the backward hemisphere by natural ensembles of particles suspended in the water column is dominated by contributions due to small particles that act as dipole scatterers. Our discussion

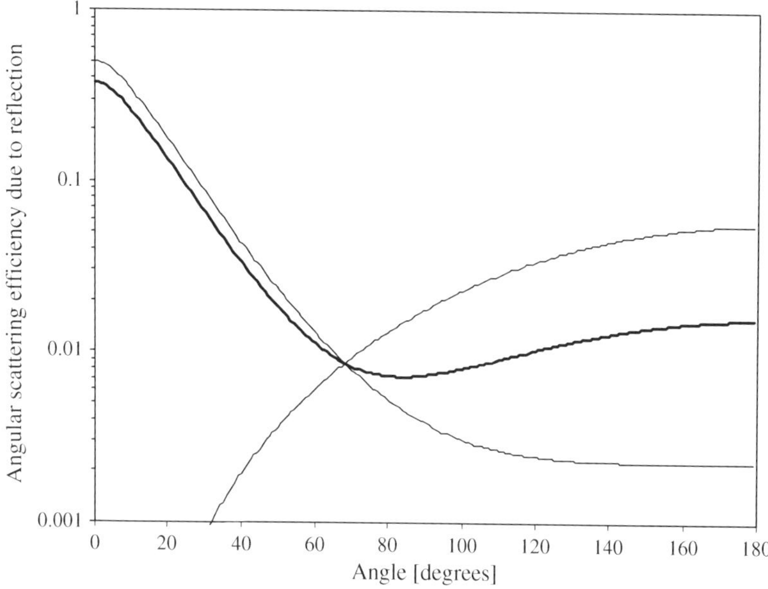

Figure 3.11. Angular scattering efficiency due to reflection for a randomly polarized beam, i.e., the scattering cross-section divided by the average geometric projected area. The thin descending curve is for a real refractive index of 1.1 and smooth particles (roughness parameter $R = 0$). The curve is almost flat as a function of angle from 100° to 180°. The thick curve is obtained for a particle with the same index and a roughness parameter $R = 0.25$. Note the initial decrease to a minimum at 90° followed by a smooth rise to 180°. The monotonically rising thin curve is for particles with perfectly diffusing surfaces (roughness parameter $R = 1$).

so far partly contradicts this piece of conventional wisdom and shows that similar scattering patterns can be obtained from the reflection of large particles. Distinguishing between the two contributions therefore becomes very difficult. This is a case where analyzing different polarization states may yield some useful information; however even this is somewhat doubtful given the results obtained by *Voss* and Fry (1984) on the measurement of the complete scattering matrix for ocean water. They found that their results could be explained by standard Mie calculations for spherical particles with a low refractive index. Their results extended from 20° to 160°. The equivalence of the spherical reflection pattern with the reflection pattern from randomly oriented arbitrary convex particle shapes would ensure that the results obtained with Mie theory could match any distribution of shapes.

Another extremely important conclusion is that light scattering in the backward hemisphere contains very little information, if any, about the shape distribution of the particles. If polarization is neglected, this is certainly true in the dipole scattering limit. Furthermore, as long as the relative refractive index is sufficiently close to 1 for the ADA to apply, this is also true in the limit of large convex particles of arbitrary shape. The only possible exception is in the direct backscatter signal at 180° where in the case of a sphere, multiple reflections can lead to large enhancements. These enhancements will not be seen for other shapes of particles.

3.3.3. Anomalous diffraction results for various particle shapes

The question therefore remains as to what angular portion of the scattering cross-section might contain information about the shape of the particles involved. What scattering features do different shapes share and what features might be unique? The second open problem is whether similar invariants as those found for the reflection term can be established for the diffractive and refractive components.

We will try to answer these questions by using the ADA. As we have already discussed it, this approximation applies to hydrosols because their refractive index relative to water is close to unity. We will inventory in this section the anomalous diffraction results for the most common shapes. Where exact solutions are not available, we will also try to obtain simple approximations that may be useful when considering large ensembles of particles. We will consider results for oriented particles and, whenever possible, for ensembles of randomly oriented particles.

We will consider two extreme shapes: disks and cylinders. As both these shapes have a single axis of symmetry, we will adopt a convention that the particle orientation is defined by the angle between the incident radiation and this axis of symmetry. For a disk, $\theta_0 = 0$ implies that the disk face is normal to the incident radiation. For a cylinder, $\theta_0 = \pi/2$ represents the case where the cylinder is perpendicular to the incident radiation. We will begin by quoting results for oriented particles and follow with the results for random orientation whenever possible.

For comparison, we will first gather together the results we have obtained for the sphere in the limit of the large particles and generalize them to apply to particles of all sizes. We have

$$\frac{\hat{s}}{4} = \frac{4\pi r^2}{4} = \pi r^2 \tag{3.82}$$

$$C_{\text{attn}} = \left(\frac{\hat{s}}{4}\right) 4\text{Re}\left[\frac{1}{2} + \frac{\exp(-i\rho)}{i\rho} + \frac{\exp(-i\rho) - 1}{(i\rho)^2}\right] \tag{3.83}$$

$$C_{\text{abs}} = \left(\frac{\hat{s}}{4}\right) 2\left[\frac{1}{2} + \frac{\exp(-4n''x)}{4n''x} + \frac{\exp(-4n''x) - 1}{(4n''x)^2}\right] \tag{3.84}$$

where $i\rho$ is defined by (3.42). In the limit of small ρ, these formulas reduce to the following form

$$\begin{aligned} C_{\text{scat}} &= C_{\text{attn}} - C_{\text{abs}} \\ &= \left(\frac{\hat{s}}{4}\right)\frac{|\rho|^2}{2} \\ &= \left(\frac{\hat{s}}{4}\right)\frac{|2(n-1)x|^2}{2} \\ &= \left(\frac{\hat{s}}{4}\right)\frac{4\left[(n'-1)^2+n''^2\right]x^2}{2} \end{aligned} \tag{3.85}$$

$$C_{\text{abs}} = \left(\frac{\hat{s}}{4}\right)\frac{2}{3}4n''x \tag{3.86}$$

The derivations of the angle-dependent scattering cross-sections assumed that the particles were large. The expressions we obtained were all in the form of the geometric cross-section times an angular component. The simplest possible improvement consists in replacing the geometric cross-section by the equivalent total cross-section while still maintaining the proportions between the terms that are appropriate to the large-particle limit. For example, the amount of light diffracted around the particle is always set equal to half of the attenuation, the appropriate proportion for large particle. The reflected and refracted terms are set by the same method. This gives the following expressions:

$$\sigma_{\text{diff}}(\theta) = \left(\frac{C_{\text{attn}}}{2}\right)2\frac{J_1^2(x\Theta)}{\Theta^2} \tag{3.87}$$

where, as before, $2\sin(\theta/2) = \Theta$, and

$$\sigma_{\text{refr}}(\theta) = \left(\frac{C_{\text{attn}}}{2} - C_{\text{abs}}\right)(1-\omega_l)\frac{8\mu^2}{(\Theta^2+4\mu^2)^2} \tag{3.88}$$

$$\sigma_{\text{scat}}^{\text{rough}}(\theta) = \omega_l\left(\frac{C_{\text{attn}}}{2}\right)\left[\frac{4}{3\pi}(\sin\theta - \theta\cos\theta)\right] \tag{3.89}$$

$$\sigma_{\text{scat}}^{\text{refl}}(\theta) = \left(\frac{C_{\text{attn}}}{2}\right)\left[\sigma_{\text{scat}}^{\text{refl},0}(\theta) + \sigma_{\text{scat}}^{\text{refl},2}(\theta)\left(1 - \frac{C_{\text{abs}}}{\hat{s}/4}\right)^2\right] \tag{3.90}$$

The last term in the square brackets in (3.90) approximates the average two-way absorption (hence the square) experienced by light reflected from the second surface of the particle. For small to moderate values of the absorption coefficient, the expression reduces to the same result as given in (3.75). Finally:

$$\sigma_{scat}^{mixed}(\theta) = (1 - R)\sigma_{scat}^{refl}(\theta) + R\sigma_{scat}^{rough}(\theta) \tag{3.91}$$

$$\sigma_{scat}^{total}(\theta) = \sigma_{diff}(\theta) + \sigma_{refr}(\theta) + \sigma_{scat}^{mixed}(\theta) \tag{3.92}$$

We will now develop equivalent expressions for other particle shapes. The first shape we will consider is the "infinite" cylinder. The size parameter of the cylinder with radius r is $x = kr = 2\pi r/\lambda$. The following are expressed for a cylinder length l with an assumption that l is much greater than the radius r.

$$C_{attn} = s\pi \mathrm{Re}\{\mathbf{H}_1(i\rho) + iJ_1(i\rho)\} \tag{3.93}$$

$$C_{abs} \approx s\left[1 - \exp\left(-\frac{\pi n'' x}{sin\theta_0}\right)\right] \tag{3.94}$$

with

$$i\rho = \frac{i2(n-1)x}{\sin\theta_0} = \frac{i2(n'-1)x + 2n''x}{\sin\theta_0} \tag{3.95}$$

$$\mu(\theta_0) = \frac{n' - 1}{\sin\theta_0} \tag{3.96}$$

$$s = 2rl\sin\theta_0 \tag{3.97}$$

where s is the projected surface area of the cylinder in the plane normal to direction of the incident light. $\mathbf{H}_1$ is the Struve function and J_1 is the first-order Bessel function (*Abramowitz* and Stegun 1964). Parameter θ_0 is the angle between the axis of symmetry of the cylinder and the direction of the incident radiation, $\theta_0 = \pi/2$ represents the case where the cylinder is perpendicular to the incident radiation. A more complete exposition of the problem can be found in *Fournier* and Evans (1996).

In the limit of small ρ, these formulas reduce to the following form

$$C_{scat} = s\frac{2|\rho|^2}{3} = s\frac{2}{3}\left|\frac{2(n-1)x}{\sin\theta_0}\right|^2 \tag{3.98}$$

$$C_{abs} = s\frac{\pi n'' x}{\sin\theta_0} \tag{3.99}$$

As noted by Fournier and Evans, the accuracy of these formulas for oriented cylinders can be improved by simply using the actual phase difference of the central light ray through the cylinder as the variable.

$$i\rho = i2x[(n^2 - \cos^2\theta_0)^{1/2} - \sin\theta_0] \tag{3.100}$$

This improvement is only useful for an oriented particle at angles near $\theta_0 = 0$. In the case of random orientation, the simpler form is perfectly adequate and allows an analytic form for some of the results:

$$\sigma_{\text{diff}}(\theta) = \left(\frac{C_{\text{attn}}}{2}\right) 2 \frac{\sin^2(x\Theta)}{\Theta^2} \tag{3.101}$$

where, as before, $2\sin(\theta/2) = \Theta$,

$$\sigma_{\text{refr}}(\theta) = \left(\frac{C_{attn}}{2} - C_{abs}\right)(1-\omega_t)\, 2 \frac{[2\mu(\theta_0)]^2}{\{\Theta^2 + [2\mu(\theta_0)]^2\}^2} \tag{3.102}$$

It is easy to see that the diffraction pattern, (3.101), for an infinite cylinder is equivalent to the pattern for an infinite slit (*van de Hulst* 1957) and therefore is always normal to the projection of the long axis of the slit (cylinder) onto a plane perpendicular to the direction of the incident light (Figure 3.12). The scattering pattern is independent of θ_0 because the cylinder is assumed to be of infinite length. It is this independence of orientation which is responsible for the preservation of the oscillations in the scattering patterns of randomly oriented cylinders seen in Figure 3.3 and Figure 3.4.

The same symmetry arguments can be used for the refractive term, (3.102). The cross-section of a cylinder whose axis is at an angle with respect to the incoming radiation is an ellipse. The result for the refractive component quoted above are based on assuming that this ellipse acts as a lens with a fixed diameter of $2r$ and a focal length that varies as $2(n'-1)/\sin\theta_0$. The refraction pattern is therefore also in a plane normal to the projection of the axis of the cylinder. Results for the refractivity terms are straightforward to derive; hence, we will not pursue further our analysis of oriented cylinders but concentrate instead on randomly oriented

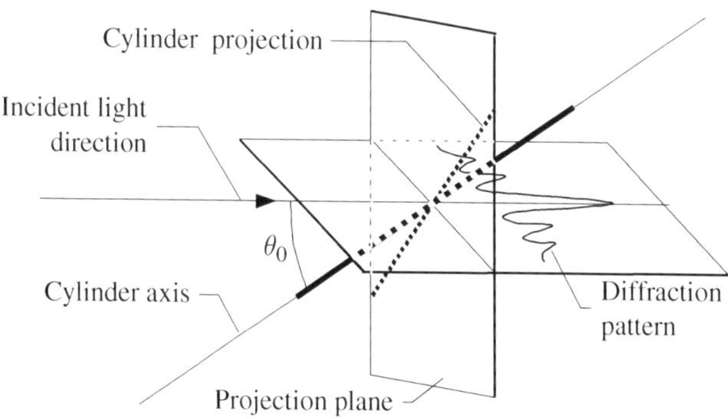

Figure 3.12. The diffraction pattern of an infinite cylinder is always perpendicular to the projection of the cylinder onto a plane perpendicular to the direction of the incident light.

ensembles. The formulas for the total attenuation and absorption cross-sections are (*Fournier* and Evans 1996):

$$C_{\text{attn}} = \frac{\hat{s}}{4}\{2+\exp(-2n''x)[\mathrm{F}(\rho')-2]\} \tag{3.103}$$

$$C_{\text{abs}} \approx \frac{\hat{s}}{4}\,[1-\exp(-4n''x)] \tag{3.104}$$

where $\rho' = 2(n'-1)x$:

$$\frac{\hat{s}}{4} = \frac{2\pi rl}{4} = \frac{\pi rl}{2} \tag{3.105}$$

and

$$\mathrm{F}(\rho') = \frac{4}{3}\rho'^2\left\{1 - \frac{\rho'\pi}{4}\times\left[J_0^2\left(\frac{\rho'}{2}\right) - \frac{2}{\rho'}J_0\left(\frac{\rho'}{2}\right)J_1\left(\frac{\rho'}{2}\right) + \left(1-\frac{2}{\rho'^2}\right)J_1^2\left(\frac{\rho'}{2}\right)\right]\right\} \tag{3.106}$$

Please note that in this formulation ρ' is real. The details of the derivation in *Fournier* and Evans (1996) explain why this is the most convenient form. In the limit of small ρ', the cross-sections become:

$$C_{\text{scat}} = \left(\frac{\hat{s}}{4}\right)\frac{4}{3}(1-2n''x)\,\rho'^2 \tag{3.107}$$

$$C_{\text{abs}} = \left(\frac{\hat{s}}{4}\right)4n''x \tag{3.108}$$

Because of the symmetry, the scattering cross-section due to diffraction is unchanged:

$$\sigma_{\text{diff}}(\theta) = \left(\frac{C_{\text{attn}}}{2}\right)2\frac{\sin^2(x\Theta)}{\Theta^2} \tag{3.109}$$

$$\sigma r_{\text{refr}}(\theta) = \left(\frac{C_{\text{attn}}}{2} - C_{\text{abs}}\right)(1-\omega_t)\,F(\mu,\theta) \tag{3.110}$$

where

$$F(\mu,\theta) = \frac{2}{\theta}\left\{\frac{2}{\theta^2/\mu^2}\left[1-\frac{1}{(1+\theta^2/\mu^2)^{1/2}}\right]-\frac{1}{(1+\theta^2/\mu^2)^{3/2}}\right\} \tag{3.111}$$

To simplify notation, we have used θ instead of $\Theta = 2\sin(\theta/2)$. Both forms are in fact equally satisfactory to the level of accuracy of the expressions. As we have shown in the previous section, the scattering due to reflection is independent of particle shape for ensembles of convex randomly oriented particles. The formulas are therefore the same as for spheres, but with the replacement of C_{attn} by the corresponding formula for randomly oriented cylinders.

One can derive a similar set of expressions for the case of oriented disks. In this case, the thickness t is assumed to be small as compared with its radius a. In the following formulas, we use the dimensionless thickness parameter of the disk

$$\tau = 2\pi \frac{t}{\lambda} \tag{3.112}$$

Once again, the orientation angle is taken as the angle between the direction of the incident radiation and the axis of symmetry. This implies that the disk is perpendicular to the incident radiation when $\theta_0 = 0$. Keeping in mind this convention, we can write down the following results.

$$C_{attn} = 2s\mathrm{Re}\{1 - \exp(-i\rho)\} \tag{3.113}$$

$$C_{abs} \approx s\left[1 - \exp\left(-\frac{2n''\tau}{\cos\theta_0}\right)\right] \tag{3.114}$$

where

$$i\rho = \frac{i(n-1)\tau}{\cos\theta_0} = \frac{i(n'-1)\tau + n''\tau}{\cos\theta_0} \tag{3.115}$$

$$s = \pi r^2 \cos\theta_0 \tag{3.116}$$

In the limit of small ρ, these formulas reduce to the following form

$$C_{attn} = s|\rho|^2 = s\left|\frac{(n-1)\tau}{\cos\theta_0}\right|^2 \tag{3.117}$$

$$C_{abs} = s\frac{2\,n''\tau}{\cos\theta_0} \tag{3.118}$$

As for cylinders, the accuracy of these formulas for oriented disks can be improved by simply using the actual phase difference of the central light ray through the disk

$$i\rho = i2\tau[(n^2 - \sin^2\theta_0)^{1/2} - \cos\theta_0] \tag{3.119}$$

This only applies to oriented disks at angles near $\theta_0 = \pi/2$. In the case of random orientation, the simpler form is here also perfectly adequate and once again allows one to derive analytic forms for some of the results.

It is interesting to note that the total attenuation cross-section for oriented thin disks shows a pattern of complete interference cancellation between the diffracted and the refracted parts. This implies that the diffracted pattern is identical to the refracted pattern, except for the irradiance lost to reflection that is not accounted for. This can be expected for flat plates: as shown by geometric optics, to first order, light passes straight through a plate. Thus, except for its magnitude and phase, the refracted pattern is identical to the incident illumination pattern on the plate. The light refracted through the plate is however subject to diffraction (i.e., the plate appears as an aperture with a phase shift of ρ' with respect to the light propagating around the particle). From Babinet's principle (section 1.7.2), the diffraction pattern of the refracted light is identical to the pattern of the light diffracted around the particle. In the case of spheres and cylinders, we neglected this diffraction of the refracted light because it was always the second-order correction of scattering due to the geometry of the particle. Indeed, the complete pattern consists of the convolution of each refracted ray with the diffraction pattern of an aperture of the size and shape of the geometric cross-section of the particle. For large particles, this diffraction pattern is very narrow and the convolution is almost indistinguishable from the pure geometric refraction pattern. In this case, the diffraction is of second order. In the case of flat plates, light goes straight through the plate, and there is no first-order refractive geometric terms. Therefore the diffraction of the refracted light dominates and becomes the first-order term. We now can make an important conclusion that in the case of scattering from thin flat plates and flat disks, the diffracted and refracted patterns are identical in shape and only differ in magnitude. This leads immediately to the following results:

$$\sigma_{\text{diff}}(\theta) = \left(\frac{C_{\text{attn}}}{2}\right) 2\frac{J_1^2(p\Theta)}{\Theta^2} \tag{3.120}$$

$$\sigma_{\text{refr}}(\theta) = \left(\frac{C_{\text{attn}}}{2} - C_{\text{abs}}\right)(1-\omega_t)\, 2\frac{J_1^2(p\Theta)}{\Theta^2} \tag{3.121}$$

$$p = \frac{r\cos\theta_0}{\sqrt{\cos^2\phi\cos^2\theta_0 + \sin^2\phi}} \tag{3.122}$$

In these expressions, $\Theta = 2\sin(\theta/2)$ as usual, ϕ is the angle of rotation around the plane perpendicular to the direction of propagation of the incident light. The projection of the disk at an angle θ_0 is an ellipse with a major axis of a and a minor axis of $a\cos\theta_0$. Parameter p represents the length of the vector from the center to the edge of the ellipse at an angle ϕ with respect to this major axis.

The formulas for the total attenuation and absorption of randomly oriented disks, obtained from direct integration, are as follows:

$$C_{\text{ext}} = \left(\frac{\hat{s}}{4}\right) 4\text{Re}\left\{\frac{1}{2} - E_3(-i\rho)\right\} \tag{3.123}$$

$$C_{\text{abs}} = \left(\frac{\hat{s}}{4}\right)[1 - 2E_3(2n''\tau)] \tag{3.124}$$

where $E_3(z)$ is the third-order exponential integral (e.g., *Abramowitz* and Stegun 1964) and

$$\rho = (n-1)\tau \tag{3.125}$$

$$\left(\frac{\hat{s}}{4}\right) = \frac{2\pi r^2}{4} = \frac{\pi r^2}{2} \tag{3.126}$$

In the limit of small values of τ, these formulas reduce to:

$$C_{\text{scat}} = \left(\frac{\hat{s}}{4}\right) 3\,|\rho|^2 = \left(\frac{\hat{s}}{4}\right) 3\,|(n-1)\tau|^2 \tag{3.127}$$

$$C_{\text{abs}} = \left(\frac{\hat{s}}{4}\right) 4n''\tau \tag{3.128}$$

Averaging the diffraction term for disks over all orientations does not lead to a simple analytic form. However, such a form can be derived using the concept of the equivalent radius of the average projected area. This radius is the radius of a disk that would have the same area as the projected area of the actual disk. It is defined as follows:

$$\pi \bar{r}^2 = \pi r^2 \cos\theta_0 \tag{3.129}$$

hence

$$\bar{r} = r\sqrt{\cos\theta_0} \tag{3.130}$$

This concept can obviously be generalized to flat thin plates of almost any shape. By almost any shape, we mean arbitrary shapes with an average aspect ratio between the minimum diameter to the maximum diameter of approximately 2 or less. The average over this equivalent radius can be carried out explicitly.

$$\sigma_{\text{diff}}(\theta) = \left(\frac{C_{\text{attn}}}{2}\right)\frac{2}{\theta^2}\left[J_1^2(r\theta) - J_0(r\theta)J_1(r\theta)\right] \tag{3.131}$$

$$\sigma_{\text{refr}}(\theta) = \left(\frac{C_{\text{attn}}}{2} - C_{\text{abs}}\right)(1-\omega_t)\frac{2}{\theta^2}\left[J_1^2(r\theta) - J_0(r\theta)J_1(r\theta)\right] \tag{3.132}$$

For the sake of simplicity, we again used θ in the equations instead of the standard form $\Theta = 2\sin(\theta/2)$. As pointed out by *van de Hulst* (1957), at the expected level of accuracy of the formulas, both forms are equivalent. The resulting angular shape of the scattering cross-sections is slightly broader than that of spheres and, as expected, all the oscillations have been almost completely damped by the averaging process.

Similar results can be obtained for oriented spheroids (*Fournier* and Evans 1991). Some results can also be obtained for ensembles of randomly oriented spheroids but at the price of some significant complexity (*Evans* and Fournier 1994).

Our ultimate purpose here was to extract general trends and overall behavior of the scattering function for typical ensembles of particles found in natural waters. The three shapes—spheres, flat plates (disks), and needles (cylinders)—which we considered so far represent extreme cases. These cases are sufficient to explore the space of possibilities, and we will therefore not derive detailed results for other intermediate shapes.

3.4. Oceanic phase function approximations

In order to be truly useful in the context of optics of natural waters, the results we obtained for single-particle scattering must now be integrated over the expected size distributions of particles found in these waters. The result of this averaging process is the volume scattering function which characterizes the angular pattern of light scattered by a unit volume of water-containing particles.

In general, the volume scattering function, β, depends explicitly on the wavelength, λ, of the incident radiation, the angle, θ, between the direction of the incident radiation and the direction of observation, as well as on the state of polarization. It depends implicitly on the composition of the ensemble of particles in terms of size, shape, and refractive index. If we combine the effect of shape, and index into an average scattering cross-section $\overline{\sigma}$ for the ensemble, we can write that

$$\beta(\vartheta, \lambda) = \int_0^\infty \frac{\overline{\sigma}(\theta, \lambda, r)}{2\pi} n(r) dr \tag{3.133}$$

where $n(r)$ is the number concentration of particles with radii in a range $(r,\ r+dr)$ per unit volume. This parameter is also referred to as the particle size distribution and is discussed in detail in a separate chapter. Note that we have divided the cross-section by 2π to account for the fact that in our previous notation, the integral over the azimuth angle φ had already been carried out. The total scattering coefficient per unit volume is obviously given by the integral of the volume scattering function over all angles.

$$b(\lambda) = \int_0^{2\pi} \int_0^{\pi} \beta(\theta, \lambda) \sin(\theta) d\theta\ d\varphi \tag{3.134}$$

As there are many practical situations where the composition of the ensemble of scattering particles is almost constant in terms of size, shape, and refractive index while the total number of particles varies (and with it the total scattering coefficient, b) over more than an order of magnitude, it is often convenient to express the scattering properties of the medium in terms of an essentially non-dimensional form (with a dimension of sr^{-1}) called the phase function $p(\theta)$:

$$p(\theta, \lambda) = \frac{\beta(\theta, \lambda)}{b(\lambda)} \tag{3.135}$$

From our previous discussion, the phase function is subject to the following normalization condition.

$$\int_0^{2\pi} \int_0^{\pi} p(\theta, \lambda) \sin(\theta) d\theta \, d\varphi = 1 \tag{3.136}$$

By simply substituting the expressions we derived in the previous sections for the various contributions to the scattering cross-section we can already derive some interesting observations. For the sake of simplicity and clarity, we will first discuss the case where none of the particles absorbs light.

The first important observation we can make is that for both the refraction and reflection contributions to the scattering cross-section, the angular behavior is independent of particle size. In both cases, the effect of the size is expressed in terms of the geometric cross-section multiplied by the scattering efficiency. Recalling that in the case of no absorption, the total scattering cross-section is equal to the total attenuation cross-section, we can write the contribution to the phase function of the refracted term $p_{\text{refr}}(\theta, \lambda)$ as follows:

$$\begin{aligned} p_{\text{refr}}(\theta, \lambda) &= \frac{1}{2\pi} \frac{\int_0^\infty \sigma_{\text{refr}}(\theta, \lambda, r) n(r) dr}{\int_0^\infty C_{\text{scat}}(\lambda, r) n(r) dr} \\ &= \frac{1}{2\pi} \frac{(1-\omega_i) \text{F}(n'-1, \theta) \int_0^\infty \frac{C_{\text{scat}}(\lambda, r)}{2} n(r) dr}{\int_0^\infty C_{\text{scat}}(\lambda, r) n(r) dr} \\ &= \frac{1}{4\pi} (1-\omega_i) \text{F}(n'-1, \theta) \end{aligned} \tag{3.137}$$

As it is implied by (3.52), for spheres the angular part is given by the following simple formula:

$$p_{\text{refr}}(\theta, \lambda) = \frac{1}{4\pi} (1-\omega_i) 2 \frac{4(n'-1)^2}{[\Theta^2 + 4(n'-1)^2]^2} \tag{3.138}$$

The derivation can obviously be carried out for cylinders, and the result will be the same as for spheres except for the use of the angular function for cylinders. In the case of disks, the refractive term is identical to the diffracted term, except for a reduction in its magnitude due to the amount of light reflected by the surfaces of the particle. Since the angular form of this term depends on the size of the particle, we cannot simplify its contribution to the phase function in the same way.

The same approach can be used for evaluating the contribution to the phase function of the light reflected by the surface of the particle. Once again, the angular part of the reflected radiation is independent of the size of the particle. This is true for all ensembles of randomly oriented convex particles, including disks. We can therefore write that:

$$p_{\text{refl}}(\theta, \lambda) = \frac{1}{4\pi}\omega_t F_{\text{refl}}(n', \theta) \tag{3.139}$$

For example, as was shown in section 3.3.1, an ensemble of randomly oriented convex particles with a rough surface will have the following angular dependence for its reflective term:

$$F_{\text{refl}}(n', \theta) = \frac{3}{4\pi}(\sin\theta - \theta\cos\theta) \tag{3.140}$$

By using the other results from section 3.3.1, an ensemble of randomly oriented convex particles with smooth surfaces will have the following angular dependence function:

$$F_{\text{refl}}(n', \theta) = \frac{\omega_{\parallel 0}\left|r_{\parallel}(\theta)\right|^2}{4\omega_t} + \frac{\omega_{\perp 0}\left|r_{\perp}(\theta)\right|^2}{4\omega_t} + \frac{\omega_{\parallel 2}\left|r_{\parallel}(\theta)\right|^2\left(1-\left|r_{\parallel}(\theta)\right|^2\right)^2}{4\omega_t} + \frac{\omega_{\perp 2}\left|r_{\perp}(\theta)\right|^2\left(1-\left|r_{\perp}(\theta)\right|^2\right)^2}{4\omega_t} \tag{3.141}$$

with

$$\omega_t = \frac{\omega_{\parallel 0}+\omega_{\perp 0}}{2} + \frac{\omega_{\parallel 2}+\omega_{\perp 2}}{2} \tag{3.142}$$

The $r_{\parallel}(\theta)$ and $r_{\perp}(\theta)$ are the Fresnel reflection coefficients [equations (3.63) and (3.64)] for the perpendicular and parallel polarization states. Terms $\omega_{\parallel j}$ and $\omega_{\perp j}$ are the integrated reflection coefficients for the front ($j = 0$) and back ($j = 2$) surfaces of the particles. These terms are defined in section 3.3.2.

We have so far managed to show that the forms for the reflective contributions to the phase function are completely independent of the particle size distribution. This conclusion also applies for any ensemble of convex randomly oriented particles. For spheres and cylinders, the refractive contribution is also to first order independent of the size distribution. This is an enormous simplification. For ensembles whose average total scattering cross-section is dominated by the contribution of large particles, the scattering in the back hemisphere can be dominated by the refraction and reflection terms.

This domination by the large-particle fraction is very often the situation that is observed in the measurements carried out in ocean waters. Our results imply that under those conditions, in the back hemisphere, the functional form of the phase function is almost completely independent of the particle size distribution. These results imply that the phase function will show a considerable amount of regularity in the backscattering angular range. According to our analysis, the main factors affecting the phase function in the back hemisphere are the value of the mean refractive index and the degree of surface roughness of the particles.

The contribution from the shape of the particle will only be noticeable at intermediate angles ($\theta \cong n' - 1$) in the forward hemisphere. For spheres and cylinders, the refractive contribution is also independent of the form of the size distribution. It only depends on particle shape and refractive index. The one exception to this conclusion are particles that are both sufficiently flat and small so that their angular diffraction pattern is broader than their refractive angular scattering pattern. In that case, since the angular pattern due to refractive scattering is extremely narrow, one must take into account the broader pattern of diffraction of the light coming through the particle itself. Accordingly, except for their amplitude, the angular patterns of the light passing through the particle and of the light diffracted around it become identical to first order.

We have now exhausted the fortunate circumstances that allowed us to compute the refractive and reflective contribution to the phase function for some types of particles without explicit reference to the size distribution. The angular distribution of both the pure diffractive term and the refractive term from sufficiently flat particles are explicitly dependent on size and require definite assumptions to be made about the size distribution.

The fact that the relative refractive index is always near unity allowed us to considerably simplify the descriptions of single-particle scattering. Similarly, the fact that the particles size distributions in water can often be approximated by an inverse power law will allow an important simplification of the results for the scattering functions of an ensemble of particles. Using an inverse power form for the particle size distribution implies the existence of some very general relationships that can be derived from first principles.

Most people are uneasy about using an inverse power relationship for a size distribution because of the infinities involved in the evaluation of parameters at both small and large sizes. In addition, we have been trained to think that a mere power law is always just a convenient approximation to some more complicated function. This assumption certainly holds when one is dealing with relatively simple systems. For truly complex systems evolving in open environments, time and time again, power laws have been found to apply over astounding ranges. For complex systems, power laws could in fact be in some sense the "exact solution." They certainly seem to be the only regular feature that has consistently been observed. It has recently been confirmed that the size distribution of living organisms follows such laws over an astounding 17 order of size magnitudes (*Schmid* et al. 2000, Marquet 2000). *Kiefer* and Berwald (1992) among others have postulated a predator/prey system to explain the inverse fourth power found in measurements of the population size distribution of the microbial planktonic community. This inverse power type of size distribution is often called a Junge distribution. More detailed discussion of experimental data supporting this distribution form and of the models leading to it is given in Chapter 5. We will use this form here to analyze the features of the overall oceanic phase functions. We start by writing down an explicit formula for the phase function of a suspension with a power-law particle size distribution:

$$p_{\text{diff}}(\theta, \lambda) = \frac{\frac{1}{4\pi}\int\limits_0^\infty C_{\text{scat}}(\lambda, r) f(\theta, \lambda, r)\left(\frac{k}{r^m}\right) dr}{\int\limits_0^\infty C_{\text{scat}}(\lambda, r)\left(\frac{k}{r^m}\right) dr} \tag{3.143}$$

where we used the inverse power approximation for the size distribution $n(r) = kr^{-m}$. Note that in anticipation of the notation of Chapter 5, symbol k denotes here the magnitude parameter of the size distribution.

Some important relations can be derived here by simply carefully considering the functional relationships of the various terms. For instance, we know from (3.18) and (3.20) that λ in $C_{scat}(\lambda, r)$ and $f(\theta, \lambda, r)$ only occurs in the non-dimensional size parameter, x, the definition of which we shall recall here for convenience:

$$x = \frac{2\pi r}{\lambda}$$

This non-dimensional dependence of the phase function on the ratio of the physical dimensions to the wavelength of the incident radiation is in fact a perfectly general property of the interaction of any wave with a particle.

We also know that C_{scat} can be written as the product of the average projected area of a particle times the scattering efficiency that is only dependent on the particle radius, r, through the size parameter, x. On changing the variable of integration to the size parameter x instead of particle radius, r, we obtain.

$$p_{\text{diff}}(\theta, \lambda) = \frac{1}{4\pi}\frac{\left(\frac{\lambda}{2\pi}\right)^{3-m}\int\limits_0^\infty \pi x^2 Q_{\text{scat}}(x) f(\theta, x)\left(\frac{k}{x^m}\right) dx}{\left(\frac{\lambda}{2\pi}\right)^{3-m}\int\limits_0^\infty \pi x^2 Q_{\text{scat}}(x)\left(\frac{k}{x^m}\right) dx} \tag{3.144}$$

In this expression for the phase function, the explicit wavelength dependence cancels out, and we come to an important conclusion: for power-law (Junge) distributions of particle sizes spanning a range of 0 to ∞, the phase function is to first order independent of wavelength. The only source of wavelength dependence will be through the variation with wavelength of the refractive index. Another important conclusion is that for a Junge distribution of inverse power m, the total scattering cross-section will vary to first order as $\lambda^{-\gamma}$, where λ is the wavelength and

$$\gamma = m - 3 \tag{3.145}$$

which has been known since 1929 in atmospheric physics as the Ångström law (e.g., *Heintzenberg* and Charlson 1996). *Twardowski* et al. (2001) note other derivations of this relationship. Calculations of the attenuation coefficient spectra by *Kishino* (1980), who integrated power-law size distributions over a limited particle radii range of 0.25 to 15 μm with a Mie kernel, yield similar results as noted by *Kitchen* et al. (1982). By substantially extending the scope of these Mie-based calculations, a relationship essentially identical to (3.145) has been also obtained by *Boss* et al. (2001)

$$m = 3 + \gamma - 0.5\exp(-6\gamma) \tag{3.146}$$

who also noted that for the particle size distribution slope, $m > 3$, typical of seawater, the difference between these two formulations is negligible.

Equation (3.145) implies that we can evaluate the value of the exponent of the power-law particle size distribution by simply measuring the total scattering coefficient as a function of wavelength and fitting the wavelength dependence of the result by a power law (*Fournier* 2000).

There is another important property of the phase function that we can derive from general considerations (*Fournier* and Forand 1994). The normalized angular distribution of the light diffracted around a particle is given by:

$$f(\theta, x) = x^2 \left[2\frac{J_1^2(x\Theta)}{(x\Theta)^2} \right] \tag{3.147}$$

where, as before, $\Theta = 2\sin(\theta/2)$.

By using the z-variable defined by (3.40), which we here recapitulate for convenience:

$$z = x\Theta$$

we can rewrite the diffraction contribution to the phase function as:

$$p_{\text{diff}}(\theta, \lambda) = \frac{1}{4\pi} \frac{\int\limits_0^\infty \pi \left(\frac{z^2}{\Theta^2}\right) Q_{\text{scat}}\left(\frac{z}{\Theta}\right) \left(\frac{z^2}{\Theta^2}\right) \left[2\frac{J_1^2(z)}{z^2}\right] \left(\frac{k\Theta^m}{z^m}\right) d\left(\frac{z}{\Theta}\right)}{\int\limits_0^\infty \pi x^2 \, Q_{\text{scat}}(x) \left(\frac{k}{x^m}\right) \, dx} \tag{3.148}$$

In the case of large particles which dominate light scattering near the forward direction, $Q_{scat} \cong 2$. By substituting this limit and simplifying the formula we obtain:

$$p_{\text{diff}}(\theta, \lambda) \approx \Theta^{m-5} k \frac{\int_0^\infty \left[\frac{J_1^2(z)}{z^2}\right] z^{2-m} dz}{\int_0^\infty \pi x^2 Q_{scat}(x) \left(\frac{k}{x^m}\right) dx} = \Theta^{m-5} k \left\{\frac{\mathbf{I}_1}{\mathbf{I}_2}\right\} \quad (3.149)$$

Note that the ratio of the two integrals with limits from 0 to ∞ is independent of the scattering angle and is merely a number. We have established the remarkable fact that for power-law size distributions with an inverse power m, the part of the phase function due to diffraction approaches the forward angle as an inverse power η of the scattering angle and that this power is directly related to m:

$$\eta = 5 - m \quad (3.150)$$

For convenience, we used in this derivation the formula for the diffraction of light around an obstacle with a circular geometric cross-section. If we used the equivalent expression for another shape with a finite cross-section, a different expression for the angular function thus obtained would have merely resulted in different values for the ratio $\mathbf{I}_1/\mathbf{I}_2$ in (3.149). The functional dependence would be the same because it only requires that the scattering efficiency approach a constant in the limit of the large-particle sizes and that the angular scattering function for each particle be properly normalized to unity. The latter condition can be rephrased as follows. The forward diffraction peak of a finite object narrows as the inverse of the square of the characteristic size parameter. Therefore, the normalized amplitude must scale as the square of the characteristic size parameter to compensate for that narrowing.

According to the above arguments, the phase function should go to infinity at $\theta = 0$. In a real experiment, any apparatus used to measure the phase function has a finite aperture. This aperture produces an angular diffraction pattern which is convolved with the true phase function of seawater. The convolution ensures that the measured phase function is always finite at $\theta = 0$. As the design of experiments improved and measurements were taken closer to the forward angle with equipment with progressively larger apertures and smaller intrinsic diffraction, the inverse power behavior was observed to continue and to stretch ever closer to $\theta = 0$. No one has yet measured at what angle the true phase function turns over and stops increasing as an inverse power. Such a result would imply the existence of a cut-off of the particle size distribution at large sizes. Note that at this time, any turnover in the slope of the phase function near $\theta = 0$ requires one to postulate a purely arbitrary cut-off of the Junge particle size distribution. The mathematical singularity at $\theta = 0$ is easily handled by simple integration, and one can thus obtain finite values for all the measurable parameters related to the phase function. Because the imposition of an unobserved cut-off in the size distribution is completely arbitrary and the problem of handling the infinities is easy, we believe that one should use the pure inverse power forms derived so far for the phase function.

We can combine the expression for the wavelength dependence of the total scattering coefficient on the exponent of the power-law size distribution with the slope of the angular dependence of the phase function near $\theta = 0$ on the same Junge exponent in order to obtain

$$\gamma = 2 - \eta \tag{3.151}$$

This equation shows clearly the remarkable fact that by measuring the logarithmic slope η of the phase function near $\theta = 0$ at one wavelength, we can compute the full spectral dependence of the total scattering coefficient fixed by the value γ of its power law. We can of course use either the wavelength dependence of the total scattering coefficient or the logarithmic slope of the phase function in the near forward angle to determine the Junge exponent of the particle size distribution.

We have systematically used the minimum number of assumptions to derive results which are as general as possible. We have unfortunately reached the end of this fruitful approach, and further progress will now require the use of more specific, and thus restrictive, approximations.

The fact that we are integrating over a broad size range allows one to make rather crude assumptions about the details of the functions involved. The only requirement is that the approximating functions have the correct asymptotic behavior. *Fournier* and Forand (1994) as well as *Forand* and Fournier (1999) used this approach to obtain an analytic approximation to the diffractive part of the phase function. We corrected some minor typographical errors in their papers.

They first approximated the scattering efficiency by the following simple expression:

$$Q_{\text{scat}}(x) \approx \frac{\frac{\rho'^2}{2}}{1 + \frac{\rho'^2}{4}} \tag{3.152}$$

This expression is correct both in the small x and in the large x limits. It rises monotonically from 0 to the asymptotic value of 2. However, the interference between diffraction and refraction that causes the series of alternating maxima and minima in Q_{scat} is neglected to first order by this expression. The integration over a broad size distribution minimizes the magnitude of an error incurred by this neglect.

To obtain an integrable expression, Fournier and Forand used a binomial form to represent the normalized angular diffraction function of the individual particle:

$$\frac{f(\theta, x)}{2\pi} \approx \frac{1}{4\pi}\frac{1+\frac{4}{3}x^2}{\left(1+\frac{\Theta^2 x^2}{3}\right)^2} \tag{3.153}$$

where $\Theta = 2\sin(\theta/2)$, as before. In this approximation, the total scattering cross-section, b, is given by:

$$\begin{aligned} b(\lambda, m, n') &= k\left(\frac{2\pi}{\lambda}\right)^{m-3}\pi\int_0^\infty Q_{\text{scat}}(x)x^{2-m}dx \\ &= k\frac{\pi}{\cos\frac{\pi m}{2}}\left[\frac{2\pi(n'-1)}{\lambda}\right]^{m-3} \end{aligned} \tag{3.154}$$

Finally they obtained the following expression for the diffractive component of the phase function

$$\begin{aligned} p_{\text{diff}}(\theta, \lambda) = \frac{1}{4\pi}\frac{1}{(1-\delta)^2\delta^\nu}\Big(&[\nu(1-\delta)-(1-\delta^\nu)] \\ &+\frac{4}{\Theta^2}[\delta(1-\delta^\nu)-\nu(1-\delta)]\Big) \end{aligned} \tag{3.155}$$

with

$$\nu = \frac{3-m}{2}, \quad \delta = \frac{\Theta^2}{3(n'-1)^2}, \quad \Theta = 2\sin(\theta/2) \tag{3.156}$$

Combining the expression for this diffractive term with the terms for refraction and reflection, we obtain a total phase function approximation that can be compared to experimentally measured volume scattering functions.

3.5. Basic experimental comparison

There are very few complete marine phase function data sets that combine measurements at both large and small angles. The data of *Petzold* (1972) stand out in this respect. It is probably the most complete data set in terms of angular coverage. A detailed comparison of this data set with the volume scattering function of spheres reveals a significant discrepancy. There is no evidence of the effect of the refractive term.

To establish this, we first compared the results of an approximation that did not include the refractive term with those of an approximation that did. The effect of the refractive term was to raise the volume scattering function around $\theta = n - 1$ above the level it had when the term was neglected. The effect of refraction then shows up on a logarithmic plot of the volume scattering function as a hump in an otherwise smooth monotonic decay as a function of angle. No experimental volume scattering function which we analyzed displayed this feature.

According to our analysis, there are two possible ways in which this refractive term can disappear. In the first case, the particle is indeed spherical, but it has a large enough absorption coefficient that all the light entering the particle is absorbed. This hypothesis would imply that the absorption and scattering coefficients for large particles is equal. In the second case, the particles are non-spherical. We have seen that in the extreme case of flat plates, the refractive term actually takes on the same form as the diffractive term. In this case, the particle does not have to absorb significantly. Most results using Mie theory to fit experimental phase functions use the first approach (i.e., introduce artificial absorption) to eliminate both the refractive hump around $\theta = n' - 1$ and the sharp resonances around the backscatter direction. Values of the refractive index obtained by this technique should be viewed with some reserve as the effect on which they are based could be explained as easily by non-sphericity of the particles. The ratio of the total absorption and scattering coefficients is the key factor that could allow one to distinguish among the two hypotheses. Accordingly, the data of Petzold favor the hypothesis of non-sphericity. In any case, in further fitting the results of experiments with the phase function, we deliberately took the pragmatic approach of setting the refractive term to zero.

As we mentioned before, the data sets of Petzold are the most complete in terms of angular coverage. Unfortunately, the data were obtained at only one wavelength. This precludes us from using it to verify the wavelength dependence of the approximations. The only data set that adequately covers both angular and wavelength range is that of *Whitlock* et al. (1981) for turbid estuary waters. Using a moored buoy, they measured phase functions simultaneously at 50 nm intervals from 450 to 800 nm. This data set covers a sufficiently wide angular and wavelength range to be an excellent test of the concepts outlined in the previous section. *Forand* and Fournier (1999) first performed a fit to each phase function using a standard non-linear simplex algorithm (*Nelder* and Mead 1965, *Press* et al. 1989) that minimized the relative least squares error. The formula used in this fit was:

$$\beta(\theta, \lambda) = A\ p(\theta, \lambda) \tag{3.157}$$

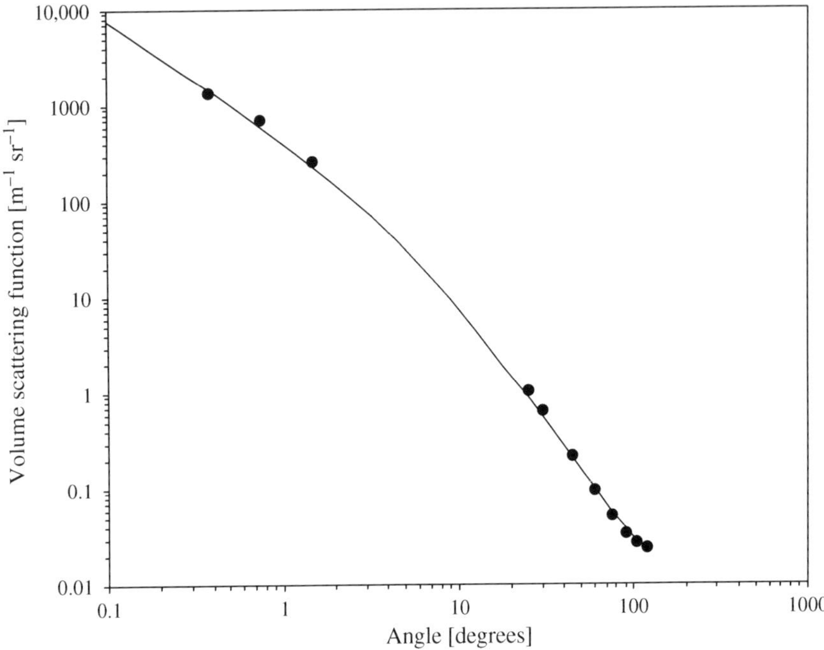

Figure 3.13. Absolute phase function data of Whitlock and colleagues (*Whitlock* et al. 1981) at a wavelength of 800 nm (dots) and the best fit using phase function (3.155) [solid line, $m = 3.77$, $n' = 1.09$, the magnitude constant, equation (3.157), $A = 6.272$].

The form of the phase function $p(\theta, \lambda)$ is assumed to follow (3.155). The parameters to fit are therefore the total scattering amplitude A, the refractive index, n', and the power of the particle size distribution m. The result of this fit at a wavelength of 800 nm is shown in Figure 3.13. The relative RMS error is less than 8% showing the close match of the functional form to the experimental phase functions. Similarly close fits of the phase functions were obtained at all other wavelengths.

Fournier and Forand also performed a fit to the entire set of phase functions using the power-law wavelength dependence whose exponent is given in equation (3.151). The following equation was used for this second fit:

$$\beta(\theta) = A_t \left(\frac{\lambda}{\lambda_r}\right)^{3-m} p(\theta) \tag{3.158}$$

where λ_r is an arbitrary reference wavelength which was chosen to be 550 nm and A_t is the total amplitude. The results of this overall fit are compared with the parameters obtained from the single wavelength fits in Figure 3.14. The solid line

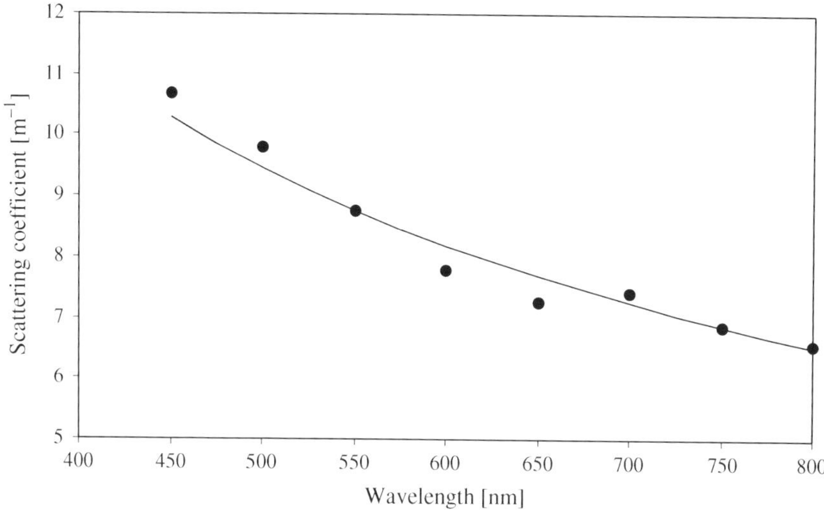

Figure 3.14. Absolute scattering coefficient as a function of wavelength derived from the data of *Whitlock* et al. (1981). The dots represent each the result of an independent fit of the experimental phase function at the respective wavelength. The solid line is the result of the fit of these results by using equation (3.158) with $m = 3.79$ and $n' = 1.08$.

is the result of the overall fit. The result of a direct least squares fit to an arbitrary power law as a function of wavelength is indistinguishable from the results of the approximation. This close agreement clearly shows that the overall formula for the volume scattering function is also the best fit to the total scattering cross-section as a function of wavelength.

As it was mentioned previously, there are very few data sets that incorporate both near forward-angle and large-angle data points. The combination of both is required to obtain reliable values of the size distribution slope m and refractive index, n'. *Forand* and Fournier (1999) therefore carefully analyzed the full data sets of *Petzold* (1972) and *Whitlock* et al. (1981). They first corrected the data to take out the scattering due to pure seawater due to the Einstein–Smoluchowski density fluctuations (section 2.4). After this correction and eliminating a single result that was an obvious outlier, they found a remarkable regularity. The refractive index of the scattering particles was centered at a mean value of 1.09, with a very tight standard deviation of 0.01. The mean value of the power-law size distribution slope, m, was 3.649 with a standard deviation of 0.123. This value of m corresponds to a variation of the total scattering coefficient as a function of wavelength of $1/\lambda^{0.649}$. The corresponding angular dependence of the phase function near $\theta = 0$ should be expressed as $1/\theta^{1..35}$. All the slopes as a function of angle near $\theta = 0$ measured directly by *Petzold* (1972) fall within the quoted standard deviation of this value.

3.6. Conclusions

In the case of scattering from particles with the small refractive index typical of the particles found in natural waters, several very general results can be derived. Probably the most important of those is the existence of a universal form for the scattering of light in the back hemisphere which depends only on the refractive index and the surface roughness of the particles. This form applies to all ensembles of randomly oriented convex particles. This universality is confirmed by a wealth of experimental data on scattering in the backward direction by oceanic waters.

Another important result is a consequence of the previous finding and states that the information about particle shape is mainly due to the refractive contribution. The maximum effect of that shape is expected to show up around the angle $\theta = n' - 1$, where n' is the real part of the refractive index. For a typical value of the refractive index $n' = 1.09$, this corresponds to a zone around 5°. This zone is notoriously neglected in most light scattering measurements. Further studies in this area would have the potential of verifying how much shape information could actually be extracted from the phase function.

With the assumption of an inverse power law with exponent m for the particle size distribution, an assumption that data and modern complexity theory supports (*Marquet* 2000), we obtain some further significant results. The total scattering coefficient will vary as a function λ^{3-m} and the volume scattering function in the near forward direction will vary as θ^{m-5}. A corollary to these statements is that a measurement of either the total scattering as a function of wavelength or the logarithmic slope of the near forward-angle scattering function will determine the exponent of the power-law size distribution. This information can then be combined with a measurement of the scattering in the back hemisphere to obtain the value of the average refractive index of the particles.

The mean refractive index and general features of the particle size distributions are thus the only features that can be reliably and easily extracted from the oceanic volume scattering functions. This function is remarkably insensitive over most of its angular range to variations in particle shape and size because it involves contributions of particles with a relative refractive index near unity. Most of the other regularities observed in the form of the volume scattering function are due to the fact that the inverse power size distribution is a relatively good approximation to the real size distribution.

3.7. Problems

A lot of marine plankton are encased in a thin shell which has a significantly higher index of refraction relative to water than that of their liquid core. Typically, the relative index of refraction of the shell is around 1.1, while the relative index of refraction of the liquid core is 1.03. If the shell is thin enough and its relative index is small enough that the fundamental assumption of anomalous diffraction (i.e., approximately straight path of light propagation through the particle) still applies, the main effect of a shell on light scattering by phytoplankton cells with shells can be modeled by using two effective indices of refraction, one for the transmitted term and one for the reflective term. Assuming a spherical particle of radius r with a shell thickness t, derive the formulae for these two effective indices.

Hint: For the transmitted term, the phase difference must be accounted for precisely, while for the reflective term the Fresnel reflection coefficient must be modeled accurately.

Chapter 4

Measurements of light scattering by particles in water

4.1. Introduction

The light scattering functions of a suspension or aerosol provide a complete description of the incoherent interaction of light with the particles. There are 16 such functions of the scattering angle which form a 4×4 Mueller matrix, $\mathbf{M}$, which describes a linear transform of the irradiance (power per unit area) and polarization of the incident light beam into the intensity (power per unit sold angle) and polarization of the scattered beam. The most researched and measured is the element $\mathbf{M}_{11}$ of this matrix, which describes the effect of the particles on the intensity of the scattered light. This element is the volume scattering function, also referred to as the Rayleigh ratio in the older literature. Note that the intensity, as referred to here, is not the intensity referred to traditionally in physics. This latter quantity is the irradiance in the terminology used in this chapter.

For particles in water, the volume scattering function, which we already defined conceptually in (1.9) and will define operationally in this chapter, is a sum of two components: the volume scattering function of particles themselves and that of pure water (seawater). The first component can itself be partitioned into contributions of individual particle classes, such as microbial particles in seawater (*Stramski* et al. 2001, *Stramski* and Mobley 1997), phytoplankton, and mineral particles, to start with. Interestingly, a similar approach has developed in the atmospheric sciences (e.g., *Levoni* et al. 1997).

We discussed in Chapter 2 the volume scattering function of water and its theoretical relationships to the wavelength of light and the scattering angle. In Chapter 3, we discussed the various theoretical models of light scattering by particles, as well as contributions to the particle volume scattering function of the effects of diffraction, refraction and absorption, as well as reflection of light by the particles. Here, we will be solely concerned with problems related to the experimental determination of the volume scattering function and the usage of such experimental data. These problems include measurement techniques, experimental errors, representative data, and methods of approximation of the volume scattering function.

In reviewing approximation methods, we will be looking at the experimental data from a black-box point of view, as opposed to the discussion involving the basic physical processes we carried out in Chapter 2 and Chapter 3. We will examine efficient and realistic methods of approximation of experimental light scattering for use in many aspects of the transfer of radiant energy in the sea. Applications of such approximation methods range from underwater visibility to remote sensing. Recent examples of such applications are a study of the effect of the form of the volume scattering function on the reflectance of the sea as a function of the solar angle (*Morel* and Gentili 1993) and a study of the effect of the form of the approximation on numerical solutions of the radiative transfer equation in seawater (*Mobley* et al. 2002).

4.2. Scattering function

4.2.1. Definitions and units

Some of the definitions given in this section have already been introduced in Chapter 2. We recapitulate these definitions here for easy reference.

The volume scattering function, also referred to in the older publications as the Rayleigh factor (e.g., *Kaye* and Havlik 1973), characterizes the angular pattern of light scattered by a volume of a medium, e.g., hydrosol or aerosol. This function, usually denoted by β, is a proportionality factor that relates the intensity of light scattered in a given direction, $\boldsymbol{\xi}$, by an infinitesimal volume $\mathrm{d}V$ of a scattering medium that is illuminated by a plane wave of irradiance E:

$$\mathrm{d}I(\boldsymbol{\xi}) = \beta(\boldsymbol{\xi}) E \,\mathrm{d}V \tag{4.1}$$

If the medium is axially symmetrical about the direction of propagation of the incident light beam, the volume scattering function, β [$\mathrm{m^{-1}\ sr^{-1}}$], is thus operationally defined (e.g., *Jerlov* 1976) as follows:

$$\beta(\theta) = \frac{\mathrm{d}I(\theta)}{E\,\mathrm{d}V} \tag{4.2}$$

where θ is the scattering angle, $\mathrm{d}I$ [$\mathrm{W\,sr^{-1}}$] the intensity of light (i.e., power per solid angle) scattered at angle θ, and E [$\mathrm{Wm^{-2}}$] the irradiance (i.e., power per area), by a plane light wave, of the scattering volume $\mathrm{d}V$ [$\mathrm{m^3}$], i.e., of a solid of intersection of the incident beam with the field of view of a detector. The geometry of a light scattering experiment aimed at the determination of the volume scattering function is schematically shown in Figure 4.1.

The scattering angle is measured between the direction of the incident beam, i.e., the direction of a vector perpendicular to the incident wave front at the scattering volume and the optical axis of the detector of the scattered light. Thus, if the detector faces the incident beam, the scattering angle equals 0. The angular resolution of the

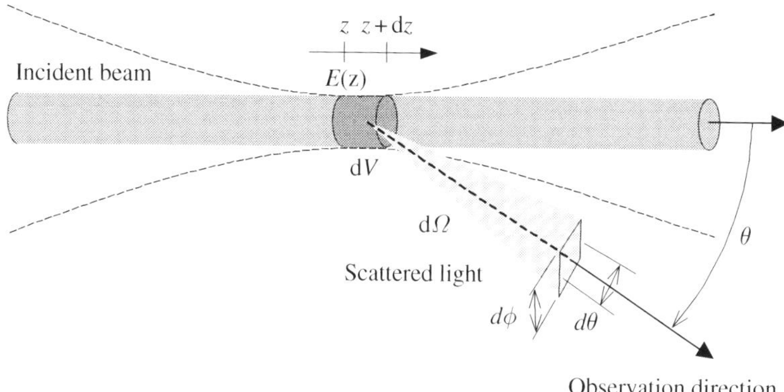

Figure 4.1. Geometry of the volume scattering function definition. The incident beam (medium shaded) is shown as a parallel beam with a circular cross-section because this is typically the simplest and most convenient arrangement. However, other beam geometries are equally applicable, e.g., a focused beam (dashed lines) that allows to significantly increase the irradiance of the scattering volume dV (heavy shading), provided that the incident wave has substantially plane wavefronts *within* the scattering volume. However, the non-parallel beam limits the angular range in which the scattered light *alone* can be observed. It also limits the scattering volume length, dz, to a small focal region. The solid angle $d\Omega$ (light shaded) contains the scattered light being observed, with $d\phi$ and $d\theta$ being the angular resolutions in the azimuthal scattering angle and the scattering angle respectively. The irradiance E is that at the scattering volume dV.

volume scattering function measurements is determined by two factors: (1) the angular field of view of the detector, which determines the range of β values sampled from the scattering volume, and (2) the acceptance angle of the detector, which determines a range of the scattering angles sampled by the detector at each point within the scattering volume. The latter quantity is a volume of the medium defined by the intersection of the incident light beam and of the field of view of the detector of the scattered light. We will discuss measurement errors applicable to typical experimental geometries of a light scattering meter later in this chapter.

Note that definition (4.2) does not require the beam to have specific geometries besides the requirement of the incident light wave being plane within the scattering volume. This requirement is needed for the specification of the scattering angle. Neither does definition (4.2) require that the irradiance distribution within a beam cross-section at the scattering volume be of a specific form.

Of course, if one wants to compare the experimental and theoretical volume scattering functions, such comparison must be made for the same irradiance distribution forms. This specifically applies to light scattering by particles illuminated by tightly focused laser beams. Such a theory for Gaussian beams has been developed for spherical particles by *Gouesbet* and Maheu (1988, see also a recent review by *Gouesbet* and Gréhan 2000), and extended to a sphere with an inclusion by *Gouesbet* et al. (2001) and to irregular, nearly spheroidal particles by *Barton* (2002). The localization principle of light (*van de Hulst* 1957) allows simplifying that complicated theory in the case of particles larger than the wavelength of light (*Sloot* et al. 1990).

It is generally accepted that natural waters are scattering media with axial symmetry, although there are mechanisms that orient at least some particles in natural water (e.g., magnetotaxis in algae, *Frankel* et al. 1997), and it has been independently argued that polarized light scattering results for seawater imply such orientation (*Kadyshevich* et al. 1976). We will discuss these mechanisms and experimental evidence of non-random orientation of living particles in Chapter 6. Unfortunately, the scarce data which hint at the effect of orientation on light scattering by particles in natural waters prevent one from making definite conclusions, besides one that the problem remains open.

In many applications, e.g., the theory of radiative transfer, one is not interested in the magnitude of the scattering function as much as in its form. In such cases, the phase function, $p(\theta)$, is used:

$$p(\theta) = \frac{\beta(\theta)}{b} \tag{4.3}$$

where the scattering coefficient, b [m^{-1}], is an integral of the scattering function over the full solid angle

$$\begin{aligned} b &= \int_{4\pi} \beta(\theta, \phi)\,\mathrm{d}\Omega \\ &= \int_0^{2\pi} \int_0^{\pi} \beta(\theta, \phi)\, \sin\theta \mathrm{d}\theta\, \mathrm{d}\phi \\ &= 2\pi \int_0^{\pi} \beta(\theta)\, \sin\theta \mathrm{d}\theta \end{aligned} \tag{4.4}$$

where ϕ is the azimuth angle, the last line of that equation applies to axially symmetrical scattering functions.

It follows from (4.3) and (4.4) that

$$\begin{aligned} \int_{4\pi} p(\theta)\,\mathrm{d}\Omega &= \int_{4\pi} \frac{\beta(\theta)}{b}\,\mathrm{d}\Omega \\ &= \frac{1}{b}\int_{4\pi} \beta(\theta)\,\mathrm{d}\Omega \\ &= 1 \end{aligned} \tag{4.5}$$

Note that other normalization conventions for the phase function are also used (e.g., *Haltrin* 1998).

The scattering coefficient, b, is one of key parameters in the radiative transfer theory. In the simplest version of that theory, applicable to single-scattering media, the scattering coefficient appears in the Lambert law, which in its integral form can be written as follows:

$$F(z) = F(z_0)\exp[-b(z - z_0)] \tag{4.6}$$

where $F(z_0)$ is the light power at an incidence plane in the medium and $F(z)$ is the attenuated flux, measured at a plane which is parallel to the incidence plane and located at a distance z away from it along the direction of the propagation of light. If the scattering of light by a slab of a scattering medium is considered, i.e., if there is a refractive index change at the incidence and exit planes of the slab, reflections at these planes have to be accounted for in order to separate their effect on light attenuation from the scattering by the medium itself. *Morel* and Bricaud (1986) discuss fine points of this law in relation to scattering, absorption, and attenuation, and of the concepts of the scattering, absorption, and attenuation coefficients.

One can also define the forward and backward scattering coefficients, which for an axially symmetric volume scattering function, β, can be defined as follows:

$$\begin{aligned} b_\mathrm{f} &= 2\pi \int_0^{\pi/2} \beta(\theta)\,\sin\theta \mathrm{d}\theta \\ b_\mathrm{b} &= 2\pi \int_{\pi/2}^{\pi} \beta(\theta)\,\sin\theta \mathrm{d}\theta \end{aligned} \tag{4.7}$$

The latter is one of the key parameters in the various models of radiative transfer used in remote sensing of the aquatic environment. In fact, the diffuse reflectance, $E_\mathrm{u}/E_\mathrm{d}$, where E_u and E_d are the upwelling and downwelling irradiances at the surface of a semi-infinite water body, can be expressed as $\sim 0.33 b_\mathrm{b}/a$, where a is the absorption coefficient (e.g., *Morel* and Prieur 1977).

The separation of the directional structure of light scattering (the phase function) and of the magnitude of light scattering (the scattering coefficient) is the basis of numerous models of radiative transfer such as the Monte Carlo and photon migration models (e.g., *Wu* et al. 1993). These and other models of radiative transfer, as specifically applied to the atmosphere–ocean system, are discussed at length in an excellent book by *Mobley* (1994).

The directional structure of light scattering is frequently summarized with the mean cosine of the scattering angle, another important parameter in the theory of radiative transfer in scattering media:

$$\begin{aligned} g = \langle\cos\theta\rangle &= \frac{\int_{4\pi} \beta(\theta)\cos\theta \,\mathrm{d}\Omega}{\int_{4\pi} \beta(\theta)\,\mathrm{d}\Omega} \\ &= \frac{1}{b}\int_0^{2\pi}\int_0^{\pi} \beta(\theta)\cos\theta\,\sin\theta \mathrm{d}\theta\,\mathrm{d}\phi \end{aligned} \tag{4.8}$$

$$= -\int_0^{2\pi}\int_1^{-1} p(\theta)\cos\theta \,\mathrm{d}\cos\theta \mathrm{d}\phi$$

$$= 2\pi\int_{-1}^{1} p(\theta)\cos\theta \,\mathrm{d}\cos\theta$$

where $p(\theta)$ is the phase function with assumed axial symmetry. The mean cosine vanishes for phase functions that are either isotropic or symmetric about the scattering angle of 90°. The case of nearly isotropic, symmetric scattering pattern, referred to as Rayleigh scattering, is applicable to particles that are very small as compared to the wavelength of light. Such 'particles' are e.g., fluctuations of the refractive index of seawater discussed in Chapter 2. The mean cosine of seawater and other natural waters is close to but not equal to unity, as discussed in section 4.4.2.4.

Another integral measure of the directional asymmetry of the scattering function is the average square of the scattering angle, $\langle\theta^2\rangle$:

$$\langle\theta^2\rangle = 2\pi\int_{-1}^{1} p(\theta)\theta^2 \, d\cos\theta \tag{4.9}$$

The average square of the scattering angle appears in the small-angle scattering models of image transmission in turbid media that rely on using the concept of the point spread function, e.g., *McLean* et al. (1998) and references therein.

4.2.2. *Single and multiple scattering*

If particles are sufficiently far away from each other, that is, if the volume concentration of the particles is low, each particle scatters light independently. The volume scattering function in this single-scattering approximation is a linear superposition of the scattering patterns of all particles in the scattering volume. The condition of single scattering can be expressed as follows (*Bohren* 1987):

$$cz(1-g) << 1 \tag{4.10}$$

where c is the attenuation coefficient of the medium [m^{-1}], z is the pathlength in the medium [m], and g is the mean cosine of the scattering angle (4.8).

If the Rayleigh scattering dominates, i.e., $g \approx 0$, which indicates a substantial symmetry of the scattering function about the scattering angle of 90°, then equation (4.57) can be transformed into the following condition:

$$z << \frac{1}{c} \tag{4.11}$$

The inverse of the attenuation coefficient, c, has the meaning of the mean pathlength between attenuation events in the medium. Thus, it follows from (4.11) that single scattering will dominate in a Rayleigh scattering medium if the pathlength is much smaller than the mean attenuation pathlength. By using a value of $g = 0.8$ to represent natural waters (e.g., *Dera* 1992), we can use (4.10) to formulate a similar, yet less restrictive condition for single scattering in natural waters:

$$z << \frac{5}{c} \tag{4.12}$$

The attenuation coefficient of clear natural waters assumes values on the order of $0.2\,\text{m}^{-1}$ (at a minimum located in the blue-green part of the visible spectrum, e.g., *Dera* 1992); thus the right side of (4.12) evaluates to 25 m. Since "much less than" can be replaced by "an order of magnitude less than," the single scattering at wavelengths near to that of the minimum of the attenuation coefficient dominates in clear natural waters at distances on the order of up to about 2.5 m.

4.2.3. Measurements of the scattering function

4.2.3.1. Instruments

A few commercial instruments (nephelometers) are available for the measurements of the volume scattering function in a significant angular range. Most of these instruments are designed to be operated in a laboratory and not on board ship or *in situ*. Brice-Phoenix instruments (a prototype of which is described by *Brice* et al. 1950) were to our knowledge the only commercial wide-angle nephelometers used in the early oceanographic investigations. Recently, an *in situ* laser diffractometer (LISST-100, Sequoia Scientific Inc.) has become available. Independently, a three-angle monochromatic nephelometer (ECO-VSF, Wet Labs Inc.) and a single-angle spectral nephelometer (HydroScat-6, HOBILabs, Inc.) have been introduced as commercial *in situ* backscattering nephelometers. These last two instruments are aimed mainly at the determination of the backscattering coefficient (as discussed later in this chapter). Aside from these modest commercial offerings, measurements of light scattering in natural waters have been traditionally performed using instruments built by researchers themselves especially in the case of *in situ* measurements. Interestingly, a nephelometer calibration survey regarding the early nephelometers (*Kratohvil* et al. 1962) implies that this does not seem to have affected the quality of the early data, as there was "no significant difference in results obtained by means of commercially available instruments and those constructed and built by individual workers."

Table A.1. of the Appendix lists a representative, although not exhaustive, selection of nephelometer designs. Instruments intended for aerosol measurements are also included in that table as their optical designs may also serve as inspirations for the development of novel nephelometers intended for measurements on hydrosols.

4.2.3.2. Difficulties

Measurements of the scattering function are made relatively difficult for several reasons. First, this function can vary about five orders of magnitude between the small angles ($\sim 0.1°$) and medium angles ($\sim 90°$). This imposes high demands on the dynamic range of the light detector system. In addition, the significant changes of the function with the scattering angle require precision alignment and accurate positioning of the key assemblies of the nephelometer relative to each other.

Second, the necessarily small sample volumes of most nephelometers may fail to contain large particles, such as marine snow (e.g., *Alldredge* and Silver 1988) whose number concentrations may be much less than 1 particle cm^{-3}. Yet, as it is evident from visual observations underwater, these large particles are likely to affect optical properties of water bodies relevant in the large-scale radiative transfer processes such as propagation of sunlight into and out of the ocean. Large particles affect the forward-scattering part of the scattering function relatively more than small particles. Thus, these particles are of importance for applications relying on the small-angle light scattering approximations, e.g., underwater imaging. It is thus surprising to find that very few attempts were undertaken to quantify the effect of large particles on light scattering by seawater (*Hou* 1997, *Hou* et al. 1997, *Carder* and Costello 1994).

Large and necessarily delicate particles of marine snow are very likely to disintegrate on sampling and handling of the sample as discussed at length in the following chapter on the particle size distributions. This makes it virtually impossible to measure light scattering properties of these particles with *in vitro* nephelometers, *In* situ nephelometers would have a better chance at measuring the scattering function of such particles if they had a reasonable chance of finding them in their sample volumes. Unfortunately, such chances are very slim for a typical *in situ* polar nephelometer design. Indeed, consider a power-law approximation for a particle size distribution (to be discussed in Chapter 5) representative of coastal waters, where concentration of these particles is likely to be substantial, i.e., $dN(D) \cong 100D^{-3}dD$ [particle cm^{-3}] (*Hou* 1997). At $D = 0.1$ cm, this yields on the average 0.01 particles per cm^3 within a diameter range of $D = 0.1$ to 0.2 cm. One would need to analyze a volume on the order of 100 cm^{-3} to *find* one of these large particles. Thus, an *in situ* nephelometer with a sample volume on the order of 1 cm^3 would on the average need to process 100 *different* sample volumes just to get one measurement representative of such large particles. As a result, measurements of the scattering function with such a nephelometer are almost guaranteed to be severely biased toward the contributions of the smaller particles. Yet, some application of the light scattering theory, e.g., in an underwater imager model, may refer to distances in water of several meters and more. This yields sample volumes on the order of 10 m^3, i.e., $\sim 10^7$ cm^3 that would contain thousands of such large particles and virtually assure their significant contribution to light scattering.

Third, the scattering function defined in (4.2) is actually an ensemble average of a statistical variable, since the homogeneous "scattering medium" implied in that definition is an idealization of the actual medium with the scatterers being randomly distributed in space. The spatial distribution of the particles can significantly change in a time comparable to a typical measurement time even in a perfectly still sample because some particles, such as bacteria, may be mobile (e.g., *Blackburn* et al. 1998), and every one is subject to a size-dependent degree to Brownian motion. In *in vitro* measurements, the sample volume is typically mixed to "assure" homogeneity. This creates convection currents that transport particles into and out of the scattering volume. All these factors may cause significant fluctuations of the nephelometer signal with time. These fluctuations should be averaged if the result is to represent the "average" medium. The fluctuations are generally caused by changes in the size distribution of particles in the sampling volume, and by changes in the orientation of those particles that are non-spherical, in an incoherent addition to the instrument-generated noise (*Boxman* et al. 1991). Incidentally, such fluctuations in the light power transmitted by a scattering medium have been utilized to determine the particle size distribution (e.g., *Shen* and Riebel 2003).

Note that too long an averaging time may cause other problems as the sample itself may change in time, especially if it contains microbial particles. For this reason, as well as to simply enable more samples to be analyzed, the measurement time should be minimized.

An experimenter's dilemma thus follows: the increase in the measurement precision with the length of the averaging period must be judiciously balanced with the decrease in the measurement accuracy caused by changes, usually irreversible, in the properties of the particles or the suspension as a whole. This dilemma applies to both *in situ* and *in vitro* measurements. In the *in situ* case, the question is how representative the snapshot measurement is of the sampling site environment. In the *in vitro* case, the very sampling, as well as irreversible changes that may occur after the sample acquisition may cause additional mis-representation of that environment. These issues apply also in the case of the particle size distribution measurements discussed in the following chapter.

The design of *in situ* instruments faces additional difficulties due to their submersion in water. The need for a robust case and thick transparent windows that can withstand water pressure at depth complicates the design. When such instrument remain submerged for a long time, the case and the window are subjected to biofouling that can significantly affect the instrument readings (*Dolphin* et al. 2001, *Barth* et al. 1997). Various means that prevent biofouling have been proposed, e.g., using the "windshield wiper" principle (*Dolphin* et al. 2001, *Ridd* and Larcombe 1994) or various anti-fouling coatings whose efficiency have been recently evaluated by *McLean* et al. (1997).

Several measurement methods of the one-dimensional (i.e., axially symmetric) volume scattering function have been devised, each with its characteristic angular range. Typically, that function is measured using different types of instruments at the small angles (0° to 5°), moderate angles (5° to 175°), and large angles (175° to 180°). This specialization is due to the angular range limitations and resolution characteristic of the various optical designs and also to practical limitations imposed on the nephelometer size. The angular ranges stated here are somewhat arbitrary but follow a historically established pattern.

Before we discuss typical designs of the scattering meter, we should address radiometric and calibration considerations of the nephėlometer design and the measurement errors. We select for this discussion the polar nephelometer design that is typically used in the moderate-angle range (5° to 175°). Although other designs may have specific design issues, we feel that the discussion centered on the polar nephelometer is sufficiently representative, given the large body of experimental data obtained with polar nephelometers.

4.2.3.3. The polar nephelometer

Typical polar nephelometer designs for oceanographic applications have been described by *Lee* and Lewis (2003), *Kullenberg* (1984), *Sugihara* et al. (1982a, 1982b), *Kullenberg* (1968), *Sasaki* et al. (1960), *Tyler* and Richardson (1958), *Atkins* and Poole (1952) among others. See also Table 1.1. for a more detailed list. This type of nephelometers has been also used for numerous non-oceanographic applications, for example by *Haller* et al. (1983), *Sherman* et al. (1968), *Pritchard* and Elliot (1960; atmospheric studies), *McIntyre* and Doderer (1959), and *Brice* et al. (1950). The design theory of a polar nephelometer and calibration procedures are discussed by *Leong* et al. (1995), *Kullenberg* (1984), *Holland* (1980), *Privoznik* et al. (1978), *Petzold* (1972), *Fry* (1974), and *Tyler* (1963).

In a polar nephelometer, the scattering angle is scanned by rotating in the scattering plane either a detector (Figure 4.2.A; e.g., *in vitro*, *Hunt* and Huffman 1973; *in situ*, *Kullenberg* 1968) or a periscope whose exit port faces a stationary detector (Figure 4.2.B; e.g., *in vitro*, *Prandke* 1980; *in situ*, *Kullenberg* 1984). Some designs (e.g., *Wyatt* and Jackson 1989) use fixed detectors aimed at specific scattering angles.

Rapid development and reduction in the prices of sufficiently sensitive detector arrays encouraged electronically scanned as opposed to mechanically scanned polar nephelometers (Figure 4.3 and Figure 4.4). Electronic scanning allows one to measure the scattering function in a fraction of the time characteristic of the mechanically scanned design. It is especially suitable for characterization of individual particles in a flow-cytometric approach (*Bartholdi* et al. 1980). That time for fast mechanically-scanned instruments is on the order of several milliseconds (*Hespel* et al. 2001, *Moser* 1974).

With suitably miniaturized detectors, the fixed-detectors design permits measurement of the scattering function in several scattering planes at once, i.e., to study two-dimensional volume scattering function, dependent not only on the scattering angle but also on the azimuth angle (*Wyatt* and Jackson 1989).

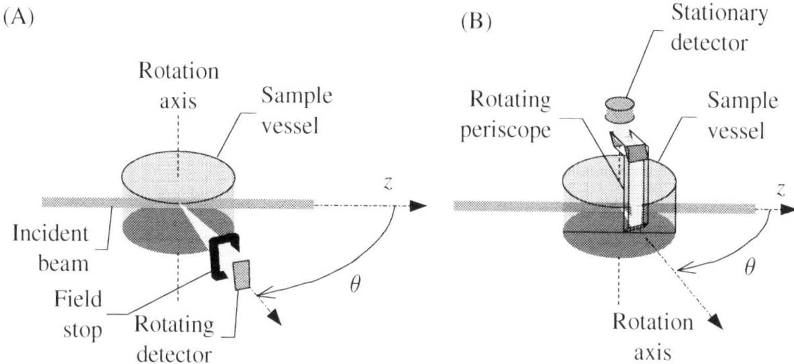

Figure 4.2. Two typical designs of the polar nephelometer. The detector size and its distance from the rotation axis defines the acceptance angle of the detector and angular resolution of the scattered light measurement. The field stop likewise defines the scattering volume (in the periscope version, the field stop is at the submerged end of the periscope inside the sample vessel). The detector aperture along with the field stop represent a radiance meter that receives the scattered light. *Panel A*: The detector rotates around the transparent sample chamber. This design makes the detector signal dependent on variations in the quality of the sample vessel wall. Such a dependence is avoided in the *in situ* version of that design which encloses the detector in a watertight housing with a transparent window. Hence, the detector receives the scattered light through the same window, independent of the scattering angle setting. The nephelometer design shown in panel B shares this advantage. If a photomultiplier is used as a detector, its sensitivity to variations in the magnetic field, induced as the detector arm rotates, may require enclosing the detector in a magnetic shield. *Panel B*: A periscope folds the light path to the detector. This enables one to use a stationary detector and reduce the overall size of the instrument but requires attention to mitigate the sensitivity of the lightpath-folding components of the rotating periscope to the polarization of the scattered light.

This direction has been taken by several research groups in the last decade who used commercial array detectors (intensified CCD) instead of discrete detector systems to record one-dimensional and two-dimensional distributions of irradiances that are produced by converting angular intensity distributions with the use of ellipsoidal and paraboloidal mirrors (e.g., *Hirst* et al. 1994). These instruments are capable of measuring two-dimensional angular optical scattering (TAOS) but to our knowledge have so far been developed for single-particle characterization.

Fast scanning nephelometers that utilize a "single" detector (either a truly single detector, like a PMT or a photodiode, or a detector array, such as a CCD) for the measurement of the scattered light in the entire angular range accessible to the instrument face the problem of very large variations in the scattered light power. Indeed, that intensity can decrease by several orders of magnitude from the forward to backward scattering direction. Solutions to this problem have included the use of logarithmic amplifiers for the light scattering signal (e.g., *Gucker* et al.

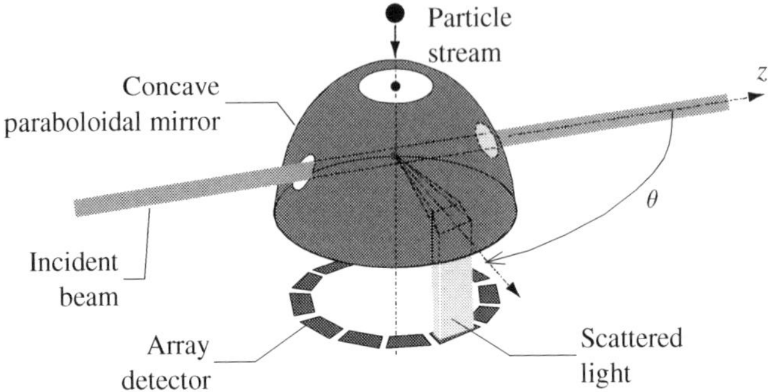

Figure 4.3. A general-angle polar nephelometer utilizing an array detector and a paraboloidal (or ellipsoidal) mirror (e.g., *Bartholdi* et al. 1980) to fold the scattered light paths. This design allows for rapid electronic scanning of the scattering angle (and the azimuth angle if a two-dimensional array detector is used). The acceptance angle (angular resolution) is defined by the sizes of the array elements and the mirror. The scattering volume must be limited to the immediate vicinity of the mirror's focal point because the quality of imaging the scattered light onto the detector plane quickly deteriorates with increasing scattering volume size. This design is thus best suited for the measurements of single-particle light scattering as shown here.

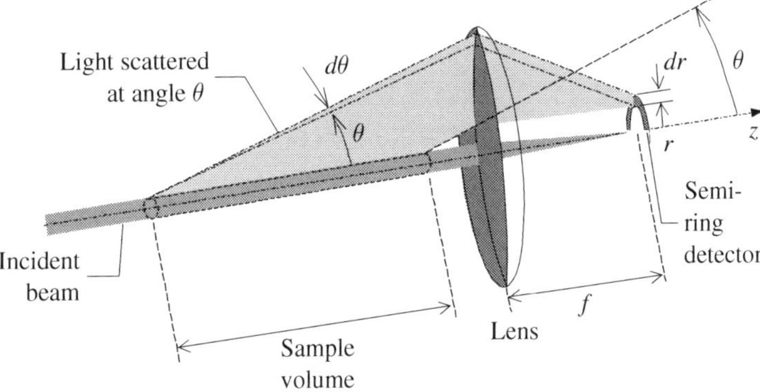

Figure 4.4. Small-angle Fourier-transform nephelometers typically use an array of concentric semi-ring or quarter-ring detectors (only one is shown for clarity), or a two-dimensional array detector, enabling fast electronic rather than slow mechanical scanning of the scattered angle. This design is standard for laser diffractometers intended for measurements of the particle size distribution (e.g., *Agrawal* and Pottsmith 2000). The lens with the focal length, f, transforms the scattering angle, θ, into a radial distance $r = f \tan\theta$ from the lens axis at the lens focal plane. The ring width, dr, controls the angular resolution. Note that the ring detector receives light scattered by the entire (applicable) scattering volume at the angle θ, as shown by light shading. For a given lens diameter, the sample volume length, and hence the volume itself, decreases with increasing scattering angle. The exiting beam power may be monitored by a detector centered at the beam focus.

1973, note that the detector system linearity range remains the ultimate limiting factor here) or variable neutral density filters (e.g., *Watson* et al. 2004).

In a nephelometer that uses an array of independent detectors for each scattering angle of a set of discrete angles, the light scattering power level and gain of each detector can be set independently.

In each case, the detector's field of view is suitably limited so that the scattered light comes from a volume of the scattering medium (scattering volume) on the order of several mm^3 to several cm^3 in the case of the multi-particle nephelometers, and much less, on the order of 10^{-3} mm for single-particle nephelometers. Likewise, the acceptance angle of the detector is also suitably limited to collect light scattered into a small range about the scattering angle.

Note that these parameters are inherently different in multi-particle and single-particle nephelometers (Figure 4.5). To start with, it is the multi-particle nephelometer that measures the volume scattering function as defined in Chapter 3 and repeated here for convenience:

$$\beta(\theta, \lambda) = \int_0^\infty \frac{\overline{\sigma}(\theta, \lambda, a)}{2\pi} N(a) da \tag{4.13}$$

where $\overline{\sigma}(\theta, \lambda, a)$ is the average differential cross-section for scattering by particles with radius a, and $N(a)$ is the number concentration of particles with radii in a range $(a, a + da)$ per unit volume, i.e., the particle size distribution. We made a simplifying assumption here that all particles in the scattering volume have the same composition, shape, and orientation, hence the sizes are the only differentiating characteristics. Note a division of the cross-section by 2π to account for the fact that in our notation, the integral over the azimuth scattering angle φ had already been carried out over an angle of 2π. In contrast to the multi-particle nephelometer, the single-particle nephelometer measures just the scattering cross-section of the particle that is, at the moment, in the sample volume of the nephelometer.

If the incident light is polychromatic, the wavelength of the incident beam is typically selected by using a filter. In measuring light scattering at the short wavelengths in the visible spectrum, the same kind of filter is sometimes placed in front of the detector because particulate and dissolved matter may fluoresce. Without the second filter, the isotropic angular pattern of fluorescence may add to, and thus distort, the angular light scattering pattern. Suitable combinations of a polarizer (incident light path) and an analyzer (scattered light path) can be used to measure the scattering functions for polarized light. We postpone the discussion of such measurements until the following section.

In designing a nephelometer, one must consider the sensitivity range of the light detector system. It is also important to eliminate multiple reflections at the interfaces between water and the sample container, as well as between the sample container and air, and to minimize the forward-scattered stray light (e.g., *Leong* et al. 1995) which, when back-reflected at the sample vessel wall, may significantly

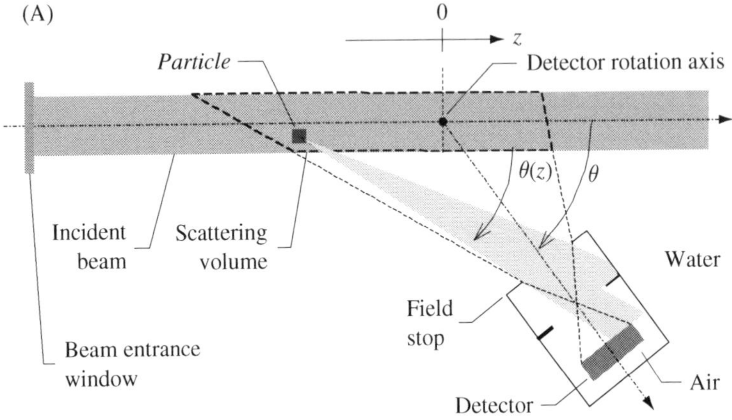

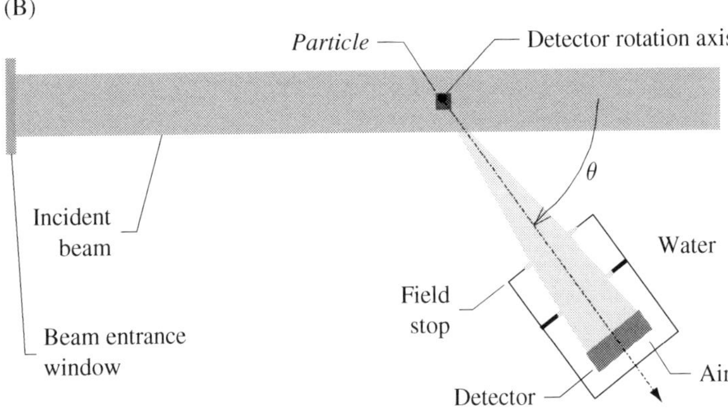

Figure 4.5. The sample volume and angular resolution are inherently different in multi-particle (*panel A*) and single-particle (*panel B*) polar nephelometers. In the multi-particle case, all particles within an angle-dependent scattering volume contribute simultaneously to the scattered light power received by the detector. The contribution of each particle, at position-dependent scattering angle, $\theta(z)$, to the scattered light power at a nominal scattering angle, θ, depends on that particle position within the angle-dependent scattering volume. In the single-particle case, particles are supplied one-at-a-time to a much smaller "scattering volume" that nominally determines only the medium contribution. They are localized mechanically or optically to provide power scattered at the nominal scattering angle. The presence of water may in fact be limited to that small "scattering volume" alone.

contribute to the power of light scattered at angles greater than 90°. Incidentally, the effect of the reflection of light at the water-container wall interface can be minimized by slightly tilting the interface with respect to the incident light beam, a trick that works well for the measurement of backscattered light (e.g., *Spicer* et al. 1999). The refraction of light at the sample container wall and air interface must be accounted for, since such refraction affects the effective acceptance solid angle of the detector of the scattered light and may affect the scattering angle itself, e.g., if the sample container has flat walls. Finally, the calibration procedure and fixtures must be considered.

4.2.3.4. Radiometric considerations

The minimum power to be sensed by a nephelometer is that of the light power scattered at 90° by pure water (seawater) when the polarization of the incident light is parallel to the scattering plane. This plane contains the direction of the incident beam and that of observation. The minimum power, F_{min} [W], can be calculated by using the following equation, resulting from the definition of the volume scattering function given by (4.1):

$$F_{\mathrm{min}} = \beta(\lambda, 90°) E V \Omega_{\mathrm{w}} \tag{4.14}$$

where $\beta(\lambda, 90°)[\mathrm{m}^{-1}\,\mathrm{sr}^{-1}]$ is the volume scattering function of pure seawater at 90° (where it assumes the minimum), Ω_{w} [sr] is the acceptance solid angle in water of the detector, and V [m^3] is the scattering volume of seawater illuminated by a laser beam of irradiance, E [$\mathrm{W\,m}^{-2}$].

We set arbitrarily the wavelength to $\lambda = 633\,\mathrm{nm}$ in air (HeNe laser). According to *Morel* (1974), we have $\beta(\lambda,\ 90°) = 7 \times 10^{-5}$. Typically, the scattering volume is on the order of $10^{-9}\ \mathrm{m}^3$ ($1\ \mathrm{mm}^3$). The detector acceptance solid angle, Ω_w, can be reasonably set at $2.4 \times 10^{-4}\,\mathrm{sr}$ (corresponding to an angular resolution of 1°). The beam irradiance, E, is on the order of $1000\,\mathrm{W\,m}^{-2}$ ($1\,\mathrm{mW\,mm}^{-2}$, typical of low-power HeNe lasers). With these parameters, one obtains $F_{\mathrm{min}} \sim 1.7 \times 10^{-14}\,\mathrm{W}$ (17 fW or 5.3×10^4 photons/s) from (4.14). Losses in the optical path due to reflection at the interfaces of optical elements and incomplete collection of light due, for example, to vignetting typically reduce this scattered light power. We neglect here a loss due to attenuation by the sample, i.e., pure water (seawater).

The relative photon (shot) noise, i.e., the coefficient of variation (equal to the standard deviation divided by the average value), can be obtained by noting that the probability of the number of photons in a photon flux is described by the Poisson probability distribution with an average number, N. The relative shot noise thus equals $\sqrt{N}/N = (\sqrt{N})^{-1}$, that is about 0.4% in the above example. However, this inherent photon flux (shot) noise is typically insignificant in comparison to the noise due to the fluctuations in the number of particles in the scattering volume during the measurement time (which may easily reach 20% of the signal), and to the signal-independent noise of the light detection system.

The detection system noise can be estimated without any reference to the scattered light signal. A simplified, back-of-the-envelope estimate of that noise can be obtained as follows. The sensitivity of a photodetector is specified by its noise equivalent power (NEP; $\mathrm{W\ Hz}^{-1/2}$). This is the light power equal to that of the detector noise contained in a 1 Hz bandwidth about a frequency at which the light power is modulated. We are referring here to the signal modulation frequency which one might want to impose on the incident light in order to differentiate the scattered light-related signal from that due to ambient light and electrical interference. This modulation allows one to detect the scattered signal within a very narrow bandwidth around the modulation frequency, in a process referred to as synchronous detection, or lock-in amplification. Incidentally, high ambient light power will degrade the measurements even with synchronous detection by adding its photon noise component to the measurement noise. We refer an interested reader to an introduction to this measurement method by *Blair* and Sydenham (1975), a concise review by *Meade* (1982), and discussions of the precision of this method (*Gualtieri* 1987, *Gillies* and Allison 1986).

For a photomultiplier (PMT)-based detection system without an external amplifier, the signal-to-noise ratio, SNR, can be obtained from the following simple formula:

$$\mathrm{SNR} = \frac{F_{\min}}{NEP\sqrt{\Delta f}} \tag{4.15}$$

PMT detectors have NEP's on the order of $10^{-15}\,\mathrm{W\,Hz^{-1/2}}$. Thus, an SNR ~ 1 results for a scattered light power of $1.7 \times 10^{-14}\,\mathrm{W}$ sensed within a modulation frequency bandwidth of about $\Delta f = 280\,\mathrm{Hz}$.

Let us also consider a photodiode-based detection system. In contrast to an expensive PMT detector system which requires a high-voltage supply on the order of 1000 V and a large volume PMT housing, a silicon- or gallium-based photodiode is a rugged solid-state low-voltage detector which affords a detection system at a much lower cost and with a smaller volume.

Good solid-state photodiodes have NEPs on the order of $10^{-15}\,\mathrm{W\,Hz^{1/2}}$ at 633 nm and 1 Hz bandwidth about a modulation frequency on the order of 1 kHz. A photodiode is linear over about 10 to 11 decades of incident light power, compared to two to three decades for a PMT. In contrast to a PMT, a photodiode can easily tolerate overexposures of many orders of magnitude. As with the PMT, a photodiode is a current generator.

The photocurrent, I, generated by a photodetector with responsivity, R (photocurrent per unit light power, P, received), can be calculated as follows:

$$I = PR \tag{4.16}$$

Silicon photodiodes have a responsivity of about $0.5\,\mathrm{A\,W^{-1}}$ at $\lambda = 633\,\mathrm{nm}$. When illuminated with $1.7 \times 10^{-14}\,\mathrm{W}$, such a photodiode would generate a current of about $0.5\,\mathrm{A\,W^{-1}} \times 1.7 \times 10^{-14}\,\mathrm{W} = 0.85 \times 10^{-14}\,\mathrm{A}$ ($\sim 10\,\mathrm{fA}$). To measure current of this magnitude, we need to amplify it significantly. This is best done with a current-to-voltage (CTV) amplifier based on a high-quality operational amplifier. Unfortunately, the CTV amplifier contributes significantly to the noise of a photodiode-based light detection system and increases the overall NEP of such a system by factor on the order of 10 and more, i.e., to $> 10^{-14}\,\mathrm{W\,Hz^{-1}}$.

This reduces the system bandwith by a factor on the order of more than 0.01. In the present example, it is $\sim$3 Hz. Note that by long integration of a photodiode-based detection system ($\sim$100 s), one can achieve an NEP on the order of 10^{-16} W Hz^{-1} (*Eppeldauer* 2000).

A meaningful discussion of light detection systems is outside the scope of this book. We do, however, note representative references for an interested reader. The CTV amplifier, i.e., transimpedance amplifier, is discussed in detail, e.g., by *Graeme* (1995) and *Graeme* et al. (1971). *Eppeldauer* (2000) and *Eppeldauer* and Hardis (1991) discuss specific design issues and circuit component selection for photodiode-based low-light detection systems.

A bandwidth of several Hz and lower can be readily achieved with a lock-in amplifier although at the expense of the measurement time, Δt [s], which equals (e.g., *Gualtieri* 1987):

$$\Delta t = \frac{1}{4\Delta f} \tag{4.17}$$

From (4.17), it follows in the present example that the measurement time with a photodiode-based light detection system is $\sim$0.1 s. Fluctuations of the scattered light power due to fluctuations in the number of the large particles in the scattering volume require signal averaging during a time interval greater than that by a factor of 10 or more. Thus, time savings realized with a more sensitive (but larger and more expensive) detector system are in part offset by the need to average the scattered light.

We will now discuss the other end of the scattered signal range, the maximum power of the scattered light. It follows from the definition of the scattering function (4.2) that aside from the measurement of the scattered light power, we also need to measure the laser beam power, which enters (4.2) in the form of the incident beam irradiance. A 1 mW laser beam incident on a PMT connected to a high-voltage power supply would cause serious problems for that photodetector. If the PMT were capable of handling this input power, a typical current of 10^{-3} A W^{-1} $\times$ 10^{-3} W $\times$ 10^{6} = 10^{6} A, would be generated. The first term in this equation is the photocathode responsivity, the second is the incident light power, and the last is the PMT gain factor. The reasonable maximum PMT anode current is typically about 0.1 mA (10^{-4} A). Thus, one needs to attenuate the laser power by a factor on the order of 10^{6} A/10^{-4} A = 10^{10}! This can be achieved by the use of stacked neutral density filters and also by a reduction of the high-voltage applied to the PMT.

A photodiode-based detector system would handle a direct beam overload much better but would not provide useful results without modifications either. A photodiode illuminated by a beam of 1 mW would generate a current of 0.5 mA The maximum signal voltage which can be provided by such a system is the CTV amplifier supply voltage, typically 10 V. The signal voltage, corresponding to a photocurrent, I, supplied by a photodiode can be obtained by multiplying I by the transimpedance of the amplifier, which is on the order of 10^9 in the case of a low-light detection system. However, by blind substitution of the input current of 0.5 mA, that results from multiplication of the incident light power (1 mW) by a representative responsivity of the photodiode (0.5 A/W) in such an equation would yield $0.5 \times 10^{-3} \times 10^9 = 5 \times 10^5$ V, which is much greater than the achievable output voltage of 10 V. By traversing this reasoning backward, we arrive at the maximum power of $10\,\mathrm{V}/(10^9 \times 0.5) = 2 \times 10^{-8}$ W that can be handled by a photodiode-based detector system. Thus, the direct beam power has to be attenuated by a factor of $2 \times 10^{-8}\,\mathrm{W}/10^{-3}\,\mathrm{W} = 2 \times 10^{-5}$, where 10^{-3} W is the direct laser beam power. This can be achieved by using a neutral density filter and/or setting an appropriate gain in the CTV amplifier.

4.2.3.5. Alignment

The alignment of a polar nephelometer ensures that the axis of the incident light beam passes through the rotation axis of the nephelometer's detector, is in the detector rotation plane, and fills the detector aperture when the latter is at a scattering angle of 0°. A minimalistic approach to the nephelometer alignment is to rely on the machining and assembly tolerances. Another approach is to provide in the nephelometer design a means for alignment of the components (e.g., *Jonasz* 1991b). The alignment method used by Jonasz relies on a fixture whose essential components are schematically depicted in Figure 4.6.

The alignment fixture consists of two transparent screens (A and B) each with an identical grid, or other means of determining the beam footprint position at each screen. These two grids need only to be roughly symmetrical about the rotation axis of the detector when the fixture is mounted in the nephelometer. The screen grid centers do not need to be aligned with the rotation plane or the desired beam axis. The set of the two screens is first oriented so that the direction from screen A to screen B indicates the desired direction of beam propagation. This orientation corresponds to the scattering angle of 0°. The beam is subsequently and arbitrarily positioned so that the center of its footprint at screen A is within the grid area. The position of the beam at screen A, say P_A, is noted.

The fixture is subsequently rotated by 180° so that screen B assumes the location of screen A and vice versa. If the beam axis passed through and were perpendicular to the detector rotation axis, which is identical with the rotation axis of the set of screens, the beam would have passed through screen A at is new location at the same position P_A. Thus, the beam can be properly aligned by making it pass through the same position at screen A in its two locations (before and after the rotation by 180°).

The alignment is facilitated by the presence of screen B, now at the former location of screen A. Position P_B of the beam at screen B (which is identical with position P_A at screen A) should be noted before attempting to align the beam. The

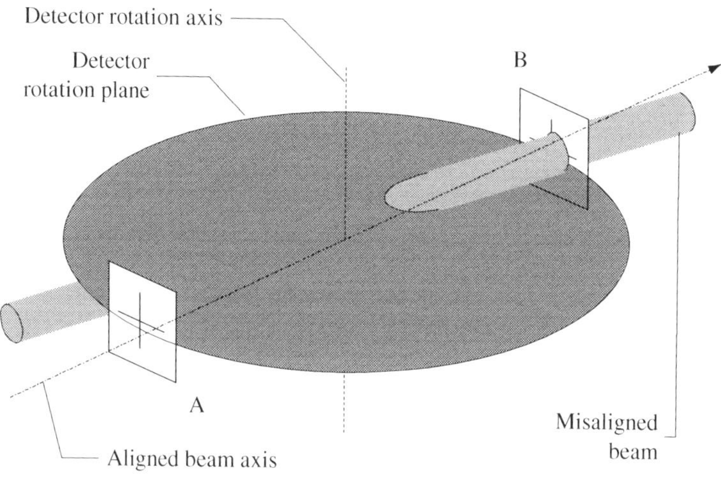

Figure 4.6. Alignment of a polar nephelometer by means of a fixture that can rotate about the axis of the detector rotation. The fixture contains two transparent screens A and B, each with identical grids of etched lines or other means to enable visual location of the beam footprint position. Refer to the text for the alignment procedure. Here, the misaligned beam propagates from left to right and from below to above the detector rotation plane.

beam should not leave this position. On completing this alignment step, the beam is made to pass through and at a right angle to the axis of rotation of the detector.

The next step is to place the beam axis in the detector rotation plane. This may require a parallel translation of the detector or the beam. Finally, the detector axis is aligned to coincide with that of the beam at a scattering angle of 0° by positioning the detector at that angle, so that it faces the beam, and adjusting the orientation of the detector axis so that the detector signal generated by the intercepted beam is maximized. As we discussed already, the beam power must be appropriately attenuated during this last alignment step.

4.2.3.6. Calibration

The geometrical extent of the scattering volume of the nephelometer changes as $1/\sin\theta$, where θ is the scattering angle (Figure 4.7). Early designs of polar nephelometers (e.g., *Tyler* and Richardson 1958) employed a stop, rotating with the detector, introduced by *Waldram* (1945). The area of this stop projected onto a plane perpendicular to the beam axis, varies as $\sin\theta$ canceling the effect of the $1/\sin\theta$ factor.

Variations in the scattering volume geometry are only a part (albeit typically a dominant one) of the total effect of the detector rotation. Another part comes from variations in the effective acceptance angle of the detector with the scattering angle. That angle varies for different elements of the scattering volume that may be located at different distance from the detector. Such variations are most

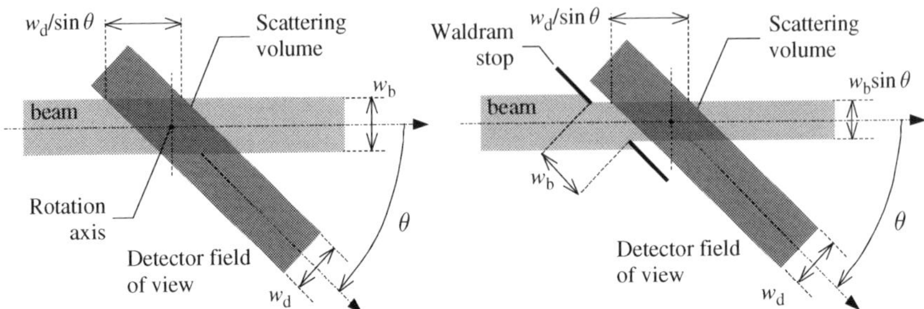

Figure 4.7. *Left:* A scattering volume (dark shaded) in an idealized nephelometer varies with the scattering angle θ as $1/\sin\theta$ because the area of the intersection of the beam with the detector field of view equals $w_b \times w_d/\sin\theta$ and the depth of this volume (in a direction perpendicular to the paper plane) does not depend on θ. *Right:* The Waldram stop, rotating with the detector, changes the beam width as $w_b \sin\theta$, which cancels the $1/\sin\theta$ factor in the dependency of the scattering volume on θ. That dependency is an important factor in the scattering function measurement. However, the Waldram stop does not compensate for variations in the solid angle subtended by the detector at various points of the scattering volume. Such variations may also affect that measurement.

pronounced at small and large scattering angles, where the scattering volume is rather elongated. Thus, more recent nephelometer designs have used a calibration procedure to compensate for these two effects of the detector rotation.

A calibration technique relying on a movable light-diffusing screen has been introduced by *Pritchard* and Elliot (1960) and improved by *Tyler* (1963). The calibration procedure discussed here is essentially that of *Pritchard* and Elliot (1960) with improvements by *Fry* (1974) and by *Jonasz* (1991b). *Kaye* and Havlik (1973) discuss the calibration of an axially symmetrical small-angle nephelometer, integrating the scattered light over the azimuth angle.

An alternative to the diffuse-screen calibration method has been reported by *Kullenberg* (1984). That method is based on the use of a fluorescent dye and allows one to determine the calibration factor for a scattering angle with just one measurement per scattering angle, as opposed to the many measurements required for each scattering angle by the diffusing-screen method. Fluorescence in most dyes is excited by short-wavelength light. Relatively few dyes can be excited by red light such as that of a HeNe laser (e.g., *Lee* et al. 1989).

During the calibration, one determines a function that accounts for systematic changes in the detector signal with the scattering angle. These changes are due to changes in the scattering volume and effective solid angle subtended by the detector at the various locations within the scattering volume. Both types of changes are independent of the polarization of the incident light. Thus, one calibration function is sufficient to calculate all polarized light volume scattering functions. It is tacitly assumed that the detector is not polarization-sensitive.

The diffuse-screen calibration is conducted by translating along the beam axis (or detector axis) a translucent screen inserted into the sample vessel filled with clear water so that scattering by water can be neglected as compared with that by the screen. The screen is positioned and oriented by using a suitable fixture so that (1) it completely intercepts the incident light beam, and (2) the illuminated area of the screen can be "seen" by the detector, i.e., the screen plane is not parallel to the detector axis. It is not necessary to know the bi-directional reflectance of the screen, but it should not vary greatly with the incidence and observation angles for reasons to be discussed later in this section.

The calibration fixture should enable one to measure the relative position of the screen, z_s, which is a linear function of the distance, z, along the beam axis (Figure 4.8). An arrangement whereby the screen travels along the axis of the detectors' field of view (*Tyler* and Austin 1964) results in a calibration procedure more complex than that described here, especially when the screen path is offset from the detector axis due, e.g., to machining and assembling tolerances.

The calibration procedure can be summarized as follows. The detector is set to a scattering angle, θ, from a range to be investigated, and the detector signal

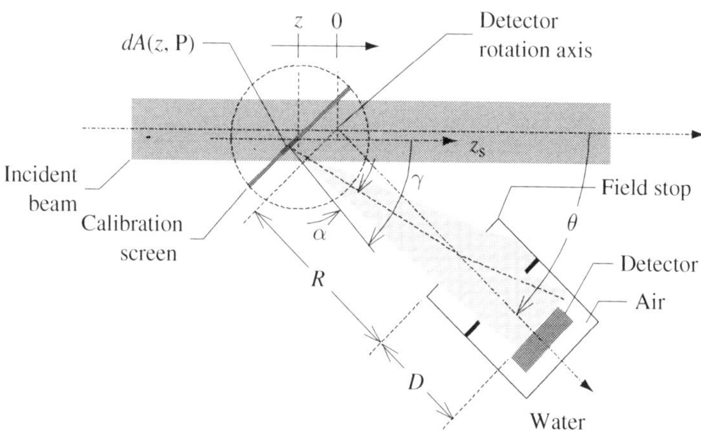

Figure 4.8. Geometry of the scattering meter calibration with a light-diffusing screen, here shown with its center at a position z. A transmission configuration, used for the scattering angle $\theta < 90°$, is shown here: the detector is on the opposite side of the screen relative to the light source. A reflection configuration, with the detector and the light source on the same side of the screen, must be used for angles greater than 90°. Small offsets of the screen travel axis (the track of the center of the screen rotation) from the center of the detector rotation must be accounted for when splicing calibration data from the transmission and reflection configurations. A baffle reducing light reflection at the inner wall of the detector housing is also shown. Note refraction at the water–air interface at the field stop/window position. Also note some vignetting of the detector by the field stop, which contributes to variations in the effective acceptance solid angle of the detector.

$dS(\theta, z_s)$ is recorded at each screen position, z_s. At each such position, the power of light scattered by the screen at a position, z_s, that reaches the detector is controlled by the effective solid angle Ω subtended by the detector at that position. This solid angle equals 0 if the illuminated area is outside the field of view of the detector. In order to cancel the effect of the beam power, the detector signal is normalized by its value measured when the center of the beam footprint at the screen is located at the detector rotation axis. This procedure is repeated for as many scattering angles as required.

The orientation of the screen is immaterial as long as (1) the illuminated area of the screen fills the field of view of the detector when the latter is oriented at a scattering angle of 90°, (2) the whole incident beam footprint is visible at the screen, and (3) changes in the bi-directional reflectivity of the screen are small within that field of view. In the reflection mode, the specular reflection angle should be avoided.

The integral, $S(\theta)$, of $dS(\theta,\ z_s)$ with respect to the screen position, z_s, can be regarded an effective product of the scattering volume and the solid angle subtended by the detector at the elements of this volume. The inverse of the normalized $S(\theta)$ and a solid angle, Ω_0, subtended by the detector at the center of the detector rotation gives the calibration function.

The axis of the screen rotation translates along axis z_s (Figure 4.8) that in general is slightly offset by a distance x_o from the beam axis. When the screen coordinate is z_{s0}, the center of the beam footprint at the screen coincides with the detector rotation axis. The path of the screen rotation axis may also make a small angle α with the beam axis. This makes the beam axis coordinate increment, dz, equal to a product of that of the screen multiplied by $\cos\alpha$. For simplicity, it is assumed here that $x_o \approx 0$ and $\cos\alpha \approx 1$. Thus, we can transform z_s (a relative screen position) into an absolute screen position, z as follows:

$$z = z_{s0} - z_s \tag{4.18}$$

The "zero" position of the screen, z_{s0}, is determined by finding the maximum of $dS(90°, z)$. If the axis of screen rotation is offset from the beam axis, this position may differ for the two calibration modes (transmission and reflection) and, more generally, for different screen orientations.

During calibration, the screen normal (Figure 4.8) is set to make an angle γ with the beam axis. This angle is positive when the normal is to be turned clockwise from the direction of the beam axis to assume direction γ. When the detector is set to an angle θ from a range of 0 to 90° (the forward-scattering range), the screen normal is at an angle γ such that the detector receives light diffusely transmitted by the screen (transmission mode). When the detector is set to an angle from a range of 90 to 180° (the backscattering range), the screen normal is at an angle $-\gamma$, and the detector receives light diffusely reflected by the screen (reflection mode).

In the following discussion, it is assumed that the field of view of the detector at the detector rotation center is larger than the diameter of the incident beam, and that the beam diameter is small compared to the distance $R+D$ from the detector rotation center to the detector aperture, where R is the radius of rotation of the field stop and D is the distance between the field stop and the detector aperture (Figure 4.8). We will first calculate the scattered light power that the detector receives from the screen as a function of the absolute position, z, of the screen expressed by (4.18). It is this power that the detector would receive from a thin slice of the scattering volume at the location of the screen. We will then integrate this signal over distance, z, along the beam axis to get a signal representative of what would have been received by the detector from the scattering volume at that scattering angle.

The light flux, dF, which the detector receives from a small screen surface element, dA, located at point P of the illuminated area of the screen (Figure 4.8), can be expressed as follows:

$$dF(\theta, z, \alpha, \gamma, \mathrm{P}) = L(\alpha, \gamma, \mathrm{P})\Omega(z, \mathrm{P}) \cos\alpha(z, \mathrm{P})\, dA(z, \mathrm{P}) \tag{4.19}$$

where θ is the scattering angle, z indicates the screen position, $\alpha(z,\mathrm{P})$ is the angle between the screen normal and the chief ray from the screen area element dA to the field stop of the detector, γ is the angle the screen normal makes with the beam axis, L is the radiance emitted by the screen, and Ω is the solid angle that the detector subtends at point P.

For a sufficiently narrow beam, α is to a reasonable degree independent of P. On the other hand, the solid angle, Ω, is a complex function of position of the screen area element, dA, because at some positions, the detector may be vignetted by the field stop. Such minor vignetting is in fact depicted in Figure 4.8.

The radiance, L, produced by the screen element, dA, is related to the bi-directional reflectance, $\rho[\alpha(z), \gamma]$, of the screen and to the irradiance of the incident beam as follows:

$$L(\alpha, \gamma, \mathrm{P}) = \frac{E_n(\mathrm{P}) \cos\gamma\, \rho[\alpha(z), \gamma]}{\cos\alpha} \tag{4.20}$$

where E_n is the irradiance of the incident beam. The bi-directional reflectance, ρ, of the screen is assumed to be independent of position P within the screen.

Irradiance E_n can be expressed as follows:

$$E_n(\mathrm{P}) = E_{n0}\, \varepsilon(\mathrm{P}) \tag{4.21}$$

where $\varepsilon(\mathrm{P})$ is the normalized, dimensionless irradiance distribution.

In the following discussion, it is assumed that $\rho[\alpha(z), \gamma]$ is a slowly varying function of α. Thus, the orientation of the screen normal, as defined by γ, should prevent the detector from viewing the screen at a specular reflection angle when operating in the reflection mode of calibration. By integrating dF with respect to z and position, P, within the screen one obtains:

$$F(\theta,\gamma) = E_{n0}\rho(\theta,\gamma)\int_{z_{\min}}^{z_{\max}}\int_{A}\varepsilon(\mathrm{P})\Omega(z,\mathrm{P})\cos\gamma\, dA(z,\mathrm{P})dz \tag{4.22}$$

Integration over z is effectively truncated to within a range of $[z_{\min}, z_{\max}]$ within which the illuminated part of the screen is seen by the detector, i.e., where $\Omega(z, \mathrm{P}) > 0$. The assumption of ρ being independent of z and P enabled us to factor out ρ from the above integral and cancel it in the following equation.

By dividing $F(\theta,\gamma)$ through $dF(\theta, z=0, \gamma)$, i.e., $dF(\theta, z=0, \gamma, \alpha, \mathrm{P})$ integrated over the illuminated area, A, of the screen, one cancels the effect of E_{n0} and ρ on F. This way, we obtain:

$$\begin{aligned} C(\theta) &= \frac{F(\theta,\gamma)}{dF(\theta, z=0,\gamma)} \\ &= \frac{1}{\Omega_0}\frac{\int_{z_{\min}}^{z_{\max}}\int_A \varepsilon(\mathrm{P})\Omega(z,\mathrm{P})\cos\gamma\, dA(z,\mathrm{P})dz}{\int_A e\varepsilon(\mathrm{P})\cos\gamma\, dA(z,\mathrm{P})} \\ &= \frac{1}{\Omega_0}\frac{\int_{z_{\min}}^{z_{\max}}\int_{A_n} \varepsilon(\mathrm{P})\Omega(z,\mathrm{P})\, dA_n(z,\mathrm{P})dz}{\int_{A_n}\varepsilon(\mathrm{P})\, dA_n(z,\mathrm{P})} \\ &= \frac{1}{\Omega_0}\frac{\int_V \varepsilon(\mathrm{P})\Omega(z,\mathrm{P})\, dV(z,\mathrm{P})}{\int_{A_n}\varepsilon(\mathrm{P})\, dA_n(z,\mathrm{P})} \end{aligned} \tag{4.23}$$

where we used equalities $dA_n = dA\cos\gamma$ and $dV = dz\, dA_n$, as well as assumed that $\Omega(z=0, \mathrm{P}) \approx \Omega(z=0) = \Omega_0$. $C(\theta)$ is the calibration function of the nephelometer.

When the screen is removed and a sample is poured into the sample vessel (Figure 4.9), the flux, dF, received by the detector from an element $dV(z, \mathrm{P})$ of the scattering volume is expressed as follows:

$$dF[\theta(z), z, \mathrm{P}] = \beta[\theta(z,\mathrm{P})]E_{n0}\varepsilon(\mathrm{P})dV(z,\mathrm{P})\,\Omega(z,\mathrm{P})e^{-cT(z,\mathrm{P})} \tag{4.24}$$

where θ is the scattering angle, β is the volume scattering function of the sample, c is the attenuation coefficient (e.g., *Dera*, 1992) of the sample, and T is the distance from the face of the beam entrance window to the detector field stop via the volume element, dV.

The total flux, $F(\theta)$, equals the integral of dF over the scattering volume, V:

$$\begin{aligned} F(\theta) &= E_{n0}\int_V \beta[\theta(z,\mathrm{P})]\varepsilon(\mathrm{P})\Omega(z,\mathrm{P})e^{-cT(z,\mathrm{P})}dV(z,\mathrm{P}) \\ &= E_{n0}\beta'(\theta)\int_V \varepsilon(\mathrm{P})\Omega(z,\mathrm{P})e^{-cT(z,\mathrm{P})}dV(z,\mathrm{P}) \end{aligned} \tag{4.25}$$

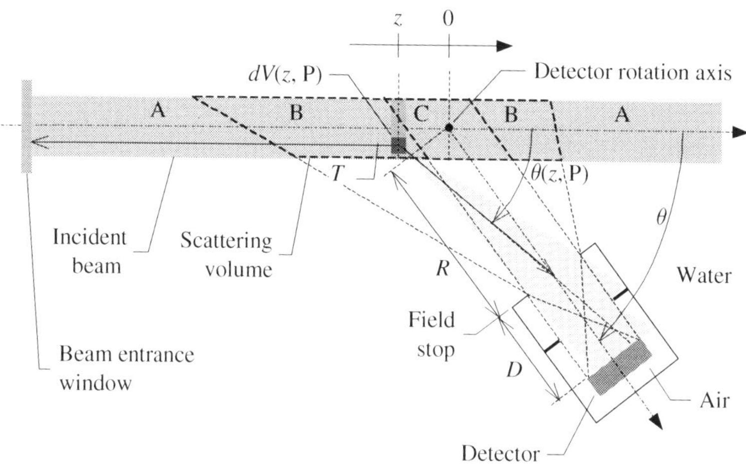

Figure 4.9. Geometry of the scattering meter measurements for calculation of the signal received from a slice of the scattering volume, here shown at a position z. The detector receives no light scattered by an element dV if that element is positioned in the region A of the beam. The detector is fully illuminated by light scattered by the element dV, if that element is in the region C, and is partly illuminated if dV is in the region B of the beam. Note refraction at the water–air interface at the field stop window which must be accounted for in calculating the solid angle subtended by the detector aperture at dV.

where we factored out the effective scattering function, β', i.e., the scattering function averaged over the scattering volume, V:

$$\beta'(\theta) = \frac{\int_V \beta[\theta(z,\mathrm{P})]\varepsilon(\mathrm{P})\,\Omega(z,\mathrm{P})e^{-cT(z,\mathrm{P})}dV(z,\mathrm{P})}{\int_V \varepsilon(\mathrm{P})\Omega(z,\mathrm{P})e^{-cT(z,\mathrm{P})}dV(z,\mathrm{P})} \tag{4.26}$$

and θ is the $\theta(z=0,\ \mathrm{P}_0)$, with P_0 located at the beam axis. This is equivalent to the following formal representation of F:

$$F(\theta) = \beta'(\theta)E_{n0}(V\Omega)' \tag{4.27}$$

where the symbol $(V\Omega)'$ represents an effective product of the scattering volume and acceptance angle of the detector:

$$(V\Omega)' = \int_V \varepsilon(\mathrm{P})\Omega(z,\mathrm{P})e^{-cT(z,\mathrm{P})}dV(z,\mathrm{P}) \tag{4.28}$$

This product corresponds to a product $dVd\Omega$ in the following form of the operational definition of the volume scattering function:

$$dF(\theta) = \beta(\theta)EdVd\Omega \tag{4.29}$$

If the effective scattering function, β', is measured with a nephelometer of a low angular resolution, the result may significantly differ from the actual volume scattering function, β, especially at scattering angles close to 0° and 180° (*Jonasz* 1990).

What we still need is a means of canceling E_{n0} in (4.25). This can be done through dividing $F(\theta)$ by the power, F_0, received by the detector at a scattering angle of 0°. F_0 equals:

$$F_0 = E_{n0} e^{-cT_0(z,\mathrm{P})} \int_{A_n} e(\mathrm{P}) dA_n \tag{4.30}$$

where $T_0 = T(z=0)$. Finally,

$$\begin{aligned} \frac{F(\theta)}{F_0} &= \beta'(\theta) \frac{\int_V e(\mathrm{P})\Omega(z,\mathrm{P}) e^{-c[T(z,\mathrm{P})-T_0]} dV(z,\mathrm{P})}{\int_{A_n} e(\mathrm{P}) dA_n} \\ &\approx \beta'(\theta) \frac{\int_V e(\mathrm{P})\Omega(z,\mathrm{P}) dV(z,\mathrm{P})}{\int_{A_n} e(\mathrm{P}) dA_n} \end{aligned} \tag{4.31}$$

because the product $c[T(z) - T_0]$ is typically much smaller than unity [note that $T_0 \approx T(z)$], so that $\exp\{-c[T(z) - T_0]\} \approx 1$ for $z \subset [z_{\min}, z_{\max}]$. This assumption is usually satisfied, except for low-resolution nephelometers and strongly attenuating media at the large scattering angles, in which case it may contribute to potentially sizable systematic error in β' (*Jonasz* 1990).

With this caveat, the fraction in the second line of (4.31) is identical with the rightmost fraction in (4.23). Hence, by combining (4.31) and (4.23), we have:

$$\beta'(\theta) \approx \frac{F(\theta)}{F(0)} \frac{1}{\Omega_0 C(\theta)} \tag{4.32}$$

By replacing power of light, F, with an electrical signal, $S = \mathrm{const} \times F$, generated by the detector system of the nephelometer, (4.32) can be reformulated as follows:

$$\begin{aligned} \beta'(\theta) &\approx \frac{S_S(\theta)}{S_S(0)} \frac{1}{\Omega_0} \frac{1}{C(\theta)} \\ &= \frac{S_S(\theta)}{S_S(0)} \frac{1}{\Omega_0} \frac{dS(\theta, z=0, \gamma)}{S(\theta, \gamma)} \end{aligned} \tag{4.33}$$

where S_S is the signal obtained with a sample in the nephelometer, while S is the calibration screen signal. The unit of the scattering function, β', is length^{-1} sr^{-1}, where the length unit is that used in calculating the integrals.

Sample functions $dS(\theta, z, \gamma)$ are shown in Figure 4.10 for a calibration experiment in the transmission mode with a nephelometer whose angular resolution is about 1°. Integrals of these functions with respect to z and normalized by $dS(z=0)$, i.e., $S(\theta, \gamma)/dS(\theta, z=0, \gamma)$, are used in the nephelometer calibration function $1/C(\theta)$. The calibration function $1/C(\theta)$ is shown in Figure 4.11. For a relatively high-resolution nephelometer, this function is similar to an expression const/sin θ that describes just the angle-dependent changes in the scattering volume geometry. The differences from that expression increase as the scattering angle becomes either small (approaching 0°) or large (approaching 180°).

The solid angle $\Omega_0 = \Omega(z=0)$ is determined from the nephelometer geometry as follows:

$$\Omega_0 \approx \frac{A_\mathrm{d}}{(R+nD)^2} \tag{4.34}$$

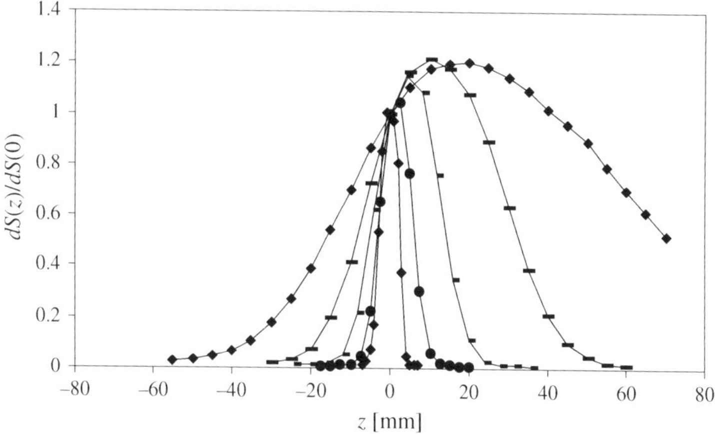

Figure 4.10. Functions $dS(z)$, normalized by their value at $z=0$, for a nephelometer with an acceptance solid angle of about 2×10^{-4} sr, i.e., the scattering angle, θ, resolution of about 1°. These functions have been measured in the transmission mode of the diffuse-screen calibration with the screen normal at an angle of $\gamma = 22.5°$. Each such function, measured for a particular scattering angle (5°, 10°, 20°, 40°, and 90° from left to right) provides a value of the calibration function when integrated over z and normalized by $dS(z=0)$. The form of function $dS(z)$ changes from one which is symmetrical about $z=0$ at a scattering angle $\theta = 90°$ to a highly asymmetrical one at $\theta = 5°$. These functions show the combined effect of changes in the area of the scattering volume cross-section perpendicular to the beam axis and in the detector acceptance angle, subtended at that cross-section, with the position of the cross-section along the beam axis.

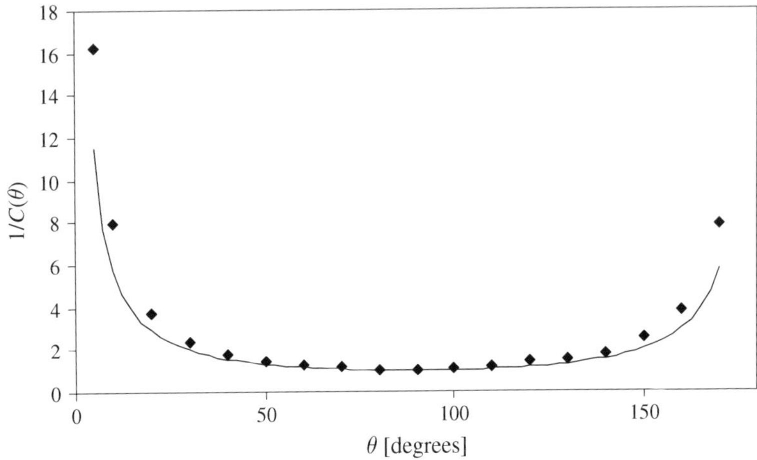

Figure 4.11. Calibration function (points) for a nephelometer with an acceptance solid angle of about 2×10^{-4} sr, i.e., the scattering angle resolution of about 1°. The solid curve represents a function const/sin θ. Differences between the data points and the curve are caused by changes in the detector acceptance angle. These differences increase as the scattering angle becomes either small or large. For the nephelometer in question, these differences increased from 0% at 90° to about 30% at both 10° and 170°.

where A_d is the detector aperture area, R is the radius of rotation of the detector field stop (Figure 4.9), D is the distance between the field stop and the detector aperture, and n is the refractive index of water.

4.2.3.7. Measurement errors

Measurement errors of the light scattering function with a polar nephelometer come from several sources: (1) finite angular resolution of the nephelometer, (2) stray light due to reflections inside and imperfections of the surfaces of the optical elements of the nephelometer, and (3) diffraction at apertures that limit the incident light beam or the diffraction of the incident beam itself, as well as (4) electrical interference and noise of the light source and detection system. Errors caused by the finite angular resolution of the nephelometer cannot be avoided as the nephelometer must admit some scattered light, i.e., must have a finite angular resolution. Likewise, those due to diffraction cannot be avoided because that latter process is fundamentally linked to the propagation of light around obstacles. Other errors can be limited by careful optical and electronic design of the nephelometer.

Effects of the finite angular resolution. Jonasz (1990) evaluated numerically errors due to a finite angular resolution of the polar nephelometer as functions of the scattering angle and of that angular resolution.

As it is evident from Figure 4.9, the nephelometer measures the scattering function, β', averaged over the scattering volume. At a given nominal scattering angle, θ, contributions to that average function depend on positions of the relevant elements of the scattering volume in relation to the detector. Such a contribution, from a dz-thick slice of the scattering volume at z, can be expressed as follows:

$$dF[\theta(z), z] = dzE_n dV(z)\, e^{-cT(z)} \int_{\theta_{\min}}^{\theta_{\max}} \beta[\theta(z)]\Omega(z, \theta)d\theta \qquad (4.35)$$

where, E_n is the irradiance $\theta_{\min}$ and $\theta_{\max}$ specify an applicable range of θ at z, this range being defined by the intersection of the field stop projection at the detector aperture and the detector aperture itself (Figure 4.12). For simplicity, we assume a factorable irradiance distribution, E_n, across the incident beam.

The contribution of a slice of the scattering volume is obviously unresolved when measuring the volume scattering function. Hence, as we discussed earlier, the latter must be regarded as an average scattering function. We can define it

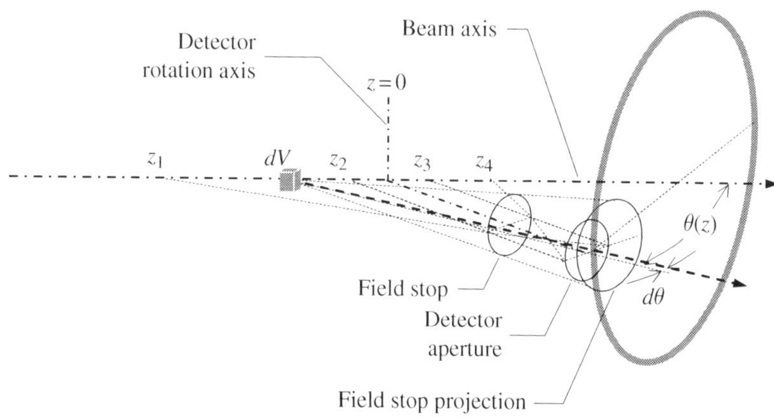

Figure 4.12. Light scattered by an element dV of the scattering volume projects the detector field stop onto the detector aperture plane. If dV is outside range $[z_1, z_4]$, i.e., in the region A of the beam in Figure 4.9, the projection misses the detector aperture and the detector receives no light. If the volume element is inside position range $[z_2, z_3]$, i.e., the region C of the beam in Figure 4.9, the entire detector aperture is illuminated. Otherwise, only a part of the detector aperture is illuminated, giving rise to vignetting of that aperture by the field stop. Refraction at the water–air interface of the field stop must be accounted for in order to correctly evaluate the vignetting effect. The detector, when illuminated, collects light from the element dV at a range of the scattering angle, θ, which depends on the nephelometer geometry and the position of the volume element. The large shaded circle represents the locus of light rays scattered by the volume element dV at $[\theta, \theta + d\theta]$.

with the following equation, looking from a slightly different perspective than that we used in the previous section:

$$\beta'(\theta') = \frac{\int_{z_1}^{z_2} e^{-cT(z)} \int_{\theta_{min}}^{\theta_{max}} \beta[\theta(z)]\Omega(z,\theta)d\theta dz}{\int_{z_1}^{z_2} e^{-cT(z)} \int_{\theta_{min}}^{\theta_{max}} \Omega(z,\theta)d\theta dz} \tag{4.36}$$

where $\theta' = \theta(z=0)$. This equation is consistent with the definition (4.26) of the average volume scattering function as measured by a polar nephelometer, if one assumes that (1) the dimensionless irradiance distribution, ε(P), is independent of position P within the beam cross-section, and (2) integration over the scattering volume, V, is replaced by an equivalent integration over distance, z, along the beam axis and over a z-dependent scattering angle range sampled by the detector aperture possibly vignetted by the field stop: θ_{min} to θ_{max}. This definition of the average volume scattering function is somewhat different than that used by *Jonasz* (1990), who did not include the attenuation factor in the denominator. As a result, we obtained somewhat different results in the backscattering angle range.

The magnitude of a contribution to the integral in the numerator of (4.36) depends on several factors: (1) the attenuation of light by the sample, as represented by $e^{-cT(z)}$, where T(z) is the distance from the face of the beam entrance window to the detector field stop via the intersection of the beam axis and the detector rotation axis, (2) averaging of the scattering function over the scattering angle within the acceptance angle $[\theta_{min}, \theta_{max}]$ of the detector (Figure 4.12). It is this averaging, which requires that the bi-directional reflectance and transmittance of the calibration screen be both weak functions of the incidence and observation angles. In the case of the actual sample, an error introduced by such averaging increases with the steepness of the scattering function as a function of the scattering angle.

The whole integral also expresses (3) averaging of the scattering function over a scattering angle range defined by the field of view of the detector at the beam, i.e., $[\theta(z_1), \theta(z_4)]$, where z_1 and z_4 are defined in Figure 4.12, and (4) modification of the light flux received by a detector aperture at $\theta(z)$ by the solid angle $\Omega(z, \theta)$ over the position range $[z_1, z_4]$ of the scattering volume element.

Consider the effect of the attenuation first. Sizeable attenuation of light at distances comparable to the light pathlength, $T(z)$, from z_1 through z (where light is scattered) and to the detector window (field stop) in the nephelometer sample space can occur for moderate values of the attenuation coefficient, c. As an example, by using (4.6) with the c substituted for b, one obtains for a representative scattering volume length on the order of 0.1 m an attenuation of the beam power by $\geq 10\%$ for $c \geq 1\ \text{m}^{-1}$. The most significant effect of the attenuation of light by the sample can be expected in the backscattering angle range, where the pathlength, T, varies most as a function of position z along the beam axis.

Let us now consider the solid angle, Ω, subtended by the detector at the various elements of the scattering volume and the effect of the angular field of view of the detector. A finite solid angle, Ω, as can be seen in Figure 4.12, causes the nephelometer detector aperture to sample light scattered into a range of angles that, for any element of the scattering volume, produces an average scattering function over that range. Contributions to that average are weighed by the solid angle subtended at the volume element by the height of an area element of the detector aperture sampling light scattered into an angular range $[\theta, \ \theta + d\theta]$. In addition to being averaged over the detector acceptance angle, the scattering function is averaged over a scattering angle range $[\theta(z_1), \ \theta(z_4)]$ (Figure 4.12) corresponding to the angular field of view of the detector. For the scattering angles of less than 90°, an element of the scattering volume that is furthest from the detector contributes the scattering function value at the smallest angle of this range. Closer elements contribute a value corresponding to a progressively larger scattering angle. The reverse is true for the scattering angles greater than 90°. These contributions are weighed by the product of the solid angle $\Omega(z)$ that the illuminated part of the detector aperture subtends at a scattering volume element at z and the volume of that element. It is also affected by the attenuation of light, $e^{-cT(z)}$, along a pathlength, T, from z_1 through z (where light is scattered) and to the detector window (field stop).

The product of the solid angle, Ω, and the volume of an element of the scattering volume is a complex function of the position, z, along the beam. It vanishes when $z < z_1$ or $z < z_4$ (Figure 4.12). In a range $z_1 \leq z \leq z_4$, this product initially increases and then decreases, as can be gathered from Figure 4.10.

The relative measurement error of the volume scattering function, defined as follows:

$$\varepsilon = \frac{\beta_{\text{avg}} - \beta_{\text{true}}}{\beta_{\text{true}}} \tag{4.37}$$

is shown in Figure 4.13 for a turbid water sample. The effect of the attenuation factor, $e^{-cT(z)}$, along a pathlength from z_1 through z (where light is scattered) and to the detector window (field stop) is relatively minor (Figure 4.13). In fact, it can only be discerned at the scale of that figure for a low-resolution nephelometer at a large scattering angle.

Grasso et al. (1995, 1997) have recently evaluated the effect of the nephelometer geometry on the scattering function in a polar nephelometer of type A (Figure 4.2) much along the same lines. In addition, they examined errors caused by reflection of the scattered light at the interfaces of the sample vessel. Such a reflection combines light scattered at an angle θ with that scattered at an angle $\theta + \pi$. Strictly speaking, there is an additional term, resulting from backscattering of light originally scattered at the angle of θ which is also reflected at the vessel–air interface, but with a high asymmetry of the scattering function; the contribution of this term to the detector signal is generally negligible.

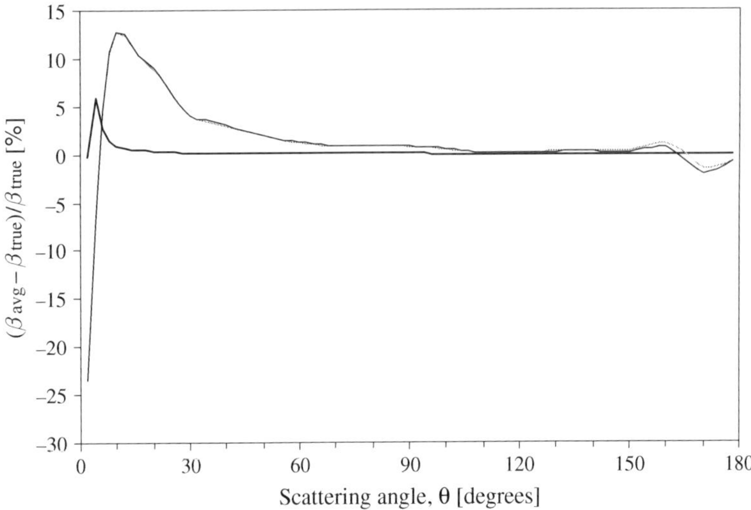

Figure 4.13. Systematic measurement error [%] of the volume scattering function with a high- (thick line) and a low-resolution (thin line) nephelometers. The scattering angle resolution of the high-resolution nephelometer is about 1°. That of the low-resolution nephelometer is about 2.2°. A scattering function Haoce, st. 11, of *Petzold* (1972) is used as the true scattering function. The attenuation coefficient of seawater is $2\,\mathrm{m}^{-1}$. The effect of the attenuation of light is relatively minor and discernible only for the low-resolution nephelometer, as can be seen from the near coincidence of the thin black and gray lines, the latter obtained for the attenuation coefficient of $0.2\,\mathrm{m}^{-1}$.

These errors are compounded by the error of the solid angle $\Omega(z = 0)$ and that of the calibration function $1/C(\theta)$. These latter errors are combined into the calibration error, $\varepsilon_{\mathrm{cal}}$, as follows:

$$\varepsilon_{\mathrm{cal}}^2 = \left[\frac{d\Omega(z=0)}{\Omega(z=0)}\right]^2 + \left[\frac{d[dS(\theta, z=0, \beta)]}{dS(\theta, z=0, \beta)}\right]^2 + \left[\frac{dS(\theta, \beta)]}{S(\theta, \beta)}\right]^2 \tag{4.38}$$

where we assumed that these errors are not correlated. The maximum error in the solid angle is on the order of 2%, based on the accuracy of the measurement of relevant variables. Relative errors in $dS(\theta,\ z=0,\ \beta)$ and $S(\theta,\ \beta)$ are of the order of 5% each. Thus, the relative calibration error is about 7.5%.

In addition, there are random errors due to noise of the detection system and to the photon shot noise of the scattered light flux itself, both discussed in section 4.2.3.4. This latter is likely to contribute only at a low incident light power and at scattering angles close to that of the scattering function minimum.

Fluctuations in the incident beam power contribute about 2% of the signal, for a typical unstabilized HeNe laser, unless they are compensated for by simultaneously measuring the beam power with a reference detector. Fluctuations in the number of particles in the scattering volume may contribute 10 to 20% of the signal. All these noise sources are uncorrelated. Thus, the relative measurement error may easily be on the order of several tens percent (e.g., *Kullenberg* 1984). This may easily dip the scattering functions of the clearest waters below the scattering function of pure water (seawater) (e.g., *Vaillancourt* et al. 2004).

Stray light. Errors related to the stray light are caused by the detector accepting light reflected at the surfaces of optical elements of the nephelometer, as well as light scattered by imperfections of the material of lenses, windows, and by the defects or contamination of the optical surfaces.

The sizeable errors that can be included in this category are potentially those due to a residual reflection of the scattered light (especially that scattered forward) at the water–glass, and glass-air interfaces of the sample container, or the sample stream (in flow-cytometric applications). Indeed, consider reflection at a water–glass interface of a round sample container of light scattered, e.g., at 5°. Unless the sample container is slightly conical (e.g., *Sasaki* et al. 1960) this reflected light will be measured by the scattered light detector when the latter is positioned at 175°, the conjugate angle in this example. Reflectivity of an interface between media with refractive indices n_1 and n_2 at normal incidence is (e.g., *Hecht* 1987) is

$$R = \left(\frac{n_2 - n_1}{n_2 + n_1} \right)^2 \tag{4.39}$$

With $n_1 \cong 1.34$ and $n_2 \cong 1.55$, we have $R = 0.0053$. Consider now the scattering function of "clear" seawater (see the average scattering functions of seawater later in this chapter). Such function yields a ratio of $\beta(5°)/\beta(175°)$ on the order of 0.2/0.0002. Thus, the reflection at the water–glass interface of light scattered at 5° would have contributed about $0.0002/(0.0053 \times 0.2) \cong 0.0002/0.001 = 20\%$ of light scattered at 175°, neglecting the attenuation of the sample by a factor of $\sim e^{-2cD}$, where c is the attenuation coefficient of the sample and D is the sample container diameter. Given that the scattering function increases little or not at all with angle in the backscattering range, and increases very rapidly with decreasing scattering angle, the situation worsens with the increasing scattering angle in the backscattering range. Therefore, optically black surfaces are frequently used to reduce this component of the stray light.

Diffraction effects. In the small-angle range, the nephelometer measures the incident beam light diffracted by the beam stop in addition to light scattered by the particles. Although the concept of the volume scattering function, β, does not extend to "volume diffraction function" at a two-dimensional obstacle, the intensity of light measured by the nephelometer detector at an angle, θ, can be generalized in the small-angle approximation to include the diffracted light as follows:

$$\beta_{\text{eff}}(\theta) = \frac{dI_{\text{scat}}(\theta) + \dfrac{dF_{\text{diff}}(\theta)}{d\Omega}}{EdV}$$
$$= \beta_{\text{scat}}(\theta) + \frac{E_{\text{diff}}(\theta)}{E}\frac{\theta}{dV_{90}}(R+D)^2\left(1-\frac{D\theta}{2w}\right) \tag{4.40}$$

where dF_{diff} is the diffracted power, $d\Omega$ is the acceptance solid angle of the detector (note that $dI = dF/d\Omega$)), E_{diff} is the distribution of the diffracted irradiance (power/area) at a plane $2R+D$ away from the beam stop (Figure 4.14), R is the rotation radius of the detector field stop, D is the distance of the detector from the field stop, w is the detector width in the scattering plane (we assume for simplicity a rectangular detector), and dV_{90} is the scattering volume at $\theta = 90°$; hence dV_{90}/θ approximates (for $\theta << 1\,\text{rad}$) the angle-dependent scattering volume, $dV(\sin\theta)$. A factor of $1 - D\theta/(2w)$ approximates the vignetting of the diffracted light by the detector field stop, where w is the detector width. This factor is replaced by 0, when it becomes negative.

Consider the case of a rectangular beam stop, with a width of $2r_{\text{bs}}$, as measured in the scattering plane. The distribution of irradiance of light diffracted by this stop at the detector plane in the scattering plane of the nephelometer with a field stopped parallel (plane wave) beam can be described by the Fresnel (near-field) diffraction approximation (e.g., *Hecht* 1987):

$$\frac{E_{\text{diff}}(\theta)}{E} = \frac{2}{4}\left\{[C(u_2) - C(u_1)]^2 + [S(u_2) - S(u_1)]^2\right\} \tag{4.41}$$

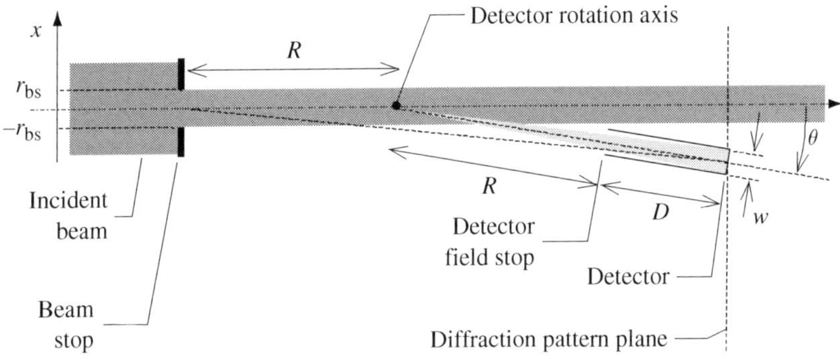

Figure 4.14. Geometry of the effect of the incident beam diffraction on the measurement of the scattering function at a small angle. As the scattering angle increases, the field stop vignettes an increasing area of the detector for the light diffracted by the beam stop. The inner surface of the detector tube of length D is assumed to be perfectly absorbing.

where 2 represents the diffraction profile maximum in a direction perpendicular to the scattering plane, and $C(u)$ and $S(u)$ are the Fresnel integrals:

$$
\begin{aligned}
C(u) &= \int_0^u \cos\frac{\pi t^2}{2}dt \cong \frac{1}{2} + f(u)\sin\frac{\pi u^2}{2} - g(u)\cos\frac{\pi u^2}{2} \\
S(u) &= \int_0^u \sin\frac{\pi t^2}{2}dt \cong \frac{1}{2} - f(u)\cos\frac{\pi u^2}{2} - g(u)\sin\frac{\pi u^2}{2}
\end{aligned}
\tag{4.42}
$$

where the approximations (e.g., *Mielenz* 1998), with

$$
\begin{aligned}
f(u) &\cong \frac{1+0.926u}{2+1.792u+3.104u^2} \\
g(u) &\cong \frac{1}{2+4.142u+3.492u^2+6.67u^3}
\end{aligned}
\tag{4.43}
$$

apply if $|u| >> 1$, i.e., in the case of typical nephelometer geometry, and where

$$
\begin{aligned}
u_1(\theta) &= \left(L_{\text{neph}}\frac{\theta}{2} - r_{bs}\right)\left(\frac{2}{\lambda L_{\text{diff}}}\right)^2 \\
u_2(\theta) &= \left(L_{\text{neph}}\frac{\theta}{2} + r_{bs}\right)\left(\frac{2}{\lambda L_{\text{diff}}}\right)^2
\end{aligned}
\tag{4.44}
$$

are the non-dimensional positions of the beam stop edges perpendicular to the scattering plane relative to the observation point, i.e., the center of the detector, $L_{\text{neph}} = R + D$ and $L_{\text{diff}} = 2R + D$ are respectively the observation point distance from the detector rotation axis and the observation point distance from the beam stop, θ is the scattering angle ($<< 1$ rad), r_{bs} is the beam stop half-width in the scattering plane (Figure 4.14), and λ is the wavelength of light in the sample.

A sample diffraction pattern for a polar nephelometer discussed earlier in this chapter is shown in Figure 4.15. Note that with finite angular resolution nephelometers, the diffraction pattern is averaged over the detector acceptance solid angle. With a quasi-monochromatic light source, the angular diffraction pattern is also averaged over the effective spectral range of the light source and the detector combined.

The diffraction component may significantly affect the measurements of the scattering function as shown approximately in Figure 4.16 for pure seawater. Note that the specific shape of the forward-scattering part of the function shown in that figure *reflects* the geometry of the nephelometer considered in that example and may differ from that for other nephelometer geometries. In addition, the sharp decline will, in the case of actual measurements, be smoothed by reflections at the detector tube wall and a contribution of small particles that are extremely difficult to remove from "pure" water.

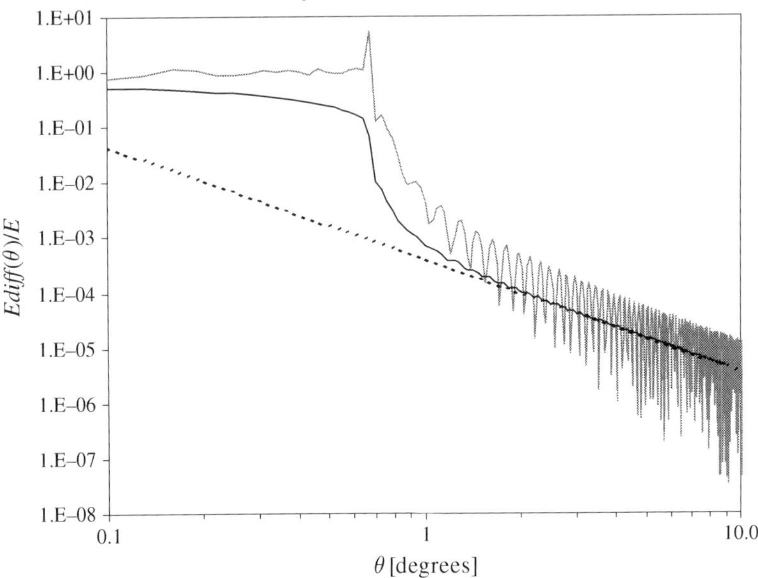

Figure 4.15. Diffraction of a parallel monochromatic beam (wavelength of 550 nm) stopped to a width of 5 mm as a function of the scattering angle of a polar nephelometer, as observed at a distance of 30 cm from the beam stop (*gray curve* – the actual diffraction pattern, *black curve* – diffraction pattern as observed with a 5 mm wide detector). The plateau on the left, marked by a sharp fall-off, represents the beam cross-section. The envelope of the smoothed diffraction pattern at angles $> \sim 1°$ decays according to a power law with a slope (here) of about -2.

4.2.3.8. Other nephelometer designs

Fast scanning polar nephelometers. The advent of flow cytometry with its requirement for single-particle light scattering measurements within milliseconds stimulated the design of rapid-scan nephelometers. The scanning is performed either mechanically, with one stationary detector (*Ulanowski* et al. 2002, *Moser* 1974, *Gucker* et al. 1973), or electronically, with several stationary detectors, each measuring light scattered into a different angular range (*Wyatt* and Jackson 1989, *Wyatt* et al. 1988, *Bartholdi* et al. 1980) and more recently with an imaging array (*Grasso* et al. 1997, 1995, *Hirst* et al. 1994). Quick measurements of the complete scattering pattern of single non-spherical particles opened the way to rapid online identification of the particles from their differential scattering cross-section itself (*Shvalov* et al. 1999, *Holler* et al. 1998, *Hirst* and Kaye 1996).

An interesting variation of the scanning polar nephelometer was introduced by Loken and colleagues (*Loken* et al. 1976) and later re-developed in a new form by *Chernyshev* et al. (1995) and *Maltsev* (2000). In this approach, the scattering pattern of a single particle is scanned by observing light scattered by the particle

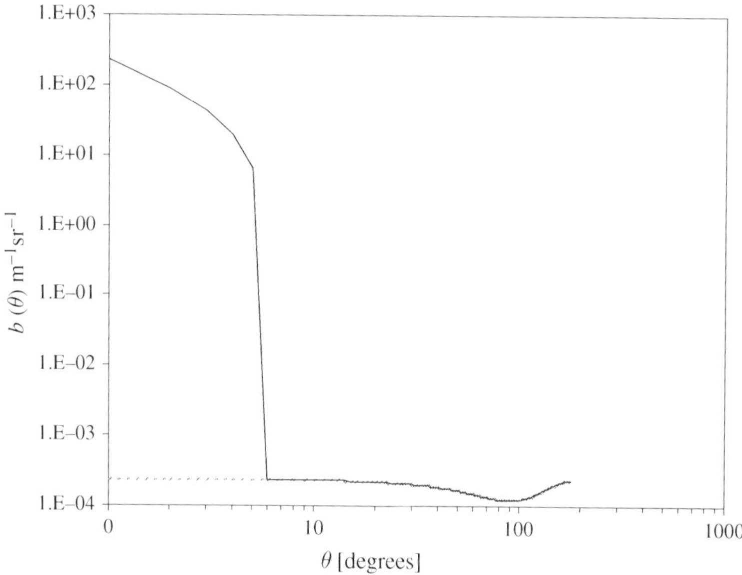

Figure 4.16. Contribution to the scattering function of diffraction of 550 nm light at the beam stop in a polar nephelometer. The base function is that of pure seawater ($S = 35$, thin dashed gray curve). The diffraction component is the smoothed Fresnel diffraction pattern shown in Figure 4.15. The parameters of the nephelometer are $R = 0.1$ m, $D = 0.1$ m, $w = 5$ mm, $r_{bs} = 2.5$ mm, refer to Figure 4.14; the small and large limits with this parameter are about 2° and 178°. The diffraction pattern is evaluated at a distance of $2R + D$. The sharp drop at ~6° is caused by the vignetting of the detector by its field stop that sets the acceptance angle. The shape of the forward-scattering part of the function *reflects* the geometry of the nephelometer considered here and may differ from that for other nephelometer geometries. In addition, the sharp decline will, in the case of actual measurements, be smoothed by the reflections at the detector tube wall and by contribution of small particles that are extremely difficult to remove from the "pure" water.

as it passes near a stationary detector. This method relies on the principles that we discussed in the analysis of the systematic error of the scattering function. Here, the particle itself is the sole "element" of the scattering volume. As it moves along the beam axis, a fixed field-of-view detector receives light scattered at a varying angle.

Simultaneous static and dynamic light scattering. Interest in the applications of dynamic light scattering for the characterization of suspended particles or macromolecules resulted in nephelometer designs permitting simultaneous measurement of the scattering function and dynamic light scattering (e.g., *Bantle* et al. 1982). Such a nephelometer has been also described and used by *Witkowski* et al. (1993)

to investigate growth of *Chlorella* cells in suspension. The dynamic light scattering was used to infer the cell structure at a submicron scale.

The small-angle nephelometer (< 5°). The small-angle range (0° to 5°) requires a nephelometer design different from that of the polar nephelometer. *Kullenberg* (1968) developed an *in situ* small-angle nephelometer, in which the scattering angle range seen by the detector is selected with a set of conical mirrors and annular stops all coaxial with the incident light beam. This instrument uses a HeNe laser as a light source. Three interchangeable sets of the mirrors and stops permit the measurements of light scattered in angular ranges of 0.5–1°, 2–2.5°, and 3–3.5°. The use of a lens system (not elaborated on by the author) permitted the measurement of the volume scattering function at angles ranging from 25 to 135°. A single-angle (0.5°) *in situ* nephelometer is also described by *Duntley* (1963). It measured light scattered from a hollow cylinder of light formed by the apertured illumination source.

Another design of the small-angle nephelometer is based on the Fourier transform of the angular field of the scattered light intensity into a two-dimensional distribution of irradiance (Figure 4.4). Such transform is performed by a convex lens. The resulting two-dimensional distribution of irradiance is located in a focal plane of the lens. It follows from geometrical optics that the light incident at the lens at an angle, θ, with respect to the optical axis is focused onto a circle of radius $f \tan \theta$ in the focal plane, where f is the focal length of the lens. Thus, light scattered at various angles from a parallel beam coaxial with the optical axis of the lens is focused at concentric circles of different radii. This design is implemented in the laser particle analyzers (see *Cornillault* 1972 for an early design of such analyzer), also referred to as laser diffractometers. We discuss these instruments at more detail in Chapter 5.

An early *in situ* nephelometer of this type is described by *Petzold* (1972). An annular field stop placed in front of a PMT allowed the scattered light to pass but obscured the center light spot. By exchanging the field stops, light scattered at 0.057–0.114°, 0.114–0.229°, and 0.229–0.458° could be measured, one stop at a time. In that instrument, the beam pathlength in water was 50 cm. *Spinrad* et al. (1978) also described a small-angle nephelometer based on a similar principle. A double-purpose *in situ* laser diffractometer and nephelometer has been recently made available commercially (*Agrawal* and Pottsmith, 2000).

In an early design of a lens-based nephelometer (*Bauer* and Morel 1967, *Bauer* and Ivanoff 1965), a photographic film was placed in the focal plane of the lens to record the radial irradiance pattern. A light stop was used to obscure the central spot, corresponding to the incident light beam.

Later, *McCluney* (1974) detected small-angle scattered light in ten angular regions simultaneously with a modulation–demodulation technique employing two masks placed at the focal plane of the lens to select the scattering angle. Each mask contained ten concentric zones of alternating opaque and transparent regions with a different period in each zone. The masks' patterns were shifted by 180° in phase. One mask was stationary, while the other rotated with a constant angular velocity. Thus, each zone periodically interrupted light scattered into its angular range. These interruptions occurred at a different frequency for each zone, so the detector signal contained superimposed waveforms at the frequencies of all zones. The signals for each zone were obtained by an analog demultiplexing circuit.

Recent designs of the Fourier-transform nephelometers use either custom photodiode arrays (*Hirleman* et al. 1984, *Dodge* 1984) or general-purpose two-dimensional detector arrays (*Conklin* et al. 1998, *Dueweke* et al. 1997) for the measurement of the irradiance pattern in the focal plane of the Fourier transform lens.

An interesting departure from attempts to separate the incident and scattered light based on their angular distributions is offered by the photorefractive and other nephelometry, which we will discuss in the remaining part of this section. *Fry* and colleagues (1992b) used a photorefractive crystal ($BaTiO_3$) to separate the scattered light from incident light at exactly 0°. Such crystals bend coherent light beams by forming a refractive index gradient in response to the electric field of the incident light wave. This process is relatively slow which enables the light scattered by a particle undergoing Brownian motion to pass through the crystal undeviated while the stationary incident beam bends away (after a delay on the order of seconds to minutes for low-beam powers on the order of mW to μW).

Interferometry-based measurements of small-angle scattering have also been attempted (*Batchelder* and Taubenblatt 1989, *Taubenblatt* and Batchelder 1990). This technique utilizes the fact that the phase of the incident light wave is modified by the scattering particle. When the scattered and incident waves are added, the effect is to shift the phase of the combined wave as compared with that of the incident wave. For particles with diameters, D, much smaller than the wavelength of the incident light, the phase shift at 0° is proportional to D^3.

Modulation transfer function (MTF)-based and point spread function-based nephelometry. The volume scattering function can also be derived from measurements of the MTF that describes the decrease, with increasing spatial frequency, of an optical system resolution. In the theory of conventional short-range imaging in air, this decrease is caused by imperfections of the optical system and by the diffraction of light. In transmission of an image through seawater and other turbid media, the scattering of light is the major factor limiting the imaging system resolution. This loss of resolution with increasing spatial frequency in the image can be determined by measuring the contrast of a test target observed through a layer of seawater of known thickness. The test target is either a sinusoidal reflectance pattern (a single spatial frequency) or a repetitive white-black bar pattern (a wide spatial frequency range). The contrast of the pattern image, i.e., the ratio of modulation of irradiance in the target to that in the image, is the MTF.

Wells (1969) developed a small-angle scattering theory which relates the MTF of a collimated light beam in seawater, here denoted by $F(\nu, R)$, where ν is the spatial frequency [cycles m^{-1}] and R [m] is distance in seawater to the scattering function of seawater, β, as follows:

$$\beta(\theta) = -2\pi \times \frac{\partial}{\partial\theta}\left\{\theta \int_0^\infty \nu[\ln F(\nu, R) - \ln F(\infty, R)] J_0(2\pi\theta\nu R) d(\nu R)\right\} \quad (4.45)$$

where J_0 is the Bessel function of the first kind (e.g., *Korn* and Korn, 1968) and νR is the angular spatial frequency [cycles rad^{-1}] (*Huang* et al. 1994, *Hodara* 1973).

A related method (*Mertens* and Phillips 1972) is based on the measurements of the beam spread function (BSF, e.g., *Mertens* and Replogle 1977) of a light scattering medium, which is a Hankel (or Fourier–Bessel) transform of the MTF:

$$\mathrm{BSF} = \frac{2\pi}{R^2} \int_0^\infty F(\nu, R) J_0(2\pi\theta\nu R) d(\nu R) \quad (4.46)$$

For readers who would like to explore this route, we note a fast algorithm for the numerical Hankel transform that has been recently published by *Magni* et al. (1992).

Radiance field inversion. *Zaneveld* (1974) proposed an algorithm, improved by *Wells* (1983), for measuring inherent optical properties (e.g., *Dera* 1992), such as the absorption coefficient and the scattering function, of a scattering and absorbing medium by measuring moments of the radiance field in that medium. In short, the equation of radiative transfer in such a medium can be expanded in spherical harmonics as follows:

$$n\frac{dL_{n-1}}{dz} + (n+1)\frac{dL_{n+1}}{dz} + (2n+1)A_n L_n = 0 \quad (4.47)$$

where L_n is the n-th moment of the radiance field, defined as follows:

$$L_n = \int L(\theta, \phi) P_n(\cos\theta) d\Omega, \quad n = 0, 1, 2, \ldots \infty \quad (4.48)$$

θ and ϕ are the elevation and azimuth angles respectively, P_n is the n-th Legendre polynomial, $d\Omega = \sin\theta d\theta d\phi$, and coefficients A_n are related to the inherent optical properties of the medium as follows:

$$A_n = c - 2\pi \int \beta(\xi) P_n(\cos\xi) d\xi, \quad n = 0, 1, 2, \ldots \infty \quad (4.49)$$

where c is the attenuation coefficient of the medium and $\beta(\xi)$ is its scattering function.

As $P_0 = 1$, it follows that $A_0 = c - b = a$, where b and a are the scattering and absorption coefficients respectively. Similarly, it can be proven that $A_\infty = c$. Coefficients A_n increase smoothly and monotonically to that latter value. Thus, with the first few (on the order of 10) values of A_n determined from the radiance field measurements and with an independent estimate of c, one can reconstruct the scattering function from the following expression:

$$\beta(\xi) = \frac{1}{4\pi}\sum_{n=0}^{\infty}(2n+1)(c - A_n)P_n(\cos\xi) \tag{4.50}$$

Doss and Wells (1992) describe the design of an instrument which measures simultaneously the first ten moments L_n of the radiance field with ten separate collectors of light. Each collector has the angular response shaped to follow a Legendre polynomial of the corresponding degree. Recent numerical and analytical evaluation of the measurement errors of the scattering function with such an instrument (*Holl* and McCormick 1995) suggests that significant errors may substantially limit practical applications of this type of nephelometer.

Speckle-based nephelometry. This method, developed recently by *Brogioli* et al. (2002), utilizes the speckle phenomenon that is usually considered a nuisance when making measurements of light scattering with a laser as a light source. Their method is based on the following relationship between the power spectrum of the speckle field generated by a scattering medium, illuminated by a coherent plane wave and the scattering function:

$$S_{\delta I}(q) \propto \beta(q) \tag{4.51}$$

where $S_{\delta I}$ is the power spectrum of the intensity fluctuation in an image of speckles at a plane that is at a distance z away from the sample, q is the amplitude of the scattered wave vector

$$q = \frac{4\pi}{\lambda}\sin\frac{\theta}{2} \tag{4.52}$$

with λ being the wavelength of the incident light in the medium surrounding the particles and θ being the scattering angle. Equation (4.51) is valid in the small-angle approximation and under the following conditions for the speckle plane distance z:

$$z < \frac{d}{\theta^*} \tag{4.53}$$

where θ^* is the scattering angle that includes a significant scattered light power, here assumed to be much less than 1 rad, d is the parallel beam diameter, and

$$z << \frac{L^2}{\lambda} \tag{4.54}$$

with L being the transverse dimension of the image sensor that is used to measure the intensity fluctuations.

4.2.4. Measurements of the scattering coefficients

The scattering coefficients can be determined by numerical integration of the scattering function determined over a sufficiently large angular range in order to minimize the truncation error, most severe in the forward-scattering range due to a high asymmetry of the function. However, this is clearly a laborious method.

A shortcut, whereby one could optically, rather than numerically, integrate the scattered light, is a desired choice here. It is also one of potentially higher accuracy than that of numerical integration of the scattering function because of the inherently greater scattered light power available to a detector of an integrating nephelometer, than that available to a detector of a nephelometer.

Beuttell and Brewer (1949) were first to propose a nephelometer that would measure the scattering coefficient of a medium via optical integration. Their design (Figure 4.17) was based on the integration of light scattered by the medium over a large range of the scattering angle, with an implicit assumption of the axial symmetry of the scattering function of the medium. *Charlson* (1993) compiled a representative bibliography of this type of nephelometer and its applications in atmospheric sciences. Additional references can be found in *Gordon* and Johnson (1985). A review of the integrating nephelometer has been recently published (*Heintzenberg* and Charlson 1996).

As it follows from Figure 4.17, the detector in an integrating nephelometer (with a Lambertian directional response characteristics, i.e., the response independent of

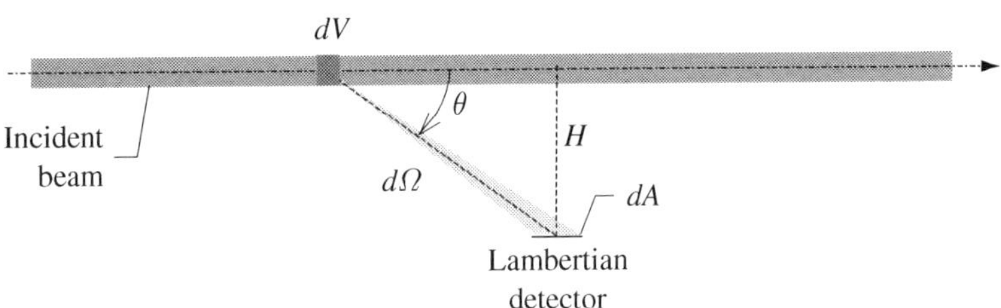

Figure 4.17. The principle of the integrating nephelometer. The distance of the detector from the beam, H, should be minimized in order to allow integration over a scattering angle range as close as possible to a range of 0 to 180° but to keep the beam length, and thus the angle-dependent attenuation suffered by light scattered by dV, reasonably small. The principle shown here also applies to a design where the roles of the detector and light source are reversed, i.e., a highly directional ("collimated") detector is viewing light scattered by the medium illuminated with a Lambertian light source that replaces the Lambertian detector we show here. This latter design, where the field of view of the detector is terminated by a light trap, is used more frequently because it affords a simpler design.

direction) receives from a beam volume element, $dV(\theta)$, the scattered light power, $dF(\theta)$:

$$
\begin{aligned}
dF(\theta) &= dI(\theta)d\Omega(\theta) \\
&= \beta(\theta)EdV(\theta)d\Omega(\theta) \\
&= \beta(\theta)E\frac{dV_{90}}{\sin\theta}\frac{dA\sin\theta}{H^2/\sin\theta} \\
&= \beta(\theta)\sin\theta EdV_{90}d\Omega_{90}
\end{aligned}
\tag{4.55}
$$

where dI is the scattered light intensity, $d\Omega$ is the angle that the detector subtends at dV, E is the beam irradiance (power/area), H is the distance of the detector from the beam, and dV_{90} and $d\Omega_{90} = dA/H^2$ are the beam volume element, dV, and the solid angle, $d\Omega$ (dA is the detector area) respectively, at a scattering angle of 90°. Thus, when the scattered power is integrated over the distance along the beam, and thus effectively over the resulting range of the scattering angle, one obtains a value that is proportional to the scattering coefficient (4.4), provided that the scattering angle range is sufficiently large. In this simple approximation, we neglected the attenuation of the scattered light along an angle-dependent path $H/\sin\theta$. As it is evident from Figure 4.17, it is the orientation of the detector, with the normal perpendicular to the beam axis, which introduces the correct $\sin\theta$ factor in the integration of the scattered light power.

Gordon and Johnson (1985) as well as *Rosen* et al. (1997) examined theoretical models of the integrating nephelometer of this type. A major disadvantage of this measurement method is a potentially sizeable error resulting from the angular truncation of the scattering function (Figure 4.18). This problem, dependent on the specific geometry of the nephelometer as well as on the scattering function shape, is of less significance for atmospheric particles. This seems to have limited the interesting concept of an integrating nephelometer to atmospheric studies, where the truncation error has been extensively investigated (see *Heintzenberg* and Charlson 1996 for references). In fact, we have been able to find just one example of the usage of such an instrument in marine studies (*Sternberg* et al. 1974).

However, by shielding the detector from the forward-scattered light, one may as easily measure the backscattering coefficient, b_b. The scattering functions of natural waters are generally much less steep in the scattering angle range of 90° to 180°, than in the range of $<90°$. Hence, the backscattering coefficient can in principle be measured with a reasonable accuracy in this range (Figure 4.19).

Given the difficulties of measuring the scattering and backscattering coefficients by optical integration, other approaches have evolved. *Forand* and Fournier (1999, see also *Forand* et al. 1993) describe a small-angle nephelometer that was used, in conjunction with the Fournier–Forand (FF) approximation to the aquatic scattering

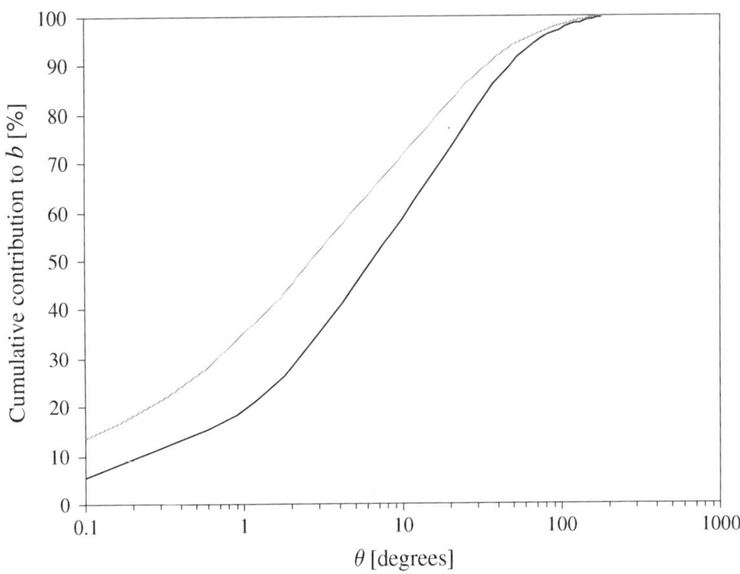

Figure 4.18. Relative contributions, $(2\pi/b)\int_0^\theta \beta(t)\sin t\,dt$ to the scattering coefficient, b, for two seawater scattering functions, β: gray curve - Atlantic, Bahamas (*Petzold* 1972, station 7, clear water), black curve - Atlantic, off New Jersey, USA (*Lee* et al. 2003, #48, turbid water). In order to keep the angular truncation error below 10%, an integrating nephelometer must collect light scattered at angles much less than 1° in both cases. The numerical integrations were carried out here with the trapezoid formula. For the gray line, it yields $b = 0.125\,\mathrm{m}^{-1}$ vs. $0.123\,\mathrm{m}^{-1}$ from a fit to the Fournier–Forand (FF) approximation (see notes in Table 4.3). For the black line, it yields $b = 16.64\,\mathrm{m}^{-1}$ vs. $16.05\,\mathrm{m}^{-1}$ from an FF fit.

function (section 4.5.2.1), to measure the spectra of the scattering coefficient in the visible. An FF approximation was fitted to the data obtained with the nephelometer, and the parameters of that fit were then used to calculate the scattering coefficient from an analytical expression.

Independently, it has been proposed to use the measurements of the scattering function at a small (about 3–45°) or large angle (120–140°) to retrieve the scattering and backscattering coefficient from an experimental relationship between those values and the respective scattering coefficients. These methods are discussed in more detail following the overview of the experimental scattering functions later in this chapter (section 4.4.2.2). An *in situ* instrument utilizes that proportionality to measure the backscattering coefficient simultaneously, with the absorption coefficient (*Dana* et al. 1998) although not in the same sample volume.

An interesting method of measuring the backscattering and absorption coefficients simultaneously, based on the two-stream approximation to the radiative

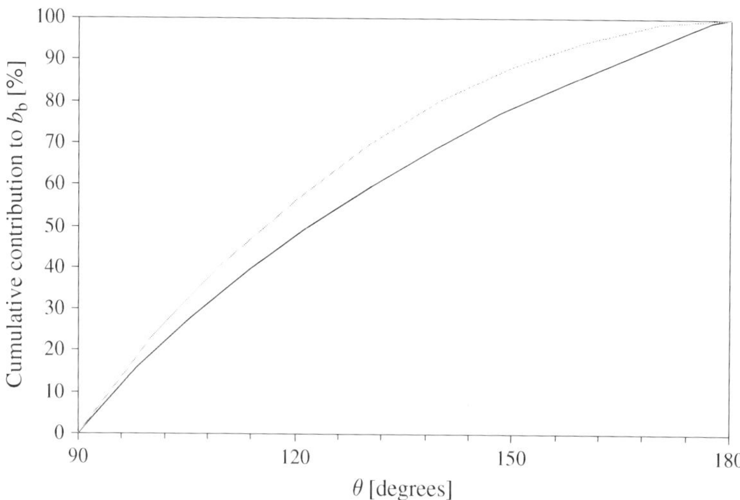

Figure 4.19. Relative contributions, $(2\pi/b_b)\int_{90}^{\theta}\beta(t)\sin t\, dt$ to the backscattering coefficient, b_b, for scattering functions, β, differing in their shape in the backscattering region: steep (black line, Atlantic off New Jersey, USA; *Lee* et al. 2003, #60, which we found to be one of the most steep in the backward direction), and flat (gray line, San Diego Harbour, *Petzold* 1972, time 20:40). In the case of the flat function, a 10% error occurs in an integrating nephelometer that truncates the scattering angle range at ~150°. In the case of the steep function, integration must include scattering angles of up to ~165° for an error of the same magnitude. The numerical integration simulating the action of an integrating nephelometer is carried out with the trapezoid formula.

transfer theory, was proposed some time ago by *Bukata* et al. (1980). The principle of that method is summarized in Figure 4.20. The proposed instrument would have consisted of two large chambers that approximated an optical medium of finite depth. The chambers would be identical, apart from the reflectivities, ρ_1, and ρ_2, of their bottoms. The upwelling irradiances, E_{u1} and E_{u2}, measured by the detector in each chamber can thus be expressed as follows:

$$
\begin{aligned}
E_{u1} &= \frac{b_b E_{d0}}{2(a+b_b)}\{1-\exp[-2(a+b_b)z]\} + \rho_1 E_{d0}\exp[-2(a+b_b)z] \\
E_{u2} &= \frac{b_b E_{d0}}{2(a+b_b)}\{1-\exp[-2(a+b_b)z]\} + \rho_2 E_{d0}\exp[-2(a+b_b)z]
\end{aligned}
\tag{4.56}
$$

where E_{d0} is the downwelling irradiance at the source, a is the absorption coefficient, b_b is the basckscattering coefficient, and z is the chamber depth. These equations (identical aside from the bottom reflectivities) can be solved for b_b and a. The proposed device would nevertheless have to be quite bulky (prototype

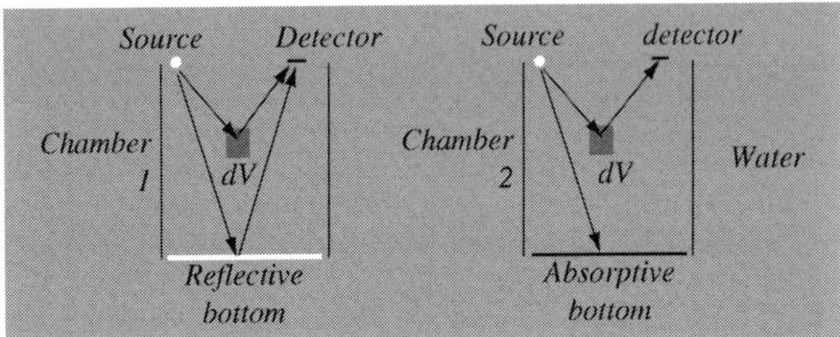

Figure 4.20. The principle of the simultaneous measurement of the backscattering and absorption coefficients (*Bukata* et al. 1980) based on a two-stream approximation to the radiative transfer theory in turbid media of finite optical depth. In such media, the reflectivity of the bottom controls the irradiance component received by the detector from the bottom. With fixed absorption and increasing backscattering coefficient, the detector output in chamber 1 decreases and that of chamber 2 increases toward a common value that represents backscattering by the water. With fixed backscattering and increasing absorption, the outputs of both chambers' detectors decrease.

chambers were 1 m high), making it difficult to deploy the instrument in the field. In addition, the instrument response was non-linear and not described sufficiently well by the simple model, although the model did reproduce the salient features of that response.

More recently, a number of algorithms to determine the backscattering coefficient and absorption based on the measurements of the irradiance reflectance, E_u/E_d, and irradiance attenuation coefficient, $K_d(z) = -1/E_d(z)dE_d(z)/dz$, were published, based on more advanced approximations to the radiative transfer theory (e.g., *Leathers* and McCormick 1997). A significant appeal of such algorithms is their inherent *in situ* character as measurements of irradiance with well-established instrumentation is much less invasive than sampling of the water for analysis either *in vitro*, or into the limited sample volumes of *in situ* instruments. The inherently large "sampling" volumes of methods based on these algorithms yield results that are representative of light scattering in large-scale radiative processes in natural water bodies, unlike those provided by the small-volume point-measurement methods. One should note that *in situ* irradiance measurements may be significantly affected by fluctuations due to focusing by the wavy surface of a water body and by passage of clouds over the measurement area.

However, at the present time, the direct, small-volume approach seems to have gained an upper hand, as recent widespread availability of commercial instruments for *in situ* measurement of the coefficients of absorption, a (with the ac-9 absorption meter, WET Labs, Philomath, Oregon, USA), and attenuation of light, c, made it possible to routinely measure spectra of the scattering coefficient of natural waters by simply subtracting the absorption from the attenuation coefficient (e.g., *Babin* et al. 2003).

The "integral" methods of measuring the scattering coefficients open the way to spectral measurements. One should, however, cautiously examine the reliability of such measurements related to cross-spectral contamination due to spectral down-conversion processes such as phytoplankton fluorescence for the small-volume measurement methods (*Vaillancourt* et al. 2004, *Bricaud* et al. 1983) and Raman scattering (e.g., *Stavn* 1993, *Stavn* and Weidemann 1992), as well as phytoplankton fluorescence (*Stavn* and Weidemann 1992) for methods based on the light field measurements.

4.3. Polarized light scattering: the scattering matrix

We are now going to take a broader look at the characterization of polarized light scattering and discuss the scattering matrix. This matrix transforms the incident light irradiance and polarization state information, assembled into a four-dimensional Stokes vector, into the scattered light Stokes vector. Its major virtue is the complete description of the interaction of light with a single particle or the incoherent interaction of light with a suspension. It is the completeness of that description which is the main reason for our interest here: even if one intends to measure solely the volume scattering function or the scattering coefficient, it is worth knowing what, at least in theory, can be gained from the knowledge of the complete description of the particle–light interaction. Theoretical and experimental evaluation of the scattering matrix has gained momentum within the past decade due both to a significant progress in the theory of light scattering by non-spherical and/or non-homogeneous particles and to improvements in measurement and computational techniques.

4.3.1. Stokes vector

The Stokes vector completely specifies a light beam: wavelength, direction, irradiance, and polarization state. Let us first consider the formal definition and its implications and then discuss how the Stokes vector can be determined experimentally.

Consider a plane light wave represented by the following electric vector:

$$\mathbf{E} = E_{\|}\mathbf{e}_{\|} + E_{\perp}\mathbf{e}_{\perp} \tag{4.57}$$

with

$$\begin{aligned} E_{\|} &= E_{0\|}e^{i\delta_{\|}}e^{i(\omega t-kz)} \\ E_{\perp} &= E_{0\perp}e^{i\delta_{\perp}}e^{i(\omega t-kz)} \end{aligned} \tag{4.58}$$

where $E_{\|}$ is the time-dependent amplitude of the electric field of the light wave polarized parallel to the scattering plane, i.e., the plane, containing the incident and observation directions, and $E_\perp$ is the time-dependent amplitude of the electric field of the light wave polarized perpendicular to the scattering plane. Vectors $\mathbf{e}_{||}$ and $\mathbf{e}_\perp$ are unit vectors for these two polarization directions. For coherent light (at the time scale of the measurements), the phase difference $\delta_{||} - \delta_\perp = \delta$ is independent of time. The phase difference δ is a random function of time in incoherent light. As usual, one uses the electric vector alone to represent electromagnetic fields (i.e., electric and magnetic fields) because the effect of the magnetic field on electrons in matter can be neglected at velocities at which these electrons are driven by the electric field (e.g., *Crawford* 1968).

The four elements of the Stokes vector are defined as follows (e.g., *Born* and Wolf 1980):

$$\begin{aligned} I &= \langle E_{||}E_{||}^* + E_\perp E_\perp^* \rangle \\ Q &= \langle E_{||}E_{||}^* - E_\perp E_\perp^* \rangle \\ U &= \langle E_{||}E_\perp^* + E_\perp E_{||}^* \rangle \\ V &= i\langle E_{||}E_\perp^* - E_\perp E_{||}^* \rangle \end{aligned} \tag{4.59}$$

or, with the use of (4.58),

$$\begin{aligned} I &= E_{0||}^2 + E_\perp^2 \\ Q &= E_{0||}^2 - E_\perp^2 \\ U &= 2E_{0||}E_{0\perp}\cos\delta \\ V &= -2E_{0||}E_{0\perp}\sin\delta \end{aligned} \tag{4.60}$$

where $\langle\rangle$ denotes the time average over an interval much larger than the wave period and the asterisk denotes the complex conjugate of a complex variable.

The quantities I, Q, U, and V all have dimension of irradiance, i.e., power per area. From (4.59), it can be seen that $(I+Q)/2$ is the irradiance of a beam component polarized in the scattering plane, and $(I-Q)/2$ is the irradiance of a beam component polarized in the orthogonal plane. In fact, another frequently used definition of the Stokes vector replaces elements I and Q with the parallel-polarized and perpendicular-polarized irradiances respectively: $\mathrm{I}_{||} = \langle E_{||}E_{||}^* \rangle = E_{0||}{}^2$ and $\mathrm{I}_\perp = \langle E_\perp E_\perp \rangle = E_{0\perp}{}^2$. The elements U and V are related to the inclination of the major axis, asymmetry parameter, and handedness of the polarization ellipse of the light beam (e.g., *Bohren* and Huffman 1983) as follows:

$$\begin{aligned} I &= c^2 \\ Q &= c^2 \cos 2\eta \cos 2\gamma \\ U &= c^2 \cos 2\eta \sin 2\gamma \\ V &= c^2 \sin 2\eta \end{aligned} \tag{4.61}$$

where $c^2 = a^2 + b^2$, with a and b being the semimajor and semiminor axes of the polarization ellipse, $0 \leq \gamma \leq \pi$ is an angle between the positive axis direction of the reference plane and the semimajor axis, and $\pi/4 \leq \eta \leq \pi/4$ is a measure of the ellipse asymmetry, defined as follows: $|\tan \eta| = \mathrm{b/a}$. The sign of V specifies the handedness of the polarization ellipse: the positive sign indicates the right-handed (clockwise) rotation of the electric vector tip, as seen by an observer looking at the light source, and the negative sign indicates a left-handed ellipse. *Bohren* and Huffman (1983) nicely summarize the history and pitfalls of the polarization ellipse handedness conventions. Since the Stokes vector is defined with respect to a reference plane, if this plane changes, as in a scattering event, the vector itself must be accordingly transformed (e.g., *Bohren* and Huffman 1983).

A variety of notations for the Stokes vector components is used in the optical literature as noted by *Bohren* and Huffman (1983). Sample Stokes vectors, with unit irradiance ($I = 1$), are listed in Table 4.1.

It follows from (4.59) that the elements of a Stokes vector fulfill the following relationship:

$$I^2 \geq Q^2 + U^2 + V^2 \tag{4.62}$$

with the equality applicable in the case of completely polarized light.

Table 4.1. Sample Stokes vectors of unpolarized and polarized light with unit irradiance.

No polarization	**Linear polarization**			**Circular polarization**	
	Arbitrary orientation angle γ	**Parallel to reference plane**	**Perpendicular to reference plane**	**Right-handed**	**Left-handed**
$\begin{bmatrix}1\\0\\0\\0\end{bmatrix}$	$\begin{bmatrix}1\\ \cos 2\gamma\\ \sin 2\gamma\\0\end{bmatrix}$	$\begin{bmatrix}1\\1\\0\\0\end{bmatrix}$	$\begin{bmatrix}1\\-1\\0\\0\end{bmatrix}$	$\begin{bmatrix}1\\0\\0\\1\end{bmatrix}$	$\begin{bmatrix}1\\0\\0\\-1\end{bmatrix}$

See text for the derivation of the Stokes vector elements for unpolarized light.

If light is completely polarized, $E_\perp = aE_\| \exp(i\delta)$, with a and δ being constants. We then have

$$\begin{aligned} I &= \langle E_\| E_\|^* \rangle (1+a^2) \\ Q &= \langle E_\| E_\|^* \rangle (1-a^2) \\ U &= \langle E_\| E_\|^* \rangle 2a\cos\delta \\ V &= \langle E_\| E_\|^* \rangle 2a\sin\delta \end{aligned} \tag{4.63}$$

which, after some algebra, yields the equality sign in (4.62). If light is not polarized, then $E_\|$ and $E_\perp$ are uncorrelated. This can be expressed as $E_\perp = aE_\| \exp[i\delta(t)]$, where $a \equiv 1$ and the phase difference $\delta(t)$ is a random function of time with the mean value of 0. Note that a must equal unity, otherwise light will be partially polarized. From (4.59), we have:

$$\begin{aligned} I &= \langle E_\| E_\|^* \rangle (1+a^2) = 2\langle E_\| E_\|^* \rangle \\ Q &= \langle E_\| E_\|^* \rangle (1-a^2) = 0 \\ U &= \langle E_\| E_\|^* \rangle 2a\,\langle \cos\delta(t) \rangle = 0 \\ V &= i\langle E_\| E_\|^* \rangle 2a\,\langle \sin\delta(t) \rangle = 0 \end{aligned} \tag{4.64}$$

because $a^2 = 1$ and both $\cos\delta(t)$ and $\sin\delta(t)$ are random variables with the means of 0. Unpolarized light is, for example, generated by thermal sources, such as an incandescent lamp.

It thus follows that a ratio:

$$\frac{\sqrt{Q^2+U^2+V^2}}{I} \tag{4.65}$$

is the measure of polarization of a light beam characterized by a Stokes vector $[I, Q, U, V]$.

4.3.2. Measuring the Stokes vector

To determine the Stokes vector of a light beam, we could start with measuring irradiances of the two linearly polarized components of the lightwave's electric vector: $I_\|$ and $I_\perp$, which would yield the I and Q parameters. We would need a linear polarizer and a detector that is insensitive to the direction of polarization of light as shown in Figure 4.21.

Following these measurements, we would still have a rather incomplete image of the polarization and would not be able to specify whether the polarization ellipse major axis is in the $(+\mathbf{E}_\|, +\mathbf{E}_\perp)$ and $(-\mathbf{E}_\|, -\mathbf{E}_\perp)$ or in $(-\mathbf{E}_\|, +\mathbf{E}_\perp)$ and

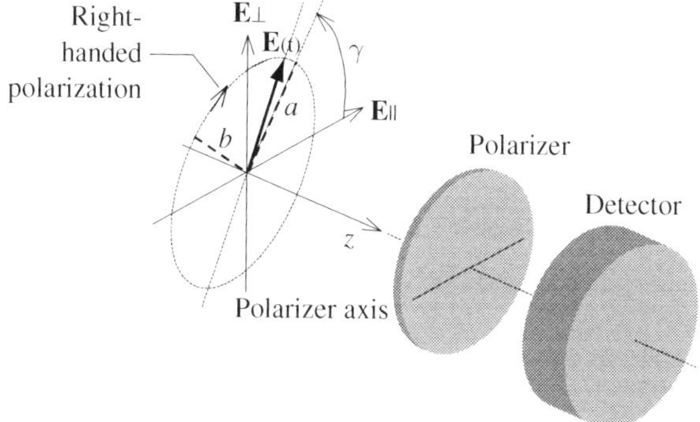

Figure 4.21. Measurement of irradiances of the linearly polarized components of an elliptically polarized light beam propagating along the z-axis. The reference frame is otherwise arbitrary unless set by the symmetry of the problem. Here the reference plane is the $\mathbf{E}_{||}z$ plane. The polarization is right-handed, as indicated by a small arrow at the polarization ellipse whose parameters are a (semimajor axis), b (semiminor axis), and γ (inclination angle). The ellipse is a projection onto a plane $\mathbf{E}_{||}\mathbf{E}_{\perp}$ of a helix traced in space by the tip of the electric vector $\mathbf{E}(t)$. The linear polarizer's axis is set for the measurement of the irradiance of the parallel-polarized component of the beam. A detector is assumed to be insensitive to the polarization state of light.

$(-\mathbf{E}_{||}, -\mathbf{E}_{\perp})$ quadrants, let alone determine the polarization ellipse parameters (a, b, γ). Additional measurements are clearly needed. From Figure 4.21, it follows that by measuring irradiance with the polarization axis of the linear polarizer rotated at $\gamma = +45°$ and then at $-45°$, with respect of the $\mathbf{E}_{||}z$ plane, we should at least be able to determine in which quadrants the major axis of the polarization ellipse is located. Rotation of the reference frame by 45° introduces a new frame with unit vectors $\mathbf{e}_{/}$ (for the $+45°$ axis) and $\mathbf{e}_{\backslash}$ (for the $-45°$ axis). The old unit vectors $\mathbf{e}_{||}$ and $\mathbf{e}_{\perp}$ are expressed in the new frame as follows:

$$\begin{aligned}\mathbf{e}_{||} &= \frac{1}{\sqrt{2}}(\mathbf{e}_{/} + \mathbf{e}_{\backslash}) \\ \mathbf{e}_{\perp} &= \frac{1}{\sqrt{2}}(\mathbf{e}_{/} - \mathbf{e}_{\backslash})\end{aligned} \tag{4.66}$$

Thus, the electric vector, $\mathbf{E}$, can now be expressed in the new reference frame as follows:

$$\begin{aligned}\mathbf{E} &= E_{||}\mathbf{e}_{||} + E_{\perp}\mathbf{e}_{\perp} \\ &= \frac{1}{\sqrt{2}}\left[(E_{||} + E_{\perp})\mathbf{e}_{/} + (E_{||} - E_{\perp})\mathbf{e}_{\perp}\right] \\ &= E_{/}\mathbf{e}_{/} + E_{\backslash}\mathbf{e}_{\backslash}\end{aligned} \tag{4.67}$$

It follows from the above equation and (4.58) that the irradiances of the beam components polarized along the new reference frame axes are:

$$\begin{aligned} I_{/} &= \langle E_{/} E_{/}^{*} \rangle \\ &= \frac{1}{2} \langle E_{0||}^{2} + 2E_{0||} E_{0\perp} \cos\delta + E_{0\perp}^{2} \rangle \\ I_{\backslash} &= \langle E_{\backslash} E_{\backslash}^{*} \rangle \\ &= \frac{1}{2} \langle E_{0||}^{2} - 2E_{0||} E_{0\perp} \cos\delta + E_{0\perp}^{2} \rangle \end{aligned} \tag{4.68}$$

where δ is the relative phase difference $\delta_{||} - \delta_{\perp}$. Rotation of the reference frame mixes the original components and introduces a phase difference. As pointed out by *Hecht* (1987), the phase difference implies that there are two non-zero orthogonal polarization components of the beam.

By subtracting the second equation in (4.68) from the first, we arrive at

$$\begin{aligned} I_{/} - I_{\backslash} &= 2 \langle E_{0||} E_{0\perp} \cos\delta \rangle \\ &= \mathrm{U} \end{aligned} \tag{4.69}$$

Incidentally, from the first of equations (4.68) and the alternative definitions of the two first components of the Stokes vector as $\mathrm{I}_{||} = E_{0||}{}^{2}$ and $\mathrm{I}_{\perp} = E_{0\perp}{}^{2}$, we also have:

$$\begin{aligned} \mathrm{U} &= 2\mathrm{I}_{/} - 2\mathrm{I}_{||} - 2\mathrm{I}_{\perp} \\ &= 2\mathrm{I}_{/} - 2\mathrm{I} \end{aligned} \tag{4.70}$$

which requires one measurement less (no need for the measurement with the linear polarizer axis at −45°).

However, even after these additional measurements, we still cannot unambiguously determine the relative phase δ because by knowing the $\cos\delta$, we cannot find whether δ is positive or negative. This prevents us from determining the handedness of the polarization ellipse. If we could delay each of the / and \ beam components' phases by $\pi/2$ (i.e., by ¼ wavelength), then the cosines (i.e., even functions of the relative phase, δ) in (4.68) would change to sines (odd functions). Thus, from the sign of the sine, we would be able to find the sign of the relative phase δ, and the measurement set would be complete. This can be easily achieved by inserting before the detector a quarter-wave plate, whose fast axis is oriented along the || axis, followed by a linear polarizer. We need to make two measurements, one with the polarizer oriented at +45°, as shown in Figure 4.22, and the other at −45°.

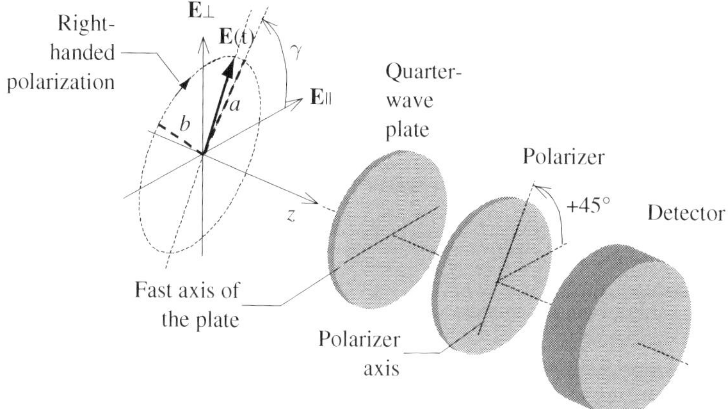

Figure 4.22. Measurement of the right-handed circularly polarized components of an elliptically polarized light beam propagating along the z-axis. The reference frame is arbitrary unless it needs to be set by the symmetry of the problem. Here the reference plane is the $\mathbf{E}_{||}z$ plane. The polarization is right-handed, as indicated by a small arrow at the polarization ellipse. Parameters of this ellipse are a (semimajor axis), b (semiminor axis), and γ (inclination angle). The ellipse is the projection onto a plane xy of a helix traced in space by the tip of the electric vector $\mathbf{E}$(t). The axes of the retarder (quarter-wave plate) and of the linear polarizer are set for the measurement of the irradiance of the right-handed component of the beam.

Incidentally, these measurements give the irradiances, I_R, and I_L, respectively of the right-handed and left-handed circularly polarized components of the beam. Indeed, a combination of a quarter-wave plate followed by a linear polarizer oriented at an angle of 45° to the fast axis of the quarter-wave plate is a circular analyzer. Note that a polarized beam of light can be expressed as a mixture of two linearly polarized orthogonal components *or* two circularly polarized components of opposite handedness (e.g., *Hecht* 1987). By orienting the linear polarizer at +45° and −45°, we can determine, respectively, the irradiance of the right-handed, I_R, and left-handed, I_L, circular polarization components.

These two latter measurements lead to the following equation

$$\begin{aligned} I_R - I_L &= -2\langle E_{0||}E_{0\perp}\sin\delta\rangle \\ &= V \end{aligned} \tag{4.71}$$

which completes the task of defining the Stokes vector parameters. Note that the parameter V can also be obtained as follows (similar to the alternative equation for U):

$$V = 2I_R - 2I \tag{4.72}$$

again reducing the number of required measurements by one.

4.3.3. The scattering matrix

As we have already noted that the volume scattering function is one of sixteen functions of the scattering angle, which are necessary for a complete description of the interaction of light with a scattering medium (e.g., *Bohren* and Huffman 1983). These 16 functions are elements of the scattering matrix, also referred to as the Mueller matrix. This matrix transforms the Stokes vector of the incident light beam into the Stokes vector of the scattered light.

The scattering matrix, **M**, with 4 × 4 elements, is defined as follows:

$$\mathbf{S}_{\mathrm{s}} = \frac{1}{(kr)^2}\mathbf{M}\mathbf{S}_{\mathrm{i}} \tag{4.73}$$

where $\mathbf{S}_x$[W m^{-2}] is the Stokes vector, with components denoted customarily by I, Q, U, and V, and where subscript x assumes values of s or i for scattered and incident light respectively, k [m^{-1}] is the wave number of the incident light, and r [m] is the distance from the scattering particle or a volume of the scattering medium to the detector of the scattered light.

Incidentally, the scattering of light is not exceptional in its calling for the use of such matrix. All linear incoherent interactions between an optical system and light can be described in the same manner. Moreover, interaction of light with an optical system composed of a series of components can be described by a Mueller matrix that is a product of Mueller matrices, with each matrix representing the effect of a component on the beam of light. This property is exploited in measuring the scattering matrix, as we will outline it later in this section.

4.3.4. The form of the scattering matrix for media with various degrees of symmetry

The scattering matrix of an isotropic but otherwise arbitrary medium has 16 independent elements. By considering symmetry of the scattering medium, it can be proven (e.g., *Perrin* 1942) that an isotropic asymmetric scattering medium has a scattering matrix with only ten independent elements:

$$\begin{bmatrix} a_1 & b_1 & -b_3 & b_5 \\ b_1 & a_2 & -b_4 & b_6 \\ b_3 & b_4 & a_3 & b_2 \\ b_5 & b_6 & -b_2 & a_4 \end{bmatrix} \tag{4.74}$$

By an isotropic symmetric scattering medium, we mean a medium, such that every spherical volume of that medium has a center of symmetry and that every plane passing through that center is a plane of symmetry. For such a medium, the number of independent elements is reduced to six:

$$\begin{bmatrix} a_1 & b_1 & 0 & 0 \\ b_1 & a_2 & 0 & 0 \\ 0 & 0 & a_3 & b_2 \\ 0 & 0 & -b_2 & a_4 \end{bmatrix} \tag{4.75}$$

Finally, for an isotropic medium made of a suspension of homogeneous spherical particles, the matrix has only four independent elements:

$$\begin{bmatrix} a_1 & b_1 & 0 & 0 \\ b_1 & a_1 & 0 & 0 \\ 0 & 0 & a_3 & b_2 \\ 0 & 0 & -b_2 & a_3 \end{bmatrix} \tag{4.76}$$

Hence, the difference $M_{11} - M_{22} = a_1 - a_2$ is a measure of the non-sphericity of the particles. If the particles are small in relation to the wavelength of light or if their refractive index is close to that of the surrounding medium, then $M_{34} = M_{43} \cong 0$.

A more detailed exposition of the effect of various symmetries of the scattering medium on the form of the Mueller matrix can be found in *van de Hulst* (1957).

4.3.5. Derivation of the scattering matrix

We will close the discussion of the scattering matrix with a brief explanation of how its form can be derived from the relationships between the electric fields of the incident and scattered light. We will follow the approach of van de Hulst. To simplify the matter, we will consider a homogeneous sphere illuminated by a polarized beam of light with an electric vector represented by $[E_{||}, E_{\perp}]$. The symmetry implies that it cannot introduce cross-polarization in the scattered light, i.e. we must have:

$$\begin{bmatrix} E_{||,s} \\ E_{\perp,s} \end{bmatrix} = \frac{\exp[-ik(r-z)]}{ikr} \begin{bmatrix} A_{11} & 0 \\ 0 & A_{22} \end{bmatrix} \begin{bmatrix} E_{||,i} \\ E_{\perp,i} \end{bmatrix} \tag{4.77}$$

Thus, with k, r, and z being constant, the relationship between the incident and scattered field vectors is:

$$\begin{bmatrix} E_{||,s} \\ E_{\perp,s} \end{bmatrix} = c \begin{bmatrix} A_{11} E_{||,i} \\ A_{22} E_{\perp,i} \end{bmatrix} \tag{4.78}$$

where $c = \exp[-ik(r-z)]/(ikr)$ is a constant. On the other hand, from (4.73) and (4.59) we have

$$\begin{bmatrix} I_s \\ Q_s \\ U_s \\ V_s \end{bmatrix} = |c|^2 \begin{bmatrix} M_{11}I_i + M_{12}Q_i + M_{13}U_i + M_{14}V_i \\ M_{21}I_i + M_{22}Q_i + M_{23}U_i + M_{24}V_i \\ M_{31}I_i + M_{32}Q_i + M_{33}U_i + M_{34}V_i \\ M_{41}I_i + M_{42}Q_i + M_{43}U_i + M_{44}V_i \end{bmatrix} \quad (4.79)$$

Consider the I_s component. From (4.59) and (4.78), we have (neglecting the constant factor):

$$\begin{aligned} I_s &= \langle E_{\|,s}E^*_{\|,s} + E_{\perp,s}E^*_{\perp,s} \rangle \\ &= A_{11}A^*_{11}\langle E_{\|,i}E^*_{\|,i}\rangle + A_{22}A^*_{22}\langle E_{\perp,i}E^*_{\perp,i}\rangle \end{aligned} \quad (4.80)$$

However, from (4.59) and (4.79), we similarly have for I_s:

$$\begin{aligned} I_s &= \langle E_{\|,s}E^*_{\|,s} + E_{\perp,s}E^*_{\perp,s} \rangle \\ &= M_{11}\langle E_{\|,i}E^*_{\|,i}\rangle + M_{11}\langle E_{\perp,i}E^*_{\perp,i}\rangle + M_{12}\langle E_{\|,i}E^*_{\|,i}\rangle - M_{12}\langle E_{\perp,i}E^*_{\perp,i}\rangle \\ &\quad + M_{13}\langle E_{\|,i}E^*_{\perp,i}\rangle + M_{13}\langle E_{\perp,i}E^*_{\|,i}\rangle iM_{14}\langle E_{\|,i}E^*_{\perp,i}\rangle - iM_{14}\langle E_{\perp,i}E^*_{\|,i}\rangle \\ &= (M_{11}+M_{12})\langle E_{\|,i}E^*_{\|,i}\rangle + (M_{11}-M_{12})\langle E_{\perp,i}E^*_{\perp,i}\rangle \\ &\quad + (M_{13}+iM_{14})\langle E_{\|,i}E^*_{\perp,i}\rangle + (M_{13}-iM_{14})\langle E_{\perp,i}E^*_{\|,i}\rangle \end{aligned} \quad (4.81)$$

By comparing (4.80) and (4.81) and noting that in the second line of (4.79) the neglected constant c has been squared ($|c|^2$), i.e., assumed the same form as it would have assumed in the second and third equations of (4.81), had it not been neglected. Thus, we can cancel it altogether and obtain

$$\begin{aligned} M_{11} + M_{12} &= |A_{11}|^2 \\ M_{11} - M_{12} &= |A_{22}|^2 \\ M_{13} + iM_{14} &= 0 \\ M_{13} - iM_{14} &= 0 \end{aligned} \quad (4.82)$$

and

$$M_{11} = \frac{|A_{11}|^2 + |A_{22}|^2}{2}$$

$$M_{12} = \frac{|A_{11}|^2 - |A_{22}|^2}{2} \tag{4.83}$$

$$M_{13} = 0$$

$$M_{14} = 0$$

For homogeneous spheres, as discussed in Chapter 3, $A_{11} = S_2$ and $A_{22} = S_1$, where S_1 and S_2 are the Mie amplitude functions. Thus, the first equation in (4.83) reduces to $M_{11} = (|S_1|^2 + |S_2|^2)/2$. This derivation paves the way for explaining other elements of the Mueller matrix in terms of the elements of the amplitude matrix **A** in (4.77), and we shall not continue this simple but tedious process.

4.3.6. Significance of the various elements of the scattering matrix

In an isotropic symmetric scattering medium, the scattering matrix has the form of (4.75). Thus, from (4.79), it can be seen that element M_{11} of the scattering matrix characterizes the scattered irradiance for unpolarized light ($Q = U = V = 0$), as does the volume scattering function, β. Thus,

$$\beta = \frac{N}{k^2} M_{11} \tag{4.84}$$

where N is the number concentration of particles [length^{-3}] and k is the wave number of the incident light in the medium surrounding the particles. All particles are assumed to be identical. Otherwise, the multiplication by N must be replaced by summation over all particles (or integration of M_{11}/k^2, weighed by the particle size distribution, over a particle size range).

Element M_{12} describes the linear cross-polarization introduced into the scattered light by particles of an isotropic symmetric scattering medium. Indeed, if $M_{12} = 0$ for such a medium, then $I_s = M_{11} I_i$ and $Q_s = M_{22} Q_i$, so the polarization of the incident linearly polarized light (see Table 4.1) is preserved. Incidentally, equations (4.83) imply that $M_{12}(\theta = 0) = M_{12}(\theta = \pi) = 0$ for spheres, as can be inferred from equations (3.13) and (3.14) which indicate that $S_1(0) = S_2(0)$ and $S_1(\pi) = S_2(\pi)$.

Element M_{41} characterizes the optical activity of the scattering medium, i.e., the degree by which the medium introduces circular polarization into the scattered light. Indeed, consider incident unpolarized light, for which the Stokes vector parameters $Q_i = U_i = V_i = 0$. A Stokes vector for light scattered by a medium with a scattering matrix with a non-zero element M_{41} will have a non-zero element $V_s = M_{41} I_i$. For an isotropic homogeneous medium of homogeneous optically-inactive spheres, element M_{41} is zero. Note that the Stokes vector for totally circularly polarized light has only I and V as non-zero elements (Table 4.1).

This can be easily verified by expressing the electric vector field components as $E_{||} = E_0 \exp(-i\phi t)$ and $E_{\perp} = E_0 \exp(-i\pi) \exp(-i\phi t)$, which describes a rotating electric vector **E**, and by using (4.60).

Element M_{34} links the linear polarization of the incident light with circular polarization of the scattered light. It was found to strongly depend on the size of the particle, as well as on the magnitude and distribution of the complex refractive index within the particle (*Bickel* et al. 1976).

4.3.7. Relationships between the elements of the scattering matrix

The 16 elements of the scattering matrix of a single particle are not independent. (e.g., *Hovenier* et al. 1986). Relationships between these elements which involve either squares of the elements or products of these elements are concisely reviewed by *Hovenier* (1999). These relationships can be used for testing the correctness of calculations of measurements of the scattering matrix (*Hovenier* and van der Mee 1996, *Fry* and Kattawar 1981). We will quote only the relationships for the scattering matrix of a suspension below (*Hovenier* 1999, *Fry* and Kattawar 1981) and refer the interested reader to one of the references cited above for single-particle relationships.

$$\begin{aligned} (M_{11} \pm M_{22})^2 - (M_{12} \pm M_{21})^2 &\geq (M_{33} \pm M_{44})^2 + (M_{34} \mp M_{43})^2 \\ (M_{11} \pm M_{12})^2 - (M_{21} \pm M_{22})^2 &\geq (M_{31} \pm M_{32})^2 + (M_{41} \pm M_{42})^2 \\ (M_{11} \pm M_{21})^2 - (M_{12} \pm M_{22})^2 &\geq (M_{13} \pm M_{23})^2 + (M_{14} \pm M_{24})^2 \end{aligned} \tag{4.85}$$

Note that for homogeneous spheres, $M_{13} = M_{14} = M_{23} = M_{24} = M_{31} = M_{32} = M_{41} = M_{42} = 0$ and $M_{11} = M_{22}$, $M_{12} = M_{21}$, $M_{33} = M_{44}$, and $M_{34} = -M_{43}$. Thus equations (4.85) become:

$$\begin{aligned} &{M_{11}}^2 - {M_{12}}^2 \geq {M_{33}}^2 + {M_{34}}^2 \\ &(M_{11} \pm M_{12})^2 - (M_{21} \pm M_{22})^2 \geq 0 \end{aligned} \tag{4.86}$$

4.3.8. Measurements

The basic principle of measurement of the scattering matrix is shown in Figure 4.23. One simply uses various combinations of linear and circular polarizers in the incident and scattered light paths at selected orientations and handedness, to determine certain linear combinations of the scattering matrix elements (see the text). These combinations can then be solved for individual elements of the scattering matrix. This basic process is rather time consuming (in older references, measurement times on the order of hours have been reported, e.g., *Kadyshevich* et al. 1971) and is not feasible if either the particle suspension changes rapidly

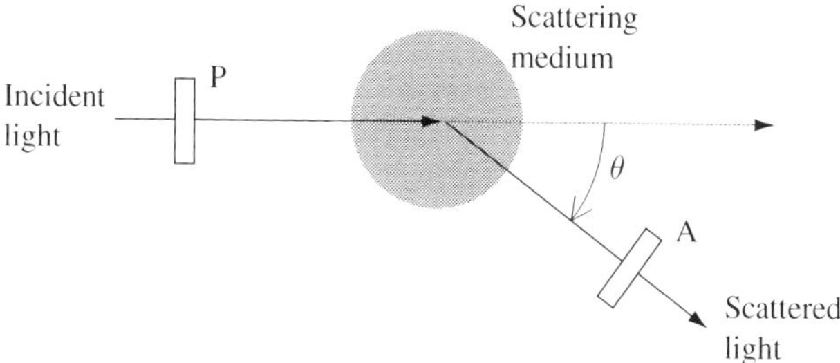

Figure 4.23. Measurement of the scattering matrix. P is a polarizer, an optical element which modifies polarization of the incident light. A is an analyzer, i.e., an optical element that selects certain polarization state of the scattered light. By using various combinations of linear and circular polarizers at selected orientations, one can determine corresponding linear combinations of the scattering matrix elements (see the text) which can be solved for individual elements of the scattering matrix.

or the allotted measurement time is very short, such as in flow-cytometric applications. However, this basic procedure gives a firm insight into the measurement process, and we shall use it to outline how individual elements of the scattering matrix of a scattering medium or a particle can be determined this way. We will then outline faster approaches.

In considering the basic measurement procedure, we should first note that, according to (4.73) and the principle (mentioned earlier) that the incoherent interaction of light with an optical system is described by a product of the Mueller matrices of components of this system, the Stokes vector of the scattered light $\mathbf{S}^{\mathrm{s}}$ can, aside from the constant $1/(kr)^2$, be expressed as follows:

$$\mathbf{S}^{\mathrm{s}} = \mathbf{M}_{\mathrm{A}}\mathbf{M}\mathbf{M}_{\mathrm{P}}\mathbf{S}^{\mathrm{i}} \tag{4.87}$$

where $\mathbf{M}_{\mathrm{A}}$, $\mathbf{M}$, and $\mathbf{M}_{\mathrm{P}}$ are the Mueller matrices of the analyzer, scattering medium, and polarizer, and $\mathbf{S}^{\mathrm{i}}$ is the Stokes vector an incident beam of unpolarized light.

Let us first consider the simplest case, no polarizer nor analyzer, and unpolarized incident light. Thus, from (4.87) we have

$$\begin{bmatrix} I_{\mathrm{u}} \\ Q_{\mathrm{u}} \\ U_{\mathrm{u}} \\ V_{\mathrm{u}} \end{bmatrix}^{s} = \begin{bmatrix} M_{11} & M_{12} & M_{13} & M_{14} \\ M_{21} & M_{22} & M_{23} & M_{24} \\ M_{31} & M_{32} & M_{33} & M_{34} \\ M_{41} & M_{42} & M_{43} & M_{44} \end{bmatrix} \begin{bmatrix} I_{\mathrm{u}} \\ 0 \\ 0 \\ 0 \end{bmatrix}^{i} \tag{4.88}$$

where the subscript ‘u’ means unpolarized light. This leads to

$$I_{\mathrm{uu}}^{\mathrm{s}} = M_{11} I_{\mathrm{u}}^{\mathrm{i}} \tag{4.89}$$

where the subscript 'uu' denotes measurements performed with no polarizer (the first u) and no analyzer (the second u). Thus,

$$M_{11} = \frac{I^{\mathrm{s}}_{\mathrm{uu}}}{I^{\mathrm{i}}_{\mathrm{u}}} \tag{4.90}$$

Let us now consider the combination of a linear polarizer parallel to the scattering plane in the incident light path and no analyzer at all in the scattered light path. The Mueller matrix of an ideal linear polarizer with such an orientation and with no attenuation is (e.g., *Bohren* and Huffman 1983):

$$\frac{1}{2}\begin{bmatrix} 1 & 1 & 0 & 0 \\ 1 & 1 & 0 & 0 \\ 0 & 0 & 0 & 0 \\ 0 & 0 & 0 & 0 \end{bmatrix} \tag{4.91}$$

The Mueller matrix corresponding to an identity transformation (no analyzer at all) is the unity matrix. It is neglected here. From (4.87), we thus have

$$\begin{aligned} \begin{bmatrix} I_{\|\mathrm{u}} \\ Q_{\|\mathrm{u}} \\ U_{\|\mathrm{u}} \\ V_{\|\mathrm{u}} \end{bmatrix}^{s} &= \begin{bmatrix} M_{11} & M_{12} & M_{13} & M_{14} \\ M_{21} & M_{22} & M_{23} & M_{24} \\ M_{31} & M_{32} & M_{33} & M_{34} \\ M_{41} & M_{42} & M_{43} & M_{44} \end{bmatrix} \frac{1}{2} \begin{bmatrix} 1 & 1 & 0 & 0 \\ 1 & 1 & 0 & 0 \\ 0 & 0 & 0 & 0 \\ 0 & 0 & 0 & 0 \end{bmatrix} \begin{bmatrix} I_{\mathrm{u}} \\ 0 \\ 0 \\ 0 \end{bmatrix}^{i} \\ &= \begin{bmatrix} M_{11} & M_{12} & M_{13} & M_{14} \\ M_{21} & M_{22} & M_{23} & M_{24} \\ M_{31} & M_{32} & M_{33} & M_{34} \\ M_{41} & M_{42} & M_{43} & M_{44} \end{bmatrix} \begin{bmatrix} I_{\mathrm{u}}/2 \\ I_{\mathrm{u}}/2 \\ 0 \\ 0 \end{bmatrix}^{i} \end{aligned} \tag{4.92}$$

The form of the Stokes vector for the parallel-polarized light (the rightmost term in the second line of the above equation) agrees with that of equations (4.59) and Table 4.1: $[I = I_{\|}, Q = I_{\|}, 0, 0]^{\mathrm{T}}$, where the superscript T stands for the transposed vector and $I_{\|} = I_{\mathrm{u}}/2$. From (4.92) we have:

$$I^{\mathrm{s}}_{\|\mathrm{u}} = \frac{M_{11} + M_{12}}{2} I^{\mathrm{i}}_{\mathrm{u}} \tag{4.93}$$

Of course we could have:

$$M_{12} = 2I^{\mathrm{s}}_{\|\mathrm{u}}/I^{\mathrm{i}}_{\mathrm{u}} - M_{11} \tag{4.94}$$

or we can perform another measurement with the linear analyzer oriented perpendicular to the scattering plane and obtain

$$I^{\mathrm{s}}_{\perp\mathrm{u}} = \frac{M_{11} - M_{12}}{2} I^{\mathrm{i}}_{\mathrm{u}} \tag{4.95}$$

This way, the M_{12} element can also be obtained as follows:

$$M_{12} = (I^{s}_{||u} - I^{s}_{\perp u})/I^{i}_{u} \tag{4.96}$$

Finally, let us consider the measurement of the M_{22} matrix element. We would first use a linear polarizer and a linear analyzer, both parallel to the scattering plane. The use of (4.91) leads to the following matrix equation:

$$\begin{bmatrix} I_{||\,||} \\ Q_{||\,||} \\ U_{||\,||} \\ V_{||\,||} \end{bmatrix}^{s} = \frac{1}{2}\begin{bmatrix} 1 & 1 & 0 & 0 \\ 1 & 1 & 0 & 0 \\ 0 & 0 & 0 & 0 \\ 0 & 0 & 0 & 0 \end{bmatrix} \begin{bmatrix} M_{11} & M_{12} & M_{13} & M_{14} \\ M_{21} & M_{22} & M_{23} & M_{24} \\ M_{31} & M_{32} & M_{33} & M_{34} \\ M_{41} & M_{42} & M_{43} & M_{44} \end{bmatrix} \times \frac{1}{2}\begin{bmatrix} 1 & 1 & 0 & 0 \\ 1 & 1 & 0 & 0 \\ 0 & 0 & 0 & 0 \\ 0 & 0 & 0 & 0 \end{bmatrix} \begin{bmatrix} \mathrm{I}_{u} \\ 0 \\ 0 \\ 0 \end{bmatrix}^{i}$$

$$= \frac{1}{2}\begin{bmatrix} M_{11}+M_{21} & M_{12}+M_{22} & 0 & 0 \\ M_{11}+M_{21} & M_{12}+M_{22} & 0 & 0 \\ 0 & 0 & 0 & 0 \\ 0 & 0 & 0 & 0 \end{bmatrix} \begin{bmatrix} I_{u}/2 \\ I_{u}/2 \\ 0 \\ 0 \end{bmatrix}^{i} \tag{4.97}$$

that yields

$$I^{s}_{||\,||} = \frac{M_{11}+M_{21}+M_{12}+M_{22}}{4} I^{i}_{u} \tag{4.98}$$

It is clear that we still need to perform a few additional measurements. Let us try a perpendicular polarizer and perpendicular analyzer, for which the Mueller matrix is as follows (e.g., *Bohren* and Huffman 1983):

$$\frac{1}{2}\begin{bmatrix} 1 & -1 & 0 & 0 \\ -1 & 1 & 0 & 0 \\ 0 & 0 & 0 & 0 \\ 0 & 0 & 0 & 0 \end{bmatrix} \tag{4.99}$$

This yields:

$$I^{s}_{\perp\perp} = \frac{M_{11}-M_{21}-M_{12}+M_{22}}{4} I^{i}_{u} \tag{4.100}$$

From (4.98) and (4.100), we have

$$I^{s}_{||\,||} + I^{s}_{\perp\,\perp} = \frac{M_{11}+M_{22}}{2} I^{i}_{u} \tag{4.101}$$

Short of using the value, we have already determined, we still need additional measurements to remove M_{11} from the equation. We could try two additional measurements with mixed orientations of the polarizer and analyzer: ($||$, $\perp$) and ($\perp$, $||$). These two measurements yield:

$$I^{s}_{|| \perp} + I^{s}_{\perp ||} = (1/2)(M_{11} - M_{22}) \tag{4.102}$$

that can be used for cross-checking of the formerly obtained M_{11} matrix element [see (4.90)]. Finally,

$$M_{22} = 2[(I^{s}_{|| ||} + 1/I^{i}_{u \perp \perp}) - (I^{s}_{|| \perp} + I^{s}_{\perp ||})] \tag{4.103}$$

Other matrix elements can be determined in a similar manner. The complete prescription for the measurement of the scattering matrix (after *Hielscher* et al. 1997, modified to be consistent with our notation) is shown in (4.104):

$$\frac{1}{I^{i}_{u}} \begin{bmatrix} I^{s}_{uu} & I^{s}_{||u} - I^{s}_{\perp u} & I^{s}_{/u} - I^{s}_{\backslash u} & I^{s}_{Ru} - I^{s}_{Lu} \\ I^{s}_{u||} - I^{s}_{u\perp} & \begin{matrix} 1[(I^{s}_{|| ||} + I^{s}_{\perp \perp}) \\ -(I^{s}_{|| \perp} + I^{s}_{\perp ||})] \end{matrix} & \begin{matrix} 1[(I^{s}_{/ ||} + I^{s}_{\backslash \perp}) \\ -(I^{s}_{/ \perp} + I^{s}_{\backslash ||})] \end{matrix} & \begin{matrix} 1[(I^{s}_{L ||} + I^{s}_{R \perp}) \\ -(I^{s}_{R \perp} + I^{s}_{L ||})] \end{matrix} \\ I^{s}_{u/} - I^{s}_{u\backslash} & \begin{matrix} 1[(I^{s}_{|| /} + I^{s}_{\perp \backslash}) \\ -(I^{s}_{|| \backslash} + I^{s}_{\perp /})] \end{matrix} & \begin{matrix} 1[(I^{s}_{//} + I^{s}_{\backslash\backslash}) \\ -(I^{s}_{/ \backslash} + I^{s}_{\backslash /})] \end{matrix} & \begin{matrix} 1[(I^{s}_{L/} + I^{s}_{R \backslash}) \\ -(I^{s}_{R\backslash} + I^{s}_{L/})] \end{matrix} \\ I^{s}_{u R} - I^{s}_{u L} & \begin{matrix} 1[(I^{s}_{|| L} + I^{s}_{\perp R}) \\ -(I^{s}_{|| R} + I^{s}_{\perp L})] \end{matrix} & \begin{matrix} 1[(I^{s}_{/L} + I^{s}_{\backslash R}) \\ -(I^{s}_{\backslash R} + I^{s}_{/ L})] \end{matrix} & \begin{matrix} 1[(I^{s}_{L L} + I^{s}_{R R}) \\ -(I^{s}_{RL} + I^{s}_{LR})] \end{matrix} \end{bmatrix} \tag{4.104}$$

Let us now discuss faster methods to obtain the Mueller matrix elements. These methods rely on an observation that the irradiance of a light beam passing through a combination of retarders and polarizers depends on the relative angles between the optical axes of these components and a reference frame. Consider an optical system composed of a retarder (say a quarter-wave plate) that introduces a phase difference (delay) δ between the polarized components of the beam oriented along the fast and slow axes of the retarder, and a liner polarizer. The irradiance, E, of the beam that passes through such a system can be described by the following equation (*Berry* et al. 1977):

$$\begin{aligned} E(\alpha, \chi, \delta) = \frac{1}{2}\left[\mathrm{I} + \left(\frac{Q}{2}\cos 2\alpha + \frac{U}{2}\sin 2\alpha\right)(1 + cos\delta)\right] \\ + \frac{1}{2}[V\sin\delta + \sin(2\alpha - 2\chi)] \\ + \frac{1}{4}[(Q\cos 2\alpha - U\sin 2\alpha)\cos 4\chi \\ + (Q\sin 2\alpha + U\cos 2\alpha)\sin 4\chi](1 - \cos\delta) \end{aligned} \tag{4.105}$$

where α is the angle that the transmission axis of the polarizer makes with the positive parallel axis of the reference frame and χ is the angle the fast axis of the retarder makes with the positive parallel axis of the reference frame. If the retarder is rotated at a constant angular velocity, ω, which is a preferable solution to rotating the polarizer as problems related to the detector polarization sensitivity can be avoided, then the angle χ can be expressed as a function of time, t:

$$\chi = \omega\, t \tag{4.106}$$

On substitution in equation (4.105), this yields a periodic function of time which can be expressed as follows:

$$\mathrm{E}(\chi) = C_0 + C_2 \cos 2\chi + C_4 \cos 4\chi + S_2 \sin 2\chi + S_4 \sin 4\chi \tag{4.107}$$

which is a Fourier series, whose coefficients can be obtained via Fourier analysis of the $\mathrm{E}(\chi)$ time series. The Stokes parameters can then be conveniently retrieved from the Fourier coefficients values by comparing these coefficients with those in (4.107). Fast measurement of the Stokes parameters opens a possibility to rapidly measure the Mueller matrix of a scattering medium or any other optical system which can be described with that matrix.

Azzam (1997) reviews this and other methods of rapid measurements of the scattering matrix. *Mujat* and Dogariu (2001), who provide a concise review of the methods of Mueller matrix measurements, note that it was *Azzam* (1978) who first proposed a method of simultaneous measurement of the whole Mueller matrix via the Fourier analysis of a time-dependent waveform obtained in similar fashion to that just described. The polarizing and analyzing optics would consist of stationary parallel polarizers and of two quarter-wave retarders synchronously rotating at angular velocities of ω and 5ω.

Thompson et al. (1980) developed a nephelometer for measuring the whole Mueller matrix of single particles and suspensions by using Pockel cells, voltage-controlled retarders, modulated at four different frequencies. The modulated signals were synchronously detected with 16 lock-in amplifiers. A similar instrument was later used by *Voss* and Fry (1984) to measure the complete scattering matrix of ocean water in under 2 minutes.

Mujat and Dogariu (2001) have recently developed a much simpler method relying on a stationary linear polarizer and two liquid crystal retarders in the polarization unit, and a single photoelastic modulator and a division of amplitude with two stationary linear polarizers in the analysis unit. That system is capable of completing the measurement of the entire Mueller matrix in about 50 ms, which opens the possibility of analyzing rapidly varying suspensions and other optical systems.

4.4. Light scattering data for natural waters

4.4.1. The volume scattering function of seawater

Since 1945, when possibly the first report on light scattering by distilled and natural waters was published (*Hulburt* 1945), hundreds of measurements of the volume scattering function have been performed *in situ* and *in vitro* in many regions of the world's ocean, seas, and lakes. The results of volume scattering function measurements were also reviewed by *Kirk* (1983a), *Jerlov* (1976), *Kullenberg* (1974), *Morel* (1973), *Jerlov* (1968, 1963), and *Duntley* (1963).

Sources of experimental data on the scattering function of seawater are listed in the Appendix section (Table A.2). Although we attempted to compile a reasonably comprehensive list, we do not claim it to be exhaustive. Extensive computer-readable data collections and a graphical atlases of volume scattering functions of seawater have been recently compiled (*Jonasz* 1996, 1992).

Selected volume scattering functions for water bodies whose optical properties span a broad range are shown in Figure 4.24. This figure illustrates the following general trends of the scattering function of seawater:

1. The scattering function is typically a smooth function of the scattering angle.
2. The magnitude of the scattering function may vary greatly from one water body to another, spanning an impressive range of about 8 decades over the experimentally accessed angle range.
3. The form of the function tends to vary less than the magnitude.
4. The scattering function is very steep in the forward-scattering range (0–90°). At scattering angles of less than 5° the log-log scale slope is on the order of 1.5 and more (Fig. 4.27). This slope typically increases to ~2.5 and more in the range of ~20 to ~90°. At even greater angles the slope generally decreases. That slope generally decreases with increasing magnitude of the function. The steep forward-scattering slope accounts for the strong asymmetry of the scattering function.
5. The minimum of the volume scattering function, if one exists, typically moves from the vicinity of 100° in clear waters toward 180° in turbid waters. A moderate-to-steep rise in the backward direction typically follows such a minimum, although in some (especially in turbid) waters, the function decreases monotonically with increasing angle.

The similarity in the form of the scattering functions prompted definitions of typical, or average, scattering functions (*Mobley* 1994, *Morel* 1973) of natural waters. This similarity is all the more striking in that it cuts across various water bodies, seasons, and wavelengths. We re-examined a large body of historical data on the scattering function and significant recent additions and propose updated versions of such "average" functions that are typical of clear and turbid natural waters. We do this by calculating geometric averages of the particle scattering functions from a data collection of scattering functions compiled by *Jonasz* (1996,

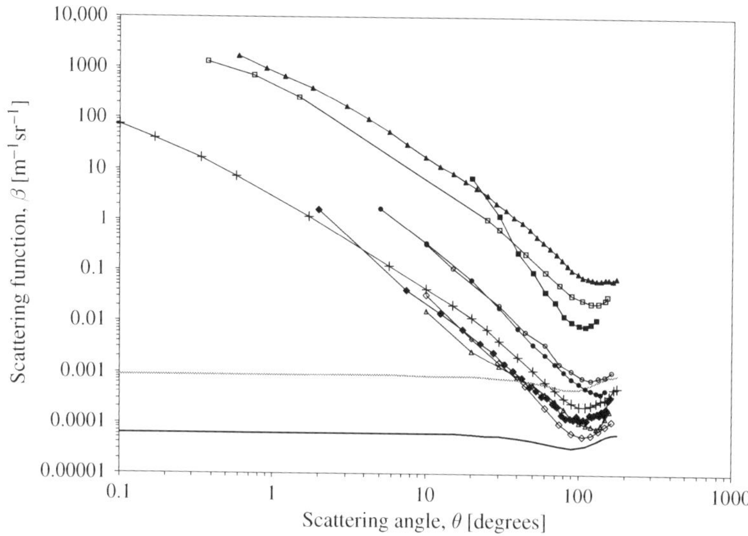

Figure 4.24. Selected volume scattering functions for water bodies whose optical properties span a broad range. From top to bottom: ▲ Northwestern Atlantic off New Jersey, USA (*Lee* et al. 2003, #48), ■ Charles River, Massachusetts, USA (*Beardsley* 1968), □ the mouth of Back River, Virginia, 29 Aug 1979, 800 nm (*Whitlock* et al. 1981), ∘ North Atlantic, 5 May 1986, 40 °N 64 °W, depth 150 m, 633 nm (*Jonasz* 1991b), • Baltic Sea, surface layer, June 1977, average of 12 samples, 633 nm (*Jonasz* and Prandke 1986), + Atlantic off Bahamas, 25 °N 78 °W, depth 1880 m, 510 nm (*Petzold* 1972), ◇ Sargasso Sea 27 °N 63 °W, depth 10 to 15 m, 655 nm (*Kullenberg* 1968), △ Drake Passage 58 °S 63 °W, depth 400 m, 655 nm (*Kullenberg* 1984), ♦ Mediterranean Sea, 28 May 1998, 35.9 °N 28.1 °E, depth 40 m, 520 nm (*Mankovsky* and Haltrin 2002a), *thick lines: gray* – pure seawater at 400 nm, equation (4.127), *black* – pure water at 700 mm, equation (4.127).

1992). By focusing on the particle scattering, we removed to the first order the effect of the wavelength on the scattering function shape brought about by the wavelength selectivity of light scattering by pure seawater. We will discuss this effect shortly. The residual wavelength selectivity is due (as we already discussed in Chapter 3) to the dispersion of the refractive index of the particles, as well as to the absorption by the particles (e.g., *Babin* et al. 2003).

In selecting data sets for the clear and turbid averages, we initially divided the data set into an open ocean and a coastal subset based on the measurement location. When analyzing a ratio of the total scattering function to the scattering function of water at the same wavelength, it became clear that the minimum of that ratio, R_{TW}, for the open ocean data sets is much more narrowly distributed $[\langle R_{TW}\rangle = 1.70,\ \sigma(R_{TW}) = 0.66]$ than that for the nominally coastal data sets. This led us to select an arbitrary value of the maximum ratio at 3, which is approximately equal to the average value of that ratio plus two standard deviations and to name all data sets with the minimum $R_{TW} < 3$ as "clear water" data sets. All other data sets were classified as "turbid water" data sets. This classification led to the derivation of the average phase functions for the clear and turbid seawater shown in Table 4.2.

We have chosen the geometric average to more equally represent all scattering functions that differ by an order of magnitude. Nevertheless, the particle phase function for clear seawater almost coincides with the "typical" particle phase function obtained by *Morel* (1973, his Table II) as shown in Figure 4.27.

The average functions for clear and turbid seawater are shown in Figure 4.25 (in log-log scale that shows the small-angle detail) and Figure 4.26 (in linear-linear scale that shows the large angle detail). These average functions have been approximated with the FF formula (*Fournier* and Forand 1994, with modifications by *Forand* and Fournier 1999), which we will discuss shortly, to yield approximation parameters shown in Table 4.3.

4.4.2. *Integral characteristics of the scattering function*

4.4.2.1. *Scattering coefficients*

The scattering coefficient, b, varies in natural waters over a wide range. The absolute minimum in the visible is that set by the scattering coefficient of water (seawater) that reaches down to $\sim 0.005\,\mathrm{m}^{-1}$ (pure water) and $\sim 0.007\,\mathrm{m}^{-1}$ (pure seawater at $S = 35$) at 700 nm. The addition of particles significantly increases that minimum and extends the range to between 0.008 and $9.3\,\mathrm{m}^{-1}$ in a wavelength range of 515 to 550 nm (*Haltrin* et al. 2003) for waters ranging from clear open ocean waters to turbid coastal areas. The upper limit of that range may approach $10\,\mathrm{m}^{-1}$ in turbid inland waters such as those of the North American Great Lakes (*Bukata* et al. 1980). A recent survey of the particle-related component of the scattering coefficient in coastal waters off Europe (*Babin* et al. 2003) extend the upper limit (at a wavelength of 555 nm) to $\sim 30\,\mathrm{m}^{-1}$.

By analyzing 101 scattering functions with the scattering coefficient in that range, Haltrin and colleagues found a relatively high correlation ($r^2 = 0.88$) between the scattering and backscattering coefficients:

$$b_b = b_{bw} + 0.00618(b - b_w) + 0.00322(b - b_w)^2 \qquad (4.108)$$

where the subscript 'w' denotes the scattering coefficient of pure seawater. Note that the scattering coefficients, as referred to in (4.108), are actually non-dimensional quantities expressed relative to the scattering coefficient of $1\,\mathrm{m}^{-1}$. Also note that in a smaller range of the scattering coefficient, the correlation between b_b and b may be significantly smaller.

The scattering coefficient of particles is generally found to increase linearly with the particle mass load, although the slope of that relationship, i.e., the particle mass-specific scattering coefficient, denoted customarily as $b_p{}^m$ was found to vary

Table 4.2. The particle scattering phase functions typical of the "clear" and "turbid" seawater and related scattering coefficients and average cosines for the particle volume scattering functions shown in Figure 4.25 and Figure 4.26.

	Clear seawater[a]	**Turbid seawater**[b]		**Clear seawater**[a]	**Turbid seawater**[b]
θ **[degrees]**	$p(\theta)$	$p(\theta)$	θ **[degrees]**	$p(\theta)$	$p(\theta)$
0.2	89.6	687.	80	0.000177	0.00289
0.5	13.9	317.	85	0.000152	0.00247
1	6.38	83.4	90	0.000118	0.00214
2	1.94	21.4	95	0.000106	0.00192
5	0.275	2.42	100	0.000113	0.00175
10	0.0700	0.630	105	0.000111	0.00159
15	0.0274	0.274	110	0.0000995	0.00148
20	0.0147	0.151	115	0.0000971	0.00139
25	0.00801	0.0799	120	0.0000939	0.00132
30	0.00480	0.0484	125	0.0000923	0.00128
35	0.00308	0.0309	130	0.0000926	0.00124
40	0.00205	0.0209	135	0.0000951	0.00116
45	0.00142	0.0149	140	0.000106	0.00122
50	0.00100	0.0110	145	0.000129	0.00129
55	0.000733	0.00829	150	0.000125	0.00130
60	0.000541	0.00642	155	0.0000985	0.00126
65	0.000400	0.00508	160	0.000120	0.000929
70	0.000305	0.00408	165	0.000174	0.000976
75	0.000231	0.00333	170	0.000213	0.00117
b [m^{-1}][c]	0.0839	0.778			
$\langle cos\theta \rangle$[d]	0.962	0.948			

[a]A geometric average of 108 data sets from a computer-readable data collection compiled by *Jonasz* (1996, 1992): *Atkins* and Poole (1952, English Channel—5 sets), *Austin* (1973, Sargasso Sea—5, Pacific—22), *Beardsley* (1968, Atlantic—2), *Gohs* et al. (1978, Baltic Sea—3), *Jonasz* (1991b, Atlantic—17), *Kullenberg* (1984, Drake Passage—9, Peru upwelling—4), *Kullenberg* and Olsen (1972, Mediterranean Sea—11), *Kullenberg* (1969, Baltic Sea—5), *Kullenberg* (1968, Sargasso Sea—3), *Mertens* and Phillips (1972, Bahamas—2), *Petzold* (1972, Bahamas—3, Pacific—2), *Matlack* (1974—9, as quoted by *Hodara* 1973), *Tyler* (1961, Pacific—3).

[b]A geometric average of 161 data sets from a data collection compiled by *Jonasz* (1996, 1992): *Atkins* and Poole (1952, English Channel—5 sets), *Beardsley* (1968, Atlantic—3), *Gohs* et al. (1978, Baltic Sea—91), *Jonasz* (1991b, Atlantic—4), *Kullenberg* (1984, Peru upwelling—3), *Kullenberg* and Olsen (1972, Mediterranean Sea—5), *Kullenberg* (1969, Baltic Sea—10), *Mertens* and Phillips (1972, Bahamas—5), *Morrison* (1970, Atlantic coast, off New York—2), *Petzold* (1972, San Diego—3), *Prandke* (1980, Atlantic—3), *Reuter* (1980b, Baltic Sea—1), *Reese* and Tucker (1973, San Diego Bay—6, as quoted by *Morel* 1973), *Tyler* (1961, Pacific—1), *Whitlock* et al. (1981, Back River, Virginia, USA—8)

[c]Calculated by fitting the Fournier-Forand approximation (FF, *Forand* and Fournier 1999, *Fournier* and Forand 1994, Table 4.3) to the respective data above. Calculations that use a successive approximation integration [trapezoid rule, angle range: variable start(variable step)end = 10°(0.1°)180°; the variable start and step of the small-angle portion is adjusted to keep the relative incremental change of the integral to less than 0.001] yield results that are significantly different, 0.0824 and 0.769, than those obtained by using the FF fitting.

[d]Calculated by the numerical integration of the volume scattering functions using the successive approximation algorithm discussed in note *c*.

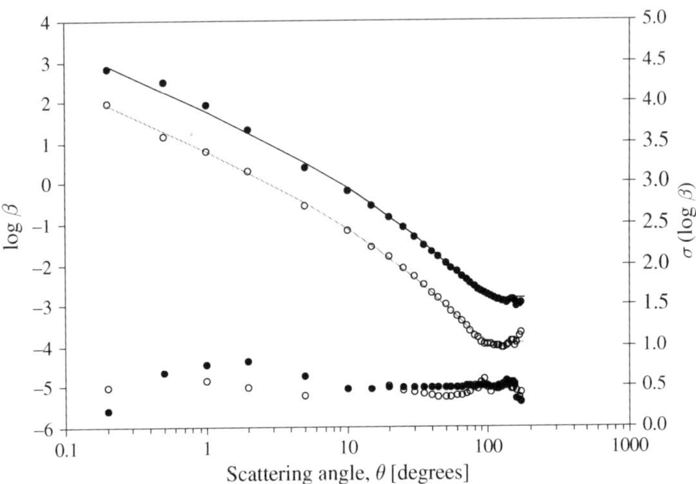

Figure 4.25. Particle volume scattering functions representative of the clear and turbid ocean waters. These functions are geometric averages of respectively 108 and 161 data sets from a computer-readable data collection compiled by *Jonasz* (1996, 1992). The data were measured by several research groups or individual researchers with both *in situ* and *in vitro* techniques, in various seasons and in various regions and depths of the ocean. See Table 4.2 for references to the original data sources and nominal locations. Points with lines represent scattering functions and their approximations (left *Y*-axis) with the Fournier-Forand (FF) function (*Fournier* and Forand 1994, *Forand* Fournier 1999): turbid seawater (• and a black line), clear seawater (○ and a gray line). Approximation parameters are listed in Table 4.3. Symbols without lines represent standard deviations of the $\log \beta$ (right *Y*-axis). These deviations are relatively stable throughout the entire angular range. Variations for $\theta < 10°$ reflect mainly the scarcity of data sets in this angular range. The volume scattering functions, scattering coefficients, and average cosines of the scattering angle corresponding to the volume scattering functions shown here are listed in Table 4.2.

between about 0.1 and $0.8\,\text{m}^2\text{g}^{-1}$ in turbid waters (*Hofmann* and Dominik 1995, *Baker* et al. 1983, *Baker* and Lavelle 1984) to about $1\,\text{m}^2\text{g}^{-1}$ in open ocean surface water (*Gordon* and Morel 1983), i.e., generally increasing with the water clarity. This is no surprise because $b_p{}^m$ should in general depend on the particle size distribution, refractive index (*Baker* and Lavelle 1984), and also on particle shape (*Jonasz* 1987c). Such a tendency, suggested by *Baker* and Lavelle (1984), would indicate that these properties of the particles are not randomly distributed.

It is unclear whether the reported variability reflects the actual conditions in the various waters because these results have been obtained, as pointed out by *Babin* et al. (2003) in a recent survey of $b_p{}^m$, with various instruments for the measurement of the scattering coefficient and various experimental protocols. These authors used uniform experimental procedures throughout the survey and

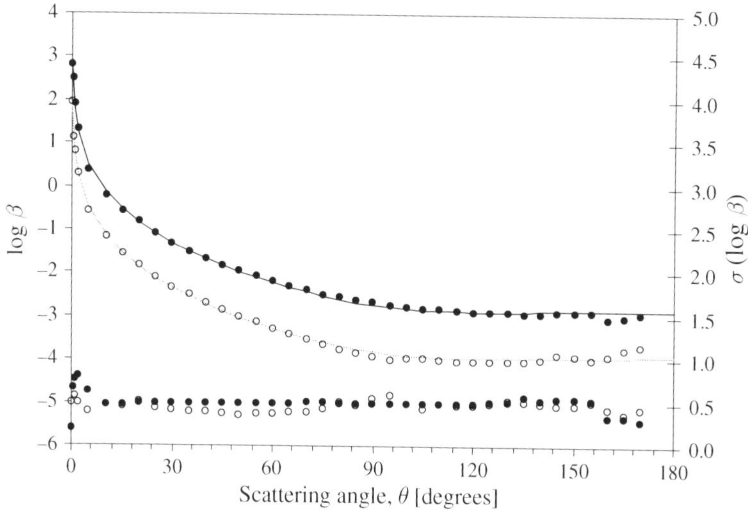

Figure 4.26. Particle volume scattering functions representative of turbid (• and a black line) and clear (○ and a gray line) ocean waters shown in Figure 4.25. Here the scattering angle axis is linear to better display the medium- and large-angle range.

obtained the average values of b_p^m at $0.5\,m^2g^{-1}$ in costal areas surrounding Europe and $1\,m^2g^{-1}$ in clear, open ocean waters, supporting the trend reported earlier.

The spectrum of the scattering coefficient of suspended particles in a range of coastal and open ocean waters is found to be nearly flat (*Babin* et al. 2003, *Gould* et al. 1999, *Barnard* et al. 1998), with a slope of $\gamma \sim 0.22$ in the power law $\lambda^{-\gamma}$, where λ is the wavelength of light relative to a wavelength of $1\,\mu m$, only slightly increasing with decreasing wavelength — much slower than would be indicated by the λ^{-1}-dependence, frequently used to characterize open ocean waters. Most spectra of the particle mass-specific scattering coefficient, b_p^m show residual spectral features due to phytoplankton, at 475 and 675 nm, especially in clear, open ocean waters, where the mineral particles contribute little to the scattering coefficient (*Babin* et al. 2003). Such a decrease of the scattering coefficient with increasing wavelength, calculated from the scattering function measurements, even for very turbid waters, has also been demonstrated by *Forand* and Fournier (1999) who used the data of *Whitlock* et al. (1981) obtained for turbid river waters. We already discussed that work at the end of Chapter 3.

This wavelength dependency is a function of the shape of the particle size distribution as well as the refractive index of the particles. As we discussed it in Chapter 3, even if the refractive index does not depend on the wavelength, the spectrum of the scattering coefficient varies as λ^{-4} for particles smaller than

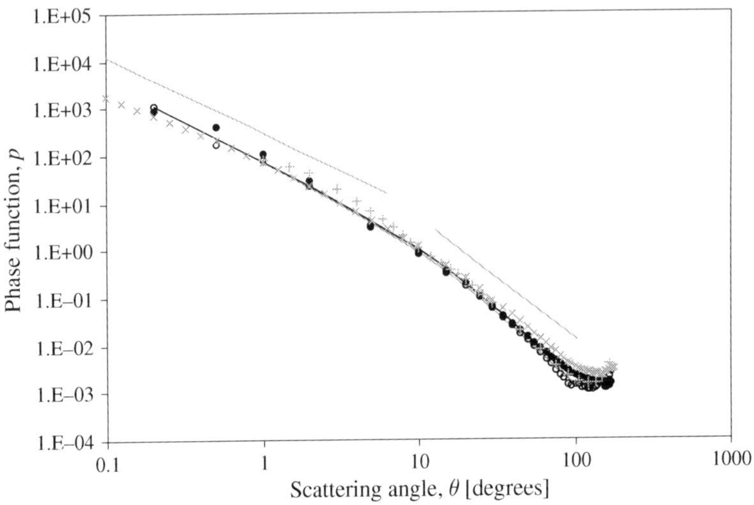

Figure 4.27. Particle phase functions representative of the clear and turbid ocean waters: turbid (•) and clear seawater (○) functions from Table 4.2, "typical" particle function of *Morel* (1973, +), "typical" particle function (turbid-to-clear waters) of *Mobley* (1994, ×) at 514 nm, based on the measurements of *Petzold* (1972) also included in the turbid and clear datasets. The almost undistinguishable lines represent the Fournier–Forand (FF) approximations for turbid (black) and clear (gray) seawater calculated by dividing the FF functions obtained with the coefficients from Table 4.3 by the respective values of the scattering coefficient, b, from Table 4.2. Note that all phase functions cross in an angular range about 10°, suggesting that $b = \text{const} \times \beta(\theta)$, where const = 1/(phase function at about 10°). The short straight gray line sections correspond each to a power-law $\sim\theta^{-S}$ with a slope, s, of 1.6 and 2.6 for the scattering angle, θ, ranges of 0.1 to ~6° and 12 to ~100° respectively.

the wavelength of light. As the particle size grows, the slope of this wavelength dependency decreases, to eventually reach 0 for particles much larger than the wavelength of light. This particle size effect alone is described by the relationship

$$\gamma = m - 3 \tag{4.109}$$

between the slope, γ, of the scattering coefficient spectrum

$$b(\lambda) \propto (\lambda/\lambda_0)^{-\gamma} \tag{4.110}$$

where λ_0 is a reference wavelength, whose role is to merely render λ dimensionless, and the slope, m, of the power-law size distribution of the particles. This relationship, discussed in Chapter 3, and recently examined for marine particles by *Boss* et al. (2001), has been known in the atmospheric optics as the Ångström law since 1929 (e.g., *Heintzenberg* and Charlson 1996).

Table 4.3. Parameters of the Fournier–Forand (FF) approximation (*Fournier* and Forand 1994) with modifications in (*Forand* and Fournier 1999) of the particle volume scattering functions shown in Figure 4.25 and Figure 4.26.

	Clear seawater		Turbid seawater	
Parameter	**Value[a]**	**Precision**	**Value[a]**	**Precision**
b	0.0839	0.001%	0.778	0.001%
n	1.098	0.001	1.073	0.001
m	3.37	0.01	3.59	0.01
approximation error	0.170		0.104	

[a]The FF approximation parameters were determined via a systematic, two successive approximation search for a minimum of the approximation error surface in three-dimensional parameter space (b, n, m). Parameter b is a magnitude factor, defined by $\beta = bp$, where p is the FF phase function defined by (4.130). Hence, b is simply the scattering coefficient, as defined by (4.4). The error is defined as a $(1/N)\sqrt{[(\log\beta_i - \langle\log\beta\rangle)/\sigma^2]}$ where N is the number of data points and σ^2 is the variance of $\log\beta$ over the whole applicable data set population. Only data up to the scattering angle of $\sim 120°$ have been used in all the fits. This was done because the FF function does not include a component representing reflection of light by particles, an effect that may be important in the backscattering angle range and also because the large-angle scattering data come typically with larger errors than those at small-to-medium angles. The error surface is that of a long narrow, slightly curved valley. The alongside profile of the valley bottom has a rather broad minimum. Thus, combinations of m and n along a substantial range of the valley bottom yield approximation errors comparable to those listed here. This is observed because an increase in the relative refractive index, n, can to some extent compensate for a decrease in the size distribution slope (*Jonasz* and Prandke 1986), as shown in Figure 4.29. Although the fitting algorithm used here is different than that used by *Forand* and Fournier (1999, quoted at the end of Chapter 3), the results are consistent: we arrived at $n = 1.082$ and $m = 3.8$ vs. 1.09 and 3.77 obtained by Forand and Fournier for a particle scattering function measured by *Whitlock* et al. (1981) at 800 nm.

The refractive index of mineral particles is typically assumed to be real in the visible, i.e., the particles are assumed to not absorb light. That of phytoplankton mimics the spectrum of a collection of pigments, most notably the chlorophylls. Yet, many common minerals that contribute to the mineral particulate pool do absorb in the visible, mostly through the presence of iron oxide that gives them distinct yellowish to red color. This is confirmed by significant, iron concentration-dependent absorption in the blue part of the visible (*Babin* and Stramski 2002). As the scattering efficiency decreases with the increasing imaginary part of the refractive index, absorption of light by the particles may reduce the degree of the wavelength dependency of the scattering coefficient.

4.4.2.2. Relationships between the scattering coefficients and scattering function at selected angles

Given the difficulties of measuring the scattering coefficients, the observation of a relative stability of the scattering function shape as measured in waters from a broad range of water bodies have led to attempts at developing simple means for estimating the scattering and backscattering coefficients by measuring light scattering at a fixed angle.

Early research indicated that the scattering coefficient may by closely proportional to the volume scattering function at 45° (*Kullenberg* 1974), with the proportionality constant ranging from 0.0031 to 0.0029. Interestingly, that range includes prepared suspensions of quartz particles in water (*Hodkinson* 1963). Data from the Baltic and Mediterranean seas yield somewhat greater values on the order of 0.0040 to 0.0048.

Small-angle scattering was also investigated, and close correlations were found between the scattering coefficient and the scattering function in an angular range of 4° to 6° (*Jonasz* 1980, *Kopelevich* and Burenkov 1971, *Mankovsky* 1971). Jonasz found the following relationship between the scattering coefficient, b [m^{-1}], and the scattering function $\beta(5°)$ [$m^{-1}sr^{-1}$]

$$b = 0.155(\pm 0.0029)\beta(5°) - 0.009(\pm 0.001) \tag{4.111}$$

in turbid coastal waters (the Gdansk Bay) with a correlation coefficient of 0.982. A single standard deviation value is given in the parentheses. Mankovsky obtained a proportionality coefficient of 0.14 for $\beta(4°)$. Note that the numerical values of the scattering coefficient in such correlations, where these values are calculated by numerical integration of the scattering function, are affected by the methods used to extrapolate the incomplete scattering function data into the small-angle range that most significantly contributes to the scattering coefficient.

It is thus encouraging to realize that numerical modeling based on the Mie theory (*Morel* 1973, *Reuter* 1980a) indicates that there should be a close correlation between the scattering coefficient and the volume scattering function in a range of ~4° to ~10°. This is also observed experimentally as shown in Figure 4.27. All phase functions shown there, of which that of Morel is based on an independent set of measurements, cross at a scattering angle of about 10°. As the phase function is defined by $p(\theta) = \beta(\theta)/b$, it follows that $b = const \times \beta(\theta)$, where, according to Figure 4.27, $const = p(\sim 10°)^{-1}$.

Such relationships are not surprising because almost all scattered light is contained within an angular range of less than 45° (e.g., Figure 4.18). According to *Petzold* (1972), who measured scattering functions in an angular range of 0.1° to 170°, between 45 and 64% of the scattering coefficient is due to light scattered at angles of less than 5°.

Similar relationships between the backscattering coefficient, b_b [m^{-1}], and the volume scattering function at an angle from a range of 90° to 180° were also investigated by *Oishi* (1990), who found a significant relationship at 120°:

$$\begin{aligned} b_{\mathrm{b}} &= 7.19\beta(120°) - 0.43 \times 10^{-4} \\ &= 2\pi \times 1.14\beta(120°) - 0.43 \times 10^{-4} \end{aligned} \quad (4.112)$$

with a correlation coefficient of $\sim$1.000. In fact, the backscattering coefficient was reasonably well correlated with the scattering function for all angles in that range.

Maffione and Dana (1997) argued that b_{b} can be estimated to within $\sim$9% by using $\beta(140°)$ as follows:

$$b_{\mathrm{b}} = 2\pi \times 1.08\beta(140°) \quad (4.113)$$

This relationship was also examined by *Haltrin* et al. (2003) who obtained a similar coefficient of $2\pi \times 1.151$ ($r^2 = 0.999$) for 869 scattering function with values at 140° ranging from $\sim 1.5 \times 10^{-4}$ to $\sim$0.04. Here, as in their evaluation of the correlation between b_{b} and b, which we mentioned earlier, a very large dynamic range contributes to the high value of r^2.

Boss and Pegau (2001) also examined the shape variability of the scattering phase function in the backward direction and concluded that, as suggested originally by *Oishi* (1990), the angle close to 120° is better suited for this purpose on theoretical grounds. By explicitly accounting for the role of light scattering by pure seawater and analyzing data not available to *Maffione* and Dana (1997), they concluded that

$$b_{\mathrm{b}} = 2\pi \times 1.1\beta(\theta_0) \quad (4.114)$$

at an angle $\theta_0 = 117° \pm 3°$ with an error of less than 4%. More recently, *Vaillancourt* et al. (2004) evaluated the relationship between b_{b} and $b(140°)$ for nine marine phytoplankton cultures and found the proportionality coefficient to be $2\pi \times (0.82 \pm 0.01)$, where 0.01 is the single standard deviation.

4.4.2.3. Smoothness

The volume scattering functions measured in natural water bodies are generally thought to be smooth functions of the scattering angle, typical of media with a wide distribution of sizes of the scattering centers. This impression may have come from the usual coarse increments of the scattering angle in these measurements. Some high-resolution measurements do show small-amplitude oscillations (*Voss* and Fry

1984, *Mankovsky* et al. 1970, *Sasaki* et al. 1960) which may indicate the presence of significant quasi-monodispersed particle populations. Low-angular frequency oscillations have been observed in some high-resolution scattering function data for suspensions of bacteria (*Cross* and Latimer 1972) but not in suspensions of unicellular algae (e.g., *Schreurs* 1996, *Burns* et al. 1976), except for cylindrical algae cells colonies (*Prochlorotrix hollandica*, *Volten* et al. 1998, *Schreurs* 1996). This latter case is not surprising: oscillations of the scattering pattern of monodisperse cylinders are preserved even for randomly oriented cylinders as we discussed it in Chapter 3.

4.4.2.4. Asymmetry

The volume scattering functions of seawater are highly asymmetric functions of the scattering angle (as referred to the scattering angle $\theta = 90°$). A rigorous measure of that asymmetry is the mean cosine of the scattering angle, defined by (4.8). The average cosine is an important parameter in the radiative transfer theory of turbid media (e.g., *Bohren* 1987). It has been reported to be typically in a range of 0.7 for the clearest natural waters, to over 0.97 for turbid, coastal waters (*Lee* et al. 2003, *Dera* 1992). We found the lower limit to be somewhat too small, as we will discuss later in this section in more detail. Our geometric average scattering functions yield average cosines of 0.89 (clear water) and 0.95 (turbid water). However, the data analyzed in the process of developing these average functions contain functions of potentially extreme asymmetry, for example those of *Beardsley* (1968) and *Mankovsky* and Haltrin (2002a), as shown in Figure 4.24, and of *Mankovsky* and Haltrin (2002b).

The high positive value of the average cosine is due to the rapid decrease of the scattering function value for natural waters by several orders of magnitude with the increasing angle from a range of 0° to 90° to a broad minimum occurring between 90° and 180°. In the backscattering range (90° to 180°), these functions usually increase, although much less than in the forward-scattering range (0–90°). The scattering angle at which the minimum of the scattering function occurs generally decreases with increasing clarity of water. In turbid waters, the minimum of the scattering function may move all the way to the scattering angle of 180°. Thus, the scattering function may essentially monotonically decrease with the increasing angle in the entire angular range.

The evaluation of the mean cosine and the backscattering ratio requires the knowledge of the scattering function in the entire scattering angle range. Unfortunately, despite the early recognition of the high asymmetry of the scattering function (e.g., *Bauer* and Morel 1967) and its role in the radiative transfer in the sea, there are but a few measurements that come sufficiently close to that ideal. In a notable exception, *Lee* et al. (2003) evaluated the mean cosine as well as other integral parameters of 60 scattering functions measured in the coastal waters of Northwestern Atlantic off New Jersey, USA, in a wide range of the scattering angle (0.6° to 177.3°), with the scattering coefficient ranging from 0.38 to 9.3 m^{-1}.

They found that the mean cosine is reasonably correlated ($r^2 = 0.991$) with the backscattering probability, $B = b_b/b$:

$$< \cos\theta >= 0.986 - 3.29\frac{b_b}{b} \tag{4.115}$$

Haltrin et al. (2003), who analyzed 874 scattering functions, obtained the following relationship:

$$< \cos\theta >= \frac{1 - 4\frac{b_b}{b}}{1.0144 + 2.6307\frac{b_b}{b} - 1.2772\left(\frac{b_b}{b}\right)^2} \tag{4.116}$$

with $r^2 > 0.82$ for $0.0022 < b_b/b < 0.146$.

Although such a relationship might have been expected because the average cosine of the scattering angle, as indicated by (4.8), should decrease with increasing backscattering, the high correlation clearly states that, despite the high asymmetry of the scattering phase functions of natural waters, the role of the backscattering angular range is not negligible.

The backscattering probability b_b/b is an asymmetry parameter of the scattering function that appears in the theory of radiative transfer in turbid medium (e.g., *Haltrin* 1998) and is relevant in remote sensing because the remotely sensed reflectance of a water body is a function of the absorption and backscattering coefficient (e.g., *Morel* and Prieur 1977). It has also been used as an indication of the refractive index of the particles (e.g., *Boss* et al. 2004).

The scattering coefficient of pure seawater has a significant role in variations of the shape of the scattering function. It was already recognized some time ago (*Morel* 1965, *Morel* and Prieur 1977) that the variability in the shape of the volume scattering function of natural waters can in large part can be explained by combining two components: the scattering function of pure water or pure seawater with the scattering function of the particles whose shape varies only a little.

Given generally incomplete measurements of the scattering function as a function of the scattering angle, other measures of the scattering function asymmetry have historically been used, most notably a ratio of the scattering function at 45° to that at 135°. This ratio equals 1 for molecular (pure water) scattering. *Morel* (1973) examined the $\beta(45°)/\beta(135°)$ ratio and found that it initially increases sharply with the scattering function magnitude to eventually flatten out at high-magnitude values. This clearly indicates that the pure seawater contribution is significant in clear waters, and the initial fast increase of the asymmetry with the magnitude of scattering is at least in part due to the increase of the relatively stable particle contribution to the scattering function (*Morel* 1965), given that the scattering function of seawater is a sum of the scattering function of pure seawater and that of the particles, and that the particle size distribution in the ocean has a relatively stable slope (we will discuss that topic in Chapter 5).

We compared the values reported by Morel with those from a computer-readable data collection compiled by *Jonasz* (1996, 1992) and additional data (*Lee* et al. 2003, *Mankovsky* and Haltrin 2002a, 2002b) and found a close correspondence of these two data sets (Figure 4.28). The ranges of the ratio $\beta(45°)/\beta(135°)$ are listed in Table 4.4. At least some part of the observed variability of this ratio, as well as $\beta(135°)/\beta(90°)$, is due to the relatively high measurement errors of the scattering function at 90° and 135°, especially in clear waters. Another part of the variability is likely due to variations of the asymmetry of the scattering function with the refractive index and slope of the particle size distribution as shown in Figure 4.29 for the FF scattering function (*Fournier* and Forand 1994, with modifications by *Forand* and Fournier 1999): the combined effect of variations in the particle size distribution slope and relative refractive index of the particles can spread the $\beta(45°)/\beta(135°)$ ratio over a range of 4.68 to 23.2 within the characteristic domain of these two parameters for seawater particle assemblies.

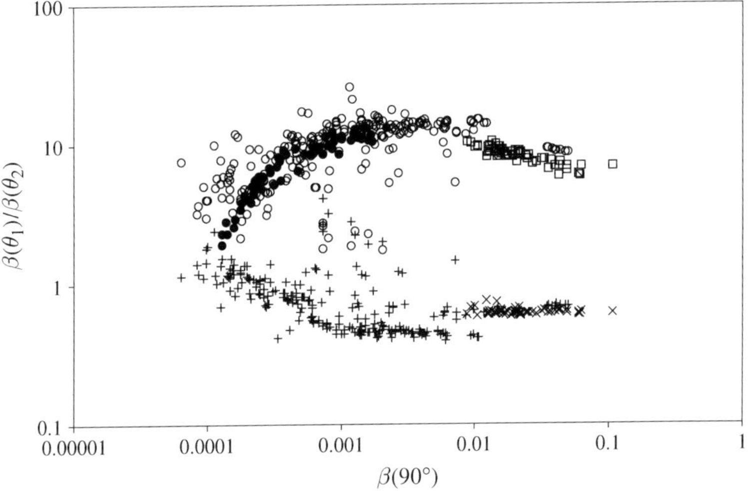

Figure 4.28. The asymmetry of the scattering function of seawater represented by the ratios $\beta(45°)/\beta(135°)$ and $\beta(135°)/\beta(90°)$ as a function of the scattering function magnitude, represented by $\beta(90°)$. Points: • $\beta(45°)/\beta(135°)$, a dark curved patch at the right: 60 points of *Morel* (1973, his Fig. 1.3) at 546 nm, English Channel, Mediterranean Sea, and the Indian Ocean, ○ $\beta(45°)/\beta(135°)$, + $\beta(135°)/\beta(90°)$, 254 data sets from a collection of data obtained by several researchers in various waters and seasons at wavelengths ranging from 366 to 850 nm) (compiled by *Jonasz* 1996, 1992), □ $\beta(45°)/\beta(135°)$, × $\beta(135°)/\beta(90°)$, 60 data sets from *Lee* et al (2003, Atlantic off New Jersey, USA). The far right-top patch (*Whitlock* et al. 1981) represent very turbid river coastal waters.

Table 4.4. Ranges of the asymmetry parameters for data shown in Figure 4.28.

	Seawater (particles + pure seawater)[a]		**Particles only[b]**	
	$\beta(45°)/\beta(135°)$	$\beta(135°)/\beta(90°)$	$\beta(45°)/\beta(135°)$	$\beta(135°)/\beta(90°)$
Minimum	1.76	0.39	1.81	0.15
Average	–	–	15.4	0.77
Standard deviation	–	–	16.5	0.91
Maximum	26.1	4.15	164	11.5

[a]Sixty data sets of *Morel* (1973, Fig. 1.3), 254 datasets from a collection of data obtained by various authors in various waters and seasons as compiled by *Jonasz* (1996, 1992, see Table 4.2 for sources).

[b]Two-hundred and twenty six data sets from the above collection.

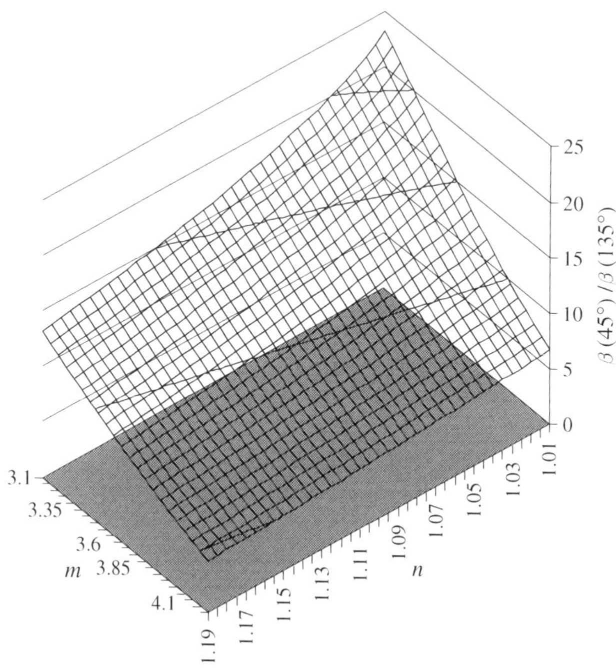

Figure 4.29. The asymmetry of the scattering function of marine particles represented by the ratio $\beta(45°)/\beta(135°)$ of the Fournier-Forand scattering function (*Fournier* and Forand 1994, *Forand* and Fournier 1999) as a function of the slope, m, of the power-law particle size distribution and the relative refractive index, n, of the particles. In the (m, n) domain representative of the marine particles, the ratio varies from 4.69 to 23.2.

Note that the FF approximation underestimates the backscattering part of the scattering function.

We also calculated the average cosine for the whole water scattering function by using 36 data sets representing a wide range of turbidity: from the clearest ocean waters (*Kullenberg* 1968) to the turbid coastal river waters (*Whitlock* et al. 1981) with angular spread that warranted such calculations and compared these values with those of the scattering coefficient. We found that the average cosine, which varied from 0.803 to 0.978, is virtually uncorrelated with turbidity. Given the wide random variability seen in Figure 4.28, this presents no surprise for this limited data set. A similar conclusion can be reached by considering the results of *Lee* et.al. (2003) for a set of 60 scattering functions measured in the coastal Northwestern Atlantic.

4.4.2.5. Wavelength dependence of light scattering

Early observations of light scattering, summarized by *Morel* (1973), indicated that the volume scattering functions of natural waters depends relatively weakly on the wavelength of light, and the degree of that dependence decreases with increasing turbidity, i.e., particle contribution to light scattering. A major part of the wavelength dependence of light scattering at mid to large angles is due to a strong, approximately $\sim\lambda^{-4}$, wavelength dependence of light scattering by pure water, as we already discussed it in Chapter 2 on theoretical grounds and about which we will shortly provide experimental data and approximations.

It has long been assumed that the particle scattering function itself depends only marginally on the wavelength of light. In fact, our discussion of the subject in Chapter 3 indicates that to the first order, the phase scattering function of the particles can be regarded as independent of the wavelength. This is confirmed by examining experimental data for $\beta(45°)/\beta(135°)$ which show virtually no correlation with the wavelength of light. If one assumes a power-law dependency, as for the scattering coefficient, the slope of such a power-law function evaluates to -0.187 ± 0.327 (1σ) for 226 datasets from a computer-readable data collection compiled by *Jonasz* (1996, 1992).

However, the $\beta(45°)/\beta(135°)$ ratio for scattering functions measured over a wide range of wavelengths at a single location [*Whitlock* et al. 1981, 450(50) 800 nm] does show a weak power-law dependence on the wavelength ($R = 0.878$, slope of 0.120 ± 0.028). Recent measurements of the visible spectra of the scattering coefficient of the particles (*Babin* et al. 2003, *Gould* et al. 1999) indicate also that the scattering coefficient of the particles may exhibit definite, albeit weak wavelength dependency.

We also observe a weak wavelength dependency for the average cosine calculated from the data of Whitlock and colleagues (Figure 4.30). Although these results represent the scattering function of the whole seawater, the contribution of pure seawater is negligible so that the data can be regarded as characteristic of the particles.

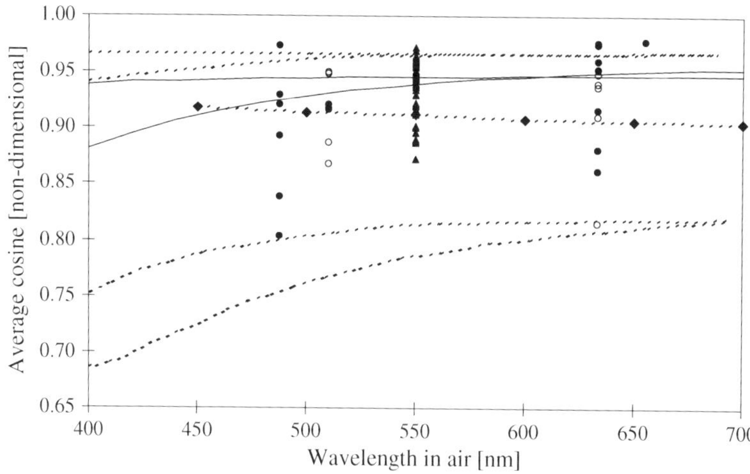

Figure 4.30. Average cosine of the scattering angle as a function of the wavelength (in air) for scattering functions of the whole seawater (particles + pure seawater). Solid lines represent the variability of the average cosine, calculated according to Eq. (4.117), for the turbid (top) and clear (bottom) particle scattering functions listed in Table 4.2. The particle scattering functions are taken to be independent of the wavelength. Points represent values of the average cosine calculated by using the Fournier-Forand approximation to scattering function data: • — turbid seawater: by using data of *Kullenberg* (1969, Baltic Sea), *Gohs* et al. (1978, Baltic Sea), *Kullenberg* and Olsen (1972, Mediterranean Sea), *Mertens* and Philips (1972, off Bahamas, the lowest point at 488 nm represents waters off Andros I.), *Petzold* (1972, San Diego harbor, CA, USA), *Reese* and Tucker (1973, San Diego Bay), ♦ and broken line, turbid river mouth waters *Whitlock* et al. (1981, Atlantic coast at Virginia, USA: the average cosine *decreases* with the wavelength!), ▲ — costal waters (*Lee* et al. 2003, Atlantic off New Jersey, USA), ○ — clear seawater: *Jonasz* (1991b, north western Atlantic), *Kullenberg* (1968, Sargasso Sea, the lowest point at 633 nm), *Kullenberg* and Olsen (1972, Mediterranean Sea), *Petzold* (1972, off Bahamas, Pacific), *Prandke* (1980, eastern equatorial Atlantic). Broken lines represent spectra calculated from Mie theory-derived tables of particle scattering (*Woźniak* 1977, also cited in *Dera* 1992): top to bottom at 400 nm: turbid waters (Gdansk Bay and Baltic Sea), typical ocean waters and Sargasso Sea.

The decrease of the average cosine of the scattering angle for light scattering by the whole seawater (particles + pure seawater) with wavelength merits some comments. As it follows from (4.8), the average cosine vanishes for a scattering function that is symmetrical about a scattering angle of 90°. The scattering function of the whole seawater is a sum of that of the particles and of pure seawater. Being symmetrical about 90°, the pure seawater scattering function does not contribute to the numerator of (4.8). However, it does contribute to the denominator, i.e., the scattering coefficient, b, and if that contribution is significant, the average cosine is reduced this way. Thus, we can express the average cosine of the whole seawater as follows:

$$\begin{aligned}\langle \cos\theta \rangle &= \frac{2\pi}{b} \int_{-1}^{1} \beta_p(\theta) \cos\theta \; d cos\theta \\ &= \langle \cos\theta \rangle_p \frac{b_p}{b}\end{aligned} \tag{4.117}$$

where b is the scattering coefficient of the whole seawater and the subscript p signifies the particles' part of a quantity.

We plotted the average cosine values calculated for a few measured scattering functions that covered a suitable angular range in Figure 4.30 along with the average cosine spectra calculated by using the average turbid and clear waters particle functions (Table 4.2) and the results of *Woźniak* (1977) derived from the Mie theory. There is a fair agreement between the present work's results and the Mie theory-derived values if these values are taken to represent the limits of the average cosine spectra. Indeed, most clear-water values calculated in this work are much higher than those predicted for the clear ocean waters from the Mie theory. Spectra that correspond to the average turbid and clear waters scattering functions (Table 4.2) more closely represent the data points calculated by using measured scattering functions.

4.4.2.6. Relative contributions of light scattering by particles, water, and turbulence

The partition of the scattering function deserves some discussion. Except for the turbulence component, most publications on the scattering of light by natural waters treat the latter as a sum of the "particles" and pure water or pure seawater. The "particles" are treated as a faceless blackbox neglecting the significant variability of the relevant characteristics within the particle population and treating it as "scatter."

Recent research indicates that such variability manifests itself significantly and predictably in the spectra of the scattering coefficient of the particles (*Babin* et al. 2003) as well as absorption (*Babin* and Stramski 2003). Although considerable effort has been expended in quantifying the light scattering of the various classes of particle present in natural waters, and is discussed later in this chapter, the prevailing blackbox approach keeps us still far from being able to provide reasonable prediction accuracy both in the forward and in the inverse problems of marine optics.

The component of the scattering function of seawater, due to "particles," is most significant at the small angles (Figure 4.24) in all natural waters. In clear open ocean waters, the light scattering by seawater itself makes a significant contribution to the volume scattering function at large angles. This contribution may be so significant that, given errors in the measurements of the scattering function, the particulate scattering function can assume negative values when scattering by pure water (seawater) is subtracted from that of the whole water (seawater). The decrease in the relative contribution of the pure seawater component with increasing magnitude of the scattering function is responsible for variations in the asymmetry of the scattering function of seawater with the function magnitude (i.e., water turbidity) as we already discussed it.

The contribution of particles to light scattering can be assessed by using an approach entirely different from the subtraction of the volume scattering function of pure seawater from the experimental volume scattering function. If seawater is illuminated with a highly monochromatic light source (e.g., a HeNe laser), the spectrum of the scattered light consists of three closely spaced peaks: the central "line" of intensity I_C and two side "lines": the Brillouin doublet (*Stone* and Pochapsky 1969, see also *Young* 1981 for an enlightening discussion of the light scattering nomenclature, including the various meanings of the Rayleigh scattering). Each of the doublet lines has intensity I_B and is symmetrically located in relation to the central line of the incident light. The wavelength spacing of the Brillouin doublet is on the order 4×10^{-3} nm at the wavelength of HeNe laser. The ratio $I_C/(2I_B)$, known as the Landau–Placzek ratio, is much smaller than unity for pure seawater. That ratio can be theoretically predicted for pure liquids and is explained as a result of the Doppler shift of the central frequency by thermally generated acoustic waves in liquid (*O'Connor* and Schlupf 1967). Particles do not give rise to the Brillouin doublet, but only to the central line at the wavelength of the incident light. *Stone* and Pochapsky (1969) measured the values of the Landau–Placzek ratio to be between 1 and 15 at a scattering angle of 90° for stored samples of seawater.

At very small angles (on the order of 0.1°), the effect of turbulence in pure seawater may become significant. The turbulence affects the spatial structure of the refractive index at a dimensional scale much larger than that of molecular density fluctuations. This effect is exerted through spatial fluctuations in temperature and salinity of seawater. It appears that the temperature fluctuations have a dominant effect (*Bogucki* et al. 1994). These fluctuations albeit very small (~ 0.01 to ~0.1 C°) have pronounced effect on the volume scattering function as predicted theoretically by *Bogucki* et al. (1998).

The effect of turbulence, which may dominate the scattering function at angles smaller than 0.1°, may increase the small-angle scattering function by several orders of magnitude. Little is known, at least in the public domain literature, about the magnitude of the effect of turbulence on light scattering by seawater *in situ*. Virtually the single experimental contribution to the knowledge of light scattering by turbulence *in situ* is the data of *Honey* and Sorensen (1970). In the context of turbulence, seawater is similar to human tissue, where discrete scattering centers are imbedded in a quasi-continuous structure of the refractive index.

4.4.3. The scattering matrix

Compared to the many investigations of the volume scattering function, there are relatively few reports on the measurements of the scattering matrix (Table A.2). This is clearly due to two factors: (1) an instrument capable of measuring the scattering matrix is much more complex than a nephelometer capable of only unpolarized light measurements, and (2) before the development of nephelometers employing electro-optical modulators in the 1970s (e.g., see a review in *Mujat* and Dogariu 2001), the measuring time of a scattering matrix tended to be substantially longer than that of the volume scattering function.

The representative investigations of the scattering matrix of seawater are discussed in this section. Selected data are shown in Figure 4.31 (upper left quadrant of the matrix), Figure 4.32 (upper right quandrant), Figure 4.33 (lower left quandrant), and Figure 4.34 (lower right quandrant). The scattering matrix elements of pure seawater (a baseline) are also shown in these figures. For the convenience of comparison, all elements (except M_{11}) are shown on the same scale.

Beardsley (1968, see the full data set in *Beardsley* 1966) was to our knowledge the first researcher who measured *in vitro* the scattering matrix in the sea and in river waters with a modified Brice-Phoenix nephelometer (*Brice* et al. 1950). The volume scattering functions, calculated from the matrix components, are similar to those reported earlier for other natural waters. The scattered light is highly polarized (linear polarization of about 40 to 70% at $\theta = 90°$). The diagonal components of the matrices, and components M_{21}, and M_{12} exceed the remaining components of the scattering matrices at a scattering angle of 30° by about two orders of magnitude for five samples of natural waters. Interestingly, elements m_{22}, m_{33}, and m_{44} of the normalized scattering matrix are greater than unity.

Shortly afterward, *Kadyshevich* et al. (1971) reported results of *in vitro* measurements of scattering matrices of seawater at a wavelength of 546 nm, in an angular range of 25 to 145°. The samples were taken about 1.5 miles offshore in the Black Sea in August and September of 1969, at depths of 3, 5, 10, and 15 m. A full set of results for one scattering matrix was obtained during 1 to 2 hours.

Components m_{13}, m_{31}, and m_{42} were negligible for all samples, m_{12} was approximately equal to m_{21}. Components m_{34}, m_{43}, and m_{23} all assumed small but significant values in the entire angular range. The non-zero values of m_{14}, m_{23}, and m_{24} indicate the presence of asymmetrical or optically active particles.

Interesting results were obtained regarding the time changes in the light scattering by a sample of seawater. The magnitude of light scattering was found to decrease by about 30% within the first 2 hours after sampling. Light scattering decreased most rapidly, to about half the original value, within the first 15 hours after sampling.

That group of researchers soon followed with an extensive set of measurements of the scattering matrix for samples of seawater from the Pacific and Atlantic Oceans, taken at depths ranging from of 0 to 2000 m (*Kadyshevich* et al. 1976). The differences in the mean values of the components for the various depth ranges

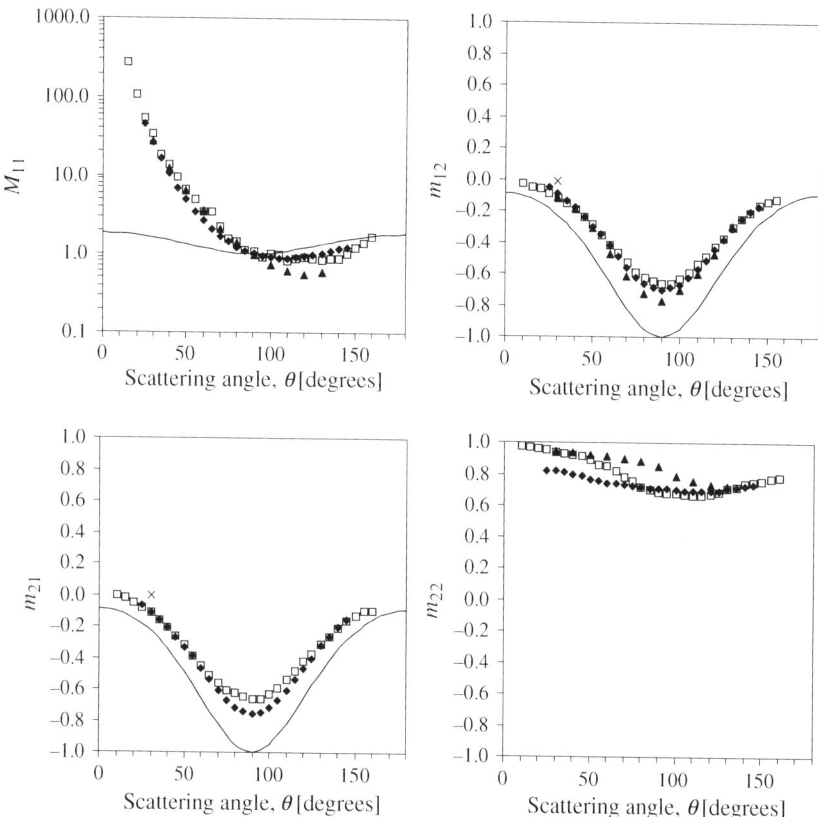

Figure 4.31. The upper left quadrant of the scattering matrix of seawater: ♦ average elements at an unspecified wavelength for the Pacific and Atlantic waters, depths of 0 to 2000 m (*Kadyshevich* et al. 1976), □ average elements at 488 nm for the Atlantic and Pacific waters, unknown depth range (*Voss* and Fry 1984), ▲ a single sample from Baltic Sea at 546 nm (*Kadyshevich* 1977). Single data points (×) represent the average of two matrices measured in the western Atlantic by *Beardsley* (1968, samples 4 and 6) at 30° and 546 nm. These averages for m_{22}, m_{33}, and m_{44} are greater than unity and have not been plotted. Thin solid lines represent the scattering matrix elements of pure seawater. Note that for seawater: m_{12} and m_{21} do not vanish at 0° and 180° due to depolarization (water molecules are anisotropic), and that $m_{22} = 1$ (it coincides with the graph frame). In all cases, the element M_{11} is normalized to unity at 90°. The remaining elements are shown as $m_{ij} = M_{ij}/M_{11}$.

were within the experimental errors, but the variances of the components exceed these errors. No systematic dependence of the components on the depth were found, except for the phase function that shows a marked decrease in asymmetry with increasing depth, as also found by *Kullenberg* (1978). The diagonal elements

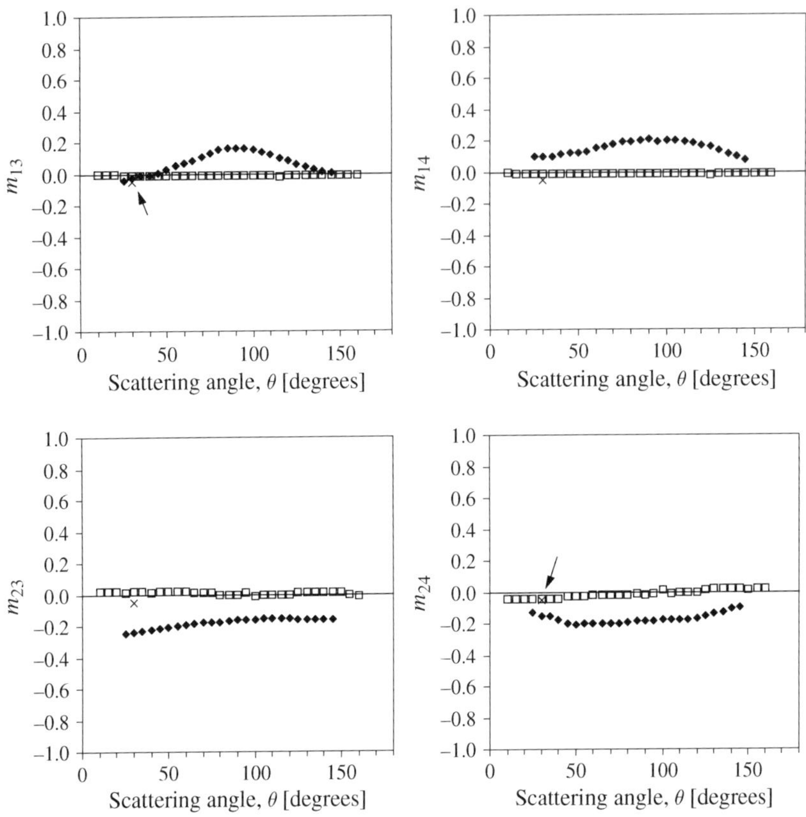

Figure 4.32. The upper right quadrant of the scattering matrix of seawater: ♦ average elements at an unspecified wavelength for the Pacific and Atlantic waters, depths of 0 to 2000 m (*Kadyshevich* et al. 1976), □ average elements at 488 nm for the Atlantic and Pacific waters, unknown depth range (*Voss* and Fry 1984). Single data points (×), pointed to by an arrow where needed, represent the average of two matrices measured in the western Atlantic by *Beardsley* (1968, samples 4 and 6) at 30° and 546 nm. Thin solid lines represent the scattering matrix elements of pure seawater. The matrix elements are shown as $m_{ij} = M_{ij}/M_{11}$.

m_{22}, m_{33}, and m_{44} appear to tend to a value of less than unity at the scattering angle of 0°.

The form of the scattering matrix (only $m31$ and $m42$ elements were close to zero) led *Kadyshevich* and colleagues to conclude that the ocean water is an anisotropic scattering medium, where the anisotropy may be caused by non-spherical particles oriented by the gravitational field as they settle through the water column. This form of the scattering matrices does not agree with more recent measurements of *Voss* and Fry (1984), reviewed further in this section.

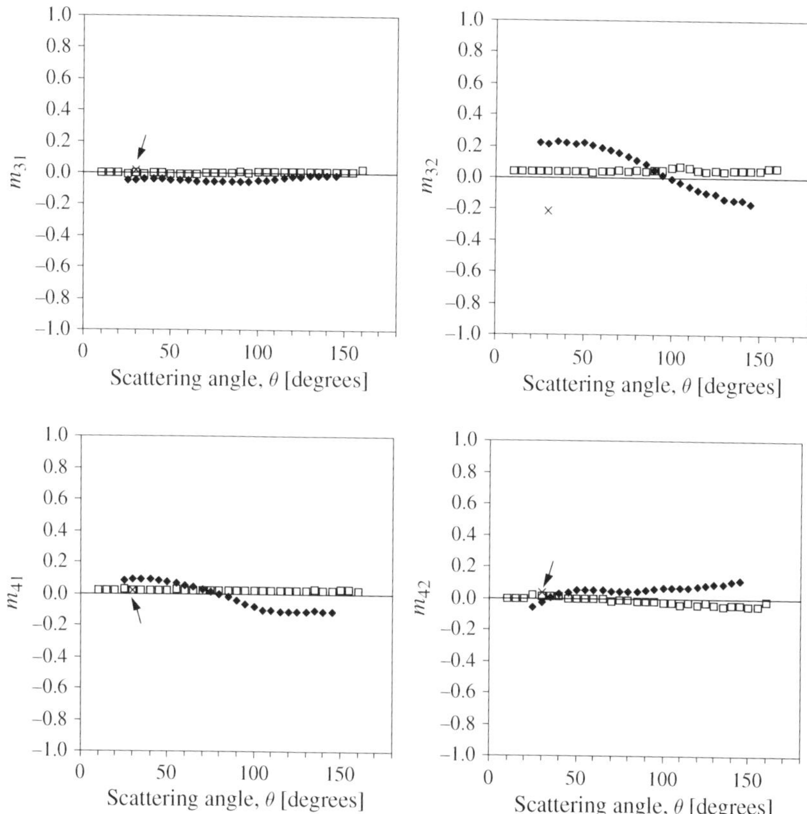

Figure 4.33. The lower left quadrant of the scattering matrix of seawater: ♦ average elements at an unspecified wavelength for the Pacific and Atlantic waters, depths of 0 to 2000 m (*Kadyshevich* et al. 1976), □ average elements at 488 nm for the Atlantic and Pacific waters, unknown depth range (*Voss* and Fry 1984). Single data points (×), pointed to by an arrow where needed, represent the average of two matrices measured in the western Atlantic by *Beardsley* (1968, samples 4 and 6) at 30° and 546 nm. Thin solid lines represent the scattering matrix elements of pure seawater. The matrix elements are shown as $m_{ij} = M_{ij}/M_{11}$.

The findings of *Padisák* et al. (2003a, 2003b) regarding the effect of the shape asymmetry of various phytoplankton cells/colonies on the orientation of these phytoplankton are interesting in this respect (see section 6.4.3.3). These findings suggest that there are cases when certain cell/colony orientations may be preferred when settling.

Kadyshevich (1977) also measured *in vitro* scattering matrices in the waters of the Baltic Sea at a wavelength of 546 nm, in an angular range of 30 to 140°. He examined 25 samples of seawater, obtained off the coast of Saremaa Island

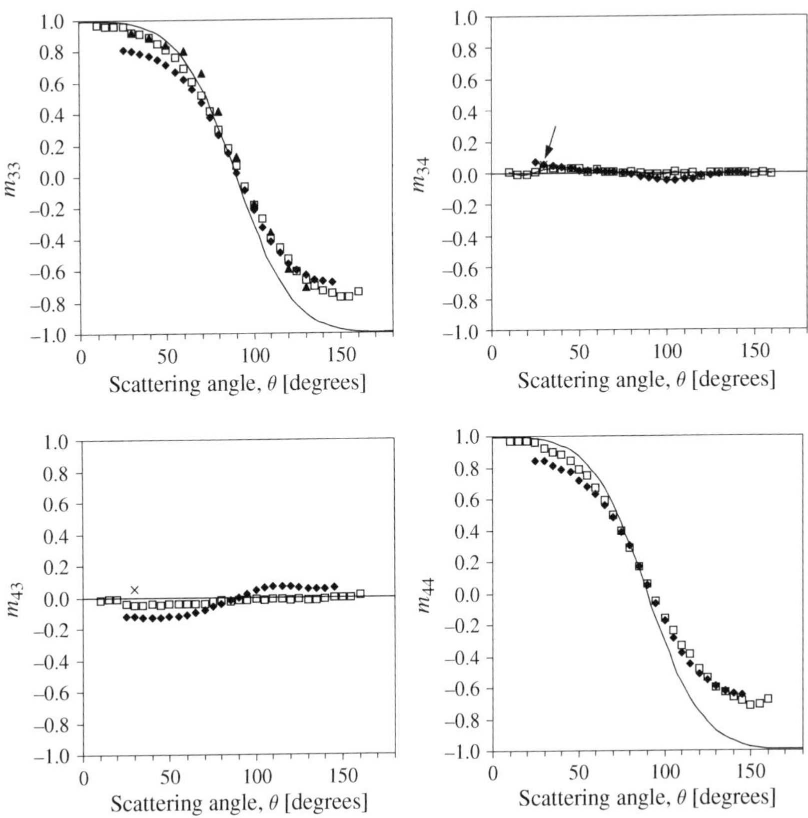

Figure 4.34. The lower right quadrant of the scattering matrix of seawater: ♦ average elements at an unspecified wavelength for the Pacific and Atlantic waters, depths of 0 to 2000 m (*Kadyshevich* et al. 1976), □ average elements at 488 nm for the Atlantic and Pacific waters, unknown depth range (*Voss* and Fry 1984), ▲ one sample from Baltic Sea at 546 nm (*Kadyshevich* 1977). Single data points (×), pointed to by an arrow where needed, represent the average of two matrices measured in the western Atlantic by *Beardsley* (1968, samples 4 and 6) at 30° and 546 nm. Thin solid lines represent the scattering matrix elements of pure seawater. The matrix elements are shown as $m_{ij} = M_{ij}/M_{11}$.

in August and September of 1973, at depths between 0 and 40 m in relatively well-mixed water body (no depth variations were noted). The matrices characteristic of these waters, in contrast to those measured by *Kadyshevich* et al. (1976) in the Atlantic and Pacific waters, were characteristic of an isotropic scattering medium: the only non-zero components were m_{11} (=1), m_{22}, m_{21}, m_{12}, m_{33}, and m_{44}.

The scattering matrices for the Baltic Sea were similar to over 200 matrices measured by *Voss* and Fry (1984) in the Atlantic and Pacific oceans in an angular range of 10 to 160°. The polarization-modulation nephelometer that they used was similar in design to that of *Thompson* et al. (1980). It permitted the measurements of the complete scattering matrix in about 2 minutes. The errors of measurements of the scattering matrix were usually less than 10%.

Contrary to earlier measurements by *Kadyshevich* et al. (1976) in those waters, and similar to those performed by *Kadyshevich* (1977) in the Baltic Sea, all non-diagonal elements of the matrices, except m_{12} and m_{21}, m_{34}, and m_{43}, were found to be equal to zero within the measurements accuracy. Thus, according to the results of Voss and Fry, seawater is essentially an isotropic scattering medium. The maximum linear polarization (m_{12}) of 60 to 80% occurred at 90°. The element m_{22} decreases from (extrapolated) unity at 0° to a minimum (0.6 to 0.8) at a scattering angle of about 100°, indicating significant non-sphericity of marine particles.

4.4.4. Volume scattering functions of various aquatic particles

The natural waters contain many species of particulate matter. Given that the interaction of light with particles is incoherent, the optical properties of water containing particles can be represented by a sum of the results of the interaction of light with each particle itself. Therefore, by knowing the optical properties, such as the scattering cross-section, of individual particles, or more realistically—particle classes—one may derive the relevant optical properties of the whole water (water + particles). This proposition seems to slowly gain recognition (e.g., *Stramski* and Mobley 1997).

The light scattering properties of many of these species have been investigated. The size distributions, refractive indices, shapes, and compositions of some of these species are discussed in the following two chapters. In this section we are concerned with measurements of the light scattering properties of individual quasi mono-disperse species of particles isolated in laboratory cultures. See Table A.3 for the summary of the data sources.

4.4.4.1. Viruses

To our knowledge, the only examination of the light scattering function of marine viruses has been reported by *Balch* et al. (2000). The functions were typical of those for particle sizes much smaller than the wavelength of light in accordance with the effective diameters of the virus particles that are on the order of 0.1 μm. Results of Balch and colleagues indicate that viruses, although abundant in seawater (on the order of 10^{12} m^{-3}, for example *Wommack* and Colwell 2000), are too small to contribute significantly to the backscattering of light by the whole seawater.

4.4.4.2. Bacteria

The interest in the optical properties of bacteria has long been fueled by potential clinical applications of light scattering as a diagnostic tool (e.g., *Wyatt* 1968). As a consequence, light scattering by numerous species of bacteria has been measured (see Table A.3 in Appendix for a representative survey) and modeled. Polarized light scattering seemed to hold promise as early experiments with identification of bacterial species and their physiological states (*Bickel* and Stafford 1981, *Bickel* et al. 1976) pointed to the m_{34} element of the scattering matrix as being especially sensitive to bacterial cell structure.

However, it soon became clear that the scattering of light by suspensions of bacteria depends in a complex manner on the size, and structure of the cells, and that random orientation of cells in suspension obscures a significant portion of information about these properties carried by the scattered light. With the development of practical applications in medical diagnostics delayed by theoretical and experimental problems, the interest in light scattering subsided. It has been revived only by the introduction of optical flow cytometers and cell sorters and realization that in dealing with single cells of known species, whose optical signatures decisively clustered in multi-parameter plots, one does not need to use realistic models to classify, identify, and count the cells.

Recent recognition of the role that bacteria may play in determining the backscattering of light by seawater (e.g., *Stramski* and Kiefer 1991) spurred several investigations in this field in marine optics as well. With few exceptions (*Kopelevich* et al. 1987, *Morel* and Bricaud 1986), these investigations concentrated on the determination of the backscattering of light by bacteria, in recognition that these small cells may significantly contribute to light scattering only in that angular range. Thus, a relatively limited selection of experimental data on light scattering functions of marine bacteria is available.

The effective diameters of bacterial cells are comparable to the wavelength of light. Thus, the scattering functions of bacteria are highly asymmetrical, decreasing by two decades within the first 6° (*Kopelevich* et al. 1987). Minima of these functions are located between 90 and 180°. The scattering function of suspension of a marine cyanobacterium was in fact found to be essentially independent of the scattering angle for angles between 120° and 155° (*Morel* and Bricaud 1986).

Weak, low-angular frequency oscillations can also be identified in some other data (Figure 4.35) e.g., *Cross* and Latimer (1972) observed distinct local minima, at about 34° and 60°, for cultures of rod-like bacteria *E. coli*. However, such minima were not found by *Lyubovtseva* and Plakhina (1976) who measured the scattering function and some matrix elements for the same species. The problem of reproducibility of measurements performed on cultures of living cells at various laboratories was discussed by *Van De Merwe* et al. (1989) who found that large discrepancies are possible and explained these discrepancies by the effects of the differences in the growth conditions of cellular cultures (Figure 4.35).

4.4.4.3. Phytoplankton

Cells of phytoplankton are much larger than those of bacteria. Plankton cells may also form macroscopic size colonies. Thus, the scattering functions are highly

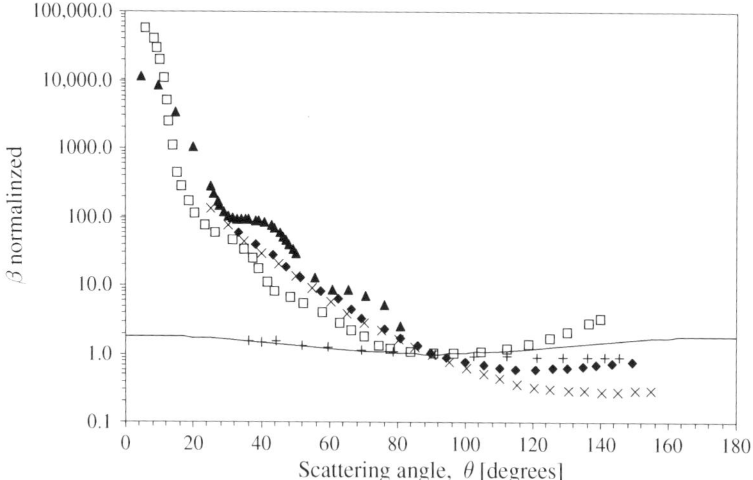

Figure 4.35. Volume scattering functions (normalized to 1 at 90°) for a marine virus (+, *Balch* et al. 2000, marine bacteriophage C2, 0.1 μm diameter, wavelength $\lambda = 514$ nm), a marine cyanobacterium (× *Morel* and Bricaud 1986, data from their Fig. 15 at $\lambda = 546$ nm; these authors also show a well-fitted Mie theory-based approximation of the scattering function for a measured size distribution of the cells peaking at 1.25 μm, and for an average relative refractive index of $1.035 - i0.001$) and for common bacteria: rod-shaped *E. coli*: ♦ *Lyubovtseva* and Plakhina (1976, 0.5 μm average diameter, 2–4 μm length, $n = 1.044$, $\lambda =$ 540 nm), ▲ *Cross* and Latimer (1972, 2.16 μm average length, 0.74 μm average diameter $n = 1.045$ for the cytoplasm, 1.10 for the cell wall—as derived from fitting a Rayleigh–Gans–Debye approximation for a shelled prolate spheroid to the angular scattering data at $\lambda = 404$ nm), and *Baccilus subtilis* which is used as a fungicide in agriculture (□ *Bickel* et al. 1976, strain UVS-42DPA, size, and refractive index not reported, $\lambda = 442$ nm). The scattering function of pure water is also shown with a solid line.

asymmetric and vary by several orders of magnitude within a measurement range extending from about 1° to typically 140°–170°.

Typically, the backscattering portions of these functions for unarmored cells are relatively flat, due to the relative refractive index of these cells being close to 1. The *Chlorella* sp., a ubiquitous freshwater and marine algae with spherical cells that have a smooth and soft cell wall has been a widely researched representative of such plankton (Figure 4.37).

The effect of a hard, silicate cell wall (such as in diatoms) or of calcite armor (as in coccolitophores) is surprisingly not too significant, as shown in Figure 4.36 where the scattering functions of some soft- and hard-walled phytoplankton cells are compared.

The scattering functions of suspensions of algae are generally smooth functions of the scattering angle even at a relatively high resolution on the order of 1°

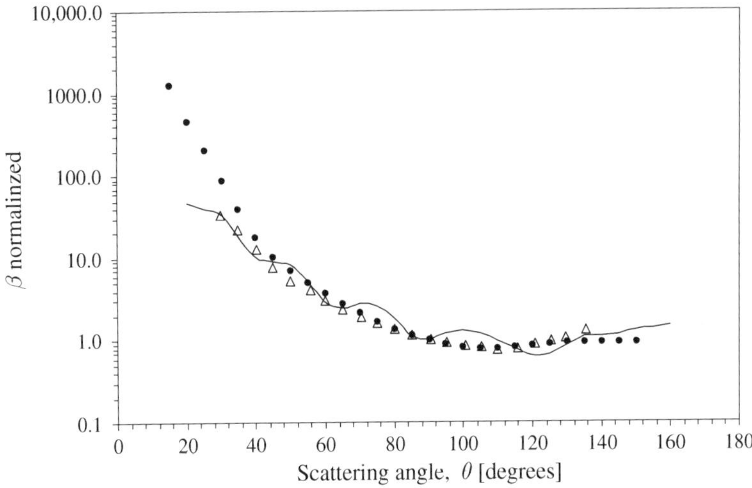

Figure 4.36. Volume scattering functions (normalized to 1 at 90°) for phytoplankton with soft and hard cells: • *C. vulgaris*, spherical soft cell wall, $D = 6\,\mu m$, $\lambda = 633\,nm$ (*Witkowski* et al. 1993)—*Prochlorotrix hollandica*, long cylinder, $D = 0.67\,\mu m$, $L = 60\,\mu m$, $\lambda = 633\,nm$ (*Volten* et al. 1998, size data from *Schreurs* 1996), △ *Thoracosphaera* with calcite armor, $D = 11\,\mu m$, $\lambda = 546\,nm$ (*Balch* et al. 1999)—*Cyclotella menegihiniana*, spherical diatom, $D = 10–30\,\mu m$, $\lambda = 633\,nm$ (Król 1998).

(*Privoznik* et al. 1978, *Burns* et al. 1976). However, some measurements (*Quinby-Hunt* et al. 1989) do show high-frequency oscillations. Definite oscillations can be seen in the scattering functions at the small angles (0.1°–19.5°, resolution of $\cong 0.2°$; *Price* et al. 1978) obtained for single algal cells ranging in size between 2 and 30 μm. In fact, it is this "fine" structure of the scattering patterns which permitted the latter authors to differentiate between species. However, it appears that the magnitude of the oscillations would be much reduced if suspensions of such cells were measured. This is supported by the results of *Kopelevich* et al. (1987) who studied, within a similar angular range (0.25°–6.5°), the optical properties of suspensions of seven marine bacteria species of various sizes (0.2 to $5 \times 1.8\,\mu m$) and shapes (spheres to cylinders) at wavelengths of 400, 500, and 700 nm.

Definite oscillations can be seen in the scattering function of *Prochlorotrix hollandica* (Figure 4.36), an "infinitely" long cylindrical algae (diameter 0f 0.67 μm, colony length:diameter $= \sim 100$). This is consistent with the fact that the random orientation of infinite cylinders does not average oscillations in the angular scattering pattern observed for individual cylinders (Figure 3.3). Interestingly, measurements by *Schreurs* (1996) of the scattering function for much larger "infinitely" cylindrical phytoplankton (*Oscillatoria agahardii*, cylinder

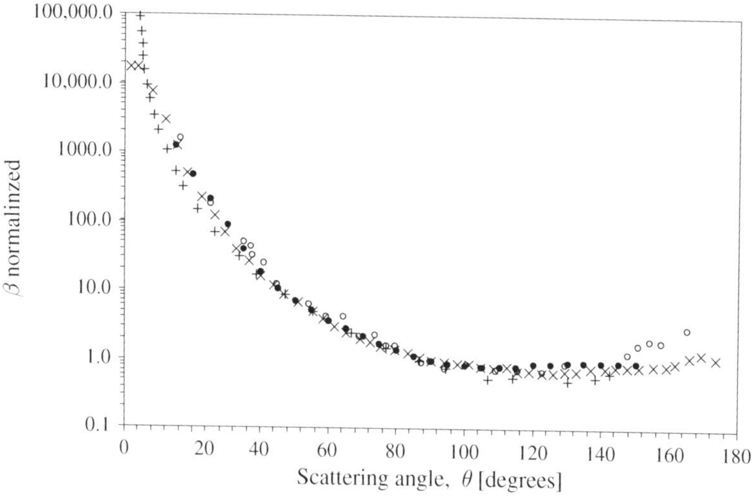

Figure 4.37. Volume scattering functions (normalized to 1 at 90°) for the *Chlorella* sp., a marine and freshwater spherical phytoplankton with a smooth soft cell wall: × *C. pyrenoidosa*, outer diameter $D = 1.7\,\mu m$, $\lambda = 633\,nm$ (*Privoznik* et al. 1978), ○ marine Chlorella off Hawaii, $D = 2\text{–}5\,\mu m$, $\lambda = 442\,nm$ (*Quinby-Hunt* et al. 1989), • *C. vulgaris*, $D = 6\,\mu m$, $\lambda = 633\,nm$ (*Witkowski* et al. 1993).

diameter, D, of $3\,\mu m$, and *O. amoena*, $D = 5.5\,\mu m$) did not show any significant oscillations. A plausible explanation of these observations can be given as follows. In the case of *P. hollandica*, the relative size parameter of the cylinder is about 4. As seen in Figure 3.4 in this relative size range, the oscillations in the angular scattering pattern have a relatively low amplitude and angular frequency, especially for a high relative refractive index (Schreurs reports an relative index of 1.235 for these cells). Such oscillations can be easily detected by an instrument with an angular resolution of $\sim 1.7°$ reported by *Volten* et al. (1998). However, the oscillation frequency increases rapidly with particle size and its amplitude decreases with increasing refractive index as seen in Figure 3.3. Thus, at a relative size of ~ 20 and ~ 37, and at a refractive index of 1.035 and 1.046, for *O. agahardii* and *O. amoena*. respectively, such oscillations in the scattering function may go undetected at the angular resolution reported.

4.4.4.4. Large aggregate particles

Despite the potential importance of these particles ($> 500\,\mu m$), frequently referred to as marine snow, in large-scale radiative transfer processes in natural waters, virtually all data on the scattering functions of aquatic particle species are relevant to relatively small particles, probably for reasons discussed in section 4.2.3.2 on the difficulties of measuring the volume scattering function. Publications by *Hou* (1997), *Hou* et al. (1997), and *Carder* and Costello (1994) appear to be the virtually single source of experimental data on light scattering by these large particles. That group measured the volume scattering functions at a wavelength of 685 nm of single particles larger than 280 μm in coastal waters of the Pacific by using an *in situ* microphotographic system. Their data refer to three relatively narrow angular ranges centered about 50, 90, and 130° respectively and represent single-particle scattering functions averaged over 5 m long sections of the instrument descent. Sample results (Figure 4.38) are shown as the Beardsley–Zaneveld approximation (*Beardsley* and Zaneveld 1969) fits to their data. The fit parameters listed in that figure caption are those listed in *Hou* (1997). A significant difference between these large-particle scattering functions and those of the smaller particles, typically measured, is the high forward slope of the function,

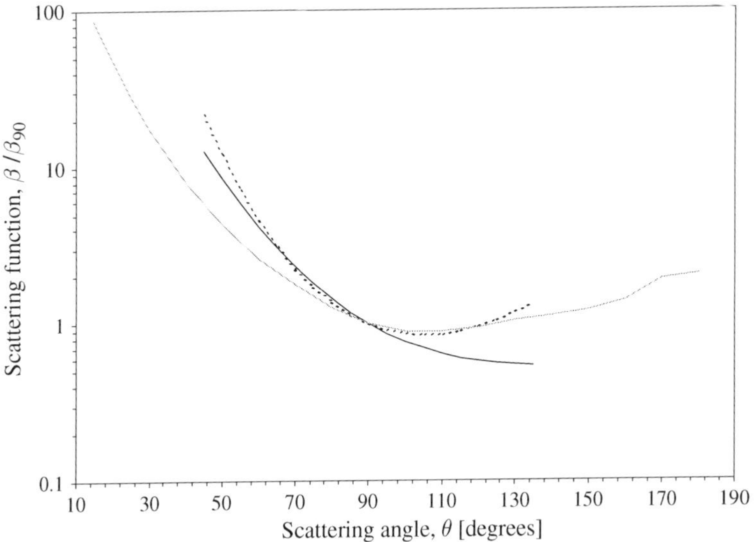

Figure 4.38. Beardsley–Zaneveld approximation (*Beardsley* and Zaneveld 1969), $\beta(\theta) = 1/[(1-F\cos\theta)^4(1+B\cos\theta)^4]$, where F and B are constants, fitted (black curves) to average single-particle volume scattering functions measured *in situ* at a diode laser wavelength of 685 nm for costal marine particles greater than 280 μm by *Hou* (1997, his Table 10; *solid black curve*—cast 4/19 10–15 m, $F = 0.82$, $B = 0.37$, *dashed black curve*—cast 4/21 0–5 m, $F = 0.96$, $B = 0.63$). Original data (not shown) are scattered over angular ranges of about 20° wide each, centered about 50, 90, and 130°. A volume scattering function of clear ocean water (*Petzold* 1972; Atlantic off Bahamas, 25 °N 78 °W, depth 1880 m, 510 ± 40 nm) is shown for comparison (*gray curve*). All functions are normalized to unity at 90°.

which translates into a significant contribution of the large particles, when present, to light scattering in natural waters.

4.4.4.5. Gas bubbles

Gas bubbles in the surface layer of the sea have become recently an object of interest as another candidate (after bacteria) for a seawater component that contributes significantly to the backscattering of light (*Terrill* et al. 1998, *Zhang* et al. 1998, *Stramski* 1994). Small bubbles are not distorted by the shear in the sea-surface layer and thus are perfect spheres.

Pure bubbles either dissolve or rise to the sea surface, and their effects on light scattering can only be brief (e.g., *Zhang* et al. 1998, *Johnson* 1986). Gas bubbles in seawater acquire a thin (monomolecular) coating of organic matter of high-molecular weight, i.e., substances that are present in seawater as dissolved organic matter (e.g., *D'Arrigo* et al. 1984, *D'Arrigo* 1984, *Johnson* and Cooke 1980). This stabilization mechanism is pertinent to small bubbles ($D <\sim 10\,\mu m$). For the larger bubbles, stabilization by adsorbed particles has been demonstrated by *Johnson* and Wangersky (1987). Such stabilization significantly extends lifetimes of bubbles in water and creates persistent bubble populations, even at depths that are not reachable by uncoated bubbles injected at the sea surface. See section 5.8.4.7 for more details on bubble populations in natural waters and their size distributions.

The scattering of light by a clean gas bubble in water can be explained exactly by the Mie theory (e.g., *Bohren* and Huffman 1983). The scattering of light by a coated bubble is likewise explained by the coated sphere theory (e.g., *Bohren* and Huffman 1983). Due to the relative refractive index of the bubble being less than unity, important modifications are observed in the light scattering properties of bubbles as compared with those of particles with the relative refractive index above unity.

First, the high-frequency oscillations (ripples) on the scattering efficiency as a function of the relative bubble size (x) disappear for bubbles. The ripples for solid spheres are due to interference between surface waves riding the sphere surface (see also section 6.4.1). Thus, as pointed by *Marston* et al. (1982), the lack of ripples for bubbles may be caused by the lack of entrapment of the internal surface waves for spheres with the relative refractive index of less than unity.

A coating of protein of under 0.1 of the bubble diameter on a bubble with a size parameter $>\sim 60$ (bubble diameter about $10\,\mu m$ for green light) has relatively little influence on the total scattering efficiency (*Zhang* et al. 1998). The backscattering efficiency, however, depends in a complex manner on the coating thickness as found by Zhang and colleagues. First, with a film thinner than 0.001 the bubble diameter, the effect is small. As the film thickens to 0.01 of the bubble diameter, the backscattering efficiency tends to increase. For yet thicker films, the backscattering efficiency decreases until it ultimately reaches a plateau for a film thickness of 0.1 of the bubble diameter.

Second, the existence of the critical angle, $\phi_C = \arcsin(n')$, where n' is the real part of refractive index of air (i.e., that of the bubble "particle") relative to that of water, and of the Brewster angle, $\phi_B = \arctan(n')$, creates special angular ranges for bubbles in liquid (Figure 4.39). The critical angle is the angle at which the incident light wave suffers total internal reflection when approaching a plane liquid–air interface from the liquid side. The Brewster angle is the angle at which the reflectance of the air–liquid surface vanishes for light polarized linearly in the incidence plane.

Marston (1979) developed a physical model of light scattering near the critical incidence angle that approximates the angular dependence of the light scattering intensity near the critical scattering angle, θ_C

$$\theta_C = \pi - 2\phi_C \tag{4.118}$$

That model, applicable for bubbles with relative sizes $x > 25$ and validated by observations of light scattering by millimeter-sized bubbles in water, describes coarse oscillations in the scattered light intensity that are similar to the Fresnel (near-field) diffraction by a knife edge (*Marston* 1979): as the scattering angle,

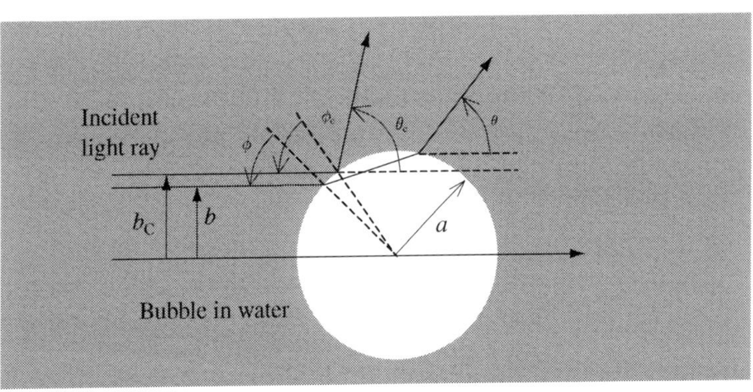

Figure 4.39. Light scattering by an air bubble in water. Total internal reflection at the water–gas surface prevents rays with impact distance $b \geq b_c$, for which the incidence angle is greater than the critical angle, ϕ_c, from penetrating into the bubble if the bubble is much larger than the wavelength of light. This latter requirement limits the tunneling of the light wave through the bubble. The existence of the critical angle creates a distinct angular scattering pattern (Figure 4.40) in the vicinity of the scattering angle $\theta_c = \pi - 2\phi_C$ corresponding to the critical angle. For air bubbles in water, $\theta_c \cong 83°$. For each refracted ray with an impact distance, b_1, such that $b_c > b_1 > 0$, and a scattering angle, θ, there exists a totally reflected ray with an impact distance, b_0, such that $a \geq b_0 > b_c$, for which the scattering angle also equals θ. Consideration of the diffraction and interference of such ray pairs leads to a physical optics model of light scattering by a bubble near the critical angle (e.g., *Marston* et al. 1982).

approaches and exceeds a value corresponding to the critical angle, the scattered light intensity decreases rapidly. For air bubbles in water, this angle is about 83°. At scattering angles smaller than θ_C (Figure 4.39), the scattered light intensity oscillates about a level slowly rising with decreasing angle until at the small scattering angles the diffraction by the bubble takes over (Figure 4.40).

The effect of the Brewster angle is to create a broad minimum of the scattered light intensity at angles around the Brewster scattering angle $\theta_B = \pi - 2\phi_B = \sim 106°$ for air bubbles in water caused by vanishing of the reflectivity of the water–gas surface for polarization parallel to the scattering plane (*Marston* et al. 1982).

The scattering patterns of clean air bubbles, as predicted by the homogeneous sphere (Mie) theory, are shown in Figure 4.40 along with those predicted by

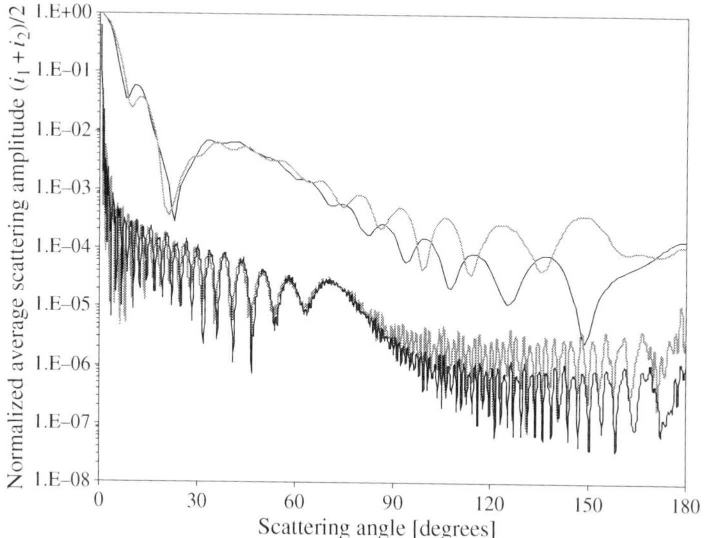

Figure 4.40. The scattering functions of small ($x = 25$, upper set of lines, angle increment 1°) and large ($x = 250$, lower set of lines, angle increment of 0.1°) air bubbles in water as calculated with the homogeneous sphere (thin lines) and a coated sphere theory (thick gray lines). The calculations were performed in double-precision arithmetic with programs developed by MJC Optical Technology that use downward recursion for the log-derivatives of the relevant Bessel functions. The refractive index of the bubble relative to seawater is taken to be 0.746. The relative refractive index of the coating is 1.20 (protein). The absolute shell thickness is the same for both bubbles and translates to a relative shell thickness (in the x-scale) of 0.0314 and 0.00314 for the small and large bubbles respectively. This corresponds to 0.1 μm (i.e., 10 times the size of large protein molecules) at a wavelength of 400 nm. A coating with the thickness approximately corresponding to a monomolecular layer of protein (10 nm) has little effect, especially for the larger bubble At a wavelength of 400 nm, the bubble diameters are ~ 3.18 μm, i.e., about the minimum stable bubble size in water (3 μm, *Zhang* et al. 1998) and ~ 31.8 μm respectively.

the coated sphere theory. *Zhang* et al. (2002) point out that the unique bump in the scattering functions of the large bubbles in water, between 60 and 80° (due to critical angle scattering at ~83°) can be used to evaluate the contribution of bubbles to the volume scattering function.

The knife-edge diffraction pattern becomes distinguishable near the scattering angle of 83° for the larger bubble. The effect of a realistic organic film (protein) at the bubble surface is relatively small at angles smaller than the critical angle but becomes relatively significant at the larger angles. The coating modifies the quasi-period of the coarse oscillations in the scattered light intensity and creates a scattering pattern that is more closely approximated by the Marston model than that of the clean bubble of the same size.

Some comments are in order regarding the calculation of the scattering pattern of a gas bubble in water by using the Mie theory. Such calculations involve recurrence relations for the logarithmic derivative of a Bessel function of an argument, $y = mx$, where x is the relative size parameter and m is the refractive index of the bubble relative to that of water. That derivative is commonly referred to as the $A_n(y)$ function, where n is the iteration index, following an algorithm established by *Deirmendijan* (1969).

Two modes of calculation of that function are possible: a fast but unstable upward recurrence (the iteration index n increases) and a slower but stable downward recurrence (n decreases from a pre-set maximum $> x$). While the upward recurrence mode generally works well for the real relative refractive index, $m > 1$, it tends to break down for $m < 1$ (Figure 4.41) . This breakdown is caused by a rapid accumulation of the rounding errors as the $A_n(y)$ approaches n corresponding to the asymptotic regime for $n >> y = mx$ (*Kattawar* and Plass 1967) for Im $m = 0$:

$$A_n(mx) \cong \frac{n+1}{mx} \tag{4.119}$$

Given a relative size parameter, x, the asymptotic regime is approached much earlier with the $m < 1$ within an x-dependent range of n required for the Mie series convergence. For $m > 1$ and moderate values of the size parameter (we tested x of up to 1000 and $m \geq 1.01$), the asymptotic regime is not reached within that range.

This breakdown of the upward recurrence algorithm requires the use of the downward recurrence in A_n for the calculation of the angular light scattering patterns for gas bubbles in liquids.

Minerals. The small size of the mineral particles, which is a prerequisite for their substantial residence time in water bodies, coupled with their large refractive index results in a somewhat lower asymmetry of the scattering function than that characteristic of phytoplankton and bacteria. An interesting result was obtained by *Lyubovtseva* and Plakhina (1976) who measured the scattering functions of

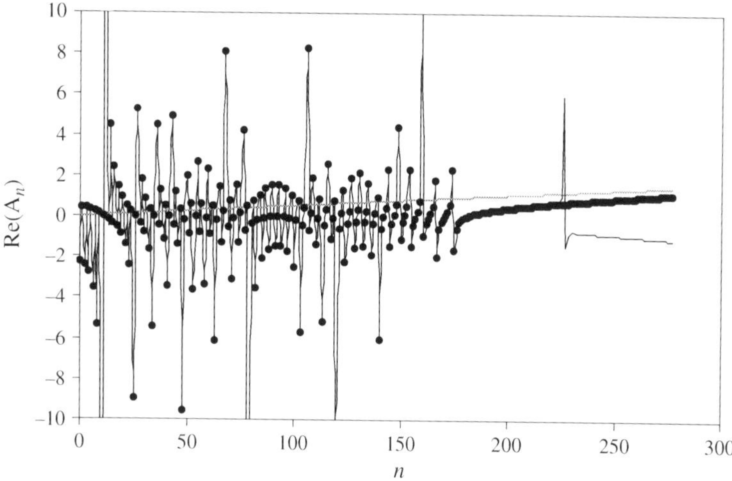

Figure 4.41. The convergence of $A_n(mx)$ function (thin black lines), used in the Mie scattering calculations, to its asymptote [thick gray line, (4.119)] for an uncoated air bubble in water (relative bubble size $x = 250$, relative refractive index $m = 0.75$). The asymptotic regime, marked by the convergence of the curve with symbols (downward recurrence algorithm), and the asymptote (gray line) is reached for $m < 1$ well below an x-dependent (*Wiscombe* 1980) value of the iteration index n required for satisfactory convergence of the Mie series (n corresponding to the far right end of the curves). Upward recurrence (thin line, no symbols) causes, unlike for $m > 1$, a divergence of the A_n from the asymptote following the last violent oscillation at $n \sim 225$. Oscillations that occur normally at the lower values of the iteration index cancel the effect of rounding errors which start to accumulate only when the A_n begins a monotonic approach to the asymptote.

montmorillonite platelets freshly dispersed in water and of a several day-old suspension. Montmorillonite platelets swell when immersed in water: their thickness can increase as much as 50-fold after several days of soaking. This swelling was reflected in the increased asymmetry of the volume scattering function as compared with that of the fresh preparation. However, Lyubovtseva and Plakhina found that the differences between the scattering functions (as well as the matrix elements as we will discuss shortly) of small mineral particles of highly diverse shapes (needles and thin plates) are relatively minor and make it difficult, e.g., to infer determination of the particle shape from the shape of the scattering function.

The relationship between the asymmetry of light scattering in the backward direction, expressed by the ratio of $\beta(140°)/\beta(90°)$, and the non-sphericity of the particles was investigated by *Gibbs* (1978). He found that between the scattering functions of water suspensions of micrometer-sized glass spheres, crushed quartz grains, and mica flakes, the scattering function of the spheres exhibited the least asymmetry, $\beta(140°)/\beta(90°) \sim 5$. The ratio $\beta(140°)/\beta(90°) \sim 22$ was found for irregular quartz grains. The mica flakes resulted in the highest asymmetry, $\beta(140°)/\beta(90°) \sim 40$. The effect of the differences in refractive index was expected to be minimal because the latter was similar for all three types of particles. Backscattering by irregular quartz grains was greater than that for spheres in direct opposition to the microwave studies of *Zerull* et al. (1977) on particle aggregates and also to the calculated scattering functions of moderately non-spherical particles. However, a recent work by *Umhauer* and Bottlinger (1991, 1990) suggests that irregular quartz grains with sizes similar to that used by Gibbs can result in a large variability of the scattering pattern.

A conclusion of *Gibbs* (1978) that the non-sphericity of particles can be determined by interpolating between the three asymmetry factors of their volume scattering function (his Fig. 4) seems unwarranted. This is because the asymmetry of the volume scattering function depends on the asymmetry of the particles in a complex, non-monotonic manner, as indicated by, for example, the numerical simulations of *Wiscombe* and Mugnai (1988).

4.4.5. The scattering matrix of various aquatic particles

4.4.5.1. Bacteria and phytoplankton

The matrices (Fig. 1.14) measured for different species (*Quinby-Hunt* et al. 1989, *Fry* and Voss, 1985, *Lyubovtseva* and Plakhina 1976) are generally quite similar to each other and to those of seawater measured by *Voss* and Fry (1984) (see section 4.4.3). The greatest differences are observed in the m_{33}, m_{34}, and m_{44} elements of the scattering matrix. In fact, it has been postulated (as we have already noted) that the sensitivity of the element m_{34} to the particle structure and shape can be used to differentiate between biological particles (*Bickel* et al. 1976), although for such differentiation to be reliable, the growth conditions for the cells must be similar (*Van De Merwe* et al. 1989). The research of *Stramski* et al. (1995), *Stramski* and Reynolds (1993), *Stramski* et al. (1988), and *Ackleson* et al. (1988a) indicates that growth conditions can significantly affect the refractive index and volume of the cells (see relevant discussion in Chapter 3).

The optical activity of marine particles has been recently examined (*Shapiro* et al. 1991, *Shapiro* et al. 1990) as indicated by a non-zero value of the m_{14} (or m_{41}) element of the scattering matrix. Significant optical activity in single cells of some dinoflagellates has been observed. This activity has a substantial diurnal variability, as m_{14} can apparently increase fourfold about midnight. The optical activity of these species is likely to be caused by helically structured chromosomes which contain substantial amount of DNA. The DNA molecules are optically active due to their own helical structure (*Oldenbourg* and Ruiz 1989).

4.4.5.2. Minerals

Early measurements of *Lyubovtseva* and Plakhina (1976) indicated that the scattering matrices of mineral particles (montmorillonite and palygorskite, Figure 4.42)

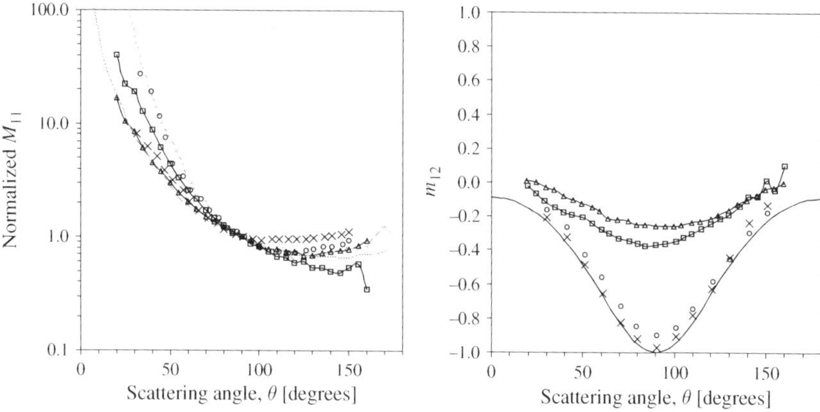

Figure 4.42. Selected elements of the scattering matrix of some minerals: ×—palygorskite and ○—montmorillonite, particle size $< 1\,\mu m$ (*Lyubovtseva* and Plakhina 1976), □—silt 3–5 μm, and △—silt 5–12 μm (*Volten* et al. 1998). The matrix element M_{11} is normalized to unity at 90°, $m_{12} = M_{12}/M_{11}$. Matrix elements M_{11} for red clay in air (solid gray curve, *Muñoz* et al. 1999) and chlorella cells (dashed gray curve, *Privoznik* et al. 1978) are shown for comparison in the left panel. Thin black curve in the right panel represents m_{11} of pure water, i.e., the Rayleigh scattering regime.

have as non-zero elements only $m_{12} = m_{21}$, m_{22}, and $m_{33} = m_{44}$ (the element $m_{11} \equiv 1$). The particle diameters were on the order of a fraction of $1\,\mu m$. Element m_{22} decreases from 1 at 30° to about 0.9 at 150°. Element m_{12} was equal to m_{21} and followed closely the angular pattern of the Rayleigh regime, except for the maximum value which was on the order of 0.8 instead of 1. Elements m_{33} and $m_{44} = m_{33}$ also followed closely the angular dependence characteristics of the Rayleigh regime. *Volten* et al. (1998) have recently measured the scattering function and linear polarization degree (the $-m_{12}$ element of the scattering matrix) of marine sediment (silt 3–5 μm and 5–12 μm fractions) at 633 nm (Figure 4.42). They found the scattering functions to have similar or lower asymmetry than that of the Petzold's turbid water function (*Petzold* 1972, San Diego harbor). The linear polarization degree was also lower than that for seawater.

Volten et al. (2001), *Muñoz* et al. (1999), and *Kuik* et al. (1991) have all measured the light scattering matrices at wavelengths of 633 and/or 441.6 nm of mineral particles (feldspar, red clay, and quartz; effective diameters on the order of several μm) in air by using an instrument employed by *Volten* et al. (1998). The results for mineral aerosols are similar to those of *Volten* et al. (1998) obtained for mineral suspensions in water. All these authors also found the maximum of the matrix element m_{34} (at about 90°) to be substantially different from zero. The maximum linear polarization degree, of mineral aerosols may even be smaller than that of marine silt suspension. It is the lowest (on the order of 0.12) for quartz aerosols. Interestingly, the degree of linear polarization for all these particles becomes negative at scattering angles of between $\sim 160°$ and $180°$. It is important to note here that the degree of linear polarization (given by $-m_{12}$, for example, *Muñoz* et al., 1999) depends on the particle size and composition in a complex manner (e.g., *Volten* et al. 2001). For particles much smaller than the wavelength of light (the Rayleigh regime), $-m_{12}$ increases from ~ 0 at the scattering angle of $0°$ and $180°$ to unity at $90°$ (Figure 4.42). However, for particles much larger than the wavelength of light (the geometric optics regime) and non-zero imaginary part, n'', of the refractive index, the degree of linear polarization, $-m_{12}$ can also be relatively large. Thus, the $-m_{12}$ is affected both by the particle size distribution and by the composition of the particles (i.e., n'').

Given the similarities between and the limited variability of the scattering matrices of the mineral particles, both *Volten* et al. (2001) and *Muñoz* et al. (1999) considered it meaningful to define average scattering matrices of mineral particles. This is of interest especially for remote sensing applications where the relevant properties of the particles are rarely known.

The limited body of experimental data regarding the angular scattering patterns of mineral particles in water seems surprising. These particles tend to dominate light scattering in both the turbid coastal waters and many inland water bodies. Many minerals that occur commonly in these water also have significant absorption in the visible, typically due to iron oxide (e.g., *Babin* and Stramski 2003). A better knowledge of the optical properties of these species would certainly contribute to, e.g., higher accuracy of the particle loads in natural waters as estimated from remote sensing.

4.5. Approximations of the volume scattering function

4.5.1. Pure water and pure seawater

The importance of an accurate functional representation for the scattering function of pure seawater cannot be understated. We have already discussed in detail the theoretical basis of that representation in Chapter 2. Here we concentrate on the practical approach and will attempt to provide simple semi-empirical approximate formulas that can be quickly used in the field to evaluate experimental results.

4.5.1.1. Scattering by density fluctuations at a molecular size scale

According to *Morel* (1974), who examined theoretically and experimentally the light scattering by pure water and pure seawater, the volume scattering function of pure water can be described by the following equation:

$$\beta(\theta) = \beta(90°)\left(1 + \frac{1-\delta}{1+\delta}\cos^2\theta\right) \tag{4.120}$$

where δ is the depolarization factor. This is essentially the scattering pattern of anisotropic particles that are much smaller than the wavelength of light, i.e., in the Rayleigh scattering regime, as we have already noted in Chapter 2. The Rayleigh scattering pattern is symmetrical about the scattering angle of 90°. Thus, the mean cosine of the scattering angle $\langle cos\theta \rangle$ vanishes.

When the unpolarized light is scattered by such an anisotropic medium, the scattered light at both 0° and 180° contains a small polarized component. The degree of linear polarization, $P_L(\theta)$, can be expressed as follows:

$$P_L = \frac{i_\perp - i_\parallel}{i_\perp + i_\parallel} = \frac{1 - \sqrt{\frac{1-\delta}{1+\delta}} \cos\theta}{1 + \sqrt{\frac{1-\delta}{1+\delta}} \cos\theta} \tag{4.121}$$

where $i_\parallel$ and $i_\perp$ are respectively the intensities of the parallel- and perpendicular-polarized components of the scattered light irradiance. The orientation of polarization is given in reference to the scattering plane. According to Morel, the best value of the depolarization factor δ is 0.09. The value of the depolarization factor reported by different authors varies significantly. However, as discussed previously in Chapter 2, the most reliable value seems to be that measured by *Farinato* and Roswell (1975) using an argon–ion laser and a narrow bandwidth detector to minimize stray light effects. According to these authors, the best value of the depolarization factor δ is 0.039. Thus, the degree of linear polarization at 0° (and 180°) of light scattered by pure water (pure seawater) is ~0.0195, i.e., there is a small perpendicular-polarization component (the value of P_L is positive).

The scattering function at 90° is expressed as follows:

$$\beta(90°) = \frac{2\pi^2 KTn^2}{\lambda^x} \frac{1}{\beta_T} \left(\frac{\partial n}{\partial p}\right)_T^2 \frac{6+6\delta}{6-7\delta} \tag{4.122}$$

where K is the Boltzmann constant, T is the absolute temperature, n is the refractive index of water at a wavelength λ in air, exponent x controls the wavelength dependency of light scattering, β_T is the isothermal compressibility of water, and p is pressure. The isothermal compressibility of seawater has been reported by *Horne* (1969) and by *Lepple* and Millero (1971) as a function of temperature, salinity, and pressure. There is an extensive literature on the refractive index of pure water and seawater. We cite important references in sections 2.4 and 6.2 where we also discuss significant differences between the different experimental data as well as between data calculated using different approximation formulas. The results of section 2.4, using the seawater index formula of *Quan* and Fry (1995) and a moderately complex evaluation procedure for pressure effects, are more accurate than those we will quote here. Here, we use the results of *Millard* and Seaver (1990) and give a formula for the calculation of the $(\partial n/\partial p)_T$ derived from their formula for the refractive index of seawater as a function of temperature, T [°C], salinity, S [ppt], and pressure, p [dbar]:

$$\left(\frac{\partial n}{\partial p}\right)_T = n_{18} + 2n_{19}p + n_{20}\lambda^{-2} + n_{21}T + n_{22}T^2 + 2n_{23}pT^2 + (n_{24} + n_{25}T + n_{26}T^2)S \tag{4.123}$$

This formula has the distinct advantage of allowing a simple evaluation of the effects of pressure on scattering. The coefficients, n_i, in (4.123) are listed in Table 6.5. The values of $(\partial n/\partial p)_T$ calculated from (4.123) are systematically lower than values reported elsewhere (Table 4.5). Since this variable enters (4.122) in the second power, such differences result in significant differences in the scattering function for seawater. For example, according to the data on the physical properties of water reported by *Kratohvil* et al. (1965) and used by *Morel* (1974) we have at $\lambda = 405\,\text{nm}$ and $T = 25°\text{C}$: $n = 1.343$, $(\partial n/\partial p)_T = 1.53 \times 10^{-10}\,\text{m}^2\text{N}^{-1}$, and $\beta_T = 4.56 \times 10^{-10}\,\text{m}^2\text{N}^{-1}$, and thus $\beta(90°) = 3.40 \times 10^{-4}\,\text{m}^{-1}$. On the other hand, by using (4.123), we have $n = 1.342$, $(\partial n/\partial p)_T = 1.48 \times 10^{-10}\,\text{m}^2\text{N}^{-1}$, and $\beta(90°) = 3.17 \times 10^{-4}\,\text{m}^{-1}$. This is less by about 7% than the previous $\beta(90°)$ value.

The effect of salinity is expressed mostly through an additional term due to the fluctuations of concentration of the salts. An NaCl solution of 0.035 [g/g], which has approximately the same Cl^- ion concentration as seawater at a salinity of 38 ppt, scatters light about 1.18 to 1.20 times as much as pure water (*Morel* 1974). Based on the experimental results for the molecular scattering of light by solutions of relevant salts and by artificial as well as natural seawater, Morel assumes that the scattering of light by seawater is greater than that by pure water by a factor of 1.3.

By using (4.122) and (4.123) with the appropriate values of the isothermal compressibility (e.g., Horne 1969), one can verify that the effect of pressure can

Table 4.5. Values of $(\partial n/\partial p)_T$ for pure water at the atmospheric pressure calculated from (4.123) and reported by other authors.

Reference	λ [nm]	T[°C]	Value from ...	
			Reference	Equation (4.123)
O'Connor and Schlupf (1967)	632.8	1	1.65×10^{-10}	1.60×10^{-10}
		25	1.47×10^{-10}	1.44×10^{-10}
Kratohvil et al. (1965)	405	25	1.53×10^{-10}	1.53×10^{-10}

be significant. For pure water, at a pressure of 1000 dbars (corresponding roughly to a depth of 1000 m in the ocean) and at a temperature of 1 °C, the pressure would decrease the scattering of light by about 13% as compared with that at atmospheric pressure.

According to Morel, the best value of the exponent, x, in (4.122) equals 4.32. This is slightly more than an exponent of 4 characteristic of a Rayleigh scattering medium. This difference is due to a reinforcement of the wavelength selectivity of $\beta(90^\circ)$ by the dispersion in n and in $(\partial n/\partial p)_T$. However, the best fit to Morel's own results (his Table 4) is obtained with an exponent of 4.24, as shown in Fig. 4.43:

$$\beta(90^\circ) = 7.47 \times 10^{-6} \lambda^{-4.24} \tag{4.124}$$

for pure water, and

$$\beta(90^\circ) = 9.68 \times 10^{-6} \lambda^{-4.24} \tag{4.125}$$

for pure seawater with a salinity of 30 to 35 ppt, where the dimension of β is $m^{-1}sr^{-1}$. In order to avoid problems with dimensions raised to non-integer powers, the wavelength (in μm) is taken to be relative to 1 μm, so that the resulting "wavelength" is non-dimensional.

Boss and Pegau (2001) developed an approximation to the *Morel* (1974) data that includes the salinity as follows:

$$\beta(90^\circ) = 1.38 \times 10^{-4} \left(\frac{\lambda}{0.5}\right)^{-4.32} (1 + 0.008108S) \tag{4.126}$$

where λ(μm) is the wavelength of light and S (ppt) is the salinity. The constant 0.008108 represents a ratio of 0.3 to 37 in the practical salinity (non-dimensional) units. The practical salinity is approximately equal to salinity expressed in ppt (for example, *Dera* 1992).

We have slightly modified the approximation of Boss and Pegau by adopting an exponent of −4.24 as previously indicated. This modification results in the following equation:

$$\beta(90^\circ) = 7.47 \times 10^{-6} \lambda^{-4.24} (1 + 0.008027S) \tag{4.127}$$

where λ is the wavelength in μm, relative to a wavelength of 1 μm that reduces to (4.124) for $S = 0$. Both approximations are shown in Figure 4.43 compared with the *Morel* (1974) data.

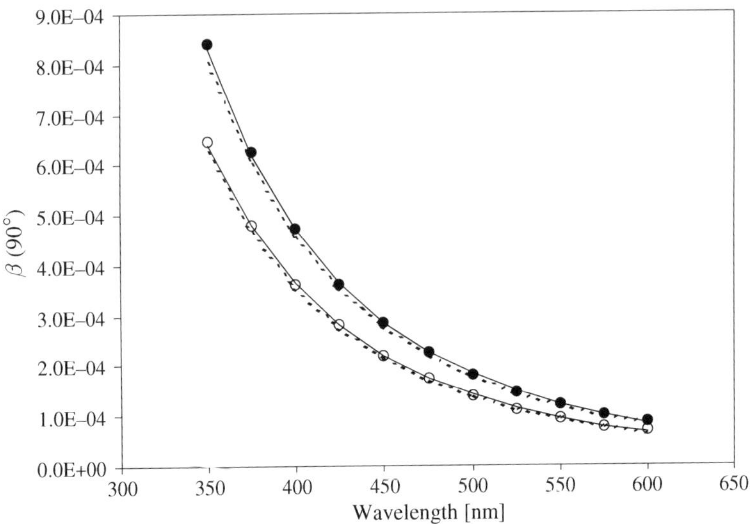

Figure 4.43. Spectra of the value of the volume scattering function at 90° of pure water (∘) and pure seawater (•, $S = 37$ ppt). Points represent the data of *Morel* (1974). Lines represent best fits as described by equations (4.127) (full lines) and (4.126) (dashed lines).

Finally, by integrating the pure water (seawater) scattering function (4.120) over the entire solid angle we obtain an expression for the scattering coefficient of pure water (e.g., *Morel* 1974) as a function of the wavelength and salinity:

$$\begin{aligned} b_{\mathrm{w}} &= \frac{8\pi}{3}\beta(90^{\circ})\frac{2+\delta}{1+\delta} \\ &= 0.00012\lambda^{-4.24}(1+0.008027S) \end{aligned} \tag{4.128}$$

where we used $\delta = 0.09$ after *Morel* (1974) and also used equation (4.127) (Figure 4.44).

The correctness of the data on the scattering of light by the pure water (pure seawater) is of utmost importance in the separation of the effects of water and particles on the scattering of light.

4.5.1.2. Scattering by micro-turbulence

The effect of turbulence on small-angle scattering is well established in atmospheric optics (e.g., *Crittenden* et al. 1978). However, as we already mentioned in section 4.4.2.6, there are few data available for seawater. The turbulence is essentially a continuation, toward the large sizes, of the molecular density fluctuations in seawater and can contribute significantly to the volume scattering function at very small angles ($< 1°$).

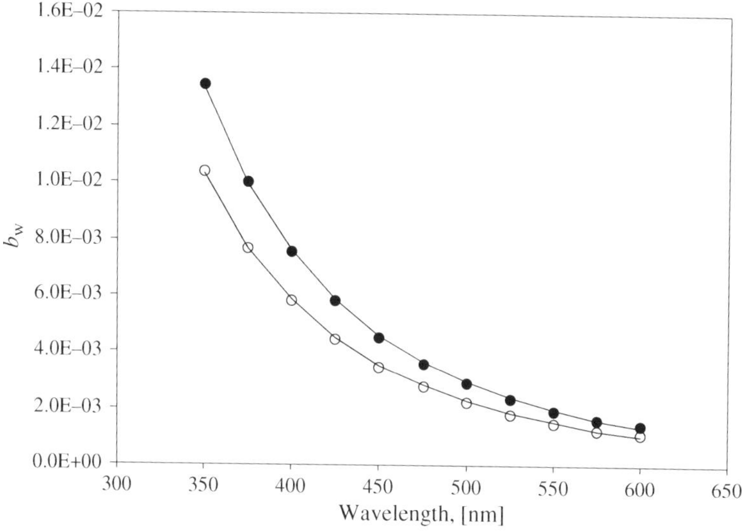

Figure 4.44. Spectra of the scattering coefficient of pure water (○) and pure seawater (●, S = 37 ppt). Points represent the data of *Morel* (1974). Lines represent best fits as described by equation (4.128).

The effect of turbulence cannot be modeled using a scattering theory based on the localized-particle model, such as the Mie theory, nor on the molecular scattering model. It requires ray-tracing through a continuous spatial distribution of the refractive index. Alternatively, a parabolic Helmholtz equation describing the propagation of light through water can be solved numerically to obtain the spatial distribution of the amplitude of the light beam (*Bogucki* et al. 1994). The effect of turbulence on the volume scattering function is most likely quite variable as suggested by numerical experiments performed by Bogucki and colleagues. Their results indicate that the component of the volume scattering function due to turbulence attains a higher magnitude in low-turbulence environments than in high-turbulence environments although at a smaller scattering angle ($\beta \sim 10^7\,\mathrm{m^{-1}sr^{-1}}$ for low-turbulence at scattering angles $<\sim 0.001°$, $\beta \sim 10^3\,\mathrm{m^{-1}sr^{-1}}$ for high-turbulence environments at angles $<\sim 1°$). A continuation of that work (*Bogucki* et al. 1998) indicates that the scattering of light by turbulence can exceed that due to marine particles at angles smaller than 0.1°.

Early calculations (*Tatarski* 1961, as quoted by *Chilton* et al. 1969) show that the scattering function due to turbulence can be approximated by:

$$\beta(\theta) = \exp\left[-\mathrm{const}\frac{(kR)^2}{2}(1-\cos\theta)\right] \qquad (4.129)$$

where k is the wavenumber of the light wave and R is the radius of the representative turbulence eddy diameter. An interesting feature of the work of Chilton et al. is the use of a semi-analytical Monte Carlo technique originally developed for modeling the propagation of high-energy cosmic-ray protons.

4.5.2. *The whole seawater and particulate matter*

The search for a universal analytical formula approximating the volume scattering function of seawater has been stimulated mainly by the needs of modeling the radiative transfer in seawater. However the usefulness of a good approximation reaches beyond this application because the reduction in the number of parameters required for the description of the function simplifies the classification and hopefully identification of the light scattering signatures of the medium (*Reynolds* and McCormick 1980). The need to reduce the large number of "degrees of freedom" of the original data also stimulated the search for a versatile approximation of the particle size distributions of marine particles that we discuss in the next chapter.

A purely statistical approach to the approximation and classification of the volume scattering function, can also be useful (*Dean* 1990, *Price* et al. 1978). The statistical approach of principal components is discussed in more detail later in this section and in the next chapter. With such an approach, *Price* et al. (1978) successfully used a statistical method of classification of multi-parameter data (cluster analysis) adapted from high-energy physics (*Ludlam* and Slansky, 1977) to identify phytoplankton species from single-particle light scattering measurements.

All approximations of the volume scattering functions reviewed in this section include powers of the scattering angle to account for a power-law dependence of the volume scattering function on the scattering angle in an angular range $\theta < 90°$ which we discussed already in Chapter 3.

4.5.2.1. *The FF function*

This two-parameter approximation for the marine phase function, derived by *Fournier* and Forand (1994), has received little attention in the marine optics community until recently, when it was recognized as one of the most accurate representation so far of real oceanic scattering functions (*Haltrin* 1997), especially in the forward-scattering region. Its slow acceptance was probably influenced by its somewhat complex form and a need for a non-linear fitting algorithm. We discussed the derivation of this function and its implications in Chapter 3. Here we provide a concise overview and concentrate on the discussion of the fitting procedure. Note that by virtue of its derivation, the FF function applies to particle scattering only. Thus, the pure seawater scattering must be subtracted from experimental light scattering data for the whole seawater for a meaningful fit. This especially applies to clear ocean waters. In turbid waters, where the contribution of the pure seawater is generally negligible, the FF function can usually be fitted to the scattering data for the whole seawater.

In its most recent form (*Forand* and Fournier 1999), the FF phase function can be expressed as follows:

$$p_{\mathrm{FF}}(\theta) = \frac{1}{4\pi}\frac{1}{(1-\delta)^2\delta^{\nu}}\Big([\nu(1-\delta)-(1-\delta^{\nu})] + \frac{4}{u^2}[\delta(1-\delta^{\nu})-\nu(1-\delta)]\Big) - \frac{1}{16\pi}\frac{1-\delta_{\pi}{}^{\nu}}{(1-\delta_{\pi})\delta_{\pi}{}^{\nu}}(3\cos^2\theta-1) \tag{4.130}$$

where

$$\nu = \frac{3-m}{2},\ \delta = \frac{u^2}{3(n-1)^2},\ \delta_{\pi} = \frac{4}{3(n-1)^2},\ u = 2\sin(\theta/2) \tag{4.131}$$

and $1 < n < 1.35$ is the refractive index of the particles relative to seawater and $3.5 < m < 5$ is the slope of the power-law size distribution:

$$f(D) = kD^{-m} \tag{4.132}$$

Note that the FF approximation for the phase function does not include the wavelength of light. Only the scattering coefficient which in this formalism is expressed as follows:

$$b(\lambda, m, n) = \frac{k2^{-m}\pi}{\cos(\pi m/2)}\left[\frac{2\pi(n-1)}{\lambda}\right]^{m-3} \tag{4.133}$$

is wavelength dependent. Finally, the normalized cumulative probability of scattering into an angular range of 0° to θ is given by the following expression:

$$w(\theta) = \frac{1}{(1-\delta)\delta^{\nu}}\left(1-\delta^{\nu+1}-\frac{u^2}{4}(1-\delta^{\nu})\right) - \frac{1}{8\pi}\frac{1-\delta_{\pi}{}^{\nu}}{(1-\delta_{\pi})\delta_{\pi}{}^{\nu}}\pi\cos\theta\sin^2\theta \tag{4.134}$$

which leads to the following expression for the backscattering probability (range of 90° to 180°):

$$w_{\mathrm{b}} = 1 - w(90°) \tag{4.135}$$

Forand and Fournier (1999) used a SIMPLEX non-linear fitting algorithm to find the parameters of the phase function. In this work, we used a simple search algorithm instead, which returns parameter values consistent with the more complex SIMPLEX algorithm. The search algorithm was developed to fit the scattering function, not the phase function, because most experimental light scattering data are available as absolute scattering functions rather than phase functions. Thus, we needed to retrieve one more parameter in addition to m and n, i.e., the scattering coefficient, b.

The algorithm performed a systematic search for a minimum of an RMS error of approximation in three-dimensional space (b, m, n). Due to the well-behaved error surface, we could use a two-step approximation. We first found the minimum error in a sparse grid of m and n that covered a realistic domain relevant to marine particles. At each grid intersection, we searched for the error minimum by using a fast converging triple-step algorithm in the b-dimension as follows. The error was evaluated at three values of the b equally spaced in the $\log b$ scale. The value of b which yielded the minimum error became the center of a new triple of b values, with the edge values spaced $1/2$ of the previous step in the $\log b$ scale. Once the minimum error was located in the coarse (m, n) grid, the grid was magnified and the second order search was repeated by using the same macro.

Figure 4.45 shows the contour map of the approximation error in the (m, n) subspace, as obtained with our algorithm for the turbid seawater listed in Table 4.2. It can be seen that the error isolines reflect the mutually compensating roles of the particle size distribution slopes and the refractive index of the particles. This compensation is based on the dependence of the scattering efficiency of particles on the refractive index. It becomes limited for the large values of the index and for sufficiently large particles. In that region, the scattering efficiency flattens out as a function of the refractive index which limits the compensational effect of changes in the refractive index (e.g., *Jonasz* and Prandke 1986).

4.5.2.2. Exponential and power-law functions

At small angles ($\theta < 1°$), the phase function of seawater has also been approximated by using an exponential function (*Chilton* et al. 1969):

$$p(\theta) = \exp[-C(1 - \cos\theta)] \tag{4.136}$$

where C is a constant. This expression approximates closely the diffraction pattern of an opaque disk up to the first zero of the pattern.

For small angles, $\theta < 1°$, equation (4.136) can be approximated by the following formula (*Chilton* et al. 1969):

$$p(\theta) = C_1 \exp(-C\theta^2) \tag{4.137}$$

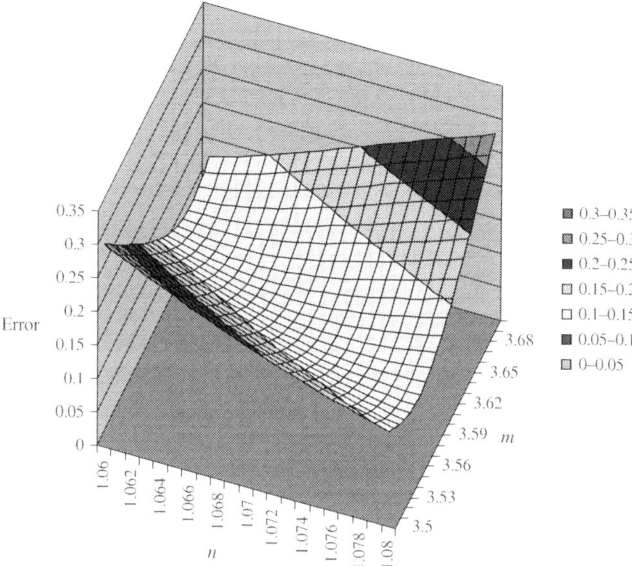

Figure 4.45. An error surface map, $\varepsilon(m, n)$, for the second-order fitting of the FF function to the turbid seawater particle scattering function listed in Table 4.2 and shown in Figure 4.25. The third parameter, the scattering coefficient, b, has already been fitted for each combination of m and n. The minimum error of 0.1036 is located at $m = 3.59$ and $n = 1.073$. The innermost contour (containing the minimum error) represents an error of 0.105, the increment between contours is 0.005. The error surface resembles a long valley: combinations of the m and n along the valley bottom yield comparable approximation error values.

and at large angles. $\theta > 1°$, the phase function was approximated by:

$$p(\theta) = C_2(1-\cos\theta)^{-b} \tag{4.138}$$

where

$$\begin{aligned} C_1 = &\frac{1}{2\pi\eta}\{1-\exp[-\eta(1-\cos\theta)]\} \\ &+\exp[-\eta(1-\cos\theta)]\,(1-\cos\theta)\frac{[(1-\cos\theta)/2]^{b-1}}{b-1} \end{aligned} \tag{4.139}$$

and

$$C_2 = C_1(1-\cos\theta)^b \exp[-\eta(1-\cos\theta)] \tag{4.140}$$

According to *Chilton* et al. (1969), the scattering functions of *Morrison* (1967) were well approximated by the choice of $b = 1.08$. However, the available small-angle scattering data were fitted by widely ranging values of η (10^3 to 10^8), indicating the need for accurate small-angle scattering function measurements, especially when evaluating imaging problems in seawater.

Morrison (1970) reviewed the following approximations to the volume scattering function:

$$\beta(\theta) = C \exp(-A\theta - B\theta^2) \tag{4.141}$$

in an angular range of 30° to 50°, with A, B, and C, being adjustable parameters, and

$$\beta(\theta) = C \exp(D\theta) \tag{4.142}$$

in an angular range of 115° to 135°, where the adjustable parameters, C and D, have different values in both equations. Morrison also tested the following log-normal approximation:

$$\ln \beta(\theta) = A + B \ln \theta + C(\ln \theta)^2 \tag{4.143}$$

with adjustable parameters, A, B, and C, which he found suitable for expressing the scattering function of seawater at the small angles.

In their study of the asymptotic optical field in the ocean, *Beardsley* and Zaneveld (1969) approximated the scattering function by using the following formula:

$$\beta(\theta) = \frac{\beta_0}{(1 - F\cos\theta)^4(1 + B\cos\theta)^4} \tag{4.144}$$

where β_0, F, and B are adjustable parameters. Parameter F controls the forward-scattering slope and parameter B controls the backscattering slope, almost independently. This approximation fits the scattering function data relatively well in the mid- and backscattering angular ranges and was used to model the light scattering functions of large particles (e.g., *Hou* 1997) and algae (*Balch* et al. 1999).

McLean and Voss (1991) approximated the phase function of seawater with an equation:

$$p(\theta) = \theta_0 \left[2\pi(\theta_0^2 + \theta^2)\right]^{-2/3} \tag{4.145}$$

where θ_0 is a free parameter. This expression was used by *Wells* (1973) because it provides a closed form solution for the MTF of a light scattering medium resulting from the Wells' small-angle radiative transfer theory. However, this seems to be the only advantage of that functional representation of the scattering function as it approximates the scattering functions of natural waters rather poorly. *DeWeert* et al. (1999) used a modification of that function as follows:

$$p(\theta) = \frac{\text{const}}{(\theta^2 + a^2)^{3/4}(\theta_0^2 + \theta^2)^{1/2}} \tag{4.146}$$

The parameter θ_0 affects the form of the function at most angles, while a fixes a finite value of the phase function at $\theta = 0°$. According to DeWeert and colleagues, a choice of $\theta_0 = 0.007$ rad and $a = 0.0005$ rad approximates the benchmark phase functions (*Petzold* 1972) very well out to angles of less than 90°.

Recently, *McLean* et al. (1998) introduced another variation of the exponential approximation in their study of the time-dependent point spread function of seawater:

$$p(\theta) = A\frac{\exp[-(\Theta/\Theta_0)^{1/2}]}{(\Theta/\Theta_0)^{3/2}\Theta_0{}^2}\theta^{-b} \tag{4.147}$$

where $\Theta = 2\sin(\theta/2)$, $\Theta_0 = 2\sin(\theta_0/2)$, A is a normalization constant equal to $\{1 - exp[-(2/\Theta_0)^{1/2}]\}^{-1}$ and θ_0 is a characteristic scattering angle. This approximation behaves as $\theta^{-3/2}$at small scattering angles in accordance with experiment. McLean and colleagues found $\theta_0 \approx 0.13$ rad (7.45 deg) to represent the average scattering function of seawater proposed by *Mobley* (1994).

A systematic investigation of the approximation errors by functions involving powers of the scattering angle has been reported (*Jonasz* 1980) for the volume scattering function in an angular range of 10° to 90°. Two power-law functions and a gamma function were used:

$$\beta(\theta) = A\theta^{-B} \tag{4.148}$$

$$\beta(\theta) = A(C + \theta)^{-B} \tag{4.149}$$

$$\beta(\theta) = A\theta^{-B}\exp(-C\theta) \tag{4.150}$$

In each equation, the adjustable constants A, B, and C have different meanings and values. The approximation errors were calculated for 88 volume scattering functions measured by Prandke in the Baltic waters (*Gohs* et al. 1978) and in the eastern equatorial Atlantic waters (*Prandke*, personal communication). The lowest average approximation errors were found for the modified power-law and gamma functions [equations (4.149) and (4.150)]. The errors and average values of the adjustable constants are listed in Table 4.6. As discussed in Chapter 3, the power-law function approximation at the small angles is consistent with the physical interpretation of the light scattering functions produced by particles with a power-law size distribution function. Note that the shape variability of the scattering functions, as expressed by the standard deviations of parameters A and C, is much smaller than that of the function magnitude (parameter a).

Table 4.6. Average errors and approximation coefficients for scattering functions investigated by *Jonasz* (1980).

Area	***N***	**Parameter**	**Power law, equation (4.148)**	**Modified power law, equation (4.149)**	**Gamma function, equation (4.150)**
Baltic Sea	39	Error [%]	5.5	1.1	1.2
		A [m^{-1}sr^{-1}]	154 ± 98	161 ± 486	121 ± 81
		B	2.67 ± 0.13	2.80 ± 0.25	2.56 ± 0.33
		C	–	1.05 ± 1.86	0.0036 ± 0.0084
Gdansk Bay	46	Error [%]	1.1	0.9	0.3
		A [m^{-1}sr^{-1}]	299 ± 174	630 ± 352	220 ± 140
		B	2.53 ± 0.05	2.69 ± 0.06	2.38 ± 0.08
		C	–	1.58 ± 0.86	0.0043 ± 0.0017
Atlantic	3	Error [%]	1.4	0.04	0.6
		A [m^{-1}sr^{-1}]	69	120	60
		B	2.62	2.74	2.55
		C	–	0.80	0.0025

All scattering function data by Prandke (*Gohs* et al. 1978: the Baltic Sea data; personal communication: Atlantic data). N denotes the number of functions examined. The approximation error is defined as $\Sigma[\beta(\theta_i) - \beta_x(\theta_i)]^2/\Sigma[\beta(\theta_i)]^2$, where i numbers data points, and the subscript "x" denotes approximated values. A single SD is quoted following the ± symbol.

Aas (1987) approximated the phase functions of turbid waters (reported by *Whitlock* et al. 1981) by using the following expressions:

$$p(\theta) = 0.092\ (1.00002 - \cos\theta)^{-0.7}, \quad 0 \le \theta \le 10°$$

$$p(\theta) = 0.0113\ (1 - \cos\theta)^{-1.7}, \quad 10 \le \theta \le 90° \tag{4.151}$$

$$p(\theta) = 0.0256 + 0.0099\cos\theta - 0.0143\sin\theta, \quad 90 \le \theta \le 180°$$

He used similar functional representations to analyze the results of the measurements reported by *Bauer* and Morel (1967):

$$p(\theta) = 0.00328\ (1.0006 - \cos\theta)^{-1.4}, \quad 0 \le \theta \le 10°$$

$$p(\theta) = 0.00224\ (1.017 - \cos\theta)^{-1.8}, 10 \le \theta \le 90° \tag{4.152}$$

$$p(\theta) = 0.00629 + 0.00272\cos\theta - 0.0411\sin\theta, 90 \le \theta \le 180°$$

The data of *Bauer* and Morel (1967) are characterized by a more pronounced forward scattering than are the functions measured by *Whitlock* et al. (1981).

Finally, *Haltrin* (1997) provides the following expansions (and their coefficients) for several experimental scattering functions in powers of the scattering angle:

$$p(\theta) = \frac{4\pi}{b} \exp\left[\sum_{n=0}^{5} (-1)^n s_n \theta^{\frac{n}{2}}\right] \tag{4.153}$$

4.5.2.3. Legendre polynomial expansion

Probably the first to use the Legendre polynomial expansion to approximate the phase function of turbid media was *Chandrasekhar* (1960). This expansion utilizes the fact that the Legendre polynomials (e.g., *Korn* and Korn 1968) represent an orthogonal basis in an infinitely dimensional space. Thus, a function of a parameter μ which varies between $+1$ to -1, such as a phase function, $p(\mu)$, where $\mu = \cos\theta$, with θ from a range of 0 to 180°, can be expressed as follows:

$$p(\mu) = \frac{1}{4\pi} \sum_{n=0}^{\infty} a_n P_n(\mu) \tag{4.154}$$

The Legendre polynomials are especially suitable for an expansion of an asymmetrical function as they all assume a value of 1 at $\mu = 1$, oscillate between $+1$ and -1, and assume alternatively values of $+1$ and -1 at $\mu = -1$, as the polynomial degree is being incremented by unity. The coefficients a_n can be easily obtained by using the orthogonality property of the Legendre polynomials:

$$\int_{-1}^{1} P_n(\mu) P_m(\mu) d\mu = \begin{cases} 0 & m \neq n \\ \dfrac{2}{2n+1} & m = n \end{cases} \tag{4.155}$$

It follows from (4.155) that on multiplying both sides of (4.154) by $P_m(\mu)$, we have:

$$a_n = \int_{-1}^{1} p(\mu) P_n(\mu) d\mu \tag{4.156}$$

The calculation of these coefficients is somewhat easier explained than done with experimental data that are typically defined on a sparse grid of μ in a sub-range of the $[-1,\ 1]$ range which rarely includes a small-angle section ($0 \leq \theta \leq 1°$, i.e., $1 \geq \mu \geq 0.9998$). Thus, we include this approximation mainly for completeness. The high asymmetry of the marine phase function also requires an inclusion of tens of terms in the sum in equation (4.154) (e.g., *McCormick* and Rinaldi 1989). The major strength of the Legendre polynomial expansion is in the theory of radiative transfer in turbid media where such an expansion provides a relatively simple approach to solving the radiative transfer equation.

The Legendre polynomials can be expressed as follows:

$$P_n(\mu) = 2^{-n} \sum_{k=0}^{n/2} (-1)^k \frac{(2n-2k)!}{k!(n-k)!(n-2k)!} \mu^{n-2k} \tag{4.157}$$

The first few Legendre polynomials are, as can be easily deduced from equation (4.157):

$$\begin{aligned} P_0(\mu) &= 1 \\ P_1(\mu) &= \mu \\ P_2(\mu) &= \frac{1}{2}(3\mu^2 - 1) \end{aligned} \tag{4.158}$$

From (4.156), (4.155), the second equation in (4.158), and the definition (4.8) of the average cosine, $\langle \mu \rangle$, we have for an axisymmetric phase function (for which the integration over the azimuth angle results in multiplication by 2π):

$$\begin{aligned} a_1 &= \int_{-1}^{1} p(\mu)\mu \, d\mu \\ &= 3\langle \mu \rangle \end{aligned} \tag{4.159}$$

4.5.2.4. Eddington functions

By retaining only the two first terms in the expansion of the phase function in terms of the Legendre polynomials (4.154), we obtain the following approximation:

$$p(\mu) = \frac{1}{4\pi} (1 + 3\langle \mu \rangle \mu) \tag{4.160}$$

that was introduced by Eddington for studies of radiative energy transfer in stellar atmospheres and in the Earth atmosphere (e.g., *Shettle* and Weinman 1970). However, given the high asymmetry of the typical phase function of seawater ($\langle m \rangle > 0.7$), this is a rather poor approximation here. Consequently, a modified Eddington approximation, referred to as delta-Eddington functions, has been frequently used to represent highly asymmetric phase functions, such as those of atmospheric aerosols and also human tissue (e.g., *Prahl* 1988):

$$p(\mu) = \frac{1}{4\pi} [2f\delta(1-\mu) + (1-f)(1 + 3\langle \mu \rangle \mu)] \tag{4.161}$$

or seawater (e.g., *McCormick* 1987)

$$p(\mu) = \frac{1}{4\pi} [2f\delta(1-\mu) + (1-f)(1 + 3\langle \mu \rangle \mu) + 5kP_2(\mu)] \tag{4.162}$$

and *McCormick* and Rinaldi (1989)

$$p(\mu) = \frac{1}{4\pi}\left[2f\delta(1-\mu) + (1-f)\sum_{n=0}^{\infty} a_n P_n(\mu)\right] \quad (4.163)$$

where f is selected from a range of 0 to 1, δ is the delta function, and k [in (4.162) is a higher-order (than that of $\langle\mu\rangle$) scattering factor. McCormick provides equations for estimating the unknown coefficients in (4.163) from the radiance field in a homogeneous turbid slab, an approach similar to that of *Zaneveld* (1974) and *Wells* (1983) for determining the scattering function of seawater.

4.5.2.5. The Henyey–Greenstein function and related approximations

Although we came across statements in the literature that the Henyey–Greenstein (HG) function was originally used to approximate the angular distribution of radiance in calculations of the radiative transfer through the atmosphere, that function was actually introduced as a phase function (*Henyey* and Greenstein, 1941). It has gained a prominent place as an approximation of the angular pattern of light scattering in such diverse fields of science as medical optics (*Jacques* et al. 1987) and marine optics, as discussed below. This function is expressed as follows:

$$p(\theta) = \frac{\gamma(1-g^2)}{4\pi(1-2g\cos\theta+g^2)^{3/2}} \quad (4.164)$$

where γg is the mean cosine of the scattering angle. The parameter γ is frequently set to 1 in that equation, making g numerically equal to the average cosine. This latter equality gives the most commonly used fitting procedure for experimental functions, defined in an angular range that is wide enough to allow meaningful calculation of the average cosine. A fitting procedure of the HG function to the light scattering function that is based on minimizing the mean-square error of the approximation is given by *Kamiuto* (1987). The HG function can obviously be expanded into a series of the Legendre polynomials (4.154). In fact, with $\gamma = 1$, we have (*Haltrin* 1997):

$$p(\theta) = \frac{1}{4\pi}\sum_{n=0}^{\infty}(2n+1)g^n P_n(\cos\theta) \quad (4.165)$$

At the minimum of the mean-square error integral over $-1 \le \cos\theta \le 1$, we have (*Kamiuto* 1987):

$$\sum_{n=1}^{\infty} n\left[(2n+1)g^{2n-1} - a_n g^{n-1}\right] = 0 \quad (4.166)$$

where a_n are coefficients of the expansion of the experimental phase function being fitted in terms of the Legendre polynomials. Equation (4.166) can be solved numerically for g by the bisection method.

The backscattering probability of the HG function can be expressed analytically (*Haltrin* 1997):

$$B = 2\pi \int_{-1}^{0} p_{\mathrm{HG}}(\theta) d\cos\theta = 2\pi \frac{1-g}{g}\left(\frac{1+g}{\sqrt{1+g^2}} - 1\right) \tag{4.167}$$

However, the HG function delivers a relatively poor approximation of the real scattering functions in both the ocean and the atmosphere. This prompted modifications of the function by including a second term (*Haltrin* 1999, *Plass* et al. 1985):

$$p(\theta) = t\frac{1-{g_1}^2}{4\pi(1-2g_1\cos\theta+{g_1}^2)^{3/2}} + (1-t)\frac{1-{g_2}^2}{4\pi(1-2g_2\cos\theta+{g_2}^2)^{3/2}} \tag{4.168}$$

where t, g_1, and g_2 are adjustable parameters. *Kattawar* (1975) discusses the procedure to calculate these parameters to fit experimental data.

A single-term HG function does not reduce to the Rayleigh phase function. *Cornette* and Shanks (1992) proposed a modification that enables the Rayleigh phase function to be attained in the limit:

$$p(\theta) = \frac{3}{2}\frac{1-g^2}{2+g^2}\left(\frac{1+\cos^2\theta}{1-2g\cos\theta+g^2}\right)^{3/2} \tag{4.169}$$

The analytical qualities of the HG approximation prompted, and continues to do so, numerous variations on the theme. *Lerner* and Summers (1982) approximated the phase function of seawater by using a function related to the HG function:

$$p(\theta) = h[\sinh^2(\theta_0/2) + \sin^2(\theta/2)]^{-(1+f)} \tag{4.170}$$

where h is a normalizing constant, θ_0 is the scattering angle below which the function deviates significantly from the asymptotic behavior at the large scattering angles, and $1+f$ is selected to adjust the function's behavior at large scattering angles. On fitting an experimental scattering function tabulated by *Jerlov* (1968), Lerner and Summers obtained the following values of the parameters:

$h = 0.44$, $\theta_0 = 2.5°$, and $f = 0.35$. If the normalization condition (4.5) for the phase function is used, the parameter h must be changed to 0.035.

Finally, *Haltrin* (1997) proposed a delta-hyperbolic approximation to the phase function that is related to the HG function:

$$p(\theta) = 2g\delta(1 - \cos\theta) + \frac{1-g}{\sqrt{2(1-\cos\mu)}} \tag{4.171}$$

where $\delta(1 - \cos\theta)$ is a Dirac delta function and g is a shape parameter. This function yields the following expression for the backscattering probability, p_b:

$$p_b = \frac{1-g}{2+\sqrt{2}} \tag{4.172}$$

4.5.2.6. Statistical methods

Principal components. The volume scattering function can be treated as a vector in an N-dimensional space, where each dimension represents a fixed scattering angle. This allows one to approximate the scattering function by expressing it as a linear combination of a set of orthogonal basis vectors in that space. One such set, referred to as principal components (for example, *Anderson* 1958), is provided by the characteristic vectors (eigenvectors) of the covariance matrix of a population of the experimental scattering functions. Although this looks similar to the well-known method of representing a vector by a set of numbers which express projections of the vector onto each basis vector, there is significant difference between these two paradigms. In the case of the principal component expansion, a vector can be represented well by just using a first few principal components.

In order to be reasonably meaningful, the method of principal components requires a fairly large data set of measured volume scattering functions as an input (the defining set). As we already mentioned, each function is considered as an N-dimensional vector, $\boldsymbol{\beta}$, with components $\beta_n = \beta(\theta_n)$, $n = 1, \ldots, N$. The base vectors, $\mathbf{f}_r$, $r = 1, \ldots, N$, fulfill the following equation

$$\mathbf{Cov}\ \mathbf{f}_r = l_r \mathbf{f}_r \tag{4.173}$$

where **Cov** is the $N \times N$ covariance matrix, of the set of experimental vectors $\boldsymbol{\beta}$, and l_r are the roots (eigenvalues) of the covariance matrix. It follows that each vector $\boldsymbol{\beta}$ from the defining set can be well approximated by the sum:

$$\boldsymbol{\beta} = \langle\boldsymbol{\beta}\rangle + \sum_{r=1}^{R} c_r \mathbf{f}_r \tag{4.174}$$

where $\langle\boldsymbol{\beta}\rangle$ is the average vector for the defining set and c_r, $r = 1 \ldots R < N$ are the best fit coefficients that, as usual, are calculated by using the orthogonality conditions of the base vectors, $\mathbf{f}_r$.

It turns out that given the form of the covariance matrix of the volume scattering function for seawater, the number of the basis vectors, R, required to account for the major portion of the variability in the scattering function is on the order of 2 to 3 (*Kopelevich* and Burenkov 1972).

We will note in advance that a similar statistical treatment also applies to the particle size distribution Unfortunately, integration of the basis vectors of the particle size distribution weighed by the single-particle scattering pattern— which, for example, can be given by the Mie theory—yields a set of vectors which are not orthogonal and cannot serve as a basis for the scattering function expansion. This prevents the inverse transformation of an expansion of a scattering function in terms of the principal components into an expansion of the particle size distribution in terms of its principal components. Interestingly, a similar route of inverting the scattering function into the particle size distribution was pursued by *Alger* (1979) who used an appropriately spaced set of narrow size distributions as the "basis" of an expansion of a size distribution.

Two-component model of Kopelevich. Kopelevich and Mezhericher (1983) proposed a two-component model of the oceanic particle scattering function. This model is based on the statistical analysis of experimental functions and on an assumption that the population of particles in seawater is composed of two subpopulations: small mineral particles with a high relative refractive index of 1.15 and a balance of large, organic particles with a low refractive index of 1.03. The mineral (high-density) particles must be small to achieve any significant residence time in the water column. The organic (low-density) particles can be much larger and still enjoy a comparable residence time. The base functions for these two populations are concentration specific. The model is summarized by the following equation, expressing a linear combination of the scattering function of seawater and of the base functions:

$$\beta(\theta,\lambda) = \beta_{\rm w}(\theta,\lambda) + \nu_{\rm s}\beta_{\rm s}(\theta,\lambda))\lambda^{-1.7} + \nu_{\rm l}\beta_{\rm l}(\theta,\lambda)\lambda^{-0.3} \tag{4.175}$$

where $\beta_w(\theta,\ \lambda)$ is the scattering function of pure seawater as defined by (4.120) and (4.125), ν_x, is the volume concentration of particles in $\mathrm{cm^3 m^{-1}}$, β_x is the scattering function of a particle fraction in $\mathrm{m^{-1}sr^{-1}\ ppm^{-1}}$, and x is either "s" for the small particle fraction or "l" for the large fraction. The wavelength (in nm) is given relative to a wavelength of 550 nm. The base functions are given in Table 4.7.

The allowed volume concentrations ranges are $0.01 \le \nu_{\rm s} \le 0.2\,\mathrm{ppm}$ and $0.01 \le \nu_{\rm l} \le 0.4\,\mathrm{ppm}$. Given a scattering function for the whole seawater, the volume fractions can be determined from the following equations:

$$\begin{aligned} \nu_{\rm s} &= -1.4\times10^{-4}\beta_1 + 10.2\beta_2 - 0.002 \\ \nu_{\rm l} &= 2.2\times10^{-2}\beta_1 - 1.2\beta_2 \end{aligned} \tag{4.176}$$

where $\beta_1 = \beta(1^\circ,\ 550\,\mathrm{nm}),\ \beta_2 = \beta(45^\circ,\ 550\,\mathrm{nm})$.

Table 4.7. The concentration-specific particle scattering functions of the two-component model of *Kopelevich* (1983).

θ [degrees]	$\beta_s(\theta)$ [$m^{-1}sr^{-1}$ ppm^{-1}]	$\beta_l(\theta)$ [$m^{-1}sr^{-1}$ ppm^{-1}]	θ [degrees]	$\beta_s(\theta)$ [$m^{-1}sr^{-1}$ ppm^{-1}]	$\beta_l(\theta)$ [$m^{-1}sr^{-1}$ ppm^{-1}]
0	5.3	140.	45	0.098	0.00062
0.5	5.3	98.	60	0.041	0.00038
1	5.2	46.	75	0.020	0.00020
1.5	5.2	26	90	0.012	0.000063
2	5.1	15	105	0.0086	0.000044
4	4.6	3.6	120	0.0074	0.000029
6	3.9	1.1	135	0.0074	0.00002
10	2.5	0.2	150	0.0075	0.00002
15	1.3	0.05	180	0.0081	0.00007
30	0.29	0.0028	–	–	–
b[$m^{-1}ppm^{-1}$]	1.34	0.312			

Mobley (1994) has recently discussed and used this model to approximate Petzold's scattering function for coastal seawater (*Petzold* 1972) as well as the spectral behavior of the scattering function in the clear/turbid waters observed by *Morel* (1973). He found a reasonable qualitative agreement, but concluded that the Kopelevich model underestimates the small-angle scattering in the Petzold function and overestimates the large-angle scattering.

4.5.2.7. Abstract multi-component models

Although these models are not technically approximations, because of their universality, we believe that it is useful to mention the development of such models, if only to illustrate the rationales. These models are similar to the statistical model of Kopelevich in that they are based on applying physical reasoning to explain the observed oceanic scattering functions. However, they lack that statistical model's generality and apply, in principle, to a specific experimental scattering function. All these models so far use the Mie theory of light scattering to provide a weighting function for the calculation of the scattering function of particles via numerical integration of the particle size and refractive index distributions.

To our knowledge, the first such model was developed for a clear-water scattering function measured by *Kullenberg* (1968) in the Sargasso Sea (*Brown* and Gordon 1973, *Gordon* and Brown 1972). An approach similar to that of Gordon and Brown was used to analyze the average scattering functions in the Baltic waters for two seasons: summer and winter (*Jonasz* and Prandke 1986).

Zaneveld et al. (1974) used a systematic search approach in fitting that scattering function (*Kullenberg* 1968) by using Mie theory. The best-fit theoretical scattering function was selected from numerous combinations of the functions from a database calculated for 40 power-law particle size distributions and for relative refractive indices of 1.02, 1.05, 1.075, 1.10, and 1.15. These authors found that the measured scattering function is best reproduced using a three-component model of the particle size–refractive index distribution. The major component, with a particle size distribution having a slope, m, of 3.5, was found to have the refractive index of 1.15 and to dominate the large particle size range. The two other components had refractive indices and slopes of 1.075 and 3.9, and 1.05 and 3.7 respectively.

4.5.2.8. Multi-component models based on actual particle species

Accumulation of data on the optical properties of particles present in natural waters opens the possibility of a "natural" approach to fitting experimental scattering functions. In that approach, the scattering function is expressed as a linear combination of the pure water (seawater) function and components, representing the contributions of particle species, either measured or calculated by using a light scattering theory and refractive index distributions of these species (*Stramski* et al. 2001, *Stramski* and Mobley 1997). Although the "forward" problem, i.e., modeling of the effect of variability of these characteristics of the particles on light scattering in natural waters is straightforward, the "inverse" problem, i.e., fitting the experimental scattering functions may be more complex as a minimization problem in multi-dimensional parameter space.

4.6. Problems

1. Errors in the volume scattering function due to reflection of scattered light at the sample container wall

Assume that a nephelometer utilizes a cylindrical glass vessel as a sample container. Light scattered at a small angle, θ, is reflected at the two interfaces created by the container (water–glass, glass–air) and measured by a detector at an angle $\theta' = 2\pi - \theta$. Calculate the contribution of that light to the scattered light measurements in an angular range of $\theta' > \pi/2$ as a function of θ' for the optical properties of clear seawater. Account for the attenuation of light by the sample. When, if at all, might one need to consider the interference between light reflected at the water–glass and glass–air interfaces?

2. Why a suspension changes color depending on the background?

A researcher was visually examining a sample of a fine suspension in a typical spectrophotometer cuvette for the presence of large contaminant particles when he noticed a change in the suspension tint when observing the cuvette contents against background of different brightness. When observed against a dark background, the suspension looked bluish, but against a bright background it looked brownish. Explain this effect. What, if any, conclusions can be reached and about which properties of the suspension, assuming that the backgrounds did not modify the daylight illumination spectrum in the laboratory?

3. *Mueller matrix of a linear polarizer*
   Prove that the Mueller matrix of a linear polarizer is described by (4.91).

4. *Measuring linearly polarized light scattering*
   A researcher would like to measure the scattering of polarized light with a nephelometer that is capable of measuring unpolarized light scattering. The construction of that nephelometer allows for insertion of polarizers and analyzers into the light path. The researcher intends to use a linear polarizer in the incident beam and a linear polarizer in front of the detector. What element of the scattering matrix, if any, will be measured with the polarizer and analyzer oriented perpendicularly to the scattering plane?

5. *Fluctuations of the scattered light*
   A nephelometer with a small scattering volume on the order of 1 mm^3 is used to measure light scattered by a sample of seawater. Water is gently mixed inside the sample container of the nephelometer. A baffled experimenter finds that the scattered light intensity, $I(\theta)$, significantly fluctuates as a function of time. He/she plans to evaluate the mean scattering intensity by averaging a series of measurements taken at $t_i = i\,dt$. What parameters of the instrument and characteristics of seawater come into play in selecting the magnitude of the time interval dt and the length of the measurement series? Discuss the effect of the response time of the detector system in the nephelometer (at a fixed scattering angle).

6. *Errors of the scattering coefficient calculation*
   Estimate the error of the scattering coefficient calculated from a scattering function measured in a limited range of the scattering angle. Use one of the Petzold's scattering function (*Petzold* 1972, also in *Mobley* 1994) determined in a wide range of the scattering angle as an "exact" complete function to evaluate the error in the total scattering coefficient, as a function of the small-angle limit in the scattering coefficient integral.

7. *Absorption meter utilizing reflective tube*
   In the discussion of the integrating nephelometer in this chapter, we said that such a nephelometer measures the scattering coefficient, i.e., integrates the volume scattering function of the medium because its detector is oriented parallel to the beam axis to achieve the correct integrand weighing of $\sin\theta$, where θ is the scattering angle. Yet, in a reflective tube-type absorption meter (Chapter 2), which is said to "integrate out" scattering, the detector is perpendicular to the beam axis. How is such an integration possible in the reflective tube case?

Chapter 5

The particle size distribution

5.1. Introduction

The particle size distribution (PSD), i.e., the relationship between the particle size and concentration, is one of the key parameters defining the interaction of light with natural waters and hence has a paramount importance for the optics of natural waters. However, the PSD is also of great importance for other fields of oceanography and limnology: biology, chemistry, and sedimentology. This wide field of interests reflects the complexity of populations of particles in natural waters, of their dynamics and interaction with other components of the environment.

The definition of an aquatic particle is somewhat vague and depends on the viewpoint of a research field. One could argue that water is nothing else but suspended particles, whose sizes range from the molecules, through fish and whales (in the case of seawater!). In the optics of natural waters, the particle size range of interest is limited in the first approximation to an interval of roughly 0.01 to 1000 μm. This range is of interest to those researchers who are concerned with the optical properties of seawater and its influence on propagation of light in the sea, as well as to those who want to use optical methods to study the particles. However, the range which may be of interest to other disciplines, such as biology and geology, may well include particles that are smaller then 0.01 μm as well as much greater than 1000 μm. Zooplankton and other small organisms have sizes on the order of several millimeters (~10 000 μm). Large aggregates (flocs) of particles in seawater, which are studied *in situ* with optical methods, and whose role in radiative transfer is little researched, may exceed centimeters in size (~100 000 μm).

We have already discussed in Chapter 1 that the effect of particles on the propagation of light in natural waters depends on the product of their individual scattering cross-sections and their number concentration, i.e., the number of particles per unit volume of seawater. It is this dependence that controls the optically important particle size range through the size dependence of the cross-section and particle number concentration. At the low end of that size range, the scattering cross-section decreases too fast for this decrease to be compensated for by a sustainable increase in the number concentration. At the upper end of this range, the decrease in the number concentration is not compensated for by the enlarged cross-section because the latter increases with particle size relatively slowly in that size range.

Why are the particle size and size distribution given so much attention here, while in the earlier chapters of this book we discussed, apart from the size, other factors which affect light scattering by marine particles, such as the shape, structure, and orientation of particles? First, the *characteristic* size of an arbitrary particle relative to the wavelength of light is the simplest, scalar parameter that can, in the first approximation, be used to describe the interaction of the particle with a light wave. Moreover, the particle size has been and remains a parameter which can be relatively simply and quickly measured. Although this conclusion reaches outside the realm of scientific reasoning and reflects the way in which science advances, given technological as well as so-called socioeconomic constraints, the simplicity and affordability of a measurement method are significant advantages. Thus, the description of interaction of light with particles, as well as many other physical and chemical processes involving suspended particles, gravitates toward the use of the particle size as a shortcut parameter. Given that the concentration of particles of a given size describes the magnitude of the effect these particles have on the interaction of light with the particle population, it is only natural to relate the number concentration of particles to the particle size, i.e., use the size distribution as a first-order-of-magnitude factor affecting light scattering by populations of aquatic particles.

The PSD should be interpreted as a particle-size-dependent average number of particles of given size in a volume of water. The stress here is on the word "average." This is immediately apparent by considering a particle population that, within given size interval, contains on average so few particles per unit volume of seawater that when we attempt to count them by examining a small volume, we can obtain either 1 or 0. Particles are localized in space, and they either are found or not within a specific volume of seawater. One also needs to realize that a particle in water, although localized in space, does not occupy a fixed position all the time. The particle may drift through the actions of sedimentation, convection, and/or Brownian motion. It may also propel itself, as is the case for motile bacteria and other organisms.

However, one can readily define a time-averaged probability density of finding within a unit volume of water a particle with size from a certain range. This enables us to define the time-averaged mean value of the particle concentration. It follows that the number we get by counting particles in a specific volume of a water sample is merely an approximation to the mean number of such particles. As we shall see shortly, this approximation gets the better the greater the particle count.

The size distribution is generally time dependent also in a more fundamental way because a particle population is subject to several competing processes that can remove and add particles as well as convert one type of particles into another.

These processes include inanimate processes of coagulation and dissolution, as well as generation of new living organisms, death and decay of dead organisms, excretion, and assimilation of particles by living organisms.

This chapter is different from the previous chapters not only in its topic but also because the main stress here is on the experimental data and their analysis. We are interested in applications of issues discussed here for the modeling of light scattering, not in pursuing research in the dynamics of particle populations in water.

We have also reversed the order in which topics are discussed in this chapter as compared with the succession of topics in the previous chapters: we discuss measurement techniques *before* discussing the results. This reflects the fact that the PSD may depend not only on the size structure of the particle population but also on the measurement technique in the sense that the various measurement techniques tend to measure various attributes of the particles that are then converted to an abstract "particle size" parameter.

We will frequently use the power-law approximation when discussing the various features of and operations on the aquatic PSD. This is a convenient "first-order-of magnitude" approximation that reflects the major feature of such size distributions: a rapid decline in the particle size concentration with the increasing particle size. Thus, frequent references to the power-law size distribution throughout this chapter should not be understood as our desire to impress on the reader a particular fitness of this approximation.

5.2. The particle size definitions and the particle shape

The concept of a particle size implies that the particle geometry can be *uniquely* characterized with a single variable. However, in order for such a characterization to be unique, the particle shape must be specified. This may become very complex if a particle is irregularly shaped (Chapter 6). Unfortunately, in the case of water-borne particles, the irregularity of the particle shape, understood as a departure of that shape from that of a sphere, appears to be the norm rather than an exception. Therefore, the particle size must be understood here as an approximate, at best, characteristic of aquatic particles.

Some of the particle size definitions (*Allen* 1990a) are used mostly in fields other than oceanography and limnology. In Table 5.1, we give a reasonable selection of size definitions after a recent work by *Jillavenkatesa* et al. (2001).

The Feret size is frequently used as a measure of the contour size in image analysis. The definition of the Martin diameter attempts to restore the effect of the second dimension. The distribution of Martin diameters for a particle population can, as opposed to the Feret diameter, account for the concavity of the particle contour.

Some definitions of particle size (maximum chord, Feret and Martin diameters) refer to a two-dimensional image of the particle, such as one that can be observed

Table 5.1. Commonly used definitions of the particle size.

Particle size type	**Definition**
Equivalent spherical diameter (ESD)	Diameter of a sphere with volume equal to that of the particle
Equivalent circular diameter (ECD)[a]	Diameter of a circle with area equal to that of the particle projection onto a plane
Maximum chord diameter	Length of the longest chord of the particle projection onto a plane
Hydrodynamic diameter[b]	Diameter of a sphere having the same settling velocity as that of the particle
Feret diameter	Length of the longest chord of the particle projection onto a plane with the chord aligned parallel to an arbitrary axis
Martin diameter	Length of a chord of the particle projection onto a plane with the chord aligned parallel to an arbitrary axis and measured at the middle point of the Feret diameter measured along a perpendicular axis

[a]The ECD is sometimes referred to as the Heywood diameter.
[b]Also referred to as the Stokes diameter.

with an optical microscope. They tend to be most informative if particles are compact and their shapes are close to spherical. These simple measures characterize a particle population increasingly less accurately as the particle shapes become more asymmetric and variable across the population. For example, the adoption of either Feret or Martin diameters results in wide size distributions for a population of identical but randomly oriented needles. These definitions apply to particles randomly oriented in three-dimensional space. Projections of the particle shapes onto a reference plane are not biased in that case. A further compression of the two-dimensional particle contours to one-dimensional projections onto an arbitrary axis is also unbiased. Such is the case when observing, with an optical microscope, particles suspended in a small volume of water where the particles are free to orient themselves (presumably) randomly. However, if these particles become preferentially oriented, for example, after having been filtered onto a membrane filter, these definitions may systematically bias the particle size estimate. This is because elongated or flat particles tend to assume positions with their largest dimensions aligned parallel to the surface.

Taylor (2002) has proposed a “digital” definition of the particle size and shape as a three-dimensional array that maps the space occupied by the particle. He also proposed a four-element particle size-shape descriptor that consists of the number of the three-dimensional volume elements and moments of their spatial distribution.

Some particle size definitions do not result from the consideration of the particle shape but are introduced as equivalent size parameters by the measuring technique. These definitions include the Stokes, or hydrodynamic diameter, and the equivalent spherical volume and cross-section diameters. The Stokes diameter is defined by the sedimentation technique of particle size analysis. The particles are allowed to settle under gravity or centrifuge acceleration, and the average settling velocity of these particles is determined. This velocity is then used to calculate an equivalent particle diameter, equal to that of a sphere that has the same density and settling velocity as those of the particle.

The equivalent spherical diameter (ESD) has gained a considerable recognition in particle size analysis of water-borne particles because of the widely used electro resistance technique (discussed later in this chapter). This technique measures the particle volume as a primary parameter. The particle size is then defined as the diameter of a sphere with volume equal to that of the particle. The equivalent cross-section diameter is introduced by optical particle sizing techniques based either on measuring light scattering or on attenuation by a particle or particle suspensions.

We will discuss in Chapter 6 the shapes of particles occurring in natural waters and methods of describing those shapes mathematically. In this section, we concentrate on the distribution of the particle size alone. Thus, we now leave aside the complexities of the particle shape following the example of a physicist (*Kerker* et al. 1979) who began an analysis of the horse by assuming that it is a sphere. Accordingly, the particle size as understood in this chapter is generally the diameter of a sphere with certain property (e.g., the volume) equal to that of the particle, unless the particle size is specifically defined otherwise.

5.3. Definition and units

The size distribution has been assigned many meanings in the literature on the topic, making it rather confusing when trying to compare the various results, even within the same field. Researchers planning to discuss size distributions would help their readers enormously by rigorously defining the size distribution that they refer to, as well as specifying the units in their size distribution graphs, this latter practice being alarmingly less common that one would have desired.

The PSD, $n(D)$ (particle number per unit particle size and suspension volume), is defined as follows:

$$dN = n(D)dD \tag{5.1}$$

where dN is the mean number concentration of particles in a size interval $[D, D+dD)$. The unit of $n(D)$ used throughout this book is $\mu m^{-1} cm^{-3}$, implying that the unit of dN is cm^{-3}, and the unit of D is 1 μm. The usage of letter n stems here from the *number* concentration of the particles. We acknowledge a conflict with the same symbol traditionally used for the refractive index. However, since we will refer little to refractive index in the present chapter, we feel that this conflict should not significantly hamper our presentation.

Symbol $n(D)$ denotes a differential size distribution, also referred to as the frequency distribution of particle sizes. It characterizes the partitioning of the total particulate material into consecutive size intervals. The differential size distribution is a convenient base form of the size distribution, and we will refer to it in short as the size distribution unless a different meaning of that term is obvious from the context.

The size structure of a particle population can also be characterized by using the cumulative distribution, $N\ (D)$ [cm^{-3}]. It is related to the frequency PSD, $n(D)$, as follows:

$$N(D) = \int_D^\infty n(D')dD' \tag{5.2}$$

Thus, $N(D)$ is simply the number concentration of particles with sizes greater than D.

In geology, the cumulative size distribution, $V(D)$, defined as mass of particles with sizes below rather than above size D, is also used (e.g., *Austin* 1998). This is a tricky thing to do in the case of the number size distribution, because the number of particles smaller than the lower limit of the size range accessible to measurement, usually unknown, may be significant.

The cumulative size distribution, $N(D)$, as defined by (5.2), is a monotonically decaying function of the particle diameter, D. Thus, the derivative of that function, as implied by (5.1), is a negative function of D. In order to keep dN positive, we therefore define formally $n(D)$ as follows:

$$n(D) = -\frac{dN(D)}{dD} \tag{5.3}$$

The PSD, $n(D)$, is a limiting case of $\Delta D \to 0$ of the following size-interval-normalized histogram:

$$h(D, \Delta D) = \left| \frac{N(D+\Delta D) - N(D)}{\Delta D} \right| \tag{5.4}$$

Hence (5.4) can serve as the first-order approximation for $n(D)$ (see section 5.6.2 for an estimate of the error that arises from using the histogram to represent a power-law size distributions). The use of the absolute histogram $H(D,\ \Delta D) = h(D,\ \Delta D)\ \Delta D$ to represent the size structure is a poor practice because the forms of such histograms for a given differential (frequency) size distribution depends on the selection of the size grid as it was pointed out by *Krumbein* (1934).

One can also define related distributions of the particle volume, $v(D)$ [$\mu m^2 cm^{-3}$], and cross-section, $c(D)$ [$\mu m\ cm^{-3}$], as follows:

$$v(D) = \frac{\pi}{6} D^3 n(D) \tag{5.5}$$

$$c(D) = \frac{\pi}{4} D^2 n(D) \tag{5.6}$$

A mass size distribution, $w(D)$, which is important in sedimentology, can also be defined by multiplying the right side of (5.5) by the (potentially size-dependent) density, ρ, of the particles.

A relevant unit of the volume distribution, $1\ \mu m^3 cm^{-3}$, equals 10^{-6} ppm. Thus, by factoring out a multiplier of 10^{-6} one converts the units of volume distribution, $v(D)$, from $\mu m^3 cm^{-3}$ to ppm. The latter unit is equivalent to the mass concentration of particles on the order of mg dm^{-3}, assuming that the density of the particles is unity (i.e., the same as that of water).

The PSD, the histogram size distribution, and the volume distribution are shown in Figure 5.1 for a typical population of marine particles in a relatively narrow size range. The complex shape of the PSD is due to biological activity.

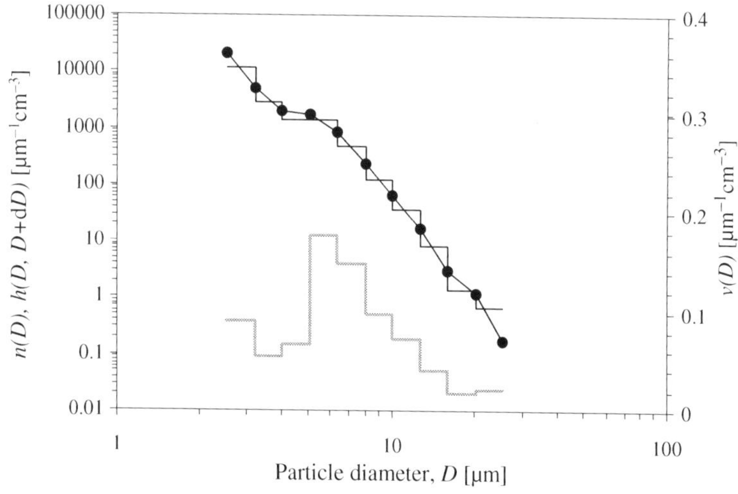

Figure 5.1. The various formats of presenting the particle size distribution: —•— number (frequency) size distribution, $n(D)$, — histogram distribution, $h(D)$, — (gray thick line) volume distribution, $v(D)$ in the histogram format, (the right y-axis). Data of M. Jonasz, obtained off Nova Scotia, at a depth of 5 m, on 15 July 1983, with a Coulter counter model ZB using a 100 μm diameter aperture. It is a part of a PSD collection compiled by

5.4. An optimum particle size grid

The optically significant size range of aquatic particles spans several decades in which the size distribution, $n(D)$, generally decreases with increasing particle size. The selection of a particle size grid at which the size distribution is to be defined, i.e., the partitioning of a particle size range into intervals, is an important decision. On one hand, such a selection may affect the quantity of information about the sample that can be extracted from the size distribution. For example, a size grid that is too sparse may prevent one from discovering the fine structure of the size distribution. Arguments based on the information content generally favor smaller size intervals. According to the sampling theorem (*Shannon* 1949) adapted to the PSD measurements, in order to resolve a feature of a size distribution, the size grid interval length must be smaller than half of the feature size scale. On the other hand, an increase in the number of the size grid points extends the measurement time. Populations of aquatic particles are dynamic and may change on time scales comparable to the measurement time. Such evolution will be discussed in a later section of this chapter in more detail. Thus, in an attempt to study a particle population at a more detailed level, one may in fact make the results less representative of that population.

An optimum size grid, i.e., one that maximizes the information content of the size distribution, can be defined only by following the analysis of the size distribution in a number of samples (*Full* et al. 1984). It follows from information theory (*Shannon* 1948) that such an optimum grid must be defined so that an equal value of the integral of a distribution is obtained within each size interval of the grid. This fixed-size grid is optimized for a set of samples and not for any individual sample. Thus, specific features of the size distributions of individual samples may still be lost if these features span narrow-size sub-ranges.

The definition of the maximum information size grid depends strongly on the application, or interest, which prompts the measurements. If one is interested in the application of the size distribution to modeling of light scattering or attenuation by the particles, the optimum grid should be evaluated for the cross-section distribution, i.e., a function proportional to $n(D)D^2$. If, on the other hand, one is interested in studying the mass flow through the population of particles, one should define the optimum grid for the volume distribution, $\sim n(D)D^3$.

It would be interesting to see what such grids might look like for featureless PSDs of particles in the open ocean waters. As we shall discuss later in this chapter, such PSDs are relatively well approximated with a power-law function $n(D) = kD^{-m}$, where the slope, $m \cong 4$. A size grid $(D_0, D_1, D_2, \ldots)$ which ensures that equal values of the integral of a moment of the size distribution, are found in each of the size-axis grid intervals $[D_0, D_1), [D_1, D_2), \ldots$ can be defined for some moments of a such a distribution. By an rth moment of the size distribution, we understand here, by analogy to the probability theory, an integral of a product of the size distribution and the size raised to an rth power. We set the grid so that each size point of that grid is greater than the previous one. Let the differential moments be defined as follows:

$$\begin{aligned} f_r(D) &= n(D)D^r \\ &= kD^{-m+r} \end{aligned} \tag{5.7}$$

Thus, the cumulative moments, $F_r(D)$, or simply moments, can be expressed as follows:

$$\begin{aligned} F_r(D) &= -\int_D^\infty kD^{-m+r}dD \\ &= -\frac{k}{-m+r+1}D^{-m+r+1} \end{aligned} \tag{5.8}$$

except when $-m+r+1=0$, as in the case of the $f_3 = n(D)\ D^{-m+3}$ for the slope, $m=4$.

The case of $-m+r+1>0$ leads to a diverging function of no interest here. Out of the remaining two choices, consider first the case of $-m+r+1<0$. Let us express the size, D, as a function of F_r, i.e., $D=D(F_r)$, as follows:

$$D = \left[-\frac{-m+r+1}{k}F_r\right]^{\frac{1}{-m+r+1}} \tag{5.9}$$

To construct the grid, we set the initial diameter, D_0, of the series of the size grid points, D_i, and work our way up the size axis. The size points are defined by a requirement that the integral of f_r over each grid size interval equals an arbitrary increment, ΔF_r. Thus, by using (5.9), we define a particle size grid as follows:

$$D_i = \left[-\frac{-m+r+1}{k}[F_r(D_0) - i\Delta F_r]\right]^{\frac{1}{-m+r+1}} \tag{5.10}$$

The grid points are related as follows:

$$\begin{aligned} \frac{D_{i+1}}{D_i} &= \left[\frac{F_r(D_0)-(i+1)\Delta F_r}{F_r(D_0)-i\Delta F_r}\right]^{\frac{1}{-m+r+1}} \\ &= \left[\frac{\dfrac{F_r(D_0)}{\Delta F_r}-(i+1)}{\dfrac{F_r(D_0)}{\Delta F_r}-i}\right]^{\frac{1}{-m+r+1}} \end{aligned} \tag{5.11}$$

Table 5.2. The size grids, D_2 and D_3, that assure the maximum information content respectively for the second moment (i.e., the particle cross-section distribution) and third moment ($\sim$ particle volume distribution) of a power-law size distribution $n(D) = 1000\,D^{-4}$ [cm^{-3}].

D_2 [μm]	$F_2(D)$ [$\mu\text{m}^2\ \text{cm}^{-3}$]	D_3 [μm]
0.500	2000	0.500
0.555	1800	0.630
0.625	1600	0.794
0.714	1400	1.000
0.833	1200	1.260
1.000	1000	1.587
1.250	800	2.000
1.666	600	2.520
2.500	400	3.174
5.000	200	4.000

$D_0 = 0.5\,\mu\text{m},\ F_r(D_0)/\Delta F_r = 0.1$. The third moment values are infinite and are not shown.

The ratio D_{i+1}/D_i depends on the ratio $F_r(D_0)/\Delta F_r$, the diameter index, i, the size distribution slope m, and the order, r, of the moment of the size distribution. For example, by selecting $D_0 = 0.5\,\mu\text{m},\ F_2(D_0)/\Delta F_2 = 0.1$ and assuming that the slope, $m = 4$, i.e., that of a typical oceanic PSD, the size grid, D_2, for the cross-section distribution, $\sim n(D)D^2$ $(r = 2)$, assumes a form shown in Table 5.2 and Figure 5.2 [where it is also compared with a size grid, D_3, for the volume size distribution, $\sim n(D)D^3,\ r = 3$]. In calculating the values of F_2, we assumed that the scale factor, k, of the frequency size distribution equals 1000, also a typical order of magnitude for the PSD in the open ocean.

The maximum D_i depends on the particle concentration. In this example, where $\Delta F_2 = F_2(D_{i+1}) - F_2(D_i) = 200\,\mu\text{m}^2\text{cm}^{-3}$ and $F_2(D_0) = 2000\,\mu\text{m}^2/\text{cm}^{-3}$, values of $F_2(D_i > 5\,\mu\text{m})$ do not support $\Delta F_2 = 200\,\mu\text{m}^2/\text{cm}^{-3}$.

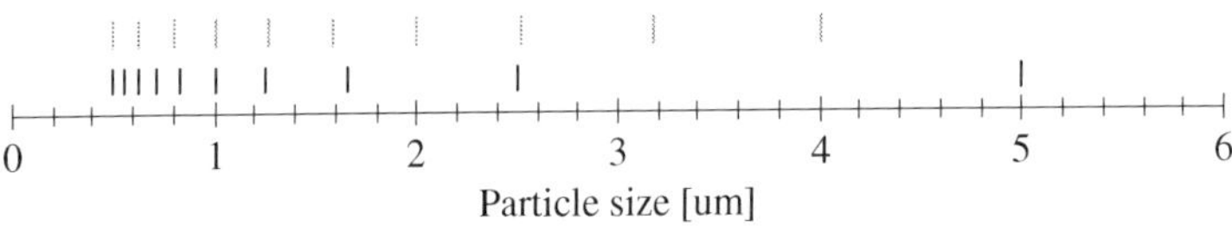

Figure 5.2. Size grids assuring the maximum information content for the second (the lower row of tics) and third (the upper row of tics) moments of a power-law size distribution $n(D) \sim D^{-4}$. Both grids start at $D_0 = 0.5\,\mu\text{m}$ and have the same number of points (tics). The third moment's grid is more evenly distributed because the integrand of that moment, $n(D)D^3 \sim D^{-1}$, decreases slower than that of the second moment, $n(D)D^2 \sim D^{-2}$. The second moment's integrand is proportional to the cross-section distribution. The third moment's integrand is proportional to the volume distribution. The size values for the second and third moments' grids and the parameters of the size distribution are listed in Table 5.2 The factor a in the grid size definition (5.13) for the third moment equals $2^{1/3}$.

When $-m+r+1=0$, the cumulative distribution moment is infinite, but its difference ΔF_0 is finite and can be expressed as follows:

$$\Delta F_0 = k\ln(D_{i+1}/D_i) \tag{5.12}$$

It follows from (5.12) that when $D_{i+1} = aD_i$, with a being a constant, then $\Delta F_0 = k \ln a$ =const for all D_i. Thus, by starting with an arbitrary value of D_0, we can define the size grid for the third moment of the size distribution (the volume distribution) as follows:

$$D_i = a^i D_0, \quad i = 0, 1, 2, \ldots \tag{5.13}$$

Although an optimum size grid for a specific set of samples may permit to retrieve the maximum information, such an optimum grid may be different for a set of samples from another water body or period. In addition, selection of a different grid for each set may make it difficult to compare size distributions between sample sets. Of course, one can interpolate and group data to express size distributions for all sets by using a common, optimum size grid, but such common grid may differ from an optimum grid for each individual set.

A practical solution to such a complex issue is to define the size grid based on the general features of the size distribution, such as the power-law decay with a slope of -4 (for marine particles, e.g., *Sheldon* and Parsons 1967b). This leads to logarithmically equal, contiguous size intervals whose widths increase according to the following equation [a particular case of (5.13) with $a = 2^{1/n}$]:

$$D_{i+1} = 2^x D_i \tag{5.14}$$

where D_i is the lower size limit of the i-th interval, and D_{i+1} is the lower size limit of the $(i+1)$-interval. The $(i+1)$-th interval encompasses particle sizes between D_{i+1} and D_i. The exponent is usually set to 1/3. As it follows from (5.14) with $x = 1/3$, the volume V_{i+1} of a sphere with a diameter D_{i+1} is twice that (V_i) of a sphere with a diameter D_i. The justification for such a selection of the particle size intervals is provided by observations that roughly equal particle volumes can be found within size intervals which are equal when expressed by using a logarithmic scale (*Sheldon* et al. 1972, *Sheldon* and Parsons 1967b) as discussed in more detail in sections 5.5.2.3 and 5.8.5.3. Thus, such a grid maximizes the quantity of information which can be extracted from the volume size distribution, $v(V)$, with a log-log slope of 0.

Other definitions of the particle size grid are also used. The *phi*-size scale, introduced by *Krumbein* (1936, see also *Tanner* 1969) is widely used in sedimentology (e.g., *Lewis* and McConchie 1994). The *phi*-transformation, a basis of this scale, is defined as follows:

$$\phi = -\log_2 D \tag{5.15}$$

where D is particle size in mm. The *phi*-size scale grid is defined for the integer values of ϕ. This logarithmic transformation converts a log-normal distribution into a normal distribution, so that probability graph paper could be used (before computers became ubiquitous) to easily plot and visualize deviations of the size distribution from log-normality.

5.5. Transforming the size distribution

5.5.1. Differentiating cumulative size distribution

The PSD $n(D)$ shown in Figure 5.1 was calculated from a cumulative size distribution, $N(D)$, by using a piecewise numerical differentiation algorithm (*Jonasz* 1983a, 1980). In that algorithm, the various size ranges are processed according to the estimated error of the particle count. For particle diameters resulting in a relative particle count error smaller than an arbitrary value of 0.1, a Lagrange interpolation polynomial (e.g., *Kreyszig* 1972, p. 652) is constructed at each triad of consecutive, diameters, and the derivative of this polynomial is assigned to the middle diameter of the triad. If the relative measurement error is greater than 0.1, a power-law approximation is calculated for each triad of consecutive data points, i.e., at (D_0, D_1, D_2), $(D_1, D_2, D_3), \ldots, (D_{M-2}, D_{M-1}, D_M)$, where M is the number of data points, and the negated value of its derivative is assigned to the middle diameter for the fully internal triads, such as (D_1, D_2, D_3), and also to the first or last value of the diameter, D_1, and D_M, for the two edge triads: (D_0, D_1, D_2) and (D_{M-2}, D_{M-1}, D_M).

In processing some of the cumulative distributions for this work, we extended this piecewise differentiation concept by fitting a second-degree polynomial with a least squares procedure, by using weights $w_i = 1/\text{var}(N_i) = 1/N_i$, where $\text{var}(x)$ is the variance of x, which follows from assuming that N_i are Poisson distributed, to all possible sets of four successive points of the cumulative distribution and evaluating the derivative at the second point of each set and also at the first point and the two last points of a set for the edge sets of four points. Note that in certain cases, the curvature of the polynomial curve fitted to the four points may result in a positive derivative value, which—when negated to obtain $n(D)$—yields a negative value. Similar situations may occurr in the piecewise differentiation discussed in the previous paragraph. In such cases, we simply set $n_i = -(N_{i+1} - N_i)/(D_{i+1} - D_i)$.

Other transformations from the cumulative to the frequency distributions have also been reported. *Jackson* et al. (1995) converted their cumulative size distributions into frequency distributions by first interpolating the cumulative distribution data with cubic splines at a fixed interval size grid to obtain $N(D_i)$, and then by calculating the derivative: $-[N(D_{i+1}) - N(D_i)]/(D_{i+1} - D_i)$. *Bush* (1951) describes an alternative procedure for converting the cumulative size distribution into the frequency distribution.

5.5.2. *Changing the size scale*

5.5.2.1. *The principles*

In comparing and/or splicing size distributions obtained with various measurement techniques, the various particle size scales might need to be reconciled. Such a need arises in the splicing case because different measurement techniques use different definitions of the particle size. For example, the resistive technique (section 5.7.1) introduces a volume-based size scale with the particle size defined as the diameter of a sphere with the volume equal to that of the particle. Optical microscopy may use a scale with the particle size being defined either as the diameter of a circle with the area equal to that of the projection of the particle or as the Feret diameter.

Let us consider a broad definition of the particle size S, which—for example—can be either the particle "diameter" , projected area, or volume. Let us consider two such particle size scales S_1 and S_2 which are related as follows:

$$S_2 = f(S_1) \tag{5.16}$$

The relationship between the two forms, n_1 and n_2, of the same PSD expressed in the size scales S_1 and S_2, respectively, is obtained (for example, *Jonasz* 1987a) by requiring that there is an equal number of particles, $dN = n(S)dS$, in the corresponding differential intervals in both size scales:

$$n_2(S_2)dS_2 = n_1(S_1)dS_1 \tag{5.17}$$

Thus,

$$n_2(S_2) = n_1(S_1)\left(\frac{dS_1}{dS_2}\right)_{S_2=f(S_1)} \tag{5.18}$$

Given the functional relationship between S_1 and S_2, and the form of $n_1(S_1)$, the $n_2(S_2)$ can be calculated from (5.18). Assume a simple case of $S_2 = aS_1$, with a being a constant factor. This transformation simply spreads or compresses the size scale depending on the value of a. From (5.18) we have $n_2(S_2) = n_2(aS_1) = (1/a)\ n(S_1)$. If the size distribution is of the power-law type $n_1(S_1) = kS_1{}^{-m}$ then $n_2(S_2) = (k/a^{1-m})S_2{}^{-m}$. Thus, given the scale transformation of $S_2 = aS_1$, the slope m of the power-law size distribution remains the same, and the magnitude term k changes by a factor of a^{1-m}. In section 5.5.2.3 we discuss practically important transformations of the particle size from the particle diameter to volume, essential in the utilization of a large body of biological data for optical modeling.

Jackson et al. (1995) provide a similar procedure for comparing the PSDs for size scales defined by the smallest sphere enclosing the entire particle and by the sphere with volume equal to that of the particle. *Austin* (1998) also describes a procedure for converting a cumulative mass size distribution, $W_1(D)$, obtained with a particle sizing method to a distribution, $W_2(D)$, which would have been obtained with another method. This procedure involves calculation of conversion factors that depend both on the particle size definitions involved and on the shape of the size distribution itself.

A method for transforming a histogram expressed as a function of one variable, for example, a histogram expressed by (5.4), into a histogram expressed as a function of another variable has also been proposed (*Falk* 1992). This method relies on the approximation of the functional relationship represented by the old histogram by a piecewise continuous function whose coefficients are calculated by using the values of that histogram. To obtain the new histogram, this function is integrated within the new histogram channel boundaries.

Although most modern particle counters allow for a fairly dense particle size grid resulting in rather detailed particle size histograms, there is a large body of older data and data acquired by low-resolution particle counters. *Lawless* (2001) developed an interesting procedure of increasing the size resolution of histograms with coarsely defined particle size grids so that, for example, histograms with different size grids can be compared. This procedure is based on the minimization of the differences between the particle count values in arbitrary sub-intervals of the original coarse size intervals with a condition that the particle count from a group of sub-intervals within an original size interval is to equal the original particle count in that interval. Thus, this interpolation method does not modify the information content of the original data as opposed to fitting an approximation to the data or smoothing of the original data. Note that the procedure of Lawless can be used for smoothing as well.

5.5.2.2. Volume-based and projected area-based diameters

One application of the particle size transformation is the comparison or splicing of size distributions determined with different techniques for the same particle population (e.g., *Jackson* et al. 1997). Such a comparison is based on deriving a functional relationship between the particle size definitions introduced by various particle sizing techniques and can be used to characterize the particle shape (*Endoh* et al. 1998, *Umhauer* and Gutsch 1997, *Inaba* and Matsumoto 1995, *Jonasz* 1987a).

Two commonly used techniques of particle size analysis are the volume-sensitive resistive particle counting (section 5.7.1) and the cross-section-sensitive particle sizing, for example, based on microscopy combined with image analysis (section 5.7.4). The particle size range of the resistive technique is roughly 0.5 to 1000 μm. The size range applicable to the image analysis technique has a lower limit of about 5 μm *in situ*.

These two particle sizing techniques introduce two scales of the particle size: (1) the volume-equivalent particle diameter (ESD) D_S i.e. the diameter of a sphere with volume, V, equal to that of the particle:

$$D_S = \left(\frac{6}{\pi}\right)^{1/3} V^{1/3} \tag{5.19}$$

and (2) the projected area-equivalent diameter (equivalent circular diameter, ECD), D_C:

$$D_C = \left(\frac{4}{\pi}\right)^{1/2} A^{1/2} \tag{5.20}$$

where A is the area of the particle projection. In the case of randomly oriented non-spherical particles, this is meant to be the orientation-averaged projection area.

These diameters are the same for a solid sphere but differ for other particle shapes. The projected area of a non-spherical particle, and thus the ECD also depends on the particle shape. Thus, one can expect a probability distribution of the projected area (and thus also of ECD) even for a monodisperse population of particles, i.e. particles which have all the same ESD. Such probability distributions have been derived for randomly oriented non-spherical particles of simple shapes (ellipsoid, cylinder, and circular cone—*Vickers* and Brown 2001, *Vickers* 1996; cube—*Brown* and Vickers 1998) and have been experimentally determined for an optical particle counter (*Butler* et al. 1989, *Chin* et al. 1988), and with an experimental device for measuring the projected areas of large, model particles (*Umhauer* and Gutsch 1997). The fact that the ECD of a non-spherical particle has a probability distribution implies that the PSD that is measured with a device sensitive to the projected area is effectively a convolution of that probability distribution with the size distribution evaluated for the mean ECD.

The mean ECD is much easier to determine. A theorem attributed to Cauchy (e.g., *Vickers* and Brown 2001) that we referred to in Chapter 3 states that the mean projected area of a convex body is ¼ of its surface area. This theorem enables one to analytically calculate the mean ECD for many simple geometric shapes and relate it to the corresponding ESD. There are similar theorems for some irregular particles, such as fractal aggregates, which we shall discuss shortly.

Let us assume that

$$D_S = p_S D_C^{q_S} \tag{5.21}$$

and note that the inverse relationship is also of the same type

$$\begin{aligned} D_C &= \left(\frac{1}{p_S}\right)^{1/q_S} D_S^{1/q_S} \\ &= p_C D_S^{q_C} \end{aligned} \tag{5.22}$$

Such empirical relationships have been established for marine particles from coastal waters off Nova Scotia (*Jonasz* 1987a) with parameters p_C and q_C in (5.22) approximately equal to 1.13 and 1.11 respectively. Relationship (5.22) was obtained for a broad range of particle types, from solid particles to aggregates.

Relationships of this type also results from the fractal aggregation theory (*Jackson* et al. 1997), i.e., apply specifically to aggregates, as we shall discuss it shortly.

By using equation (5.18), we obtain:

$$n_C(D_C) = n_S[D_S(D_C)]p_S q_S D_C^{\,q_S-1} \tag{5.23}$$

and, likewise

$$n_S(D_S) = n_C[D_C(D_S)]p_C q_C D_S^{\,q_C-1} \tag{5.24}$$

Let us now assume that $n_S(D_S) = k_S D_S^{\,-ms}$. Then, $n_C(D_C) = k_C D_S^{\,-mc}$ with the k and m parameters linked by the following relationships, obtained from (5.23):

$$\begin{aligned} k_2 &= k_1 p_1^{\,1-m_1} q_1 \\ m_2 &= q_1(m_1 - 1) + 1 \end{aligned} \tag{5.25}$$

where indices 1 and 2 signify S and C or C and S respectively.

A typical slope, m_S, of the power-law size distribution $n_S(D_S)$ in seawater is about 4 (section 5.8.5.3). Thus, given the value of $q_S = 1/q_C = 0.9$ referred to earlier in this section, the corresponding slope m_C of distribution $n_C(D_C)$ evaluates to $q_S(m_S - 1) + 1 = 3.7$. For more accurate estimation, please refer to the discussion of data shown in Figure 5.33 in section 5.8.5.3 on the power-law approximation.

A power-law transformation between the ESD and ECD can also be explained in terms of the basic properties of fractal aggregates such as the fractal dimension and the component particle size. By using the arguments of *Jackson* et al. (1997) we can—as will be shown shortly—derive the following relationship between the ESD and ECD size scales:

$$D_S = D_0^{\frac{1-d}{3}} D_C^{\frac{d}{3}} \tag{5.26}$$

and its inverse form:

$$D_C = D_0^{\frac{d-3}{d}} D_S^{\frac{3}{d}} \tag{5.27}$$

where d is the three-dimensional fractal dimension of the aggregate, D_g is the aggregate diameter of gyration, and D_0 is the primary particle diameter, assumed to be the same for all particles forming the aggregate.

We will shortly discuss a method of determining the fractal dimension, d, of an aggregate, based on the comparison of the size distribution slopes, and another, based on the measurement of the settling velocity of the particles in section 5.7.11. A reader interested in this topic is encouraged to consult a review by *Bushell* et al. (2002).

Note that the fractal dimension, in reference to the determination of the projected area of a fractal aggregate can be defined by using the *covering set* method as follows (e.g., *Bushell* et al. 2002):

$$d_2 = -\lim_{r \to 0} \frac{\log N_r}{\log r} \tag{5.28}$$

where d_2 is the two-dimensional fractal dimension and N_r is the number of image pixels (*covering set*) with the side, r, needed to fully cover the projection of the aggregate. If the three-dimensional fractal dimension $d < 2$ (geometrically transparent aggregate), then $d_2 = d$. Otherwise, $d_2 = 2$ (geometrically opaque aggregate) and $2 \leq d \leq 3$.

In general, for a fractal aggregate $d < 3$, while for a solid particle $d = 3$. The three-dimensional fractal dimension d is usually denoted by D_3 in the literature. Here we prefer to use the symbol D for the particle diameter and the lowercase d for the three-dimensional fractal dimension. For porous aquatic aggregates, d may vary from 1.26 (highly porous) to 2.59 (less porous) (*de Boer* and Stone 1999, *Li* et al. 1998, *Chen* and Eisma 1995, *Kilps* et al. 1994, *Logan* and Wilkinson 1990). In most cases reported for aquatic particles, the fractal dimension, d, is less than 2, although a range of 2 to 3 has been reported for sediment aggregates near the seabed in an estuary (*Winterwerp* et al. 2002).

The fractal dimension of an aggregate depends on the manner in which aggregates are formed (*Logan* and Kilps 1995) and on the probability of attachment of component particles (e.g., *Logan* and Wilkinson 1990). If the probability of attachment of a component particle to an aggregate is high, highly tenuous (porous) aggregates with low fractal dimension, d, are formed. If the attachment probability is low, component particles can penetrate an aggregate. This promotes the formation of dense aggregates with high fractal dimension, d (close to 3).

We assume here that d is independent of the particle size. However, marine aggregates are formed by various processes and from a diverse population of particles (e.g., *Jackson* and Burd 1998). Each of these processes is most efficient in a specific particle size range. Thus, the fractal dimension may depend on the particle size (*Logan* and Wilkinson 1990), and the formulas given here are the first-order approximation in that respect. Indeed, the small and large particles have been assigned d of 2.7 ± 0.4 and 1.6 ± 0.4 respectively (*Martinis* and Risović 1998), based on the slopes of the power-law approximations for the large- and small-angle ranges of the volume scattering function.

Bushell and Amal (1998) have pointed out the confusion regarding the degree to which the polydispersity of the component particles affects the fractal properties of aggregates. Their own simulations for diffusion-limited aggregates seem to indicate that the polydispersity effects are limited as far as the mass and geometric properties of aggregates are concerned but may have significant effect on light scattering by the aggregates.

Equations (5.26) and (5.27) are based on properties of fractal aggregates (e.g., *Jullien* and Botet 1987) and on certain simplifying assumptions (*Jackson* et al. 1997) that we shall briefly review now. Assume that a fractal aggregate with dimension d is made of N primary particles, each with diameter D_0. Thus, if the particles just touch each other (i.e., do not penetrate into each other), the ESD of such an aggregate can be calculated from:

$$\begin{aligned} D_S &= \left(\frac{6}{\pi}\right)^{1/3} V^{1/3} \\ &= N^{1/3} D_0 \end{aligned} \tag{5.29}$$

For a fractal aggregate, the number of component particles, N, is proportional to the power of d of the gyration diameter D_g (e.g., *Jullien* and Botet 1987):

$$N = k_f \left(\frac{D_g}{D_0}\right)^d \tag{5.30}$$

where k_f is a proportionality constant, referred to as the fractal prefactor. *Brasil* et al. (1999) and *Oh* and Sorensen (1997) point out that both the fractal dimension and prefactor must be known in order to correctly define the fractal properties of an aggregate. Both the fractal dimension and prefactor were found to depend on the overlap of the component particles (*Brasil* et al. 1999, *Oh* and Sorensen 1997). Indeed, the aggregate should become more "solid" with the increasing overlap. As pointed by *Oh* and Sorensen (1997) the overlap in three-dimensional aggregates is difficult to determine from two-dimensional images (projections) of these aggregates because the component particles frequently seem to overlap one another in the two-dimensional projection of an aggregate, even if they only touch one another in the aggregate.

With these caveats, the ESD of the aggregate, i.e., D_S, can thus be expressed with the following equation:

$$\begin{aligned} D_S &= \left[k_f \left(\frac{D_g}{D_0}\right)^d\right]^{1/3} D_0 \\ &= k_f^{\,1/3} D_0^{\,1-d/3} D_g^{\,d/3} \end{aligned} \tag{5.31}$$

Jackson et al. (1997) propose to evaluate the constant k_f as follows. For a solid sphere, i.e., a single-particle "aggregate" with a diameter $D_S = D_0$, i.e., for the three-dimensional fractal dimension $d = 3$, we have:

$$
\begin{aligned}
D_g &= 2\left[\frac{1}{V}\int_0^{r_0} r^2 dV(r)\right]^{1/2} \\
&= 2\left[\frac{4\pi r_0^{\ 3}}{3}\int_0^{r_0} r^2 4\pi r^2 dr\right]^{1/2} \\
&= \left(\frac{3}{5}\right)^{1/2} D_0
\end{aligned}
\tag{5.32}
$$

where r is the radial coordinate. Hence, we can set the coefficient $k_f^{\ 1/3}$ to

$$k_f^{\ 1/3} = \left(\frac{3}{5}\right)^{-d/6} \tag{5.33}$$

in order for equation (5.31) to be valid for the solid sphere case. We are still missing the relationship between the gyration diameter and the ECD, D_C. *Jackson* et al. (1997) suggest using the solid sphere analogy again:

$$D_C = \left(\frac{3}{5}\right)^{-1/2} D_g \tag{5.34}$$

This leads to an equation of the form $D_S = pD_C^{\ q}$ with

$$
\begin{aligned}
p &= D_0^{1-d/3} \\
q &= \frac{d}{3}
\end{aligned}
\tag{5.35}
$$

Interestingly, the slope parameter is $q \cong 0.8$ with $d \cong 2.6$ (for compact aggregates). This is not far from a value of 0.9 determined by *Jonasz* (1987a) for coastal marine particles. Further, as we shall see in section 5.8.5.3, the parameter q (representing the q_S) evaluates to 0.92 based on the averages of the slopes of experimental PSDs obtained with the volume- and projected area-sensitive methods. This would yield the "average" fractal dimension d of about 2.75.

The ECD of an aggregate has also been directly related to the number of the primary particles in an aggregate (e.g., *Köylü* et al. 1995 and references therein):

$$N = k_a\left(\frac{D_C}{D_0}\right)^{2a} \tag{5.36}$$

where the parameters k_a and a have each been evaluated to ~ 1.1 (1.16 and 1.1 respectively) for soot aggregates with $d \sim 1.8$, i.e., within the range of the fractal dimension characteristics of marine aggregates as we shall discuss shortly. The exponent a equals d/d_p, where d_p is the fractal dimension of the two-dimensional projection of the aggregate (*Oh* and Sorensen 1997).

The parameters k_a and a depend on the overlap, γ, between the component particles, defined as follows

$$\gamma = 1 - \frac{l}{D_0} \tag{5.37}$$

where l is the distance between the component particles' centers (assumed constant for the entire aggregate). The overlap factor of 0 corresponds to the component particles just touching each other. In the other extreme, 1 would correspond to the particles coalescing into one particle. *Oh* and Sorensen (1997) also investigated the case with $l > D_0$ where g assumes negative values. Although in flame research, which motivated their study, this case seems to be devoid of practical importance, it may be quite important in aquatic research. In natural waters, solid (opaque) particles may be bound into an aggregate by transparent strands of polysaccharides, also known as transparent exopolymer particles (TEP, e.g., *Passow* 2002, see Figure 6.38 for sample TEP particles). Thus, component particles of an aquatic aggregate may indeed be separated by distances greater than the sum of their "radii."

Results of numerical studies assign the ranges of 1.08 to 1.14 (*Brasil* et al. 1999) and ~1 to 1.1 (*Oh* and Sorensen 1997) to parameter a, and the ranges of 1.1 to 1.44 (*Brasil* et al. 1999) and ~1 to 2.2 (*Oh* and Sorensen 1997) to parameter k_a for the overlap parameter values ranging from 0 to 0.33 (*Brasil* et al. 1999) and −3 to 0.5 (*Oh* and Sorensen 1997).

On the other hand, *Batz*-Sohn (2003) has obtained the following "experimental" formula relating D_C/D_0 to N by fitting results of numerical experiments in which fractal aggregates were constructed by adding primary particles at the aggregate perimeter:

$$\left(\frac{D_C}{D_0}\right)^2 = N(1 - 0.085 \ln N)(1 - 1.1\gamma) \tag{5.38}$$

where γ is the overlap factor (that he calls the penetration parameter) from a range of 0 through 0.4.

Let us now return to the size distribution conversions. From (5.18) and (5.23) we have

$$n_C(D_C) = n_S[D_S(D_C)] \left(\frac{3}{5}\right)^{\frac{3-d}{6}} \frac{d}{3} D_0^{\frac{1-d}{3}} D_C^{\frac{d-3}{3}} \tag{5.39}$$

$$n_S(D_S) = n_C[D_C(D_S)] \left(\frac{3}{5}\right)^{\frac{d-3}{2d}} \frac{3}{d} D_0^{\frac{d-3}{d}} D_S^{\frac{3-d}{d}} \tag{5.40}$$

As an example, consider again a power-law size distribution, $n_S(D_S) = k_s D_S^{-mS}$, obtained with a resistive sizing technique (the ESD size scale). This distribution, when expressed in the ECD size scale characteristic of the image analysis, technique remains also a power-law distribution, $n_C(D_C) = k_C D_C^{-mC}$, but with modified coefficients. The relationships between these coefficients are as follows:

$$k_C = k_S \left(\frac{3}{5}\right)^{(1-m_S)\frac{3-d}{6}} \frac{d}{3} D_0^{(1-m_S)\frac{1-d}{3}}$$

$$m_C = (m_S - 1)\frac{d}{3} + 1 \tag{5.41}$$

$$k_S = k_C \left(\frac{3}{5}\right)^{(1-m_C)\frac{d-3}{2d}} \frac{3}{d} D_0^{(1-m_C)\frac{d-3}{d}}$$

$$m_S = (m_C - 1)\frac{3}{d} + 1 \tag{5.42}$$

If $d < 3$, then the slope of the size distribution obtained with image analysis (expressed as a function of D_C) for the same population would be somewhat less than that of the size distribution obtained with a resistive technique. For example, with $m_S = 4$ and $d = 1.8$, $m_C = 2.8$. Indeed, as we will see in section 5.8.5.3, the slopes of the size distributions obtained with the *in vitro*-resistive sizing technique are significantly higher than those obtained with *in situ* microscopy, although the size scale transformation is not the only reason for this discrepancy as we will discuss in a section on particle breakage (5.7.1.5).

Calculation of the scale coefficients, k_S and k_C, requires the knowledge of both the fractal dimension, d, and the component particle diameter, D_0. The latter parameter is assumed to be constant for all aggregates in a population. Although this is a reasonable assumption for fractal aggregates, such as soot, for which key relationships used here were developed, it may not be so in the case of aggregates occurring in natural waters. Thus, given the complexity of aggregates in natural water that rarely are made of fixed-size components, D_0 can be regarded at best as an indication of the primary particle size.

In splicing PSDs obtained with different sizing techniques, the size ranges of the distributions must overlap in order to solve for the fit parameters, d and D_0. Note that the extent of the overlap is affected by the transformation of the particle size to a common scale, for example, D_S, and may be different from an apparent overlap indicated by the particle diameter ranges expressed in their individual scales (D_S and D_C). The fractal dimension, d, can in principle be determined by comparing the slopes of the size distributions obtained with the two techniques (*Jiang* and Logan 1996, *Logan* and Kilps 1995). *Logan* and Kilps (1995) point out that such a determination is more accurate if the cumulative size distributions are used, because these distributions are smoother than the frequency size distributions. Once the three-dimensional fractal dimension, d, is known, then the component particle diameter, D_0, can be determined by requiring that the magnitude of the two size distributions is the same in the overlap region. *Jackson* et al.

(1997), who spliced *in vitro* Coulter counter and *in situ* microphotography-derived size distributions, minimized the root mean-square (RMS) fit error for the entire size overlap region as a function of both d and D_0. They note problems with such splicing that result from the RMS error having a shallow minimum region and were forced to use additional constraints to complete the splice in inconclusive cases. The unconstrained fits resulted in the ranges of the fractal dimension, d, and the component particle diameter, D_0, of 1.45–2.9 and 1–19.5 μm respectively for the particle size range of 20 to 300 μm. The constrained fits resulted in narrower ranges of 2.26–2.36 and 3.38–9.55 respectively.

In earlier work by *Jackson* et al. (1995), there was no overlap between the size distributions determined by resistive particle sizing and microphotography. In that case, the fit parameters were derived by simply assuming that the two distributions should have the same slopes and overlap, if extrapolated, when transformed to a common size scale. That evaluation resulted in the fractal dimension $d = 2.3$.

One should again keep in mind that the transformation of the size scale is not the only factor that causes the slope of the PSD obtained with a resistive technique to be greater than that of the size distribution obtained by image analysis for the same population. Another important factor may be breakage of aggregates during sampling and resistive sizing. We discuss this problem in section 5.7.1.5.

5.5.2.3. Biovolume and biomass spectra

By using (5.18), the size distribution, $n(D)$, can be expressed in terms of the particle volume, V, rather than the diameter, D. Following the notation of that equation, the new particle size, $D_2 = V$, is expressed as a function of the old particle size $D_1 = D$ (assuming a spherical particle):

$$V(D) = \frac{\pi}{6} D^3 \tag{5.43}$$

with the inverse function

$$D = \left(\frac{6}{\pi}\right)^{1/3} V^{1/3} \tag{5.44}$$

By using (5.18) we have:

$$\begin{aligned} n_V(V) &= n_D[D(V)] \frac{dD}{dV} \\ &= n_D[D(V)] \frac{1}{3} \left(\frac{6}{\pi}\right)^{1/3} V^{-2/3} \end{aligned} \tag{5.45}$$

In the simple case of the power-law size distribution, $n_D(D) = kD^{-m}$,

$$n_V(V) = k \frac{1}{3} \left(\frac{6}{\pi}\right)^{(1-m)/3} V^{-(2+m)/3} \tag{5.46}$$

Thus, the slope of the PSD expressed as a function of the particle volume rather than its linear size is significantly reduced. If, for example, $n_D(D) = kD^{-4}$, then $n_V(V) = k(\pi/18)V^{-2}$.

The biovolume distribution, $v(V)$, typically used by marine biologists, is expressed as follows:

$$v(V) = Vn_V(V) \tag{5.47}$$

If $n_V(V) \propto V^{-2}$ (i.e., $n_D(D) \propto D^{-4}$), then the biovolume distribution $v(V) \propto V^{-1}$, i.e., it has a power-law form with a log-log slope of approximately -1, (e.g., *Gaedke* 1992—lake, *Quinones* et al. 2003–ocean). We will shortly examine the significance of this finding. Here, for completeness, we give a formula for converting the biovolume distribution, $v(V)$, into the PSD, $n(D)$:

$$n(D) = v[V(D)]3D \tag{5.48}$$

and

$$n(D) = n[V(D)]\frac{\pi}{2}D^2 \tag{5.49}$$

The "normalized" biomass spectrum (NBS), b_N, introduced by *Platt* and Denman (1978, 1977; see a methodological discussion by *Blanco* et al. 1994), is also frequently used in biology. Note that in the biological literature, the biomass spectrum, b_N, is usually denoted by β, which is in conflict with the traditional notation for the volume scattering function. The normalized biomass spectrum, b_N, is defined as follows:

$$b_N(W) = \frac{dB(W)}{dW} \tag{5.50}$$

where dB is the biomass concentration in mass, W, increments of constant length in the logarithmic scale:

$$dB(W) = \int_W^{cW} n(W')W'dW' \tag{5.51}$$

where the length factor, c, is a constant tending to unity. If the size grid is chosen to be the octave grid, i.e., $W_{i+1} = 2W_i$, which implies that $\Delta W_i = W_{i+1} - W_i = W_i$,

$$b_N(W) = \frac{dB(W)}{W} \cong \Delta N(W) \tag{5.52}$$

i.e., the normalized biomass spectrum approximates the numerical abundance of organisms, ΔN, in size class, ΔW, as pointed out by *Platt* and Denman (1977; note that $\Delta B = \langle W \rangle \Delta N$, where $\langle W \rangle$ is the mean mass of organisms in size class ΔW).

Note that the multiplication in the upper limit of the integral in (5.51) can be converted to an addition by changing the linear size scale to the logarithmic size scale as follows:

$$dB(W) = \int_{\ln W}^{\ln W + d\ln W} n(W')W'(W'd\ln W')$$

$$= \int_{\ln W}^{\ln W + d\ln W} n(W')W'^2 d\ln W' \qquad (5.53)$$

$$= n(W')W'^2 d\ln W'$$

where we substituted W for $\ln W$ and $d\ln W$ for $\ln c$ in the upper limit of the integral in (5.51). Hence,

$$\frac{dB(W)}{d\ln W} = n(W)W^2 \qquad (5.54)$$

and we can use the calculus chain rule to relate the normalized biomass spectrum and the weight distribution of the organisms from (5.50) and (5.54):

$$b_N(W) = \frac{dB(W)}{d\ln W}\frac{d\ln W}{dW}$$
$$= n(W)W \qquad (5.55)$$

If $n(D) \propto D^{-4}$, then $n(W) \propto W^{-2}$ [as it follows from (5.46) because $W = \rho V$, where ρ is the particle material density], and we have $b_N(W) \propto W^{-1}$. An important comment is due here. Consider equation (5.54). In the present example, it implies that $dB/d\ln W \propto W^0$, i.e., the $dB/d\ln W$ is independent of W. Thus, the W^{-1} dependence of b_N is purely spurious (*Blanco* et al. 1994, *Prothero* 1986). This is because both b_N and W contain the dependence on W. See, for example, *Kenney* (1982) for an illuminating discussion of such spurious correlations. Therefore, the relationship $b_N(W) \propto W^{-1}$ can only be understood as an indication that $dB/d\ln W$ is independent of W, which in turn implies that the biomass concentration in equal logarithmic weight/size intervals, see (5.53), is nearly constant as postulated by *Sheldon* et al. (1972). Conversely, the constancy of biomass per equally sized logarithmic intervals of mass, implied by biological arguments discussed later in this chapter, leads to the log-log slope of –4 of the number size distribution, $n(D)$.

For completeness, we give a formula for converting the normalized biomass distributions into size distributions that follows from (5.18) and (5.55):

$$n(D) = b_N[W(D)]3\mathrm{D}^{-1} \qquad (5.56)$$

where we assumed that the density of the organism is independent of the organism size.

5.6. Uncertainty of the PSD measurements

It is an unfortunate practice that the uncertainty of the PSD is rarely reported. This makes it difficult to assess the significance of the results and forces the user of these data to take results at face value. In this section, we discuss the sources and significance of these uncertainties and provide a recipe for evaluation of their magnitudes. We concentrate on the discussion of uncertainties that apply in general to the size analysis of marine particles. Uncertainties that are specific for a size analysis technique are discussed separately in each technique's section.

5.6.1. Sampling and sample handling

An important source of uncertainty in the measurement of the size distribution is just the act of sampling water. Sampling is a disturbance whose effect is to change the *ambient* turbulence field. As a consequence, particles that were in equilibrium with the ambient turbulence field may break into smaller fragments or may aggregate. The time scale of these processes (e.g., *Milligan* 1995) is comparable with the typical transfer time between sampling and analysis. *Eisma* et al. (1983), who used a diver-operated filter-equipped sampler to collect small samples of seawater, report that aggregates visible to the eye easily disintegrated into clouds of small particles when the sample was disturbed or shocked. Special devices have been described for remote sampling (*Lunau* et al. 2003, 2004), sampling of particle aggregates by divers (*Droppo* et al. 1996, *Ten Brinke* 1994), and for use in waters with high concentration of particles (*Woodward* and Walling 1992). *Droppo* et al. (1996) describe both a sampler for preserving the structures of delicate flocs as well as a procedure of stabilization of the flocs in agarose for examination of the flocs with several microscopic techniques. *Stoderegger* and Herndl (1999) also note that although there is no "significant difference on gently shaking the beakers before counting," yet "intensive stirring altered the particle size spectrum significantly." We discuss the effects of sampling further in a section on the breakage of particle aggregates.

Sampling with Niskin bottles (*Gibbs* and Konwar 1983) and pumps (*Gibbs* 1981) is likely to cause breakage of large particle aggregates (100–300 μm) and a reduction in the maximum particle size by an order of magnitude. Incomplete extraction of large, quickly settling particles is likely with Niskin bottles or other samplers that have an outlet not at the lowest point within the sampler volume (*Gardner* 1977). Interestingly, some 20 years following that observation, similar sampler designs are still being marketed. To minimize the effect of removing water from a sampler through a narrow spigot, *Gibbs* et al. (1989) sampled water from within an opened Niskin bottle after the bottle was recovered.

Sample transport and storage has been noted to modify the suspended particle content (*Alldredge* and Silver 1988, p. 44). Sample handling and preparation for analysis may introduce other problems. In optical microscopy, particles are analyzed typically by either allowing them to settle onto the slide of an inverted microscope or collecting them onto a filter and analysis with an upright optical microscope. *de Boer* and Stone (1999), who used optical microscopy coupled with image analysis for investigation of freshwater aggregates, found significant differences between the two approaches to a point of recommending the use of the filtering method. Indeed, the settling method was found not to assure a sufficient collection of low-density particle and yielded blurred outlines of the particles. This was specifically important in their study aimed at the determination of fractal dimensions of the particles.

Although no significant statistical difference between the average cell volume of solid particles measured using an optical microscope and a Coulter counter was observed (*Montesinos* et al. 1983, *Jonasz* 1983a), the sample preparation for the electron microscopy may result in a significant particle shrinkage. Such shrinkage is usually evident in the images of the cells as wrinkled or collapsed surfaces. In the study of Montesinos and colleagues, the mean cell volumes of 11 microorganisms obtained using a Coulter counter were found to be about 2.5 times higher on the average than the volumes obtained using the electron microscopes. The shrinkage was found to be dependent on the microorganism (but not on its size) and on the drying method, critical point drying being worse than air drying.

5.6.2. *Insufficient sample volume*

The limited volume of water used during particle size analysis contributes to the uncertainty due to undersampling. As we will discuss later in this chapter, the concentration of marine particles greater than size D decreases typically as D^{-3}. Representative concentration of these particles at a diameter of $10\,\mu\text{m}$ is on the order of 10 particles/cm^3. Thus, a sample of few cubic centimeters is likely to yield several tens of particles. However, at a diameter of $100\,\mu\text{m}$, the particle concentration is about $10 \times (100/10)^{-3} = 0.01$ particles/cm^3, and a single sample of several cubic centimeters is quite likely to yield 0 particles. This is directly supported by experimental evidence: it is difficult to catch a small fish with a 10-l Niskin-type water sampler, although one of us has done so once. That event's probability may have actually been strongly biased by the fish's curiosity because the mean concentration of particles with that size (ESD $\sim 2 \times 10^4\,\mu\text{m}$) is extremely low: 10^{-8} particles per cm^3. The rarity of large particles in seawater prompted the sampling of very large water volumes with *in situ* filtration devices (e.g., *Bishop* et al. 1977) in experiments aimed at the determination of the vertical particle flux in the ocean. The sampling time is sufficiently large in these cases to potentially cause substantial grazing by zooplankton and other feeders. Therefore very-large-volume samplers, as well as sediment traps, which are deployed for long periods frequently utilize means to prevent grazing and other disturbances to the samples collected on the filters.

Sampling insufficient volumes at low particle concentration may introduce artifacts into the size distribution if the sampled volume cannot be adjusted on demand, for example, in the image analysis of particles deposited by filtration on a membrane filter, or image of a fixed sample volume taken with *in situ* imaging. Let $\Delta D = (D + \Delta D) - D$ be a size range. According to the definition of the size distribution, we have

$$n(D) = \lim_{\Delta D \to 0} \Delta N / \Delta D \tag{5.57}$$

where ΔN is the average number concentration of particles with sizes in a range ΔD. In practice, $n(D)$ is sometimes taken to be the value of $\Delta N / \Delta D$ itself. Assume that the number concentration of particles is so small, that when we repeatedly sample a volume V_S we obtain either 0 or 1 particle. By sampling r volumes of water, we would count M particles:

$$M = \sum_{j=1}^{r} \delta_j, \quad \delta_j = 0, 1 \tag{5.58}$$

The minimum acceptable value of M is of course unity. Thus, we must examine a series of volumes V_S until we find at least one particle. The number concentration of particles, ΔN, in a size range ΔD is then expressed as follows:

$$\Delta N = \frac{M}{rV_S} \tag{5.59}$$

As already hinted, we set the estimated $n(D)$ value to be equal to $\Delta N / \Delta D = M/(rV_S \Delta D)$.

Now, let a particle size grid $D_0,\ D_0, \ldots D_{i+1}$ be defined by the condition expressed in equation (5.13), and some arbitrary starting value of D_0. As already stated, such a grid is frequently used in natural water research because it provides roughly the same order of magnitude of particle volume within each size interval. Thus, $\Delta D_i = (D_{i+1} - D_i) = D_i(a-1)$, and

$$\frac{\Delta N_i}{\Delta D_i} = \frac{1}{r_i V_S (a-1)} D_i^{-1} \tag{5.60}$$

where a is a constant, and the unity in the numerator stands for the smallest acceptable value of the total particle count M. The variable r_i presumably increases with i, because for a typical PSD in natural waters, we need to examine more volumes, V_S, of water as the number of particles decreases with increasing particle size in order to keep the counting precision at a given level. Let us check whether this increase of the sampling volume on demand preserves the size distribution shape for a power-law size distribution $n(D) = kD^{-m}$. We have

$$\Delta N_i = \int_{D_i}^{D_{i+1}} n(D)dD$$
$$= \frac{k}{-m+1}(a^{-m+1} - 1)D_i^{-m+1} \tag{5.61}$$

The right side of equation (5.61) must be equal to $M/(r_i V_S)$ with $M = 1$. This provides us with an expression for $1/r_i V_S$ in Equation (5.60) that, on substitution in equation (5.60), yields:

$$\frac{\Delta N_i}{\Delta D_i} = \frac{k}{-m+1} \frac{a^{-m+1} - 1}{a-1} D_i^{-m} \tag{5.62}$$

i.e., the slope of the size distribution is correct. The value of the scale factor is correct only in the limit of $a \to 1$, i.e., the width of the size interval $\Delta D_i \to 0$ [use $\lim_{a\to 1}(a^{-m+1} - 1)/(a-1) = -m+1$]. Thus, in the variable-sampling-volume approach, applicable to particle sizing methods that permit real-time adjustments to the total sampled volume, there is no minimum-concentration limit.

However, the minimum-concentration limit applies to a situation where a fixed sample volume is analyzed, such as in image analysis of particles on membrane filters, or *in situ* microphotography. In that case, we must replace $r_i V_S$ in Eq. (5.60) by a constant. Thus, the minimum-concentration size distribution estimate is expressed as

$$\frac{\Delta N_i}{\Delta D_i} = \frac{1}{V_F(a-1)} D_i^{-1}$$
$$= \text{const} \times D_i^{-1} \tag{5.63}$$

where V_F is the fixed volume of water analyzed. Such a size distribution is a power-law distribution with a slope of -1. The evidence for approaching such minimum-concentration size distributions was presented by *Jackson* et al. (1997).

Incidentally, equation (5.62) enables us to assess the quality of approximation of $n(D)$ by histogram $h(D_i, \Delta D_i) = \Delta N_i/\Delta D_i$ [Eq. (5.4)]. We already know that the slope of that approximation is correct. However, the magnitude factor, k', of that estimate is not correct, as can be seen from the following equation, derived from (5.62) by setting $\Delta N_i/\Delta D_i = k' D_i^{-m}$:

$$k' = \frac{k}{-m+1} \frac{a^{-m+1} - 1}{a-1} \tag{5.64}$$

Consider typical values: $m = 4$ and $a = 2^{1/3}$. Then $k' \cong 0.64\,k$. This scale factor k is modified because we assign the $\Delta N_i/\Delta D_i$ value to a particle size D_i which is the lower bound of interval ΔD_i, while that value really corresponds to some particle size within that interval. If we knew the slope of the size distribution, we could calculate the correct particle size, D_i. Indeed, by comparing $k'D_i^{-m} = kD_{i,\text{correct}}{}^{-m}$, the correct particle size can be calculated as follows:

$$D_{i,\text{correct}} = \sqrt[-m]{\frac{k}{-m+1}\frac{a^{-m+1}-1}{a-1}}D_i \qquad (5.65)$$

For the sample values of m and a, discussed above, the $D_{i,\text{correct}} \cong 1.12D_i$.

This topic is also related to the random distribution of particles within the sample when the analysis utilizes only a sub-sample volume. We discuss these aspects of the problem in section 5.6.4.

5.6.3. *Sampling from a spatial distribution of particles in a water body*

Strictly speaking, the effects of sampling from a spatial distribution of particles are not errors. These effects represent an inherent property of particle dispersions in natural waters (as well as in other fluids): particles are not uniformly distributed throughout a water body. We specifically refrained from using the word "randomly" here. Light, water density distribution, convection currents, and turbulence within the water body may each contribute to the non-uniformity of this spatial distribution. Effects of the large-scale turbulence on the distribution of particles in natural waters have been well documented by satellite measurements of the distribution of chlorophyll and other pigments (e.g., *Gordon* et al. 1983). Even on a smaller scale, large variability has been observed in coastal waters (e.g., *Holm*-Hansen and Mitchell 1991). Such variability is also likely to exist in high-turbulence near-bottom layer of the sea. Large fluctuations of particle concentration on the scale of centimeters can be easily inferred from the *in situ* microphotographs of the particles (e.g., *Kranck* and Milligan 1992).

At the small spatial scale, which is of interest here, the distribution of small particles may be affected by small-scale turbulence, and also by the availability of nutrients that may be released from living and dead organic particles. Such clouds of nutrients may, for example, attract swarms of motile chemotactic bacteria (*Blackburn* et al. 1998). The spatial distributions of particles tend to change at a time scale of minutes and may readily cause substantial sample-to-sample variations. This topic has also been discussed in connection with the measurement errors of the volume scattering function in Chapter 4.

When making multi-instrument measurements, for example in developing/verifying a light scattering model, the same volume of seawater should ideally be analyzed at the same time by all the instruments. In an experiment that involves analysis of several different, by necessity, sub-samples of a sample, this is virtually impossible. In the multi-instrument measurements, the effect of turbulence on the uniformity of the spatial distribution of particles at the scale of the sampled volume of seawater may also modify the particles themselves (section 5.6.1).

A step in the right direction would be to make single-particle measurements of various particle properties as the particle passes through the sensing zone. However, this latter approach is likely to introduce other problems caused by modifications of at least some particles by, for example, forcing them to flow through the limited space of the sensing zone that introduces substantial shear stresses. Techniques which enable measurement of single-particle parameters *in situ* have been demonstrated (*Wang* and Hencken 1986, *Holve* and Self 1979a, 1979b) but have not gained much following in the natural water optics community.

5.6.4. Random distribution of particles within sample: sample vs. population

One of the major sources of uncertainties in the measurement of the size distribution is the random spatial distribution of particles within a homogenized sample. It is widely assumed that mixing homogenizes the spatial distribution of particles inside a sample. However, the issue of homogenization may not be as simple as one would like it to be as we will discuss shortly.

Experimental results shown in Figure 5.3 indicate that the probability of finding N particles in volume V of such a sample is approximately governed by the Poisson probability distribution:

$$p(N) = \exp(-\langle N\rangle)\frac{\langle N\rangle^N}{N!} \tag{5.66}$$

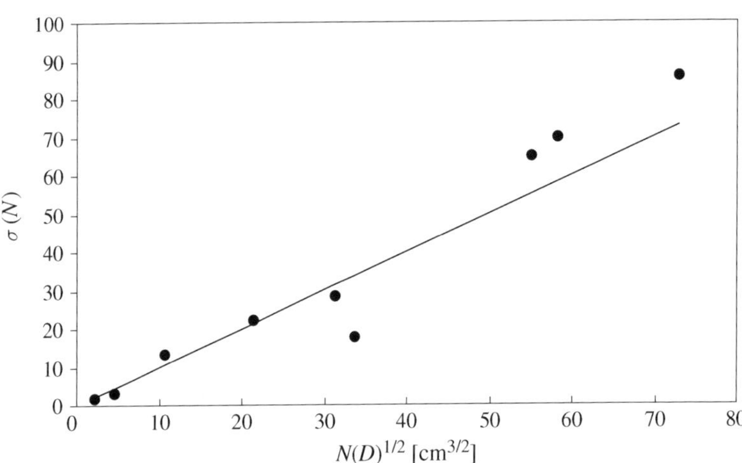

Figure 5.3. Random spatial distribution of particles within a sample volume cause Poisson-distributed random errors in the particle number concentration as it follows from the approximate equality between the standard deviation of the particle concentration, N, and the square root of N. This is a property of the Poisson probability distribution. Symbols: • experimental data obtained with a Coulter counter model ZB, — regression line with a slope of 1.2 ± 0.11, $r^2 = 0.94$, $n = 9$.

where $\langle N \rangle$ is the mean number of particles in volume V. The standard deviation, σ, of the Poisson distribution equals $\langle N \rangle^{-1/2}$. Note that for $\langle N \rangle$ greater than about 100, the Poisson probability distribution is well approximated by a Gaussian distribution with the variance set equal to the mean.

If the cumulative PSD, $N(D)$, at a diameter D is being determined by counting particles greater than a diameter D, within r volumes, V, sub-sampled from the original sample, then $N(D)$ is expressed as follows:

$$N(D) = \frac{1}{rV} \sum_{i=1}^{r} N_i(D) \tag{5.67}$$

where i numbers the sub-samples and N_i is the ith count of particles with diameters greater than D. Each N_i is an independent statistical variable distributed according to the Poisson probability distribution. Thus, the precision of the cumulative size distribution (as indicated by the variance), due to the particle counting precision, $N(D)$, is determined as follows:

$$\text{var}(N) = \frac{1}{rV} N(D) \tag{5.68}$$

where r is the number of sample volumes processed, each having a volume of V. For $rV = 1$ it reduces to $\text{var}(N) = N$. Thus, with increasing particle concentration, the relative precision of N decreases as $\sqrt{N}/N$. One should not overlook the fact that this uncertainty of N is but one of several components of the overall precision of N, which includes also method-dependent components.

Returning to the effect of the particle counting precision, an evaluation of the standard deviation or variance of $n(D)$ depends on the manner in which $n(D)$ is calculated. Let us approximate $n(D)$ by (5.4). Then the variance of the approximation, $\text{var}[h(D, \Delta D)]$, can be expressed as follows:

$$\begin{aligned} \text{var}(h) &= \frac{1}{(\Delta D)^2} \{\text{var}[N(D + \Delta D)] + var[N(D)] \\ &\quad - 2\text{cov}[N(D + \Delta D), N(D)]\} \\ &= \frac{1}{(\Delta D)^2} \{\text{var}[N(D + \Delta D)] + \text{var}[N(D)]\} \end{aligned} \tag{5.69}$$

where $\text{cov}(x, y)$ is the covariance function, here a function of the diameter lag ΔD. This equation follows from an observation that the counting noise is not correlated, i.e., $\text{cov}[N(D + \Delta D), N(D)]$ vanishes. If ΔD is small ($\lim \Delta D \to 0$), then $\text{var}[N(D + \Delta D)] \cong \text{var}[N(D)]$. Thus, for a reasonably small ΔD we have

$$\text{var}(h) \cong \frac{2\text{var}[N(D)]}{(\Delta D)^2} \tag{5.70}$$

It follows that for this simple differentiation method, the variance of the derivative is approximately proportional to var[$N(D)$]. From the increase of the uncertainty of $n(D)$ as compared with that of $N(D)$, it also follows that smoothing of the cumulative size distribution $N(D)$ data is highly recommended *before* attempting to differentiate it in order to obtain $n(D)$.

The manner of sample mixing is usually specified in rather general terms ("gently mixed" and the like). However, recent theoretical and experimental evidence suggests that the mode of mixing a sample may be significant. Mixing creates a turbulence field within the sample that is supposed to homogenize the spatial distribution of particles. This widely held assumption has been questioned. For example, *Squires* and Eaton (1991) show by numerical simulations, consistent with experimental evidence cited by these authors, that an initially homogeneous spatial distribution of particles can be preferentially concentrated by turbulence in regions of low vorticity and high strain rate of the fluid flow. Particle concentrations on the order of more than one magnitude greater than the mean concentration can be observed for particles with sufficiently high inertia relative to the fluid flow acceleration. If the assumption of a random distribution of particles within the sample was used, particle concentrations of an order of magnitude greater than the mean would be extremely improbable. Simulations by *Squires* and Eaton (1991) lead to probability distributions of the particle number concentration which resemble power-law distributions that (1) assign much higher values than the Poisson distribution to probabilities of small particle concentration and (2) decay much slower at high particle concentrations than the Poisson distribution. Perhaps "outliers" in Figure 5.3 indicate such preferential concentration of particles by mixing.

When sampling from a particle population, the sample-derived characteristics are distributed with a certain probability distribution about the population values of these characteristics. The problem lies in how large should the sample/sub-sample be in order to make the results of a particle size analysis significant. *Blanco* et al. (1994) used Monte Carlo simulations to illustrate the pitfalls of undersampling or the choice of the size grid for determination of the size distribution histogram.

The average diameter is a frequently given integral characteristic of a particle population. The error of the sample mean diameter (and other functions of the particle size) has been investigated theoretically for narrow log-normal PSDs with geometric standard deviation, σ_g, of less than 2 (*Masuda* and Iinoya 1971) and recently confirmed by numerical experiments (*Masuda* and Gotoh 1999). According to that theory, by specifying the desired probability, P, that the relative error of the average value $\langle y \rangle$ of function y of the particle diameter be within a range of $\langle y \rangle - \delta$ to $\langle y \rangle + \delta$, where δ is an arbitrary value, one sets the minimum total number, N_T, of the particles that must be measured.

Masuda and colleagues studied functions y of the particle diameter, D, defined as follows

$$y = \kappa D^{\alpha} \tag{5.71}$$

For example, if $y = D$, the diameter itself, then $\kappa = 1$ and $\alpha = 1$.

The minimum total number of the particles required for y to be given with an error of less than δ can be calculated from the following expression:

$$\ln N_T = -2 \ln \delta + \ln \omega \tag{5.72}$$

where

$$\omega = u^2 \alpha^2 \sigma_g^2 (2c^2 \sigma_g^2 + 1) \tag{5.73}$$

$$c = \beta + \alpha/2 \tag{5.74}$$

and β represents the form of the size distribution, i.e., $\beta = 0$ for the number distribution, $\beta = 2$ for the cross-section distribution, and $\beta = 3$ for the volume distribution, and σ_g is the geometric standard deviation, i.e., $\sigma(\ln D)$. The variable u is defined as follows:

$$\begin{aligned} \Phi(|u|) &= \int_{|u|}^{\infty} \exp(-\frac{\varphi^2}{2}) d\varphi \\ &= \frac{1-P}{2} \end{aligned} \tag{5.75}$$

where symbol Φ is simply the cumulative normal distribution. For example, with $P = 0.95$ we have $u = 1.96$.

For example, the mean volume diameter, $\langle D_V \rangle (\alpha = 3)$, defined as follows:

$$\langle D_V \rangle = \left[\frac{\int D^3 n(\ln D) d \ln D}{\int n(\ln D) d \ln D} \right]^{1/3} \tag{5.76}$$

can be determined with a relative error < 0.05 with a probability $P = 0.95 (u = 1.96)$ when a minimum of about 6100 particles are measured in a population with a log-normal size distribution, $n(\ln D)$, having $\sigma_g = 1.6$ (*Masuda* and Gotoh 1999).

The range of the geometric standard deviation, σ_g, assumed by *Masuda* and Gotoh (1999) is within the range representative of log-normal components of the marine PSDs examined by *Jonasz* and Fournier (1996). However, distributions wider than those are likely. Errors of particle counts were examined for such distributions by *Wedd* (2001) by assuming a Poisson probability distribution of the particle count. Wedd concluded that the Masuda–Iinoya theory (*Masuda* and Innoya 1971) underestimates the total number of particles required for a given accuracy of the particle count in the 90 to 99.9 percentile range of the particle size when $\sigma_g > 1.5$.

An alternative approach, based on the Kolmogorov–Smirnov (KS) statistics, was proposed by *Vigneau* et al. (2000). That statistics, and related test of the goodness-of-fit by a theoretical *continuous* size distribution function to experimental data, applies to the normalized cumulative size distribution, $\underline{N}_m$, $m = 1, 2, \ldots, M$, defined as a staircase function for a discrete set of the particle size measurements, arranged in an ascending series $D_1 < D_2 < \ldots < D_M$. This function, varying in steps of $1/M$, is defined as follows:

$$\underline{N}_m = N(D_m) = \frac{m}{M} \tag{5.77}$$

at the particle size grid that is defined by the sample of the particles enumerated.

As the procedure indicates, this definition of the cumulative size distribution is best applicable to microscopy or in flow cytometry, where particles are sized individually, and thus *each* particle can define a particle size, D_m. Note that such a procedure does not prevent one from regrouping the data into size intervals and presenting them in a manner discussed earlier in this chapter (section 5.3).

The confidence half-band for the function $\underline{N}_m$ is expressed as follows:

$$\Delta\underline{N}_{m;\alpha,M} = \sqrt{-\frac{1}{2M}\ln\frac{\alpha}{2}} \tag{5.78}$$

where $1-\alpha$ is the confidence level.

Given this particle size-independent band, one can calculate the upper and lower confidence limits of the cumulative size distribution determined from a specific number of particle size measurements:

$$\begin{aligned} \underline{N}_{m,-} &= \min(1, \underline{N}_m + \Delta\underline{N}_{m;\alpha,M}) \\ \underline{N}_{m,+} &= \max(0, \underline{N}_m - \Delta\underline{N}_{m;\alpha,M}) \end{aligned} \tag{5.79}$$

Note (*Vigneau* et al. 2000) that the determination of the confidence interval, i.e., $2\underline{N}_{m;\ \alpha,\ M}$, loses its significance near the beginning and end of the particle size range D_1 to D_M. This is the consequence of the definitions of the upper and lower confidence limits.

5.6.5. *Sample aging and instrumental drift*

The effect of random spatial distribution may be obscured by systematic time trends in the number concentration of particles in the sample. *Jackson* et al. (1995), *McCave* (1983), *Kadyshevich* et al. (1971) all point to similar problems. A systematic assessment of time-dependent changes in the PSD (*Jantschik* et al. 1992) suggests that short-term variations (time <120 min) in the size distribution at particle diameters <20 μm are on the order of 10 to 20%, with much greater variations occurring at the large diameters than at the small diameters (on the order of 1 μm). Storage of samples for up to 6 h in that study caused variations of 20 to 40% for particle diameters <20 μm.

5.6.6. Modifications of particles by preservation

Preservation of live cells has been reported to cause method-dependent variations of the particle size. *Lee* and Fuhrman (1987) observed that cells examined with an epifluorescence microscope may shrink as the preservative is added to the sample, although not as much as in the case of the sample preparation for scanning electron microscopy (SEM). A definite reduction in cell size of flagellates and ciliates in cultures, caused by addition of formalin (buffered, a low concentration of 0.2%), was also observed when using a Coulter counter (*Boyd* and Johnson 1995, *Morel* and Ahn 1991). *Verity* et al. (1992) quote cell volume reduction on the order of 29% when using 0.5% glutaraldehyde as a preservative. This is consistent with a later study by *Montagnes* et al. (1994), who also point out that the results are species dependent, ranging from 40 to 60%.

5.6.7. Method-dependent uncertainties

Each particle size analysis method introduces additional method-dependent uncertainties. For example, in the zone methods, such as flow cytometry, the particles are subjected to a high-shear flow. The high-shear forces in that flow field may disrupt particles that have survived sampling and preparatory sample handling. In addition, the particle shape also introduces uncertainties that depend on the particle sizing technique and may be difficult to qualify. These uncertainties are discussed in more detail in the following sections on the methods of particle size analysis.

5.7. Methods of PSD measurements

The various methods used to examine particles in different sub-ranges of the optically important particle size range have been reviewed by several authors (*Kaye* 1999, *Black* et al. 1996, *McCave* and Syvitski 1991, *Farrow* and Warren 1993, *Barth* and Sun 1991, *Allen* 1990a, *Bunville* 1984, *Stockham* and Fochtman 1979). Many short and accessible reviews of the particle sizing methods have been published (e.g., *Kaye* and Trottier 1995).

McCave and Syvitski (1991) and *Allen* (1990b) discuss the various particle size definitions and list the sources of particle size standards. *Dean* (1990) reviews the sources and specifications of zone-sensing particle size analyzers such as resistive and optical single-particle counters (flow cytometers).

Many of the particle size analysis methods originated in response to demands of technologies and sciences not related to aquatic sciences. For example, the development of the resistive particle counting was stimulated by medical applications, such as blood cell counting. As a consequence, particle sizing instruments are optimized for those "core" applications. This may limit their application potential in aquatic sciences as comprehensively discussed by *Peeters* et al. (1989). Although recently an optimized flow cytometer has been developed for oceanographic applications (e.g., *Dubelaar* et al. 1999), there seems to be little chance that the situation will soon change, as the oceanographic particle sizing market constitutes a minor fraction of the overall particle sizing market (*Dubellar* and Jonker 2000).

Each particle sizing method introduces a method-specific definition of particle size which makes it difficult to compare results obtained with different sizing methods (*Jackson* et al. 1997, 1995). Even spherical particles with narrow size distributions may be sized differently by the various methods (*Yoshida* et al. 2003). Many attempts were made to compare size distributions of various particle populations of both natural and industrial origins obtained with different particle sizing methods. *Singer* (1986) and *Singer* et al. (1988) compared the performance of sedimentation-based particle size analyzers, a resistive particle counter, and a laser diffractometer. The size distributions obtained with these instruments were similar when the suspensions contained particles with sizes in a relatively narrow range. However, the differences between the results obtained using the various techniques were much greater for polydisperse and polymodal suspensions.

Allen (1990b) compared the performance of 27 particle sizing instruments of different types and from different manufacturers. The particle sizing methods employed in those instruments ranged from sedimentation (photocentrifuges), through "light blocking" and light scattering to laser diffractometry. Silica powder with particle sizes between 0.3 and 3 μm (the mean size of about 1 μm) was used to provide test samples. The differences between the results obtained with the different instruments reached 50% (averaged over the particle size range). These differences were due to the differences in the particle sizing methods used, but also to errors in the software supplied with the instruments. The reproducibility was much better: the maximum standard deviations averaged over the particle size range were less than 4%.

An intercomparison study of particle size analysis methods (resistive particle counting, x-ray sedimentation, laser diffraction, microscopy + image analysis, laser-beam scanning, i.e., "time-of-transition," photosedimentation, and sedimentation tubes, and sieve), aimed at typical marine geology applications, was performed by *Syvitski* et al. (1991b). The authors concluded that "all instruments are associated with some bias" and left it to the readers to consider the merits of the various methods. Recently *Eisma* et al. (1996) compared *in situ* methods of size measurements for aggregates (flocs).

A similarly comprehensive "round-robin" investigation of the differences between the various methods of particle size analysis that additionally involved several instruments of different makes in most method groups was carried out by *Naito* et al. 1998. The methods used in this investigation included resistive particle counting, laser diffraction, x-ray sedimentation, photosedimentation, and optical scanning (i.e., "time-of-transition"). Although the particle shape affected particle sizing by all methods, the effect was especially pronounced for strongly non-spherical particles (fibers and flakes). In the case of the optical particle sizing methods (laser diffraction, photosedimentation, and scanning), the variability of the effective particle area/shape with the particle orientation significantly widened the particle size range, although this widening was method specific and was most pronounced for the photosedimentation method. The x-ray sedimentation method was found to be least affected by the particle shape.

Similar conclusions were formulated by *Endoh* et al. (1998), who compared the sedimentation and laser diffraction technique for fibers and flakes. These latter authors also explained significant differences between these two methods in the case of flakes.

Differences in the PSDs obtained with the various particle size analysis methods have been well established through such comparisons. On one hand, these differences are a nuisance for those who want to create an overview of the size distribution properties and their variability and have to resort to results obtained by various measurement methods. On the other hand, these differences can be used to study the secondary characteristics of the particles, such as particle shape (*Kaye* et al. 1999, *Naito* et al. 1998, *Endoh* et al. 1998, *Inaba* and Matsumoto 1995, *Jonasz* 1987b).

5.7.1. Resistive single-particle counters

5.7.1.1. Technology

The resistive particle counting technique was commercially developed by Coulter Electronics, Inc. to automate tedious blood cell counting. The potential of this technique in marine sciences was soon recognized (e.g., *Sheldon* and Parsons 1967a). Since then, many researchers have used the technique of resistive particle counting to analyze hundreds of seawater samples. Applications and peculiarities of the resistive sizing technique have been reviewed by several authors (*Lines* 1992, *Kachel* 1990, 1982). Oceanographic applications and procedures for resistive particle counting were also reviewed (*Milligan* and Kranck 1991, *Kranck* and Milligan 1979, *Sheldon* and Parsons, 1967a). *Kachel*'s (1990) review is of particular interest in regard of the sensing technology basis and the development history of hydrodynamic focusing: a technique that limits variations in the system response as a function of the particle size. The resistive particle counters were also available from Particle Data, Inc. (the Elzone system, see *Karuhn* and Berg 1984 for a review).

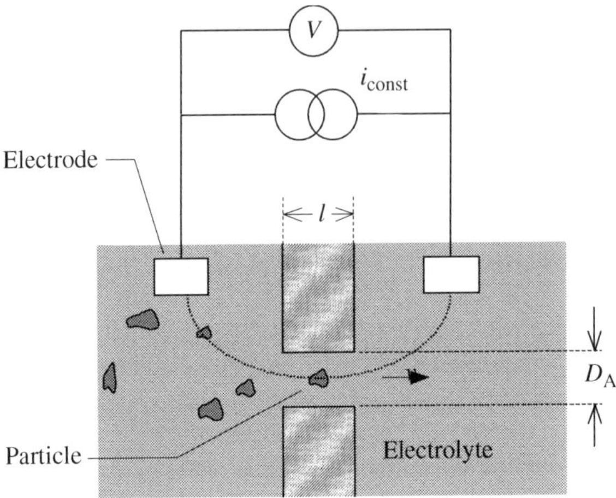

Figure 5.4. The principle of the resistive particle counting. The electrolyte closes a constant-current loop (dashed arc) between two electrodes through a small cylindrical aperture in a dielectric membrane. Particles and the electrolyte flow through the aperture at a constant velocity, v, on the order of 1 m/s. When a particle passes through the aperture, the voltage, V, between the two electrodes rises due to a momentarily increased resistance of the current path. If the particle size is between ~2 and ~60% of the aperture diameter, D_A (on the order of between 10 and 1000 μm), the pulse amplitude is approximately proportional to the particle volume, although the particle shape, structure, orientation, and its path within the aperture may also be of influence. This proportionality is established by noting the pulse height generated by quasi-monodisperse particles of known size. The voltage pulses above a threshold or within multiple voltage ranges are counted within a preset time period, yielding a concentration of particles greater than the corresponding particle size, or the concentrations within the corresponding size bins, respectively. A ratio of the aperture length, l, to the diameter, D_A, determines the electric field distribution within the aperture as well as the fluid flow field. The particle concentration is typically kept low to minimize the coincidence, i.e., simultaneous passage of two or more particles through the aperture that can cause the particle concentration to be underestimated. If a particle suspension cannot be diluted, the coincidence effects may need to be compensated for.

The resistive technique relies on sizing and counting electrical resistance pulses, each ideally caused by a single particle as it is drawn, along with the electrolyte in which it is suspended, through a small aperture connecting two chambers filled with electrolyte (Figure 5.4). The fluid (and particle) velocity in the aperture depends on the aperture size and the pressure applied and may be as high as 3 m s^{-1} (e.g., *Kubitschek* 1962). The resistance is determined by measuring the voltage, at constant current, between two platinum electrodes, submerged each in one of the two chambers. The height of the resistance pulse caused by the passage of the particle through the aperture is approximately proportional to the particle volume, V (e.g., *Jackson* et al. 1995):

$$\Delta\rho = \frac{3}{2}\xi\frac{V}{\frac{3/2}{1-\rho/\rho_p} - \frac{1}{\eta}\left(\frac{D}{D_A}\right)^2} \tag{5.80}$$

where $\xi = 16R\pi^{-2}D_A^{-4}$, ρ and ρ_p are the resistivities of the electrolyte and particle respectively, D and D_A are the diameters of the particle and aperture respectively, and η is the shape factor. The shape factor is on the order of unity. Thus, if particles are small as compared with the aperture diameter, i.e., $D/D_A << 1$, then (5.80) reduces to:

$$\Delta\rho \cong \xi V(1-\rho/\rho_p) \tag{5.81}$$

Finally, if particles are nearly non-conductive, $\rho/\rho_p \cong 0$, we have:

$$\Delta\rho \cong \xi V \tag{5.82}$$

According to *Kachel* (1990), one should be concerned with the effect of the resistivity of the particles, ρ_p, if it is less than 100ρ.

As the resistive method is sensitive to the particle volume, this method defines the particle size as the diameter of a sphere with volume equal to that of the particle, i.e., ESD. The volume of the particle-containing electrolyte that is drawn through the aperture is measured, permitting the determination of the particle concentration. The relationship between the counter response and the particle size is established by using well-defined, nearly monodisperse particles (particle size standards, e.g., *Sheldon* and Parsons 1967a, *Mercer* 1966). Particle size standards used today are almost exclusively polystyrene latex particles. In the past, pollens were frequently used as standards for the Coulter counter. This raises a question about the reliability of the particle size calibrations performed using those latter particles that are reported to swell when immersed in water (e.g., *van Hout* and Katz 2004, ~10% increase in the particle size) .

With the resistive technique, particles are measured and counted in a relatively well-defined zone. The ratio of the volume of that zone to that of the aperture ranges from nearly unity (*Wynn* and Hounslow 1997) to 4 (*Allen* 1990a), depending on the length-to-diameter ratio of the aperture. The longer the aperture, the closer is the sensing zone volume to that of the aperture because the edge effects of the electric field distribution around the aperture are less significant (*Kachel* 1982).

Apertures with diameters ranging from 12 to 1000 μm are used, each permitting the particle size analysis in a nominal particle size range on the order of 2 to 40% of the aperture diameter. *Cowan* and Harfield (1990) provide evidence that this range may in fact be as large as 2 to 60%. However, when delicate particles are to be analyzed, the aperture diameter in relation to the particle diameter may need to be selected so that this ratio is low in order to minimize breakage of the particles by the flow shear stresses near and inside the aperture (see section 5.7.1.5).

There may exist significant differences between the PSDs measured with a Coulter counter and an Elzone counter for the same samples. *Longhurst* et al. (1992) indicate that the distributions in the sub-micron particle size range measured with an Elzone counter may be higher by a factor of 1.2 than those measured with a Coulter counter. *Jackson* et al. (1995) report a factor of 1.38. The different designs of the two systems are suggested to be the cause of such differences.

We should mention extensions and modifications of the resistive particle counting technology. In the early 1970s, a potential sensing configuration was introduced by *Leif* and Thomas (1973) and improved by *Thomas* et al. (1974). In this configuration, the current was supplied to the orifice by one set of electrodes, and the particle-produced pulse was measured by using another electrode set. *Thomas* et al. were also able to determine the shape of non-spherical particles (human erythrocytes) by using their sensor.

A similar sensor configuration has been examined more recently by *Zhang* et al. (2003), with the specific aim of determining the particle shape. In addition, instead of an aperture with the electrodes located outside it, all the electrodes were deposited on the inner surface of a sensing tube. This eliminated particle recirculation errors, caused by trapping some particles in a turbulence field outside the orifice as in conventional Coulter counter sensors. Theoretical analysis and a large-scale model of the sensor indicates that the particle aspect ratio can be obtained with this sensor configuration.

Finally, some resistive particle counters combine resistive particle sizing with optical characterization through measurement of the fluorescence as we mention in section 5.7.2.1.

5.7.1.2. Coincidence

The length of an aperture is a compromise between the particle sizing accuracy and the ability of a particle counter to analyze suspensions of high particle concentrations (*Kubitschek* 1962). Given the velocity of the electrolyte flow in the aperture, and the speed of response of the electronics, the aperture length determines the accuracy of both particle sizing and counting. If this length is too small, the particle passage time may be insufficient for a pulse generated by the particle to fully develop. If this happens, the particle volume will be underestimated (e.g., *Kachel* 1990). Moreover, with substantially asymmetric (needle-like) particles, whose length is greater than that of the aperture, the pulse height may not accurately represent the total volume of the particle because a portion of the volume will be outside the sensing zone at some time during the particle passage.

On the other hand, if the aperture is too long, the sensing zone volume may be sufficiently large for the probability of two or more particles being simultaneously in that volume (coincide) to be significant at a given particle concentration.

The particles simultaneously present in the sensing volume will be counted depending on their sizes and positions in the sensing zone. Two major coincidence types are worth mentioning. In the first type, which we prefer to name the "particle-hiding" rather than the "horizontal" coincidence (*Princen* and Kwolek 1965), the largest or first particle, depending on the instrument's trigger setup (*Kachel* 1990), is counted and sized, and the other particles are "hidden" to the sensor. In the second type, which we prefer to name the "volume-summing" rather than the "vertical" coincidence (*Princen* and Kwolek 1965), the volumes of all particles in the sensing zone are summed and the particles are counted as a single, non-existing particle, larger than any of its "components." Note that this latter type of coincidence conserves the total particle volume as opposed to the "particle-hiding" coincidence. In proposing the new terminology, we hope to clear the confusion associated with the traditionally used terms.

The two coincidence types are schematically represented in Figure 5.5. These schematics refer to ideal pulse shapes, being best approximated by those generated by particles traversing the aperture at the axis. Pulses due to off-axis particles (e.g., *Kachel* 1982) or to non-spherical particles rotating during the passage (e.g., *Berge* et al. 1989) have more complicated shapes and may feature several intermediary maxima, an unnecessary complication, given that modern instruments allow rejection of such complicated pulse shapes. The method of detecting the pulse maximum is important, because it is the pulse maximum, not the pulse integral, which contains the particle volume information.

Coincidence has two effects: (1) the measured concentration of all particles is reduced by both particle-hiding and volume-summing as compared with the true concentration, and (2) the concentration of large particles may be increased by the volume-summing at the expense of that of the small particles. The phenomenon of coincidence is common to all particle counters that are based on the fixed-volume zone-sensing, i.e., the electrical resistance and optical flow cytometers, although differences exist, due to the different characteristics of the particles that are sensed by different sensor types. It is thus worth examining the effect of coincidence in more detail.

A sample volume, V_S, analyzed by a resistive sensor, passes through the sensing zone with volume, v_Z. This passage can be visualized as a sequence of the zone volumes, $v_1, v_2, \ldots, v_N$, all equal to v_Z, with $V_S = \Sigma_{i=1\ldots N}\ v_i$. In the simplest approach, for each zone volume, v_i, which contains one or more than one particle, the sensor will produce a single pulse. This pulse may represent (1) the size of one of the particles that are simultaneously in the zone volume (the "particle-hiding" interaction—the simpler case) or (2) the sum of volumes of all particles in the zone volume (the "volume-summing" interaction—the less simple case).

Consider the coincidence effect in counting particles from a monodisperse particle population with a single-threshold particle counter. In this case, particles are counted only when the counter's signal threshold is set below the pulse height produced by each particle. Given the average number concentration of the

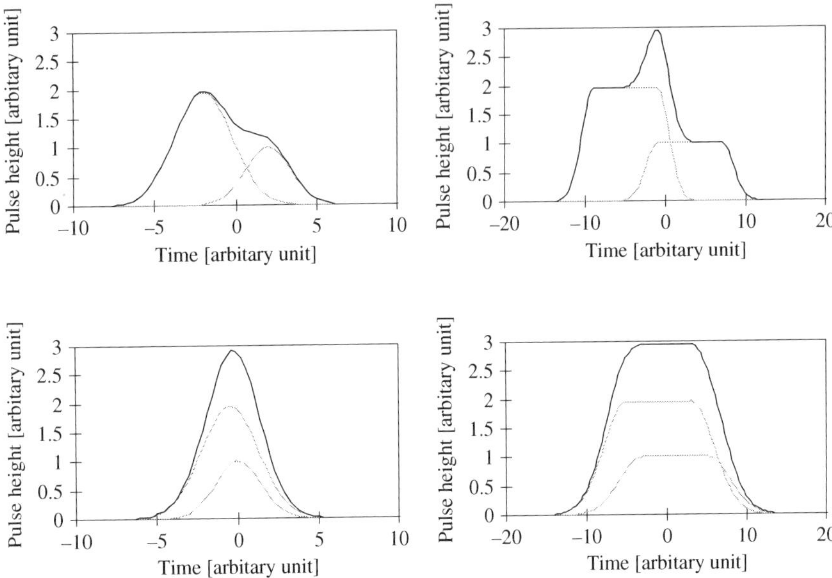

Figure 5.5. Schematic representation of the two major coincidence types "particle-hiding" ("horizontal') and "volume-summing" ("vertical") (top and bottom rows respectively) in short and long apertures (left and right columns respectively) with two-particle coincidence. The gray curves represent pulses from the individual particles, the black curves represent the signal as seen by the pulse processing electronics of the particle counter. In this simulation, the response time of the counting electronics is assumed to be negligible compared to the pulse rise time. The time intervals between the pulse maxima indicate the distances between the particles along the aperture axis. It is assumed that the particle velocity is such that the pulse has time to fully develop in every case. The pulses for long apertures have distinct "top-hat" profile as opposed to the "bell-shaped" short-aperture profiles. This schematics shows that, aside from the distance between and the relative pulse heights of the particles, the coincidence type depends also on the pulse height detection method used by the particle counter electronics. Indeed, if the magnitude of a pulse is determined by that part of the pulse profile where the pulse amplitude *does not increase* within a pre-defined time, say on the order of 10 units, the top-right panel represents the particle-hiding case, because only the pulse magnitude for the first particle is measured and the subsequent brief increase of the pulse amplitude due to the addition of the second particle's pulse is ignored. Otherwise the top-right panel represents a case of the "volume-summing" coincidence.

particles, N [cm^{-3}], in the sample, the total true number, N_S, of particles present in V_S thus equals:

$$N_S = N_Z L \tag{5.83}$$

where

$$N_Z = N v_Z \tag{5.84}$$

is the average number of particles in the zone volume, and

$$L = \frac{V_S}{v_Z} \tag{5.85}$$

is the number of the sensing zone volumes in the sample volume. The parameter L is also the limiting number of particles in volume V_S that the zone sensor can count. Indeed, due to coincidence, two or more particles being simultaneously in the sensing volume produce a single pulse, so the maximum count that the instrument can produce is the number of zone volumes in the sample volume.

The number of particles actually counted, N_S', in the sample volume is equal to or smaller than the true number, N_S. Indeed, N_S' equals the number of zone volumes containing at least one particle. This number is simply the number, L, of zone volumes in the sample volume less the number of zone volumes containing no particles, L_0:

$$N_S' = L - L_0 \tag{5.86}$$

By assuming that the probability, $p(k)$, of finding k, particles in the sensing zone volume, which on the average contains N_Z, particles, is Poisson distributed, we have:

$$p(k) = \exp(-N_Z)\frac{N_Z{}^k}{k!} \tag{5.87}$$

However, from the definition of the probability, $p(k)$, as the ratio of the number of desired outcomes (the number of findings of k particles in the sensing zone volume) to the total number of outcomes (i.e., the number of zone volumes analyzed, L), we have

$$p(k) = \frac{L_k}{L} \tag{5.88}$$

where L_k is the number of zone volumes containing k particles. Hence, from (5.87) and (5.88) with $k = 0$, we have $L_0 = L\exp(-N_Z)$ and:

$$\begin{aligned} N_S' &= L - L\exp(-N_Z) \\ &= L\left[1 - \exp\left(-\frac{N_S}{L}\right)\right] \end{aligned} \tag{5.89}$$

From the above equation, the actual number of particles in the sample volume, V_S, can be simply obtained as follows:

$$N_S = -L\ln\left(1 - \frac{N_S'}{L}\right) \tag{5.90}$$

By limiting ourselves to particle concentrations such that $N_S/L << 1$, we can obtain the following approximate expression for the coincidence-reduced count, $N_S{}'$:

$$N_S{}' \cong N_S - \frac{N_S^2}{2L} \tag{5.91}$$

From the above equation, the approximate actual number of particles in the sample volume, V_S, is as follows:

$$\begin{aligned} N_S &= L\left(1 - \sqrt{1 - 2\frac{N_S{}'}{L}}\right) \\ &\approx N_S{}' + \frac{N_S{}'^2}{2L} \end{aligned} \tag{5.92}$$

Note that the second row of the above equation is an approximation of an approximation, hence its accuracy is limited, as experienced by *Princen* and Kwolek (1965).

In order to apply these formulas, we need to know L, i.e., the ratio of the sample volume to the sensing zone volume. In the case of the resistive particle counting, due to edge effects in the electric field distribution around the aperture (*Kachel* 1982), the sensing zone volume is generally greater than the aperture volume by as much as four times in short apertures (*Allen* 1990a). In long apertures, the edge effects are less significant, and the zone volume is much closer to the geometric volume of the aperture. For example, *Wynn* and Hounslow (1997) obtained a nearly unity ratio of the sensing zone volume (0.00087 μl) to that of the aperture itself (0.00082 μl) for an Elzone particle counter aperture with a 60 μm diameter and 290 μm length.

The parameter L can be obtained, for example, by fitting formula (5.89) to experimental data of $N_S{}' = f(C)$ obtained by measuring $N_S{}'$ for a series of suspensions. Such a series can, for example, be obtained from a high-concentration suspension, by diluting it to a varying degree as represented by a fractional concentration C. *Wynn* and Hounslow (1997) discuss the merits of the various approaches to preparing such a series of suspensions. For the purpose of the fitting, it is convenient to rewrite (5.89) as follows

$$N_S{}' = L\left[1 - \exp\left(-\frac{N_{SH}C}{L}\right)\right] \tag{5.93}$$

where N_{SH} is the number of particles in the sample volume in the high-concentration suspension and C is the fractional concentration ($C < 1$). Note that N_{SH} needs not to be known *a priori*. Indeed, by starting from the (sufficiently) high-C end of the suspension sequence and working our way down the concentration scale, we shall soon arrive at $C = C_0$ such that for the $C < C_0$, the relationship $N_S' = f(C)$ becomes linear to within the measurement accuracy. In that concentration range, we have $N_S' = N_{SH}C$. Thus, one can readily determine N_{SH} with an accuracy depending on the particle count N_S', as stated by (5.68), and the accuracy of C, and obtain L by fitting (5.93) to $N_S' = f(C)$ for the entire suspension sequence.

Wales and Wilson (1961, 1962) obtained the following coincidence formula

$$N_S' \cong \frac{N_S}{1 + \dfrac{N_S}{2L}} \tag{5.94}$$

that yields too high a coincidence-reduced particle count as pointed out by *Princen* and Kwolek (1965).

In a polydispersed particle population, both types of coincidence may play a role. While the "particle-hiding" ("horizontal") coincidence will merely reduce the particle count across the size spectrum, the "volume-summing" ("vertical") interaction will transfer a portion of that count from the small-to the large-particle size range.

The relative importance of these coincidence types is practically significant, because the "volume-summing" coincidence is more difficult to model and correct than is the "particle-hiding" coincidence. *Wales* and Wilson (1961, 1962) "believed" that the "volume-summing" coincidence is less important. On the other hand, *Princen* and Kwolek (1965) advanced a contradicting view, based on the claim of the manufacturer of the Coulter counter that the device responds only to the particle volume, but without much experimental evidence. Experiments with counting nearly monodisperse particles at high concentrations clearly show (e.g., *Wynn* and Hounslow 1997) the presence of doublets and, to a much smaller degree, triplets, which suggests that the effect may indeed be significant.

Few particles simultaneously present in the zone volume (a multiplet) can be regarded as a "porous" non-spherical particle Although with resistive particle sizing, the particle volume is the primary factor, other factors (particle shape, orientation, position, and structure) cannot be ruled out as we will shortly explain. Experiments (e.g., *Eckhoff* 1969) indicate that the orientation of a porous particle may have a significant influence on the pulse height. By extrapolation, these experiments suggest that a pulse resulting from a multiplet whose component particles are in close proximity and so oriented that the projection of the multiplet onto a plane perpendicular to the aperture axis, i.e., the direction of the sensing current flow, contains no "holes" (i.e., direct pathways for the ionic current) may be quite different (greater) than one from the multiplet oriented otherwise. However, because of the requirement that the component particles be in close proximity to each other and in specific spatial configuration(s), the probability of such a multiplet configuration, leading to the volume-summing coincidence, is less likely than that of finding several particles anywhere inside the zone volume.

An interesting observation, in the coincidence context, was made by *Gibbs* (1982a) who studied breakage of flocs by an aperture of a resistive particle counter (see section 5.7.2.4). He noted that the signals from the flocs, which broke on entry to the aperture and traversed the aperture as a string of small fragments, were not registered by the counter according to the unbroken flocs' volumes, resulting in a count loss in the particle size range corresponding to the unbroken flocs. This would suggest that the "volume-summing" coincidence is less likely an event than the "particle-hiding" coincidence. In the case of resistive particle sizing, there is also the signal magnitude argument. Indeed, in that case, the pulse height is roughly proportional to the particle volume, i.e., $\sim D^3$. For example, a particle with twice the size of another particle will produce a pulse with a height that is eight times, i.e., nearly an order of magnitude, greater than that produced by the smaller particle.

The "volume-summing" coincidence has been modeled for a single-mode log-normal PSD by *Princen* and Kwolek (1965). They calculated the probabilities of the various orders of particle multiplets within the sensing zone and formulated a system of several equations involving non-linear functions of the unknown mean size and logarithmic standard deviation of the original size distribution which need to be solved for these two latter parameters by using parameters calculated from the measured size distribution.

A different approach was taken by *Wynn* and Hounslow (1997) who analyzed the problem by dividing the sensing zone volume into much smaller elements so that the probability of finding two or more particles in each element is negligible. This concept enabled them to analyze the probability distribution of the total volume of particles in the sensing volume by convolving, for all volume elements, the probability distributions of a particle being within an element. They proposed and demonstrated the results of a procedure for correcting the effects of coincidence on the PSD of calibration particles.

Here, we propose a simple solution to the problem of coincidence. This solution, based on the assumption of "particle-hiding" for polydisperse particle populations as the prevailing mode of coincidence, is an extension of the solution for monodisperse particle populations expressed by (5.89). Specifically, we postulate that the number of particles of a certain size, D, counted in the sample volume equals the number of the sensing zone volumes containing at least one particle of that size less the fraction of that number of zone volumes that *also* contain one or more particles of any size greater than D. The assumption of the "particle-hiding" coincidence frees us from having to consider details of the spatial distribution of particles inside the sensing zone.

Consider a particle size interval $D + dD$ of a polydisperse particle population. Regardless of the nature of coincidence, the observed count of particles in that size interval is smaller or equal the actual count, although, since these particles are less numerous than particles of all sizes, the potential reduction of the observed count will be limited as compared with that for all particles regardless of the size. Incidentally, since particles within the size interval $D+dD$ are also size distributed, although at a scale unrecognizable by the instrument, the larger particles in that interval can in principle hide the smaller particles. Generally, through coincidence, one or more particles of sizes $D' > D$ will reduce the count of particles of size D if they coincide with those latter particles.

Thus, the number of particles counted in a size range from D to $D+dD$ should be reduced more than is suggested by (5.86). Specifically, the reduction should be brought about by the number of sensing zones that contain both at least one particle of size D and at least one particle $D' > D$. For simplicity, and to hint at the actual numerical procedure for simulating and correcting the coincidence, let the PSD be expressed as a histogram $H(D_i)$, $i=0, \ldots i_{\max}$, where $H(D_i)=n(D_i)dD$. Given a size D_i, the number of particles counted by an instrument can thus be expressed, by extension of (5.89), as follows:

$$\begin{aligned} H_S{}'(D_i) &= L - L_0(D_i) - [L - L_0(D_i)]\left\{1 - \exp\left[-\frac{1}{L}\sum_{i+1}^{i_{\max}} H_S(D_j)\right]\right\} \\ &= [L - L_0(D_i)]\exp\left[-\frac{1}{L}\sum_{i+1}^{i_{\max}} H_S(D_j)\right] \\ &= L\left\{1 - \exp\left[-\frac{H_S(D_i)}{L}\right]\right\}\exp\left[-\frac{1}{L}\sum_{i+1}^{i_{\max}} H_S(D_j)\right] \end{aligned} \tag{5.95}$$

where the subscript S refers H to the sample volume and the term enclosed in {} in the first line of (5.95) expresses the probability, p, of finding any particle with the size greater than D_i in the sensing zone. That probability equals the fraction of the $L-L_0(D_i)$ zone volumes which contains at least one particle with size D_i and at least one particle with size greater than D_i. The term $L - L_0(D_i)$, corresponding to the term $L - L_0$ in the monodisperse coincidence derivation, represents the number of zone volumes that contain at least one particle with size D_i. From these simple premises, we have obtained a result that can be interpreted as the monodisperse coincidence model applied to each particle size interval separately and weighed by a factor equal to the probability of having no particles greater than or equal to size D_{i+1} in the sensing volume.

We note that the assumption that the probability of finding a particle of a given size is independent of that for a particle of another size may not always be true for natural waters that, for example, contain chemotactic motile bacteria. Experimental evidence discussed in section 5.6.3 indicates that such bacteria tend to concentrate around the sources of food, i.e., decomposing detrital particles.

The simulated effect of coincidence according to the simple polydisperse coincidence model expressed by (5.95) is compared in Figure 5.6 with the actual measurements (as digitized from Fig. 4 of *Wynn* and Hounslow 1997). It is clear that most of the coincidence can be explained by our simple model that relies on assuming the prevalence of the "particle-hiding" coincidence.

A "single-formula" coincidence correction cannot be given with this model, nor for that matter with any other polydisperse coincidence model advanced previously. In the present case, this cannot be done because of the summation of the size distribution histogram over the particle sizes. Nevertheless, the correction can be easily implemented as a simple numerical procedure if the measured PSD includes so low a particle concentration value for the largest measured particle size that the coincidence at this particle size is negligible (in the example shown in Figure 5.6 it is the last value, at $D = 10\,\mu\text{m}$). This procedure involves these steps: (1) use as is the $\Delta N_{\max} = n(D_{i\_\max})\Delta D$ value corresponding to the particle size interval with the negligible coincidence, presumably the one for the largest particle size $D_{i\_\max}$, (2) for each smaller particle size, use (5.95) with the sum

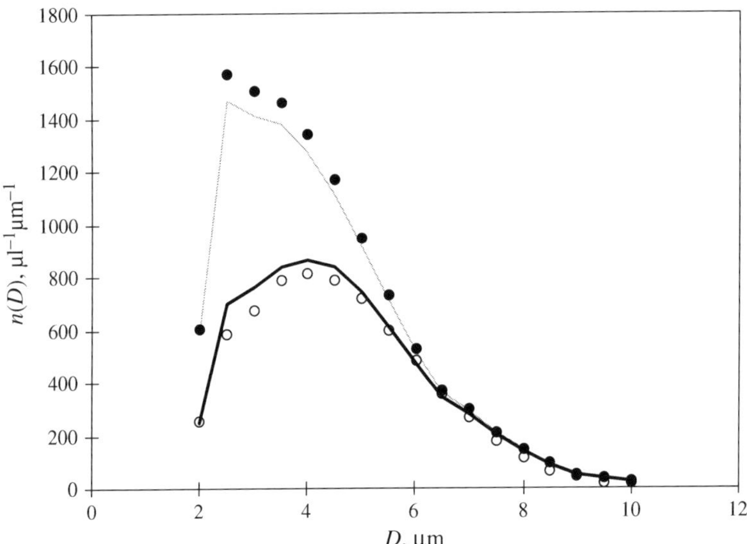

Figure 5.6. A simulation of the effect of coincidence on the particle size distribution (PSD) according to this work's polydisperse coincidence model which assumes only the "particle-hiding" coincidence (—, Equation 5.95) compared to the high-concentration PSD (○, fractional concentration $C = 0.191$, high coincidence) and low-concentration PSD (●, $C = 0.01$, negligible coincidence) data of *Wynn* and Hounslow (1997, as digitized from their Fig. 4). The real PSD is approximately that measured at $C = 0.01$. The monodisperse coincidence model (—), Equation 5.89, is clearly inadequate in simulating the coincidence effect on the PSD.

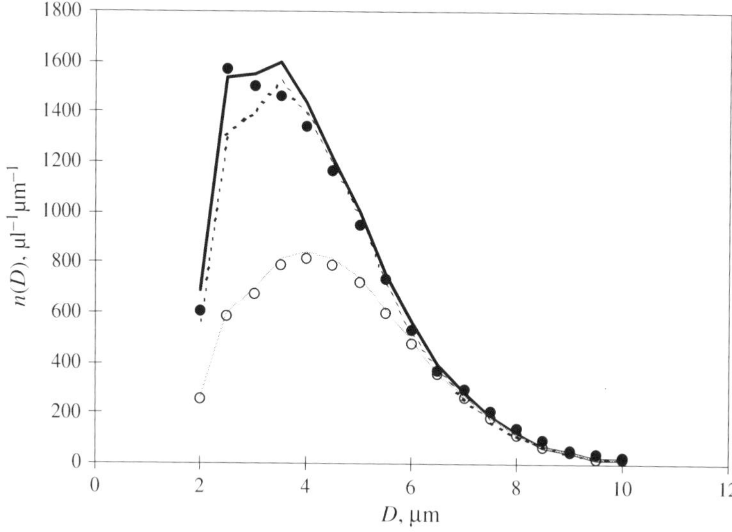

Figure 5.7. This work's simple polydisperse coincidence model (—, Equation 5.95) substantially retrieves the actual PSD (•, as approximated by that at a low fractional concentration, $C = 0.01$), similarly to the result of a conceptually more complex model of *Wynn* and Hounslow (1997, ---), when applied to the high-concentration, high-coincidence PSD (○, $C = 0.191$). In contrast, a simple monodisperse coincidence correction (-, the first line of Equation 5.92) fails substantially as can be seen from the closeness of its result to the high-coincidence PSD (○). All experimental data (symbols), as well as the *Wynn* and Hounslow model (thin dashed line) are shown here as digitized from their Fig. 4.

calculated with the particle concentrations $\Delta N(D_j)$ obtained for $D_j > D_i$. Results of application of the coincidence correction based on this procedure are shown in Figure 5.7 by using the sample data shown in Figure 5.6 along with the results of a more complex procedure of *Wynn* and Hounslow (1997).

5.7.1.3. Effects of the particle shape and trajectory inside the aperture

The theory of the resistive particle sizing technique implies that if a particle is spherical and small as compared with the aperture diameter, then the relative change in the resistance of the aperture caused by the presence of the particle in that aperture is proportional to the ratio of the particle volume to that of the aperture [equation (5.82), for example, *Kachel* 1990, *Hurley* 1970]. The proportionality constant for non-conducting particles is 3/2.

If a particle is non-spherical, the proportionality "constant" depends on the particle shape. For prolate spheroids, the proportionality "constant" increases nearly linearly with the semiaxis ratio $r = b/c$, where b is the minor semiaxis and a the major semiaxis, and assumes values from 1.02 at $r = 0.1$ to 1.5 (3/2) at $r = 1$ (sphere). For a disk with its face perpendicular to the aperture axis, the "constant" may assume a significantly higher value (3.5—*Kachel* 1990, 2.8—*Lloyd* 1982). Experiments confirm the effect of the particle shape on the particle size as determined with the Coulter counter. The response of this counter was found to vary up to about 30% due to the effect of the particle shape (*Golibersuch* 1973). Large-scale model studies (*Boyd* and Johnson 1995) indicate an orientation effect on the order of +20 to −25% for elongated particles that pass through the aperture oriented perpendicular and parallel to the aperture axis respectively. The shape effects may play a significant role in distorting the estimated size of elastic soft particles that may be deformed by the flow shear field across the aperture because the zone counters (resistive and optical) are generally calibrated by using solid, non-deformable particles (e.g., *Chalmers* et al. 1999).

The particle trajectory can also affect the measured particle size. First, the fluid velocity across the aperture cross-section is not constant (slowest at the wall and fastest at the axis, e.g., *Kachel* and Menke 1979). Thus, the transition time of a particle may vary depending on its trajectory. Due to the differences in the transition times, pulse heights produced by particles traversing the aperture along different routes may differ for short apertures, where the pulses from the fastest particle may not have a sufficient time to fully develop.

Second, due to the edge effects of the electric field distribution inside and around the aperture, the length of the sensing zone varies across the aperture cross-section, assuming the greatest value at the axis of the aperture. This may cause a reduction in the height of a pulse from a particle traversing the aperture near the aperture wall as compared with that of the same particle traversing the aperture along the axis. Experiments with multiple passages of the same particle through an aperture (*Boyd* and Johnson 1995) yield an estimate of the trajectory-dependent pulse height variations of about ±15% (see also *Kachel* 1990).

Third, the pulse height depends on the boundary effects as implied by experiments performed by using a Coulter-type instrument with a long aperture (diameter-to-length ratio of 1:20, *Berge* et al. 1990, 1989). A particle was drawn into the aperture, and the direction of flow through the aperture was reversed when the particle was about to exit the aperture. This method permitted multiple passages of the aperture by *the same* particle. Rotation of a non-spherical particle resulted in a systematic modulation of the top of the pulse. Such modulation may affect the particle size determination because the resistive counters measure the peak height of the pulse caused by a particle. In that study, the boundary effects accounted for about 25% error in the particle volume for a 7 μm particle traversing a 27 μm diameter capillary. Variations in the distance of a particle from the aperture wall may cause signal changes on the order of several percent in the particle volume, especially when the particle size is comparable with the aperture diameter (*Berge* et al. 1990, 1989). *Berge* et al. (1990) pointed out that this latter error is considerably less than that predicted by theory (*Smythe* 1972, 1964). Other experimental results (*Kachel* 1990) indicate that non-axial passage of a small particle through an aperture can cause changes of up to 80% in the measured particle size.

Generally, the trajectory-dependent effects increase with the decrease of the particle size, because small particles can assume pathways that are closer to the aperture edge than those available for the large particles (*Boyd* and Johnson 1995). These effects on one hand prompted the development of "pulse-editing" circuitry as a means of rejection of the off-axis pulse as well as devices stabilizing the radial position of particles in the aperture of resistive particle counter by hydrodynamic focusing (*Merkus* et al. 1990, *Kachel* 1990, *Kachel* and Menke 1979, *Spielman* and Goren 1968). Such stabilization is essential in the optical single-particle counters (section 5.7.2). Note that electronic pulse editing invalidates the measurement of the particle concentration and can be used essentially only when one is interested in the determination of the mean particle diameter in quasi-monodisperse population.

In the hydrodynamic focusing technique, particles are injected axially into the aperture from a capillary located a small distance away from the aperture. The electrolyte outside the injection capillary is particle free. The flow through the aperture is laminar, so that a stream of electrolyte with the particles does not mix with the clean electrolyte sheath through the aperture. The flow pattern of the aperture maps one-to-one the directions in the hemisphere in front of the aperture and centered at the aperture axis onto positions within the aperture cross-section (Figure 5.8). Thus particles injected into the flow in front of the aperture at a

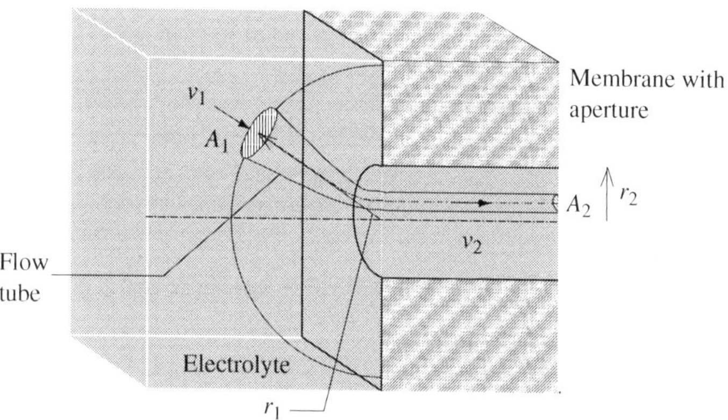

Figure 5.8. Hydrodynamic focusing (after *Kachel* 1990). The flow in front of and inside most apertures is laminar. Thus, the area $2\pi r_1^2$ of a hemisphere near the aperture with radius r_1 closed to the aperture centered at the aperture axis must map one-to-one onto the area of the aperture cross-section πr_2^2. It follows that the ratio of a cross-sectional area, A_1, of a flow tube at the hemisphere to the cross-sectional area, A_2, of that tube inside the aperture equals $2r_1^2/r_2^2$. By assuming the uniform flow velocity inside the aperture, the ratio of flow velocities v_1/v_2 equals cross sectional area, A_2, of that tube inside the aperture equals $r_2^2/2r_1^2$ the principle of conservation of mass in a non-compressible fluid. Thus, a parcel of fluid entering the aperture is stretched along the axis of the flow tube on entering the aperture.

specific angular position traverse the aperture at a corresponding cross-section position. The flow is laminar, at least reasonably far from the aperture wall, so that particles stay the initial course throughout the aperture. *Spielman* and Goren (1968) note that the injection point should be within a few millimeters away from the aperture for the focusing effect to work.

The width of the size distribution of monodisperse particles, analyzed with a hydrodynamically focused aperture, can be reduced by about 20 to 50% as compared with that characteristic of a standard aperture (*Goransson* 1990). Spielman and Goren (1968) in their pioneering paper on the subject found even greater improvement: the width of the size distribution, as measured by the standard deviation, was reduced by a factor of about 3.5 by hydrodynamic focusing. The size distribution may be less broadened with small apertures (diameter of 30 μm; *Walstra* and Oortwign 1969—although it is not clear whether in their case, this improvement was due to the combination of the counter model *and* the specific aperture they used).

Eddies that form behind the aperture may recirculate within the sensing zone particles that have been ejected from the aperture. Such recirculation may also cause some counting errors, especially at high particle concentrations (*Kachel* 1990). The recirculating particles are generally at the edge of the sensing zone, so that they must be relatively large to generate substantial pulses. The pulse shape that a recirculating particle produces is different (long, low-slope leading edge) than that of the symmetric pulses generated by particles passing through the aperture. Thus, pulses generated by recirculating particles can be rejected by evaluating the pulse shape electronically. If not rejected, the effect of recirculating particles may be especially persistent with wide PSDs. Such distributions represent populations with considerable fractions of large particles capable of producing significant pulses even when recirculating near the exit of the aperture. Note that pulse-editing invalidates the particle concentration measurement.

5.7.1.4. Effects of the particle composition and structure

The theory of the resistive zone technique states that the resistive pulse caused by the passage of a particle through the aperture depends on the ratio of the resistivity of the electrolyte to that of the particle material (e.g., *Kachel* 1990). Experiments with a large-scale model of the Coulter-type aperture and with the scale models of the various particles indicate that the material from which the particle is made has, under favorable experimental conditions, little if any effect on the measurement results (*Eckhoff* 1969). In those experiments, a high-conductivity gold-plated particle caused substantially the same signal as that caused by an insulating particle of the same volume. This similarity is expected to be due to different mechanisms of conduction of electricity in metals and in the electrolyte. The insensitivity of the counter to the particle composition holds, however, only below a certain, particle material-dependent, value of the electric field strength inside the aperture (*Kachel* 1990, *van der Plaats* and Herps 1983—experiments performed using a Coulter counter; *Eckhoff* 1969—large-scale model experiments). Above that value, the conducting particle may indeed act as a conductor causing partial reversal of the pulse polarity and a significant reduction in the measured particle size.

The average conductivity of a particle suspended in seawater may be similar to that of seawater if the particle is a loosely bound aggregate. However, many particles are living cells with a low-conductivity cell membrane. The sizes of such cells determined using a Coulter counter and an optical microscope were found equivalent to those determined using an optical microscope (*Montesinos* et al. 1983, *Jonasz* 1983a). On the other hand, *Boyd* and Johnson (1995) noted the effect of the particle resistivity (if different from that of the calibration particles used). Similarly, *O'Hern* et al. (1988) attributed significant and systematic differences between the size distributions determined in the coastal ocean waters in a diameter range of 10 to 100 μm by using a Coulter counter and by using an *in situ* Fraunhofer-type holographic camera system to differences between the conductivities of the calibration particles and those sampled from seawater. Preservation-related changes in the electrical resistivity of the phytoplankton cell wall have been suggested as a source of variations in the cell volume determined with the Coulter counter with time after the fixation of the cells (*Boyd* and Johnson 1995). Those authors also quote earlier research on this subject that indicates particle resistivity-specific effects. Interestingly, as pointed out by Boyd and Johnson, the low electrical resistivity may also be a factor in underestimating the number concentration of aggregates when measured with a resistive particle counter. We discuss in section 5.7.2.4 another, probably the more significant, reason for such underestimation: the particle breakage.

The size distributions determined by *O'Hern* et al. (1988) with the Coulter counter indicated concentrations of the particles which were lower by about an order of magnitude than those determined with the holographic camera. The differential Coulter counter size distributions were approximated with a power-law function having a slope of −4. With such a size distribution, the difference in concentration of one order of magnitude is equivalent to a change in the particle size by a factor of about 1.8. This factor, consistent with graphically reported data of O'Hern and colleagues, is much smaller than a factor of 10 attributed to the size calibration error by those authors. Although the influence of the differences in composition and structure between the calibration particles and measured particles cannot be dismissed, the differences in concentration are more likely to be due to the breakage of the particles, especially since the analyses were performed in the coastal zone.

The porosity, the distribution of the solid material within the aggregate, and the changing shape of the aggregate as it passes through the aperture can all significantly affect the determination of the aggregate size. *McCave* (1983) quoted an estimate that the sizes of coalesced aggregates may be up to 25% less than those measured. The measured sizes, in turn, will generally be smaller than those determined microphotographically in suspension.

Experiments with model particles indicate that the response of the instrument is a complicated function of the particle porosity. This response is somewhere between the response to a sphere with volume equal to the volume of the solid content of the particle and to a sphere with volume equal to the total volume of the particle, including interstitial liquid. For example, *Eckhoff* (1969) found that a hole through a particle significantly reduces the height of the resistivity pulse when that hole is oriented along the axis of the measuring aperture (i.e., allows the aperture current to pass unhindered), but causes only a slight reduction in the pulse height when the hole is oriented perpendicularly to the aperture axis. In the same experiment, when particles of the same volume and shape (cylinder) but with different length were measured, the shorter particle generated a greater pulse, another indication that the electric field within the aperture is not uniform.

Pores in the actual particle (as opposed to those in their macroscopic models) may have a size distribution and may interconnect to form a complex network. Some pores may be so small that they are not wetted by the electrolyte. Experiments with "real" particles of porous carbon and silica confirm that the resistance method used with a porous particle generates a response which is between that for a solid particle with the volume equal to the envelope volume of the particle and that for the solid-equivalent volume particle (*van der Plaats* and Herps 1984). *Horák* et al. (1982) systematically investigated the effect of porosity on the apparent particle size determined by a Coulter counter by using porous beads of cellulose and polyglycidylmethacrylate. By filling the pores with water-immiscible solvents or by coating the particles with polymer films, they found that the porosity of particles (as the pore volume fraction) on the order of 80% may lead to underestimation of the particle size by as much as about 40%.

Jackson and colleagues (1995) give an approximate treatment of the effect of porosity on the resistivity, ρ_{p}, of a particle passing through an aperture with diameter, d_{A}. Based on the results of *Archie* (1942) relating the resistivity of the porous particle vs. that of the surrounding electrolyte, ρ, to the solid volume fraction $V_{\mathrm{s}}/V_{\mathrm{p}}$ of the particle:

$$\rho/\rho_{\mathrm{p}} = \left(1 - \frac{V_s}{V_p}\right)^m \tag{5.96}$$

where $m \sim 2$, which we quote after *Jackson* and colleagues, they arrived at the following result:

$$\Delta\rho \cong \xi V(1 - \rho/\rho_p) \tag{5.97}$$

where $\xi = 16\rho/(\pi^2 d_A)$.

5.7.1.5. Particle breakage

Fragile particles, such as flocs, may break during the passage through the aperture of a Coulter counter, where high shear flow exists (*Kachel* and Menke

1979, *Kachel* 1990). Experiments and visual observations of the flow performed by Kachel demonstrate that this shear can align and deform (stretch) elastic red blood cells (erythrocytes) as shown in Figure 5.10.

As pointed out by *Kachel* (2001), breakage of an aggregate, the flow dynamics, and pulse generation by fragments of the broken aggregate all depend on several factors: the aperture geometry, flow conditions (especially at the aperture's entrance), properties of the aggregate itself (including the magnitude of forces binding the component particles), and the trigger voltage as well as the dead time of the signal processing electronics. It is essential to acknowledge the role of the trigger voltage setting and the method of detecting the pulse peak height as no particle that produces a pulse smaller than the trigger voltage receives the attention of a resistive particle counter.

Kachel (2001) suggested that the behavior of a stream of cells flowing into a narrow aperture could be regarded as a first-order model of the breakage of a loosely coupled aggregate (Figure 5.9). As the fluid flow lines converge at the aperture and a non-uniform flow velocity profile develops across the aperture cross-section, and also because the volume of liquid is conserved, significant forces arise. These forces first align the aggregate's long axis along the flow lines and then attempt to separate parts of the aggregate into a string of particles as the volume of fluid containing the aggregate is stretched from a compact shape into a thin filament. The tearing forces are caused by the high axial gradient of the flow velocity at the aperture entrance and high radial acceleration there (*Sonntag* and Russel 1987). Due to this velocity gradient, the tip of the aggregate which is close to the aperture is accelerated earlier than the far tip of that aggregate so that the aggregate may be torn apart and its components separated into a string of independent particles. These forces are so significant that they deform erythrocyte cells from their native disk shape into elongated ellipsoids, as it is evident from the top panel of Figure 5.10.

The high probability of particle breakup in a resistive particle counter aperture forced *Sonntag* and Russel (1986), who studied the effect of the shear flow on the properties of polystyrene flocs, to use a small-angle nephelometer instead. The mechanism postulated here is consistent with these observations as well as with the microscopic observations of *Gibbs* (1982a) reported earlier in section 5.7.1.2. In fact, Sonntag and Russel observed the deformation and eventual breakup of flocs of 0.14 μm polystyrene latex flowing through a 200 μm aperture. They concluded that the shape of the flocs changes from "spherical," far from the aperture, to "linear" inside it. Sonntag and Russel also found that the average number of flocs was reduced by an order of magnitude according to a relationship $\propto E^{-1}$ for flow strain rates, E, ranging from 20 000 to 200 000 s^{-1}. The velocity gradient is on the order of $\sim$50 000 s^{-1} in the aperture of a particle counter, such as Coulter counter, as estimated by *Sonntag* and Russel (1986). Thus, the shear stresses near the aperture are likely to break many if not most flocs.

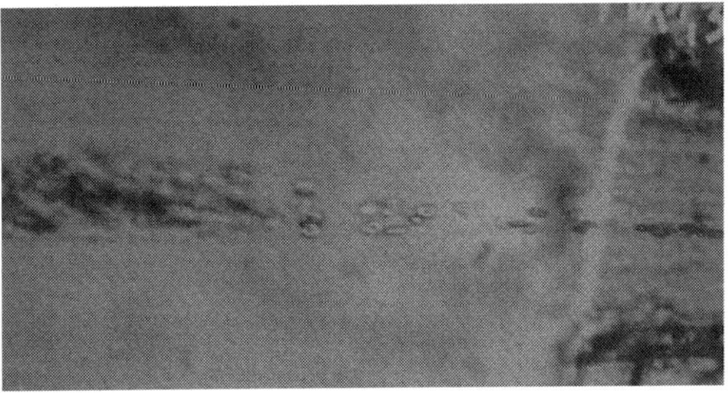

Figure 5.9. Due to hydrodynamic focusing (Figure 5.8), a wide stream of liquid with human erythrocyte cells leaving an injection capillary (top) far away from an aperture is stretched to a wire-thin filament at the axis of the aperture (bottom; the aperture is on the far right). A concentrated cloud of cells (which can be thought of as a very loosely bound "aggregate") far from the aperture is stretched by forces arising because of the flow velocity gradient so that cells ("aggregate components") become well separated inside the aperture. These forces may also be breaking flocs as they enter a particle counter aperture (Kachel 1982, Fig. 25a and d, reprinted by permission; *photographs*: courtesy of V. Kachel).

The extension and breakage of flocs in shear flows was demonstrated by *Lim* et al. (2002) who simulated time histories of flocs breaking in shear flow at velocity gradients of up to $1000\,s^{-1}$. Results of these simulations clearly show that the flocs initially extend and finally break into a long string of small fragments and individual particles. The gradient of $1000\,s^{-1}$ is much less than those expected to occur near the aperture of a resistive particle counter as indicated above.

Gibbs (1982a) reported observations with an optical microscope of the behavior of artificially generated flocs of kaolinite and natural sediment particles at the aperture entrance of an Elzone counter and stated that flocs broke on entry and traversed the aperture as a string of component particles. He speculated that the

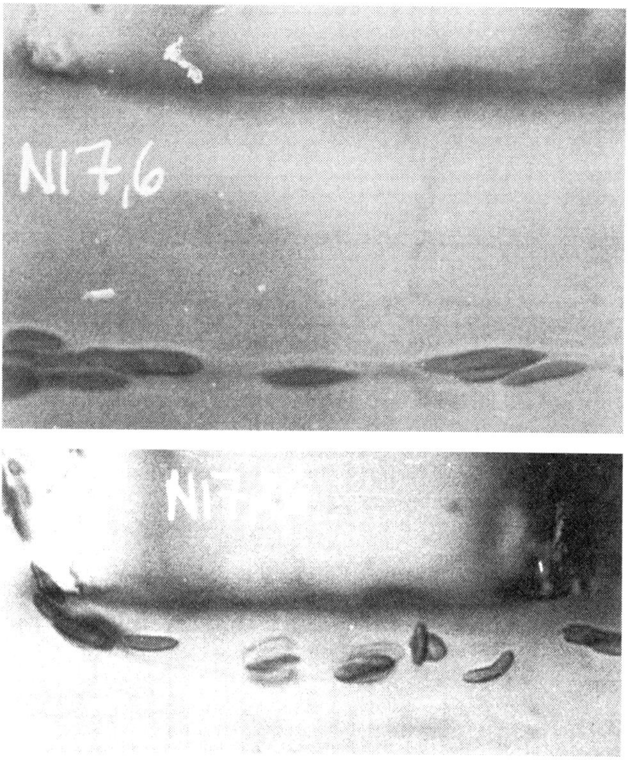

Figure 5.10. Deformation of elastic particles (human erythrocyte cells) flowing (from left to right) through an aperture: on axis of the aperture (top, stretching) and close to the aperture wall (bottom, bending and rotation). The aperture wall is the dark horizontal shadow in the upper part of each picture (Kachel 1990, Fig. 21a, b; *photographs*: courtesy of V. Kachel).

signal generated by the string either was below the detection limit of the aperture or was of wrong shape for the pulse detection system to handle.

Whatever the cause, the result was a substantial reduction of particle counts (as compared with the optical microscopy data) at particle sizes well below the upper limit of the sizing range for an aperture. In these experiments, flocs with sizes as small as 12 μm were broken when passing through a relatively large aperture of 380 μm diameter. The smaller the aperture diameter, the smaller was the diameter of the flocs that were broken during passage. The breakage rate was little influenced by the aperture diameter. These observations are important for multi-aperture measurements with aggregated particles: the large-size end of the size distribution measured with the small aperture should most likely be underestimated as compared with the small size end of a large aperture in an overlap region. On the other hand, *Matthews* and Rhodes (1970), who experimented with the effect of the aperture diameter (between 30 and 70 μm) on the measurement of the size distribution of a weakly flocculated suspension, concluded that "coincident passage of discrete components of a floc will register the same pulse as the intact aggregate."

McCave (1983) quotes experimental evidence that some aggregates (kaolinite: *Hunt* 1982) break just before entering the aperture. In their early experiments, *Kranck* and Milligan (1979) found no increase in the numbers of fine particles in natural samples re-counted with a Coulter counter. However, further experimentation by these researchers and comparisons of the Coulter counter data with those obtained with *in situ* microphotography (e.g., *Kranck* and Milligan 1988) revealed extensive differences which could be explained by breakage of aggregates. The problem is especially important in coastal waters, where large fragile aggregates of particles (flocs) are abundant.

Whether components of a broken aggregate are sized by a particle counter as one large particle or as several independent small particles is a complex matter involving, as already mentioned, the effect of the relative positions of the particles, aperture geometry, and properties of the electronics of the counter. In long apertures (length/diameter > 1.25; *Kachel* 2001), with homogeneous electric field, particles produce relatively long trapezoidal pulses. Here, several particles in close proximity and (probably) specific orientations that preclude direct ionic current path through the aggregate can produce one pulse that is a sum of the individual pulse amplitudes (maximum heights), provided that all particles are inside the homogeneous field region, as there can be a significant overlap of all component pulses at their amplitudes. As we discussed it in the coincidence section (5.7.1.2), if particles are not sufficiently close or appropriately oriented, or not all particles in a group are within the homogeneous field, the amplitude of a composite pulse registered by the counter is likely to be smaller than the sum of the individual pulse amplitudes and greater than the pulse amplitude of the smallest particle, even if each particle alone produces a pulse that exceeds the trigger level. If all aggregate fragments produce pulses below the trigger level, no particles will be counted at all: the broken aggregate is ignored by the counter. This might explain why in the breakage experiments of *Gibbs* (1982a) a resistive particle counter failed to register any aggregate particles with sizes that were observed microscopically.

In short apertures (length/diameter < 1.25; *Kachel* 2001), particles produce short bell-shaped pulses. This limits the potential for the overlap of pulses at their amplitudes. Close-by particles will still contribute to one large complex pulse, but its amplitude will generally be smaller than the sum of amplitudes of the component pulses. If particles are further apart, the resulting composite pulse amplitude will be smaller, and the pulse may even degenerate into a sequence of small individual pulses. Such a sequence may escape detection altogether if the height of the tallest peak falls below the trigger voltage.

Substantial experimental evidence based on comparing data obtained with *in situ* microphotographic cameras and with the Coulter counter indicates that *in situ* particle sizes may be an order of magnitude greater than those measured by the Coulter counter (*Chen* et al. 1994, *Kranck* and Milligan 1992, 1991, 1988 *Eisma* et al. 1991a). This led to a hypothesis that the PSDs measured with a Coulter counter represent building blocks of marine flocs. *Eisma* et al. (1991b) found no relationship between the size distributions determined using the *in vitro* methods and the *in situ* microphotographic system. However, the results obtained *in vitro* were not arbitrary and showed definite trends. Thus, the particle size determined with the Coulter counter could indicate fragility of the flocs, as suggested by *Gibbs* (1982a). Other *in situ* analysis methods also point to similar conclusions. *Bale* and Morris (1991—laser diffractometer) and *O'Hern* et al. (1988—holographic camera) both found that the particle size measured with the Coulter counter in coastal waters is significantly smaller than that determined *in situ*.

In the open ocean waters, where the low particle concentration implies a low rate of particle aggregation, an order of magnitude less flocs are formed than in the coastal waters (*Courp* et al. 1993, *Honjo* et al. 1984). In addition, the size of the flocs in the open waters is significantly smaller than that in coastal waters. *Jantschik* et al. (1992) compared bimodal PSDs evaluated in an open ocean subsurface nepheloid layer with a Coulter counter (using a 70 μm aperture) and with a Galai CIS optical scanning (aperture-less) particle counter (section 5.7.3). They did find that the Coulter counter reported somewhat smaller modal diameters than the scanning counter, but particles as large as 30 μm were counted with both counters alike. Thus, in the open ocean waters, the *in vitro* methods of the Coulter counter type may provide more realistic data, because the size distribution may have a substantial contribution from the single-particle population (*Eisma* et al. 1991b).

It would seem that *in situ* methods present an ideal solution to the breakage problem in resistive particle counters and in optical single-particle counters to be discussed shortly. However, as we point out in section 5.7.4, these methods are not free from artifacts which may affect the measurement of the size distribution in a manner similar to that of the coincidence and particle breakage.

5.7.1.6. Errors: closing comments

In evaluating the errors of the particle size measurements with a Counter counter, one should not overlook the fact that the counting error due to random distribution of particles are combined with errors due to the inaccuracies of the sample volume measurement and current/threshold setting in the counter. These two latter error sources affect the particle concentration, N, measurement, where [V] is the unit of volume, in different ways. Consider first the current/threshold setting errors. In a Coulter counter, and other resistive counters that are sensitive to the particle volume, V, the peak voltage of a pulse, v_{peak}, caused by a particle is expressed as follows:

$$v_{\text{peak}} \propto IAV \tag{5.98}$$

where I is current and A is the voltage amplification used. Hence, the particle diameter can be expressed as a function of the peak voltage and instrument settings as follows:

$$D = 3\sqrt{\frac{v_{\text{peak}}}{\kappa IA}} \tag{5.99}$$

where κ is proportionality coefficient (the instrumental scale factor). Hence, by assuming that errors of the threshold and current settings (the latter combined with the instrumental factor and amplification into one term) are small and independent, we can use the error propagation formula to obtain the following expression for the relative variance of the particle diameter, $\text{var}D/D^2$, caused by these two error sources:

$$\frac{\text{var}D}{D^2} = \frac{1}{9}\left[\frac{\text{var}v_{\text{peak}}}{v_{\text{peak}}{}^2} + \frac{\text{var}(\kappa IA)}{(\kappa IA)^2}\right] \tag{5.100}$$

Thus, based on the data from the Coulter specification, the total relative error of the particle diameter, $\varepsilon = \sigma(D)/D$, where $\sigma(D)$ is the standard deviation of D, amounts to about 0.005, assuming that the voltage threshold relative error (relative error of the bin limit voltage in multichannel instruments) is the same as that of the current setting error of about ± 0.005. The situation complicates itself from here on because the diameter error affects the particle count measurement in a manner dependent on the local slope of the PSD. For natural waters, the cumulative size distribution is approximately expressed (section 5.8.5.3) as the $-M$th power of the particle size, where $M = m - 1$, and m is the local log-log slope of the frequency size distribution, $n(D)$. Thus, the diameter error translates into the diameter-related particle concentration error, $\text{var}_{\text{D}}N$, as follows:

$$\begin{aligned} \text{var}_{\text{D}}N &= \left(\frac{dN}{dD}\right)^2 \text{var}D \\ &= \left(KMD^{-M-1}\right)^2_{D=\langle D\rangle} \text{var}D \\ &= \langle N\rangle^2 M^2 \frac{\text{var}D}{\langle D\rangle^2} \end{aligned} \tag{5.101}$$

The error, $\text{var}_{\text{V}}N$, in the cumulative particle concentration, N, due to the measurement of the sample volume, V, can be expressed as follows:

$$\begin{aligned} \text{var}_{\text{V}}N &= \left(\frac{dN}{dV}\right)^2 \text{var}V \\ &= \left(-\frac{N'}{V^2}\right)^2_{V=\langle V\rangle} \text{var}V \end{aligned} \tag{5.102}$$

$$= \langle N' \rangle^2 \frac{\text{var}V}{\langle V \rangle^2}$$

where N' is the actual particle count, not the particle concentration N. The overall error of N is thus:

$$\begin{aligned} \text{var}N &= \text{var}_D N + \text{var}_V N \\ &= \langle N \rangle^2 M^2 \frac{\text{var}D}{\langle D \rangle^2} + \langle N' \rangle^2 \frac{\text{var}V}{\langle V \rangle^2} \end{aligned} \tag{5.103}$$

By assuming $M = m - 1 = 3$ as the first approximation (section 5.8.5.3), we can expect the concentration error, $\varepsilon = \sigma(N)/\langle N \rangle$, due to the diameter error to be on the order of 0.015. When this error is combined in an RMS manner with the volume measurement error in the Coulter counter (of about 0.005), one obtains the total relative instrumental error, ε_I, of the particle concentration on the order of $\sim$0.016. Note that in the determination of $\text{var}_D N$, with M being strictly the local slope of $N(D)$ at D, one would need to use an iterative approach to find both the slope M and $\text{var}_D N$. The simplifying assumption of a diameter range definition of the slope M extending beyond the local slope at D makes it possible to determine the variance of N in one operation.

The relative counting error, ε_P, *increases* to a magnitude of 0.016 as the particle count *decreases* to 4000, according to the following equation:

$$\varepsilon_\text{P} = \left(\sqrt{N'} \right)^{-1} \tag{5.104}$$

due to random fluctuations of the particle concentration, i.e., the relative error for a Poisson-distributed random variable. Note that ε_P can be also expressed as $1/\sqrt{N} = \sigma(N)/N$ because in $N = N'/V$, the volume is just a constant from the point of view of the counting error, so it cancels in the expression $\sigma(N)/N$. Thus, the total variance of the particle concentration, N, measurements with a resistive particle counter is expressed in the first approximation, by using the assumption of the independence of the counting and instrumental errors, as follows:

$$\text{var}N = \text{var}_\text{P} N + \left(\langle N \rangle^2 M^2 \frac{\text{var}D}{\langle D \rangle^2} + \langle N' \rangle^2 \frac{\text{var}V}{\langle V \rangle^2} \right) \tag{5.105}$$

where the bracket groups the instrumental component.

As we have already stated, at particle counts higher than about 4000, the instrumental error sources dominate the variance. At the lower particle counts, the counting error source dominates.

5.7.2. Optical single-particle counters and flow cytometers

5.7.2.1. Technology

Optical methods of particle counting and characterization have long been used (e.g., *Mullaney* and Dean 1969) in medicine (FACS and EPICS instruments of Beckman Coulter Corp. , CyFlow and PAS instruments of Partec GmbH, Germany, MicroCyte instrument of Optoflow AS, Norway, Technicon H-series instruments of Bayer Corp., Diagnostics Div., USA) and for particle counting in industry (HIAC particle counters, Pacific Instruments Corp.). A distinction is usually made between optical particle counters and flow cytometers, with optical counters "simply" counting and measuring the particle size, and flow cytometers characterizing numerous optical properties of the particles, such as fluorescence at several wavelengths and scattering into one or more angular ranges. An optical particle counter thus returns the PSD alone, while a flow cytometer can provide extensive multidimensional data on each particle in the sampled population. Although the primary use of these data is to classify particles into sub-populations based on two or more dimensional point graphs, for example, the fluorescence vs. forward scattering, such data can be converted into frequency distributions, much like the PSD.

Davey and Kell (1996), *Davey* et al. (1993), *Steen* (1990), *Melamed* et al. (1990), and *Steinkamp* (1984) all review the history and technology of the optical flow cytometers, generally in the context of medical applications, with Davey and Kell referencing many applications of flow cytometers in the aquatic sciences. *Steen* (2000) reviews the flow cytometry of bacteria. Finally, *Knollenberg* and Veal (1992) review the technology and compare performances of optical particle counters for use both with hydrosols and with aerosols.

Legendre et al. (2001), *Dubelaar* and Jonker (2000), *Olson* et al. (1993), *Phinney* and Cucci (1989), and *Yentsch* et al. (1983) reviewed applications of commercial flow cytometers in aquatic research. *Dubelaar* and Jonker (2000) also reviewed a recent selection of commercially available flow cytometers and their characteristics.

The operation of a flow cytometer relies on two basic technologies: (1) hydrodynamic focusing of a thin filament of sample fluid, which contains particles, at the center of a particle-free fluid sheath in order to maintain the position of that filament in the region of minimum variability and maximum intensity of the illuminating light beam, and (2) rapid measurement of the optical effects of the cell passage through the focus of the incident light beam (light scattering, fluorescence, etc.). The flow of sample and sheath is kept in the laminar regime. This minimizes mixing of the two fluids and limits the radial position of the particles to within the thin filament. The beam focus is made highly elliptical, with the long axis of the ellipse perpendicular to the flow axis. This reduces effects of the radial distribution of irradiance within the beam focus on the signal generated by particles crossing the beam at various radial positions within the filament. The principle of the optical flow cytometer is schematically illustrated in Figure 5.11. It is worth noting that micromachining techniques have enabled

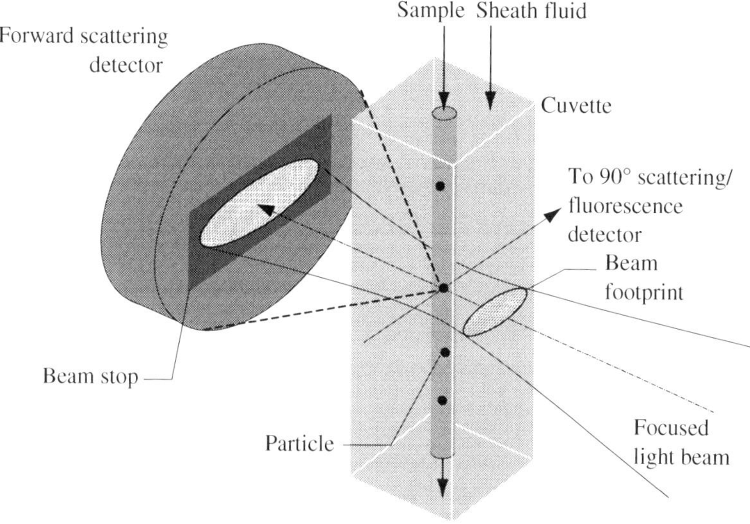

Figure 5.11. Schematic diagram of a typical optical flow cytometer. Such an instrument is essentially a single-particle nephelometer with special beam focusing and sample handling to assure high and uniform irradiance at the scattering volume. The sample handling, based on hydrodynamic focusing aligns as well as (at least partially) orients particles along the flow direction in a narrow filament at the midpoint of the elliptical cross-section of the focused light beam. The long axis of the beam focus is perpendicular to the sample filament axis to limit variation of irradiance across the sample filament. The short axis of the beam focus ellipse is typically much greater than the expected maximum particle size. Therefore, the height of the forward scattering signal pulse is a measure of the particle size. This may cause problems while analyzing cylindrical colonies of phytoplankton whose length is greater than that of the short axis of the beam focus. If beam footprint is less than the particle length along the flow axis, the signal pulse length can be measured instead, and the particles can be sized with the time-of-flight technique. Measurement of the pulse shape (slit-scanning) makes it possible to resolve the particle structure along the flow axis. Additional detectors at different angles can be used to measure the light scattering and/or fluorescence of the particle. The forward-scattering detector can be replaced with a multi-element detector that makes it possible to resolve a two-dimensional scattering pattern of a particle.

the development of miniaturized "on-chip" flow cytometers (*Lin* and Lee 2003, *Altendorf* et al. 1996).

As opposed to the resistive particle sizing technique, the fluid need not be conductive. In fact, there need not be any fluid at all, and flow cytometers have been developed and are widely used for application in aerosol characterization (e.g., *Pinnick* and Auvermann 1979). Optical flow cytometry is also more versatile as it can provide information not only about the particle size but also about its composition via auto- or induced fluorescence, as well as about the particle shape, via the measurement of the two-dimensional angular scattering pattern, or so-called slit scanning (to be discussed shortly). These latter capabilities are described in more detail in Chapter 6 which is focused on the particle shape and composition characterization. Some flow cytometers merge the optical characterization and resistive particle sizing (*Steinkamp* 1984, *Kachel* et al. 1977), or even enable imaging of aquatic particles (*Sieracki* et al. 1998) along with resistive particle sizing (*Wietzorrek* et al. 1999, 1994, *Hüller* et al. 1994, *Kachel* et al. 1980, 1979).

A flow cytometer, in addition to light scattering, typically measures the fluorescence of the particle, either inherent or that of fluorescent marker molecules sensitized to attach themselves to specific chemical binding sites on the surface of the particle. Although much of the development in that area has been done in relation to human cells, attempts have been made to extend this immunomarker technique to algal cells (e.g., *Vrieling* et al. 1996).

The particle characterization measurements are routinely performed on thousands of particles per second. Some flow cytometers are equipped with particle sorters which can be used to separate populations of particles based on their optical properties (*Lindmo* et al. 1990, *Steinkamp* 1984, *Steinkamp* et al. 1973). An interesting alternative to flow-cytometric sorting was recently demonstrated. This latter technique relies on the use of forces exerted on particles by light, much as those used in optical particle trapping and manipulation by focused light beams, pioneered by *Ashkin* et al. (1987). However, here the forces are exerted by strong light intensity gradients within a three-dimensional interference pattern (*MacDonald* et al. 2003) . Particles flowing through the interference pattern are separated based on their refractive indices, with an efficiency approaching 100% and throughputs exceeding those of the flow-cytometer-based sorting.

The sizes of particles that can be analyzed by both optical particle counters and flow cytometers range from ~ 0.2 to $\sim 10\,\mu m$ (commercial instruments), with the upper limit extended to $\sim 1000\,\mu m$ in custom-build or limited-series flow cytometers specialized for phytoplankton research. With the use of nucleic acid-specific stains and the measurement of fluorescence of these stains, flow cytometers can be used to detect and count viruses in natural waters (size between 0.02 and $0.2\,\mu m$; *Marie* et al. 1999).

The inherent possibility of multi-parameter measurements in optical flow cytometers as opposed to just the measurement of light scattering in the basic optical particle counter makes it essential to eliminate coincidence rather than correct it. This limits the particle concentration and thus the number of particles to be analyzed to between 100 and $1000\,s^{-1}$.

When the laser beam thickness at the focus is smaller than the particle dimension along the flow axis, the light scattered by various parts of the particle can be analyzed. Alternatively, the beam can be focused only moderately, but the field of view of the detector is limited by a narrow slit aperture. These variations of the flow cytometry is termed “slit scanning” (*Wheeless* 1990). With this technique, the time interval between the leading and trailing edges of the light scattering pulse generated by the particle passing through the beam focus can be used to size large particles as the flow velocity of the particle is known. This technique has also been used in aquatic science application (*Peeters* et al. 1989) for sizing elongated particles, such as cylindrical colonies of phytoplankton.

Olson et al. (1985) reported probably the first measurements at sea by using a commercial optical flow cytometer. These measurements included evaluation of light scattering and auto-fluorescence of individual cells and allowed these authors to discriminate between different groups of phytoplankton, enumerate them, and detect changes in the average pigment fluorescence resulting from changes in population structure. Since then flow cytometry has become a technique widely used in oceanography, both in land- and ship-based operations (*Legendre* et al. 2001, *Cavender*-Bares et al. 2001, *Hofstraat* et al. 1994, *Olson* et al. 1993).

Commercial flow cytometers tend to have relatively inflexible designs, optimized for the core medical/industrial applications and are relatively expensive. This prompted the development of customized inexpensive "home-made" instruments which has been described in relative detail in both the laser-based (*Shapiro* 2003) and arc-based (*Steen* 1986, *Steen* and Lindmo 1978) configurations. The two latter authors describe an inexpensive flow cytometer build around a microscope and featuring a simple flow system in which the sample is ejected from a nozzle directed at an oblique angle at a microscope slide. That arc lamp-based instrument has a particle size detection limit of 0.2 μm.

Some features of commercial instruments make it difficult to use these instruments in marine sciences (*Olson* et al. 1993, *Cunningham* 1990, *Dubelaar* et al. 1989, *Phinney* and Cucci 1989). Specifically, these instruments have a limited resolution, report the peak height or integrate the pulse generated by a particle, and can handle a small sample volume, insufficient for measurements of the large (i.e., rare) marine particle. The commercial instruments also feature high velocity gradients in the flow cell that can disrupt fragile particles. We discuss some evidence of particle breakage in flow cytometers further in section 5.7.2.4.

These shortcomings spurred the development of innovative flow cytometers designed specifically for phytoplankton analysis. *Cunningham* (1990) developed a simple, low-cost phytoplankton flow cytometer based on an air-cooled argon laser. This flow cytometer uses a flow cell with low velocity gradient and is capable of resolving single bacteria (*Synechococcus sp.*) with the diameter on the order of 1 μm. *Dubelaar* et al. (1989) and *Peeters* et al. (1989) discuss the design of a plankton flow cytometer having a particle size range of 1 to 500 μm, and a large dynamic range of 1 to 400 000. Such a dynamic range is achieved via the fast digitization of the signal pulse and numerical integration of the digitized pulse profile. That instrument can also sustain flow rates spanning about four orders of magnitude. In contrast, commercial flow cytometers can handle particles with sizes in the range of 1 to 25 μm and have a dynamic range on the order of 1 to 1000. The plankton flow cytometer permits sizing a particle via the time-of-flight measurement. The flow cell of that instrument is characterized by a low velocity gradient that imparts low shear forces on the sample preventing the breakage of phytoplankton colonies. The funnel-shaped flow cell with a square outlet of 1×1 mm is connected to a syringe-driven pump that permits accurate determination of particle concentration.

Another solution to the flow rate limitation has been the double-sheath flow cytometer (*Cavender*-Bares et al. 1998, *Eisert* et al. 1975) in which the sample flow can potentially be surrounded by two sheaths of clear liquid. In this system, the smallest cells (such as *Prochlorococcus*) are injected through the low-flow rate small-diameter inlet and the diameter of the focused illumination beam is reduced. If the analysis of large and commensurately rare cells (several micrometers in diameter) is desired, the cells are injected through the larger inlet that is normally used for the first sheath flow, and the size of the beam focus is enlarged. *Cavender*-Bares et al. (1998) achieved flow rates in a five orders of magnitude range with his system.

Despite the complexity of the flow-cytometer technology, several attempts succeeded in adapting it for *in situ* measurements (*Olson* et al. 2001, *Olson* and Sosik 2000, *Dubelaar* and Gerritzen 2000, *Dubelaar* et al. 1999) .

As compared with the sheath flow-based flow cytometers, the HIAC optical particle counter is a relatively simple instrument. Its operation is based on the measurement of the attenuation of light by a particle ("light blockage," "light obscuration") passing though a narrow, physically restricted sensing zone. It is assumed that the attenuation is proportional to the particle cross-sectional area. Note that this type of measurement can also be performed with some flow cytometers discussed above (Axial Light Loss—*Stewart* et al. 1989, *Steinkamp* 1984). Several types of sensors for the HIAC counter are available, permitting particle size analysis in the particle size range on the order of 1 to 1000 μm using multiple sensors. The dynamic particle size range attainable with a single sensor is on the order of 1:30. This instrument has been used relatively infrequently in oceanographic applications, probably because of its limited resolution (2–8 size channels), although recently new instrument generations with a much higher number of size intervals have been available. The HIAC particle counter was used to analyze the spatial and time distribution of marine particles (*Claustre* et al. 2000, *Kahru* et al. 1991, *Pugh* 1978), the size distribution of marine sediments (*Jonasz* 1987b), and phytoplankton cultures (*Stramski* et al. 2002, *Sciandra* et al. 2000). *Akers* et al. (1992) describe a modification of the standard 6-channel HIAC counter whose output was analyzed by a 1024-channel pulse height analyzer.

The "light blockage" principle has also been used in enumerating and sizing of particles as large as ~0.02 to ~2 cm, i.e., roughly, in the zooplankton size range, by using instruments referred to as optical plankton counters, both *in situ* (*Herman* et al. 1993, *Herman* 1992) and *in vitro* (*Beaulieu* et al. 1999). Recent designs of the optical zooplankton counter allow for simultaneous evaluation of the size of these large particles flowing not one-after-another, as in a small-particle flow cytometer, but side-by-side through a wide channel (*Herman* et al. 2004).

5.7.2.2. *Particle characterization techniques*

Several approaches to particle sizing with optical particle counters and flow cytometers have evolved. The simplest approach is to use the intensity of light scattering at small angle, sometimes referred to as "forward" light scattering ("forward angle light scattering," FALS), measured with a single detector, in order to predict the size of the particles. This minimizes the effect of the particle composition on particle sizing, as the "forward" light scattering is mainly due to diffraction of light at the geometric particle cross-section. In practice, the intensity of light scattered into a small solid angle about the direction close to the direction of propagation of the beam is typically measured (e.g., *Bouvier* et al. 2001—aquatic bacteria, *Cavender*-Bares et al. 2001—marine bacteria and small phytoplankton, *Robertson* et al. 1998, *Allman* et al. 1993—bacteria). Note that in the case of certain bacterial cultures, no correlation between the FALS and cell size has been found (e.g., *López-Amorós* et al. 1994–*Escherichia coli*) because the scattering of light by living cells is, in addition to the cell size, influenced by the cell shape and structure (see section 6.4.3.2).

The FALS intensity is proportional to the integral of the angular scattering pattern of the particle (differential scattering cross-section) within that solid angle. When both the scattering angle and the solid angle are small, this value generally increases as the sixth power of particle size for particles that are small as compared with the wavelength and as the second power of particle size for particles that are much larger than the wavelength. The integrated scattering intensity oscillates with a decreasing amplitude as the particle size increases between these two size extremes. Both the amplitude and frequency of these oscillations decrease with the refractive index. This has been clearly demonstrated in early experiments (e.g., *Mullaney* et al. 1976). A more recent study for aerosols with refractive indices 1.33 (water in air) and 1.5 (e.g., *Barnard* and Harrison 1988) shows that these oscillations can be minimized, and thus the resolution of the particle counter can be increased by optimizing the angular range of the scattered light collected by the instrument according to the refractive index of the particles to be sized. We are not aware of a similar optimization study for low-index particles, but experimental results for many picoplankton species (0.2 to 2 μm) indicate that the forward scattering signal is tightly correlated with the cell volume, as determined with a Coulter counter (*Olson* et al. 1989). The shape effects may play a significant role in determining the accuracy of the estimated size of elastic soft particles that may be deformed by the flow shear field across the aperture because the optical flow cytometers (and resistive particle counters) are generally calibrated by using solid, non-deformable particles (e.g., *Chalmers* et al. 1999).

We should note that in the case of particles much larger than the wavelength of light, one should expect the effect of variations of the projected area with particle orientation for non-spherical particles. Indeed, for a disk with negligible thickness, the projected varies between 0 and the disk area, depending on the disk orientation. It can be shown that the resulting probability distribution of the equivalent projection area diameter, D_A, for randomly oriented monodisperse population of such disks is as follows:

$$p(D_A) = 2\frac{D_A}{D^2} \tag{5.106}$$

resulting in an average value of D_A of:

$$\langle D_A \rangle = \frac{2}{3}D \tag{5.107}$$

Interesting experiments that quantify the effect of the particle shape on the widening of the PSD have been performed by *Umhauer* and Gutsch (1997), *Butler* et al. (1989), and *Chin* et al. (1988). *Umhauer* and Gutsch developed an experimental instrument based on "light blockage" that consisted of three identical sensors oriented at three orthogonal directions and having a common sensing volume. The instrument was designed to measure projected areas of relatively large particles ($\sim$1 mm) and was equipped with an automated feed system to enable as much as 2000 measurements on the same "model" particle. Resulting distributions of the projected area for cubes, parallelepipeds, and aggregates of spheres compared very well with numerical simulations for these model particles. By averaging the three measurements for a particle, *Umhauer* and Gutsch were able to obtain an excellent proportionality between the instrument output and the ECD.

Butler et al. (1989) and *Chin* et al. (1988) conducted similar experiments with a HIAC optical particle counter, albeit at a much lower repetition rate, for particle shapes ranging from a flake to a nearly spherical particle with diameters $\sim$400 μm. By repeatedly measuring the sensor output for the same particle dropped through the sensing zone at (presumably) random orientation, they found that particle orientation can result in a relative standard deviation ranging from 2% (spheroidal particle) to 7% (elongated particle) and substantially larger value for the flake (17%).

An extension of this technique is the use of multiangle measurements to simultaneously determine the particle size and refractive index . In practice, as with the "forward scattering" approach, particles can be sized "exactly" only if they are spherical and homogenous. In so-equipped flow cytometers, the angular scattering pattern of a particle or of a suitable subset in a limited angular range is measured. From these data, with the use Mie theory of light scattering by homogeneous sphere (e.g., *Bohren* and Huffman 1983), one can deduce (invert these data into) the particle size and refractive index. This approach has led to the development of a successful commercial flow cytometer (*Tycko* et al. 1985) optimized for the determination of size and hemoglobin content of red blood cells. These cells are not spherical in their natural state. Thus, just before the analysis they are iso-volumetrically "sphered" by applying a proprietary reagent. The inversion algorithm was based on measuring light scattering intensity at just two narrow angular ranges centered at two near-forward angles (3 and 5.5°). The particle size and refractive index was then obtained by solving a set of two non-linear equations expressing these intensities as functions of the size and refractive index in a limited domain that assured the solution uniqueness.

That technique was also used to determine the size and refractive index of several marine phytoplankton species (*Ackleson* and Spinrad 1988) and has been extended to utilize the scattering pattern in a wider angular range (*Shvalov* et al. 1999). *Green* et al. (2003b) used a similar method to measure the size distributions of components of marine suspensions: phytoplankton, detritus, and minerals with a flow cytometer. These techniques use the Mie theory applicable to the scattering of light by a spherical particle illuminated by a plane wave with constant intensity at any points at the wavefront. Unfortunately, in contrast to the clinical applications (*Tycko* et al. 1985), aquatic applications of these techniques suffer from the violations of the assumptions of the sphericity and homogeneity of the particles.

In addition, with most flow cytometers using laser light sources, the distribution of irradiance across the beam focus tends to be uniform only in the first approximation, and the usage of the Mie theory here is approximate. The problem of the scattering of light by a sphere anywhere in a Gaussian beam has been solved by using the Fraunhofer approximation (*Chevaillier* et al. 1986) and exactly (*Gouesbet* and Maheu 1988). Recently a solution was obtained for non-Gaussian beams (*Lock* and Hodges 1996). These solutions can also be used in the case of *in situ* optical particle sizing which typically gives the instrument less control over the particle location than for *in vitro* flow cytometers. Alternatively, the particle size and size distribution can be derived from raw data by deconvolving these data with the use of the instrument's response function to monodisperse particles (e.g., *Holve* and Self 1979a).

In contrast, the "slit scanning" particle sizing technique allows determining the particle size not from the measurement of the angular scattering pattern but from the time-of-flight of the particle through the beam. This approach also applies to non-spherical particles, without a need for a model of light scattering by a particular cell shape. In addition, this technique can be used to determine the cell structure. However, "slit scanning" becomes inefficient for particles with sizes much smaller than the slit width (thickness), which limits the minimum particle size to ~1 μm. A variation of "slit scanning" that relies on creating a set of interference fringes in the sensing zone has been used to characterize yet smaller particles (*Wheeless* 1990).

The sizing of non-spherical particles presents several problems. First, there is a problem of orienting such particles in the sample filament. Large long particles are relatively well oriented by the hydrodynamic focusing phenomenon or asymmetric flow nozzles (e.g., *Kachel* 1990). However, particles with sizes comparable with the filament diameter are more likely to arrive at the sensing zone randomly oriented. Second, even if the particle orientation were known, the particle shape needs to be known *a priori* for a successful conversion of the light scattering pattern information into particle size and refractive index. *Bottlinger* and Umhauer (1989), who analyzed the variability of the light scattering signal from mineral aerosol particles (quartz and limestone, size between 15 and 120 μm) oscillating in an electrodynamic trap and attempted to use these data to deconvolve the particle shape effect on the widening of the PSDs, modestly note that "... the effort required ... is relatively high."

Auto-florescence of living cells is roughly proportional to the amount of fluorescing dye, which in turn depends on the cell volume. This led to the use of fluorescence signal for phytoplankton cell sizing and discrimination via flow cytometry (*Chisholm* 1992). This approach, however, is applicable only to selected particle types (phytoplankton, bacteria, and viruses) and even there may generate data which depend on the particle history. It also may produce results which are markedly different than those obtained by using the forward light scattering as a measure of the particle size as it follows from, for example, the results of *Li* (1994, 1990).

5.7.2.3. Intercomparisons

Tests of early light scattering-based single-particle counters, whose responses were based on the measurement of light scattered by a particle (*Whitby* and Vomela 1967), revealed that the response for non-absorbing particles was similar for all counters and that the results were similar to those obtained with a microscope. However, all counters tested underestimated the size of light-absorbing particles (India ink) by a factor ranging from 2 to 5. Each of the three instruments evaluated measured light scattered by individual particles into a different range of scattering angles: a small solid angle around 90°, around 45°, and (the widest range of the three) an annular solid angle between 24 and 57° off the optical axis of the incident beam. The size resolution was the highest in the case of the instrument having the widest angular range. *Mullaney* et al. (1969) found that the forward scattering signal obtained for polystyrene beads by integrating light scattered in an angular region between 0.5° and 2° from the beam axis is only up to 15% non-linear when compared with the volume measured with a Coulter counter. As mentioned in the previous section, the forward scattering signal produced by a flow cytometer has been found to be "tightly" correlated with the Coulter counter volume (*Olson* et al. 1983).

The size distributions of platelet-like kaolinite particles, measured using a HIAC counter, were found to yield a model particle size which, at 10 μm, is about 1.2 times greater than that obtained with a Coulter counter (*Jonasz* 1987b). If the particles have a refractive index as high as kaolinite (1.2 relative to water), this difference is due mostly to the non-sphericity of the particles. The relative refractive index of phytoplankton cells is much smaller (on the order of 1.02). In that case, especially for nearly spherical cells, the difference between the PSD obtained using these two types of particle counters can be used to determine the refractive index of the particles (*Jonasz* 1986).

5.7.2.4. *Particle breakage*

Few data are available regarding the particle breakage in optical single-particle counters. This might result from the main application of these instruments to the characterization of single cells. *Peeters* et al. (1989) studied the breakage of filamentous colonies of phytoplankton cells with a 250 μm flow cell and found that such colonies start to break at flow velocities above 1 m s^{-1}. When the filaments were analyzed with a commercial flow cytometer (FACS II) nozzle (80 mm diameter, 10 m s^{-1} flow), over 90% of the filaments broke. However, *Pelssers* et al. (1990b), who studied aggregation of suspensions with an optical particle counter which they built themselves (*Pelssers* et al. 1990a), found that their hydrodynamic focusing flow system did not affect multiplets of 0.69 mm polystyrene latex containing up to seven particles. In extrapolating results such as these, one needs to take into account that the binding forces that hold together constituent particles may be different in other cases, and what truly matters in these evaluations is the magnitude of the hydrodynamic forces acting on the particles. In this respect, a valuable feature of the work of *Pelssers* et al. (1990b) in this area is the formulation of the necessary framework for the evaluation of these forces at the various critical points of flow in a hydrodynamic focusing system.

Gibbs (1982b) studied the effect of a flow cytometer (HIAC) with an aperture of 60 μm on the breakage of mineral, artificially generated flocs from kaolinite and from a sample of natural sediment. In both cases, flocs greater than about 20 μm were broken. The instrument detected only a small fraction of the flocs that decreased rapidly with the floc size. On the other hand, *Akers* et al. (1992) have found, by circulating a sample of flocculated suspension, that the process of floc breakage in a size range of 6.5 to ~100 μm is more gradual.

5.7.2.5. *Errors: closing remarks*

Like in the case of resistive particle counters, the overall error of the PSD obtained with an optical particle counter or flow cytometer can be derived by using the approach outlined in section 5.7.1.6, adapted to account for instrumental error specific to the optical particle counting and sizing technology.

5.7.3. *Scanning particle counters*

5.7.3.1. *Technology*

Scanning particle counters do not subject particles to high shear forces in flow. Such counters measure the PSD in a relatively large sample cuvette or directly *in situ*. Several types of scanning particle counters have been used in oceanography. These instruments size and count particles by measuring either

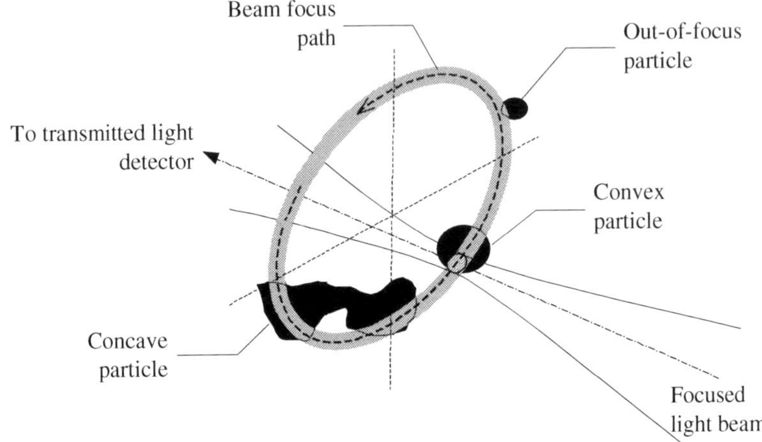

Figure 5.12. The principle of a scanning particle counter. Counters that operate in transmission mode (as shown), scattering, and in reflection mode have been developed. A tightly focused laser beam ($\sim 1\,\mu m$) rapidly scans (~ 1 to $\sim 10\,ms^{-1}$) a circular track with a diameter on the order of few millimeters. In the transmission mode, a temporary reduction in the beam power as light is scattered/absorbed by a particle in the track is measured. The duration of that light power reduction is converted to the length of the particle chord covered by the track. The finite size and non-uniform irradiance profile of the beam focus is accounted for by the signal pulse deconvolution. A similar approach is taken in the reflection (backscattering) mode. Large elongated or concave particles may generate several pulses that may be interpreted as due to several particles. In instruments utilizing the light scattering measurements, the beam focus is broader and the light scattered by the whole particle is the basis of particle sizing. In all cases, the shape and duration of a signal pulse is used to reject pulses which come from out-of-focus particles.

the "forward" scattering (PC-2000 instrument of Spectrex, USA) or the time-of-transition technique in the transmission mode (Galai CIS instrument of Galai Productions Ltd., Israel), or backscattering mode (Par-tec instrument of Lasentec Inc., USA). The operating principle of a generic scanning particle counter that uses the time-of-transition technique is shown in Figure 5.12. The scanning particle counters that use the magnitude of light scattering to determine the particle size implement principles used in optical single-particle counters, except that the light beam moves in respect to the particle.

The PC-2000 counter scans a volume of suspension along a circular track, few millimeters in diameter, with a focused laser beam. The size of a particle is determined by measuring the intensity of light forward scattered by the particle. This intensity is assumed to be proportional to the geometrical cross-section of the particle. The incident light which traverses the sample without being scattered is blocked by an opaque light trap. This instrument can analyze particles with sizes between about 0.5 and 100 μm. This size range is divided into 16 bins. A particle illuminated by a laser beam focus with diameter of ~130 μm produces a narrow pulse whose width depends on the time the beam footprint needs to traverse the particle. The out-of-focus particles produce much wider pulses, because the diameter of the beam footprint increases outside the focus zone. These wide pulses are rejected by the signal processing system. The traveling sensing zone of about 2 cm long covers a volume of 1 cm^3 in about 15 seconds. Only quite dilute suspensions (< 1000 cm^{-3}) can be analyzed. This limitation is intended to reduce the coincidence error, which is caused by interpreting the overlapping signals from two or more particles in the focus zone of the laser beam as a signal from one larger particle. This instrument was used to analyze samples of seawater from the Indian Ocean (*Chung* 1982), St. Louis Bay, Mississippi (*Sydor* and Arnone 1997), and from the Mississipi river (*Smart* et al. 1985).

The *in vitro* CIS instrument sizes a particle by measuring the time-of-transition of a focused laser beam across the particle (*Aharonson* et al. 1986). A tightly focused laser beam (1.2 μm diameter) covers a circular track with a diameter of 600 μm by moving with a fixed linear (tangential) velocity. The beam intensity is measured as a function of time. When the beam focus intercepts a particle, the beam intensity decreases momentarily due to attenuation of light by the particle, generating a negative pulse. The particle size can be estimated from the width of this pulse by deconvolving the pulse shape with the known (Gaussian) radial profile of the beam focus. The instrument's electronics evaluates the pulse shape and rejects those pulses which come from off-beam track and off-focus particle.

Weiner et al. (1998) discuss the effect of the various factors which can affect the accuracy of the particle size determination with the time-of-transit technique. For a beam focus diameter of 1.2 μm, the smallest particle which can be resolved has a diameter of about 0.5 μm. The particle size accuracy for the small particles (< 10 μm) depends on the accuracy with which the beam focus diameter is known. This requires minimization of light scattering at the optics and the sample container surfaces because it increases the beam diameter at the focus. The upper size limit is essentially set by the settling velocity of the particle and/or the particle size in relation to the circular beam track radius. The instrument's electronics evaluates the pulse shape to select only those pulses with the steepest leading and trailing edges. This nominally eliminates pulses from particles which were not traversed along their medians and from out-of-focus particles.

For the time-of-transition technique to work well in a particle size range, the scanning velocity, focus radius, scan track radius, and the particle size range and velocity of the particles relative to the beam focus must be optimized. To the first order of magnitude, operation of this instrument does not depend on the optical properties of a particle. However, the electronic system that evaluates the pulse shape must rely on finite threshold values just to distinguish a pulse from the noise. This may complicate working with highly elongated particles or with large fluffy particles which contain high-contrast component particles in a low-contrast base and introduces a possibility of counting these small components instead the whole particle. This kind of particle sizing technique is probably most reliable when used with high-contrast particles with well-defined compact outlines.

In concentrated suspensions, the scattering of light by the particles broadens the beam focus and reduces the maximum beam power at the focal point. In such suspensions, the backscattering mode of the time-of-transition is advantageous: the pathlength through the suspension can be reduced to several tens of micrometers, i.e., almost at the probe window, yet the probe minimally hinders the mixing or flow of the suspension, especially if the probe window is oriented at a small angle with respect to the suspension flow axis. Such a technique, referred to as the Focused Beam Reflectance Measurement—FBRM), is implemented in the Par-tec instruments (elliptical focus $0.8\,\mu m \times 2\,\mu m$, covering a circular track at a velocity of $\sim 2\,m\,s^{-1}$). A small cylindrical probe (25 mm diameter) can be easily inserted in the environment (*Law* and Bale 1998, *Law* et al. 1997). The small size of the probe minimizes the insertion effect. The particle size range of this instrument is about 1 to $1000\,\mu m$. *Heath* et al. (2002) and *Ruf* et al. (2000) have recently reviewed the principles and applications of the FBRM technique. These authors also discussed problems in the interpretation of the results of measurements performed with that instrument.

In the Par-tec instrument that implements the FBRM technique, the beam focus plane is relatively close to the probe window ($<<1\,mm$); hence, in flowing suspensions, the effect of the probe window orientation in relation to the particle flow may be significant in suspensions with high particle concentration ($\sim$30%—*Barrett* and Glennon 1999). The focal plane location has a significant effect on the measured PSD (see Figure 5.13). *Law* et al. (1997) also found significant effect ($\sim$50%) of the focal plane position, within a range of 0 to 4 mm away from the instrument window, on the mean particle diameter measured with a Par-tec instrument. The magnitude and direction of the observed changes were strongly dependent on the particle concentration.

The scanning particle counters cannot measure the concentration of the particles directly in such a precise way as can resistive particle counters or optical flow cytometers. The scanning counter measures merely the frequency of pulses, each representing a chord of a particle. Although it is clear that such a frequency should be related to the particle concentration, significant asynchronous variations in the rate of counting chords of various lengths have been observed as the concentration of particles changes (e.g., *Heath* et al. 2002). At higher concentrations ($>\sim 1\%$ w/v), the effect of the instrument's dead time becomes significant and, if compensated for, results in an essentially linear relationship between the total counts and particle concentration (*Heath* et al. 2002). The "sample volume" of these counters is relatively small as compared with other particle counters. Thus, at low concentrations, typical of some natural waters, the acquisition of sufficient count statistics may take significant time (minutes) as pointed out by *Law* et al. (1997).

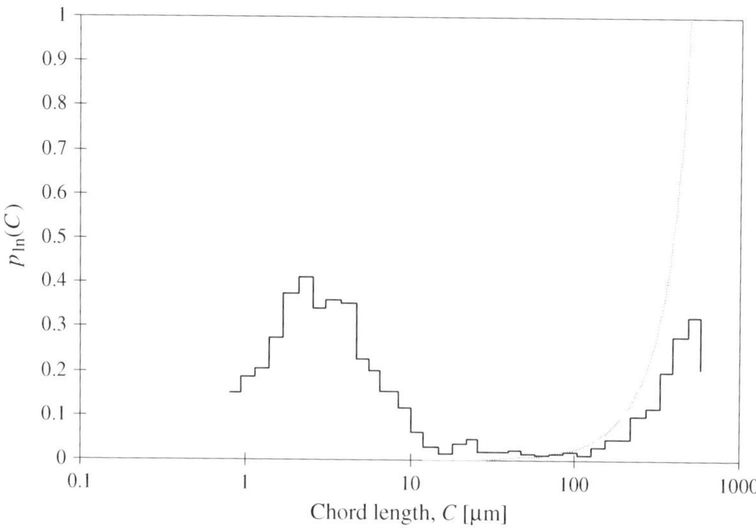

Figure 5.13. Theoretical (gray curve) and experimental (black curve) chord length distributions (CLDs) defined as $p_{\ln}(C) = dP(C)/d(\ln C)$ for a monodisperse population of SiC spheres with diameter D of 630 μm in water. The peak between about 1 and 10 μm in the experimental distribution (digitized from Fig. 10 of *Ruf* et al. 2000) is not noise but real data resulting from spheres that are away from the focal plane of the counter's beam. The experimental results are obtained for a single particle moved across the field of view of the sensor at four planes located increasingly away from the sensor. In the CLD corresponding to the plane closest to the focal plane, only the far right peak is present, consistent with the theoretical CLD.

Finally, there are the effects of the particle material, particle shape, and the structure (smoothness) of the particle surface. Given that most of the 1- to 1000-μm particle size range is in the geometric optics domain, the backscattering signal is by definition dependent on the bi-directional particle surface reflectivity. This introduces the effects of three-dimensional shape of the particle into consideration. Thus, transparent spherical particles with smooth surfaces, such as oil droplets and transparent beads, are likely to be less accurately sized (*Ruf* et al. 2000, *Sparks* and Dobbs 1993). On the other hand, opaque particles with rough surfaces and similar shapes are more likely to produce size distributions that, when appropriately scaled, are comparable with those obtained by other particle sizing methods (*Heath* et al. 2002).

5.7.3.2. Chord length distribution

The scanning particle counters that are based on the time-of-flight principle, of measuring the particle size determine essentially the chord length distribution

(CLD) of the silhouettes of particle. These distributions are replicable characteristics of the suspensions and as such can be used to differentiate one suspension from another, similar to the multi-parameter scatter plots generated by optical flow cytometers for inhomogeneous and/or non-spherical particles. However, the CLDs do differ from the corresponding PSDs. Thus, for applications in modeling of light scattering by a suspension and in any other application that requires the PSD, the CLD must be converted to the size distribution.

Note that even with monodisperse particle populations, the scanning counter will yield a CLD with a finite width. Consider the simplest case of spherical particles with a diameter D. A chord can be drawn across the particle's silhouette at any distance from the silhouette center. Thus, the chord length varies between 0 and the sphere diameter. The CLD, normalized to become a probability distribution, p, of chord length, C, can be written as follows (e.g., *Ruf* et al. 2000):

$$\begin{aligned} p_{\ln}(C) &= \frac{dP(C)}{d(\ln C)} \\ &= \frac{C^2}{D\sqrt{D^2 - C^2}} \end{aligned} \tag{5.108}$$

with $0 \leq C < D$ where P is the probability of C being within a range of C to $C + dC$, and where we used the $dP/d(\ln C)$ format of the data presented by *Ruf* et al. (2000) for easy comparison with their results. This probability distribution increases monotonically from 0 at $C = 0$, to infinity at $C = D$. This idealized distribution assumes the beam focus of the scanning particle counter has a negligible width as compared with that of the particle. Given that the beam diverges away from the focus, this implies that the depth of field of the counter is finite, and the chord definition measured across a particle worsens when the latter is away from the focal plane. Indeed, experimental results of Ruf and colleagues who obtained CLDs while moving a single spherical particle across the counter field of view clearly indicate these effects (Figure 5.13). With identical non-spherical particles, the broadening of the CLD is even stronger because the effect of the particle orientation becomes significant. This is evident from simulations performed for ellipsoidal and cuboid particles by *Ruf* et al. (2000).

Results obtained by Ruf and colleagues for actual suspensions of polydisperse spherical particles confirm that these effects are reinforced by the suspension polydispersity. The researchers also note that the high refractive index of the particles improves the particle size definition as compared to that obtained for low-index particles when operating a scanning counter in the backscattering mode.

This problem is not unique to the scanning particle counters as all particle sizing methods share it to some degree. Such "broadening" of the particle size may be due to the non-sphericity of the particles as observed in measuring variations of the intensity of light scattered by a non-spherical particle oscillating across the sensing zone of an optical particle counter (*Umhauer* and Bottlinger 1991) and/or due to the non-uniform sensing field within the sensing zone of a particle counter, as observed for monodisperse spheres analyzed with a resistive particle counter without hydrodynamic focusing (e.g., *Spielman* and Goren 1968).

For a PSD, $n(D)$, of spheres, the CLD, i.e., $n(C)$, can be evaluated from the following expression (*Wynn* 2003):

$$n(C) \propto \int_0^\infty \frac{C}{\sqrt{D^2 - C^2}} n(D)dD \tag{5.109}$$

Note that independent of the underlying normalized PSD, useful information can be obtained about the ratios of its moments from the various ratios of the moments of the normalized CLD. *Ruan* et al. (1988) derived the following relationship for spheres:

$$\langle D^k \rangle = \frac{k\langle x^2 \rangle}{2} \langle C^{k-2} \rangle \tag{5.110}$$

with $k > 1$ being the moment order.

For example, the ratio of the average volume to average diameter of the spheres $\langle D^3 \rangle / \langle D^2 \rangle$ can be obtained as follows (*Ruan* et al. 1988; note that in the convention used by these authors, $\langle C^0 \rangle = 1$):

$$\frac{\langle D^3 \rangle}{\langle D^2 \rangle} = \frac{3}{2} \langle C \rangle \tag{5.111}$$

Several methods have been developed for obtaining the PSDs from the observed CLDs. This problem has also been encountered in other particle sizing applications as demonstrated by the solution obtained by *Ruan* et al. (1988) for determining the size distribution of bubbles in foams. We mention here two such methods, both reducing the problem of sizing three-dimensional particles to sizing their two-dimensional silhouettes, developed specifically for the scanning counter's applications (1) an analytical inversion method of *Wynn* (2003) and (2) an iterative method of *Langston* et al. (2001) .

Given the complexity of and stability problems with the inversion methods, empirical methods have been developed to calculate from CLD a suspension characteristic that is comparable to a PSD. Some researchers used various CLD-weighing schemes (*Heath* et al. 2002) that include weighing of the CLD by the square and cube of the chord length, C, i.e., calculating the second and third moment of the CLD. The square weighing is reported by *Heath* et al. (2002) to produce the CLDs, with the peak located at a size comparable to that obtained with laser diffractometry. Note that the PSD obtained by the latter instrument is the second moment of the number concentration-based PSD. *Law* et al. (1997)

simply rescaled the chord length scale to fit the sizes of the "calibration" particles measured by the FBRM instrument and with optical microscopy and used it to measure predominantly monomodal size distributions. Law et al. do however caution that this method may not be suitable for complex multimodal PSDs.

5.7.3.3. Applications

Application of a time-of-transition scanning particle counter (Galai CIS 1, Galai, Inc.) in oceanography has been recently evaluated (*Ratmeyer* et al. 1999, *Jantschik* et al. 1992, *Tsai* and Rau, 1992). *Jantschik* et al. (1992) found that the particle size determined by the Galai counter was somewhat greater than that determined by the Coulter counter. In a bimodal PSD, the peak of the first mode appeared at 3.5 μm in the size distribution measured with the Coulter counter and at 7 μm in the distribution measured with the Galai counter. The peak of the second mode in the Galai size distribution was difficult to evaluate because of the statistical fluctuations in the particle concentration at that size range. Such fluctuations, also noted by *Tsai* and Rau (1992), are due to the small "sample volume," on the order of 1 mm^3, of the Galai counter and cause a low reproducibility of the results, since the probability of detecting a particle in single-particle counting is proportional to the volume of the liquid analyzed. *Tsai* and Rau (1992) compared several size distributions obtained with the Galai counter and with an optical microscope and found that the distributions "generally followed each other in shape quite closely." In a later paper, *Tsai* (1996) noted that when a liquid-flow system was used with a Galai counter (CIS-1), the results differed from those obtained with an optical microscope. *Bohling* (2005), who reviewed much of the literature of the subject, added the sample handling (operator influence; acting mostly via time delay between the sample insertion and analysis) and concentration effects to factors that affect reproducibility of the Galai-type counters.

Law and Bale (1998) and *Law* et al. (1997) used the backscattering time-of-transition Partec instrument for limited *in situ* experiments in coastal waters off the UK. Prior to its use in the field, the instrument was evaluated in-laboratory with suspensions prepared using monodispersed calibration particles; the Part-ec instrument consistently overestimated sizes of calibration particles below about 150 μm and underestimated sizes of the larger particles. Some calibration particles (latex and glass) were not sized accurately. The instrument was also operated by these authors *in situ* in the coastal waters and was found to generate meaningful results. However, results obtained in these experiments were compared with the PSDs obtained by other methods.

Bale et al. (2002) applied the particle sizing algorithms developed by *Law* et al. (1997) to observe with the FBRM technique *in vitro* changes in the median particle size caused by cyclic flow condition simulation of tidal conditions in a shallow estuary. They used a separate optical backscattering sensor to determine the concentration of particles during the experiment.

5.7.4. In situ imaging

In situ imaging systems can analyze unperturbed particles in their native environment and to avoid certain complexities of other particle-characterization systems. Several underwater photographic camera systems have been built to examine suspended particles *in situ* (*Lunven* et al. 2003, *Knowles* and Wells 1998, *Hou* 1997, *Baier* 1996, *Baier* et al. 1996, *Ratmeyer* and Wefer 1996, *MacIntyre* et al. 1995, *Costello* et al. 1995, *Maffione* et al. 1993, *Heffler* et al. 1991, *Costello* et al. 1991, *Eisma* et al. 1990, *Johnson* and Wagnersky 1985, *Honjo* et al. 1984, *Edgerton* et al. 1981).

A "sheet-of-light" structured illumination is generally used to limit the thickness of the illuminated volume of water. Such illumination helps to avoid problems with the depth of focus of the camera and to avoid shadowing of one particle by the others in the line of sight. In some underwater camera systems (Benthos camera, e.g., *Kranck* and Milligan 1992), the thickness of the illuminated volume of water is limited mechanically by positioning the camera window and the illuminator window close to each other. Flashed illumination is used to freeze the motion of the particles relative to the camera.

The "sheet-of-light" illumination, as well as a limited-depth of field at high magnification, prevents the use of this method, as is, for the determination of the three-dimensional spatial distribution of particles, unless it is used in a stereophotography configuration. The latter produces image pairs whose processing is difficult to automate for the particle size and position determinations. However, a relatively simple modification enhances the single-camera imaging method to particle size and position measurements (*Pereira* and Gharib 2004). This modification employs the fact that when a camera is focused at a plane, a blurred image of an out-of-focus particle is recorded at the image plane, as opposed to a sharp image for an in-focus particle. By obstructing most of the lens area except of two or three apertures positioned near the lens rim, one converts this defocused image into a set of two or three images of the out-of-focus particle. From the positions of the defocused images and the system geometry, one can determine the position of a particle in three-dimensional space. The particle size can be determined by measuring the defocused image brightness and compensating, where necessary, for the illumination beam intensity distribution within the measurement volume.

In one of the most comprehensive *in situ* microphotography system to date (*Eisma* et al. 1990), three cameras are used, each working at a different magnification. That microphotographic system is capable of measuring the size distribution of suspended particles between about 4 and 644 μm. The problem of handling a potentially large number of photographs is somewhat reduced by using an automated particle size analysis system based on a CCD camera and an image analysis algorithm. The image analysis-based particle sizing algorithm rejects the out-of-focus particle images and compensates for the imaging noise introduced by the photographic emulsion and by the image digitization.

Processing of images obtained by microphotography may present problems related to overlapping images of particles that are comparable to the "volume-summing" coincidence in resistive particle counters. As pointed out by *Kaye* et al. (1997), although a trained human observer can relatively easily identify and correct such problems when observing particles under the microscope rather than at a photograph, the automated image analysis system may be less clever.

The processing of fuzzy images of particles found in natural waters is no trivial task. Results tend to be quite sensitive to the selection of an edge detection algorithm and to the adopted definition of the particle size. Insufficient resolution and sensitivity of the imaging system can cause artifacts when processing images for the size distribution of the particles (*Costello* et al. 1994). These artifacts are similar to those created by breakage of particles in single-particle counters and may arise if the sensitivity of the system is too low to faithfully image a particle as an entity. This is of particular importance to aggregates composed of opaque grains embedded in a nearly transparent matrix. If the nearly transparent areas of the aggregate are poorly imaged, they may simply "dissolve" when the gray-scale image of the whole particle is converted to a binary (black-white) image, a step required in determining the particle contour. The binary image may in this case degenerate into seemingly separate small opaque particles. When the size distribution is evaluated, such particles are counted as separate entities with sizes smaller than that of the parent particle.

Jackson et al. (1995) provide a concise outline of the particle sizing with image analysis. With subtleties of the particle image identification aside, the particle size may be obtained by determining the circular equivalent diameter, D_{I}, of the particle image, i.e., the diameter of a circle with an area equal to that of the particle image. The area of the particle image is determined by simply counting pixels which constitute the particle image. Assume that there are N_{I} of such pixels, each with an area, A_{X}, which is scaled to the particle size by considering the imaging system magnification.

Thus, the circular equivalent diameter of the particle can be expressed as follows:

$$D_{\mathrm{I}} = 2\sqrt{\frac{N_{\mathrm{I}} A_{\mathrm{X}}}{\pi}} \tag{5.112}$$

5.7.5. Holography

Holography overcomes the problem of the limited depth of field of conventional microphotography and can obtain microphotographs of particles with constant size resolution in a field of very large depth (several centimeters). Thus, *in situ* spatial distribution of particles and their orientations can be determined. An early application of *in situ* holography was the determination of the settling velocity of particles by taking time-delayed holograms of particles (e.g., *Costello* et al. 1989).

Note that simple imaging with a single camera has recently also been used for this purpose (section 5.7.4).

Holograms can be taken *in situ*, avoiding problems caused by sampling. In addition, the holographic technique allows one to determine the three-dimensional shape of the particles and records the three-dimensional positions of particles within the sample volume simultaneously. However, these advantages are significantly diminished by difficulties in automating the particle sizing available with this technique. *Malkiel* et al. (1999) quote 100 to 400 h for a complete analysis of a hologram with 5000 to 20 000 particles even with considerable computer assistance. This clearly implies that routine analyses using the holographic sizing technique remain elusive.

A simple holographic recording system (Figure 5.14) is based on the in-line or Fraunhofer holography in which the direct coherent (laser) beam is superposed with the light scattered by the particles in a see-through configuration (*Craig* et al. 2000, *Nebrensky* et al. 2000, *Katz* et al. 1999, *Costello* et al. 1989, *O'Hern* et al. 1988, *Carder* et al. 1982, *Carder* and Meyers 1979, *Carder* 1978, *Stewart* et al. 1973, *Sokolov* et al. 1971, *Thompson* et al. 1967, *Knox* 1966). Such systems work well with marine particles because these particles are relatively transparent, permitting transmission of the undiffracted wave through the particle material.

Practical in-laboratory systems resolve particles with diameters down to about 5 μm. This limit is typically increased to about 10–15 μm in the case of *in situ* holographic systems, although details as small as 3 μm can be resolved (*Malkiel* et al. 1999) . The lower limit of the particle size, D_{min}, is imposed by the requirement that a sufficient angular extent of the wave diffracted by the particle is recorded. An interesting method of increasing the resolution of a holographic system by capturing a large portion of the wavefront with a small imaging detector has been recently proposed by *Liu* et al. (2002). In a standard system, the minimum size resolution can be expressed (e.g., *Malkiel* et al. 1999) as follows:

$$D_{min} > \sqrt{Z\lambda} \tag{5.113}$$

where Z is the distance from a particle to the holographic plate and λ is the wavelength of light.

Thus, critical parameters for an in-line holographic system are the size and distance of the recording plate from the particle in relation to the wavelength of light and spatial resolution of the recording system. A lens system may be used to magnify the "image" of the sampling volume at the film, so as to permit recording of a required range of the spatial frequencies of the particle image in the hologram. The particle images are reconstructed by using laser illumination and may be magnified with an optical system coupled to a TV camera. Particle images can be sized as seen on the video monitor screen, with a help of a reticle (*Thompson* et al. 1967) or digitized and processed with image analysis software (*Costello* et al. 1989). The former method imposes a significant

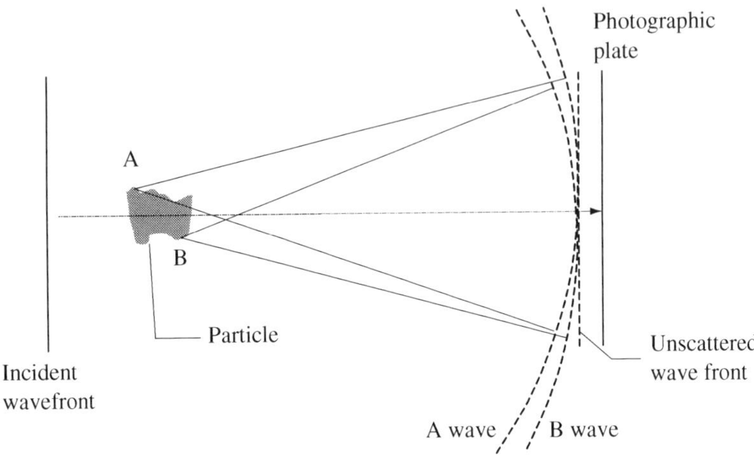

Figure 5.14. The principle of in-line (Fraunhofer) holographic microscopy. The particles must be tenuous, i.e., their refractive index relative to that of the surrounding medium must be close to unity and nearly transparent so as not to affect the incident wavefront significantly. When a particle is illuminated by an incident wave front (here shown as a plane wavefront) emitted by a coherent light source, various elements (e.g., A and B) of the particle scatter spherical waves. The wavefronts scattered by all elements of the particle (here only those scattered by the A and B elements are shown) interfere with the unscattered wavefront in space. A holographic plate records a plane section of the interference pattern, creating a hologram. The coherence length/time of such a light source must be sufficiently large to allow such interference pattern to be fixed during the recording process. This practically limits the light sources to relatively high power lasers. Due to a high depth of field (dependent on the wavelength and particle size), the hologram "freezes" a three-dimensional field of particles over a large volume (on the order of $1000\,cm^3$) and can be "played-back" for the analysis of the entire volume, plane by plane. The object can also be reconstructed by digital processing of the recorded hologram (e.g., *Xu* et al. 2002). The scene volume can contain on the order of 10^5 particles (*Malkiel* et al. 1999). Higher particle concentrations may result in the hologram deterioration through multiple scattering.

strain on the observer's eye because of the speckle noise. This noise must be effectively filtered from images of particles reconstructed from holograms for a successful application of the automatic evaluation of the particle image contour (*Costello* et al. 1989). Despite partial or complete automation, the analysis of a hologram is still a significant bottleneck as it can take hundreds of hours (*Malkiel* et al. 1999).

The holograms of particles have been generally photographed on film and used to generate three-dimensional images of particles during the image analysis phase. Recently, the holograms of three-dimensional objects have been recorded by using

CCD cameras and the three-dimensional images of the objects reproduced by the computer analysis of digital images in what is referred to as "digital holography" (*Schnars* and Jüptner 2002, *Xu* et al. 2002). The three-dimensional trajectories of particles can be traced in time by capturing a time series of holograms with a fast CCD camera, which by appropriate processing (*Xu* et al. 2003) can be compressed into a single hologram.

The relatively long hologram recording period of the original photography-based systems, which is determined by the power of the illuminating laser beam, required the use of a high-viscosity isotonic solution to slow down particles settling in an *in situ* sediment trap employing a holographic camera (*Costello* et al. 1989). Recently a comprehensive holographic *in situ* system has been reported (*Craig* et al. 2000, *Nebrensky* et al. 2000) intended for monitoring the populations of particles in the ocean. The system uses both in-line and off-axis holography and allows examining a relatively large volume of seawater (up to $10^3\,cm^3$). It has a working particle size range of few micrometers to tens of centimeters. The off-axis system uses two coherent beams and light "reflected" by the particles. In contrast to the in-line system, it is not limited by the requirement of the reasonable transparency of the particles. Here, the maximum particle size is limited only by the recording camera field of view, the coherence length of illumination light, and its power. However, the lower particle size limit of the off-axis system is restricted to about 30–40 μm by the speckle pattern in the image and by the optical system.

Alternatively, the angular scattering patterns of individual particles can be reconstructed (*Stigliani* et al. 1970) and used to size immobilized particles by applying an inversion algorithm based on a scattering theory. An automated approach to holographic particle sizing was reported by *Tschudi* et al. (1974). In that approach, the laser light scattered by a particle was simultaneously analyzed by six holographic matched filters. The holographic filters were generated using the Mie theory for six different particle sizes. The output waves of these filters were analyzed using diaphragmed photodetectors registering the intensity of the correlation signals between the angular pattern of the scattered light and the spatial transmittance pattern of the filter. The correlation signal transmitted by each filter is the highest when the particle size is the same as that for which the filter has been generated.

Malkiel et al. (1999) used a holographic imaging system mounted on a remotely controlled submersible platform to gather comprehensive data on the size distribution (10–3500 μm), spatial distribution, orientation, and motion of plankton *in situ*. *Hobson* et al. (2000) demonstrated a spatial resolution better than 20 μm for a large 75-cm deep volume of $2500\,cm^3$. Thus, technology is available for assessing the orientation of particles *in situ* as well as variations of the small-scale distribution of particles in seawater. The particle orientation, further discussed in Chapter 6, is of significant importance in modeling polarized light scattering by seawater (as discussed in Chapter 3).

5.7.6. Optical, epifluorescence, and scanning optical microscopy

Optical microscopy, one of the basic tools of microbiology, has been long used to gather a wealth of information about small particles. Its significance in the determination of particle morphology is universally recognized, especially in marine biology. However, the major disadvantage of this method remains its low throughput which severely limits the statistical significance of the results. Indeed, even when combined with image analysis, a typical optical microscopy-based particle size analysis, system can examine a fraction of the number of particles that can be analyzed by a resistive particle counter or a flow cytometer. Even with the use of image analysis software, other factors limit the throughput. Indeed, the size of the field of view at the maximum resolution is typically on the order of $100 \times 100\,\mu m^2$. At a low aquatic particle concentration on the order of 1000 particles greater than $1\,\mu m$ per cm^3, one can expect 100 particles in that field after having settled particles from a typical sample volume of $1000\,cm^3$ onto a microscope slide's $10 \times 10\,mm^2$ area. After having identified all the particles and measured them, the slide must be *mechanically* moved to a new position so that another field of view can be analyzed. This may take on the order of 0.1 s on top of the image analysis time. Thus, on the order of several hundred particles can be analyzed per second, compared to tens of thousands in the case of flow cytometry.

The maximum resolution of an optical microscope is on the order of the wavelength of light. This resolution is limited by the diffraction of light at the lenses of a microscope, which—even for an otherwise perfect lens—blurs the image of a point source, that is, a detail of an object. Due to diffraction, the distribution of light intensity in the image plane follows a complex, bell-shaped function, whose first minimum determines the radius of the blur (Airy) circle,. The significant diameter of the blur circle equals (e.g., *Born* and Wolf, 1980):

$$D_{\text{Airy}} = \frac{1.22\lambda}{n \sin u} \tag{5.114}$$

where λ is the wavelength of light, n is the refractive index in the object space, and u is the half-angle of the angular field of view of the lens system. The denominator in (5.114) is the numerical aperture of the lens system.

If two point sources are brought close together, the intensity patterns in their images, as observed through a microscope, will overlap. According to the Rayleigh criterion, such close points cannot be resolved if the maximum of the intensity pattern of the image of one point coincides with the first minimum of the intensity pattern of the image of the other point. This happens, when the distance between the two points equals the radius of the Airy circle and leads to about 26% contrast in the image of the two points (e.g., *Jonkman* et al. 2003). According to (5.114), that radius for a high-power microscope objective with a numerical aperture on the order of 1 is about $0.6\,\lambda$, that is $0.34\,\mu m$ in blue-green light ($\lambda = 0.55\,\mu m$), where the human eye is most sensitive. A high numerical aperture of 1.3, and the use of blue light ($\lambda = 0.4\,\mu m$) with a video camera as an image detector instead of the human eye which is less sensitive at that wavelength, would increase the resolution to $0.19\,\mu m$. Recently, a number of improvements in the resolution of optical microscopy resolution have been reported (*Stelzer* 2002, *Fedosseev* et al. 2005), bringing the resolution limit down to $0.1\,\mu m$.

The depth of field, z, of an optical microscope is expressed as follows:

$$z = \frac{\lambda}{4n \sin u} \tag{5.115}$$

that is a fraction of $1\,\mu m$. The object structures outside the depth of field blur the image of the structures which are within the depth of field. The small depth of field implies that the focal plane adjustment is critical. Factors that affect the precision of positioning of that plane, for example vibration, may significantly reduce the resolution.

The epifluorescence microscopy improves the accuracy of the size measurement (*Inoue* 1989, *Weiss* et al. 1989, *Sieracki* et al. 1985, *Fuhrman* 1981). In an epifluorescence microscope, the fluorescence of an object is viewed against a dark field; thus the inherent contrast of the image is improved as compared with that of conventional microscope. The use of low-noise image acquisition devices such as a cooled CCD, can additionally and significantly improve the quality of the image observed with the epifluorescence microscope (*Viles* and Sieracki 1992). An image analysis system can improve the contrast further by selectively stretching the contrast scale. Such a system can also automate the process of particle sizing and counting. Densitometric analysis of images of aquatic bacteria obtained by an epifluorescent microscope permit the determination of the bacterial cell DNA content (*Loferer*-Krößbacher et al. 1999) that enables a relatively rapid determination of the bacterial biomass.

A range of image analysis systems that can be easily coupled to conventional and fluorescent microscopes are commercially available. In addition, specialized image analysis systems have been developed by the researchers themselves. The latter range from semi-automated systems in which the operator outlines the contour of the particle (*Krambeck* et al. 1981) to fully automated systems which utilize image enhancement techniques and optimized automated edge detection algorithms (*Bloem* et al. 1995, *Viles* and Sieracki 1992, *Sieracki* et al. 1989). Typically, the edge-locating algorithms assume uniform distribution of the fluorescence intensity (image brightness) throughout the particle. Such an assumption may be well founded in the case of homogenous particles, but not in the case of structured, naturally occurring particles. The accuracy of the particle size measurement depends on the algorithm of edge detection used in the image analysis. *Sieracki* et al. (1989) found that the method utilizing the minimum of the second derivative of the image brightness profile to locate the edge of the particle is the most accurate on average. The problems with the automated detection of particles in microscopic images are similar to those experienced by processing of images obtained by *in situ* microphotography (section 5.7.4).

Optical microscopy has been extremely useful in the analysis of the particle morphology. Coupled with image analysis, it allows a greater insight into the nature of the particles and details of the size distribution in comparison with other methods as pointed out by *Kaye* et al. (1997). These researchers compared the size distributions of well-defined mineral powders by image analysis-aided microscopy (IAM) and laser diffraction (section 5.7.9) and pointed out that the inversion algorithm applied in a laser diffractometer software can hide details of the particle population characteristics.

However, optical microscopy has been less successful in providing statistical data on the populations of marine particles. As a result, PSD data obtained with optical transmission or epifluorescent microscopy before the advent of computerized image analysis are relatively rare. Representative data obtained by manual cell evaluation and sizing method include those for marine bacteria (*Stramski* and Kiefer 1990), phytoplankton (*Furuya* and Marumo 1983, *Takahashi* and Bienfang 1983), mineral particles (*Gurgul* 1993, *Lisitzyn* 1972—quoted by *Lal* and Lerman, 1975, *Jerlov* 1961—quoted by *Jerlov* 1976, *Krey* 1961). The study of *Takahashi* and Bienfang (1983) is particularly comprehensive. These authors examined the taxonomic and size composition of marine phytoplankton with cell sizes $>2\,\mu m$ in the Pacific waters off Hawaii. Samples of seawater were filtered using Nitex screens (20, 40, and $212\,\mu m$) and Nuclepore filters (0.4, 1, 3, and $5\,\mu m$). Cell dimensions, volumes total cell number, and total cell volumes calculated from data obtained using a settling chamber and an inverted optical microscope are tabulated for 23 species of marine phytoplankton with cell volumes ranging from 0.066 to $267\,000\,\mu m^3$. The majority of phytoplankton biomass was found to be in the $<3\,\mu m$ size range.

Wellershaus et al. (1973) and *Lenz* (1972) used optical microscopes in conjunction with an image analysis system to investigate the wide-range size distribution of marine particles. A semi-automatic Zeiss particle counter was used to analyze photographs of particles obtained using an optical microscope. In that counter, a light spot is projected onto an image of a particle in a photograph. The spot diameter was adjusted according to the size of the image, and served as a measure of the particle size, after accounting for the magnification of the photograph.

The advent of computers and electronic imaging devices has considerably improved the process, introducing what can be called image-analysis-aided microscopy. *Sieracki* and Viles (1992) and *Sieracki* et al. (1985) used an epifluorescence microscope combined with a computerized image analysis system to determine with much less manual effort size distributions of marine bacteria with diameters less than $1\,\mu m$. However, the limited depth of focus of conventional optical microscopy [equation (5.115)] may lead to problems in estimating the particle size and also the fractal dimensions of the particles (e.g., *de Boer* and Stone 1999) when the particle sizes span a large size range.

Optical microscopy provides much detail about particles at a very low rate, while single-particle counting (flow cytometry) provides low-detail information at a high rate. Thus, since the early days of flow cytometry, there were attempts to combine both particle analysis techniques in one instrument. Initially, images of particles observed in flow were recorded on film (*Kachel* et al. 1979, *Kay* et al. 1979). Later, an electronic imaging device was used (*Wietzorrek* et al. 1999, 1994) in a flow cytometer designed specifically for plankton analysis (*Dubelaar* et al. 1989). That system could image particles in a size range of 3 to 50 μm. *Sieracki* et al. (1998) and *Sieracki* and Sieracki (1997) developed an in-flow imaging system with a particle size range of 3–100 μm. This system utilized a custom-designed optical element which expanded the depth of focus from 75 to 300 μm. This system permits linking the fluorescence-based particle size data to images of particles, affording comprehensive characterization of the particles. An in-flow imaging system based on the resistive sizing technology with fluorescence measurement capability and a particle size range of 100–2000 μm was also built (*Hüller* et al. 1994).

Microphotometric analysis enabled measurements of the absorption spectra of individual marine particles, immobilized in gels (*Iturriaga* and Siegel 1989, 1988) and by optical trapping (*Sonek* et al. 1995). In this latter method of particle immobilization, questions have been raised about the damage by the trapping beam to biological cells (e.g., *König* et al. 1995).

Scanning optical microscopy, also referred to as confocal scanning microscopy (CSM), has about the same resolution as conventional microscopy but enables one to obtain three-dimensional images of a particle through optical sectioning. Two examples of CSM are provided by a confocal fluorescence microscope and two-photon fluorescence microscope (e.g., *Jonkman* et al. 2003, *Webb* 1996, *Kino* and Corle 1989). In the first case, the object is illuminated by a (laser) beam focused to a diffraction-limited spot. Although the fluorescence of the object is induced throughout the entire volume of intersection of the object with the focused beam, the detector received the fluorescent light only from the focal volume. This limitation is typically implemented by placing a pinhole at the back-focal plane of the beam focusing lens that via the use of a beamsplitter is shared by the illumination and detection paths. In the second case, the source of fluorescence is localized by illuminating the object with focused beam at about twice the actual excitation wavelength in a short intense pulse. The illumination conditions are such that the intensity at the beam focus is sufficiently high that the probability of two photons being delivered to a fluorophore molecule within its relaxation time is significant. This enables excitation, via a quick absorption of two successive photons of low energy, of fluorescence requiring twice as high an excitation energy. As conditions for such excitation are met only within the immediate vicinity of the beam focus, this is then the only source of the fluorescent light. It follows that two-photon microscopy does not require the use of a pinhole to spatially localize the fluorescence in the object volume.

CSM enables the determination of three-dimensional shapes and structures of aggregates (e.g., *Nagel* and Ay 2000, *Thill* et al. 1998, *Schmidt* and Bottlinger 1996), as exemplified by studies of *Cowen* and Holloway (1996) on marine and *Liss* et al. (1996) on riverine aggregates. A drawback is a low scanning velocity, requiring the particle(s) to be immobilized, although fast, video-rate CSM arrangements have been reported (e.g., *Vesely* and Boyde 1996).

5.7.7. Scanning electron microscopy

5.7.7.1. Technology

The scanning electron microscope resolves details one to two order orders of magnitude smaller than those analyzable with optical microscopy, i.e., on the order of 0.02 to 0.002 μm Such a relatively minor improvement, despite a much smaller wavelength of the electrons (on the order 0.05 nm as compared with 500 nm for optical microscopy) is due to intense scattering of electrons penetrating a thick sample.

In fact, it is the scattering by the sample of accelerated electrons which is the basis of operation of the scanning electron microscope. The magnitude of the scattered electron flux is measured by a detector located close to the sample as the electron beam scans the sample in raster fashion. The electron beam, sample, and detector are enclosed in a vacuum chamber. The scanning beam is essentially an electric current in vacuum and must be shorted by the sample to ground in order to avoid sample charging which may distort the raster scan. For this reason, the sample must be coated with a conductive substance, usually gold or carbon. The latter coating is applied if the sample chemical composition is to be analyzed by using X-ray spectroscopy. The resulting signal is displayed on a video screen synchronously with the location of the scanning beam, and a real-time image of the sample is formed. The scattered intensity varies with the nature of the sample and its three-dimensional geometry; thus the sample's image has a definite three-dimensional appearance. The sample support in a scanning electron microscope can be tilted, permitting oblique views of the sample. If such means are not sufficient to visualize the three-dimensional shape of the sample, stereomicrophotographs can be made, and the sample shape can be mapped using the same technique which is applied to the aerial stereophotogrammetry used for terrain mapping.

As noted by *Krambeck* et al. (1981), it is quite difficult to automate image analysis aimed at counting and estimating the volume of the particles with the SEM images because the common edge detecting techniques have problems with intrinsic features of SEM-generated pictures, such as the light-shadow effect. These latter authors used a semi-automated image analysis system in which the operator outlined the particle edge by using a computer-controlled digitizer.

Some scanning electron microscopes are equipped with an X-ray elemental analysis accessory. *Jambers* et al. (1996, 1995) review these methods of elemental analysis as applicable to environmental particles. As already mentioned, carbon is used as conducting coating in this case. With this accessory, the concentrations of elements with atomic weights greater than that of carbon can be determined. Unfortunately, carbon coating reduces imaging resolution by an order or magnitude from that achievable with gold coating.

5.7.7.2. Breakage and other modifications of the particles

Sample preparation required in SEM may significantly modify the particles in addition to all previously discussed alterations due to sample handling . To start with, particles have to be dried. Drying does not disturb solid particles (mineral), but those with high organic and water content (phytoplankton cells and organic-based aggregates) may collapse, and their shapes may be severely altered. In addition, the sample has to be placed in a vacuum chamber in order to be coated with a conductive layer of gold or carbon to prevent charging of the sample surface. The usage of low-energy (< 1 keV) electron beams avoids the charging of the surface, although the image resolution is compromised (*Kino* and Corle 1989). Evacuation of air to permit the coating, and the following rapid increase of the air pressure to atmospheric level when the coating chamber is opened, may complete the damage of the particles.

The shrinkage can usually be recognized in the images of cells if their original morphology is known, because it results in wrinkled or collapsed cell surfaces. Such shrinkage can account for a reduction by a factor of 2 in the particle size, as found for cultures of *Escherichia coli* cells examined in dried state by using a SEM and examined as wet cells by using phase contrast microscopy (*Trueba* and Woldringh 1980). A shrinkage by a factor of about 1.5 has been reported for particles with ESD of about 0.5 μm (*Fuhrman* 1981) in samples collected near the shore and examined within 4 h with an SEM and an epifluorescence microscope. Interestingly, no shrinkage was found in offshore samples which were preserved and examined in the same manner 3 weeks after collection. In a comprehensive study of 11 microorganisms from cultures examined with a SEM and a Coulter counter, cells shrunk by a factor of 1.4 on the average, but also as much as 1.9 times (*Montesinos* et al. 1983). The shrinkage was found to be dependent on the species (but not on the microorganism's size) and on the drying method, with the critical point drying being worse than the air drying. Even "hard" particles such as pollen are reported to shrink during preparation for SEM analysis (e.g., *van Hout* and Katz 2004 – $\sim 10\%$ size reduction).

5.7.8. Transmission electron microscopy (TEM)

TEM offers a significantly enhanced resolution (0.0001 μm), about one to two orders of magnitude higher than that of the SEM. However, due to the complex process of sample preparation and time-consuming analysis, this technique has been rarely utilized in the determination of the size distribution of marine particles (*Kim* et al. 1995, *Wells* and Goldberg 1992, *Harris* 1977). This technique has nevertheless been extensively used to study particle morphology (e.g., *Johnson* and Sieburth 1982—chroococcoid cyanobacteria, *Liss* et al. 1996—flocs in river water). Applications of TEM analysis reach beyond simple particle morphology. Indeed, densitometric analysis of TEM-derived images of bacterial cells were used to determine the dry weight of the cells (*Loferer*-Krößbacher et al. 1998).

5.7.9. Laser diffractometry

This technique and the dynamic light scattering technique, discussed in the following section 5.7.10, differ principally from the single-particle measurement techniques discussed previously because they measure the combined effects of many particles. On one hand, this amplifies the measured signal to a value which is much greater than that from any single particle in a population and makes the measurements much easier than in the single-particle techniques. However, this advantage comes at a significant price because the contributions of individual particles have to be recovered from the combined signal. Problems with such recovery add to those related to recovering the particle size from the light scattering properties of the particles. The recovery of the individual contributions of the particle from the combined signal is mathematically expressed by solving an integral equation which is known to have solution instability problems. As a result, the solution is very sensitive to small errors in the equation's functional coefficients, and the success of solving this equation relies on assumptions concerning the PSD.

5.7.9.1. Technology

The major appeal of laser diffractometry to a marine scientist is in the non-contact, non-disrupting characterization of the particles possible with this technique. The laser diffractometer can also be adapted to *in situ* usage (*Bale* and Morris 1991) to characterize large flocs, which would likely break on sampling. An *in situ* version of a laser diffractometer has been available commercially for some time (LISST-100, Sequois Scientific Inc., USA, see their website for comprehensive technical publications regarding laser diffractometry). That instrument has been recently reviewed by *Gartner* et al. (2001). As an added benefit, the LISST instrument allows the researchers an access to the results of measurements of the small-angle scattering function, which is the primary variable used to determine the PSD with a laser diffractometer.

In a laser diffractometer (e.g., *Agrawal* et al. 1991), a laser beam illuminates a sample space containing the particles (Figure 5.15; see also Figure 4.4). The angular distribution of light scattered by the suspension is determined within a

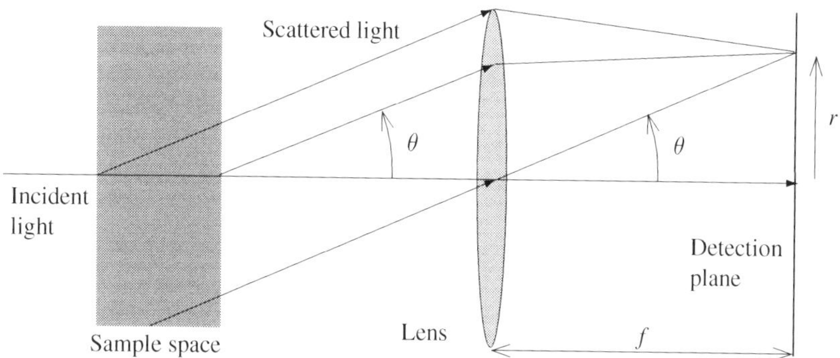

Figure 5.15. The principle of the laser (Fraunhofer) diffractometry is similar to that of the design of some small-angle nephelometers (Figure 4.4): a convex lens transforms the angular scattering pattern of a sample into a two-dimensional spatial pattern in the focal (detection) plane. Light scattered at an angle, θ, is focused at a ring with a radius $r = f \tan\theta$. If a time-averaged spatial pattern is considered, the scattering patterns of all particles in an effective volume of the sample space add incoherently at the detection plane. The effective volume depends on the scattering angle θ, distance between the sample space and the lens, and the lens diameter. Laser diffractometers typically use a segmented ring detector, a photodiode array with several active elements shaped as concentric rings centered about the optical axis of the system. Some instruments use imaging arrays (CCD) as the detectors. The scattered pattern has to be inverted into the PSD, an ill-posed mathematical problem.

relatively narrow-angle cone about the direction of the incident laser beam by using a lens and a detector positioned in the focal plane of the lens. In that configuration, a lens system performs the Fourier transform of the angular distribution of the scattered light into the radial distribution of irradiance in the focal plane. The irradiance distribution is detected typically with independent photodiodes shaped into several concentric segments. The segmented detector is positioned so that its axis of symmetry is collinear with that of the incident beam. Thus, light scattered by the particles into consecutive angular ranges is collected by the consecutive segments of the detector.

The focal length of the lens determines the angular range from which the scattered light is collected by the detector's segments. Thus, the focal length of the lens determines the particle size range of the instrument, because the angular width of the diffraction pattern of a particle is nearly inversely proportional to the particle size as mentioned in Chapter 1. The min-to-max size ratio for a lens is on the order of 1:100. Commercial laser diffractometers can measure PSDs in a range of about 0.1 to 1000 μm. With a typical design, this size range is covered by using a set of different lenses that may collect the scattered light simultaneously within several angular ranges. Some instrument designs enable the use of a single lens for nearly that entire range. Another way is to use different (fixed) lenses for the various angular ranges, as—for example, in a Coulter LS series (Beckman Coulter Inc.) (e.g., *Loizeau* et al. 1994). The size range of one lens is divided into about 15 coarse intervals, whose width increases with the particle size. Since these ranges are relatively coarse, narrow-peak PSDs may be distorted (*Agrawal* et al. 1991, *Burkholz* and Polke 1984). By merging size distributions obtained with different lenses for the same material, one may introduce additional errors (*Singer* et al. 1988).

Gianinoni et al. (2003) introduced a modified laser diffractometer design, which employs a focused, instead of a parallel (as in the standard laser diffractometer) illumination beam. The focused beam is blocked by a mask positioned in front of the optical window/lens that collects the scattered light. This design eliminates the contribution from the scattering of light at the imperfections of the detector window that are no longer illuminated by the incident beam as in the parallel beam design. The converging beam design has been developed with industrial aerosol applications in mind, where particles tend to stick to the receiver window. However, that design may be of interest for in-water research as well, for in long-term deployment of an instrument underwater, a biofilm usually develops on the optical surfaces. Note that the converging beam arrangement introduces a constraint on the beam convergency and the particle diameter range due to the fact that the particle position in the converging beam affects its contribution to the angular scattering pattern of the suspension, transformed into the radial pattern by the receiver lens.

We have already mentioned that all particles contribute simultaneously to the measured light scattering in a laser diffractometer. The complex problem of characterization of particles by inverting the overall scattering pattern of the suspension is simplified by a series of approximations. First, the effect on the scattering pattern of the particle size is maximized by selecting, in a typical laser diffractometer, the small-angle range of the scattering angles. Note that when such a limitation is adopted, the size range of the particles is accordingly limited from below, as small particles tend to contribute to light scattering at the large scattering angles. This, in fact, is the reason behind using separate detectors for several scattering angle ranges in some instruments.

Second, the overall scattering pattern is expressed by an integral equation, the Fredholm equation of the first kind, which expresses the angular scattering pattern, $\beta(\theta)$, as a function of the PSD, $f(D)$, as an unknown term under the integral:

$$\beta(\theta) = \int_0^\infty \beta_1(\theta, D, n) f(D) dD \tag{5.116}$$

where θ is the scattering angle, $\beta_1(\theta,\ D,\ n)$, i.e., the single-particle angular scattering pattern, D is the particle size, and n is the refractive index of the particles (potentially size dependent).

In order to solve such an equation, its kernel, $\beta_1(\theta, D, n)$, i.e., the single-particle angular scattering pattern, must be known. The interpretation of the kernel as the single-particle pattern implies the single-scattering regime, i.e., a moderate particle concentration. This $\beta_1(\theta, D, n)$ is generally approximated by the light scattering pattern of a sphere. In early instruments (e.g., *Cornillault* 1972), the Fraunhofer diffraction approximation to this pattern was typically used. The Fraunhofer approximation limits the particle size analysis to a minimum particle size of about 10λ, where λ is the wavelength of light. The Mie theory (e.g., *Bohren* and Huffman 1983) has also been used, as a user-selectable alternative to Fraunhofer diffraction, in the recent versions of the instruments, in order to interpret scattering patterns of particles with sizes below that which is amenable to analysis by using diffraction theory. In fact, by reviewing the product literature, one gets an impression that the invocation of the Mie theory serves as a seal of approval, implying that the particle sizing results obtained this way are "correct." Unfortunately, the reality is different: the Mie theory does apply only to light scattering by homogeneous spheres and is only an approximation (and sometimes a poor one) for light scattering by particles of other shapes. Neglecting that fact may lead to significant problem when non-spherical particles are sized, as we will discuss in more detail shortly.

As opposed to Mie theory, the Fraunhofer approximation does not require the refractive index of the particles to be defined . In this latter case, the need for the specification of the refractive index can introduce large discrepancies between the size distributions obtained with the same instrument on which both modes of analysis were available (*Loizeau* et al. 1994) . With particle populations containing particles of different composition (and thus refractive index), significant complications arise if the refractive index–size distribution is not known. *Kusters* et al. (1991) point to a potential problem with laser diffractometers that utilize the "exact," i.e., Mie-based form of the single-particle angular pattern, β_1 in equation (5.116) for the determination of the PSD: if the refractive index of measured particles relative to the suspension medium is not the same as that assumed in the calculation of β_1, serious errors may result.

Problems which are introduced by the many approximations involved in simplifying the mathematics of light diffraction in a laser diffractometer are compounded by the fact that the solution of the Fredholm equation is very sensitive to errors introduced by the measurements of the angular scattering pattern and depends to some extent on the inversion algorithm. This, in part, is the source of artifacts commonly observed in particle size analysis with laser diffractometers (*Loizeau* et al. 1994, *Agrawal* et al. 1991, *McCave* et al. 1986).

It is frequently claimed that a laser diffractometer does not need, in principle, to be calibrated because the instrument's response can be predicted with the diffraction or Mie theory. However, even for spherical particle, in which case the diffraction and Mie theory can be regarded as exact, calibration may be required. This is because the accuracy of measurement of the angular scattering pattern—the basis of the PSD determination with laser diffractometry—depends on several factors, such as (1) the light beam quality, (2) its alignment relative to the lens and the detector, (3) accuracy of the positioning of the detector relative to the lens focal plane, and (4) variations in the sensitivity to the scattered light (including the amplifier gains) across the segments of the detector (*Dodge* 1984), not accounting for factors affecting the delivery of a representative particle population to the instrument's sensing zone. The alignment issues are of particular importance for samples containing large particles whose angular scattering pattern is a very steep function of the scattering angle. The alignment problems are especially pronounced in instrument designs with removable lenses. Overall, from the manufacturer's own accounts, these factors limit the reproducibility of the particle size determination for spherical monodisperse particles to several percent.

5.7.9.2. Effects of the particle shape

The particle shape may have a significant role in the standard-design laser diffractometers that employ concentric ring detectors. As *Trainer* (2001) concludes: "... The use of spherical particle scattering models to invert scattering distributions from non-spherical particles produces significant errors in the PSD. Only non-spherical scattering models should be used to invert scattering distributions from non-spherical particles...."

Mühlenweg and *Hirleman* (1998) point out that the particle shape deviation from spherical into an elliptical shape can cause the inversion software of such laser diffractometers to misinterpret the size distribution of the particles and produce a widened, bimodal distribution. *Heffels* (1995) also noted that the angular scattering pattern of monosized particles with sharp edges averaged over the particle orientation has a smooth profile that would be interpreted by the inversion software of the typical laser diffractometer as a pattern produced by a size distribution of spheres. Such effects have also been experimentally observed by *Gabas* et al. (1994). The size distribution of equivalent spheres due to randomly oriented spheroids, as determined with a laser diffractometer, has been recently derived and used to correct the size distributions obtained for ellipsoids through laser diffraction (*Matsuyama* et al. 2000). Such effects have also been noted in intercomparison of the laser diffraction results of the PSD measurements with those obtained with other methods (*Naito* et al. 1998, *Endoh* et al. 1998).

If the detector of a laser diffractometer enables the resolution of the azimuthal scattering angle in addition to the scattering angle, for example, when each ring detector segment is further subdivided into individually addressable wedges, the effects of the particle shape may be discerned. *Heffels* (1995) demonstrated that the shape information is encoded in a relatively small number of wedges as long as the radius of the wedged ring detector contains the second and third maximum of the diffraction pattern. The strong central lobe of the diffraction pattern contains little information about the particle shape. Heffels also points out that the magnitude of the shape effects may be comparable with the effect of the refractive index of the particles. Some laser diffractometers use a CCD camera as a detector, hence the light scattering patterns recorded by these instruments may be used to obtain indications of the particle shape. The shape effects are sufficiently significant to warrant potential application in particle shape recognition. For example, *Yamamoto* et al. (2002) suggested the application of the particle shape effects as measured with laser diffractometers equipped with wedge-segmented ring detectors for the identification of phytoplankton.

5.7.9.3. Intercomparisons

Comparison between PSDs obtained with particle sizing instruments whose operation is based on different sizing techniques is made difficult by the fact that each technique may sense a different particle size. In addition, different algorithms may be used to process the data (even with instruments operating on the same principles). Finally, each instrument may divide the particle size range into a different set of size intervals.

Comparisons of the size distributions obtained using a Malvern Laser diffractometer and a Coulter counter (*Agrawal* et al. 1991, *McCave* et al. 1986) reveal a moderate similarity between results obtained with these two particle sizing techniques. This similarity depends on the lens used in the diffractometer, the particle size range, and on the number of the particle size classes used in the inversion algorithm of the diffractometer. The size distribution of particles smaller than $2\,\mu m$ tends to be underestimated in the diffractometers by about 20 to 70%. The multiangle Coulter LS-100 system, performed better in the Fraunhofer diffraction mode: 40 to 70% of particle population below $2\,\mu m$ was detected, but much worse in the Mie scattering mode: 2 to 17% (*Loizeau* et al. 1994). Interestingly, this contradicts the assumption that the Fraunhofer model does not resolve well particles with sizes below about 10λ, which would in this case correspond to sizes below about $7\,\mu m$.

Burkholz and Polke (1984) discussed operating principles, designs, and compared the performances of several laser diffractometers manufactured by Cilas, Malvern, and Leeds & Northrup with the performance of a sedimentation and an image analysis particle size analyzers. *Loizeau* et al. (1994) compared Coulter and Malvern diffractometers. The repeatability of the size distributions analyses on the same sample with a single diffractometer or various diffractometers was generally within several percent. A greater variability was caused by sub-sampling. However, significant differences between the size distributions obtained using the various sizing techniques were observed, especially in the case of platelet-shaped particles. The diffractometer results were between those obtained with the image-analysis-aided microscopy and the sedimentation method (*Burkholz* and Polke 1984). The effect of sample concentration on the PSDs obtained with some diffractometers was found to be significant. The finer the particles, the lower the concentration required in order to limit the measurement errors. However, since the scattered light intensity decreases with both the particle size and concentration, the performance of the instrument also decreases with the particle concentration. Interesting results have been obtained by *Dur* et al. (2004) who compared results for the various forms of the size distribution obtained for the mineral components of soil with a laser diffractometer and the image analysis-aided TEM. Although the number concentration PSDs seem to have agreed pretty well, the area and volume distributions have shown increasing deviations, due to weighing by the square of the particle size (area distribution) and by the cube of the particle size (volume distribution).

The particle shape effects on the PSD obtained with a laser diffractometer have also been documented by comparing the results with other particle size analysis techniques (*Naito* et al. 1998, *Endoh* et al. 1998). In fact, Endoh et al. indicated that the differences between results obtained with these two techniques could be used for the very evaluation of the particle shape. Specifically, they advanced a theoretical model suggesting that the particle flatness could be determined from the ratio of the median diameter determined by laser diffraction to that determined by sedimentation.

5.7.10. Dynamic light scattering (photon correlation spectroscopy)

5.7.10.1. Technology

The technique of dynamic light scattering (*Pecora* 2000, *Santos* and Castanho 1996, *Finsy* 1994, *Finsy* et al. 1992, *Berne* and Pecora 1976), also referred to as photon correlation spectroscopy, is widely used in polymer research and technology to determine the size of suspended particles by measuring their mobility (diffusion) in suspension. This technique is non-invasive and in principle needs no calibration besides accurate measurements and maintenance of the sample temperature. The particle size can be evaluated from the first principles governing the processes of diffusion, which depend on the particle size and temperature of the suspension solvent.

As shown in Figure 5.16, this technique relies on the effect on time-dependent light scattering of random motion of suspended particles (Brownian motion) that depends on the particle size. The instrument simply measures rapid time variations (on the scale of microseconds) of the scattered light intensity due to interference between light waves scattered by diffusing particles illuminated with a coherent light beam. The particle size is derived from that time-dependent light scattering data.

The principle of dynamic light scattering for monodisperse particles can be summarized as follows (e.g., *Finsy* 1994). Assume that we illuminate a dilute suspension with a collimated steady monochromatic light source and at time t_0 we begin to measure the time-dependent irradiance of light scattered by monodisperse spherical particles at a given angle. The time-dependence of that irradiance results from the time-dependent interference of light waves scattered by each particle at that angle. This interference is time-dependent because, as the particles are pushed

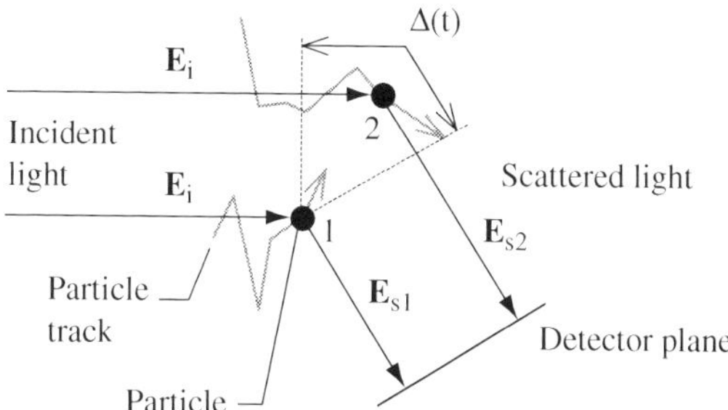

Figure 5.16. The principle of the dynamic light scattering is based on the theory of light scattering and of Brownian motion. In this schematic diagram for the homodyne configuration of a dynamic light scattering experiment, a coherent incident plane wave of wavelength λ is scattered by particles that drift within the sample space tossed around by molecules of the surrounding medium. This motion introduces time-dependent phase differences, $\Delta(t)$, between waves scattered by the particles. As a consequence, the scattered light irradiance at a specific angle, θ, $I(t,\ \theta) = \langle|\mathbf{E}_{s1} + \mathbf{E}_{s2}|^2\rangle$, randomly fluctuates in time. The scattering angle-specific autocorrelation function, $\Gamma(\tau) = \langle I(t)I(t+\tau)\rangle$, of that time-dependent irradiance decays exponentially with the time delay, τ, as $\exp(-q^2\delta\tau)$, where $q = (4\pi/\lambda)\sin(\theta/2)$, and δ is the temperature-dependent diffusion coefficient of the particles in the surrounding medium. This implies that the temperature of the sample must be known and maintained during the experiment. As the diffusion coefficient is simply related to the particle size, D, the measurement of the autocorrelation function thus enables the determination of the particle diameter. Heterodyne configurations (with large, virtually immobile reference particles, or via an interferometric setup) are also possible.

by the fluid molecules away from their positions at time t_0, the conditions for the interference of waves scattered by these particles change. Note the essential requirements fulfilled by the dilution of the sample: single scattering dominates and particles can diffuse essentially independently of each other.

The time rate of change in the scattered light intensity is related to the particle size-dependent diffusion rate: small particles diffuse faster than large ones. The changes in the intensity of light scattered by the suspension are random because particles assume random positions when they diffuse. Close to time t_0, the instantaneous scattered light intensity, $I(t_0 + \tau)$, where τ is the time delay, is pretty much the same as that at t_0, i.e., $I(t_0)$. Thus, for a small value of the time delay, τ, the instantaneous light scattering intensities are relatively well correlated with the initial intensity. As τ increases and particles drift further apart, this correlation decays until after a long time delay, at a scale set by the diffusion rate of the particles, the scattered light intensity is no more correlated with the initial intensity. Note that the concentration of the particles cannot be so small that the significant source of fluctuations in the scattered light irradiance comes from the fluctuation of the particle concentration inside the scattering volume, a volume of the suspension that is both illuminated by the light source and "observed" by the detector.

The rate at which the correlation between fluctuations of the light intensity scattered at an angle θ [the intensity autocorrelation function, $\Gamma(\tau, \theta) = \langle I(t, \theta)\, I(t+\tau, \theta)\rangle$] decays with the time delay, τ, is thus a measure of the rate of diffusion of the particles and indirectly a measure of the particle size. This size is the hydrodynamic particle diameter, i.e., the diameter of a sphere which has a diffusion coefficient equal to that of the particle.

Another view point on dynamic light scattering is offered by the Doppler-shift perspective. Indeed, each particle that moves in suspension is a moving source of the scattered light. This causes the wavelength of the scattered light, as observed by the detector, to change according to the particle velocity relative to the detector. Thus, a suspension illuminated with monochromatic light will scatter slightly polychromatic light. Hence, another name for this technique is quasi-elastic light scattering.

The larger the particles, the slower is the decay of the autocorrelation function. For large molecules (molecular weight $\sim 25\,000$) the characteristic decay time is on the order of microseconds and increases to many milliseconds for micrometer-sized particles (e.g., *Stramski* et al. 1992c). This sets the practical limits of dynamic light scattering to particles not larger than about 5 μm (e.g., *Stramski* and Sedlák 1994).

The autocorrelation function for the irradiance of light scattered by a monodisperse suspension of particles with the diameter D decays as an exponential:

$$\Gamma(D, \theta) \propto \exp(-2\gamma\tau) \tag{5.117}$$

where the factor of 2 indicates that we are concerned not with the electric fields but with the irradiance which is the square of the electric field, and

$$\gamma = \Delta(D, T)q^2 \tag{5.118}$$

with Δ being the translational diffusion coefficient of the particle that depends on the particle size and temperature of the suspension medium and q is the magnitude of the scattering vector:

$$q = \frac{4n\pi}{\lambda}\sin\frac{\theta}{2} \tag{5.119}$$

The scattering vector is the vector difference of the incident and scattered wave vectors, each with a magnitude of $2\pi n/\lambda$, where n is the refractive index of the suspension medium and λ is the wavelength of light in vacuum.

The translational diffusion coefficient is typically related to the particle size with the Stokes–Einstein equation:

$$\Delta = \frac{kT}{3\pi\eta D} \tag{5.120}$$

where k is the Boltzmann constant, η is the dynamic viscosity of the suspension medium (see also section 5.7.11), and T is that medium's absolute temperature. Thus, it follows that D is the hydrodynamic diameter of the particle.

Complications arise when the suspension is polydisperse because each of the different particle size classes sets a different particle diffusion rate and thus a rate of decay of the relevant monodisperse component of the autocorrelation function. The autocorrelation function for the polydisperse suspension is an integral of the particle size-dependent autocorrelation function weighed by the unknown PSD:

$$\Gamma_E = \int_0^{\infty} f(\gamma)\exp(-\gamma\tau)d\gamma \tag{5.121}$$

where we added a subscript E to indicate that the autocorrelation function refers here to the electric field, not the irradiance.

The distribution, $f(\gamma)$, of the decay rate, γ, and hence the PSD, via equations (5.118) and (5.120), can be obtained from (5.121) by the inverse Laplace transform. That procedure is unfortunately similar to the solution of the integral Fredholm equation in the laser diffractometry case in that it is also an ill-posed mathematical problem, making the solution strongly dependent on the noise in the experimentally determined variable on the left side of the equation. We discuss other problems related to the solution of equation (5.121), namely the scattering angle-dependent weighing of $f(\gamma)$, briefly in the following section 5.7.10.3 in the context of aquatic applications of the dynamic light scattering technique.

Particle shape can influence the particle size estimates obtained with the dynamic light scattering, and—consequently—the PSD measured with this technique. First, the particle shape affects the determination of the hydrodynamic diameter, whose specific meaning here is the diameter of a sphere with a diffusion coefficient equal to that of the particles. Second, the particle shape may affect the autocorrelation function that for non-spherical particles becomes a function of the translation *and* rotational diffusion coefficient. Such an additional "degree-of-freedom" of the particle diffusion causes the autocorrelation function Γ to decay faster than in the case of the translation alone. Third, the angular weighing factors—in the case of polydisperse suspensions—may differ from those for spherical particles. Hence the distribution of the decay rate, g, obtained by assuming that the particles are spherical may be different from the actual distribution. *Santos* and Castanho (1996) show the application of the particle shape effect for the determination of the aspect ratio of the tobacco mosaic virus.

5.7.10.2. Enhancements

The dynamic light scattering technique has been enhanced by measuring the temporal fluctuations of coherent light scattered at two orthogonal polarization (e.g., *Pitter* et al. 1999, *Bates* et al. 1997). This modified dynamic light scattering is referred to as polarization fluctuation spectroscopy (PFS). The intensities of scattered light measured at two orthogonal polarization are perfectly correlated for homogeneous spheres because the intensities of light scattered at two polarizations is only a function of the scattering angle. However, this is not the case for non-spherical particles and, due to the random rotation of such non-spherical particles in suspension, this cross-correlation is reduced. This enables a qualitative determination of the particle shape in a suspension of particles with a concentration sufficiently small to avoid substantial multiple scattering. *Kusmartseva* and Smith (2001) used this approach to develop a non-sphere detection method based on the same principles. Recently, a quantitative version of this technique has been reported, allowing simultaneous determination of the particle size and aspect ratio (*Walker* et al. 2004) .

An interesting variation of the dynamic light scattering method is direct tracking of the random movement of small particles. This principle has been applied to both single particles in three dimensions (*Garbow* et al. 1997, *Schätzel* et al. 1992) and to particle suspensions in two dimensions (*Nakroshis* et al. 2003, *Finder* et al. 2005). This approach enables the determination of the diffusion coefficient and thus the particle size and size distribution by utilizing a simple relationship between the mean square displacement $\langle L^2 \rangle$ of a particle undergoing Brownian motion and time, t:

$$\langle L^2 \rangle = \frac{4kT}{6\pi\eta D} t \tag{5.122}$$

where k is the Boltzmann constant, η is the dynamic viscosity of the suspension medium (see also section 5.7.11), T is the absolute temperature of that medium, and D is the hydrodynamic diameter of the particle.

5.7.10.3. Applications

In aquatic particle sizing applications, the technique of dynamic light scattering has been used to determine the average sizes of live and preserved cultures of marine bacteria (*Stramski* and Sedlák 1994, *Stramski* et al. 1992c, *Paul* and Jeffrey 1984). The particle sizes were found to be in general agreement with those determined using epifluorescence microscopy. The size of live bacteria determined using the dynamic light scattering technique was several times smaller than that of killed bacteria (*Paul* and Jeffrey 1984) due to the high cell motility of the species examined. In fact, the swimming velocity of bacteria can be deduced from dynamic light scattering measurements (e.g., *Boon* et al. 1974) .

An early attempt (*Paul* and Jeffrey 1984) to determine the PSD of natural samples of natural waters essentially failed, because of the diversity of the particle sizes present in the samples. As we have already mentioned (section 5.7.10.1), if a particle suspension is polydisperse, the autocorrelation function of the scattered light intensity is an integral of particle size-dependent autocorrelation functions. The retrieval of the PSD thus requires solving an integral equation, which is very sensitive to the measurement errors and has an algorithm-dependent solution. The thus obtained "PSD" is weighed by the particle size-dependent scattering cross-section. For small particles, with an angular scattering pattern nearly symmetric about 90°, this cross-section increases rapidly and monotonically with the particle size at any scattering angle. As a result, the apparent average particle size as determined using dynamic light scattering is overestimated (*Stramski* et al. 1992c). Thus, the "size distribution" must be further processed to retrieve the actual PSD. If the form of the PSD is known a *priori*, then such distribution can be calculated from the dynamic light scattering data. *Thomas* (1987) shows the procedure for a log-normal PSD. *Stramski* and Sedlák (1994; see also a minor erratum in *Stramski* and Sedlák 1995) have applied a similar technique to determine the slope of the PSD in seawater in a range of 0.1 to 8 μm. They analyzed seawater samples in which particle size was limited to between 0.1 and 8 μm by filtering through Nuclepore filters. The samples were preserved in order to ensure that the only motility of the particles was that due to the Brownian motion.

An interesting application of the dynamic light scattering technique in studies of structure of algal cells has been recently reported by *Witkowski* et al. (1993). They identified distinct features in the mobility spectra of the scattering centers that were attributable to cell components.

Another interesting application of dynamic light scattering has been reported by *Glatter* et al. (1990), who used dynamic light scattering in conjunction with electrophoresis to study conformational changes in human blood platelets due to stress. The platelets carry electric charge on their surfaces, and this charge is affected by the conformational changes. These changes also affect the hydrodynamic mobility of the platelets. Since particles suspended in seawater also carry electric charges (e.g., *Neihoff* and Loeb 1972, 1974) which affect the process of flocculation, for example, this technique could potentially be utilized to study the flocculation of small marine particles. Finally, we mention the technique of electrorotation that enables the determination of the dielectric constant of the particle (cell) as well as the dispersion of the dielectric constant of the particle-suspension medium (*Gimsa* 1999, *Gimsa* et al. 1995). Modern versions of that technique, which originated at the end of the XIX century (*Gimsa* et al. 1995), utilize dynamic light scattering for the measurement of the rotation velocity of the particle(s) in a rotating electric field, the effect brought about by interaction of the dipole moment induced in the particle and the external electric field.

5.7.11. Sedimentation techniques

The sedimentation techniques of PSD measurements utilize a significant dependence of the settling velocity of particles (in still medium) on the hydrodynamic particle size. As we shall see shortly, the settling velocity of a particle also depends on the particle shape, hence on the particle orientation for non-spherical particles, as well as on the difference between the medium and particle density.

In the classical application of the sedimentation technique, measurements are typically performed in a settling tube. This assures that the medium is still (e.g., *Syvitski* et al. 1991a). Particles are introduced at the top of the tube, and their arrival at the bottom of the tube is monitored with a variety of methods (e.g., a balance or attenuation measurement). The settling tube geometry (diameter and length) does influence the accuracy of the measurement of the sedimentation velocity (*Syvitski* et al. 1991a, *Gibbs* 1972) with large-diameter, long tubes yielding an improved accuracy. If the sedimentation rate is measured by a detector measuring the reduction in light scattering or attenuation vs. time as the particles settle out, the refractive index of the particles needs to be also known.

For small particles, the sedimentation time tends to be excessively long or infinite. Indeed, under the action of gravitation alone, particles smaller than about 0.1 μm typically stay in suspension indefinitely due to Brownian motion, even if they are denser than the suspension medium. Note that much larger particles with density nearly equal to that of the suspension medium, such as soft phytoplankton cells, may also stay in suspension indefinitely at the time scale of a sedimentation experiment because of the negligible density difference (e.g., *Smayda* 1970).

The sedimentation process can be speeded up by creating a much higher acceleration than that of gravity. This is typically done by spinning the sample at a high angular velocity. In photosedimentation disk centrifuges (e.g., *McCave* and Syvitski 1991, *Weiner* et al. 1991, *Middelberg* et al. 1990), the sample container has a form of a shallow cylinder with transparent top and bottom, much like a Petri dish. The cylinder is rotated about its axis, and the progress of sedimentation of the particles, either injected at the rotation axis or uniformly dispersed in the container, can be monitored by measuring at a fixed radial distance from the axis the attenuation of the sample. This technique is widely used in particle-processing industries. Unfortunately, it requires a fairly high concentration of the particles to be sufficiently accurate. This might be the main reason for which it has not been used for the analysis of natural water samples to our knowledge.

Another variation of the sedimentation technique is implemented as the sedimentation field fractionation (e.g., *Beckett* et al. 1988). This is one of the many variations of the field flow fractionation (FFF) , pioneered by J. C. Giddings (*Giddings* et. al. 1974), where the acceleration perpendicular to the flow direction is used to provide the force sedimenting sub-micron particles across a non-linear (parabolic) profile of the carrier fluid velocity in a channel. When a suspension of particles with various sizes is injected at the beginning of the channel, the sedimenting particles fall off toward a low velocity region. The fastest sedimenting particles fall off at first and arrive at the outlet of the channel last. In contrast, the slowest sedimenting particles arrive first at the outlet. Thus, the particles are fractionated according to the sedimentation rate, i.e., the particle size, given of course the same density for all size fractions. Note that the same idea can be used to fractionate particles with the humble sedimentation tube.

The FFF has been widely used to separate the various size fractions of submicron particles (colloids) in aquatic research (*Contado* et al. 1997, *Beckett* et al. 1988, *Vaillancourt* and Balch 2000). The optical attenuation or scattering is frequently used to determine the "size distribution" of the suspension. However, results of such analyses are generally reported as qualitative, because of the unknown weighing of the attenuation or scattering signal by the optical properties of the particles. *Wyatt* (1998) pointed out that the multiangle light scattering detection can be of significant help in the unambiguous determination of the PSD this way.

Sedimentation techniques have been widely used in the analysis of the particle size of sediment at mass particle concentrations on the order of several g dm^{-3} (e.g., *Syvitski* et al. 1991a, *Singer* et al. 1988). Note that at high concentrations, problems with interpretation of the settling velocity data arise because the sedimentation rate is affected by collective settling of the particles, also referred to as hindered settling. Indeed, independent sedimentation, as implied in the sedimentation techniques, is valid in the limit of vanishing particle concentration (volume fraction $\leq\sim 0.005$; e.g., *Hassen* and Davis 1989). If the particle concentration is high, the particles do not settle independently.

In collective settling, due to perturbation of the flow around a particle by another particle settling nearby, particles in (temporary) close proximity settle faster than would have been implied from their hydrodynamic sizes (*Hassen* and Davis 1989)—this incidentally leads to an increased settling velocity of even loosely bound aggregates as compared with that of their component particles. The resulting fluctuations in the settling velocity of a concentrated suspension may lead to errors in the hydrodynamic diameters of the particles, and thus in the PSD of the suspension measured with sedimentation techniques.

Complications aside, particles are minimally perturbed with such a settling tube technique, especially when it is applied *in situ*. This advantage led to an attempt at such *in situ* analyses of the PSDs near the sea bottom in the Nova Scotian Rise area by using a remotely controlled optical settling tube (*Spinrad* et al. 1989a). That settling tube measured variations in the attenuation of light as a function of time at a section near the tube bottom. To cut off the flux of particles from above, the tube is covered at time $t = 0$. In laboratory experiments with well-defined particles of known density, such a settling tube yielded size distributions in agreement with the Coulter counter measurements

(*Zaneveld* et al. 1982). Less accurate results were obtained for samples of marine sediment. With this device, the PSD is calculated from the time profile of the attenuation of light across a settling tube with a diameter of 5 cm and a height of 22.5 cm.

5.7.11.1. Settling velocity of a solid particle

The calculations of the PSD with a settling tube require the knowledge of the relationship between the settling velocity and particle size, here the hydrodynamic diameter. The settling velocity, v_s, is established when the driving force equals the hydrodynamic drag force (e.g., *Baird* and Emsley 1999):

$$gV(\rho - \rho_w) = C_D \eta v_s \tag{5.123}$$

where g is the gravity acceleration, V is the particle volume, ρ is the density of the particle, and ρ_w is the density of water/seawater, C_D is the drag coefficient, and η is the dynamic viscosity of water/seawater. Note that equation (5.123) can be used to determine the density of a particle by letting it settle in fluids of differing densities, ρ_w (e.g., *van Hout* and Katz 2004).

The Stokes settling law for spheres falling in still viscous medium has been widely used as a working approximation to that relationship since the particle shape is typically unknown and varies from one particle to another. For a sphere, $C_D = 3\pi D$, hence

$$v_s = \frac{g}{18\nu_w}\frac{\rho - \rho_w}{\rho_w}D^2 \tag{5.124}$$

where v_s is the settling velocity, g is the gravity acceleration, $v_w = \eta_\omega/\rho_w$ is the kinematic viscosity of water/seawater. The temperature and salinity effects are accounted for in the density of the medium. The dynamic viscosity of pure water can be calculated as follows (after *Jumars* et al. 1993, who cite *Korson* et al. 1969, error of $< 1\%$ over a temperature, T, range of 0 to 40 °C):

$$\log_{10}\eta_w = \frac{\sum_{i=0}^{2} a_{0i}T^i}{\sum_{i=0}^{1} a_{1i}T^i} \tag{5.125}$$

where T is the temperature in °C and η is in g cm^{-1} s^{-1} [i.e., the poise; to obtain the viscosity in SI units, subtract 1.0 from the right side of (5.125)]. The values of the coefficients are shown in Table 5.3. *Jumars* et al. 1993 also reference a more accurate approximation of *Sengers* and Watson (1986).

Table 5.3. Coefficients of equation (5.125) for the dynamic viscosity of pure water.

Coefficient	Value	Coefficient	Value
a_{00}	−1.57095E+02	a_{10}	8.993E+01
a_{01}	−3.09695E+00	a_{11}	1.0E+00
a_{02}	−1.827E−3	–	–

Dera (1992) gives the following, internationally accepted equation for the density of seawater, in kg m^{-3}, as a function of temperature, T, salinity, S, and pressure, p:

$$\rho_{sw} = \rho_0 \frac{1}{1 - \frac{p}{K}} \tag{5.126}$$

with ρ_0, the density of seawater at the sea level ($p = 0$) being defined as follows after *Jumars* et al. (1993), who cite *Bigg* (1967) and *Millero* and Poisson (1981):

$$\rho_0 = \rho_w + A_1 S + B_1 S^{3/2} + C_1 S^2 \tag{5.127}$$

where S is the practical salinity, and the density of pure water, ρ_w, is approximated as follows

$$\rho_w = \sum_{i=0}^{5} a_i T^i \tag{5.128}$$

and A_1, B_1, and C_1 are each also expressed by a polynomial in T [°C], with the coefficients given in Table 5.4.

Table 5.4. Coefficients of polynomials in T [°C] relevant to equation (5.127) for the density of seawater (water) in kg m^{-3} at sea level (after *Dera* 1992).

Coefficient	ρ_0			
	ρ_w	A_1	B_1	C_1
a_0	9.99842594E+02	8.24493E−01	−5.72466E−03	4.8314E−04
a_1	6.793952E−02	−4.0899E−03	1.0227E−04	–
a_2	−9.095290E−03	7.6438E−05	−1.6546E−06	–
a_3	1.001685E−04	−8.2467E−07	–	–
a_4	−1.120083E−06	5.3875E−09	–	–
a_5	6.536332E−09	–	–	–

Finally, the linear secant bulk modulus, K [b] is approximated as follows:

$$K = K_{20} + A_{20}S + B_{20}S^{3/2} + P(K_{21} + A_{21}S + B_{21}S^{3/2}) + P^2(K_{22} + A_{22}S) \quad (5.129)$$

where the coefficients are each expressed as a polynomial in T [°C] (Table 5.5 through Table 5.7) *Dera* (1992) gives a set of check values from which we cite the following: ρ_{sw} of 999.96675 kg m^{-3} at $T = 5\,°C$, $S = 0$ psu, and $P = 0$ b, with $K = 20\,337.80375$, and ρ_{sw} of 1062.53817 kg m^{-3} at $T = 25\,°C$, $S = 35$ psu, and $p = 1000$ b, with $K = 27\,108.94504$.

Table 5.5. Coefficients of polynomials in T [°C] relevant to equation (5.129) for the linear secant bulk modulus of seawater (after *Dera* 1992).

Coefficient	K_{20}	A_{20}	B_{20}
a_0	1.965221E+04	5.46746E+01	7.944E–02
a_1	1.484206E+02	–6.03459E–01	1.6483E–02
a_2	–2.327105E+00	1.09987E–02	–5.3009E–04
a_3	1.360477E–02	–6.1670E–05	–
a_4	–5.155288E–05	–	–

Table 5.6. Coefficients of polynomials in T [°C] relevant to equation (5.129) for the linear secant bulk modulus of seawater (after *Dera* 1992).

Coefficient	K_{21}	A_{21}	B_{21}
a_0	3.239908E+00	2.2838E–03	1.91075E–04
a_1	1.43713E–03	–1.0981E–05	–
a_2	1.16092E–04	–1.6078E–06	–
a_3	–5.77905E–07	–	–

Table 5.7. Coefficients of polynomials in T [°C] relevant to equation (5.129) for the linear secant bulk modulus of seawater (after *Dera* 1992).

Coefficient	K_{22}	A_{22}
a_0	8.50935E–05	–9.9348E–07
a_1	–6.12293E–06	2.0816E–08
a_2	5.2787E–08	9.1697E–10

The factor $g/(18\nu)$ for seawater at 10 °C equals ~4 × 10^5^ $m^{-1}s^{-1}$ (the acceleration of gravity equals ~9.81 m s^{-2}, the kinematic viscosity of seawater equals ~1.35 × 10^{-6} m^2 s^{-1}, e.g., *Jumars* et al. 1993). Thus, a 10 μm quartz sphere, with a relative density difference of ~1.6, would settle in seawater at a velocity of ~62 μm s^{-1}, i.e., ~5.3 m day^{-1}. This implies an analysis time on the order of 24 hr for particles in seawater under the sole effect of gravity, as indeed indicated by *Spinrad* et al. (1989a).

The Stokes law implies laminar flow around the sphere, i.e., applies in the limit of small particle size (~0.1 to ~10 μm) and relative density difference (<0.1). A more rigorous definition of the Stokes law application range involves the Reynolds number:

$$\mathrm{Re} = \frac{v_s D}{\nu} \tag{5.130}$$

By using this dimensionless number, we can state that the Stokes law applies to a range of Re << 1. In the example of a 10 μm quartz sphere in seawater, Re ~ 0.0015.

In the large particle size limit, the Stokes law was extended by Oseen (e.g., *Brun*-Cottan 1986, note errors in the Oseen formula in that paper—*Brun*-Cottan, personal communication) to larger particle sizes and densities, i.e., a quasi-laminar flow regime:

$$v_s = \frac{4\nu}{3D}\left|-1+\sqrt{1+\frac{\rho-\rho_w}{\rho_w}\frac{g}{12\nu^2}D^3}\right| \tag{5.131}$$

Gibbs et al. (1971) derived an empirical formula for the settling velocity of spheres by using data obtained for large glass spheres (diameter range of 50 to 6000 μm) with the density in a range of 2.24 to 2.755:

$$v_s = \frac{-3\eta+[9\eta^2+gD^2\rho_w(\rho-\rho_w)(3.869\times10^{-5}+0.024801D)]^{1/2}}{\rho_w(0.00011607+0.074405D)} \tag{5.132}$$

where the unit of v_s is m s^{-1}, η is the dynamic viscosity of water/seawater in N s m^{-2}, g is expressed in m s^{-2}, densities are expressed in kg m^{-3}, and the sphere diameter is expressed in m. *Gibbs* et al. (1971) also include tables of the settling velocities of spheres, calculated according to this formula for selected particle diameters, water temperatures, and salinities. Sample results obtained with the formulas of Stokes, Oseen, and Gibbs et al. are shown in Figure 5.17.

Brun-Cottan (1986) cites results of a model study (*Boido* 1947) on the settling velocity (in horizontal orientation) for plates

$$v_s = \frac{g}{7\nu}\frac{\rho-\rho_w}{\rho_w}t\langle D\rangle \tag{5.133}$$

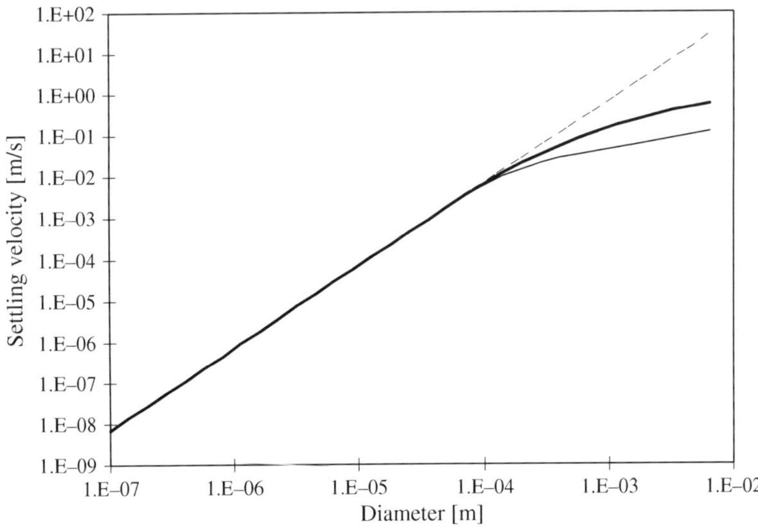

Figure 5.17. Comparison of the settling velocities of solid spheres with density of $2650\,\mathrm{kg\,m^{-3}}$ in water at $10\,°\mathrm{C}$ calculated with the theoretical formulas of Stokes (dashed line) and Oseen (thin solid line), as well as with an empirical formula of *Gibbs* et al. (1971) (thick solid line).

where t is the plate thickness and $\langle D \rangle$ is the average ECD, and for long needles:

$$v_s = \frac{g}{4\nu}\frac{\rho - \rho_w}{\rho_w}\frac{D^2}{0.4 + 2.7\frac{D}{L}} \tag{5.134}$$

where D is needle diameter and L is its length.

Baird and Emsley (1999) in their discussion of phytoplankton sinking cite analytical results for randomly oriented prolate and oblate spheroids (*Clift* et al. 1978):

$$C_{\mathrm{D,prolate}} = 6\pi r \frac{\sqrt{a^2 - 1}}{\ln(a + \sqrt{a^2 - 1})} \tag{5.135}$$

$$C_{\mathrm{D,oblate}} = 6\pi R \frac{\sqrt{1 - a^2}}{a\cos(a)} \tag{5.136}$$

where $r < R$ are the two orthogonal radii of the spheroid and $a = r/R$. See *Baird* and Emsley (1999) or *Clift* et al. (1978) for more complex results for axial and cross-axial orientations of the spheroids.

5.7.11.2. Settling velocity of an aggregate

In the case of aggregates, the settling velocity depends on the aggregate porosity that, for fractal aggregates, changes with the particle size and on the process of the aggregate creation. Small aggregates in natural waters have, on the average, greater density (e.g., *Dyer* and Manning 1999) than do large aggregates. In passing, we note that although the strength of an aggregate increases with the aggregate size, bonds between component particles weaken (*Al* Ani et al. 1991). Thus, a large aggregate may be disrupted more easily than a small one. This indeed favors particle size analysis techniques that are least invasive.

One approach in modeling the settling of aggregates is to simply extend the Stokes approximation by including a porosity term that reduces the effective density difference of the aggregate (e.g., *Bushell* et al. 2002):

$$v_s = \frac{g}{18\nu}\frac{\rho - \rho_w}{\rho_w}(1 - P)D^2 \tag{5.137}$$

where $1 - P$ is aggregate porosity. This is equivalent to assuming that the aggregate is an impermeable porous sphere, very much like the assumption taken by the proverbial physicist describing the shape of horse as a sphere in the first approximation, as amusingly told by *Kerker* et al. (1979).

Experiments suggest convincingly that the shape and permeability effects on the settling velocity of aggregates cause the latter to be substantially greater than that predicted with the Stokes approximation (5.137) (e.g., *Johnson* et al. 1996). Thus, alternative formulations have been proposed, which relate the settling velocity of fractal aggregates to their fractal dimension, d, by a power-law expression of the type (e.g., *Bushell* et al. 2002, *Johnson* et al. 1996):

$$v_s \propto L^{d-1} \tag{5.138}$$

where L is the aggregate size and d is its fractal dimension. Incidentally, relationships of these types can also be used to determine the fractal dimension of the aggregates.

The complexity of the aggregates and variability of the aggregate properties from one to another particle makes the dependence of the settling velocity on the aggregate size a statistical relationship. The average settling velocity tends nevertheless to systematically vary with particle size (e.g., *Hill* et al. 1998). However, the variability of the settling velocity for a given particle size is significant (two to three orders of magnitude, e.g., *Dyer* and Manning 1999), making it problematic the usage of the size dependence of the settling velocity to determine the size distribution of aggregates with a settling tube.

Sedimentation technique have nevertheless been frequently used to characterize aggregates (flocs) in natural waters *in situ* (*Dyer* and Manning 1999, *van Leussen* 1999, *Hill* et al. 1998, *Syvitski* and Hutton 1996, *Syvitski* et al. 1995, *Fennessy* et al. 1994, *Costello* et al. 1989, *Carder* et al. 1982). These methods rely on measuring the settling distance of specific particles, identified on two microphotographs of the same volume of seawater taken with predetermined delay. *Eisma* et al. (1997) has recently provided a review of methods of measuring settling velocity of flocs. Such measurements provide information about the relative density of these particles (i.e., "water content"), as well as the spatial distribution of component particles inside an aggregate, which determine optical properties of the aggregate. We will discuss this topic later in Chapter 6 devoted to the composition and structure of marine particles.

The settling velocity of phytoplankton cells aggregates has also been expressed by a power law (*Jackson* 1990):

$$v_s = a\left(\frac{D}{D_0}\right)^b \tag{5.139}$$

where $a = 0.00248\,\mathrm{m\ s^{-1}}$, $b = 1.17$, the cell diameter D is expressed in cm, and $D_0 = 1\,\mathrm{cm}$ renders the ratio D/D_0 non-dimensional.

5.7.12. Particle size separation with filters and screens

The determination of the PSD by sieving is a common practice in powder technology. This technique has also been utilized in oceanography to determine a coarse PSD by filtration through a cascade of filters with decreasing nominal pore sizes (*Takahashi* and Bienfang 1983, *Azam* and Hodson 1977). However, theoretical and experimental evidence has been accumulated regarding inherently poor size selectivity of the filtration and screening process (*Lee* et al. 1995, *Logan* et al. 1994, *Logan* 1993, *Johnson* and Wangersky 1983, *Sheldon* 1972, *Sheldon* and Sutcliffe 1969). Specifically, a filter of stated pore size collects a significant fraction of particles smaller than that pore size. Thus, this technique should be discouraged when the size distribution data are to be used in optical modeling.

5.8. Aquatic PSD data

5.8.1. Overview

In this section, we will explore the key features of the size distribution of marine particles. We will go beyond the size range which traditionally interests optical oceanographers and limnologists and look at a much larger size range, from colloids (0.01 μm) to large aggregates (100 000 μm = 10 cm). This size range spans seven orders of magnitude. Although particles at the extremes of this range may make negligible contributions to the optical properties of bulk seawater, they may either be directly observable with optical techniques (and thus need models of their optical properties) or be immediate precursors of particles which contribute to the optical properties of seawater.

As noted in earlier reviews (*Stramski* and Kiefer 1991, *Simpson* 1982, *Jerlov* 1976, *Kullenberg* 1974), virtually all such size distributions reported in the scientific literature indicate a rapid decrease of the number concentration of particles greater than a given size with increasing particle size. In the open ocean waters, the cumulative size distribution decreases approximately by a factor of 1/1000 for each 10-fold increase in the particle size. This tendency transcends the limited size ranges of any particle size analysis method, as indicated by generalized abundance data of the various biological species of marine particles (*Stramski* and Kiefer 1991) and combined technique measurements (*Jackson* et al. 1997).

Prior to a wide recognition of the fragility of many suspended particles, especially that of aggregates (flocs), a large body of knowledge of the PSDs in seawater was accumulated by sampling seawater and analyzing samples *in vitro*, mostly with the resistive particle sizing technique (section 5.7.1). If such data originate from a coastal region or estuary, where aggregates are common, the data acquired for the unprocessed samples probably do not represent the naturally occurring particle populations, although the results may still properly account for small aggregates with sizes too small to be broken up (see section 5.7.1.5) in an *in vitro* particle size analysis (e.g., *McCave* 1984). Measurements of the PSD with the Coulter technique led to the hypothesis that essentially equal volumes of particles can be found in logarithmically equal size intervals, from bacteria to whales (*Sheldon* et al. 1972). This is corroborated by measurements of the mass (and volume, if the density is relatively constant) size distributions (e.g., *Syvitski* et al. 1995).

5.8.2. Similarities

A colleague of one of the authors once remarked, after seeing numerous log-log graphs of the size distributions of marine particles, which that author produced from data obtained with a Coulter counter, that something must be wrong with that instrument because they all look very much alike. A physicist by training, he readily appreciated the complexity of marine particle populations and swiftly concluded that the complex processes in question cannot produce size distributions so remarkably similar to each other.

One could argue that he might have been partially right given the breakage of particles by the Coulter counter and other types of particle counters that restrict the sample flow to a narrow filament that passes through the sensing zone, now well documented (we discussed particle breakage in zone-type particle counters in sections 5.7.1.5 and 5.7.2.4). However, even the breakage argument does not explain the similarities between the size distributions of the constituent particles. Other researchers have also been baffled by the similarities between the size distributions (e.g., *Cavender*-Bares et al. 2001).

Data obtained with different techniques and different areas and seasons, covering a broad size range from colloids and viruses (0.01 μm) to zooplankton

(3 000 μm), do indeed paint a remarkably simple picture of these extremely complex particle systems in terms of the global PSDs. An interested reader may want to examine more closely the many sources of data on the PSD in natural waters listed in Table A.4 and Table A.5.

Although an extended discussion of reasons for the similarity of the shapes of the PSD in natural waters is outside the scope of this book, we briefly discuss arguments that explain a remarkable shape uniformity of the PSD of aquatic particles in section 5.8.5.3. This forthcoming discussion is not only stimulated by our curiosity as physicists in the processes that shape the PSD and in similarities between these complex particle populations and other complex systems where energy/mass is dissipated across wide size scales. It is also prompted by more practical considerations. Indeed, models of particle populations may potentially become a part of a unified model of light scattering by marine particles. Such a model could take as an input not the size distribution of the particles but some more basic properties and external conditions of a water body.

The data suggest that most of the variability in the PSD is due to the total particle concentration, not to the form of the PSD (*Jonasz* and Zalewski 1978). This is consistent with observations of a relatively high correlation between the bulk occuring inherent optical properties (such as scattering and attenuation) of naturally occurring suspensions and the particle concentration. Such a high correlation has been widely used (e.g., *Richardson* 1987) to determine the suspended particle concentration from the attenuation or scattering by natural waters.

5.8.3. Variability

Regional variations in the magnitude of the PSD, and lesser variations in its slope (shape), as well as variations with depth were established quite early (*Sheldon* et al. 1972), with various regions/depths having "characteristic" distributions. The PSDs in clear, open ocean waters, and in the mid-depths of the ocean, are generally found to be featureless. In coastal waters, especially those rich in phytoplankton, and in the euphotic zone of the open ocean, distinct features can be found in the PSDs (*Hood* et al. 1991, *Furuya* and Marumo 1983, *Jonasz* 1983a, 1980, *Reuter* 1980b, *Sheldon* et al. 1972). Such features are typically associated with the phytoplankton populations. Sharp variations in the PSDs across an oceanic front, due to variations in the phytoplankton composition, have been documented (*Kahru* et al. 1991). Resuspension of sediment in the nepheloid layer, near the ocean floor, can also result in a structured PSD (*Richardson* 1987, *McCave* 1984, 1983). The largest gradient in the composition and concentration of suspended particles was found near the bottom of the euphotic zone, between 50 and 100 m in the open ocean (*Bishop* et al. 1977).

5.8.4. *Size distributions of particle species in natural waters*

5.8.4.1. *Colloids*

Colloids (customarily assigned a size range of 0.001 to 1 μm) , also referred to as sub-micron particles, constitute a significant component of the particulate matter in natural waters (e.g., *Wells* 1998). Colloids constitute 30 to 50% of the total organic carbon pool in seawater (*Wells* 1998) and provide both a reservoir and a source of material for particle aggregation and disaggregation processes (e.g., *Kerner* et al. 2003 – river water, *Chin* et al. 1998, *Kepkay* 1994—seawater). The colloids size range includes viruses, but we discuss viruses separately (see section 5.8.4.2) mainly because viruses have a different role than colloids in the aquatic food chain.

The recent evaluations of the significant abundance of colloids in marine waters are in sharp contrast to the results obtained for stored GEOSECS samples with SEM (*Lambert* et al. 1981) who did not find many particles with sizes smaller than about 0.2 μm, although they ruled out intermethod differences by running duplicate TEM/SEM analyses.

Studies of the colloids are hampered by the smallness of the colloidal particles. In fact, despite the significance of colloids for aquatic ecosystems, few results regarding the colloid size distributions are available. These few results have been obtained with the TEM (e.g., *Wells* and Goldberg 1993, 1992, 1991, *Harris* 1977—marine colloids), FFF techniques (e.g., *Vaillancourt* and Balch 2000—marine colloids, *Contado* et al. 1997, *Beckett* et al. 1992—inland waters), and resistive particle sizing (e.g., *Chin* et al. 1998, *Longhurst* et al. 1992, *Koike* et al. 1990—marine colloids). *Beckett* et al. (1997) found the qualitative size distributions for clay samples, as determined with FFF to favorably compare with the results of TEM.

The size range of colloid particles straddles the lower limit of the resistive size technique. TEM offers a much higher size resolution but is difficult to use routinely, and the sample preparation required for that technique may modify the sizes and shapes of the particles. FFF techniques offer an interesting but so far largely qualitative alternative, as the determination of the absolute PSD [$n(D)$] with these techniques is difficult (*Vaillancourt* and Balch 2000, see also comments in *Wyatt* 1998 on improvements possible with the use of multiangle light scattering detection for the FFF). This reduces the application of the FFF techniques to the determination of the modal particle size and fractionation of the particles.

As we already noted, few and disparate data are available for the size distribution of aquatic colloids (Figure 5.18) The size distribution of these particles has been approximated with the power-law function (see section 5.8.5.3) with a widely variable slope ($m = 2.65$ to 12—the low values corresponds to a wide size range, while the high values correspond to a narrow size range section of the PSD, Table 5.8 and Table 5.9). The high-resolution PSDs show definite peaked structures (Figure 5.18) with an envelope that seems to be similar to that of a

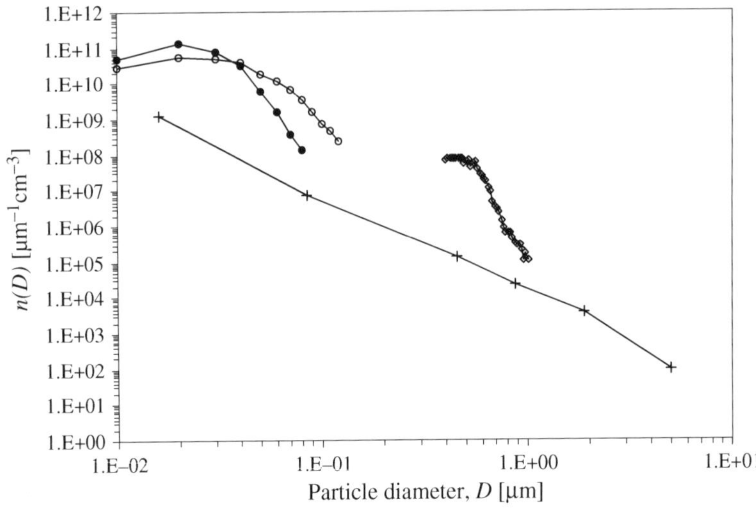

Figure 5.18. Sample PSDs of aquatic colloids. Note the structure apparent from the high-resolution size distribution measurements (open and solid circles—*Wells* and Goldberg 1992, Fig. 5b and c, transmission electron microscopy (TEM), open diamonds—*Longhurst* et al. 1992, Fig. 1, 27 Sep 1990 Elzone counter, coastal western Atlantic off Halifax, NS, Canada, 10 m). Qualitative size distributions obtained with field flow fractionation (e.g., *Vaillancourt* and Balch (2000) also show similar peaked structure in the 0.001 to 1 μm diameter range. The slope of a power-law "envelope" (not shown) of the high-resolution PSDs is similar that of the wide range PSD obtained by *Harris* (1977; crosses, an average of six PSDs from the Gulf of Mexico, TEM, 600 to 3600 m). The data of *Wells* and Goldberg (1992) are presented as functions of the minimum linear diameter.

wide size range PSDs obtained by *Harris* (1977). Qualitative size distributions obtained with FFF techniques show similarly peaked structures (e.g., *Vaillancourt* and Balch 2000—seawater, *Contado* et al. 1997—river water) in the 0.001 to 1 μm diameter range and just below 1 μm (*Yamasaki* et al. 1998), especially in productive waters.

5.8.4.2. Viruses

Viruses position themselves in the particle size range of colloids (0.001 to 1 μm), but their size distributions are much more narrow (few tens of nm) than those of the inanimate colloids. Despite significant interest in the role of viruses in aquatic (e.g., a review by *Wommack* and Colwell 2000, *Bratbak* et al. 1990) and other ecosystems (virus databases: National Institute of Health, USA, Dr. Cornelia Büchen-Osmond, http:/www.ncbi.nml.nih.gov, and the School of Biosciences of the Australian National University, Canberrea, Australia,

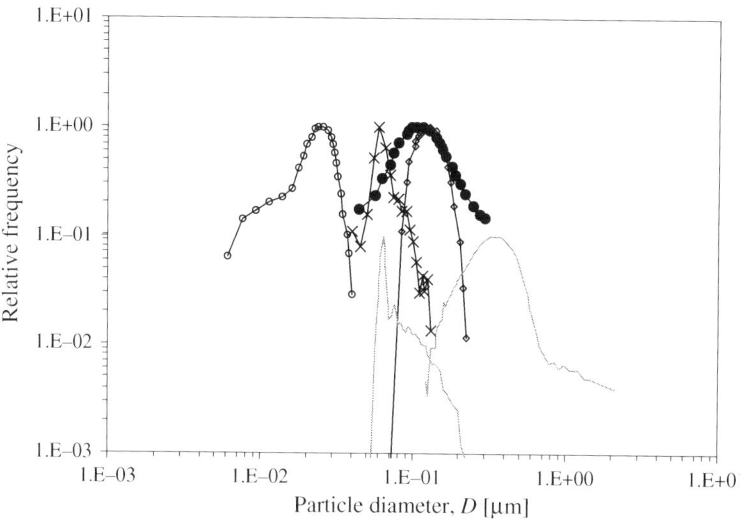

Figure 5.19. Sample PSDs of viruses (bacteriophages): solid circles—C2 virus, *Siphoviridae* (*Balch* et al. 2000, Fig. 4), open circles—MS2 virus, *Leviviridae* (*Balch* et al. 2000, Fig. 4), open diamonds—T4 virus, *Myoviridae* (*Balch* et al. 2000, Fig. 4), crosses—*Stramski* et al. (2001, Fig. 9, VIRU, after *Ackerman* and Dubow 1987, transmission electron microscopy). The data of *Balch* et al. (2000) were obtained with flow field fractionation (FFF). The size distributions of monomeric and aggregated adenoviruses (left and right gray curves respectively, *Bondoc* and Fitzpatrick 1998, Fig. 1b and d, photosedimentation disk centrifuge) are shown for comparison, scaled to 0.1 × actual frequency for convenience. Both FFF and photosedimentation techniques yield the hydrodynamic particle diameter.

http://life.anu.edu.au/viruses/), much attention is devoted to enumeration and sizing of viruses rather than to the determination of the size distribution of virus particles.

A few qualitative data (Figure 5.19), obtained with FFF (*Balch* et al. 2000—marine viruses, *Alonso* et al. 2002—sparse histogram data, not shown), TEM (*Ackermann* and Dubow 1987; as shown by *Stramski* and Kiefer 1991—a bacteriophage) and disk centrifuge photosedimentation (*Bondoc* and Fitzpatrick 1998—adenoviruses), indicate that the size distributions of single virus particles are rather narrow (several tens of nm) and center around few tens of nanometers. Bondoc and Fitzpatrick documented the effects of aggregation of adenovirus particles on the PSD (Figure 5.19). These effects may range from the appearance of distinct multiplet peaks to a very wide size distribution that does not represent at all the PSD of the virus particle monomers. The comparison of the relative widths of the aggregated virus data of Bondoc and Fitzpatrick with those of the data reported by *Balch* et al. (2000) would suggest significant contribution of aggregates in samples represented by these latter data.

Wommack and Colwell (2000) compiled and tabulated abundances and sizes of the viruses [viral capsids (heads)]. The abundances range from 0.003×10^6 to $14.9 \times 10^6\,\mathrm{cm}^{-3}$ in the open ocean waters, with capsid diameters ranging from 30 to 60 nm. In coastal waters, the abundance increases to between 0.005×10^6 and $460 \times 10^6\,\mathrm{cm}^{-3}$ with capsid size in the same size range. Based on these data, the $\Delta N/\Delta D$ at an open ocean North Atlantic site evaluates to $14.9 \times 10^6/0.03\,\mathrm{cm}^{-3}\,\mu\mathrm{m}^{-1} = 5 \times 10^8\,\mathrm{cm}^{-3}\,\mu\mathrm{m}^{-1}$ (*Bergh* et al. 1989) and at a location in the coastal zone of the Arctic Ocean to a value of $130 \times 10^6/0.02\,\mathrm{cm}^{-3}\,\mu\mathrm{m}^{-1} = 6.5 \times 10^9\,\mathrm{cm}^{-3}\,\mu\mathrm{m}^{-1}$ (*Maranger* et al. 1994). Most aquatic viruses are smaller than ~60 nm (*Bergh* et al. 1989), larger viruses (~100 to 150 nm) make < 10% of the population (*Bratbak* et al. 1992). These latter authors also report virus-like particles with capsids as large as 400 nm and tails as long as ~2 μm.

The concentration of viruses tends to covary with the concentration of their "prey," mostly bacteria (e.g., *Maranger* and Bird 1995), both in freshwater (except the data from 22 lakes in Quebec, Canada) and in marine environments in the enphotic zone:

$$\log N_V = 0.95 + 1.01 \log N_B \qquad (5.140)$$

with $r^2 = 0.72$ for 176 samples, with $N_V[\mathrm{cm}^{-3}]$ representing the total concentration of the particles. Chlorophyll concentration was found to be a marginally better predictor of the virus concentration ($r^2 = 77$ for 101 samples):

$$\log N_V = 7.08 + 0.80 \log chl\ a \qquad (5.141)$$

where *chl a* [μg dm^3] is chlorophyll *a* concentration.

5.8.4.3. Bacteria

Bacteria occur in natural waters in two populations: "attached" and "free-living." The "attached" population is of less importance for modeling of the optical properties of natural waters, hence we concentrate on the discussion of the "free-living" bacteria. We note, however, the choice of habitat (a surface or the water body) is not permanent. Bacteria can oscillate between the two modes of living. The free-living bacteria account for the majority of bacteria living in seawater (*Azam* and Hodson 1977).

Due to the considerable role of bacteria in aquatic ecosystems, the bacterial populations have been extensively researched. See a recent concise note by *Azam* and Worden 2004, as well as comprehensive reviews by *Li* et al. (2004), *del Giorgio* and Cole (1999), and *Cole* et al. (1988). The concentration of free-living bacteria with sizes between 0.2 and 2 μm is reported to be on the order of 10^3 to $10^7\,\mathrm{cm}^{-3}$ (*Cho* and Azam 1990, *Olson* et al. 1985, *Johnson* and Sieburth 1982), making these organisms a significant contributor to the optical properties of natural waters (e.g., *Morel* and Ahn 1991, *Stramski* and Kiefer 1991, *Kopelevich* et al.

1987). Among the species, *Prochlorococcus* is the dominant contributor to the concentration of bacteria, being as much as 30 times more abundant in surface waters than *Synechococcus* (e.g., *Binder* et al. 1996). The concentrations of *Synechococcus* vary from 10^4 to 10^5 cells per cm^3 in the rich waters of the Arabian Sea, and off the coast of Peru, to $10^3\,cm^{-3}$ north of the Gulf Stream (e.g., *Waterbury* et al. 1979).

The single-cell PSDs of bacteria are relatively narrow, with the full-width-at-half-maximum on the order of several percent. Sample single-cell size distributions obtained for cultured and naturally occurring aquatic bacteria are shown in (Figure 5.20) and Figure 5.21. Please refer to Table A.5 for a list of sources of information on the size distributions of aquatic bacteria. Although, from a global view point, a species-specific size distribution can be formally defined (e.g., *Rinaldo* et al. 2002), the variability of individual realizations of a species

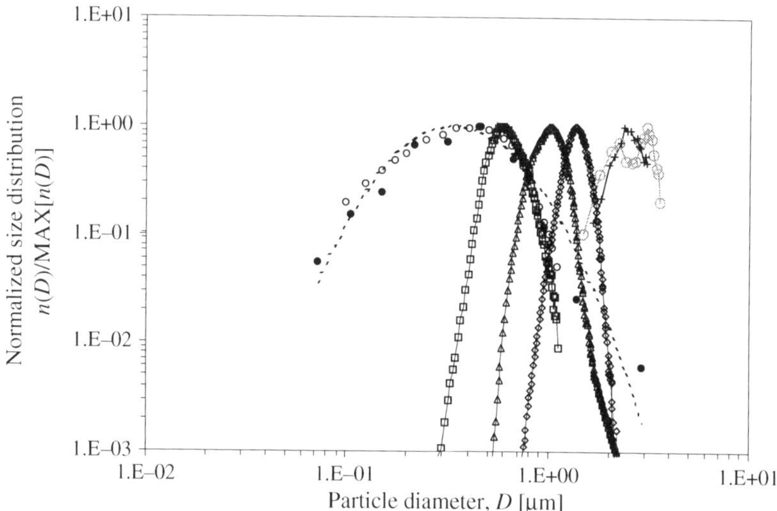

Figure 5.20. Sample PSDs of aquatic bacteria. *Freshwater bacteria* (rightmost open circles and pluses—*Chromatium* spp., Lakes Villar and Cisco near Barcelona, Spain, *Montesinos* et al. 1983). *Marine bacteria*: solid circles and dashed line—data and log-normal fit [un-normalized-fit coefficients: $n_{max} = 1.849 \times 10^8\,cm^{-3}\,\mu m^{-1}$, $D_{peak} = 0.343\,\mu m$, $\sigma = 0.4$, equation (5.169)] for a culture of free-living cylindrical heterotrophic bacteria isolated from seawater off Bermuda by Dr. J. Fuhrman (*Jonasz* et al. 1997), open circles—elongated heterotrophic, free-living bacteria culture created from coastal samples off San Diego, CA, USA (*Stramski* and Kiefer 1990), open squares—*Prochlorococcus*, and triangles—*Synechococcus* sp. (*Morel* et al. 1993; as reported by *Stramski* et al. 2001), open diamonds—a mixture of *Synechocystis* and *Anacystis marina* (*Ahn* et al. 1992, as reported by *Stramski* et al. 2001). All three latter species have nearly spherical cells. In all cases, the diameter is the equivalent spherical diameter.

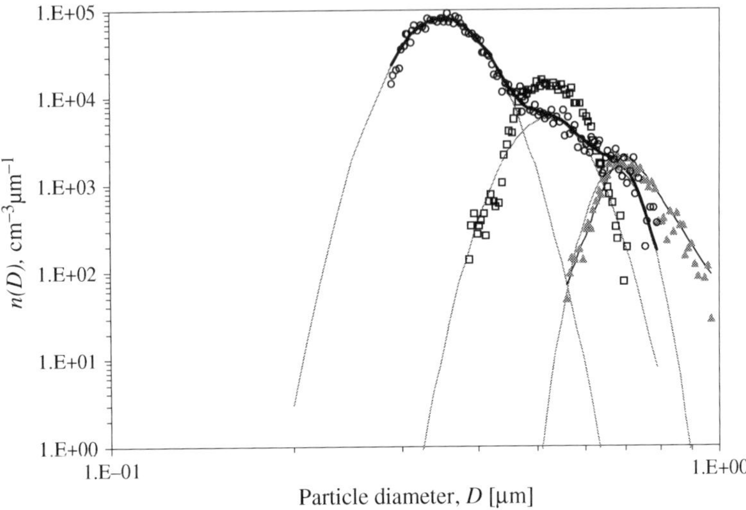

Figure 5.21. Sample PSDs of marine bacteria from the Southern Sargasso Sea as determined with a flow cytometer *in vitro* [*Cavender*-Bares et al. 2001, Fig. 5E, numerical $v(V)$ data kindly provided by Dr. Cavender-Bares were converted to the $n(D)$ format by using equation (5.48)]: *bacteria* {open circles—data, thick solid black curve and the three thin gray curves indicate respectively the sum of and the individual log-normal components, as isolated by the LND software (MJC Optical Technology) using the least-squares fit, equations (5.177) and (5.178), coefficients (B_0, B_1, B_2) of the components are respectively (−10.35, −67.18, −73.96), (−3.38, −50.55, −89.61), and (−3.06, −72.96, −214.1), left-to-right}, *Prochlorococcus* {open squares} and *Synechococcus* {full triangles—data, thin black rightmost curve—a trial-and-error hyperbolic fit, equation (5.164), $\alpha = 15.11$, $\beta = 4.061$, $\delta = 0.02$, $D_0 = 0.69$, $k = 3.6$}. Note that the peaks of the two rightmost log-normal components of bacteria nearly coincide with those of *Prochlorococcus* (open squares) and *Synechococcus* (gray full triangles). In all cases, the diameter is the equivalent spherical diameter.

PSD is significant (mean size varying by a factor of up to ~10, Figs. 2b and c in *Rinaldo* et al. 2002), similarly as in the case of phytoplankton (see section 5.8.4.4).

Although we found the log-normal distribution (see section 5.8.5.6) to represent relatively well the size distribution of heterotrophic bacteria (Figure 5.20), it fared poorer in the case of the bacterial cultures. We noted a similar tendency in the case of the gamma distribution (see section 5.8.5.8), advocated by *Ulloa* et al. (1992), who determined the size distributions using epifluorescence microscopy combined with image analysis for samples treated with 2% formaldehyde. They obtained the average values of a and b coefficients of the gamma distributions,

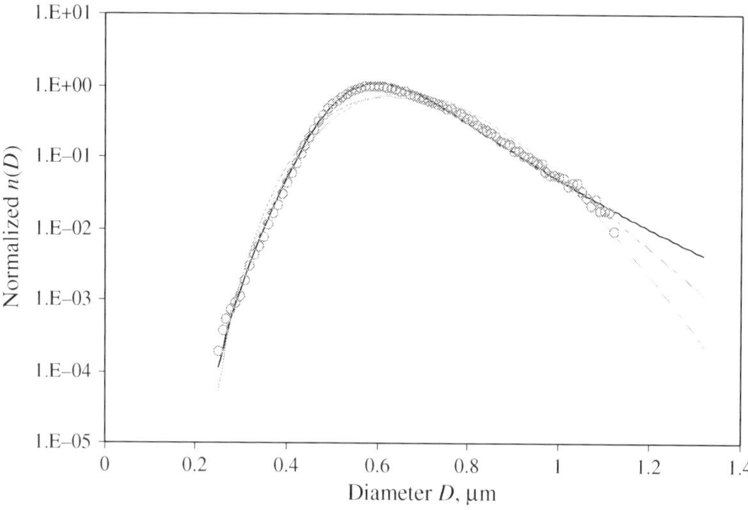

Figure 5.22. Comparison of selected approximations of the experimental size distribution of *Prochlorococcus* (open gray circles – *Morel* et al. 1993; as reported by *Stramski* et al. 2001): hyperbolic distribution [trial-and-error fit, black line overlaying most data points, eq. (5.163), $\alpha = 11.7$, $\beta = 2.27$, $\delta = 0.105$, $D_0 = 0.57$, $k = 1.25$], log-normal distribution [least squares fit, gray line, eq. (5.178), $B_0 = 1.262$, $B_1 = -10.75$, $B_2 = -26.13$], and gamma distribution [least squares fit, light gray line, eq. (5.189), $c = 6.25 \times 10^{15}$, $a = 25.7$, $b = -39.3$].

equation (5.189), of 7.26 and 14.5 respectively, as calculated from 32 original data sets. We found their results to show a weak correlation ($r^2 = 0.866$, $n = 32$):

$$b = 1.69 + 1.76a \tag{5.142}$$

We found the hyperbolic distribution (see section 5.8.5.4) to best approximate the sample size distribution of some cultured and naturally occurring populations (e.g., see a fit to the *Prochlorococcus* size distribution in Figure 5.22, and a fit to the *Synechococcus* size distribution in Figure 5.21). However, a reasonable fit was also obtained with a sum of log-normal components (see section 5.8.5.6) for the size distribution of naturally occurring bacteria (Figure 5.21).

Aggregates of bacteria are frequently observed, both in natural water samples (e.g., *Albertano* et al. 1997) and in cultures (e.g., *Jonasz* et al. 1997). The size of the aggregates may easily exceed the size of the component cells by more than an order of magnitude.

As opposed to viruses whose size and shape is largely dictated by molecular assembly rules, the size of a given species of bacteria depends on the habitat. For example, *Cole* et al. (1993) found the average cell diameter in anoxic lakes to be systematically greater by a factor of 1.3 to over 2 than that in oxic lakes.

Some observations (*Binder* et al. 1996) indicate that the mean size—as measured via forward light scattering with a flow cytometer—may increase with depth in the ocean by as much as 2.5 times (for *Prochlorococcus*) in the surface layer. A similar but much smaller increase has also been observed for *Synechococcus* and heterotrophic bacteria (*Binder* et al. 1996). The size of bacterial cells is also mediated by predation (e.g., *Jürgens* and Matz 2002). It is interesting to note in this context that the size distributions of cultures of heterotrophic marine bacteria isolated from different water bodies (Figure 5.20) seem remarkably similar.

Given the motility of some bacterial cells, the average concentration in a sample may not be indicative of those in smaller volumes within it (e.g., *Blackburn* et al. 1998). A recent review of the effects and modes of avoidance of predation on aquatic bacteria populations (*Jürgens* and Matz 2002) indicates that the typical swimming speed of bacteria may reach $140\,\mu\mathrm{m\,s^{-1}}$. The active movement of bacteria may also interfere with measurements of the dynamic light scattering (e.g., *Boon* et al. 1974, see also section 5.7.10). Some cells are chemotactic and may actively seek the sources of nutrients (*Blackburn* et al. 1998, *Bowen* et al. 1993). Simulations by *Bowen* et al. (1993) predicted bacterial concentrations orders of magnitude greater within a several μm-thick layer surrounding the cell, than those in ambient seawater. Interestingly, this raises a question of the effective size of phytoplankton cells as determined by optical methods.

Utilization by bacteria of products of phytoplankton cells or of their decay makes the concentration of bacteria covary with that of phytoplankton both in freshwater and in marine environments ($r^2 \sim 0.6$, e.g., *Cole* 2000, *Brett* et al. 1999, *Naganuma* 1997, Cole et al. 1988, *Bird* and Kalff 1984) although in specific cases this relationship may be weak (e.g., *Li* et al. 2004, *Tada* et al. 1998) due to the variety of the coupling paths between bacteria and phytoplankton. The utilization of the phytoplankton products by bacteria appears to depend also on the availability of nutrients such as N and P (*Le* et al. 1994).

Cole et al. (1988) analyzed the results published to-date for 54 freshwater and marine ecosystems and obtained the following correlation ($r^2 = 0.6$) between the bacterial, P_B, and phytoplankton production, P_P, both in $\mu\mathrm{gC\ l^{-1}d^{-1}}$:

$$\log P_\mathrm{B} = -0.483 + 0.814 \log P_\mathrm{P} \tag{5.143}$$

and, in terms of the number concentration, N_B [$\mathrm{cm^{-3}}$] of bacteria vs. chlorophyll *a* concentration in $\mathrm{mg\,m^{-3}}$:

$$\log N_\mathrm{B} = 5.97 + 0.53 \log[chl\_a] \tag{5.144}$$

with $r^2 = 0.75$. The slope of their regression is similar to those obtained by other researchers they quote (*Aizaki* et al. 1981, slope = 0.63, Japanese lakes; *Bird* and Kalff 1984, slope = 0.57, Quebec lakes) but differs from that reported in a literature review by *Bird* and Kalff (1984, slope range of 0.78 to 0.84).

Li et al. (2004) have recently analyzed a large dataset (>13 000 data points) of paired measurements of the chlorophyll concentration and bacterial concentration (excluding *Synechococcus*) in the ocean. They found that these variables were log-normally distributed with geometric medians of 0.28 mgChl m^{-1} and $4.5 \times 10^{11}\,m^{-3}$. The maximum chlorophyll-specific bacterial concentration of $\sim 7 \times 10^{12}\,m^{-3}\,mgChl^{-1}$ was established in that study, indicating equal—at that limit—contributions of bacteria and phytoplankton to the marine carbon pool. The slope of the power-law relationship between the chlorophyll concentration (independent variable) and bacterial concentration was 0.46 ± 0.09 for the whole dataset. Li et al. also noted a substantial effect of temperature on the ratio of the bacterial concentration to the phytoplankton chlorophyll concentration, with the ratio increasing with temperature.

5.8.4.4. Phytoplankton

Phytoplankton span an extensive size range of ~1 to ~1000 μm populated by some 4000 marine species and on the order of 15 000 freshwater species (*Falkowski* et al. 2003, *Sournia* et al. 1991). This range has been customarily subdivided into pico- (<2 μm), nano- (2 to 20 μm), and microplankton (>20 μm).

The size distributions of individual plankton species are relatively narrow, with full-widths-at-half-maxima on the order several tens percent, similar to those of bacteria. This similarity, extending to other components of aquatic ecosystems, such as zooplankton and fish, is captured in the model of the size structure of aquatic ecosystems (*Thiebaux* and Dickie 1993) referred to in more detail in section 5.8.5.6. Sample size distributions of aquatic phytoplankton species are shown in Figure 5.23.

Please refer to Table A.5 for sources of information on the size distribution of phytoplankton. The cell size ranges representative of phytoplankton species are available in many textbooks on the subject (e.g., *Pickett*-Heaps 1975, *Drebes* 1974). A concise summary for the species typical of the eastern Atlantic can be found in *Kronfeld* (1988). The size distributions of individual species of phytoplankton can be well approximated (Figure 5.24.) with the log-normal distribution (see section 5.8.5.6), or the hyperbolic distribution (see section 5.8.5.4).

Phytoplankton can exist not only in the single-cell state but also as colonies of single cells. The sizes of these colonies can reach into mm to cm size range, dwarfing the size ranges of the individual cells. A less spectacular example is shown in Figure 5.23.

The size distribution of phytoplankton cells, like that of bacteria, depends on the habitat conditions and the cell life phase. In early experiments using flow-cytometric analysis of cultured population of *Dunaliela tertiolecta*, *Ackleson* et al. (1988a) showed reversible doubling of the mean cell volume. Similar effects of the diurnal cycle, temperature (Figure 5.25), and nutrient availability were discovered by *Stramski* et al. (2002), *Reynolds* et al. (1997), and *Stramski* and Reynolds (1993) for *Thalassiosira pseudonana*. Similar results were obtained

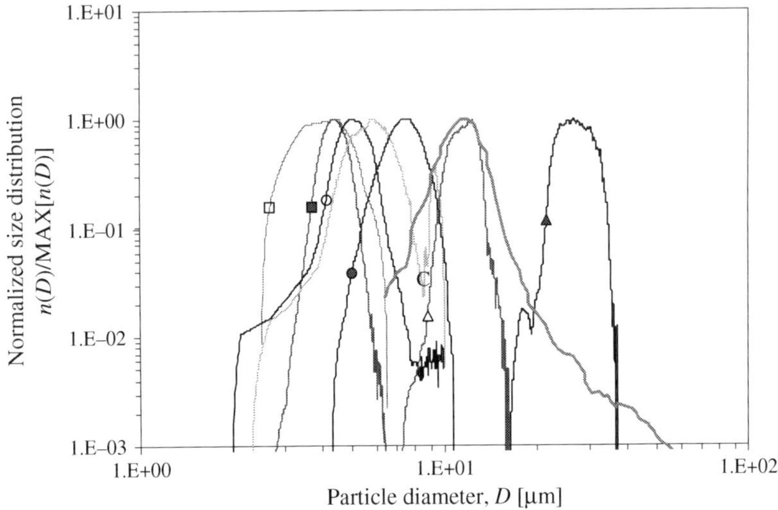

Figure 5.23. Sample PSDs of marine phytoplankton. Left-to-right (at the maxima): *Pavlova lutheri* ($D_{avg} = 4.26\,\mu m$, open square, *Bricaud* et al. 1988), *Isochrysis galbana* ($D_{avg} = 4.45\,\mu m$, solid square, *Ahn* et al. 1992), *Porphiridium cruentum* ($D_{avg} = 5.22\,\mu m$, open circle, *Bricaud* et al. 1988), *Prymnesium parvum* ($D_{avg} = 6.41\,\mu m$, light gray line without a symbol, *Bricaud* et al. 1988), *Dunaliella tertiolecta* ($D_{avg} = 7.59\,\mu m$, solid circle, *Stramski* and Reynolds 1993), *Hymenomonas elongata* ($D_{avg} = 11.77\,\mu m$, *Ahn* et al. 1992), *Prorocentrum micans* ($D_{avg} = 27.64$, *Ahn* et al. 1992). All the above size distributions correspond to those shown in Fig. 9 of *Stramski* et al. (2001, numerical data: courtesy of D. Stramski). A wide range size distribution of colony-forming *Microcystis aeruginosa* (thick gray line with no symbol, *Dubelaar* and van der Reijden 1995, calculated from a three-point moving average of the volume size distribution shown in their Fig. 4b) is shown for comparison. The average diameters are as listed in Table 1 of *Stramski* et al. (2001). The particle diameter is here the equivalent spherical diameter.

for *Nannochloris* sp. (*DuRand* and Olson 1998) where diel variations of the cell diameter (~2 to ~3 μm) on the order of 50% were observed. Thus, one can expect significant intraspecies modifications of the PSD of a phytoplankton species.

The composite size distributions of the phytoplankton in aquatic ecosystems (e.g., *Stramski* et al. 2001) can be approximated by a power law kD^{-m}, with the slope m of about 4 [which represents the "constant biomass density" hypothesis (see section 5.8.5.3)] as shown in Figure 5.32. In fact, given limited resources and size-dependent rate of their assimilation by living organisms (e.g., *Cyr* 2000, *Chisholm* 1992), phytoplankton, as well as other components of aquatic and terrestrial ecosystems, have limited maximum abundances that are inversely proportional to the organism volume (mass) (*Belgrano* et al. 2002, *Agusti* and Kalff

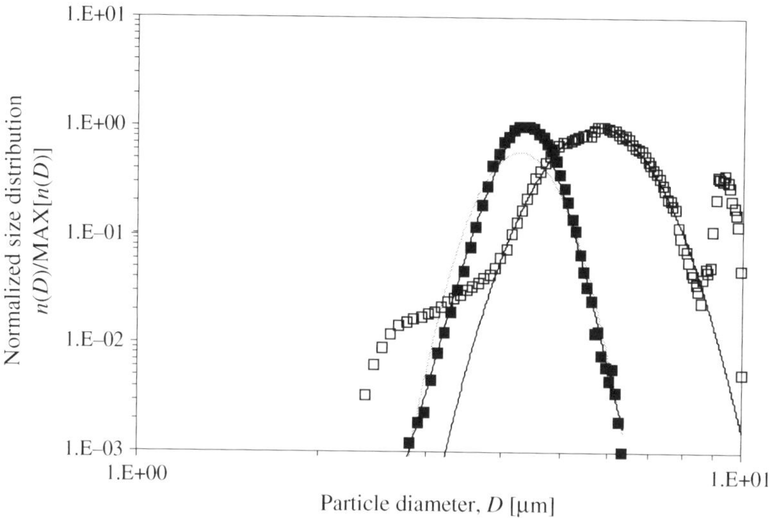

Figure 5.24. PSDs of marine phytoplankton can be approximated either by the hyperbolic function [equation (5.163)] as in the case of *Isochrysis galbana* (experimental data of *Ahn* et al. 1992—*solid squares*; trial-and-error fit—*black curve*, $\alpha = 22.56$, $\beta = -1.56$, $\delta = 0.056$, $D_0 = 4.42$ µm, $k = 1.27$) or by the log-normal function [equations (5.177) and (5.178)] as in the case of *Prymnesium parvum* (experimental data of *Bricaud* et al. 1988—*open circles*, log-normal fit—*black curve* of the central mode, $D_{min} = 4$ µm, $D_{max} = 8$ µm, $B_0 = -28.13$, $B_1 = 74.03$, and $B_2 = -48.73$). Note that the least-squares log-normal fit (light gray curve) does not work well for *I. galbana*. All the above size distributions correspond to those shown in Fig. 9 of *Stramski* et al. (2001, numerical data: courtesy of D. Stramski) except that not all data points are shown here for clarity. The particle diameter is the equivalent spherical diameter.

1989, *Witek* and Krajewska-Soltys 1989, *Agusti* et al. 1987, *Duarte* et al. 1987). *Sabetta* et al. (2005) points out that the shape of the size distribution of phytoplankton (in the 5 to 1000 µm size range in their study) is poorly correlated with the species composition. The relative invariance of the phytoplankton PSD, is—they argue—a consequence of substitution of one species for another of the same size range when the environmental conditions or simply the species succession so dictates.

Sample *in situ* PSDs of marine and lacustrine phytoplankton are shown in Figure 5.23 and Figure 5.26. For other representative size distributions of phytoplankton and other aquatic organisms in wider size ranges, please see Figure 5.32. It is clear that the size distribution of phytoplankton as a community is a linear combination of the size distributions of individual species, with magnitudes conforming to the "maximum biomass density" conjecture.

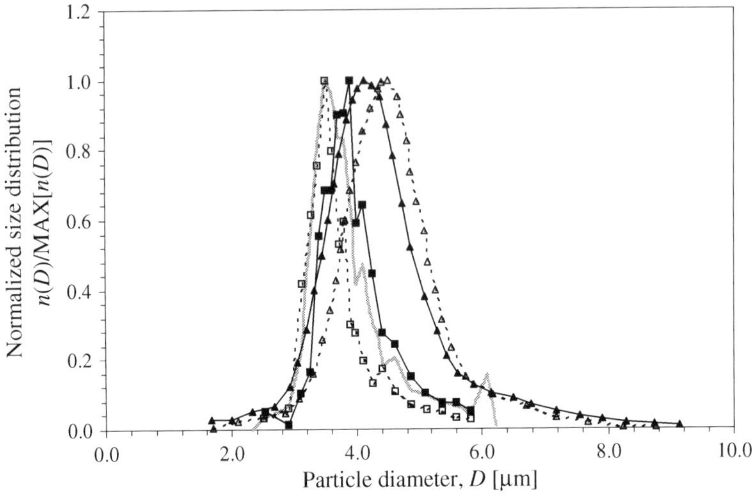

Figure 5.25. The size distribution of a phytoplankton species depends to a significant degree on the habitat conditions as shown by this comparison of the "model" size distribution of *Thalassiosira pseudonana* (thick gray line, *Stramski* et al. 2001) with size distributions of the same species obtained in different habitat conditions: time of day—05:00 hours (open squares and dashed line) and 10:00 hours (solid squares and line, both reported by *Stramski* and Reynolds 1993, their Fig. 3), and temperature: 7°C (open triangles and dashed line) and 25°C (solid triangles and line, both reported by *Stramski* et al. 2002, their Fig. 4). The changes are not random. The particle diameter is the equivalent to the spherical diameter.

In general contrast with the shapes of marine bacteria and viral capsids, the shapes of phytoplankton cells and their colonies range widely from nearly spherical (e.g., *Chlorella* sp.) to extremely elongated needle-shaped cells (e.g., *Nitzschia* sp.) and colonies (e.g., *Anabena flosaquae*). We discuss this topic in more detail in Chapter 6. This diversity of particle shapes implies caution when applying the size distributions based on the ESD to optical modeling. In the case of non-spherical shapes, the size distributions would need to be converted to the projection area-based form as described in section 5.5.2.2.

5.8.4.5. Detritus and aggregates

Detritus is commonly understood as the non-living organic component of the aquatic particle population. Detrital particles range from single, few nanometer-sized grains of organic matter generated via spontaneous aggregation from the dissolved organic matter (e.g., *Chin* et al. 1998) to complex organo-mineral aggregates (flocs) in the millimeter size range (e.g., *Riebesell* 1991).

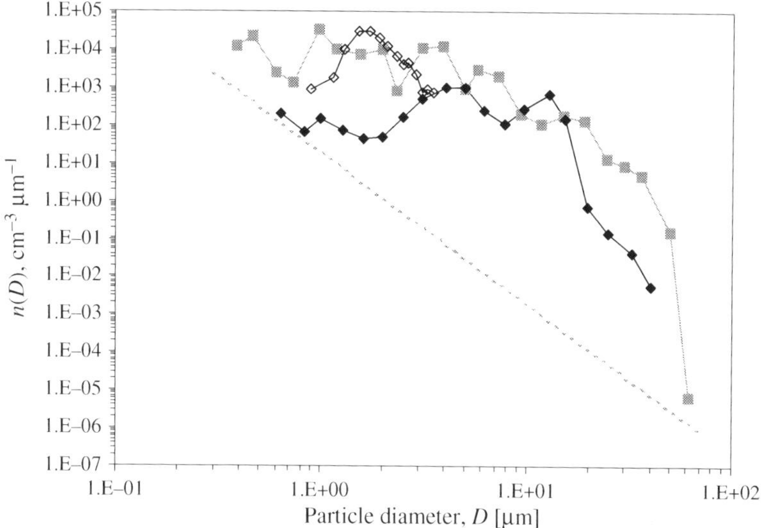

Figure 5.26. Sample absolute size distributions of lacustrine and marine phytoplankton: Lake Arendsee, Northern Germany (gray solid squares and line, *Tittel* et al. 1998, converted from the biomass spectrum of the autotrophs in their Fig. 2e), coastal western Atlantic, surface, spring 1997 (black open diamonds and line, *Green* et al. 2003a, Fig. 2A, spring), and Gerlache Strait, Antarctica, 5 m, December 1995 (solid black squares and line, *Rodríguez* et al. 2002; the peak at about 12 µm represents large flagellate *Cryptomonas*). The particle diameter is the equivalent spherical diameter. The dashed gray line represents a $n(D) \sim D^{-4}$. The wavy size distributions illustrate the effect of combining narrow distributions of the individual species with absolute magnitudes conforming to the "maximum biomass density" conjecture.

Given their vast size range, a global view size distribution of these particles can only be obtained by combining data obtained with several particle sizing techniques. Due to the extremely delicate nature of the large aggregates and commensurate problems with sampling (see section 5.6.1), reliable data on the size distribution of these particle can only be obtained by *in situ* microphotography.

This does not mean that all aggregates are separated on sampling into constituent particles. *Eisma* et al. (1983) note that even pronounced sonification did not appear to disaggregate all particles. Only complete removal of organic matter by wet oxidation (by H_2O_2) achieved that goal.

Until recently, the only method of identifying detrital particles was through optical microscopy (e.g., *Lenz* 1972). However, the availability of fluorescent stains identifying live and dead organic matter (e.g., *Green* et al. 2003a, 2003b) opens a possibility for routine determinations of the size distribution of these particles with an optical flow cytometer, albeit in a limited size range and with reservations due to a number of fit factors and corrections in their method. Nevertheless, given the large settling velocity in water of single-mineral particles with sizes exceeding several tens of micrometers, one can safely conclude that particles in that size range (except in high-energy surf zones) are all either living phytoplankton, zooplankton, or detrital aggregates—which can incorporate mineral grains. At the small end of the size range, the detrital particles with nanometer sizes in water are likely to be mostly generated by aggregation of dissolved organic matter (e.g., *Chin* et al. 1998), i.e., they are organic in nature.

In fact, the "live-dead" dichotomy is somewhat fuzzy, given that detrital aggregates may contain live bacteria and phytoplankton cells. Many observations *in situ* point to a wide variety of particles included into aggregates. *Kranck* et al. (1992) found in a quantitative study that wet oxidization of sampled suspensions in a wide range of high-sediment load aquatic environments (Nith River, NS, Canada, Amazon River, Brazil, and San Francisco Bay, USA) produces volume size distributions $dV/dlogD$ that have a sharp cut-off at a particle diameter that is about the same as the small-size cut-off of *in situ* PSD determined via microphotography. This, they conclude, indicates that almost all particles examined *in situ* were aggregates. It also illustrates that the aggregates consists of an organo-mineral mixture, with the organic part playing an important role as a "glue" holding mineral grains together.

The aggregates exist in all parts of the whole size range of detrital particles, from nanoparticles (e.g., *Wells* and Goldberg 1992) through the centimeter size range, where they are referred to as marine snow (e.g., *Alldredge* and Silver 1988). An important class of aquatic aggregates are the transparent exopolymer particles (TEP) (*Passow* 2002) that were first identified by *Wiebe* and Pomeroy (1972). These "invisible" particles, formed from strands of polysaccharides exuded by phytoplankton cells, facilitate the aggregation of particles in water. Indeed, it is well documented (e.g., *Passow* et al. 2001, *Passow* and Alldredge 1994, *Kranck* and Milligan 1988) that the abundance of aggregates increases significantly near the ends of phytoplankton blooms. DNA-specific stains such as DAPI permitted to identify another class of detritus particles: the DAPI yellow particles (DYP, e.g., *Mostajir* et al. 1998, 1995a, 1995b) generally found in the $< 10\,\mu m$ size range. The origin of these particles is unclear.

The existence of large and fragile detrital particles (marine snow) which can reach several millimeter in size has been well known for a long time through direct observations either by divers or from submersibles (*Nishizawa* et al. 1954) and more recently *Syvitski* et al. (1983) and through *in vitro* analysis of carefully sampled seawater (e.g., *Eisma* et al. 1983). However, early reports of such particles were either qualitative or focused on the biology and chemistry of these particles (see a review by *Alldredge* and Silver 1988) and did not receive much attention by the marine optics community. Only the recent proliferation of *in situ* microphotography combined with automated image analysis permitted accumulation of quantitative data on the size distribution and other properties of such particles and allowed a preliminary assessment of their optical effects (*Hou* et al. 1997, *Carder* and Costello 1994). The potential significance of these large particles to underwater visibility, remote sensing, and to the radiation balance of the Earth is yet to be fully assessed.

As we have already alluded to, the form and composition of aquatic aggregates are very diverse, ranging from all-mineral flocs to complex organo-mineral flocs with rich bacterial population and other living organisms (e.g., *Simon* et al. 2002, *Alldredge* and Silver 1988). A study of aggregation of diatoms in the open sea, conducted with a scanning electron microscope (*Buck* and Chavez 1994), revealed that diatom aggregates may constitute on the order to 10% of the autotrophic biomass. Considerable chemical gradients occur near the flocs. For example, the concentrations of ammonia in marine snow exceed by several orders of magnitude that in the surrounding volume of seawater. These gradients may serve as beacons for chemotactic bacteria looking for food (*Blackburn* et al. 1998). A settling floc may leave a long trail of dissolved organic substances that can be used by zooplankton to actively pursue the sinking particle (*Kiørboe* 2001).

The available data on the concentration of the large particles of marine snow indicate a wide range of that concentration: from 0.001 to $1\,\text{cm}^{-3}$ in the upper few tens of meters of the ocean and from 1×10^{-6} to $0.001\,\text{cm}^{-3}$ in the deep ocean (*Alldredge* and Silver 1988). The interested reader may want to consult *Simon* et al. (2002) who compiled aggregate abundance data in many water bodies. The few data on the size distributions of the large particles, derived from *in situ* microphotography (*Chen* et al. 1994, *Courp* et al. 1993, *Kranck* et al. 1992, *Kranck* and Milligan 1992, *Eisma* et al. 1991a), follow generally the power-law trend of the PSD (Figure 5.27), but its slope ($m \approx 1.3$) is much smaller than that of the *in vitro* data and the single-particle trend of the generalized size distribution ($m \approx 4$, e.g., *Stramski* and Kiefer 1991) which represent marine microorganisms.

Jackson and Burd (1998) recently reviewed aggregation in the marine environment and its effect on the size distribution of marine particles. Note that the process of aggregation is not unique to the marine environment and has considerable industrial and analytical significance. For example, clinical immunoassay instruments employ chemically induced aggregation at the size scale of protein molecules (10 nm) to detect antigen proteins in human blood (e.g., *Singer* et al. 1973).

The mechanism of aggregation varies depending on the particle size range (*McCave* 1984, *Hunt* 1982, 1980). However, all aggregation processes have this in common: before particles can aggregate they have to come within a small distance of each other, i.e., they have to collide. Small aggregates ($\sim 1\,\mu\text{m}$) are mostly formed via Brownian coagulation. Large aggregates of particles are formed through two processes important in this size range: shear coagulation and differential settling. With shear coagulation, the collisions are caused by the non-uniform distribution of the velocity of fluid: particles within the layers of fluid

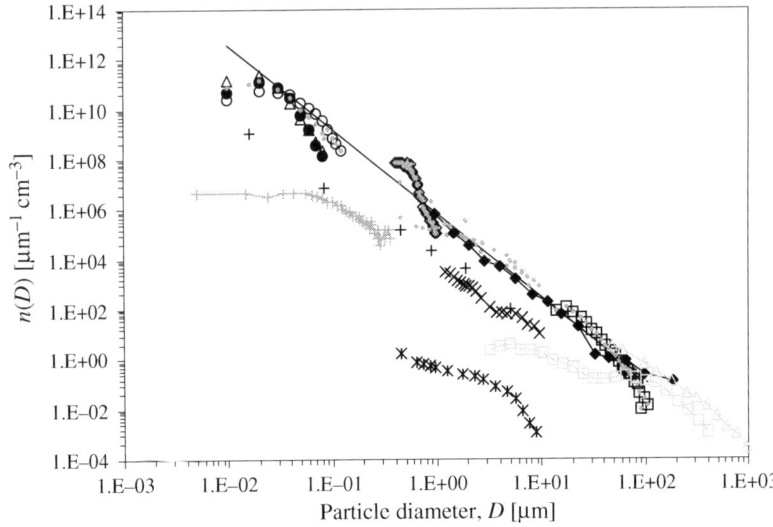

Figure 5.27. Composite PSD (PSD) of marine detritus in coastal waters (solid line, no symbols, $n(D) = kD^{-m}$, $k = 7.15 \times 10^5$, $m = 3.36 \pm 0.06$, with 1 SD shown following the $\pm$ symbol) based on data from several oceanic areas: coastal eastern Pacific (open and solid circles and open triangles, *Wells* and Goldberg 1992, Fig. 5, TEM), coastal western Atlantic (open diamonds, *Longhurst* et al. 1992, Fig. 1, 27 Sep 1990, Elzone counter, off Halifax, 10 m), Gulf of Mexico (pluses—*Harris* 1977, an average of six PSDs from depths between 600 and 3600 m), coastal western Atlantic (x's, *Green* et al. 2003a Fig. 3a, spring), GEOSEC station 306, 3879 m (crossed x's—bottom of panel, *Lambert* et al. 1981, Fig. 8, organics, SEM), and Baltic Sea (open squares, *Lenz* 1972, optical microscopy). The data used to calculate the power-law approximation are shown as gray dots, overlying the original data where not re-scaled. A randomly chosen PSD of transparent exopolymer particles (Baltic Sea, solid diamonds and line, *Mari* and Burd 1998, Fig. 9, panel 27/09) is very close ($m = 3.40 \pm 0.1$, $D = 0.97$ to 44.5) to the composite detrital PSD, as is a PSD obtained *in situ* with microphotography (Amazon River, Brazil, *Kranck* et al. 1992, gray open triangles and line) and the large-size end of the riverine colloid PSD (light gray pluses, *Kim* et al. 1995, arbitrary scale). The shapes of all these PSDs are markedly different from the small-size part of some *in situ* microphotography PSDs, for example, the data of *Courp* et al. (1993, Fig. 3, northwestern Mediterranean Sea, gray open squares and line, $m = 1.05 \pm 0.09$ for $D = 3.17$ to 128 μm), although the large-size part of the latter PSD has a similar slope. In all cases, the particle diameter is the equivalent circular diameter.

moving with different velocities move relatively to each other and have a chance to collide. In differential settling coagulation, the differences in particle velocity are caused by the particles themselves, as particles of different sizes and forms have different settling velocity in the gravitational field.

The particle collision probability increases with the square of the particle concentration (e.g., *Jackson* 1995) and strongly depends on the surface properties of the particles. With the large particles, the chances of forming an aggregate upon collision depend on the surface stickiness. Such stickiness is provided, for example, by glue-like substances excreted by aging phytoplankton cells (*Alldredge* et al. 1993). The glue can also be produced by adsorption of dissolved substances present in seawater on the surfaces of the particles (see references in *Logan* et al. 1994). Large aggregates with hundreds of primary particles (*Buck* and Chavez 1994) are generally formed when a phytoplankton bloom decays. At that time, the concentration of the plankton cells is high, and the aging cells excrete sticky substances aiding in their aggregation.

The process of particle aggregation has been observed with an *in situ* camera (*Milligan* 1995, *Eisma* and Li 1993, *Eisma* et al. 1983, *Kranck* and Milligan 1988). It has also been successfully modeled (*Li* et al. 2004, *Stemmann* et al. 2004a, 2004b, *Jackson* 1995, *Jackson* and Lochmann 1993) numerically for the open ocean environment. Unfortunately, despite the importance of these particles, there are but a handful of their size distributions (e.g., *Green* et al. 2003a, *Lenz* 1972—detritus, *Carrias* et al. 2002, *Mari* and Burd 1998, *Mari* and Kiørboe 1996, *Passow* and Alldredge 1994—TEP, *Wells* and Goldberg 1992—mostly organic marine colloids). Size distributions have also been determined for aggregates *in situ* that are likely to contain a significant fraction of organic components (e.g., *Courp* et al. 1993, *Kranck* et al. 1992, *Kranck* and Milligan 1992, *Eisma* et al. 1990). It seems nevertheless that more interest is expressed in the simple enumeration of these particles as can be judged from a sizeable table of abundances of detrital aggregates compiled by *Simon* et al. (2002). Representative sample data (Figure 5.27) show that the size distribution of these particle is likely to follow a power law in a wide size range, from tens of nm to hundreds of μm.

There is a striking difference in the slope of the composite power-law approximation of the size distributions of detrital particles obtained with *in vitro* ($m \sim 3.4$) and that of the "small-size" range of PSDs of aquatic particles obtained with *in situ* methods ($m \sim 1.3$) (Figure 5.27 and Figure 5.28). This difference is significant, for if one ventured to extrapolate a size distribution obtained with *in situ* imaging into the realm of the colloids, there should be hardly any of these in water, contrary to experimental evidence. The difference between the *in* vitro and *in situ* data is certainly likely to partly result from disaggregation of some particles upon sampling (5.6.1) and during the measurement (see section 5.7.2.4 in respect of the particle breakage by flow cytometers). The low number of small particles in the *in situ* images could also result from shading of these particles by the larger ones, as each such particle obscures part of the field of view represented by the "in-focus" viewing pyramid defined by that particle's projection outline. *Wells* and Goldberg (1992) point out that the tapering of their colloid size distribution at the small-size ends may be due to incomplete recovery of the smallest particles

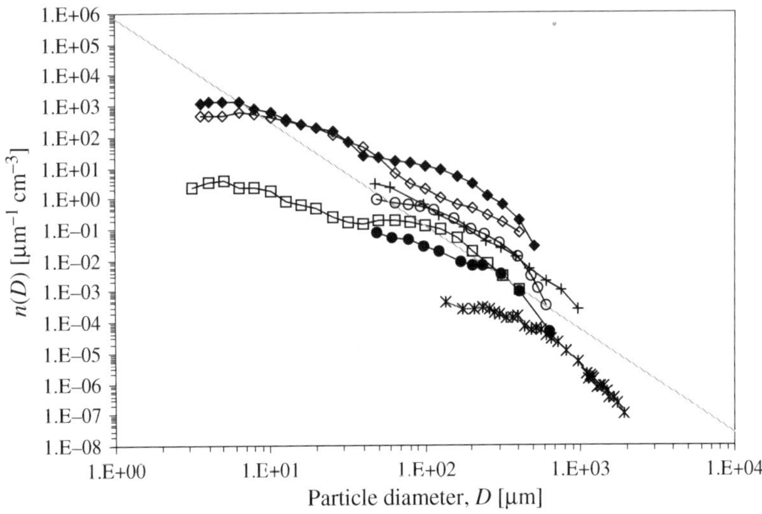

Figure 5.28. The slopes of size distributions obtained *in situ* (lines with symbols), as those obtained with *in vitro* methods (Figure 5.27), vary markedly throughout the respective size ranges, with the average of slopes in the small- and large-size ranges being close to that of the composite PSD of detrital particles from Figure 5.27 (gray line, no symbols). Open squares—western Mediterranean (*Courp* et al. 1993, Fig. 3), open diamonds—Schelde River estuary and solid diamonds—Shelde River at Temse, the Netherlands (*Eisma* et al. 1990, Fig. 7)—all three PSDs have been obtained with the same microphotography instrument and procedure, crossed x's—Monterey Bay, northeastern Pacific (*Maffione* et al. 1993, Fig. 8, 50 m, particle size from their backscattering cross-section *in situ*), solid circles—Skagitt Bay, northeastern Pacific, open circles—San Francisco Bay, and pluses—Amazon River (*Kranck* et al. 1992, *in situ* microphotography). In all cases, the particle diameter is the equivalent circular diameter.

from the samples. Whatever the case, Figure 5.27 and Figure 5.28 illustrate that each size particle size analysis method leads to PSDs with a similar variation in the slope: small slope at the small-size end of its range and large slope at the large-size end. In fact, even with the different magnification results obtained simultaneously with the same instrument—here an *in situ* camera system—one observes a similar effect (e.g., *Eisma* et al. 1990, their Fig. 6, 1:10 vs. 1:1 magnification data).

On the other hand, the observed "waviness" of the size distribution of strictly biological particles (from viruses to macroplankton) is a consequence of combining the "monodisperse" size distributions of the various contributing particle species. It is not feasible to estimate from the data available how much of this waviness in the case of the detrital particles is induced by the particle size analysis methods and how much is due to the natural variability of the size distribution of the contributing particle species. Its existence only implies that the "composite" size distribution we postulate in Figure 5.27, as well as other composite PSDs in the literature, should be treated with caution, as a global scale first-order approximation.

5.8.4.6. Minerals

Single-grain mineral particles may either appear as individual particles in water or be incorporated into aquatic aggregates. In either capacity, due to their high refractive index, mineral particles may contribute significantly to the optical properties of natural waters, provided that the size distribution of these particles has a sufficient magnitude.

As with detritus, the size distribution data specifically related to mineral particles are few. Sample size distributions of aquatic mineral suspensions as well as suspensions that are likely dominated by mineral particles are shown in Figure 5.29. Given the scarcity of reports on the size distribution of "certified" mineral particles, it is difficult to comment on the shape of their representative size distribution.

The sample distributions and much research that is specifically related to mineral components of aquatic suspensions concentrate on the size distributions of single-grain mineral particles (e.g., *Ratmeyer* et al. 1999) as building blocks of aquatic mineral-related suspensions. Such distributions are usually assessed for samples that were disaggregated by wet combustion (e.g., by leaching with H_2O_2) and sonification. The size distributions of single grains of these particles are formed generally through various fragmentation processes that yield size distributions conforming to the Weibull form (e.g., *Brown* and Wohletz 1995).

A power-law approximation has also been used to represent the size distribution of mineral particles (e.g., *Schoonmaker* et al. 1998—coastal ocean zone). Schoonmaker et al. derived the power-law size distribution slope by fitting Mie-based calculations to experimental values of the scattering coefficient of the suspensions. The use of the power-law approximation in this context has been recently criticized by *Stavn* and Keen (2004) who found significant discrepancies between the mass concentrations of mineral particles derived with a numerical model of particle resuspension utilizing PSDs measured for the sediment, and the power-law approximations derived from fitting the scattering coefficient.

Single-grain mineral particles enter natural waters through several pathways: washing of mineral debris by rain, rivers, erosion of the sea coast by waves, and fallout of aerosols generated by wind in arid areas of the continents. Mineral aerosols can travel thousands of kilometers over the ocean, such as the mineral dust stirred by storms over African or Asian deserts (e.g., *Prospero* 1996). In the close vicinity of a dust source, such as the Sahara (e.g., *Ratmeyer* et al. 1999), mineral aerosol fallout is a significant source of the particulate load of natural waters.

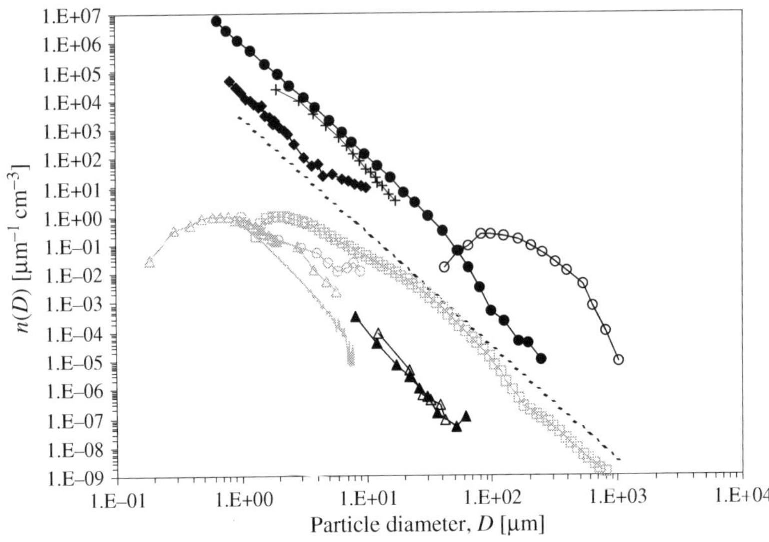

Figure 5.29. Sample PSDs (PSDs) of minerals in water. *Black symbols* represent absolutely scaled data: solid circles—PSD of disaggregated mineral particles measured with a Coulter counter and open circles—PSD of *in situ* flocs measured with *in situ* microphotography, San Francisco Bay [both PSDs calculated from the data of *Kranck* and Milligan 1992 as shown in their Fig. 4, Coulter counter (CC), spherical equivalent diameter (ESD)], aeolian PSDs of heavy ($\rho_{avg} = 4.53\,g\,cm^{-3}$, solid triangles) and light ($\rho_{avg} = 2.65\,g\,cm^{-3}$, open triangles) minerals measured off Bahamas *in situ* at a depth of 30 m [*Carder* et al. 1986, *in situ* holography, circular equivalent diameter (ECD)], pluses—particles in an Antarctic fjord with a high mineral load (Ezcurra Inlet, King George Island, *Jonasz* 1983b, CC, ESD). Gray symbols and lines represent relative data (each set is normalized to unity at the maximum) related to minerals in water: curve with no symbols—the size distribution of particles from Asian soil sample taken at Dunhuang, China, along a route of dust storms, as measured for an ultrasonicated suspension in seawater (*Stramski* et al. 2004, CC, ESD), open triangles—aluminosilicates in a stored GEOSECS sample from station 58, 4424 m (*Lambert* et al. 1981, SEM, ECD), open circles—Ezcurra Inlet, King George Island, Antarctic (*Gurgul* 1993, optical microscopy, Feret diameter), open squares—Po River, Italy (*Vignati* et al. 2003, laser diffractometry, ECD). The dashed line represents a power-law distribution, equation (5.156), with a slope $m = 4$.

We shall also note that a significant source of mineral particles are blooms of coccolitophores, phytoplankton species whose cells are clad in calcite plates. The size distributions of these mineral particles closely reflect the quasi-monodisperse size distribution of the relevant phytoplankton species, for example, *Emiliania huxleyi*.

5.8.4.7. Gas bubbles

Similarly to minerals in water, gas bubbles also exhibit significant influence on the optical properties of water (e.g., *Stramski* and Tęgowski 2001, *Zhang* et al. 1998). Gas bubbles are generated in water mostly by turbulent motion of water, although biological sources, in the water column or at the seabed and underwater gas vents (where they exist), also contribute. Breaking waves are a significant source of air bubbles in the open waters of large inland water bodies and in the oceans (e.g., *Terrill* et al. 2001, *Lamarre* and Melville 1991, *Baldy* 1988). It has also been postulated that the bubbles injected into the waterbody by breaking waves generate a significant population of small micrometer-sized second-generation bubbles upon bursting at the water surface (*Leifer* et al. 2000, *Johnson* 1986).

In addition to those transitory bubbles, persistent bubble populations exist at concentrations as high as 10^6 bubbles m^{-3} also in quiescent waters (e.g., *Terrill* et al. 2001, *O'Hern* et al. 1988, *Mulhearn* 1981) at depths of down to several tens of meters (e.g., *Medwin* 1977). Uncoated bubbles cannot exist at such concentrations so deep because they have insufficient lifetimes, limited by either their rising to the surface or dissolution (e.g., *Johnson* and Wangersky 1987). Two meachanisms of stabilization of the persistent bubbles have been proposed: (1) stabilization of small bubbles ($D < 10\,\mu m$) by adsorbed surfactants (e.g., *D'Arrigo* 1984, *D'Arrigo* et al. 1984) and (2) stabilization of the larger bubbles by layers of particles (*Johnson* and Wangersky 1987). Adsorption of particles that may outweigh the buoyancy of the bubble alone would also explain, in part, the significant concentration of bubbles at relatively large depths (*Mulhearn* 1981).

Bubbles are efficient scatterers of both light (see section 4.4.4.5) and acoustic waves. The acoustic properties of gas bubbles are sufficiently different from those of solid particles to make acoustic measurements of bubble populations a method of choice (e.g., *Terrill* and Melville 2000, *Vagle* and Farmer 1992), although scattering of sound by gas vacuoles in phytoplankton cells (*Sandler* et al. 1992) may interfere with the measurements of the free bubble populations.

Mulhearn (1981) cites observed size distributions of bubbles (Medwin 1977, 1970) that have a piecewise power-law approximation, equation (5.156): one with a slope of about $m = 2$ for bubble diameters between 30 and $120\,\mu m$ and a larger slope of $m = 4$ for the diameter between 120 and $400\,\mu m$. As seen from Figure 5.30, where sample size distributions of air bubbles generated by breaking waves are shown, such a change in the size distribution slope can in some cases be conveniently described by an exponential distribution [equation (5.195)] or log-normal distribution [equation (5.178)]. Note several-orders-of-magnitude variations in the bubble concentration across inhomogeneous spatial bubble distributions (bubble clouds) as reported by *Vagle* and Farmer (1992).

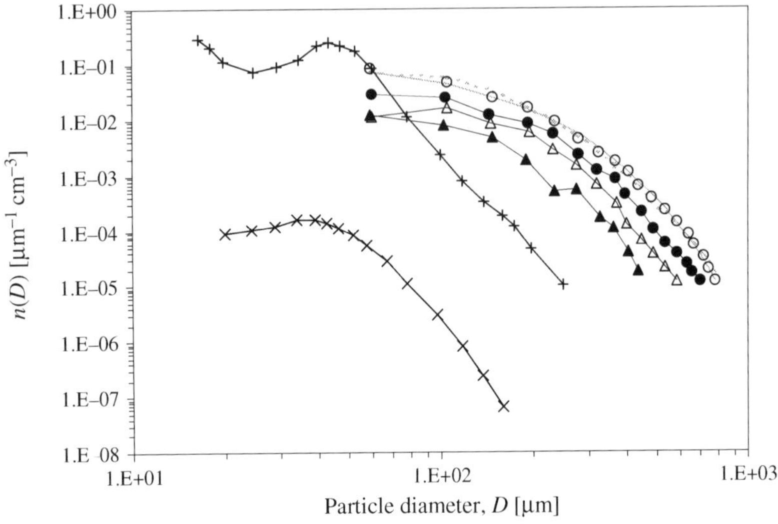

Figure 5.30. Sample size distributions of bubbles in seawater entrained by breaking waves (data of *Terrill* et al. 2001, their Fig. 1a) at depths of 0.7 m (open circles—data, solid gray line—exponential fit, equation (5.195), $a = 0.1541$, $c = 0.01211$, $r^2 = 0.998$, dashed gray line—log-normal fit, equation (5.178), $B_0 = -12.67$, $B_1 = 12.45$, $B_2 = -3.360$), 1.3 m (solid circles), 2.2 m (open triangles), and 4.1 m (solid triangles) and in a bubble cloud at the wind speed of 11 m s^{-1} off California (*Vagle* and Farmer 1992, calculated from data shown in their Fig. 11) at 0.5 m (pluses) and 7.3 m (x′s).

5.8.5. Approximations

Approximations of the PSD attempt to reduce the number of variables required for a "complete" description of the distribution. Without doubt, the general concern here is one of the "closeness" of the approximation to the experimental data. Based on this criterion, the fitness of an approximation can be judged on the compromise it makes between a reasonable representation of the data and the number of fit parameters required.

However, one should not lose sight of another important point. The experimental data are the outcome of physical/biological processes that shape the size distribution. Thus, one should (primarily) strive to arrive at an approximation that not only is close to the data but also models the cooperation of these processes. One obvious example is that of the size distribution of phytoplankton. At the level of individual species, one may want to look for an approximating function that reflects the life cycle of the species' cells and decays rapidly beyond a (generally narrow) size range characteristics of the species. However, at the level of the phytoplankton community, a reasonable approximation is one that represents well the sum of contributions of the various species.

5.8.5.1. Goodness-of-fit tests

Fitting an approximation to PSD data involves the selection of an approximating function that reflects the physical (biological) model of the process governing the particle population. If the underlying process is unknown, or if several processes participate in shaping the population to an unknown degree each, statistical methods can be used in selecting one (or more) of the candidate functions.

Statisticians developed several methods (goodness-of-fit tests) for testing whether an experimental distribution can be approximated by a theoretical distribution function. These methods typically involve calculating a test value, referred to as a statistic, which represents the difference between the experimental and theoretical distributions and comparing the result with a reference value based on probability theory. That reference value is a function of the probability (confidence level) that the theoretical distribution is the right choice, i.e., the observed difference is a result of random sampling of the data from a global data population. One should note that testing whether a size distribution model is applicable to a data set can readily be turned into curve fitting by minimizing the statistic.

The goodness-of-fit tests generally involve some form of weighing the data points $(x, y)_i$, where i is the data point index. In a limited case, applicable to the PSD measurements, such weights usually reflect the errors associated with the y coordinate—the PSD in our case. The x coordinate—the diameter in our case—is assumed to be known accurately. For those readers interested in the more general case, we suggest consulting the following references: *Press* et al. 1989, *Cecchi* 1991, *Irvin* and Quickenden 1983.

If errors are associated only with y, then one typically uses as the weights the values of $1/\sigma_i^2$, where σ_i^2 is the variance of y_i. The weighing is of significant importance especially with steep functions of y, such as the PSDs typical of natural waters. Indeed, the weights of $w_i = 1/\sigma_i^2 = 1/y_i$ result from a Poisson process, characteristic of the size distribution measurement. Thus, such weighing greatly favors the smallest particle size of the measured size range. However, such a fit may not always be relevant to optical modeling. For example, if one were interested in modeling the small-angle volume scattering function, the weighing should reflect the contribution of a given particle size to that function in addition to accounting for the measurement errors of the size distribution.

We give here a brief summary of two representative tests, the χ^2 and KS test, and refer the reader to one of the many textbooks or online resources on statistics (e.g., the NIST/SEMATECH website) for more in-depth reading.

The χ^2 (chi-square) goodness-of-fit test is widely used for curve fitting and is also applicable to testing the normalized histogram size distribution, i.e., probability distribution of the distribution of particle counts across a grid of size bins. In that latter capacity, this test requires that the particle count vs. size data be grouped into discrete bins each defined by a size range D_i to D_{i+1} and containing $h_i = \Delta N_i$ counts of particles with sizes within the bin size range in the sample volume examined. There is a minimum limit of the number concentration (approximately five counts) per bin below which the test becomes unreliable. The test is also sensitive to the selection of the size grid used for the definition of the bins.

In order to carry out such a test, one must calculate the test statistics, χ_l^2:

$$\chi_l^2 = \sum_{i=1}^{M} \frac{(h_i - g_i)^2}{g_i} \tag{5.145}$$

where g_i is the value obtained for the i-th bin from a theoretical model of the size distribution, M is the number of bins, and l is the number of degrees of freedom, which is calculated as follows:

$$l = M - k \tag{5.146}$$

where k is the number of fit parameters of the model. The resulting value of χ_l^2 is either compared to a reference value, $\chi_{l,\alpha}^2$, for a confidence level, α (say 0.05), listed in a table in one of the many statistics textbooks, or calculated from the χ^2 probability distribution (e.g., *Press* et al. 1989). If $\chi_l^2 > \chi_{l,\alpha}^2$, then the model may be a poor choice, because the probability that $\chi_l^2 > \chi_{l,\alpha}^2$ is small (it equals the confidence level). Alternatively, that probability can be evaluated directly as described in *Press* et al. (1989). Notation χ^2 does not mean χ raised to the second power, rather it is indicative of the prescription for the calculation of the statistics.

Equation (5.145) follows from a more general definition of the χ^2 statistics:

$$\chi_l^2 = \sum_{i=1}^{M} \frac{(h_i - g_i)^2}{\mathrm{var}(h_i)} \tag{5.147}$$

where var(x) is the variance of x, because in particle counting, the probability distribution of h_i for each i is the Poisson distribution. Then, $\mathrm{var}(h_i) = \langle h_i \rangle \cong g_i$, where a model-derived value of g_i serves as an estimate of the unknown mean value of h_i, i.e., $\langle h_i \rangle$. Certainly, if one has a better estimate of the measurement error in ΔN_i, it is that estimate which should be used instead of g_i in the denominator of each term in (5.145). For example, if the error is reduced by making r measurements (with r being possibly different for each ith measurement), then each g_i's should be multiplied by a factor $1/\sqrt{r_i}$. In either form, the χ^2 statistics can be used to determine the fit parameters through least-squares minimization (e.g., *Press* et al. 1989) that in the linear fit cases (straight-line and polynomial fits, and their derivatives) yields the parameters directly. In the case of a non-linear fit, the minimization is an iterative process.

As a rough guide, if $\chi^2 \leq l$, i.e., $\chi^2/l \leq 1$ (χ^2/l is simply χ^2 per degree of freedom), then the size distribution model is acceptable from the curve-fitting standpoint. Indeed, if the number of the model parameters estimated from the data was 0, then each term of the sum in (5.145) would be roughly at most 1, as indicated by (5.147), since the numerator, g_i, approximates var(h_i) when deviations from the expected value, g_i, are random. Hence $\chi^2 \sim l$. If $\chi^2/l > 1$, one should consider using another size distribution model. Why should one use the number of degrees of freedom, l, rather than the number of data, M, as a reference? Each parameter, estimated from the data, makes the model function fit the data better and reduces the value of the χ^2. Indeed, an nth degree polynomial ($n+1$ fit parameters) passes exactly through $n+1$ data points. Thus, as each fit parameter effectively "nulls" one term in (5.145), one should compare χ^2 not to M but rather to l in order to account for the commensurate reduction in χ^2 brought about by introducing and estimating $M-l$ model parameters.

The χ^2 test can be also applied to fitting an $n(D)$ size distribution model to size distribution data, $n_{\mathrm{E}}(D)$. Terms h_i in (5.145) can then be interpreted as the mean values, $\langle h_i \rangle$, of the experimental distribution. In the absence of repeated counting per particle diameter in the data set, we still have $\mathrm{var}[n(D_i)] \propto \mathrm{var}(\langle h \rangle_i)$ (see section 5.6.3) and thus $\mathrm{var}(\langle h_i \rangle) = \langle h_i \rangle \cong g_i$. Note that we still use the model yardstick ($\langle h_i \rangle \cong g_i$) to estimate the fit errors.

Many researchers use the coefficient of determination, r^2, as an indicator of the goodness-of-fit in the case of a linearized curve fit. This coefficient is a measure of correlation between the variables in question. In the case of the linear relationship between these variables, it is simply the square of the correlation coefficient. The coefficient of determination is defined, in general, as follows:

$$r^2 = \frac{\sum_{i=1}^{M} \left(y_i^{(r)} - \langle y \rangle \right)^2}{\sum_{i=1}^{M} \left(y_i - \langle y \rangle \right)^2} \tag{5.148}$$

where $y_i^{(r)}$ are obtained by applying the regression; here $y_i^{(r)} = a + bx_i$, y_i are the input data and $\langle y \rangle$ is the mean y. Thus, r^2 is a ratio of the sum of residuals accounted for by the regression to the total sum of residuals.

In reference to fitting the power-law to log-transformed PSD data, this practice has been recently given firmer foundations prompted by applications of the power-law process in reliability research (e.g., *Gaudoin* et al. 2003). In that simple test, the data are understood to support the hypothesis of a power law, $n = kD^{-m}$, if the r^2 evaluated using a linear regression of $\log n$ vs. $\log D$ exceeds a critical value. We reproduce a table of the critical r^2 values as functions of the sample size and the confidence level after *Gaudoin* et al. (2003) in the Appendix.

An alternative goodness-of-fit test, designed for use with cumulative distributions, is the KS test (e.g., *Vigneau* et al. 2000 and an online resource: search the web for NIST/SEMATECH). Also note that this test does not provide means of readily estimating by a minimization procedure the fit parameters of the theoretical distribution such as those provided by the χ^2 test.

In the KS test, a statistical variable, the KS distance, d_{KS}, is calculated for the cumulative experimental, $N_{\mathrm{E}}(D_i)$ and theoretical $N_{\mathrm{T}}(D_i)$ size distributions as follows:

$$d_{\mathrm{KS}} = \max |N_{\mathrm{E}}(D_i) - N_{\mathrm{T}}(D_i)| \tag{5.149}$$

where $i = 1, \ldots M$ is the index of an observed value and the corresponding evaluation of the theoretical distribution $N_{\mathrm{E}}(D_i)$ is a step function defined by (5.77) and $N_{\mathrm{T}}(D_i)$ is the corresponding value of the theoretical cumulative distribution. If $d_{\mathrm{KS}} > d_{\mathrm{KS},\alpha}$, where α is the confidence level (e.g., 5%), then the hypothesis that the theoretical distribution fits the experimental one is rejected. The values of $d_{\mathrm{KS},\alpha}$ are provided by many statistical data-processing software packages. Note that various formulations of the KS test exist in the literature. It is thus important to ensure that the values of $d_{\mathrm{KS},\alpha}$ correspond to the test formulation at hand.

Alternatively, one can evaluate the probability, P, that $d_{\mathrm{KS}} > d_{\mathrm{KS},\alpha}$, as follows

$$P = 2\tilde{p}(1 - \tilde{p}^3 + \tilde{p}^8 - \ldots) \tag{5.150}$$

where, as suggested by *Vigneau* et al. (2000),

$$\tilde{p} = \exp(-Md_{\mathrm{KS}}^2 - d_{\mathrm{KS}}) \tag{5.151}$$

If $P < \alpha$, then the hypothesis that the experimental cumulative size distribution is represented by the theoretical size distribution is rejected.

Another method is to use the Kullback–Liebler (*KL*) distance (e.g., *Lwin* 2003, *Koh* et al. 1989, *Martin* 1970), defined as follows:

$$d_{\mathrm{KL}} = \sum_i p_i \ln \frac{p_i}{q_i} \tag{5.152}$$

where p_i is the experimental PSD value for size D_i, expressed as probability, i.e., $(N_{i+1} - N_i)/N_{\mathrm{T}}$, with N_{T} being the total number of particles in the size range D_1 to D_M and q_i comes from a corresponding set of values of the theoretical size distribution. Note that d_{KL} is the relative entropy of a probability distribution (*Qian* 2001).

5.8.5.2. Effects of the logarithmic transformation

Some approximations to the PSD can be transformed to a polynomial by taking the logarithm of n and/or of D. A simple example is the power-law approximation (section 5.8.5.3). This modifies the weights factors that are set as $1/\mathrm{var}(h_i)$ in (5.147). In general, the size distribution is transformed from h_i to $y_i = f(h_i)$, then (in the limit of the small variance) we have

$$\mathrm{var}(y) = \left(\left. \frac{dy}{dh} \right|_{h=\langle h \rangle} \right)^2 \mathrm{var}(h) \tag{5.153}$$

In the power-law example, $y = f(h_i) = \ln h_i$, where we used the natural logarithm for the sake of simplicity. Thus, in the first approximation:

$$\begin{aligned} \mathrm{var}(y_i) &\cong \frac{1}{\langle h_i \rangle^2} \mathrm{var}(h_i) \\ &= \frac{1}{\langle h_i \rangle} \end{aligned} \tag{5.154}$$

where in the second line we used the assumption that the probability distribution of h_i is the Poisson distribution. This weighing implies that only the first few data points matter in the fitting of a power-law to aquatic PSDs.

In fitting an approximation to log-transformed size distribution, the appropriate weights can be set to $w = 1$, if the particle count is sufficiently high. Indeed, in the limit of small variance, we have

$$\begin{aligned} \mathrm{var}(\ln N) &= \left(\frac{d \ln N}{dN} \right)^2 \mathrm{var} N \\ &= \frac{1}{\langle N \rangle^2} \left(\mathrm{var_P}\ N + \langle N \rangle^2 M^2 \frac{\mathrm{var} D}{\langle D \rangle^2} + \langle N' \rangle^2 \frac{\mathrm{var} V}{\langle V \rangle^2} \right) \\ &= \frac{\mathrm{var_P}\ N}{\langle N \rangle^2} + M^2 \frac{\mathrm{var} D}{\langle D \rangle^2} + \mathrm{var} V \\ &= \frac{1}{\langle N \rangle} + M^2 \frac{\mathrm{var} D}{\langle D \rangle^2} + \mathrm{var} V \end{aligned} \tag{5.155}$$

where we used equation (5.105), applicable to resistive particle sizing [with M being the power-law slope of the cumulative PSD, D being the particle diameter, and V being the sample volume] in the second line of (5.155) and the equality $\mathrm{var_P}(N) = \langle N \rangle$, valid for the Poisson-distributed N, in the last line of that equation. Hence, var(ln$N$) for large N ($>\sim 4000$ for the Coulter counter, as evaluated in section 5.7.1.6) is determined mostly by instrumental errors (in particle diameter, D, and sample volume, V) and not by the counting error. The instrumental error contribution to the variance of ln$N$ is just a constant here, so the weight, w, is also approximately constant and can be simply set to 1. Note that for low maximum particle count, the contribution from the counting error should be included in the expression for the variance, and also for the weights which then cease to be unity. Note that since $n(D)$ is a scaled difference of two independent random numbers $N(D + dD)$ and $N(D)$, we have $\mathrm{var}[n(D)] = \mathrm{var}[N(D + dD)] + var[N(D)]$ (see Eq. 5.69). Hence the results discussed here apply also to $n(D)$.

We should note that the logarithmic transformation of the size distribution biases the fit parameters depending on the magnitude of the distribution. Indeed, the fitting process always uses a weighing scheme, whether it is set knowingly by the researcher, or left to itself (unknowingly?, when the weights are simply ignored, which by default sets them all to 1). As an illustration, consider fitting an approximation to a cumulative size distribution, $N(D)$, and to its logarithmic transform, $\ln N(D)$, and set all weights to 1 in the first case. After taking the logarithm of $N(D)$ and neglecting the contribution from instrumental errors, the weights are transformed to $N^2(D)$. Thus, if we elect to use the weights of 1 in the second case, except when N is sufficiently large, the fit parameters are bound to be different from those obtained by fitting the approximation to the original data, $N(D)$.

5.8.5.3. The power-law function

The power-law function, probably used for the first time by *Bader* (1970) in the context of the size distribution of marine particles as measured with a Coulter counter, can be expressed as follows:

$$n(D) = k\left(\frac{D}{D_0}\right)^{-m} \tag{5.156}$$

where k [$\mathrm{cm^{-3}\,\mu m^{-1}}$] and m [non-dimensional] are constants. To date, it has probably been the most frequently used approximation to the size distribution of particles in natural waters. Parameter $D_0 \equiv 1$ [L], where [L] is the dimension of D, usually omitted, renders the ratio D/D_0 non-dimensional. The role of D_0 is sometimes neglected which forces one to assign to k a dimension of volume^{-1} × length^{m} (*Lerman* et al. 1977).

An interesting point concerning the dimensional homogeneity of empirical equations, such as (5.156), has been raised by *Prothero* (1986). By rendering the D non-dimensional, we created a dimensionally homogeneous equation (pleasing to a physicist) but have not increased our understanding of the underlying phenomenon (a power-law relationship between the particle size and number concentration). Equation (5.156), as applied to a particular data set, is an empirical equation that emphasizes correlation between the particle size and concentration. We sweep aside models of the PSD for large ranges of the particle size (to be discussed shortly) that justify the power law as an approximation of choice here, because this approximation is generally applied to a small sub-range of the particle size, where other models may provide a better approximation. As Prothero points out, an empirical equation, even if made dimensionally homogeneous, is not rendered the *proper* equation by such homogenization. Thus, an empirical equation should not be understood as a justification for a model of the dependency of the particle concentration on the particle size. Such a justification must come from the relevant science (we would have preferred to say "physics," but the complex problem at hand calls for a more inclusive wording). This becomes clear when one realizes that the power law, as related to the PSD data obtained with resistive counters, is frequently applied to two, or even three sub-ranges, of the particle size (e.g., *Jonasz* 1983a). Thus, the result of making equations such as (5.156) dimensionally homogeneous is fundamentally different from doing the same while deriving, for example, an equation that relates the distance traveled by an object falling in the Earth gravity field to the travel time.

The power law can be simply fitted to experimental data by taking a logarithm of both sides of (5.156). This transforms that equation into a linear equation:

$$\log n(D) = \log k - m \log D \tag{5.157}$$

Coefficients log$k$ and m can thus be simply found by using the linear least-squares regression, where the weights $w = 1/\text{var}[n(D)]$ can be set all equal to 1 for reasons discussed in section 5.8.5.2. Estimates of the variance of the cumulative and frequency size distributions are given in sections 5.7.1.6 and 5.6.4 respectively. In section 5.7.1.6 we specifically refer to resistive particle counters, but the estimates given there can be readily adapted to optical particle counters and flow cytometers.

Sample approximations are shown in Figure 5.31 for a size distribution from the open ocean surface waters. Most observed size distributions have a two-slope form, with the slope in the small-size range being smaller than that of the large-size range (e.g., *Zalewski* 1977—the Baltic Sea). Note that this conclusion has been reached based on a limited particle size range: *Gordon* et al. (1972) observed three-slope form of the size distribution—with the slopes in the smallest and the largest size ranges being both larger than that of the middle size range, when the lower size limit was reduced.

By integrating (5.156) from D to ∞ one readily obtains a power law for the cumulative size distribution:

$$\begin{aligned} N(D) &= \int_D^\infty n(D')dD' \\ &= \frac{k}{m-1} D^{-m+1} \end{aligned} \tag{5.158}$$

assuming that $m > 1$. We shall see shortly that this assumption is justified in view of the range of values of the slope m of the PSDs of natural waters.

The power-law size distribution diverges as the particle size, D, tends to 0. This should be viewed as an indication that the underlying model cannot be extrapolated to infinitely small particle size.

Diverse arguments have been advanced in support of the suitability of the power-law function for the representation of the size distributions of particles in natural waters. Biological arguments (*Rinaldo* et al. 2002, *Camacho* 2001,

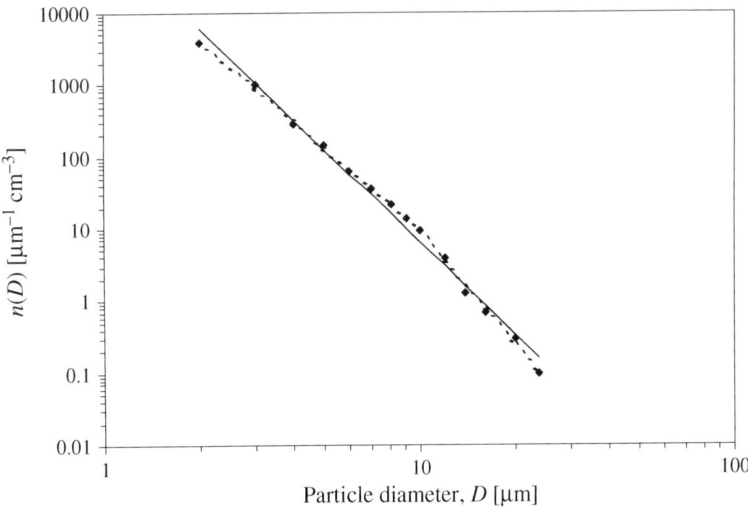

Figure 5.31. A PSD measured in the surface waters of the mid-Atlantic (*Jonasz* 1992: JONATL78.P08, column 4) as approximated using a single-segment and a two-segment power law. Single-segment approximation (χ^2 per degree of freedom = 19.6): $FD(D) = 1.17 \times 10^5 D^{-4.25}$. Two-segment approximation (χ^2 per degree of freedom = 0.73): $FD(D) = 5.75 \times 10^4 D^{-3.77}$ ($D = 2$ to $10\,\mu$m) and $FD(D) = 1.16 \times 10^6 D^{-5.11}$ ($D = 10$ to $24\,\mu$m). All fit parameters were obtained via the logarithmic transform. The weights were set to unity in each case.

Camacho and Solé 2001, *Kiefer* and Berwald 1992, *Platt* and Denman 1978, 1977, *Sheldon* et al. 1977) are based on the fact that in the aquatic food chains, in general, a large predator feeds on a small prey. Thus, the rate of production of biomass by a species, and consequently the number concentration of the species organisms, must be proportional to a negative power of the species' organism size for the ecosystem to sustain itself. Interestingly, predation is not a limiting constraint here, as plants across a vast size range (approximately six orders of magnitude) of aquatic and terrestrial genera exhibit population densities inversely proportional to a power of the body mass (e.g., *Belgrano* et al. 2002, *Cyr* et al. 1997b). This has been interpreted as a limitation of the population density of organisms by their energy requirements.

An interesting experiment in this respect was performed by *Stramski* et al. (1992a) who examined the effect of grazing by a ciliate on the size distribution of a laboratory culture of a marine cyanobacteria *Synechococcus*. They observed that the initially peaked size distribution, representing the monoculture of *Synechococcus*, evolved with time to a monotonically decreasing, power-law-like size distribution.

Borgmann (1987) compared various biological PSD (biomass spectra) models and points out that they provide similar results, largely independent of their complexity. Note that such models refer to the ecosystem as a whole. Indeed, it has been postulated (*Thiebaux* and Dickie 1993, *Boudreau* et al. 1991), in agreement with experimental data, that the size distribution of the individual species in the ecosystem is described by a log-normal function. We discuss this topic in more detail in section 5.8.5.6.

One should not lose sight of the fact that the size distribution of aquatic particles is composed of not only clearly defined size distributions of live ecosystem components, such as bacteria, phytoplankton, zooplankton, and fish, but also of single-grain mineral particles and aggregates of all of some or the above species. In fact, random aggregation has been also implicated in reference to both animate and inanimate particles (*Stemmann* et al. 2004a, 2004b, *Camacho* 2001—general aggregation models, *Jackson* 1995—aggregation in phytoplankton bloom, *Hunt* 1982—Brownian [$m = 2.5$], and shear [$m = 4$] aggregation, and differential [$m = 4.5$] settling aggregation).

Exclusively inanimate aggregation processes also generate the PSD of aerosols that has also been approximated by a power-law function (e.g., *Li* et al. 2004, *Hunt* 1982). In fact, a PSD of that type is frequently referred to in the oceanographic literature as the Junge distribution, after *Junge* (1963) who used this form to represent the PSDs of atmospheric aerosols.

Power laws have been widely used to represent results of processes as disparate as creation of craters on cosmic bodies (*Baldwin* 1965—sizes of craters on Mars) to socioeconomics (*Gabaix* and Ioannides 2004—sizes of cities, *Pareto* 1896—income distribution), language (*Zipf* 1949—usage of words), and computer science (*Mitzenmacher* 2004 — file size distribution). These laws are also inherent in fractal geometry and are an indication of the self-similarity of a system at various size scales. *Vidondo* et al. (1997) extensively review the applications of power laws and their implications in aquatic biology.

After Vilfredo Pareto (1848–1923) and George Kingsley Zipf (1902–1950), the power laws have been frequently referred to as the Pareto and/or Zipf laws. *Schmidt* and Housen (1995), who discussed the application of dimensional analysis to solving problems related to the creation of explosion or impact craters, point out that power laws can routinely be obtained through dimensional analysis of complex relationships in nature and comment that these laws "are often not derivable solely from mathematical and physical arguments."

The power law has also been implicated in turbulent diffusion in the atmosphere (the law of "4/3," *Richardson* 1926) and in the sea (*Ozmidov* 1967). One can generalize that the power law applies to processes in which the energy is transferred throughout a physical system in a diffuse, random manner. The populations of particles in natural waters are but one example of such a system. On the other hand, *Cavender*-Bares et al. (2001) advocated a network analogy, with the biological systems being shaped by a network of interactions as in the food web.

In section 5.8.5.6, we postulate that the sum of log-normal functions is a more general approximation of the marine size distribution. In fact, the power-law function is a special case of the log-normal function. The approximation of the size distribution with a sum of log-normal components focuses on the details of the distribution and does not conflict with the global, approximate representation of the size distribution of suspended particles by a power-law function.

The coefficients of the power-law approximation to the size distribution reported by various authors are shown in Table 5.8 for volume-sensitive particle sizing methods and in Table 5.9 for other particle sizing methods (mostly projected area sensitive). In that table, each slope usually represents a relatively narrow range of the particle size. Thus, that slope can be though of as a *local* slope of the particle distribution characteristic of the respective size range. The data contained in Table 5.8 and Table 5.9 are plotted in Figure 5.33, which illustrates a weak (but quite discernible in the actual size distribution data) tendency of the *local* slope of the power-law approximation to the PSD to increase with the particle diameter in the range of 1 to 1000 μm. Such a tendency is accommodated naturally in more complex approximations, such as the log-normal distribution (section 5.8.5.6) or the gamma distribution (section 5.8.5.8).

It is tempting to explore in this context the formalism introduced in section 5.5.2.2 for relating the equivalent spherical, D_S, and circular, D_C, particle diameters by (again!) a power-law relationship of the type:

$$D_S = p_S D_C{}^{q_S} \tag{5.159}$$

By using the average slope data for the volume-sensitive and projected area-sensitive methods (the caption of Figure 5.33) we obtain $q_S = 0.92$ according to (5.25). This yields (see section 5.5.2.2) an estimate of the three-dimensional fractal dimension of the aquatic particles of ~ 2.75.

The high slopes in the sub-micron particles' PSDs reported in Table 5.8, Table 5.9, and Figure 5.33 would indicate (in the log-normal decomposition approach that we discuss in section 5.8.5.6) that the size distribution of these particles is the smallest log-normal component of the global size distribution. In line with the population dynamics model advanced by *Thiebaux* and Dickie (1993) (see section 5.8.5.6), this may be the ultimate "prey" population of aquatic particles that provide material for the generation of larger particles (e.g., *Chin* et al. 1998).

As we noted in the caption of Figure 5.33, the average slope for the data from Table 5.8 and Table 5.9 is about 4. The slope of 4 was used as a representative slope by *Stramski* and Kiefer (1991) in their discussion of the role of the particles in light scattering by seawater. These latter authors approximated the size distribution of phytoplankton, reported by *Takahashi* and Bienfang (1983) for the tropical

Table 5.8. The coefficients, k and m, of the power-law approximations to the size distribution of particles in natural waters. The data were obtained with volume-sensitive methods.

D [μm]		k [$cm^{-3} μm^{-1}$]		m [non-dimensional]		Data count	Reference
Minimum	Maximum	Minimum	Maximum	Minimum	Maximum		
0.4	1			11.1	12.1	2	*Longhurst* et al. (1992)
0.65	1			7			*Gordon* et al. (1972)
1	10	1.18E+05	1.47E+05	2.8	3.4		*Gordon* and Brown (1972)[A]
1	10	3.30E+04	5.50E+04	3.4	4.1		*Gordon* and Brown (1972)[B]
1	10	1.05E+04	2.80E+04	4	5		*Gordon* and Brown (1972)[C]
1	20			3.64	4.08		*Richardson* (1987)
1	5			3.99			*Mari* and Burd (1998)
1	5			3.25	3.66	10	*Brun*-Cottan (1971)[A]
1	5			3.30	3.67	10	*Brun*-Cottan (1971)[B]
1	20			3.56	4.08	5	*Brun*-Cottan (1971)[C]
1	10,000			3.5	3.61		*Jackson* et al. (1997)
1.26	3.5			3.26	3.84	12	*McCave* (1983)[A]
1.26	32			3.78	4.24	25	*McCave* (1983)[B]
1.5	15			3.5	4.5		*Brun*-Cottan (1976)
1.5	22			4.41	5.49	81	*Spinrad* et al. (1989b)
1.9	7.5	1.51E+05		2.81	3.95	6	*Bader* (1970)
2	5			3.6	4.5		*Reuter* (1980b)
2	6			2.76	3.18	23	*Kitchen* and Zaneveld (1990)
2	6			2.83	3.45	32	*Kitchen* and Zaneveld (1990)
2	6			2.69	3.29	17	*Kitchen* and Zaneveld (1990)

(*Continued*)

Table 5.8. Continued

D [μm]		k [$cm^{-3}μm^{-1}$]		m [non-dimensional]		Data count	Reference
Minimum	Maximum	Minimum	Maximum	Minimum	Maximum		
2	6			2.64	3.24	17	*Kitchen* and Zaneveld (1990)
2	6			3.2	3.68	14	*Kitchen* and Zaneveld (1990)
2	6			3.3	3.96	63	*Kitchen* and Zaneveld (1990)
2	6			3.56	4.28	28	*Kitchen* and Zaneveld (1990)
2	7.5	1.00E+05	1.00E+05	2.7	3.7	160	*Jonasz* (1983a)
2	8.6	3.47E+04		3.41		21	*Jonasz* and Prandke (1986)[A]
2	9.6	3.72E+04		2.73		12	*Jonasz* and Prandke (1986)[B]
2	32			3.55		155	*Bradtke* (2004)
2.26	14			3.73	4.29	53	*Lerman* et al. (1977)
2	19			2.7	3.2	2	*Stoderegger* and Herndl (1999)
2.32	22.3			3.43	4.23	12	*McCave* (1985)
3	20	5.54E+05		4.66		83	*Jonasz* (1983b)
3.5	32			5.08	6.40		*McCave* (1983)
5	20			3.38	5.24	10	*Brun*-Cottan (1971)[A]
5	20			4.01	5.02	10	*Brun*-Cottan (1971)[B]
5	60			4.95			*Mari* and Burd (1998)
5.1	14.9	2.51E+06		3.70	5.39	5	*Bader* (1970)
6	16			5	5.82	23	*Kitchen* and Zaneveld (1990)[A]
6	16			5.01	5.83	32	*Kitchen* and Zaneveld (1990)[B]
6	16			4.9	5.42	17	*Kitchen* and Zaneveld (1990)[C]
6	16			4.88	6.16	17	*Kitchen* and Zaneveld (1990)[D]
6	16			4.43	5.75	14	*Kitchen* and Zaneveld (1990)[E]
6	16			4.52	5.4	63	*Kitchen* and Zaneveld (1990)[F]

6	16		4.41	5.51	28	*Kitchen* and Zaneveld (1990)[G]
7.5	32	7.50E+11	4.4	5.6	160	*Jonasz* (1983a)
9.6	32	4.45E+05	4.86		12	*Jonasz* and Prandke (1986)[A]
8.6	32	4.06E+05	4.56		21	*Jonasz* and Prandke (1986)[B]
9.2	12.8		9.95	12.5	2	*O'Hern* et al. (1988)
9.4	23		3.83	5.91	2	*O'Hern* et al. (1988)
13.7	21.7		4.34	4.41	2	*O'Hern* et al. (1988)
17.3	59.4		4.02	6.68	6	*McCave* (1985)

Abbreviations: CC = Coulter counter, EZ = Elzone counter.
Bader (1970)—Abaco Bight, Bahamas, Atlantic, CC.
Bradtke (2004)—Gdansk Bay, Baltic Sea, measurements performed throughout a year, CC.
Brun-Cottan (1971):
A—Northwestern Mediterranean, 0 to ~500 m, June 1969, CC.
B—Northwestern Mediterranean, 0 to ~500 m, November 1969, CC.
C—Northwestern Mediterranean, 0 to ~500 m, June and November 1969, CC.
Brun-Cottan (1976)—Western Mediterranean, 300–900 m, CC.
Gordon and Brown (1972):
A—Atlantic, Bahama Banks, CC.
B—Sargasso Sea, surface, CC.
C—Sargasso Sea, below mixed layer, CC.
Gordon et al. (1972)—Atlantic, Bahamas, CC.
Jackson et al. (1997)—Monterey Bay, CA, USA, Pacific, volume sensitive.
Jonasz (1983a)—Baltic Sea, CC.
Jonasz (1983a)—Baltic Sea, CC.
Jonasz (1983b)—Ezcurra Inlet, King George Island, Antarctic, 10 m.
Jonasz and Prandke (1986)
A—Baltic Sea, June 1977, CC.
B—Baltic Sea, March 1976, CC.

(Continued)

Table 5.8 Continued

Kitchen and Zaneveld (1990):

A—North Pacific central gyre north, 1–70 m, mixed layer, CC.
B—North Pacific central gyre south, 1–70 m, mixed layer, CC.
C—North Pacific central gyre north, 41–80 m, attenuation max, CC.
D—North Pacific central gyre south, 41–80 m, attenuation max, CC.
E—North Pacific central gyre, 81–90 m, CC.
F—North Pacific central gyre, 90–120 m, CC.
G—North Pacific central gyre, 121–130 m, CC.

Lerman et al. (1977)—Equatorial North Atlantic, 29 to 5111 m.
Longhurst et al. (1992)—Coastal western Atlantic, off Halifax, NS, Canada, 10 m, Elzone.
Mari and Burd (1998)—Kattegat, EZ, fit to the average of all observations.
McCave (1983)—Atlantic, Nova Scotian Rise, nepheloid layer, "new suspension," CC, average ±SD.
McCave (1983):

A—Atlantic, Scotia Rise, nepheloid layer, "new suspension," CC, average ±SD.
B—Atlantic, Scotia Rise, nepheloid layer, "old suspension," CC, average ±SD.

McCave (1985)—Atlantic, Nova Scotian Rise, 20 to 1000 m above bottom, CC, average ±SD.
O'Hern et al. (1988)—Coastal Pacific, off Catalina Island, 2–32 m, CC.
Reuter (1980b)—Eckenforde Bay, Kiel Fjord, Baltic Sea, CC.
Richardson (1987)—Atlantic, Iceland Rise, 150–2500 m, CC, average ±SD, CC.
Spinrad et al. (1989b)—Pacific, off Peru, surface to 400 m, average ±SD, CC.
Stoderegger and Herndl (1999)—Coastal North Sea, EZ.
The data were obtained with volume-sensitive methods. Data details are appended to the table.

Table 5.9 The coefficients, k and m, of the power-law approximations to the size distribution of particles in natural waters. The data were obtained with projected area-sensitive and other non-volume sensitive methods.

D [μm]		k [$cm^{-3}μm^{-1}$]		m [non− −dimensional]		Data count	Reference
Minimum	Maximum	Minimum	Maximum	Minimum	Maximum		
0.02	2	9.69E+06		2.65		6	*Harris* (1977)
0.04	0.1			5.98	7.16	3	*Wells* and Goldberg (1992)
0.06	0.3			2.6		1	*Kim* et al. (1995)
0.1	5			4.35			*Stramski* and Sedlák (1994)
0.1	100			4.5		4	*Lal* and Lerman (1975)[A]
0.1	100			3.76		1	*Lal* and Lerman (1975)[B]
0.1	100			4		1	*Lal* and Lerman (1975)[C]
0.47	2445			4		1	*Gaedke* (1992)
0.5	5			2.39	3.09	34	*Atteia* and Kozel (1997)
0.5	7.3			4.01	4.17	16	*Cavender*-Bares et al. (2001)[A]
0.5	7.3			4.18	4.27	17	*Cavender*-Bares et al. (2001)[B]
0.5	7.3			4.23	4.31	6	*Cavender*-Bares et al. (2001)[C]
0.5	7.3			4.15	4.31	9	*Cavender*-Bares et al. (2001)[D]
1	50			2.9			*Spinrad* et al. (1989a)[A]
1	80			3.39	3.59		*Spinrad* et al. (1989a)[B]
1	200			2.99			*Mari* and Burd (1998) TEP
1	10,000			2.96	3		*Jackson* et al. (1997)
2	5			4.44		5	*Harris* (1977)
2	16			2	3		*Kullenberg* (1970)[A]
2	16			1.5			*Kullenberg* (1970)[B]
2.1	32.2			2.75		1	*Wellershaus* et al. (1973)

(Continued)

Table 5.9 Continued

D [μm]		*k* [$cm^{-3}\mu m^{-1}$]		*m* [non – –dimensional]		Data count	Reference
Minimum	Maximum	Minimum	Maximum	Minimum	Maximum		
2.8	9.1			4.03		5	*Wellershaus* et al. (1973)
3	100			1.87	2.39	14	*Worm* and Søndergaard (1998)
3.5	70			2.90	3.23	2	*Spinrad* et al. (1989a)
5	60			3.59	6.13	34	*Atteia* and Kozel (1997)
10	40			3.1	3.9	6	*Chung* (1982)
10.31	127.3			3.77	3.77	5	*O'Hern* et al. (1988)
17.2	104.6			5.28		1	*Lenz* (1972)
64.0	156.6			5.24	7.80	5	*Bishop* et al. (1978)[A]
63.3	146.0			5.16	7.37	3	*Bishop* et al. (1978)[B]
70	318			4.29	4.38	2	*Spinrad* et al. (1989a)
116.5	214.5			4.04	5.56	4	*Bishop* et al. (1978)[C]
537.8	1769.0			4.85	5.75	4	*Bishop* et al. (1978)[D]
				3.5	3.9		*Forand* and Fournier (1999)[A]
				3.55	3.75		*Forand* and Fournier (1999)[B]
				3.22	4.3		*Cyr* et al. (1997a, 1997b)
0.012	152			3.4	4.1		*Twardowski* et al. (2001)

Abbreviations: DLS = dynamic light scattering, LVFS = large volume filtration system, OM – optical microscopy, ST = settling tube, TEM = transmission electron microscope, TEP = transparent exopolymer particles, TOT = time- of-transition particle counter.

Atteia and Kozel (1997) karstic aquifier, Bied, Switzerland, TOT (data not used in the slope vs. particle size regression), *Bishop* et al. (1978):

A—Southeast and equatorial Atlantic, foraminifera, LVFS Dec 73 to May 74, 400 m, OM.

B—Southeast and equatorial Atlantic, foraminifera fragments, LVFS Dec 73 to May 74, 400 m, OM.

C—Southeast and equatorial Atlantic, foraminifera, LVFS + OM, Dec 73 to May 74, 400 m, OM.

D—Southeast and equatorial Atlantic, foraminifera fragments, LVFS + OM, Dec 73 to May 74, 400 m, OM.

Cavender-Bares et al. (2001):
A—North Sargasso Sea, FC, average ± SD, FC, bacteria and phytoplankton only.
B—South Sargasso Sea, FC, average ± SD, FC, bacteria and phytoplankton only.
C—Gulf Stream, FC, average ± SD, FC, bacteria and phytoplankton only.
D—Western Atlantic, off Cape Cod, FC, average ± SD, FC, bacteria and phytoplankton only.
Chung (1982)—Indian Ocean, GEOSEC station 453, 7–4135 m, stored samples, Spectrex counter.
Cyr et al. (1997a)—A summary of biological PSD data for 18 lakes worldwide.
Forand and Fournier (1999):
A—Cabot Strait, western coastal Atlantic, 0 to 320 m, Oct. 1989, inversion of small-angle scattering.
B—Haro Strait, eastern coastal Pacific, 0 to 300 m, Jan. 1992, inversion of small-angle scattering.
Gaedke (1992)—Lake Constance, Central Europe, biogenic particles only, CC, OM.
Harris (1977)—Gulf of Mexico, 600 to 3600 m, TEM.
Jackson et al. (1997)—Monterey Bay, CA, USA, Pacific, projected area sensitive.
Kim et al. (1995)—River water (Water of Leith, Dunedin, New Zealand), TEM.
Kullenberg (1970)—Sargasso Sea, surface, VSF fitting.
Kullenberg (1970)—Baltic Sea, 10 m, VSF fitting.
Lal and Lerman (1975):
A—Indian Ocean, diatoms + foraminifera, OM.
B—Pacific, Sverdrup's data for foraminifera in sediment, OM.
C—Mid Atlantic Ridge, Black's data for coccoliths, foraminiferas and forms fragments in sediment, OM.
Lenz (1972)—Western Baltic Sea, Kiel Bight, average of 80 samples from depths 5 to 25 m, autumn/spring 1960/1961.
Mari and Burd (1998)—Kattegat, TEP, OM, fit to the average of all observations.
O'Hern et al. (1988)—Pacific, off Catalina Island, 2–32 m, *in situ* holography.
Spinrad et al. (1989a)—Atlantic, Scotia Rise, at the bottom, ST.
Stramski and Sedlák (1994)—Pacific, off California, USA, clear surface water, DLS, CC.
Twardowski et al. (2001)—Gulf of California, April 1999, backscattering ratio fitting.
Wellershaus et al. (1973):
A—Atlantic Ocean, off Portugal, average (?) of four samples, May 1971, 10–200 m, OM, particles settled in a plankton microscope.
B—Indian Ocean, average (?) of 5 samples: Dec 1964 – Jan 1965, 700 to 5000 m, OM, particles on a membrane filter
Wells and Goldberg (1992)—Coastal eastern Pacific, off San Diego, surface, TEM.
Worm and Sondergard (1998)—Lake Frederiksborg, Denmark, summer 1995, TEP, OM, filtered samples.
The data were obtained with projected area-sensitive and other non-volume-sensitive methods. Data details are appended to the table.

Pacific waters, by using a power-law function with a slope, $m = 3.86$ or 4.37, depending on the inclusion of a data point.

Note that *Wells* and Goldberg (1992), who extended down to 20–30 nm the particle size range in which the power-law approximation applies, obtained much greater slopes m in the size range of 20–100 nm, although the number concentration of the smallest particles seems to decline. As noted by these authors, this decline in the rate of the increase of the PSD could be due to incomplete extraction or retention for analysis of the smallest particles.

The slope of ~4 for many power-law approximations of the size distributions of particles in natural waters has a special significance for the distribution of particle volumes, as will be obvious from the following derivation. Consider a particle size grid that is defined by D_{i+1} =const D_i. The volumes of particles with sizes within a grid interval (D_i, D_{i+1}) are calculated as follows:

$$\begin{aligned} V(D) &= \int_{D_i}^{D_{i+1}} \frac{\pi}{6} D^3 k D^{-4} dD \\ &= \frac{\pi}{6} k \ln \frac{D_{i+1}}{D_i} \\ &= \frac{\pi}{6} k \ln const \end{aligned} \tag{5.160}$$

Thus, the total volume of particles within each such size interval is constant, a hypothesis advanced by *Sheldon* et al. (1972). Given the oscillations of actual particle size spectra about the power-law trend line (Figure 5.32), this hypothesis should be regarded as referring to the global trend line and not to the individual components of the PSD which, as we discussed it earlier in this section, are better approximated by log-normal functions (section 5.8.5.6).

An extension of the power-law approximation has been proposed by *Lawler* (1997). In this approximation, the slope coefficient, m, is not a constant as in the pure power law, but varies as a function of D as follows:

$$m(D) = m_0 \log D \tag{5.161}$$

where m_0 [non-dimensional] is a constant. *Ceronio* and Haarhoff (2005) discuss this approximation in more detail and provide a least-squares-based approximation procedure. They also proposed an extension of this approximation as follows:

$$m(D) = m_0 + m_1 \log D \tag{5.162}$$

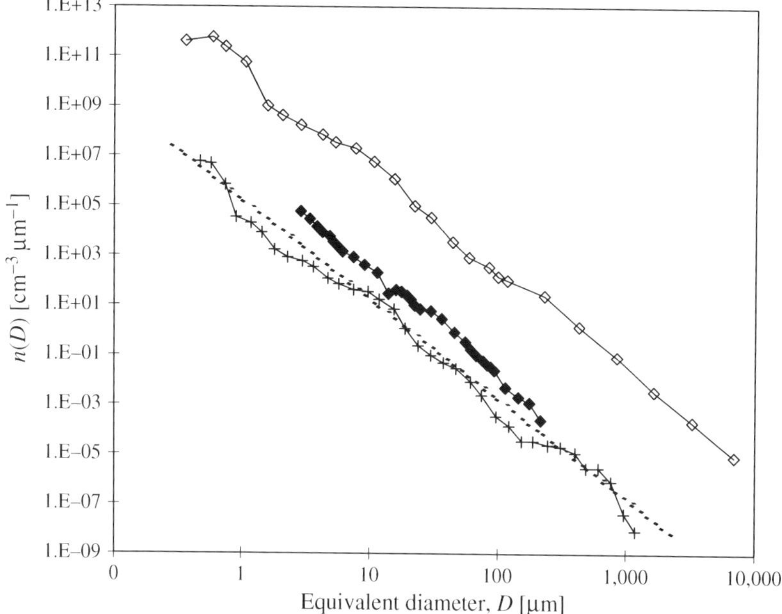

Figure 5.32. Large size range PSDs follow, on average, a power law with slope of about −4 (dashed line), although a distinct waviness is clearly visible at smaller size scales. The data shown are digitized from *Gaedke* (1992, her Fig. 3, a large alpine Lake Constance, +), *Jackson* and Burd (1998, Monterey Bay, CA, USA, ♦), and *Quinones* et al. (2003, Sargasso Sea, ◇). The data of Gaedke were transformed from the normalized biomass spectra (NBS), $b_N(C)$ to $n(D)$, where C is the carbon mass per organism and D is the equivalent spherical diameter (ESD). The size scale was transformed from carbon (C) per organism to ESD, by first converting C (pg) to volume (μm^3) by using an average ($V = 0.141/C$) of the conversion factors listed by Gaedke and then converting the volume to ESD. The data of Quinones et al. were converted to $n(D)$ from the NBS form. The size scale was transformed to ESD from the organism volume scale. The Jackson and Burd data were converted to $n(D)$ from $n(D_{opt})$, where D_{opt} is the equivalent optical diameter according to (5.40) with $c = 1$ and $d = 1.8$.

This is consistent with the results presented in Figure 5.33 except that the value of the slope m_1, for the D_C case, is substantially lower (0.59) than the sample value of ~1.3 given by Ceronio and Haarhoff for the case of (5.161) for drinking water obtained through South African treatment plants.

5.8.5.4. Hyperbolic distribution

The hyperbolic distribution, pioneered by *Barndorff*-Nielsen (1978, 1977), is closely related to the power-law distribution (section 5.8.5.3). The hyperbolic

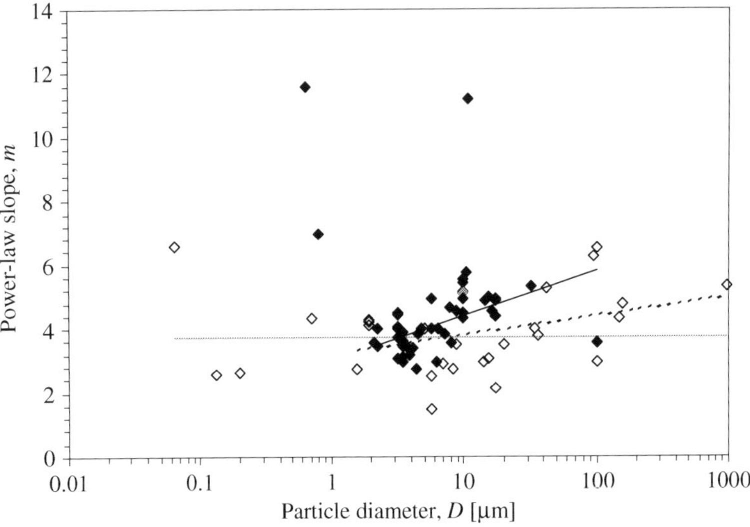

Figure 5.33. The relationship between the average slope, m, of the power-law approximation for the size distribution of marine particles and the particle diameter, D, for volume-sensitive measurement methods ($D = D_S$, solid symbols and line) and for projected area-sensitive methods ($D = D_C$, open symbols and dashed line), all based on data from Table 5.8 and Table 5.9. Each data point refers to a geometric average of endpoints of a finite diameter range and represents typically an average slope of the power-law approximation to the size distribution in that diameter range. Sub-micron particle data are excluded from the regression data groups. The volume-sensitive regression group does not include a lone high-slope data point at about 10 μm (gray solid symbol, *O'Hern* et al. 1988) that is markedly different from all other data reported for this size range. Data of *Brun*-Cottan (1971), *Atteia* and Kozel (1997), and *Bradtke* (2004) were not available at the time of the calculations and have not been used in the regression. These data conform to the trend. Regression lines: $m = m_S = 3.03 + 1.38\ \log D_S$, $r^2 = 0.31$ and $m = m_C = 3.23 + 0.59\ \log D_C$, $r^2 = 0.14$. Average slopes are $m_S = 4.17 \pm 0.85$ and $m_C = 3.91 \pm 1.14$ excluding sub-micron size range in each case. One standard deviation is given following the $\pm$ sign. The long horizontal gray line represents the average of slopes ($m_C = 3.72 \pm 0.05$) for unspecified diameter ranges that were obtained mostly through fitting the optical properties of the particle populations (the last few entries in Table 5.9). A weak tendency of the slope to increase with the particle diameter, apart from the sub-micron diameter range, reflects a slight but distinct curvature of the log-log plots of the size distribution of aquatic particles, accounted for by multi-segment power-law approximations, as shown in Figure 5.31.

distribution is a finite distribution defined by two asymptotic power-law functions, one with a positive and the other with a negative exponent. Thus, the hyperbolic distribution has a peak near the particle diameter at which these two power-law functions intersect. The equation of the resulting distribution has this form (*Bagnold* and Barndorff-Nielsen 1980):

$$\log n(D) = -\alpha\sqrt{\delta^2 + (\log D - \log D_0)^2} + \beta((\log D - \log D_0) + k \tag{5.163}$$

where $\alpha, \beta, \delta, D_0$, and k are fit parameters. Parameter k is the scale factor, D_0 is the diameter at which the two asymptotic power laws intersect. The slopes of these asymptotes, both of the form $y = kx^{-m}$, are:

$$m_1 = \alpha + \beta$$
$$-m_2 = \alpha - \beta \tag{5.164}$$

Parameter δ controls the radius of curvature of the transition between the two asymptotes, the smaller the value, the sharper the turn from one to the other asymptote. *Bagnold* and Barndorff-Nielsen (1980) give the following geometrical interpretation for that parameter: a derived parameter ξ, defined as follows:

$$\xi = \delta\sqrt{m_1 m_2} \tag{5.165}$$

is the distance, Δn, between the intersection of the two asymptotes and the maximum value of $n(D)$.

This distribution has been used to approximate experimental size distributions of a variety of particle ensembles (e.g., *Durst* and Macagno 1986) including marine sediment (e.g., *Bagnold* and Barndorff-Nielsen 1980). We found it to approximate some size distributions of marine bacteria better (Figure 5.22) than either the log-normal or gamma distribution. A major disadvantage discouraging its wider use in the approximation of the size distribution of aquatic particles is the need to use a non-linear minimization process to obtain the fit parameters (see *Christiansen* and Hartmann 1991, as well as *Durst* and Macagno 1986 for details on the fitting programs).

5.8.5.5. Normal (Gaussian) function

The size distribution of a phytoplankton species has a well-defined mean diameter and diminishes for both larger and smaller diameters than that mean diameter. This fits the description of a Gaussian function. Indeed, a combination of Gaussian functions:

$$n(D) = \sum_i a_i \exp[-t(D - D_{0i})^2] \tag{5.166}$$

where i numbers the function used, fits experimental data on phytoplankton components of marine size distributions *relatively well* (*Jonasz* and Prandke 1986,

Jonasz 1983a, *Jonasz* 1980). Coefficients a, t, and D_0 were found to range respectively from about 410 to 800 $\text{cm}^{-3}\mu\text{m}^{-1}$, 0.6 to 0.7 μm^{-2}, and 6 to 6.2 μm (values typical of several surface water samples in the Baltic, June 1977). Components of this type were associated with phytoplankton species. Otherwise only the power-law term was observed. *Kullenberg* (1970) attempted to fit his experimental volume scattering functions by using a Gaussian approximation to a hypothetical PSD. However, his data were too few to differentiate convincingly between the Gaussian and the power-law approximation.

The major conceptual disadvantage of the Gaussian function as a functional representation of the size distribution is the fact that it allows unrealistic negative particle diameters. This disadvantage is removed by replacing the Gaussian function with the log-normal function (*Jonasz* and Fournier 1996).

Pawlak and Kopec (1998) note that the Gaussian (or normal) distribution represents but one case of a class of φ-normal probability distribution functions with the cumulative distribution, $F_\varphi(D)$, defined as follows:

$$F_\varphi(D) = P[f(D) < \varphi(D)] = F\left[\frac{\varphi(D) - \mu}{\sigma}\right] \tag{5.167}$$

where $F(x)$ is the cumulative normal probability distribution of the standardized statistical variable $x = [\varphi(D) - \mu]/\sigma$, where the mean value and standard deviation of $\varphi(D)$ are respectively μ and σ. The function $\varphi(D)$, defined in a range (D_0, ∞) with $D_0 \geq 0$, with $\lim_{D\to D0+} \varphi(D) = -\infty$ and $\lim_{D\to\infty} \varphi(D) = +\infty$ must be differentiable in that range. If $\varphi(D) = D$, we obtain the normal distribution of D. If $\varphi(D) = \ln D$ we obtain the log-normal distribution (section 5.8.5.6). *Pawlak* and Kopec (1998) used the following function $\varphi(D)$ to approximate the size distribution of *Scenedesmus obliquus*:

$$\varphi(D) = D - a\frac{D-b}{(D-b)^2 + c} \tag{5.168}$$

where a, b, and c are the fit parameters and $c > 0$ and the second term can be regarded as a correction to the normal distribution case, $\varphi(D) = D$.

5.8.5.6. The log-normal function

Like the power-law distribution discussed in section 5.8.5.3, the log-normal probability distribution has applications in diverse areas, ranging from business (*Shimizu* and Crow 1988) to oceanography (*Campbell* 1995). *Limpert* et al. (2001) review applications of the log-normal distribution in various sciences. The log-normal distribution is generally the result of a process, which can be mathematically characterized by a product of many random variables, for example the process of fragmentation. Indeed, a fragmentation process with the probability of fragmentation independent of the particle size leads to the log-normal function

(*Shimizu* and Crow 1988, *Middleton* 1970) as originally found by Kolmogorov in 1941 (cited by *Tenchov* and Yanev 1986). If the probability of fragmentation is proportional to the particle size, the Weibull distribution (section 5.8.5.10) results. However, the difference between a log-normal distribution and a Weibull distribution may be made quite small by the appropriate selection of the distribution parameters (*Tenchov* and Yanev 1986). Thus, it may be difficult to discern at the measurement precision characteristic of the particle size analysis techniques applicable to aquatic particles. *Aitchinson* and Brown (1957, Section 10.2) summarize applications of the log-normal distribution in the approximation of the PSD. *Crow* (1988) discusses applications of the log-normal distribution to model the size distribution of atmospheric particles. *Heintzenberg* (1994) discusses the properties of the log-normal distribution as applicable to PSD approximations and calculations.

The use of log-normal distributions for approximating the PSD has important advantages:

(1) there is no need to "break" the power-law approximation to reflect changes in the log-log slope of the size distribution;
(2) the log-normal approximation assumes finite values for all particle diameters, except in the limit $B_2 > 0$, when the log-normal function becomes a power-law function;
(3) mass and area distributions resulting from the log-normal distribution are also log-normal (*Kerker* 1969);
(4) the power-law function is a limiting case of the log-normal function, so it is naturally included as an approximation of the size distribution.

We should also mention certain pitfalls of choosing the log-normal distribution after *Halley* and Inchausti (2002), who cite comments by *Mandelbrot* (1997) (incidentally these problems also relate to other "long-tailed" distributions):

(1) extreme sensitivity of all moments of the distribution to even small departures from log-normality, which makes the calculation of moments from the parameters of a fitted log-normal distribution unreliable
(2) slow convergence of the approximate values of the moments to their asymptotic value. This feature is due to an extremely "long tail" of the log-normal distribution at the large-size end of the particle size scale. We have experienced this problem ourselves (*Jonasz* and Fournier 1996) when trying to evaluate correlations between the total particle surface and volume and the peak diameter of the distribution as fitted to several hundred size distributions of marine particles.

The log-normal size distribution of the zero-th order can be expressed as follows (e.g., *Ross* 1978, *Casperson* 1977):

$$n(D) = n_{\max} \exp\left[-\frac{(\ln D - \ln D_{peak})^2}{2\sigma^2}\right] \tag{5.169}$$

where $n_{\max}$ is expressed as follows:

$$n_{\max} = \frac{N_{\text{tot}}}{\sqrt{2\pi}\sigma D_{\text{peak}} \exp\frac{\sigma^2}{2}} \tag{5.170}$$

N_{tot} is the total number of particles [equal to unity in the case of $n(D)$ being the probability distribution], D_{peak} is the particle diameter corresponding to the peak of the size distribution, and σ is the standard deviation of $\ln D$, i.e., the geometric standard deviation of D.

The width parameter, σ, is related to the ratio of the maximum and minimum diameters, $D_{\max}$ and $D_{\min}$, of the full-width-at-half-maximum of the log-normal $n(D)$ through the following equation (*Jonasz* and Fournier 1996):

$$\frac{D_{\max}}{D_{\min}} = \exp(2f\sigma) \tag{5.171}$$

where

$$\begin{aligned} f &= \sqrt{2\ln 2} \\ D_{\min} &= D_{\text{peak}} \exp(-f\sigma) \\ D_{\max} &= D_{\text{peak}} \exp(f\sigma) \end{aligned} \tag{5.172}$$

In contrast to the power-law distribution, the log-normal size distribution yields a finite total number of particles, N_{tot}, total particle cross-sectional area, A_{tot}, and volume, V_{tot} (*Heintzenberg* 1994):

$$\begin{aligned} A_{\text{tot}} = \pi &\left[\frac{1}{2}D_{\text{peak}} \exp(2\sigma^2)\right]^2 N_{\text{tot}} \\ &\times \exp\left\{-\frac{1}{2\sigma^2}\left[(\ln D_{\max} \exp(2\sigma^2)) - \ln D_{\max}\right]^2\right\} \end{aligned} \tag{5.173}$$

and

$$\begin{aligned} V_{\text{tot}} = \frac{4}{3}\pi &\left[\frac{1}{2}D_{\text{peak}} \exp(3\sigma^2)\right]^3 N_{\text{tot}} \\ &\times \exp\left\{-\frac{1}{2\sigma^2}\left[(\ln D_{\max} \exp(3\sigma^2)) - \ln D_{\max}\right]^2\right\} \end{aligned} \tag{5.174}$$

Equations (5.173) and (5.174) apply to the –1-th order generalized log-normal particle size distribution (see, e.g., *Casperson* 1977), not to the zero-th order mostly discussed in this section.

Note that successful applications of these formulas, as also pointed out by *Heintzenberg* (1994), require that the parameters of the log-normal distribution be evaluated for a particle size range that contributes significantly to the total projected area and volume.

The log-normal function was postulated to approximate the size distribution of marine particles in samples of seawater from several GEOSECS stations in the Atlantic and Pacific, at depths ranging from 286 to 5474 m (*Lambert* et al. 1981). The particles were collected on 0.4 μm Nuclepore filters and analyzed with a scanning electron microscope. Portions of the filters were coated with carbon and examined using a scanning electron microscope equipped with an X-ray elemental analysis accessory, which enabled the determination of species-specific PSDs. A total of between 100 and 500 particles were analyzed for each sample in a diameter range of 0.2 to 10 μm. The diameter of a particle is taken to be the diameter of a circle with an area equal to that of the particle. The peak diameter was between 1 and 2.5 μm, and the width parameter σ of the size distribution was found to be in a range of 0.5 to 0.7. The quality of the log-normal approximation could in some cases be significantly improved by eliminating the extreme data points in the tails of the distribution. This might be due to the presence in the size distribution of other modes, due to particle populations marginally overlapping in size with the main particle size range.

The log-normal function was also found to approximate well the size distributions of non-spherical clay particles (*Jonasz* 1987b) measured using a Coulter counter, model ZBI with a 100 μm aperture, and using an HIAC particle counter, model 320 with a CMH-150 particle size sensor.

The cell size distributions reported in the literature frequently appear to be log-normal at visual inspection (see Table A.5 for sources of the relevant PSD data). Interestingly, mathematical models of cell growth and evolution of isolated cell populations do not lead to a log-normal distribution of cell sizes (e.g., *Tyson* and Hannsgen 1985). However, the agreement between the models and the experimental data is questionable. In fact, Tyson and Hannsgen note that cell size distributions with log-normal size distribution have been reported (*ibid. Scherbaum* and Rasch 1957, *Collins* and Richmond 1962). Analysis of the biomass spectrum in aquatic ecosystems composed of organism groups linked via a prey–predator relationship led to the log-normal function as a natural descriptor of the contribution of an organism to the total ecosystem biomas spectrum (*Thiebaux* and Dickie 1993, *Boudreau* et al. 1991). The derivation of the log-normal form of the size distributions of the individual organism groups was based on the fact that the production, $P(w)$, is proportional to a power, b, of the body mass, W (allometric relationship):

$$P(W) = aW^b \tag{5.175}$$

The observation of (1) a persistent curvature of the size distribution of marine particles, when plotted in a log-log scale, (2) the multimodal appearance of many such size distributions, as well as (3) previous suggestions in the literature that complex size distributions of geological material can be well modeled by a sum of log-normal functions (*van Andel* 1973) led us to develop an automated algorithm of the decomposition of a marine PSD into a sum of log-normal components (*Jonasz* and Fournier 1999, 1996). In that work, we postulated that the size distribution of marine particles is essentially a linear combination of a cascade of log-normal components, according to the following equation:

$$n(D) = \sum_{1}^{k_{\max}} n_k\ (D) \tag{5.176}$$

where index k numbers the log-normal components $n_k(D)$. Each of these components is approximated with a zero-th order log-normal distribution function:

$$n_k\ (D) = n_{\max,k}\ \exp\left[-\frac{(\ln D - \ln D_{peak,k})^2}{2\sigma_k{}^2}\right] \tag{5.177}$$

where $n_{\max,k}$ is the maximum value of the component, $D_{\text{peak,k}}$ [μm] is the peak diameter, and σ_k is the width parameter. By taking the logarithm of both sides of (5.177) and performing simple algebraic transformations, one obtains (we omit the component index for simplicity):

$$\log n(D) = B_0 + B_1 \log D + B_2(\log D)^2 \tag{5.178}$$

where:

$$\begin{aligned} B_0 &= \log n_{\max} - (\log D_{\text{peak}})^2 \frac{\ln 10}{2\sigma^2} \\ B_1 &= \log D_{\text{peak}} \frac{\ln 10}{\sigma^2} \\ B_2 &= -\frac{\ln 10}{2\sigma^2} \end{aligned} \tag{5.179}$$

Note that (5.178) reduces to a power law when B_2 vanishes. Equations (5.179) can be solved for $n_{\max}$, D_{peak}, and σ to yield:

$$\begin{aligned} \log n_{\max} &= B_0 - \frac{B_1^2}{4B_2} \\ \log D_{\text{peak}} &= -\frac{B_1}{2B} \\ \sigma^2 &= -\frac{\ln 10}{2B_2} \end{aligned} \tag{5.180}$$

We consider only those functions defined by (5.177) which fulfill the condition of $B_2 < 0$. This ensures that the extremum of the function is a maximum. According to the assumption about the size distribution being a cascade of log-normal components, each log-normal component dominates in a particular size interval. Thus, if that size interval is somehow identified, one could determine the parameters of the respective log-normal component, for example, by using the least-squares fitting procedure for the log-log transformed original data.

In the algorithm of *Jonasz* and Fournier (1996) for fitting a sum of log-normal functions to PSD data, the size interval dominated by a particular log-normal function is identified by repeatedly scanning the size distribution with a window whose width is systematically varied. During each scan set with a fixed window width, the log-normal function is fitted to the data from within the window. Since the number of data points in a size distribution is usually quite moderate, the quality of the log-normal fit for all realistic window widths and locations can be assessed. Once a log-normal component is found, it is subtracted from the PSD data, and the modified data serve as the input for the next round of scans with this window width. If the PSD value (data point) from which a component value at that particle size has been subtracted falls below a preset limit, that data point is removed from the set. Thus, normally the number of data points decreases during this procedure which is terminated if there is no sufficient data left or, less likely, if no components have been found. For each of the next set of scans, the window width is incremented by unity, until the maximum allowed window width. Each set of scans may result in a set of log-normal components which approximate the original data with a specific accuracy. The algorithm is completed by selecting a set of log-normal fits based on either the minimum of the approximation error or other criterion set by the user, for example, the approximation error and the number of components.

Important comments are in order here. First, although the PSD may be better approximated with several, "interpretable" components, the extent of this interpretability is limited by the number of degrees of freedom of the data set, because each log-normal component reduces the number of degrees of freedom by 3. Second, each new component increases the number of parameters required for the description of the data set. From the purely numerical perspective, this seems to be counterproductive—we noted earlier that one goal of the approximation is to reduce the number of parameters. Indeed, given a sufficiently large number of components one simply exchanges the original data (the primary set of parameters) by an equally numerous set of the fit parameters. However, once the "interpretability" aspect is acknowledged, the advantage of that exchange should become clear, as there is usually little interpretability in the original data set. We also discuss this aspect of the PSD analysis later in this section.

Sample results are shown in Figure 5.34 and Figure 5.35. Log-normal components identified by the algorithm just described range in shape from the power-law-like function to Gaussian-normal-like function. Note that the failure

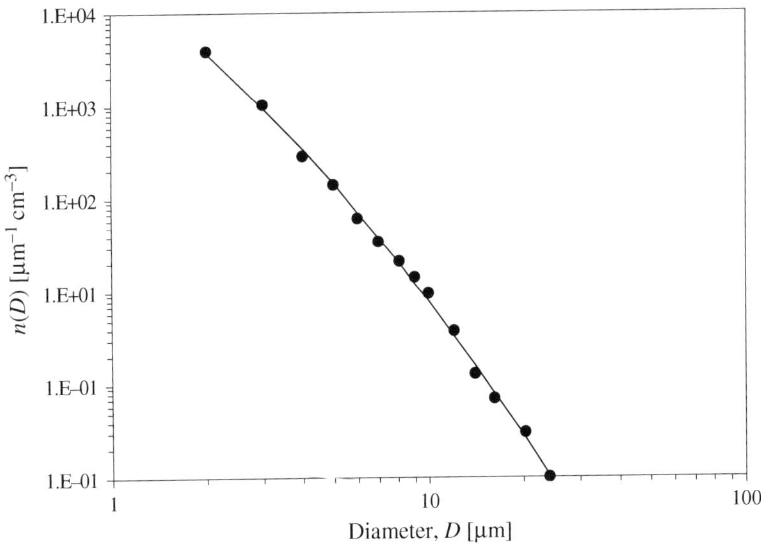

Figure 5.34. A log-normal approximation of the size distribution of Figure 5.31: $\log n(D) = 4.41 - 2.50 \log D - 1.02 (\log D)^2$ (χ^2 per degree of freedom = 0.24). The fit parameters were obtained via the logarithmic transform. All weights were set to unity. The χ^2 was calculated by assuming only the counting error.

to account for the instrumental error (see section 5.7.1.6) in evaluating the χ^2 results in the χ^2 value in excess of 200 per degree of freedom (!) in the case of data shown in Figure 5.35. that include very high particle count values.

The statistics of and correlations between the parameters of 853 log-normal components of the 412 PSDs determined using the Coulter technique by different researchers in different areas and seasons are shown in Table 5.10. The average values of coefficients B_0, B_1, and B_2 represent a geometrically averaged component characteristics, i.e., $n_{avg}(D)$ such that $n_{avg}(D) = [n_1(D)\ n_2(D) \ldots\ n_m(D)]^{1/m}$. The average values of n_{max}, D_{peak}, and σ do not have simple meanings and are given here for the sake of completeness only. The two sets of averages do not yield the same function of the diameter, D, because they are related via non-linear functions (5.179) and (5.180). The average error of approximation of the size distribution with the sum of log-normal components was 0.057 ± 0.030. The number of components per size distribution varied from 1 to 6, with an average of 2.18 ± 1.22. The value of 1 standard deviation (SD) is shown following the $\pm$ sign.

Significant correlations exist between the D_{peak} and n_{max}, as well as between D_{peak} and the width parameter, σ, of the component, as can be seen in Figure 5.36

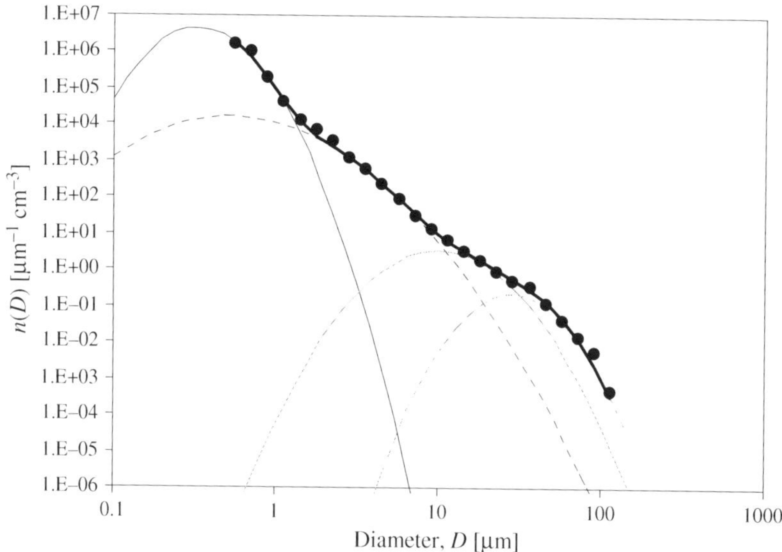

Figure 5.35. A multi-component log-normal approximation (thick black curve) of a PSD measured in the Northwest Atlantic waters (•, unpublished data: courtesy of K. Kranck and T. Milligan, file KRAATL86.P08 in *Jonasz* 1992). The approximation coefficients from equation (5.178): first component (thin solid curve) $B_{01} = 4.05$, $B_{11} = -1.16$, $B_{21} = -2.12$, second component (dashed curve) $B_{02} = -17.04$, $B_{12} = 22.52$, $B_{22} = -7.75$, third component (gray solid curve) $B_{03} = 4.97$, $B_{13} = -7.04$, $B_{23} = -7.30$, fourth component (gray dashed curve) $B_{04} = -4.27$, $B_{14} = -9.60$, $B_{24} = -4.81$. The first log-normal component removes the greatest amount of the approximation error. Each of the following components removes a progressively smaller amount of that error (χ^2 per degree of freedom = 0.021). The fit parameters were obtained via the logarithmic transform. All fitting weights were set to unity. The χ^2 was calculated by assuming both the counting and instrumental errors.

and Figure 5.37. The equations of the approximating lines (see also the correlation coefficients in Table 5.10) shown in these two figures are respectively:

$$\ln n_{\max} = (8.070 \pm 2.799) - (2.446 \pm 0.032) \ln D_{\text{peak}} \tag{5.181}$$

and

$$\sigma = (0.626 \pm 0.186) - (0.111 \pm 0.002) \ln D_{\text{peak}} \tag{5.182}$$

where the value of 1 SD of the respective parameter is shown following each $\pm$ sign.

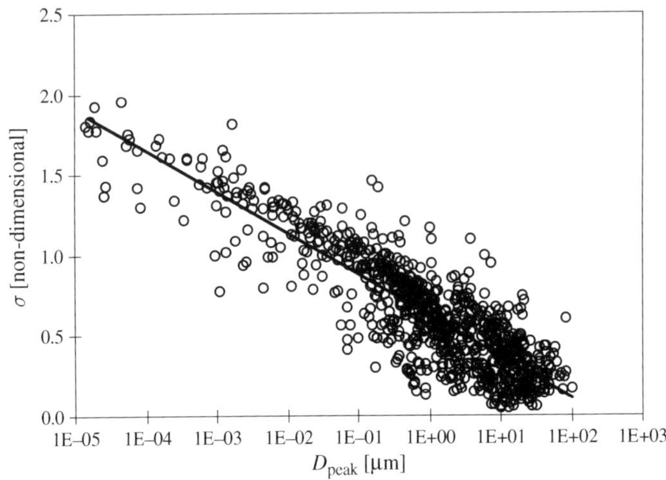

Figure 5.36. Relationship between σ and D_{peak} for 853 log-normal components of 412 particle size distributions (*Jonasz* and Fournier 1996) measured in various waters and seasons by several researchers (as compiled by *Jonasz* 1992). Approximation line equation: $\sigma = (0.626 \pm 0.186) - (0.111 \pm 0.002) \ln D_{peak}$, with 1 SD shown following each $\pm$ sign ($r^2 = 0.760$).

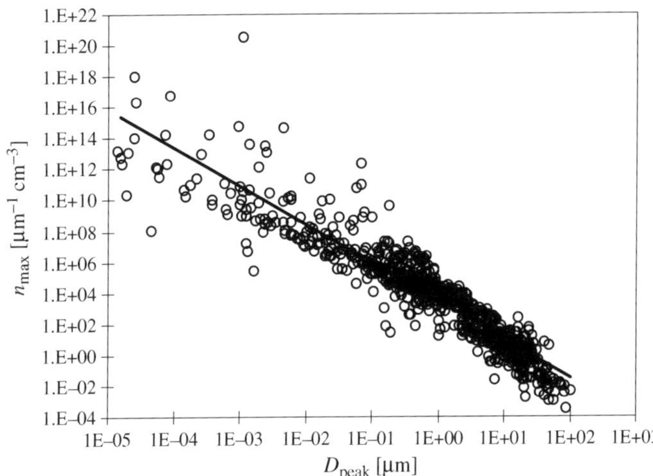

Figure 5.37. Relationship between n_{max} and D_{peak} for 853 log-normal components of 412 particle size distributions (*Jonasz* and Fournier 1996) measured in various waters and seasons by several researchers (as compiled by *Jonasz* 1992). The range of n_{max} is limited to keep the number of decades on the n_{max}-axis manageable. The data points not shown conform to the general trend. Approximation line equation: $\ln n_{max} = (8.070 \pm 2.799) - (2.446 \pm 0.032) \ln D_{peak}$, with 1 SD shown following each $\pm$ sign ($r^2 = 0.873$).

Table 5.10. Correlations, expressed using r^2, between the parameters of log-normal components of marine particle size distributions measured with a Coulter counter (*Jonasz* and Fournier 1996—412 size distributions measured by various authors in various seasons and areas of the world ocean).

Parameter	**ln n_{max}**	**ln D_{peak}**	**σ**
ln n_{max}	1.000	0.873	0.478
ln D_{peak}	0.873	1.000	0.760
σ	0.478	0.760	1.000

Parameter	**B_0**	**B_1**	**B_2**
B_0	1.000	0.963	0.840
B_1	0.963	1.000	0.942
B_2	0.840	0.942	1.000

The correlations between the coefficients, B_0, B_1, and B_2 are greater than those between the parameters n_{max}, D_{peak}, and σ because of the smoothing effect of the logarithmic transform.

The log-normal components, which range in shape from the power-law-like function to Gaussian-normal-like function, may be interpreted as the size distributions of the various classes of marine particles, for example, populations of various phytoplankton species. Indeed, *Jonasz* and Fournier (1996) found two "standard" components (Figure 5.38 and Table 5.11) in 412 size distributions measured in various seasons and regions of the world ocean by different researchers. *Bradtke* (2004) who used the algorithm just described to analyze 970 PSDs measured with a Coulter counter in the coastal waters of the Baltic Sea (Gdańsk Bay) also noted the existence of several "standard" components and linked other transitional components to the occurrences of phytoplankton species.

Although "standard" components of the marine size distribution can be generated by other techniques, for example, by the method of characteristic vectors (see section 5.8.5.11), these components lack the physical and biological interpretation possible for the log-normal components. The existence of "standard" components as identified by *Bradtke* (2004) as well as *Jonasz* and Fournier (1996) with this algorithm is remarkable, as the decomposition algorithm approaches the task from a purely numerical perspective. It would have certainly been desirable to enhance it so that such "standard" log-normal components attributable to various particle species would be searched for rather than arbitrary log-normal functions whose sum happens to minimize the approximation error of the PSD.

This approach would be particularly attractive from the biological point of view, given the use of log-normal functions to represent ecosystems in natural waters (*Thiebaux* and Dickie 1993, *Boudreau* et al. 1991). However, intraspecies

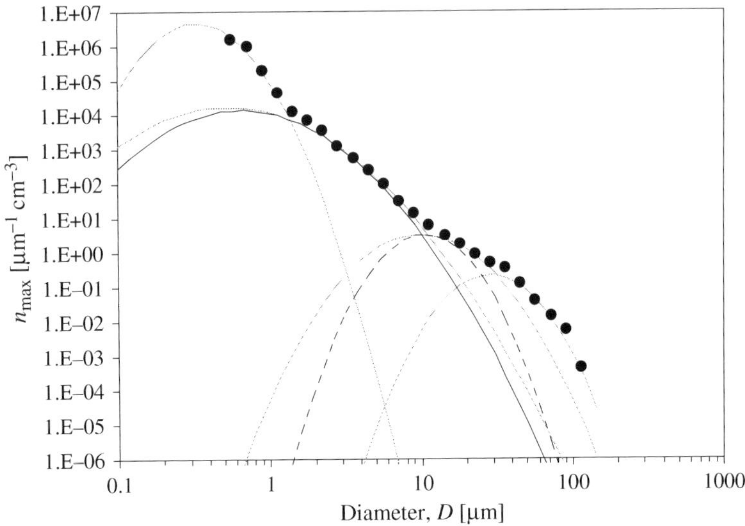

Figure 5.38. "Standard" components (Table 5.11) of the marine particle size distribution identified by *Jonasz* and Fournier (1996), who analyzed 412 particle size distributions measured in various seasons and regions of the world ocean by different researchers (as compiled by *Jonasz* 1992): first component (solid black curve) $B_{01} = 4.038$, $B_{11} = -0.9511$, $B_{21} = -2.542$, second component (dashed curve) $B_{02} = -8.447$, $B_{12} = 17.55$, $B_{22} = -8.595$, compared with the size distribution from Figure 5.35 (symbols) and its retrieved components (gray solid lines).

Table 5.11. Parameters of two "standard" components of the marine size particle distribution (*Jonasz* and Fournier 1996).

Parameter	**Component 1**	**Component 2**
n_{max} [$\mu m^{-1} cm^{-1}$]	13400	3.28
D_{peak} [μm]	0.65	10.5
σ	0.673	0.366
B_0	4.038	−8.447
B_1	−0.9511	17.55
B_2	−2.542	−8.595

variability of the PSDs "characteristic" for a particle species (see examples in section 5.8.4.4.) may make such a decomposition difficult especially when the differences between the approximation errors of a size distribution with various component sets are relatively small.

Jonasz and Fournier (1996) found that only about 5% of the components of the marine size distribution could be well approximated with a power law, which is represented with a straight line on a plot of $\log n(D)$ vs. $\log D$. Interestingly, this conclusion does not rule out the first-order representation of the PSD of marine particles by a power-law function in a large range of particle diameter. Such a function would simply be an envelope of the sum of log-normal components.

5.8.5.7. Theoretical models of the size distribution of living cells

In a growing cell population (exponential growth phase), each parent cell divides into two daughter cells. Thus, the cell number grows exponentially, hence the name of that growth phase. If the only property of the population in that growth phase that changes is the cell number, the cells are in a steady-state growth phase. This situation was considered by *Collins* and Richmond (1962), who derived a general equation for the probability distribution of sizes of living cells in steady-state exponential growth (after *Koppes* et al. 1987):

$$\frac{dD}{dt} = \frac{g}{p_{exi}(D)}[2P_{new}(D) - P_{exi}(D) - P_{div}(D)] \tag{5.183}$$

where g is the cell growth rate, defined as ln2 times the number of cell doublings per hr, P_{new}, P_{exi}, P_{div} are the cumulative probability distribution of newborn, existing, and dividing cells, and $p_{exi}(D) = dP_{exi}(D)/dD$. In the original formulation of the Collins–Richmond equation, the probability distributions of the cell size were measured and the growth rate, g, was deduced from (5.183). In the reverse approach (e.g., *Koppes* et al. 1987), the P_{exi} was determined from (5.183) by assuming certain forms for the dependency of the growth rate on the cell size. Note that the determination of the probability distribution, P_{exi}, from (5.183) requires the knowledge of P_{new} and P_{div}.

Bell and Anderson (1967) derived a general equation for the size distribution of cells in a culture in an exponential growth phase. They assumed that the growth rate $g(V)$ is proportional to the cell volume, V, i.e., $g(V) = g_1 V$ (an exponential growth model) and that the probability $p(V)$ of cell division, when the cell volume is V, is proportional to the difference between the cell volume, V, and the volume at which the division should occur (twice the minimum cell volume, V_0, of the cell just after a division), $p(V) = p_1(V - 2V_0)$. Under these conditions, the size distribution of the cells is expressed with two equations, each valid for a different size range:

$$n(V) = n(2V_0)\left(\frac{V}{2V_0}\right)^{\gamma-2} \exp\left[-\gamma\left(\frac{V}{2V_0} - 1\right)\right] \tag{5.184}$$

if $V \geq 2V_0$, and

$$n(V) = 4n(2V_0)\left(\frac{V}{2V_0}\right)^2 \gamma \exp(-\gamma) \int_1^{V/V_0} (x-1)x^{\gamma} \exp(-\gamma x)\frac{dx}{x} \tag{5.185}$$

if $V < 2V_0$, where in both cases $\gamma = 2p_1 V_0/g_1$. The partitioning of the entire size range is related to the fact that the cells divide roughly after having grown to a volume $2V_0$.

Tyson and Hannsgen (1985) also considered the exponential growth phase and arrived at the following equation for the size distribution of cells:

$$n(V) = \frac{2}{V_0}\frac{q-1}{2-q}\left[\left(\frac{V}{V_0}\right)^{-q} - \left(\frac{V}{V_0}\right)^2\right] \tag{5.186}$$

for $V_0 \leq V \leq 2V_0$ and

$$n(V) = \frac{2^{2-q}-1}{2V_0}\frac{q-1}{2-q}\left(\frac{V}{2V_0}\right)^{-q} \tag{5.187}$$

for $2V_0 \leq V$, where in both cases, V_0 is the cell volume at birth, and q is a positive constant on the order of 1 to 10. They have also assumed that the growth rate is proportional to the cell size V.

Campbell and Yentsch (1989) also considered the exponential growth of a cell population and assumed a constant growth rate of cell volumes. These authors did not attempt to derive the size distribution from the dynamics of cell population. Rather they derived an approximation formula for the distribution of cell volumes which can be used to fit an experimental distribution characterized by the average cell volume V_m and the standard deviation σ of the volume:

$$n(V) = \frac{2\mu - 1}{r} \exp\left[-\frac{1}{2}\left(\frac{\sigma\mu}{r}\right)^2\right] \exp\left[-\frac{\mu(V - V_0)}{r}\right][F(x) - F(y)] \tag{5.188}$$

where r is the linear rate of growth of the cell volume, μ is the exponential rate of cell division, F is the standard cumulative normal distribution of probability, and $x = (V' - V_m)/\sigma$, $y = (V' - 2V_m)/(2\sigma)$, and $V' = V_0 - \sigma^2\mu/r$.

5.8.5.8. Gamma function

The gamma distribution function is defined as an extension of the gamma probability distribution (e.g., *Korn* and Korn 1968):

$$n(D) = cD^a \exp(-bD) \tag{5.189}$$

where c and $b > 0$, and

$$a = 1 + \frac{\langle D \rangle^2}{\sigma^2(D)}$$
$$b = \frac{\langle D \rangle}{\sigma^2(D)} \tag{5.190}$$

with $\langle D \rangle$ being the average diameter relative to $1\,\mu m$ and $\sigma^2(D)$ is diameter variance. The total number of particles, N_{tot}, is here finite and equal to (e.g., *Risović* 1993):

$$N_{tot} = c\frac{\Gamma(a+1)}{b^{a+1}} \tag{5.191}$$

where $\Gamma(x)$ is the integral of the gamma function

$$\Gamma(x) = \int_0^\infty t^{x-1} \exp(-t)dt \tag{5.192}$$

also referred to as the gamma function itself (e.g., *Korn* and Korn 1968).

A sample fit to experimental data is shown in Figure 5.39.

The gamma function (5.189) and its linear combinations have been shown to approximate well the size distributions of marine bacteria (*Risović* 1993). The gamma function has also been reported to represent the size distributions of marine bacteria better than the log-normal function, according to the χ^2 test (*Ulloa* et al. 1992).

Jonasz (1980) fitted the gamma approximation to over 80 size distributions of marine particles in a range of about 2 to $40\,\mu m$ and found that this approximation fits the experimental data worse than either the power law or the characteristic vectors.

The gamma function has been used to describe the size distribution of PSDs in sedimentology (*Kranck* 1993, 1987, 1986 *Kranck* and Milligan 1991). According to the derivation of *Kranck* and Milligan (1985), the volume size distribution, $v(D)$, of mineral sediment particles can be approximated by a gamma function written as follows:

$$v(D) = QD^h \exp[-Kv_s(D)] \tag{5.193}$$

where Q [ppm] is the scale factor, h is the slope of the volume distribution of the original size distribution from which sedimentation occurs, parameter K [s cm^{-1}] depends on the turbulence of the water body, and v_s [cm s^{-1}] is the sedimentation velocity of the particles with diameter D [μm].

Term D^h describes the volume distribution whose number size distribution is expressed with a power law: $h = m + 4$, where m is the slope of the frequency size

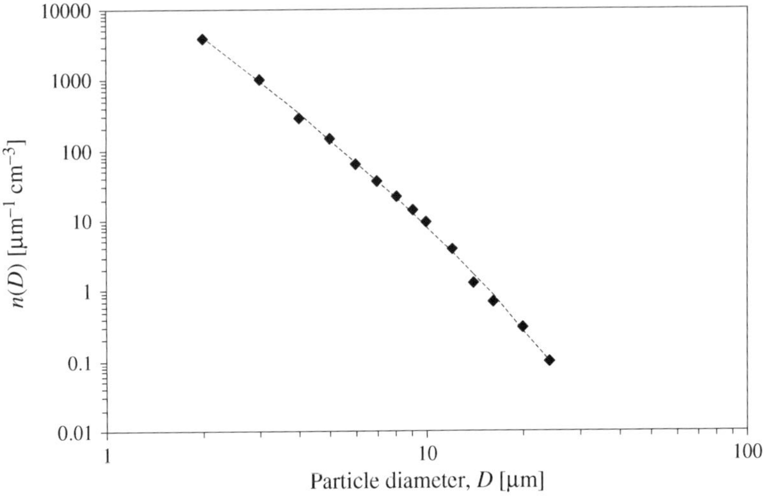

Figure 5.39. A particle size distribution measured with a Coulter counter in the surface waters of the mid-Atlantic (solid diamonds, *Jonasz* 1992—file JONATL78.P08, column 4) is approximated using the gamma distribution [dashed curve, equation (5.194), $c = 4.992 \times 10^4$, $a = -3.312$, $b = 0.1099$]. This gamma approximation yields a χ^2 per degree of freedom = 0.094. In comparison, a log-normal approximation shown in Figure 5.34 yields χ^2 per degree of freedom= 0.24. The fit parameters were obtained via a logarithmic transform with the weights set to unity. The χ^2 was calculated by assuming only the counting error.

distribution expressed with the power-law function (5.156). The exponential term, $\exp(-Kv_s)$, describes the loss of particles from the suspension due to settling (see section 5.7.11).

The slope coefficient, h, and the cut-off parameter, K, range roughly between 0.2 and 0.3 and 10 and 25 respectively for processed samples of river sediments (Indus river, *Kranck* 1987). The processing was aimed at removing organic matter and de-flocculation of the particles, so that the single-grain mineral size distribution could be measured. These parameters assume respectively the ranges of 1.32 to 3.56 and between 1.45 and 33 for *in situ* size distributions obtained with a camera in coastal waters (San Francisco Bay, *Kranck* and Milligan 1992).

A modified gamma function was proposed by *Risović* (1993):

$$n(D) = cD^a \exp(-bD^g) \tag{5.194}$$

where, a, b, c, and g are the fit parameters, for the approximation of the global size distributions, i.e., the sums of the size distribution of individual species of particles in seawater as well as the size distribution of mineral particles. The modified gamma function was also used to approximate the size distribution of aerosol particles in the calculations of light scattering by atmospheric aerosols (*Deirmendijan* 1969).

5.8.5.9. The exponential function

The modified gamma function becomes an exponential function if the slope parameter a in (5.189) vanishes:

$$N(D) = c\exp(-bD) \tag{5.195}$$

It has also been used to approximate the size distributions of marine particles. This function has a finite value at the particle diameter of 0. The value of the exponent b is reported to be in a range of 1.40 to 2.37 for several distributions measured in a diameter range of 2.4 to 15 μm using a Coulter counter in the Pacific surface waters north of Hawaii (*Sugihara* and Tsuda 1979) and 0.27 to 0.85 for PSDs measured in a diameter range of 0.5 to 60 μm with a Galai CIS particle counter in the waters of karstic aquifiers in Switzerland (*Atteia* and Kozel 1997). The correlation coefficient of the relationship $\log N$ vs. D was between 0.965 and 1.0. Recently, *Syvitski* et al. (1995) have shown that the reverse cumulative size distribution of particles $N'(D)$ (the number of particles with diameters smaller than D) can be well represented by an exponential function (5.195).

A combination of two exponential functions was postulated to represent the size distribution of marine particles (*Zuur* and Nyffeler 1992):

$$n(D) = \frac{1-a}{\langle D\rangle_1}\exp\left(-\frac{D}{\langle D\rangle_1}\right) + \frac{a}{\langle D\rangle_2}\exp\left(-\frac{D}{\langle D\rangle_2}\right) \tag{5.196}$$

where a is the relative contribution of the second component to the size distribution and $\langle D\rangle_1$ and $\langle D\rangle_2$ are the average particle diameters of the two components. Zuur and Nyffeler found that the values of $a = 0.0015$, $\langle D\rangle_1 = 0.8\,\mu m$, and $\langle D\rangle_2 = 2\,\mu m$ represented well typical size distributions in the northeast Atlantic waters measured by *Nyffeler* and Ruch (1989) with the Coulter counter.

We found the exponential function to fit well a sample size distribution of air bubbles entrained by breaking waves (calculated by using the data of *Terrill* et al. 2001) as shown in Figure 5.40.

5.8.5.10. Weibull and Rosin–Rammler functions

The Weibull distribution (*Weibull* 1939, see also *Kondolf* and Adhikari 2000) was introduced on an empirical basis in order to describe the size distribution of a particle population formed by fragmentation (crushing). This distribution, as

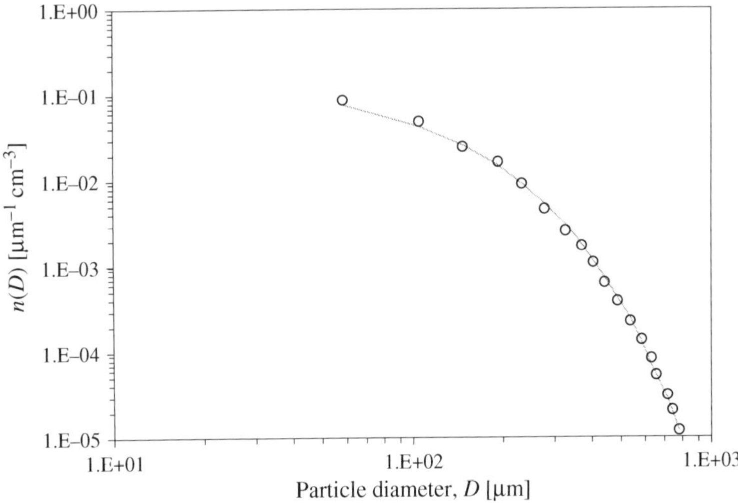

Figure 5.40. Sample exponential fit to the size distributions of bubbles in seawater entrained by breaking wave zone (data of *Terrill* et al. 2001, their Fig. 1a) at a depth of 0.7 m (open circles—data, solid gray line—exponential fit, equation (5.195), $c = 0.1541$, $b = 0.01211$, χ^2 per degree of freedom $= 3.3 \times 10^{-4}$). The fit parameters were obtained via the logarithmic transformation of the data. All weights were set to unity. The χ^2 was calculated by assuming only the counting error.

formulated by *Tenchov* and Yanev (1986), can be written as follows:

$$n(D) = n_0 \frac{c}{b} \left(\frac{D - D_0}{b} \right)^{c-1} \exp\left[-\left(\frac{D - D_0}{b} \right)^{c} \right] \tag{5.197}$$

where n_0 is the particle concentration (scale) factor with a dimension of number of particles per unit volume, the non-dimensional parameter c is the "dimensionality" parameter, D_0 is the smallest particle diameter of the distribution, and b, with the unit of the particle diameter, is related to the extent of fragmentation. As is the case for the log-normal and modified gamma distributions, the Weibull distribution yields a finite total number of particles. If $c = 1$, this distribution becomes an exponential distribution.

The key properties of the Weibull distribution are (e.g., *Tenchov* and Yanev 1986):

$$\begin{aligned} D_{\text{peak}} &= D_0 + b \left(\frac{c-1}{c} \right)^{1/c} \\ n(D_{\text{peak}}) &= n_0 \, \frac{c}{b} \left(\frac{c-1}{c} \right)^{(c-1)/b} \exp\left(-\frac{c-1}{c} \right) \end{aligned} \tag{5.198}$$

and

$$\langle D \rangle = D_0 + b\Gamma\left(\frac{1}{c}+1\right)$$
$$\sigma^2(D) = b^2\left[\Gamma\left(\frac{2}{c}+1\right) - \Gamma^2\left(\frac{1}{c}+1\right)\right] \tag{5.199}$$

where D_{peak} is the value of D, corresponding to the distribution peak, and $\Gamma(x)$ is the gamma function.

The Weibull distribution is a result of a random fragmentation process where the probability of splitting a particle into fragments depends on the particle size. If that probability is independent of the particle size, the log-normal size distribution results (see section 5.8.5.6). Both distributions can be made quite similar, and the experimental errors may prevent one from reaching a definite conclusion on which one is a better fit to experimental data (*Tenchov* and Yanev 1986).

Recently, *Brown* and Wohletz (1995) demonstrated that the Weibull distribution arises naturally as a consequence of the fragmentation process being fractal. Fragmentation of a solid particle is initiated by generation of a fractal crack tree. In aggregates, the fragmentation process is a cleavage of bonds between the component particles that can also be proven to lead to the Weibull distribution (*Tenchov* and Yanev 1986). According to the derivation of Brown and Wohletz, the exponent, c, is related to the three-dimensional fractal dimension, d, describing the fragmentation process as follows:

$$c - 1 = \frac{d}{3} \tag{5.200}$$

Brown and Wohletz (1995) also show that the Weibull distribution is a derivative of the Rosin–Rammler distribution (*Rosin* and Rammler 1933) that has been used, in particular, for describing the size distribution of fragments in coal processing and in geology (*Kittleman* 1964). The Rosin–Rammler distribution is a cumulative distribution expressed as follows (e.g., *Brown* and Wohletz 1995):

$$N(D) = N_T \exp\left[-\left(\frac{D}{D_0}\right)^r\right] \tag{5.201}$$

where $N(D)$ is the number of particles with sizes greater than D, N_T is the total number of particles, r is a dimensionless constant, and D_0 is related to the average particle diameter. Note that simple fitting of this distribution to experimental data, to avoid using a direct non-linear fitting process, requires taking double logarithms which greatly reduces the effect the features of the data set may have on the values of the fit parameters.

In marine applications, the Weibull distribution, in its integral (Rosin–Rammler) form;

$$N(D) = N_0 \left\{ 1 - \exp\left[-b \left(\frac{D}{D_0} \right)^c \right] \right\} \tag{5.202}$$

where N_0 is the scale parameter, and b and c are the shape parameters, has been used to approximate over 50 cumulative size distributions measured using a Coulter counter (model A, 100 μm aperture) in the eastern equatorial Pacific, off the Galapagos Islands (*Carder* et al. 1971). The best fits were obtained by using a multi-segment Weibull distribution, with the parameters estimated separately for several (on the order of 3) consecutive particle size ranges. The shape parameter, c, varied from about 0.3 to 1.45, increasing with the decreasing particle size.

A theoretical analysis of the process of dissolution of particles, as they settle through the water column, and of the breakage of the partially dissolved particles into smaller fragments, whose number is a power function of the particle size (*Lal* and Lerman 1975), also led to the Weibull distribution.

5.8.5.11. The principal component (characteristic vectors) method

The size distribution can be treated as a vector in an n-dimensional space, where each dimension represents a fixed particle diameter. Such an approach permits the use of the well-known statistical method of principal components (e.g., *Anderson* 1958) to approximate any vector from a population of such vectors by a relatively few *characteristic vectors* of the covariance matrix of the population. We have discussed this method in section 4.5.2.6 in relation to the approximation of the volume scattering functions of natural waters.

The method of principal components requires a fairly large data set of PSDs to be obtained before any of the size distributions can be approximated. As mentioned, each distribution, represented by n data points is considered as an n-dimensional vector $\mathbf{n}$ with components $n_i = n(D_i)$, $i = 1$ to r. The $n \times n$ covariance matrix **Cov** of the set of vectors $\mathbf{n}_j$, $j = 1$ to q, is calculated along with the characteristic vectors (eigenvectors) $\mathbf{f}_k$, $k = 1 \ldots r$, with components $f_{ki} = f(D_i), i = 1$ to r, which are defined by the following equation:

$$\mathbf{Cov}\ \mathbf{f}_k = l_k \mathbf{f}_k \tag{5.203}$$

where l_k are the roots of matrix **Cov**. The characteristics vectors form an orthogonal basis in the r-dimensional space. Thus, each vector $\mathbf{n}$ from the analyzed set can be expressed as a linear combination of the characteristic vectors as follows:

$$\mathbf{n}_j(D_i) = \langle \mathbf{n}(D_i) \rangle + \sum_{k=1}^{q} c_k \mathbf{f}_k \tag{5.204}$$

where $\langle \mathbf{n}(D_i) \rangle$ is the average vector for the considered set and c_k are the best fit coefficients. Since the characteristic vectors are orthogonal and can be normalized, the fitting procedure is very simple.

The covariance matrix of the PSD in seawater is such that q on the order of 2 to 3 is sufficient to account for most of the variability in the PSD. The first three principal components for a set of 102 PSDs from the Baltic Sea waters are shown in Figure 5.41 (after *Jonasz* 1983a). An excellent approximation can be obtained for a *typical* size distribution of such a set (Figure 5.42). In fact, in the case shown in Figure 5.42, the first component alone approximates the size distribution almost as well as the first three components. In that set, the first component alone accounts for about 97% of the total variance of the size distribution. However, in the case of a distribution less typical of the set, the approximation worsens considerably (Figure 5.43).

In an extensive study of this approximation method, *Bradtke* (2004) noted that due to the high steepness of the aquatic PSD, the use of the characteristic vectors of the covariance matrix of $n(D)$ leads to approximations that are good in the small-size part of the particle size range analyzed and poor in the large-size part of that range. She also examined two other approaches that (1) use the covariance matrix of $\log n(D)$ to improve the approximation of the PSD in the large-size part at the cost of worsening the approximation in the small-size part, and (2) use the covariance matrix of standardized $n(D)$, i.e., $n_S(D) = [n(D) - \langle n(D) \rangle]/\mathrm{var}[n(D)]$,

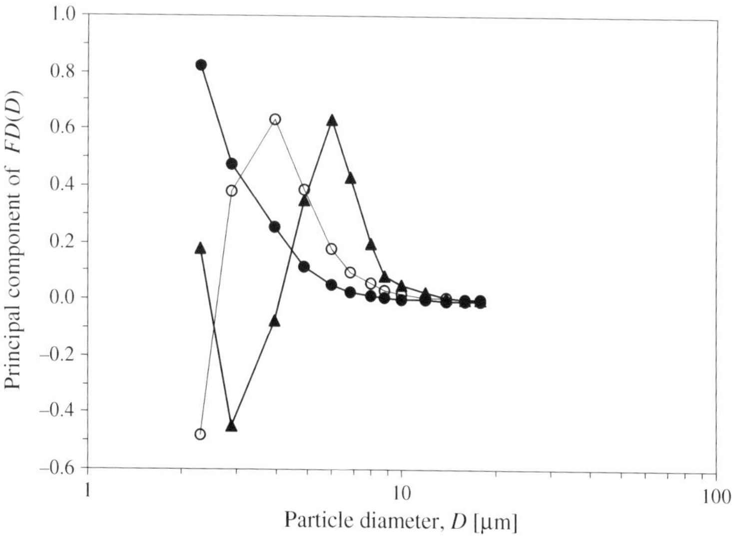

Figure 5.41. The first three principal components (normalized to unit length) of the covariance matrix of 102 particle size distributions measured in the Baltic waters (*Jonasz* 1983a).

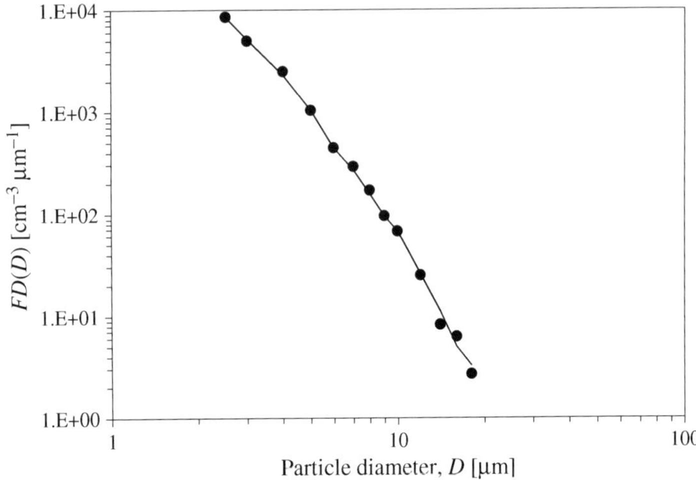

Figure 5.42. Typical size distribution of the set whose principal components are shown in Figure 5.41, as approximated (*Jonasz* 1983a) using the three first principal components (χ^2 per degree of freedom = 0.090). All weights in fitting were set to unity. The χ^2 was calculated by assuming only the counting error.

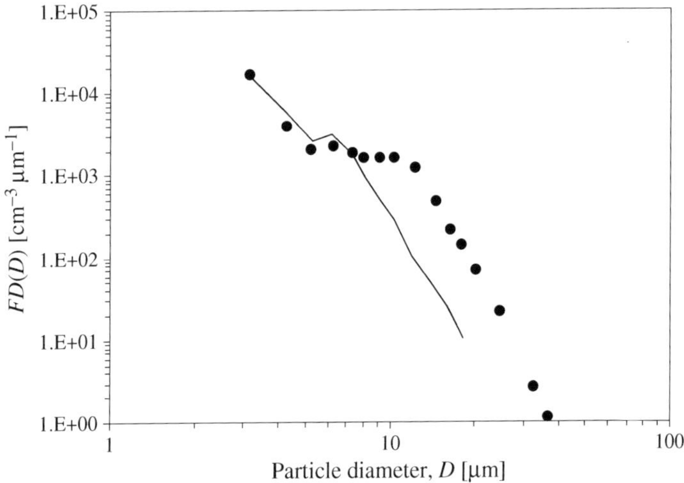

Figure 5.43. A size distribution which is less typical of a set whose principal components are shown in Figure 5.41, as approximated using the three first principal components (*Jonasz* 1983a). In this case, the approximation does not reproduce the phytoplankton-related feature with the modal particle diameter of about 20 μm(χ^2 per degree of freedom = 6.2). All weights in fitting were set to unity. The χ^2 was calculated by assuming only the counting error.

that is equivalent to the correlation matrix of $n(D)$. This latter approach yielded the lowest approximation error of the PSD over the whole examined size range.

The principal component analysis was applied to large collections of the size distribution data (*Bradtke* 2004—970 PSDs measured in the Gdańsk Bay, Baltic Sea, *Poulet* et al. 1986—over 1200 PSDs measured in the waters of the Gulf of St. Lawrence, *Jonasz* 1983a, 1980—over 160 PSDs measured in the Baltic waters, *Zalewski* 1977—Baltic waters, *Kitchen* et al. 1975—263 PSDs measured off the coast of Oregon).

In the study of *Kitchen* et al. (1975), the first principal component accounted for 74% of the variance of the PSD in the data set, the second component accounted for 18%, and the third for only 3% of the total variance. Since the first two principal components accounted for about 92% of the total variance, a PSD from the data set can be relatively accurately expanded using these principal components. As noted by Kitchen et al., the first expansion coefficient corresponds to the total particle volume, the second coefficient indicates which size interval contains the largest proportion of the total volume. Both expansion coefficients were highly correlated with the particulate carbon and chlorophyll concentrations, indicating that a large proportion of the particulate volume was composed of phytoplankton.

It is worth noting that the orthogonality of the characteristic vectors is not preserved by a transformation of the size distribution into the scattering function that uses expansion (5.204) of the size distribution. Thus, the expansion coefficients c_k in (5.204) are not the expansion coefficients of the volume scattering function into its own set characteristic vectors, i.e.,

$$\int_0^{\infty} f_k(D)\beta_1(\theta, D, \eta)dD \neq g_k(\theta) \tag{5.205}$$

where η denotes (only here) the relative refractive index of the particles, which potentially is also particle size dependent and $g_k(\theta), k = 1, \ldots, K$ is the set of the characteristic vectors of the covariance of the scattering function. Hence the expansion coefficients c_k, equation (5.204), cannot be used for the inversion of the volume scattering function into the PSD.

5.8.5.12. Average PSD

In the first approximation, expansion (5.204) retains only the first principal component. In the study of PSD in the Baltic (*Jonasz* 1983a, 1980), that component accounts for about 97% of the size distribution variance. The first principal component is almost identical to the average **n**. Thus, a simple representation of the PSD (*Jonasz* 1983a, 1980) is possible:

$$\mathbf{n}_j(D_i) = a\,\langle \mathbf{n}(D_i)\rangle \tag{5.206}$$

where a is the fit parameter.

This simple approximation is supported by experimental evidence (*Jonasz* and Zalewski 1978) that, excluding the areas or seasons of high biological activity, most variability in the size distribution of marine particles, is due to variations in the total particle concentration and not to variations in the shape of the PSD. Equation (5.206) is in fact the basis of many approximately linear empirical relationships involving suspended particles, for example, the attenuation coefficient vs. the particle concentration.

5.9. Problems

1. Probability distribution of the equivalent projection area of a thin disk

Prove that the probability distribution of the equivalent projection area diameter of a randomly oriented monodisperse population of disks with negligible thickness is expressed by equation (5.106) and that the average ECD is expressed by equation (5.112).

2. Calculation of the frequency size distribution from experimental data

You were handed a table of particle counts vs. particle diameter. Can this table be converted to the frequency (number) size distribution of the particles as defined by (5.1). If not, what additional data you would need?

3. Converting volume size distribution into number size distribution

1. Derive equation (5.48) for converting the volume size distribution $v(V)$ into the number size distribution $n(D)$.
2. You found a relevant size distribution in the literature and want to compare it with your results. Unfortunately, the literature data were expressed as *dV/dlogD* [ppm] vs. D [μm]. Derive a conversion from that size distribution format to $n(D) = dN/dD$ [$\text{cm}^{-3}\mu\text{m}$] vs. D [μm].

4. Estimating power-law fit parameters and χ^2

For a frequency size distribution, $n(D)$, which can be reasonably approximated with a power-law, calculate the fit parameters for $\log n(D)$ vs. $\log D$ by (1) setting all regression weights to unity, i.e., assuming that instrumental errors dominate the variance of $n(D)$, and (2) setting the weights by using equation (5.70) for the variance of $n(D)$ with the numerator set to $2N(D)$, where N is the cumulative size distribution, i.e., assuming that the counting error dominates the variance of $n(D)$. The sample frequency size distribution can, for example, be simulated by generating an exact (mean) power-law cumulative size distribution with parameters representative of aquatic particles, and then introducing the Poisson-distributed noise by sampling the "actual" value of $N(D)$ for each particle diameter, D, from a Poisson distribution with the mean equal to the exact value of the cumulative size distribution. Note that when the mean value, $N(D)$, exceeds about 100, the Poisson probability distribution is well approximated by a Gaussian distribution with the variance set to the mean value.

Compare the magnitudes of the χ^2 obtained via equation (5.147) for the fit parameters obtain with equal weights and the logarithmically-transformed weights (use 5.154).

Chapter 6

Refractive indices and morphologies of aquatic particles

6.1. The refractive index: introductory remarks

As we discussed in more detail in section 1.5.1, the complex refractive index of a medium (such as that of a particle)

$$n = n' - in'' \tag{6.1}$$

characterizes the effect of that medium on the phase (through n') and on the amplitude (through n'') of an electromagnetic wave. The imaginary part of the refractive index determines a reduction of the amplitude of electromagnetic wave (light), and consequently a reduction of its power, *inside* optically homogenous medium, as described by the Beer–Lambert law:

$$F(x) = F_0 e^{-ax} \tag{6.2}$$

where F_0 is the light power at a start plane in the medium (at $x = 0$), $F(x)$ is the light power at a distance x from that plane, and a, typically measured in m^{-1}, is the absorption coefficient. The effect of an interface, such as a cuvette window, is not included here. Equations equivalent to (6.2) apply also to inhomogeneous media which scatter as well as absorb light; however, the coefficient of absorption is then replaced by the scattering coefficient, b, which represents the effect of light scattering alone, or by the attenuation coefficient, c, which represents the additive effects of absorption and scattering ($c = a + b$).

The absorption coefficient, a, is related to the imaginary part of the refractive index as follows (see section 1.5.1):

$$a = \frac{4\pi}{\lambda} n'' \tag{6.3}$$

where λ is the wavelength of the wave in vacuum.

The real and imaginary parts of the refractive index cannot be set arbitrarily. These functions of the wavelength (or wave frequency) are related through a Kramers–Kronig (KK) relationship (e.g., *Bohren* and Huffman 1983, *McKellar* et al. 1982, *Hale* and Querry 1973, *van de Hulst* 1957):

$$n'(\lambda) = 1 + \frac{2\lambda^2}{\pi} P \int_0^\infty n''(\lambda_i) \frac{d\lambda_i}{\lambda_i(\lambda^2 - \lambda_i^2)} \tag{6.4}$$

where P indicates the principal Cauchy value of the integral, λ_i is simply the integration variable. It should be noted that the KK relationship is actually the precise mathematical expression of the principle of causality (i.e., a signal of finite speed cannot cause a disturbance at a point before it propagates to that point).

Thus, if the real part of the refractive index is different than 1, the medium *must* absorb light, i.e., n'' must be non-zero at some wavelengths. It also follows that the contribution of the imaginary part of the refractive index to the integral is more heavily weighed for wavelengths λ_i close to λ. Thus, as noted by *Harvey* et al. (1998), when evaluating the integral in (6.4) for a given wavelength, λ, one needs to know precisely the values of n'' mostly in the vicinity of λ. Other values may be known approximately. Nevertheless, the evaluation of the real refractive index of a substance from the KK relationship is prone to problems and may yield rather imprecise values in a narrow wavelength interval as shown in Figure 6.1, although—when used to obtain the refractive index values in a large wavelength range—that relationship may still provide a valuable insight into the spectrum of the refractive index. Finally, if the term containing the integral in (6.4) is small, the real and imaginary parts of the refractive index can be related using an approximate theory of Helmholtz and Ketteler (section 6.3.2).

We note that spectral attenuation of light by a suspension can be transformed to a function, assuming complex values, which fulfills the KK relationships (e.g., *Ku* and Felske, 1986). In the case of a suspension of homogeneous spheres, such a function is related to the complex refractive index of the particles through Mie theory (e.g., *Bohren* and Huffman 1983). Thus, the KK analysis can also be used to determine the refractive index of suspended particles through the use of Mie theory.

Since the refractive index of a medium depends on the density of electrons in that medium (e.g., *Bohren* and Huffman 1983), the index increases with the density of matter. The real parts of the refractive index of minerals which commonly occur in seawater can be expressed as (*Carder* et al. 1974, $r^2 = 0.89$ for eight common minerals, coefficients are adjusted to represent the absolute refractive index of minerals):

$$n' = 0.293\,\rho + 0.804 \tag{6.5}$$

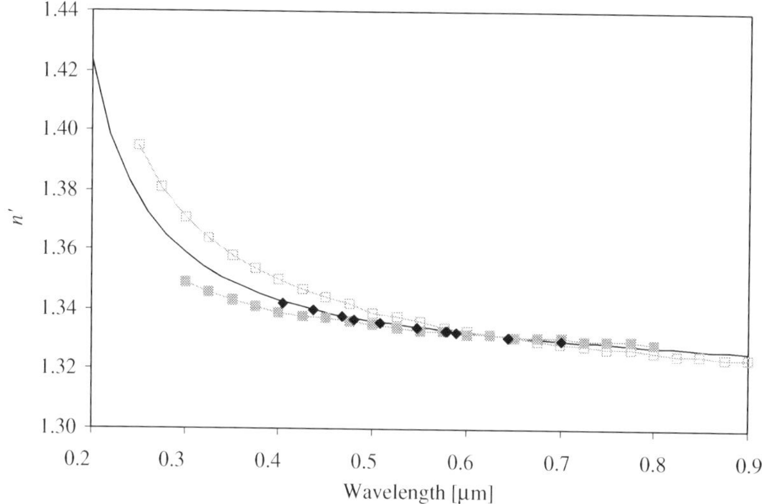

Figure 6.1. The real part, n', of the refractive index of water determined from the Kramers-Kronig analysis may yield refractive index data that are not sufficiently precise for a small wavelength range, as shown here by using the results of *Hale* and Querry (1973, *solid gray squares*, $T = 25°C$) and *Querry* et al. (1991—*open gray squares*, $T = 25$ to $30°C$, these authors used a more accurate spectrum of n'' was used as compared than that used by Hale and Querry 1973) for pure water vs. an internationally approved approximation (*Harvey* et al. 1998—*black curve* with no symbols) that is reported to be accurate to better than 1.5×10^{-5} in a wavelength range of 0.4 to 0.7 μm at ambient pressure. Data published by *Austin* and Halikas (1976, as reproduced by *Quan* and Fry 1995—*solid black diamonds*) for $T = 25°C$ and $S = 0$ ppt are shown for comparison.

where ρ [g cm^{-3}] is the density of the mineral. *Woźniak* and Stramski (2004) have extended this correlation to 29 common minerals (see also section 6.4.3.4) and obtained the following regression equation ($r^2 = 0.77$) (Figure 6.2):

$$n' = 0.198\rho + 1.034 \quad (6.6)$$

Such relationships should be regarded as large-range trends. Indeed, when the density range of a data set of *Woźniak* and Stramski (2004) is limited to be less than 3.3 g cm^{-3}, r^2 drops to an insignificant value of 0.31.

Based on the approach of *Aas* (1996), a similar relationship is found for organic matter in various states of hydration (*Babin* et al. 2003):

$$n' = 1 + 0.45(\rho - 1) \quad (6.7)$$

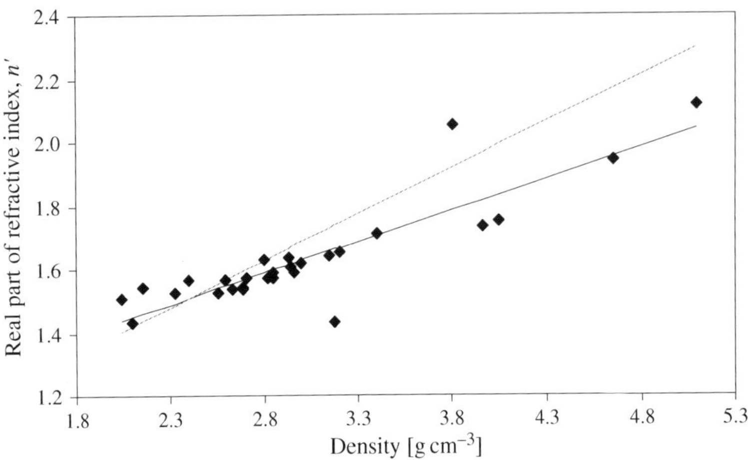

Figure 6.2. The real part, n', of the refractive index of minerals as a function of the mineral density. Data ($n = 29$ points) are taken from *Woźniak* and Stramski (2004). The corresponding regression line (black line, $r^2 = 0.77$) equation is specified in (6.6). The gray line represents equation (6.5) from *Carder* et al. (1974; $r^2 = 0.89$, $n = 8$). The original data refer to the refractive index relative to that of water, here assumed to be 1.34.

where ρ [$\mathrm{g\,cm^{-3}}$] is the density of the organic matter in its hydrated state. We should also mention that a relation of type (6.7), rewritten to use the well-known refractive index increment, $n_c = dn'/dC_\mathrm{p} \sim 0.185\,\mathrm{cm^3\,g^{-1}}$ of aqueous solutions of proteins (e.g., *Barer* and Joseph 1954), where C_p is the mass concentration of protein, in grams per $\mathrm{cm^3}$ of the solution, has been used to determine the protein content of cells:

$$C_\mathrm{p} = \frac{n' - n'_w}{n_c} \tag{6.8}$$

with n_w being the refractive index of the solvent, i.e., water.

For water in the visible-light spectral range (0.4 to 0.7 μm), the imaginary part of the refractive index, n'', is in a range of 10^{-10} to 3×10^{-8}, and the absorption coefficient a is in a range of $3 \times 10^{-9}\,\mathrm{\mu m^{-1}}$ to $5 \times 10^{-7}\,\mathrm{\mu m^{-1}}$. Thus, there is no appreciable absorption at distances comparable to the dimensions of marine particles (0.1 μm to, say, 1000 μm), and the imaginary part of the refractive index of water is generally neglected in evaluating light scattering by particles in water. If the refractive index of water, n_w, is expressed using (6.1) with $n_\mathrm{w}'' = 0$ and that of the particle is expressed as

$$n_\mathrm{p} = n'_\mathrm{p} - in''_\mathrm{p} \tag{6.9}$$

then the refractive index of the particle relative to water is, in the first approximation, expressed as:

$$n = \frac{n'_{\mathrm{p}}}{n'_{\mathrm{w}}} - i\frac{n''_{\mathrm{p}}}{n'_{\mathrm{w}}} = n' - in'' \tag{6.10}$$

The relative refractive index in (6.10) is the key parameter used in light scattering theories to describe the interaction of the particle with a plane light wave. However, in order to calculate the absorption of light inside a particle with equations (6.2) and (6.3), one needs to use in (6.3) n''_{p} rather than n'' just obtained.

6.2. Refractive index of water and seawater

6.2.1. Dependence on wavelength, temperature, and salinity

In many instances, the refractive index of the material of a particle is known relative to air. The calculation of the refractive index of the particle relative to that of water or seawater thus requires the knowledge of the refractive index spectrum of water or seawater. A comprehensive review and tabulated values of the refractive index of water can be found in *Querry* et al. (1991: n' and n'' in a wavelength range of 10^{-7} to 1.23 μm) and in *Hale* and Querry (1973: n' and n'' between 0.2 μm to 0.8 μm). Note that these values of n' in the visible are relatively imprecise, likely due to the KK analysis they used. Tables of the refractive index for seawater can be found in *Jerlov* (1976), *Austin* and Halikas [1976—*Quan* and Fry (1995) reproduce some of their data], *Horne* (1969), *Cox* (1965: a summary of early measurements by *Utterback* et al. 1934), and *Riley* (1975: data of *Matthäus* 1974). Extensive bibliographies of the refractive index measurements of pure water are given by *Thormählen* et al. (1985) and *Schiebener* et al. (1990).

The real part of the refractive index of water decreases with increasing wavelength in the visible (Figure 6.3) which is a region of *normal dispersion* (e.g., *Bohren* and Huffman 1983) for water. The dependence of the refractive index, n', on wavelength, λ, of light is well approximated in such a region by the Cauchy equation (e.g., *Jenkins* and White 2001):

$$n'(\lambda) = P + \frac{Q}{\lambda^2} + \frac{R}{\lambda^4} \tag{6.11}$$

where P, Q, and R are expansion coefficients. These coefficients are listed in Table 6.1 for the two top curves in Figure 6.3.

At a fixed wavelength, the real part, n', of the refractive index of seawater decreases approximately quadratically with increasing temperature (Figure 6.4),

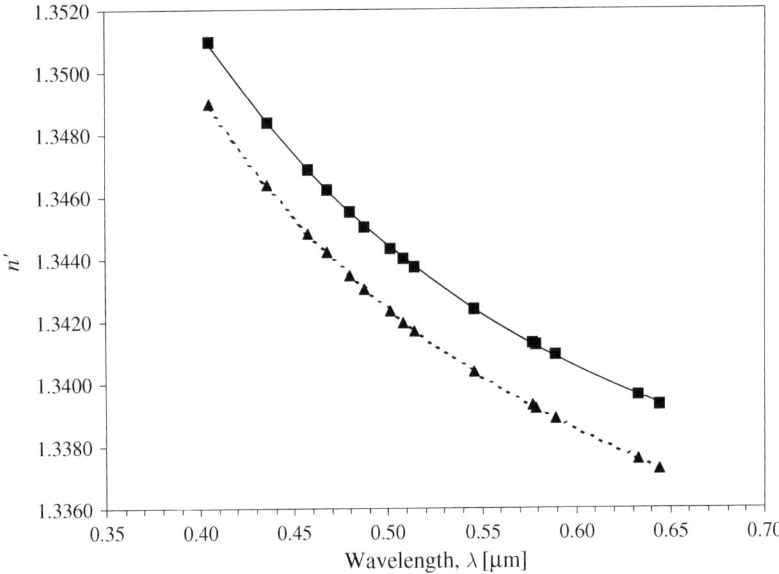

Figure 6.3. The real part, n', of the refractive index of seawater vs. the wavelength of light (in air) based on the data of *Matthäus* (1974) as cited by *Riley* (1975) (symbols). *Solid squares*—temperature, $T = 0°C$ and salinity, $S = 35$ ppt, *dashed curve*—a Cauchy approximation: $n' = 1.330989 + 0.003587/\lambda^2 - 5.488E{-}05/\lambda^4$ to those data. *Solid triangles*—$T = 25°C$, $S = 35$ ppt, *solid curve*—a Cauchy approximation: $n' = 1.329011 + 0.003562/\lambda^2 - 5.182E{-}05/\lambda^4$.

Table 6.1. Sample coefficients of the Cauchy equation for water at a salinity $S = 35$ ppt and at two temperatures.

Parameter	**$T = 0°C$**	**$T = 25°C$**
P	1.330989	1.329011
Q	0.003587	0.003562
R	−5.488E−05	−5.182E−05

To calculate these coefficients, we used the data of *Matthäus* (1974) as cited by *Riley* (1975). The coefficients are to be used with the wavelength (in air) measured in μm.

at constant salinity and increases essentially linearly with the salinity of seawater (Figure 6.5) at constant temperature. Such relationships are the basis of optical methods of the determination of density and salinity of seawater (e.g., *Alford* et al. 2006, *Seaver* 1987).

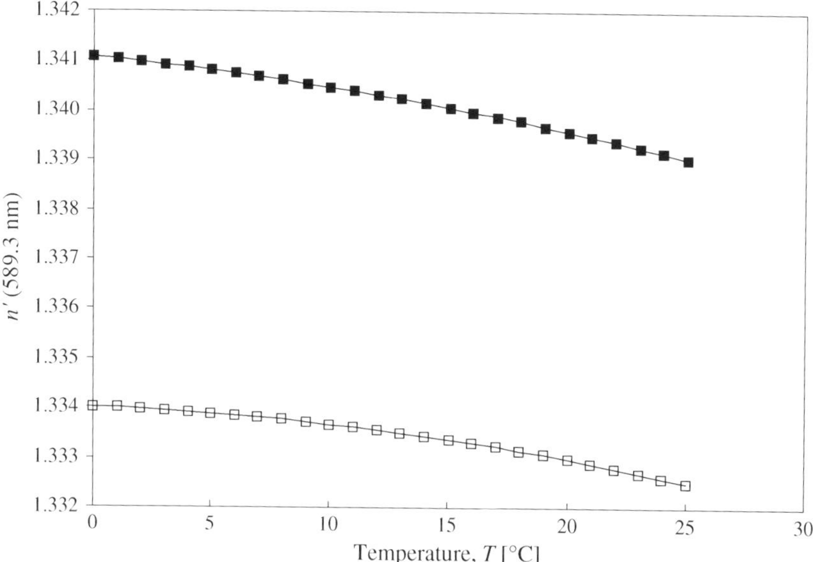

Figure 6.4. The real part, n', of the refractive index of seawater vs. temperature, based on the data of *Matthäus* (1974) as cited by *Riley* (1975) (symbols). The wavelength of light (in air) $= 0.5893\,\mu$m. *Solid squares*: salinity, $S = 36$ ppt, equation of the approximation curve: $n' = 1.34108 - 4.6696\text{E}{-}5T - 1.3964\text{E}{-}06T^2$. *Open squares*: $S = 0$ ppt, equation of the approximation curve: $n' = 1.33402 - 1.7653\text{E}{-}5T - 1.7053\text{E}{-}06T^2$.

6.2.2. *Real part of the refractive index*

Eisenberg (1965) developed a semi-empirical formula for the refractive index of pure water (see section 2.4) for $404\,\text{nm} < l < 706\,\text{nm}$, $0°\text{C} < T < 60°\text{C}$, and $0\,\text{dbar} \le p < 11000\,\text{dbar}$, which we repeat here for easy reference:

$$n'(\lambda, T, p) = \left[\frac{1 + 2f(n', \lambda, T, p)}{1 - f(n', \lambda, T, p)}\right]^{1/2} \tag{6.12}$$

where

$$f(n', \lambda, T, p) = A_t(\lambda)\left[\frac{\rho_w(T, p)}{\rho_w(t = 4°C, p)}\right]^{B(\lambda)} \exp[-C(\lambda)t] \tag{6.13}$$

where the wavelength of light in vacuum, λ is in nm, temperature T is in °C, and ρ is the pure water density given by equation (5.128) at the ambient pressure, and by equation:

$$\rho_w(p) = \rho_w \frac{1}{1 - \frac{p}{K}} \tag{6.14}$$

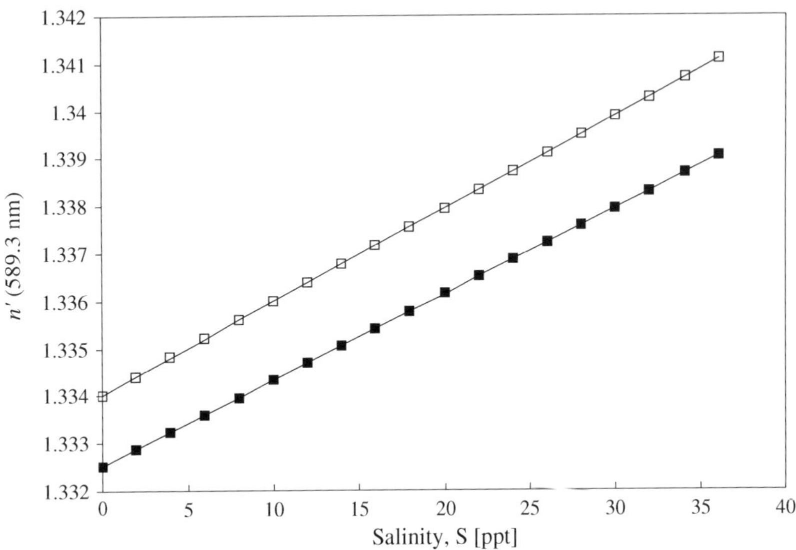

Figure 6.5. The real part, n', of the refractive index of seawater vs. salinity, S, based on the data *Matthäus* (1974) as cited by *Riley* (1975) (symbols). The wavelength of light (in air) = 0.5893 μm. Open squares: temperature of 0°C, equation of the approximation line: $n' = 1.3340 + 0.000196S$. Solid squares: temperature of 25°C, equation of the approximation line: $n_r = 1.3325 + 0.000181S$.

for pressure p, where K is the bulk modulus of water (at the salinity $S = 0$, see section 5.7.11.1). In Table 6.2, we give expansion coefficients for the wavelength-dependent parameters A_t, B, and C as polynomials in λ:

$$x(\lambda) = \sum_i a_i \lambda^i \tag{6.15}$$

where x is either A, B, or C, obtained by fitting the Eisenberg's data (his Table IIa). Sample test values are $n' = 1.3343126$ and 1.3341604 for $T = 3$ and 8°C at $\lambda = 589.26$ nm (sodium line) at the ambient pressure. See the Eisenberg's paper for more test values. Note that the pressure, p, is the total pressure applied to water. In the case of a natural water body, this is the sum of the atmospheric and hydrostatic pressures.

Based on the approximation of the Lorentz–Lorenz formula,

$$f(n) = \frac{1}{\rho_m} \frac{n^2 - 1}{n^2 + 2} \tag{6.16}$$

where ρ_{m} is the molar density of the liquid, other empirical formulas have been developed. Some have become the basis of internationally accepted

Table 6.2. Coefficients of expansions of wavelength-only-dependent constants A_t, B, and C (6.13) for the refractive index of pure water in polynomials in λ [nm].

Coefficient	A_τ	B	$C \times 10^5$
a_0	4.0673510E−01	3.4802E−01	−1.4755588E+01
a_1	−1.6413543E−03	4.2904E−03	2.3424461E−01
a_2	6.0236998E−06	−1.4150E−05	−1.0742454E−03
a_3	−1.2347793E−08	2.3757E−08	2.6418845E−06
a_4	1.4614739E−11	−2.0099E−11	−3.6719528E−09
a_5	−9.3804771E−15	6.8254E−15	2.7163151E−12
a_6	2.5358086E−18	–	−8.3313233E−16

The coefficients listed provide an approximation with an accuracy better than 2.39E−07 for A_t, 1.68E−05 for B, and 0.00031 for $C \times 10^5$.

approximations of the refractive index of pure water over substantial ranges of the wavelength of light, temperature, and pressure. In particular, *Schiebener* et al. (1990) developed such an approximation by analyzing the most reliable data sampled from an extensive database (>3000 datapoints) reported, along with a similar approximation formula, by *Thormählen* et al. (1985). The formula of *Schiebener* et al., adopted by the International Association of Properties of Water and Steam (IAPWS) in 1991, has been recently improved by *Harvey* et al. (1998) who used an updated equation of state for water and converted historical data to the new temperature scale (ITS-90, e.g., *Preston*-Thomas 1990a, 1990b). This new formula, adopted by IAPWS in September 1997 (*IAPWS* 1997), differs from the formula of *Schiebener* et al. (1990) merely by the values of the expansion coefficients. We cite it here after *Harvey* et al. (1998):

$$\begin{aligned} f(n) &= \frac{1}{\rho'}\frac{n^2-1}{n^2+2} \\ &= n_0 + n_1\rho' + n_2\rho'^2 \\ &\quad + n_3T' + n_4\lambda'^2T' \\ &\quad + \frac{n_5}{\lambda'^2} + \frac{n_6}{\lambda'^2 - \lambda'^2_{UV}} + \frac{n_7}{\lambda'^2 - \lambda'^2_{IR}} \end{aligned} \tag{6.17}$$

where $T' = T/(273.15\,\text{K})$, with T[K] being the absolute temperature, $\rho' = \rho/(1000\,\text{kg m}^{-3})$, with $\rho(T, p)$ [kg m^{-3}] being the water density (dependent on the temperature, T, and pressure, p), and $\lambda' = \lambda/(0.589\,\mu\text{m})$, with λ [μm] being the wavelength of light in vacuum. The values of the coefficients are listed in Table 6.3 The density of water, ρ, can be approximated as explained earlier in

Table 6.3. Coefficients relevant to equation (6.17) for the refractive index of pure water (*Harvey* et al. 1998, *IAPWS* 1997).

Coefficient	*Value*	**Coefficient**	*Value*
n_0	2.44257733E−01	n_5	1.58920570E−03
n_1	9.74634476E−03	n_6	2.45934259E−03
n_2	−1.66626219E−02	n_7	9.00704920E−01
n_3	−3.73234996E−03	λ'_{UV}	2.29202E−01
n_4	2.68678472E−04	λ'_{IR}	5.432937E+00

The coefficients apply for the state parameters being expressed in relative scales: $T' = T/(273.15\,\mathrm{K})$, with T [K] being the absolute temperature, $\rho' = \rho/(1000\,\mathrm{kg\,m^{-3}})$, with $\rho(T, p)$ [$\mathrm{kg\,m^{-3}}$] being the water density (dependent on the temperature, T, and pressure, p), and $\lambda' = \lambda/(0.589\,\mu\mathrm{m})$, with λ [μm] being the wavelength of light in vacuum.

this section in reference to the Eisenberg's equation for the refractive index. It follows that the refractive index, n, can be obtained from the following equation:

$$n = \left[\frac{1 + 2f(n)\rho'}{1 - f(n)\rho'}\right]^{1/2} \tag{6.18}$$

with an accuracy that varies depending on the wavelength, temperature, and pressure ranges (as stated in *Harvey* et al. 1998). In the visible (0.4 to 0.7 μm), at the ambient pressure, these formulas approximate the refractive index of water with an accuracy better than 1.5×10^{-5}. Harvey et al. provide a set of check values for testing implementations of their formula. We quote here a sample of such check values: $n = 1.394527$ at $\lambda = 0.2265\,\mu\mathrm{m}$, $T = 0°\mathrm{C}$, and pressure $p = 0.1\,\mathrm{MPa}$, and $n = 1.324202$ at $\lambda = 1.01398\,\mu\mathrm{m}$, $T = 100°\mathrm{C}$, and pressure $p = 100\,\mathrm{MPa}$. A sample spectrum of the refractive index according to this approximation is shown in Figure 6.6.

We will now cite several empirical formulas for the refractive index of seawater (that includes the case of pure water with a salinity of 0) as a function of the wavelength of light, temperature, salinity, and pressure.

McNeil (1977) gave the following expression for the refractive index of seawater with the values of the coefficients listed in Table 6.4, based on fitting a polynomial expression to experimental data of *Austin* and Halikas (1976):

$$\begin{aligned} n'(\lambda, T_c, S_w) = n_0 + n_1\lambda^{-2} + n_2\lambda^{-4} & \\ + n_3T^2 + n_4S(n_5 + n_6T) & \\ + n_7p(n_8 + n_9S)(n_{10} + n_{11}T) & \end{aligned} \tag{6.19}$$

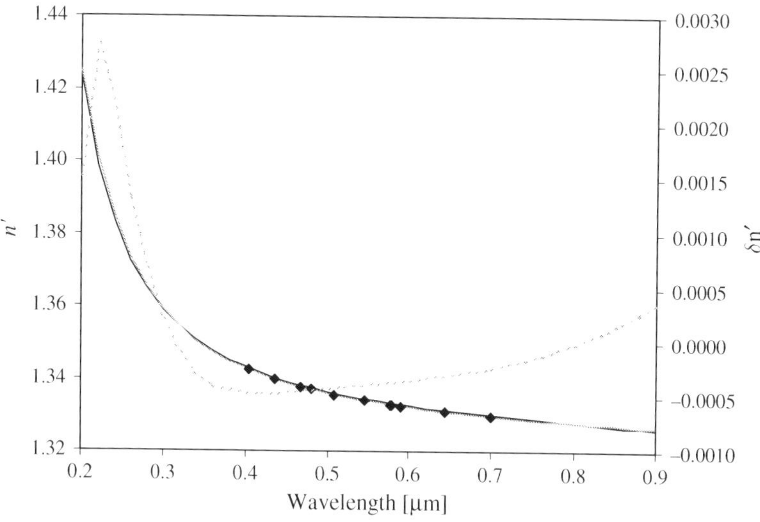

Figure 6.6. The real part, n', of the refractive index of water at $T = 25°C$ calculated from equations (6.17) and (6.18) representing an internationally approved approximation (*Harvey* et al. 1998—*black curve with no symbols*) that is reported to be accurate to better than 1.5×10^{-5} in the wavelength range of 0.4 to 0.7 μm at the ambient pressure. For comparison, an approximation of *Quan* and Fry (1995—*solid gray curve*), equation (6.24), is also shown. The difference, $\delta n'$, between the Quan and Fry and the Harvey et al. approximations is shown with a *dashed gray curve* (right y-axis). The data published by *Austin* and Halikas (1976, as reproduced by *Quan* and Fry (1995—*solid black diamonds*) for $T = 25°C$ and $S = 0$ ppt, which served as the basis of the approximation of Quan and Fry, are shown for comparison.

Table 6.4. Coefficients of equation (6.19) (*McNeil* 1977) for the refractive index of seawater as a function of the wavelength, λ [nm], temperature, T [°C], salinity, S [ppt], and pressure, p [kg-force cm^2].

Coefficient	Value	Coefficient	Value
n_0	1.3247E+00	n_6	−2.0E−02
n_1	3.3E+03	n_7	1.45E−05
n_2	−3.2E+07	n_8	1.021E+00
n_3	−2.5E−06	n_9	−6.0E − 04
n_4	4.0E−05	n_{10}	1.0E+00
n_5	5.0E+00	n_{11}	−4.5E−03

where the wavelength λ is in nm, temperature T is in °C, salinity S is in ppt, and pressure p is in kg-force cm^2 (=9.8068 dbar). This approximation generates a value of $n = 1.35355$ at $T = 10°\text{C}$, $S = 35\,\text{ppt}$, and pressure, $p = 700\,\text{kg-force cm}^{-2}$ at a wavelength of 500 nm. *McNeil* (1997) gives a set of additional check values.

An increase in pressure by 1 dbar corresponds roughly to an increase of the depth in water by 1 m; thus, the value of pressure in decibars corresponds approximately to the depth in water. *McNeil* (1997) gives the following simplified formula for the conversion between the pressure [dbar] and depth, z [m]:

$$p(z) = 1.01991z \tag{6.20}$$

at a temperature of 0°C and salinity of 35 ppt. A more accurate depth-to-pressure conversion algorithm has been developed by *Saunders* (1981). That conversion accounts for the variation of the gravity acceleration [assumed constant in (6.20)] with latitude. We cite the Saunders conversion here in its inverted form:

$$p(z) = \frac{1}{4.42 \times 10^{-6}} \left[(1-F) - \sqrt{(1-F)^2 - 8.84 \times 10^{-6} z} \right] \tag{6.21}$$

where z is in meters, and

$$F = 0.00592 + 0.00525 \sin^2\left(\frac{\varphi \pi}{180}\right) \tag{6.22}$$

where φ [°] is the latitude. A check value is $p = 7500.007$ dbar at $\varphi = 30°$ for $z = 7321.45$ m. McNeil's formula, equation (6.20), yields a value of $p = 7467.2$ for this depth. A yet more accurate and commensurately complicated conversion is given by *Fofonoff* and Millard (1983).

The pressure dependence in the McNeil formula is not accurate enough to be used in evaluating the derivative of n' as a function of pressure in the calculation of light scattering by water from first principles (see section 2.4). Also, *Quan* and Fry (1995) note that errors in approximating the empirical data at $p = 0$ with (6.19) may exceed the experimental errors of the original data ($\delta n \sim 3 \times 10^{-5}$) by more than an order of magnitude ($-4 \times 10^{-4} < \delta n < 8 \times 10^{-4}$). This points to some significant problems with the McNeil formula.

Millard and Seaver (1990) developed an approximation formula with 27 expansion coefficients for the real part of the refractive index as a function of the wavelength of light λ [μm], salinity, S [ppt], temperature, T [°C], and pressure, p [dbar]. They used the empirical data sets of *Stanley* (1970), *Mehu* and Johannin-Gilles (1968), *Waxler* et al. (1964), and *Tilton* and Taylor (1938). This approximation is applicable in the following domain: $0.5 < \lambda < 0.7\,\mu\text{m}$, $0 < T < 30°\text{C}$, $0 < S < 43\,\text{ppt}$, and $0 < p < 11\,000\,\text{dbar}$, with an accuracy varying from 0.4 ppm for fresh water at atmospheric pressure to 80 ppm for seawater at high pressures. This formula is expressed as follows:

$$\begin{aligned} n(T, p, S, \lambda) = {} & n_0 + n_1\lambda^2 + n_2\lambda^{-2} + n_3\lambda^{-4} + n_4\lambda^{-6} + n_5 T \\ & + n_6 T^2 + n_7 T^3 + n_8 T^4 + (n_9 T + n_{10} T^2 + n_{11} T^3)\lambda \\ & + (n_{12} + n_{13}\lambda^{-2} + n_{14} T + n_{15} T^2 + n_{16} T^3)S + n_{17} ST\lambda \\ & + n_{18} p + n_{19} p^2 + n_{20} p\lambda^{-2} + (n_{21} T + n_{22} T^2)p + \\ & + n_{23} p^2 T^2 + (n_{24} + n_{25} T + n_{26} T^2)pS \end{aligned} \quad (6.23)$$

The coefficients appearing in (6.23) are listed in Table 6.5 The following test values can be used to check the implementation of this formula (more are listed by *Millard* and Seaver 1990) at a wavelength of 589.26 nm: $n' = 1.332503$ at $T = 25°\text{C}$, $S = 0\,\text{ppt}$, $p = 0\,\text{dbar}$, $n' = 1.338838$ at $T = 25°\text{C}$, $S = 35\,\text{ppt}$, $p = 0\,\text{dbar}$, and $n' = 1.341631$ at $T = 25°\text{C}$, $S = 35\,\text{ppt}$, $p = 2000\,\text{dbar}$.

Quan and Fry (1995) developed the following simple expression by fitting a polynomial approximation to the experimental data of *Austin* and Halikas (1976) at a (hydrostatic) pressure of 0 in the following domain: $0°\text{C} < T < 30°\text{C}$, $0\,\text{ppt} < S < 35\,\text{ppt}$, and $400\,\text{nm} < \lambda < 700\,\text{nm}$ (a selection of the data is quoted by Quan and Fry in their paper):

$$\begin{aligned} n'(\lambda, T, S) = {} & n_0 + (n_1 + n_2 T + n_3 T^2)S + n_4 T^2 \\ & + (n_5 + n_6 S + n_7 T)\lambda^{-1} + n_8\lambda^{-2} + n_9\lambda^{-3} \end{aligned} \quad (6.24)$$

The values of the coefficients are listed in Table 6.6 The reader may want to use the following test values: $n' = 1.3427041$, at $T = 25°\text{C}$ and $S = 0\,\text{ppt}$, and $n' = 1.3493724$ at $T = 25°\text{C}$ and $S = 35\,\text{ppt}$, both at $\lambda = 400\,\text{nm}$, when implementing the Quan and Fry approximation.

Quan and Fry (1995) also evaluated the quality of approximation of the empirical data of *Austin* and Halikas (1976) by the McNeil empirical equation (6.19) (*McNeil* 1977) and by an approximation of *Matthäus* (1974, which Quan and Fry also quote) and found that both yield significant errors.

Huibers (1997) compared—for pure water at room temperature ($T = 25°\text{C}$)—the approximation of *Schiebener* et al. (1990), which was refitted by *Harvey* et al. (1998) to become the current IAPWS approximation and a simpler formula (6.24) of *Quan* and Fry (1995). He found the latter formula not only to be of comparable accuracy to that of Schiebner et al., but also to apply over a greater wavelength range ($200 < \lambda < 1100\,\text{nm}$) than that originally stated by Quan and Fry. We show the *Quan* and Fry (1995) approximation for pure water in Figure 6.6 and the *Harvey* et al. (1998), i.e., the IAPWS approximation [equations (6.17) and (6.18)],

Table 6.5. The coefficients of equation (6.23) (*Millard* and Seaver 1990) for the refractive index of seawater as a function of the wavelength of light λ [μm], salinity, S [ppt], temperature, T [°C] of seawater, and the pressure, p [dbar].

Coefficient	**Value**	**Coefficient**	**Value**	**Coefficient**	**Value**
n_0	1.3280657E+00	n_9	1.0508621E−05	n_{18}	1.5868383E−06
n_1	−4.5536802E−03	n_{10}	2.1282248E−07	n_{19}	−1.5740740E−11
n_2	2.5471707E−03	n_{11}	−1.7058810E−09	n_{20}	1.0712063E−08
n_3	7.5019660E−06	n_{12}	1.9029121E−04	n_{21}	−9.4834486E−09
n_4	2.8026320E−06	n_{13}	2.4239607E−06	n_{22}	1.0100326E−10
n_5	−5.2883907E−06	n_{14}	−7.3960297E−07	n_{23}	5.8085198E−15
n_6	−3.0738272E−06	n_{15}	8.9818478E−09	n_{24}	−1.1177517E−09
n_7	3.0124687E−08	n_{16}	1.2078804E−10	n_{25}	5.7311268E−11
n_8	−2.0883178E−10	n_{17}	−3.5894950E−07	n_{26}	−1.5460458E−12

Table 6.6. Coefficients of equation (6.24) (*Quan* and Fry 1995) for the refractive index of seawater as a function of temperature, T [°C], salinity, S [ppt], and wavelength λ [nm] at ambient pressure.

Coefficient	Value	Coefficient	Value
n_0	1.31405E+00	n_5	1.5868E+01
n_1	1.779E−04	n_6	1.155E−02
n_2	−1.05E−06	n_7	−4.23E−03
n_3	1.6E−08	n_8	−4.382E+03
n_4	−2.02E−06	n_9	1.1455E+06

along with the difference $dn' = n'_{\mathrm{QF1995}} - n'_{\mathrm{HV1998}}$. That difference in the visible (400–700 nm) ranges from −0.00048 to −0.00023. We also show the differences between the various approximations for the refractive index of seawater and the IAPWS approximation (*Harvey* et al. 1998) in Figure 6.7.

Which approximation for the refractive index of water/seawater to use? When dealing with fresh water at non-zero pressures, the IAPWS approximation (*Harvey* et al. 1998, IAPWS 1997) is the most comprehensive, yet undoubtedly the most complex due to the use of a complicated formulation for the water density. If the effect of pressure is of no concern, then the approximation of *Quan* and Fry (1995) is a reasonable substitute. The latter approximation is also sufficient for the upper 100 m or so of the ocean as the effect of pressure is relatively minor there. Indeed, in the upper 100 m of the ocean at low latitudes, the temperature changes with depth by ~ -15°C (from $\sim$ 25 to $\sim$ 10°C, e.g., *Dera* 1992), the pressure changes by +100 dbars, while the salinity is almost constant. Given these ranges, the refractive index at 500 nm changes with depth in the upper layer of the tropical ocean from about 1.342393 to about 1.344011, i.e., increases by $\sim$ 0.00162, according to the *Millard* and Seaver (1990) approximation. Most of this change is due to a drop in the seawater temperature: $\delta n' = 1.343859_{T=10°\mathrm{C},\ p=0\,\mathrm{dbar}} - 1.342394_{T=25°\mathrm{C},\ p=0\,\mathrm{dbar}} = \sim 0.00147$. The depth must increase by additional $\sim$1000 m in order for the pressure effect to begin dominating the change in the refractive index with depth. If the pressure (i.e., water depth) is so significant, the *Millard* and Seaver (1990) approximation should be used for seawater.

6.2.3. Imaginary part of the refractive index

The imaginary part, n_w'', of the refractive index of water assumes a local minimum at about 0.42 μm (Figure 6.8). The values of n_w'' are derived from the absorption data by using (6.3). The measurement of the absorption of light by pure water is a challenging task, as shown by the discrepancy of a small sample of relevant data plotted in Figure 6.8. We discuss the problems related to the determination of the absorption of light by water in section 2.5 and note that the significant part of research related to the measurement of absorption by water

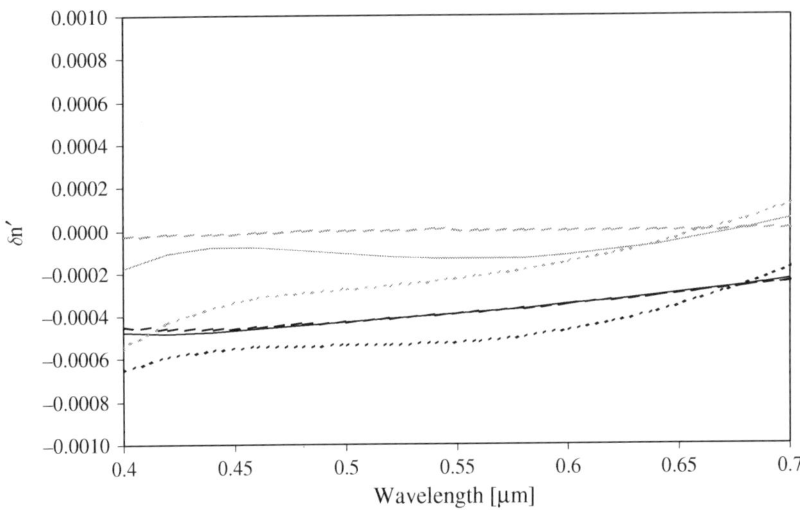

Figure 6.7. Differences of the real part, $\delta n'$, of the refractive index of water as calculated by various approximation formulas. The *black solid curve* represents $\delta n' = n'_{\mathrm{QF1995}} - n'_{\mathrm{H1998}}$ where QF1995 denotes the *Quan* and Fry (1995) approximation [equation (6.24)] and H1998 denotes an internationally approved approximation (*Harvey* et al. 1998) [equations (6.17) and (6.18)], the *black short dashed curve* represents $\delta n' = n'_{\mathrm{M1977}} - n'_{\mathrm{H1998}}$, where M1977 denotes the approximation of *McNeil* (1977) [equation (6.19)], the *black long dashed curve* (almost identical with the black solid curve) represents $\delta n' = n'_{\mathrm{MS1990}} - n'_{\mathrm{H1998}}$, where MS1990 denotes the approximation of *Millard* and Seaver (1990) [equation (6.23)], the *gray solid curve* represents $n'_{\mathrm{QF1995}} - n'_{\mathrm{M1977}}$ at $T = 25°\mathrm{C}$, $S = 0\,\mathrm{ppt}$, $p = 1\,\mathrm{bar}$, the *gray short dashed curve* represents that difference at $T = 25°\mathrm{C}$, $S = 35\,\mathrm{ppt}$, $p = 1\,\mathrm{bar}$, and the *gray long dashed curve* represents $n'_{\mathrm{MS1990}} - n'_{\mathrm{H1998}}$ at $T = 25°\mathrm{C}$, $S = 0\,\mathrm{ppt}$, $p = 1\,\mathrm{bar}$.

has been devoted to devising ingenious methods of purifying water, which is one of the best solvents, and to devising measurement methods which minimize the effect of light scattering by particulate contaminants.

Results of many measurements of the absorption of light by seawater and by pure water (Figure 6.9 shows a representative sample) indicate absorption higher than that obtained by an analysis of *in situ* long-pathlength irradiance attenuation data (*Smith* and Baker 1981). This latter analysis yields absorption data that are also higher than the most accurate to-date results of *Pope* and Fry (1997).

6.3. Refractive indices of particles

The complex refractive index of aquatic particles spans a relatively large range as shown in Table 6.7 More detailed data for the various particle species are compiled in Table A.6 and Table A.7.

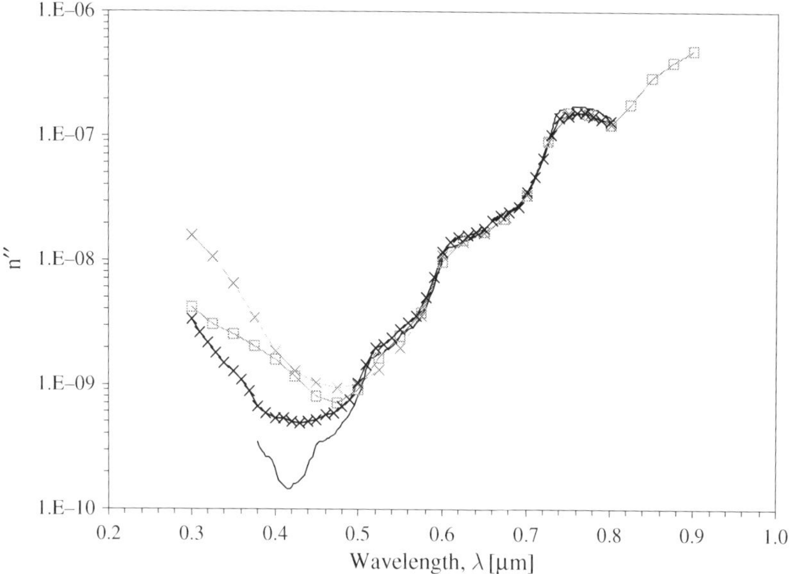

Figure 6.8. The most accurate to-date spectrum of the imaginary part of the refractive index, n," of pure water (*black solid curve with no symbols*) as calculated from the absorption data of *Pope* and Fry (1997:380 to 727.5 nm) and *Kou* et al. (1993: 728 to 800 nm) at 22°C. For comparison, some older data are shown: *Hale* and Query (1973: *gray x's*), *Smith* and Baker (1981: *black x's*) and *Querry* et al. (1991: *gray open squares*).

The determination of the refractive index of a particle can be much more difficult than that of determining the refractive index of the particle material in bulk. Although the approach from the bulk position is a reasonable alternative if one knows the composition of the particle material, some marine particles do not exist in bulk matter form, for example, a phytoplankton cell. Also, the bulk refractive index may lose meaning on approaching a particle size on the order of below 0.1 nm (*Bohren* and Huffman 1983). It is difficult to judge how important this problem might be in the case of aquatic particles, because there are no data on the presence of such small particles in seawater. Some of the smallest particles in seawater, the viruses, have sizes on the order of tens of nanometers (e.g., *Bratbak* et al. 1990, see also section 5.8.4.2). Even smaller colloidal particles with sizes of about 5 nm were found in large quantities in seawater (e.g., *Wells* and Goldberg 1992, see also section 5.8.4.1).

If the material is a mixture, as in the case of a phytoplankton cell, its refractive index can be calculated by applying a mixing rule (section 6.3.1.1). Such calculations entail an assumption that the refraction of light in an inhomogeneous particle can be described with one of the molecular refraction models. Although

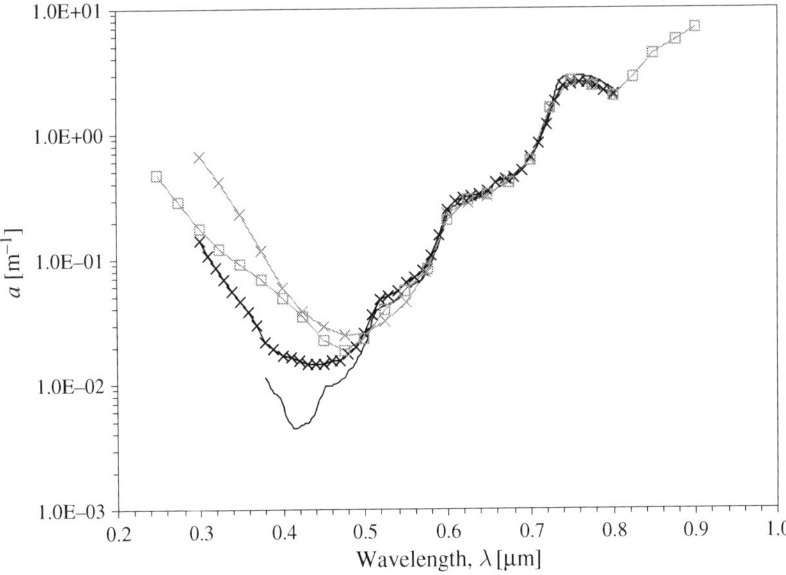

Figure 6.9. For convenience, we show here the absorption coefficient, a, for water. Note that this figure is essentially the same as Figure 6.8 because the absorption coefficient is related to the imaginary part of the refractive index, n,″ by equation (1.23). As in Figure 6.8, the absorption coefficient of pure water, as measured by *Pope* and Fry (1997: 380 to 727.5 nm) and *Kou* et al. (1993: 728 to 800 nm) at 22°C is shown by the black solid line with no symbols. For comparison, older data are shown, as calculated from the refractive index results of *Hale* and Query (1973: *gray x's*) and *Querry* et al. (1991: *gray open squares*), and as reported by *Smith* and Baker (1981: *black x's*). Note the absence in the data of Hale and Querry of a shoulder at about 0.52 μm. That shoulder appears in the Smith and Baker's data as well as in some other contemporary data, for example those of *Querry* et al. (1978: not shown).

these calculations provide results generally consistent with those obtained using other methods, this approach must be regarded as a rough approximation, since the intra-particle components frequently form granules that are much larger than the molecular size, violating some of the assumptions used in the development of the mixing rules.

Thus, methods derived from theories of the interaction of light with particulate matter remain one of the few feasible alternatives in determining an *effective* refractive index for many kinds of aquatic (as well as atmospheric) particles. In the following sections, we will both discuss literature data on the refractive index of suspended particles and review methods of determining the refractive index. We feel that the understanding of a method and thus a firm grasp of its limitations is an important component of the assessment of the refractive index values one encounters in a publication. Thus, as in the previous few chapters,

Table 6.7. Ranges of the refractive index estimates for the major types of aquatic particles in the visible.

Particle type	**Refractive index relative to water**	
	n'	n''
Minerals	1.07[a] to 1.58[a]	$< 1 \times 10^{-7}$[b] to 2×10^{-2}[c]
Detritus	1.04[d] to 1.07[e]	7×10^{-5} to 6×10^{-4}[d]
Oil[d]	1.10 to 1.12	2×10^{-5} to 9×10^{-3}
Phytoplankton	1.02 to 1.09	9×10^{-4} to 5×10^{-2}
Bacteria	1.03 to 1.07	1×10^{-4} to 3×10^{-3}
Viruses	1.04[f]	–

[a]*Woźniak* and Stramski (2004).
[b]Quartz.
[c]Saharan dust (*Patterson* et al. 1977). Albeit $n'' \sim 10^{-5}$ for many common minerals (that appear white in powdered state), there are notable exceptions: amorphous carbon ($n'' \sim 1$, *Gillespie* and Lindberg 1992), iron oxides (Fe_3O_4 : $n'' \sim 0.4$, Fe_2O_3 : $n'' \sim 0.5$ in a range of 350 to 450 nm, falling to $n'' \sim 0.03$ at 650 nm, *Gillespie* and Lindberg 1992), and manganese oxide (MnO_2, $n'' \sim 0.1$, *Gillespie* and Lindberg 1992).
[d]*Stramski* et al. (2004a), the refractive index of detritus depends on the state of hydration, see (6.7).
[e]*Green* et al. (2003b).
[f]*Stramski* and Kiefer (1991).
For detail, please refer to Table A.6 and Table A.7.

we put a deliberate stress on the discussion of the measurement and calculation techniques. We list the refractive index estimates of aquatic particles (as well as those obtained for relevant atmospheric particles) in Table A.7.

We included the estimates of the refractive index of atmospheric particles because particles found in the atmosphere usually end up in seawater (see also section 5.8.4.6). In fact, *Windom* (1969) estimated that some 75% of inorganic particles in the ocean come from the atmosphere (*Twardowski* et al. 2001). Single-grain iron-rich mineral particles (densities on the order of 4 to 5 g cm^{-3}!) as large as 20 μm were found in seawater during a major aeolian input event to the Sargasso Sea (June 1980, *Carder* et al. 1986). Such particles are typical of the Saharan desert, which was identified as a probable source of the aeolian input. Fly-ash particles were found in the deep waters of the Sargasso Sea (*Deuser* et al. 1983). Similarly, the deserts and arid areas of Asia yield significant contribution to the particulate load of the Pacific Ocean.

6.3.1. The average refractive index

Data reviewed in the following sections refer to an *average* refractive index of aquatic particles. Such an average refractive index represents well a population of homogeneous particles made of the same material. In many cases, aquatic particle populations are materially inhomogeneous. There are several types of inhomogeneity. We discuss two major types: (1) an *inhomogeneous particle* case: particles in a population are inhomogeneous but are all identical as far as their material composition is concerned, (2) an *inhomogeneous population* case: particles in a population are homogeneous, but the population contains particles of different material.

6.3.1.1. An inhomogeneous particle

An inhomogeneous particle can be treated in the first approximation as a multi-component solution. The average refractive index of a solution is necessarily a function of the refractive indices of the components. Each component contributes its own molecular refraction, modified by the presence of the molecules of all the other components of the solution. Several approximations to the refractive index of a mixture have been developed (e.g., *Aas* 1996, 1981). The most widely used approximations (mixing rules) include the rules of Gladstone–Dale, Lorenz–Lorentz, Bruggeman, and Maxwell Garnett.

We should note that these rules have been developed for solutions, i.e., they refer to the molecular size scale of the solution components. Components of a phytoplankton cell, for example, lipid granules, may have sizes that are several order of magnitude greater than that. Accounting for the effect of each of those components on the interaction of light with the cell by using the average or effective refractive index is a rough approximation because each such component contributes uniquely to the light scattering properties of the whole cell. Yet, an approximation of this nature may still be a valuable shortcut in many applications that deal with suspensions of inhomogeneous particles, because the subtle influences of the particle components on the light scattering properties of a particle tend to average out in a suspension of many cells.

The Gladstone–Dale rule (e.g., *Aas* 1996) models the refractive index of a mixture of materials (a solution), n'_{m}, as a volume-average refractive index:

$$n_{\mathrm{m}} = 1 + \frac{1}{V_{\mathrm{m}}} \sum_{j} V_j (n_j - 1) \tag{6.25}$$

where the subscript j numbers the components of the solution. Equation (6.25) asserts that the contribution of a component to the refractive index of the mixture is proportional to the partial volume of that component (V_j/V_{m}) and allows for the volume of the mixture, V_{m}, to be different from the sum of volumes of the components.

The molecular refractivity rule of Lorentz and Lorenz (e.g., *Aas* 1996) is given by:

$$\frac{n_{\mathrm{m}}^2 - 1}{n_{\mathrm{m}}^2 + 2} = \frac{1}{V_{\mathrm{m}}} \sum_{j} V_j \frac{n_j^2 - 1}{n_j^2 + 2} \tag{6.26}$$

where the meanings of the symbols are the same as those in (6.25). Equation (6.26) accounts for the modification of the electric field of the incoming light wave, as seen by a molecule of the solution, by the presence of the molecular neighbors of the molecule.

The Bruggeman rule (e.g., *Chýlek* et al. 2000, 1988) is presented here in a generalized form to account for more than two components of the mixture:

$$\sum_j f_j \frac{n_j^2 - n_m^2}{n_j^2 + 2n_m^2} = 0 \tag{6.27}$$

where $f_j = V_j/V_m$ is the volume *fraction* of the *j*-th component.

The Maxwell–Garnett *rule* (e.g., *Chýlek* et al. 2000, 1988) for a two-component mixture is expressed as follows (*Chýlek* et al. 1988):

$$n_m^2 = n_0^2 \frac{n_1^2 + 2n_0^2 + 2f_1(n_1^2 - n_0^2)}{n_1^2 + 2n_0^2 - f_1(n_1^2 - n_0^2)} \tag{6.28}$$

In the case of phytoplankton, the differences in the refractive index of a mixture, due to the choice of a mixing rule, are not too large. Changes in the water content of phytoplankton cause changes of the refractive index of a cell which are almost two orders of magnitude greater than the differences resulting from the choice of a mixing rule (*Aas* 1981).

Yet, due to the sensitivity of the angular light scattering of a single particle to the refractive index of the particle, small differences in the refractive index due to the use of the various mixing rules may translate into large differences in the scattering pattern of the particle. In particular, the Bruggeman rule may yield much better results than the widely used volume mixing rule (*Chýlek* et al. 1988). This latter conclusion is based on a comparison, by *Chýlek* et al., of measured and calculated angular scattering patterns of electromagnetic waves by homogeneous acrylic spheres (size parameter $x = \pi D/\lambda = 2\pi$) and by spheres with size-distributed water inclusions (1.76 and 2.7% by volume) at a microwave wavelength ($\lambda = 3.1835$ cm). The mode radius of inclusions was 0.02 cm, corresponding to $|n_a x_c| = 0.32$, where x_c is the size parameter of inclusions and n_a is the refractive index of water with respect to acrylic. The refractive indices of the sphere materials were $1.686 - i0.007$ (acrylic) and $7.70 - i2.48$ (water inclusions). The measured angular scattering pattern results were compared with those calculated using Mie theory for refractive indices derived with various mixing rules. The refractive indices calculated using the mixing rules of Bruggeman, Maxwell–Garnett, and of Chýlek–Srivastava (*Chýlek* et al. 1988) resulted in the closest fits to the experimental data by the calculated functions, with errors on the same order of magnitude as the measurement errors. The fit obtained for the refractive index calculated using the volume mixing rule (Gladstone–Dale) was one of the worse, resulting in an error greater by an order of magnitude than the measurement error.

Unfortunately, to our knowledge, no similar study has been done by using relative refractive indices similar to those of aquatic particles.

As we already hinted, the applicability of mixing rules is of particular interest in the case of phytoplankton cells that are inhomogeneous in a highly organized way. For example, a phytoplankton cell may contain a thin, high refractive index shell and a low refractive index core. In the first approximation, these cells can be modeled by a core-shell structure, as in a study of *Quinby*-Hunt et al. (1989), who examined the effect of the cell structure of nearly spherical green algae *Chlorella* on polarized light scattering at 441 nm. They found that the structured spherical cell model (inner sphere of $n = 1.08 - i0.05$ and a 60 nm thick coating with $n = 1.13 - i0.04$) yields more realistic angular scattering patterns than does the average refractive index. The refractive index parts could be varied only within a narrow range of ± 0.005 to assure a reasonable fit between the measured and predicted angular scattering patterns. The best-fit average refractive index model ($n = 1.085 - i0.048$) resulted in the predicted backscattering being greater by a factor of ~ 3 than that observed. Similar differences between observed and predicted patterns were found for the M_{34} element of the scattering matrix (see section 4.3.3 for definition). Note that the estimates of the imaginary part of the refractive index are nearly an order of magnitude greater than those obtained by other researchers (e.g., see section 6.3.2.4).

Zaneveld and Kitchen (1995) and *Kitchen* and Zaneveld (1992) both compared the use of an average refractive index for modeling of the volume scattering function (VSF) of phytoplankton populations vs. the use of two- and three-layer concentric cell models. Light scattering by three-layer cell model was calculated by using the approach of *Mueller* (1974) based on the theory described by *Kerker* (1969). Zaneveld and Kitchen echo the earlier findings *Quinby*-Hunt et al. (1989) by noting that the use of a structured cell model avoids the need for high refractive index particles when modeling a realistic marine VSF, which indicates the inadequacy of the effective refractive index for phytoplankton as a replacement for a more detailed specification of the actual cell structure.

6.3.1.2. Inhomogeneous populations of homogeneous and inhomogeneous particles

Studies of the optical properties of atmospheric aerosols indicate that the use of an average refractive index of aerosol composed of populations of particles having different refractive indices may result in significant errors in the absorption coefficient (by a factor of 10) and in the angular scattering pattern (*Gillespie* et al. 1978). This conclusion is based on simulations, using Mie theory, of the optical properties of aerosols represented by either two-particle populations with an absorbing ($n = 1.8 - i0.5$) and a non-absorbing ($n = 1.5$) fraction, each having a different size distribution, or a single population of particles with an average refractive index ($n = 1.5 - i0.005$). The average refractive index was calculated using the volume mixing rule of Gladstone and Dale (6.25). The size distribution of the average model was a sum of the size distributions of the two-component model.

The adequacy of the "equivalent" homogenous model for modeling of the VSF by non-homogeneous particles was also examined by *Mita* and Isono (1980). They compared the theoretical angular light scattering patterns for coated spheres with size distributions and refractive indices representative of the atmospheric aerosols with patterns calculated for the "equivalent" homogeneous spheres. The effective (average) complex refractive index of the homogeneous spheres was found to depend not only on the refractive indices of the components of the coated spheres but also on the size distribution of the particles. The VSF was well reproduced by using an effective refractive index for angles less than 90°. At angles >90°, significant deviations from the functions calculated using the true representations of the particles were observed.

Similar conclusions were reached in studies that used parameters representative of marine particles. *Kitchen* and Zaneveld (1992) compared layered and homogeneous sphere models to evaluate the effect of the phytoplankton cell structure on the VSF and attenuation by particles representative of the Pacific central gyre at 660 nm, with the particle size distribution (PSD) in a range from 0.25 to 14 μm. Results obtained with the examined models were compared with attenuation measurements performed in the same waters. Soft ($n = 1.02 - i0.005$) and hard ($n = 1.09 - i0.005$) homogenous particles yielded 29 and 114% of the measured attenuation respectively. Three-layered models (from outer to inner layer: $n_0 = 1.15$ to 1.20, $n_1 = 1.02 - i0.005$, $n_3 = 1.09$) yielded between 57 and 75% of the attenuation. The absorption of light by suspensions of various model particles was essentially insensitive to the particle structure model as variations of less than 2.5% (except the hard-particle model) were observed. The three-layered sphere models produced as much or more backscattering and near-forward scattering as did the homogeneous hard sphere model. Only the three-layered model reproduced the slope of the sample VSF, for the scattering angles of between 0 and 90° (*Petzold* 1972), measured in oligotrophic waters typical of the study area. Variations in the models caused significant variations in the shape of the function but little variations in the magnitude of that function.

The study of *Kitchen* and Zaneveld (1992) was later extended by *Zaneveld* and Kitchen (1995) who examined the effect of the stratification of a phytoplankton cell on the VSF, as well as on the light attenuation and absorption spectrum of phytoplankton suspension, all near the absorption maxima ($\lambda_{max} \pm 30$ nm) of chlorophyll. The PSD in a diameter range of 0.2 to 30 μm was derived from the results of *Kitchen* and Zaneveld (1990). Three-layered models yielded the least variations in the scattering function (average of 9%) and the two-layered models yielded the greatest variations (average of 24%, with a maximum variation of 71% in the backscattering region). The homogenous sphere model resulted in an intermediate variability of the scattering function. The absorption and attenuation coefficients were little influenced by the model of the internal structure of the cell. A variability of about 15% was observed in the magnitude of the attenuation spectra, with the two-layered model yielding the lowest estimates.

6.3.2. *Refractive index of particles from integral optical properties of suspension*

The concept of the refractive index naturally differentiates the optics of suspensions from the optics of solutions. Indeed, in the case of the solution (the scattering centers have molecular size), one cannot use the concept of the refractive index to characterize the scattering center itself. In that case, the refractive index refers only to the solution as a collection of these centers. In contrast, for a suspension (the scattering centers are much larger than molecules), the refractive index can be used to characterize the material of the scattering centers (particles). A particle in a suspension can be assigned a meaningful refractive index because it typically contains a sufficiently large number of molecules so that it can be treated as a chunk of particle material.

Although one can formally define the refractive index of a suspension (e.g., *van de Hulst* 1957), such an index is different from that of the refractive index of the particle material and cannot be used in a theory of light scattering by particles in order to calculate optical properties of the suspension. Therefore, relationships between the refractive index and composition of solutions (here water and seawater) that we developed in section 1.5.1 cannot be applied in the case of a suspension. Indeed, the determination of the refractive index of the particle material typically involves two steps: (1) converting the measurable optical properties of the suspension (such as the attenuation coefficient) to optical properties of particles of that suspension (such as the attenuation efficiency) and (2) inverting (solving for) optical properties of the particles to obtain the refractive index with the help of a theory of light scattering. This second step usually leads to relationships that can only be solved numerically.

In this section, we will identify the key optical properties of the particles (optical efficiencies) and relate them to the bulk optical properties of the suspension. In the next section, we will recapitulate a theory of light scattering (the anomalous diffraction theory, discussed in more detail in section 3.3.1) that provides simple, analytical approximations to the optical efficiencies of the particle as functions of the refractive index of the particle material, so that the second step in the process of the determination of the refractive index of particles can be performed.

We recall that the optical efficiency (for absorption, scattering, or both, section 1.6.1), Q_y, is the ratio of light power, ΔF_y removed by a particle (due to absorption: $y = a$, scattering: $y = b$, or both: $y = c$), to the light power, F_i, incident on the projected area of the particle, P, where the projection is performed onto a plane perpendicular to the direction of propagation of the incident light beam:

$$Q_y = \frac{\Delta F_y}{F_i} = \frac{\sigma_y E}{PE} = \frac{\sigma_y}{P} \tag{6.29}$$

where σ_y is the absorption, scattering, or attenuation cross-section of the particle and E is the irradiance (i.e., power per unit area) of the incident light beam.

It thus follows [(6.100) in Problem 2] that, with a suspension containing N identical particles per unit volume, the optical efficiency of the particle can be determined from the following equation:

$$Q_y = \frac{c_y}{NP} \tag{6.30}$$

where c_y is either the coefficient of absorption ($y = a$), scattering ($y = b$), or attenuation ($y = c$) of the suspension, used in the Beer–Lambert law (6.2). Note that (6.30) applies to suspensions only in the low concentration range, as defined by the prevalence of the single scattering process (see section 4.2.2).

If the particles are all of the same material but are size distributed, we may define a size averaged optical efficiency of the particles in the suspension:

$$\langle Q_y \rangle = \frac{c_y}{\int_0^\infty f(D)P(D)dD} \tag{6.31}$$

where $f(D)$ is the (frequency) PSD and D is the particle "diameter." In fact, in this latter case, a simpler approach in trying to determine the refractive index of particles in suspension is to use the optical coefficient of the suspension directly as the key optical property that can be expressed with a scattering theory as a function of the refractive index of the particle material.

In some cases, the optical coefficients of the suspension may be given in the mass concentration-specific form, $c_y{}^*$ (e.g., *Babin* and Stramski 2004). In that case we have:

$$Q_y = \frac{M_p c_y{}^*}{P} \tag{6.32}$$

where M_p is the mass of the particle. This equation applies to a suspension of identical particles. For polydisperse suspensions, we obtain:

$$\langle Q_y \rangle = \frac{\rho \int_0^\infty f(D)V(D)dD}{\int_0^\infty f(D)P(D)dD} c_y{}^* \tag{6.33}$$

where ρ is the particle density and V is the particle volume.

6.3.2.1. Anomalous diffraction approximation (ADA) revisited

The anomalous diffraction theory (*van de Hulst* 1957, see section 3.3.1) provides relatively simple, analytical expressions for the optical efficiencies of scattering, absorption, and attenuation of light by a particle with the real part of the refractive index having the real part close to that of the surrounding medium. This theory has been recently enhanced (e.g., *Yang* et al. 2004a, 2004b) to enable surprisingly accurate evaluations of the optical efficiencies of tenuous particles, despite the neglect of the edge effect (see section 3.3.1).

According to the simpler, original theory of van de Hulst, as applied to homogeneous spheres (*van de Hulst* 1957), the optical efficiencies are expressed as follows:

$$Q_c = 2 - 4e^{-\rho' \tan\beta} \frac{\cos\beta}{\rho'} \sin(\rho' - \beta) - 4e^{-\rho' \tan\beta} \left(\frac{\cos\beta}{\rho'}\right)^2 \cos(\rho' - 2\beta) + 4\left(\frac{\cos\beta}{\rho'}\right)^2 \cos 2\beta \quad (6.34)$$

$$Q_a = 1 + 2\frac{e^{-4xn''}}{4xn''} + 2\frac{e^{-4xn''} - 1}{(4xn'')^2} \quad (6.35)$$

$$Q_b = Q_c - Q_a \quad (6.36)$$

The parameter x is a dimensionless particle size, defined as follows:

$$x = \frac{\pi D}{\lambda} \quad (6.37)$$

where D is the particle diameter, λ is the wavelength in the medium surrounding the particle. The angle β is defined as follows

$$\tan\beta = \frac{n''}{n' - 1} \quad (6.38)$$

where n'' is the imaginary part and n' is the real part of the refractive index of the particle relative to that of the surrounding medium. Finally, the parameter ρ' is the phase shift of the light wave that would have been introduced by the slab of the particle material with the thickness equal to the sphere diameter:

$$\rho' = 2x(n' - 1) \quad (6.39)$$

If the particle does not absorb light ($n'' = 0$), then (6.34) simplifies to:

$$Q_c = Q_b = 2 - 4\frac{\sin\rho'}{\rho'} + 4\frac{1-\cos\rho'}{\rho'^2} \tag{6.40}$$

Many aquatic particles, such as certain bacterial and phytoplankton cells, can be more reasonably modeled by a coated sphere. *Quirantes* and Bernard (2004) have recently provided a relatively simple extension of the anomalous diffraction approximation (see sections 3.3.1 and 6.3.2.1) for coated spheres with the core/coating diameters D_1 and D_2 and refractive indices n_1 and n_2 respectively:

$$\begin{aligned} Q_c = {} & 2 - 4\frac{\cos\beta_1}{\rho'_1} z e^{-z\rho_1' \tan\beta_1} \sin(z\rho'_1 - \beta_1) \\ & - 4\left(\frac{\cos\beta_1}{\rho'_1}\right)^2 \left[e^{-z\rho'_1 \tan\beta_1}\cos(z\rho'_1 - 2\beta_1) - \cos 2\beta_1\right] \\ & - 4\frac{\cos\beta_2}{\rho'_2}\left[e^{-\rho'_2\tan\beta_2}\sin(\rho'_2 - \beta_2) - z e^{-z\rho'_2 \tan\beta_2}\sin(z\rho'_2 - \beta_2)\right] \\ & - 4\left(\frac{\cos\beta_2}{\rho'_2}\right)^2\left[e^{-\rho'_2\tan\beta_2}\sin(\rho'_2 - 2\beta_2) - z e^{-z\rho'_2\tan\beta_2}\sin(z\rho'_2 - 2\beta_2)\right] \end{aligned} \tag{6.41}$$

$$Q_a = 1 + 2\left[\frac{z e^{-z\rho''_1}}{\rho''_1} + \frac{e^{-z\rho''_1} - 1}{(\rho''_1)^2} + \frac{e^{-\rho''_2} - z e^{-z\rho''_2}}{\rho''_2} + \frac{e^{-\rho''_2} - e^{-z\rho''_2}}{(\rho''_2)^2}\right] \tag{6.42}$$

where indices 1 and 2 refer, as already stated, to the parameters of the core and the coating respectively, and

$$z = (1 - q^2)^{1/2} \tag{6.43}$$

$$q = D_1/D_2$$

and

$$\rho'_1 = 2x(n'_2 - 1) \tag{6.44}$$

$$\rho'_2 = 2x[qn'_1 + (1-q)n'_2 - 1] \tag{6.45}$$

$$\tan\beta_2 = \frac{n''_2}{n'_2 - 1}$$

with $x = \pi D_2/\lambda$, and

$$\rho''_1 = 4xn''_2 \tag{6.46}$$

$$\rho'_2 = 4x[qn''_1 + (1-q)n''_2] \tag{6.47}$$

$$\tan\beta_2 = \frac{qn''_1 + (1-q)n''_2}{qn'_1 + (1-q)n'_2 - 1} \tag{6.48}$$

ADA expressions have been developed for other particle morphologies: hollow spheres (*Aas* 1984), finite cylinders (*Liu* et al. 1998, *Aas* 1984), infinite cylinders (section 3.3.3, *Fournier* and Evans 1996) and disks (*Aas* 1984; this work, section 3.3.3), oriented spheroids (*Fournier* and Evans 1991). Some results for randomly oriented spheroids have been obtained by *Evans* and Fournier (1994) at the price of some significant complexity. *Yang* et al. (2004b) discuss a statistical approach to ADA, whereby the particle shape and orientation is accounted for by allowing pathlengths (and phase delays) along the various pathways of light rays through the particle to be randomly distributed.

Typical graphs of the optical efficiencies of homogeneous spheres as functions of the real and imaginary parts of the refractive index are shown in Figure 6.10 for $n'' = 0$ (no absorption) and in Figure 6.11 for $n'' = 0.005$. This presentation format is chosen because we are aiming here at the use of the anomalous diffraction theory for the determination of the refractive index of the particles; the usual presentation of these curves uses the relative particle size, x, as the independent variable.

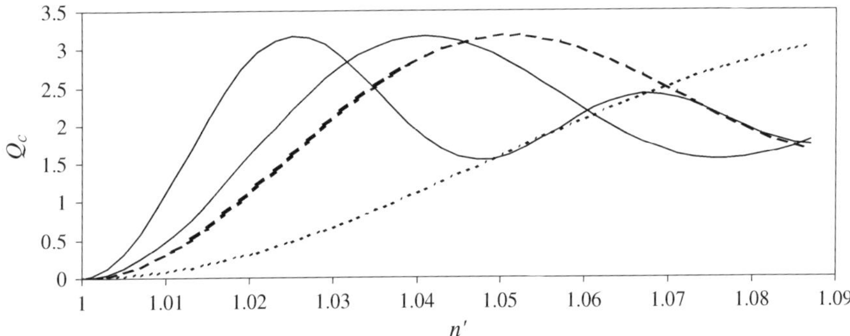

Figure 6.10. Optical efficiency, Q_c, of a non-absorbing homogeneous sphere, the first-order-of-magnitude approximation for a phytoplankton cell, at wavelengths away from those of the absorption maxima of photosynthetic pigments, as a function of the real part of the refractive index, n', for the dimensionless particle size $x = \pi D/\lambda = 10$, 20, 40, and 80 (from bottom to top at $n' = 1.02$, dashing is used merely to help the curve identification), where D is the cell diameter. The refractive index of the cell material is given relative to that of the surrounding medium. It can be seen that for n'' in the range shown, typical of low-index aquatic particles, $n' = f(Q_c,\ x)$ becomes a multi-valued function of Q_c for a sufficiently large particle. Thus, the determination of the refractive index by solving $Q_c(n';\ x) = Q_{c,\ \mathrm{exp}}$ for n', where $Q_{c,\ \mathrm{exp}}$ is an experimental value, is not possible for the large particles without limiting the refractive index range through independent constraints.

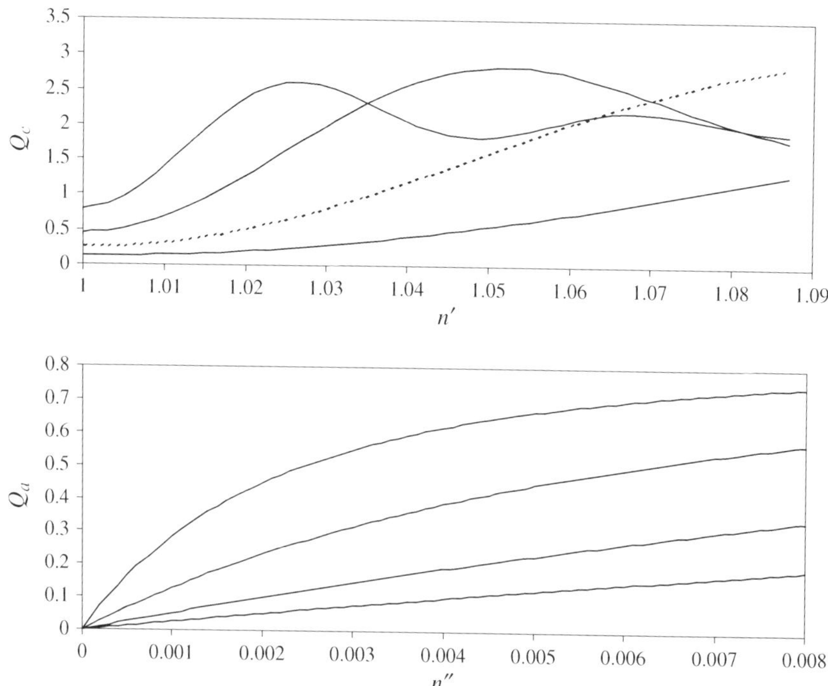

Figure 6.11. Optical efficiencies for attenuation (*upper panel*) and absorption (*lower panel*) of an absorbing homogeneous sphere (the first-order-of-magnitude approximation for a phytoplankton cell at wavelengths near those of the absorption maxima of photosynthetic pigments) as functions of the real (n') and imaginary (n'') parts of the refractive index, for the dimensionless particle size $x = \pi D/\lambda = 10, 20, 40$, and 80 (from bottom to top at $n' = 1.02$ and $n'' = 0.001$; dashing is used merely for the curve identification). D is the cell diameter. The attenuation efficiency, Q_c (upper panel), is calculated for $n'' = 0.005$. The refractive index of the cell material is given relative to that of the surrounding medium. The $n'' = f(Q_a, x)$ is single valued in the entire range of x, but $n' = f(Q_c, x)$ becomes a multi-valued function of Q_c for a sufficiently large particle. Thus, the determination of the real part, n', of the refractive index by solving $Q_c(n'; x) = Q_{c,\,\mathrm{exp}}$ for n', where $Q_{c,\,\mathrm{exp}}$ is an experimental value, is not possible for the large particles without limiting the refractive index range through independent constraints.

Incidentally, although $x = \pi D/\lambda$, the format $Q(x)$ only approximately represents the wavelength spectra of the optical efficiencies of a fixed-size particle, because the refractive index is itself a function of the wavelength. In many cases, the refractive indices of aquatic particles have relatively low dispersion in the visible, for example, those of most phytoplankton components except pigments near their absorption bands (e.g., *Aas* 1981). However, even with the low dispersion of protein (protein does not appreciably absorb light in the visible), the adoption of a wavelength-independent refractive index of the cell may cause significant errors in the calculation of a spectrum of the optical efficiency Q_c of a non-absorbing particle (Figure 6.12).

If the particle size is known, and the particles can be reasonably approximated by a homogeneous sphere, it would seem that we should easily be able to determine the real (n') and imaginary (n'') parts of the refractive index by solving a set of two equations: $Q_c(x, n', n'') = Q_{c,\,\mathrm{exp}}$ and $Q_a(x, n'') = Q_{a,\,\mathrm{exp}}$, where the

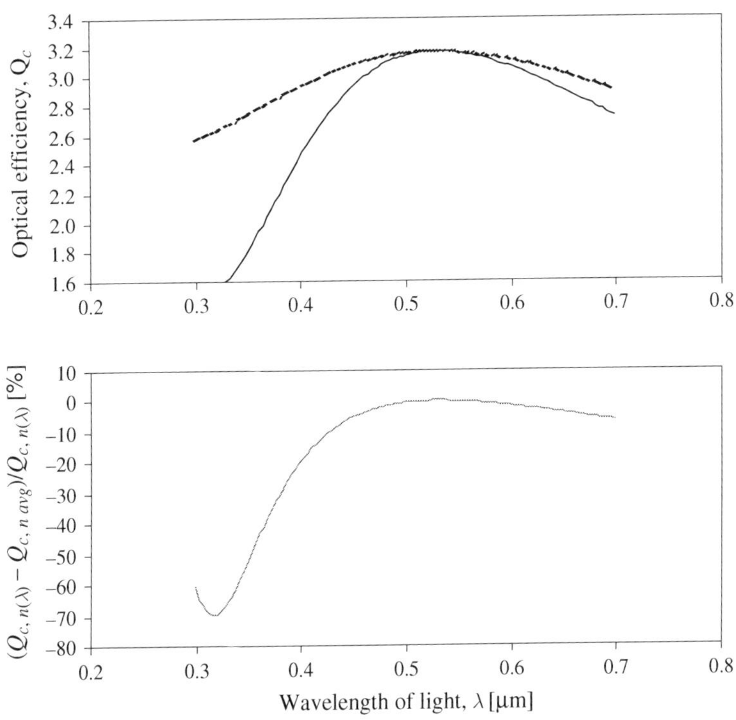

Figure 6.12. The type of wavelength dependency of the refractive index of a particle may significantly affect the calculation of the optical efficiency spectrum for light attenuation, as shown in this example of a sphere of 0.65 (v/v) protein solution (a phytoplankton cell model) with the diameter of 5 μm in seawater at $T = 25°\mathrm{C}$ and $S = 35.5$ ppt. *Upper panel: solid curve* Q_c for the refractive index of protein defined as $n = n'(\lambda) - i0$, where $n'(\lambda) = 1.578 + 0.00753/\lambda^2$ (after *Aas* 1996, the refractive index of the protein solution is calculated with the Gladstone–Dale mixing rule), *dashed curve*—an average refractive index of solution in the visible, $n = n_{\mathrm{avg}} = 1.433 - i0$. *Bottom panel*: relative difference between the attenuation efficiencies. The refractive index of seawater, is approximated by $n_{\mathrm{w}} = n'_{\mathrm{w}}(\lambda) - 0i$, where $n'_{\mathrm{w}}(\lambda) = 1.329011 + 0.003562/\lambda^2 - 0.00005182/\lambda^4$. As the imaginary part of the refractive index of protein is essentially 0 in the visible, $Q_c = Q_b$.

subscript "exp" denotes the measured value. However, as much as this is possible for n'', because $n'' = f(Q;\ x)$ is a single-valued function of n'', the task becomes complicated for n', because $n' = f(Q_c;\ x,\ n'')$ may be a multi-valued function of n' for particles within a certain range of x (Figure 6.10 and Figure 6.11).

6.3.2.2. Real refractive index from backscattering and absorption

Twardowski et al. (2001) used Mie theory to develop a relationship between the real part of the refractive index, n', the particle backscattering, and the shape of the particle-specific attenuation spectrum of marine suspensions. This relationship can be summarized as follows.

$$n' = 1 + {b'_{\mathrm{p}}}^{0.5377+0.4867\gamma^2}(1.4676 + 2.2950\gamma^2 + 2.3113\gamma^4) \qquad (6.49)$$

where $b'_{\mathrm{p}} = b_{\mathrm{bp}}/b_{\mathrm{p}}$ is the backscattering ratio of the particles in suspension and γ is the slope of the particle-specific power-law attenuation spectrum of the suspension. In fact, the parameter γ is here a proxy for the slope, m, of the power-law size distribution of the particles, which these authors assumed to be representative of the marine particle suspensions. This substitution of γ for m is possible due to a relationship between γ and m (see section 3.4).

6.3.2.3. Determinations of the real refractive index

Carder et al. (1972) used the theory of anomalous diffraction of *van de Hulst* (1957) to estimate the real refractive indices of nearly spherical, soft-walled phytoplankton cells (*Isochrysis galbana*, see Table A.6). The modal diameter ($\sim 4\,\mu$m) of a narrow PSD of these cells, measured using a Coulter counter model A, places these cells in the univalued region of $n' = f(Q_b;\ x,\ n'' = 0)$. The scattering coefficient, from which Q_{b} can be determined was itself *inferred* from the VSF $\beta(45°)$ of the suspension of cells, according to an assumption that $\beta(45°)$ is approximately proportional to the scattering coefficient (better correlation between the VSF and the scattering coefficient was found for the scattering angle of 6°, *Kopelevich* and Burenkov 1971). The coefficient of proportionality was determined using polystyrene latex. The scattering efficiency Q_{b} [equation (6.36)] of the cells at wavelengths of 546 and 578 nm was calculated from the measurements of the "scattering coefficient," the cell diameter, and cell number concentration. The cells were assumed not to absorb light at the two wavelengths examined. A similar method was used by *Kopelevich* et al. (1987) to determine the real refractive indices of isolated heterotrophic marine bacteria (see Table A.6) by comparing the experimental values of the attenuation efficiency of the cells (obtained through spectral attenuation measurements) with values calculated with (6.40).

Zaneveld and Pak (1971) determined an average real refractive index of naturally occurring marine particles (see Table A.6) also by using the anomalous diffraction theory (*van de Hulst* 1957) to express a ratio of the scattering coefficients at two different wavelengths, $b(436\,\mathrm{nm})/b(546\,\mathrm{nm})$, as a function of the slope of the power-law size distribution (section 5.8.5.3) and of the refractive index of the particles. Like *Carder* et al. (1972), *Zaneveld* and Pak used the VSF, at 45°, as a proxy for the scattering coefficient. The correlation between b and $\beta(45°)$ (see also section 4.4.2.2) is, in the case of aquatic particles, the consequence of a relative stability of the shape of the size distribution (*Jonasz* and Zalewski 1978) and of the composition of the particles. Thus, in the ratio $b(436\,\text{nm})/b(546\,\text{nm})$, the unknown proportionality coefficient, b_0, in $b = b_0\ \beta(45°)$, is nearly canceled because b and $\beta(45°)$ depend only weakly on the wavelength of light. The refractive index of the particles was estimated by comparing experimental and theoretical values of the ratio, $b(436\,\text{nm})/b(546\,\text{nm})$.

In a somewhat different approach, *Jonasz* (1986) determined both the refractive index and non-sphericity of spheroidal cells of *Rhodomonas* sp. (major axis of 14.6 μm, minor axis of 9 μm) (see Table A.6) by comparing the size distributions of these cells obtained using a HIAC particle counter (see section 5.7.2), with a distribution calculated for a HIAC-type particle counter from a PSD measured with a Coulter counter (see section 5.7.1). The HIAC counter sizes particles according to the attenuation of (white) light they cause and is sensitive to the product of the projected area of the particles (hence the particle non-sphericity, see also section 5.7.2.2) and the attenuation efficiency (hence, the refractive index). In contrast, the Coulter counter sizes a particle according to the change it causes in the electrical resistance of the measurement zone of the counter. This non-optical method is sensitive to the particle volume. The response of the HIAC counter was simulated by using a simplified form of the attenuation efficiency (we corrected a minor misprint in *Jonasz* 1986)

$$Q_c = \begin{cases} \rho'/3 & \text{for} \quad \rho' < 6 \\ 2 & \text{for} \quad \rho' > 6 \end{cases} \tag{6.50}$$

for non-spherical particles, based on the experimental results of *Hodkinson* (1963). We have $\rho' = 2x(n' - 1)$, so that the approximated attenuation efficiency is a function of the real part, n', of the refractive index of the particles relative to that of water. The real part was varied in the calculation of the simulated size distribution characteristic of the HIAC counter, which used a Coulter counter size distribution as an input, until the best fit to the distribution measured with the HIAC counter was achieved. The relatively high refractive index of this phytoplankton species ($n' = 1.08$) so obtained is consistent with the high fat content of the cells, almost four times as high as that expected for *Isochrysis galbana*, whose refractive index was estimated at about 1.03 (*Carder* et al. 1972).

6.3.2.4. The BMS method for complex refractive index

Bricaud and Morel (1986) and *Stramski* et al. (1988) developed a comprehensive method (BMS) of determining the spectra of the complex refractive index of phytoplankton cells. This method, originally based on the anomalous diffraction theory (*van de Hulst* 1957) of homogeneous spheres and on the Ketteler–Helmholtz theory of anomalous dispersion, has more recently been augmented with the use of the Mie theory of light scattering (*Stramski* et al. 1993, *Stramski* and Kiefer 1990). Numerous studies of the refractive index of phytoplankton and its diel variability have been made with this method (e.g., *Stramski* et al. 2002, *DuRand* and Olson 1998, *Reynolds* et al. 1997, *Stramski* et al. 1995, *Stramski* and Reynolds 1993, *Stramski* et al. 1993, *Ahn* et al. 1992, *Morel* and Ahn 1990). Values of the refractive indices of phytoplankton obtained with this method are listed in Table A.6. Similar methods have been used to determine the refractive index of mineral particles [*Marzo* et al. 2004, gypsum in the infrared (IR)].

The BMS method uses an approximate Helmholtz–Ketteler theory of the refractive index of matter. In that theory, matter is treated as a set of oscillators, which can be excited by an electromagnetic wave. If the incident wave frequency is sufficiently close to one of the resonance frequencies of the oscillators, the oscillators having that resonance frequency absorb energy from the wave. According to the Helmholtz–Ketteler theory, the real part, n', and the imaginary part, n'', of the refractive index, n, are expressed by the following equations:

$$n' = 1 + \varepsilon - \varepsilon \sum_j \frac{\kappa_j \nu_j}{1 + \nu_j^2} \tag{6.51}$$

$$= 1 + \varepsilon - \delta n'$$

$$n'' = \varepsilon \sum_j \frac{\kappa_j}{1 + \nu_j^2} \tag{6.52}$$

where $\varepsilon << 1$ is a constant, and

$$\nu = \frac{2}{\gamma}(k - k_0) \tag{6.53}$$

where $k_0 = 2\pi/\lambda_0$ is the wave number corresponding to an oscillator's resonance, $k = 2\pi/\lambda$ is the wave number of the incident wave, κ is the oscillator strength, and γ is the damping constant of the oscillator. The oscillator parameters k_0 and γ, as well as the product $\kappa\varepsilon$, can be determined from a component of the spectrum of n'' corresponding to that oscillator. The constant ε is determined from the measurements of light attenuation by the cells.

The determination of the spectrum of the complex refractive index of particles begins with the measurement of the absorption spectrum of a suspension of cells. The imaginary part, n''_{cm}, of the refractive index of the *cell material* is calculated from the absorption coefficient of the suspension, using (6.30) (with $y = a$) and (6.35). The $n''_{cm}(\lambda)$ determined this way represents an "average particle," if particles in the suspension are size- and/or refractive index-distributed. The spectrum of n''_{cm} is then decomposed by using several oscillators, whose resonance wave numbers represent the absorption peaks of pigments known to be present in the cells. Parameters of the oscillators are determined by trial-and-error or by using a numerical algorithm (e.g., *Hoepffner* and Sathyendranath 1991). In reference to the use of such an algorithm, Hoepfner and Sathyendranath note that the Helmholtz–Ketteler model of the interaction of light with matter describes well a collection of oscillators that do not interact with each other. In that case, the spectral shape of an absorption band is described by the Lorentz function. In photosynthetic pigments, strong vibrational interaction may occur between the molecules. In such a case, the spectral shape of an absorption band, and thus the shape of the corresponding n'' peak, is better described by a Gaussian function, not by the Lorentz function. The form of the function has a direct bearing on numerical algorithms for extracting the oscillators' parameters from the absorption spectra.

Having determined the oscillator parameters, one can generate the wavelength-dependent term of the real part, n', of the refractive index. The unknown, wavelength-independent term ε is determined by fitting a spectrum of the attenuation efficiency, calculated by using the anomalous diffraction theory as a function of ε, to the experimental spectrum (see Figure 6.13). The determination of ε, which involved guesswork in the original presentation of the BMS method, was simplified by *Stramski* et al. (1988). They noted that the term $\delta n'$ [see (6.51)] varying in the case of phytoplankton between -0.01 and $+0.01$ assumes a value of 0 usually for at least one wavelength in the visible. Thus, at each such wavelength, λ_ε, the real part, n', of the refractive index equals just $1+\varepsilon$, as it follows from (6.51). Consequently, after λ_ε is located in the spectrum of n''_{cm}, one can iteratively solve the equation:

$$Q_c(\varepsilon) = Q_{c,\exp}(\lambda = \lambda_\varepsilon) \tag{6.54}$$

where $Q_{c,\exp}$ is the experimental value of Q_c at λ_ε (Figure 6.13). This method of determining ε is usable when the particle size, x, is sufficiently small so that Q_c, which is also a function of x, depends significantly on ε. For the large x and/or n', $Q_c(\varepsilon)$ may even become multi-valued (Figure 6.13). This limits the cell size and n' ranges in which one can use this method. If there are several wavelengths at which $\delta n' = 0$, the accuracy of the determination of ε can be improved by simultaneously solving (6.54) for ε at all these wavelengths. We simplified the discussion by assuming that all particles are of the same size. If they are not, then the optical efficiencies, as well as the refractive index, must be treated as particle size averages (*Stramski* et al. 1988, *Bricaud* and Morel 1986). Typical spectra of the refractive index of phytoplankton, obtained with this method, are shown in Figure 6.14. The complex refractive indices of phytoplankton, obtained by various researchers using this method, are compiled in Table A.6.

The BMS method, through its use of the anomalous diffraction approximation for spheres, is based on an assumption that the phytoplankton cells are

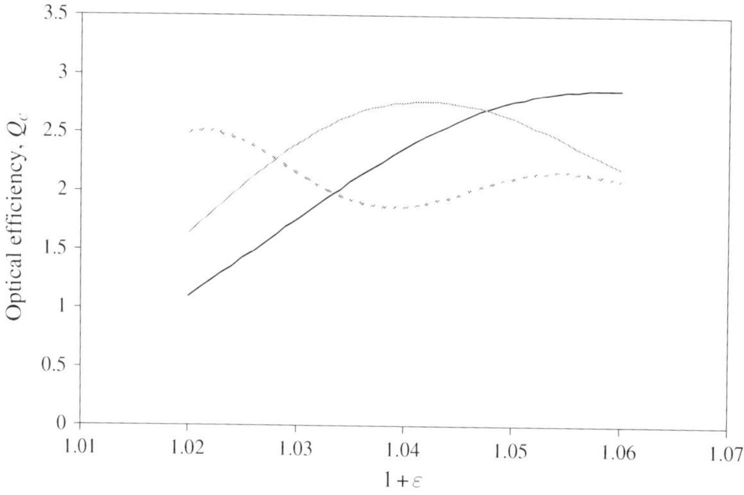

Figure 6.13. Optical efficiency of a phytoplankton cell, here approximated by a homogeneous sphere, as a function of the wavelength independent term, ε, real part of the refractive index, $n' = 1 + \varepsilon - \delta n'(\lambda)$ (relative to seawater) at a wavelength, λ_ε, where the wavelength-dependent part, $\delta n'$, vanishes. The resulting equation $Q_c(n') = Q_{c,\,\exp}(\lambda = \lambda_\varepsilon)$, is solved for ε in the last key step of the Bricaud–Morel–Stramski method (*Bricaud* and Morel 1986, *Stramski* et al. 1988) of calculating the spectrum of the complex refractive index of phytoplankton from their attenuation and absorption spectra. The results shown here were obtained for $\lambda_\varepsilon = 440\,\text{nm}$. The *black curve* represents Q_c for a $5\,\mu\text{m}$ cell, here $\varepsilon = f(Q_c)$ is single valued, yet the value of ε may be less accurate for the upper part of the refractive index range. The *solid and dashed gray curves* represent Q_c for a 7 and $14\,\mu\text{m}$ cells respectively. In these two latter cases, $\varepsilon = f(Q_c)$ is multi-valued, and solving for ε requires additional constraints.

homogeneous spheres. In reality, phytoplankton cells are frequently non-spherical. Many phytoplankton species whose refractive indices are listed in Table A.6 have spheroidal cells. The most non-spherical cells studied with this method (*Bacillariophycae* diatoms) have a ratio of the largest to the smallest dimension of about 2.5. If the size of non-spherical cell is determined with a method (e.g., resistive particle sizing method) yielding the volume-equivalent spherical diameter, the average geometrical cross-section of the cells may be underestimated. This may lead to an overestimation of the real part of the refractive index (*Jonasz* and Prandke 1986) because a refractive index higher than the actual index would then be needed to reproduce the experimental value of Q_c. Another effect of the particle shape is in the potentially different dependence of the optical efficiencies on the phase shift parameter, δ.

The phytoplankton cells may not only be non-spherical but also inhomogeneous as they contain vacuoles and fat globules. The light-absorbing pigments of the

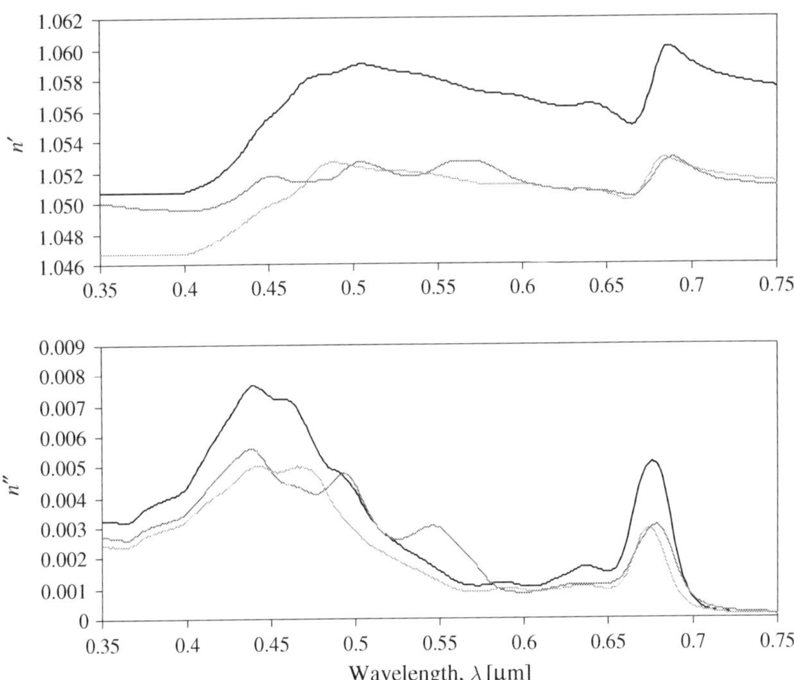

Figure 6.14. The real part, n', and the imaginary part, n'', of the refractive index of selected phytoplankton according to *Stramski* et al. (2001; that reference lists only selected values of the refractive index; the full spectra, courtesy of D. Stramski, are not shown there). *Black curve: Isochrysis galbana* (based on data from *Ahn* et al. 1992), *light gray curve: Emiliania huxleyi* (based on data from Ahn et al. 1992), *dark gray curve*: generic phycocyanin-rich picoplankton *Synechocystis* sp. The spectra reflect the major features of the optics of phytoplankton cells, determined by absorption of photosynthetic pigments near 0.44 μm and 0.675 μm.

cell are concentrated in chloroplasts which may be aligned along the cell wall, as in the case of "spherical" *Chlorella* cells (e.g., *Quinby*-Hunt et al. 1989). Some phytoplankton cells, such as coccolithophores, are encased in a calcite shell, which may modify light scattering by the cell. Indeed, with the following coated sphere model of a coccolithophore, based on data from section 6.4.3.3 – shell thickness = 0.1 μm, refractive index = 1.19, core diameter = 16 μm, refractive index = 1.015, the scattering efficiency, Q_b, of a calcite shelled cell evaluates at ~2.68 by using a coated sphere calculator (MJC Optical Technology) that implements the *Toon* and Ackerman (1981) algorithm, while that of a naked cell (core alone) evaluates at ~2.09. Interestingly, experimental results do not seem to support a conclusion that the calcite shell of a coccolithophorid, *E. huxleyi*, contributes significantly to light scattering by the cell (*Volten* et al. 1998). Finally, the results of the measurements of light absorption by the cells may also be misinterpreted if the effect of concentration of chlorophyll in the chloroplasts is not accounted for (*Haardt* and Maske 1987).

Interesting conclusions regarding the effect of the particle shape can be drawn from a modeling study of *Stramski* and Piskozub (2003) who compared the scattering corrections required to properly measure the absorption of light by suspensions of cells with a spectrophotometer. The cuvette with a sample is placed at the entry port of an integrating sphere in order to collect as much of the light scattered by the sample as possible. Not all the scattered light is collected, however, and absorption spectra measured in such a way indicate significant absorption at a wavelength (typically 750 nm) beyond the red absorption peak of chlorophyll where phytoplankton is known not to absorb light. A standard method of correcting the spectra is to subtract the absorption value at 750 nm from the spectrum. *Stramski* and Piskozub used Monte Carlo ray tracing to evaluate the contribution of the scattered light to the measured absorption of light by cells with two different morphologies: a small "spherical" *Synechococcus* (a cyanobacterium) and a thick pillbox-shaped *Thallassisira pseudonana* (a centric diatom). Evaluation of the scattering correction requires the knowledge of the VSF of the suspension (see section 4.2.1) in addition to the knowledge of the measurement system geometry. The VSF, when evaluated from Mie theory (homogeneous spheres), provided a passable correction for the "spherical" cells but not for the pillbox-shaped cells.

Although the theoretical and experimental spectra of the optical cross-sections for absorption and scattering agree well, also in the case of non-spherical algae (*Morel* and Bricaud 1986; *Bricaud* et al. 1988), this is not necessarily a proof that the refractive index of the cells is determined correctly because, for example, effects of cell structure have not been accounted for, and the "refractive index" so determined may be a classification parameter rather than the actual refractive index. However, the review of the refractive index results obtained using that and many other techniques leads to a conclusion that, apart from the extreme cases and for the lack of a more reliable approach, the effects of the particle shape on the optical efficiencies can be neglected while estimating the refractive index of the moderately non-spherical plankton cells, especially given the accuracy of the measurements of absorption and scattering by the suspensions.

Measurements of absorption of light by small particles. The measurements of the absorption spectra of suspensions play a central role in the method of determining the complex refractive index of phytoplankton cells. Some comments are in order here.

The absorption spectrum of a particle may be significantly different from that of the cellular material in bulk. First, the absorption peaks may be flattened, a phenomenon called the *package effect* (*Morel* and Bricaud 1981, *Duysens* 1956). The package effect is caused by the dependence of the absorption efficiency on the particle size and refractive index. Thus, a successful determination of the absorption coefficient of the suspended particles *must* involve a theory of interaction of light with particulate matter.

Second, light is attenuated by suspensions through absorption *and* scattering. Thus, in order to measure light absorption by a suspension one must somehow collect the scattered light, adjust the measurements results to compensate for the part of the attenuation caused by light scattering, or reduce light scattering by immersing the cells in medium of the same effective refractive index as practiced by *Barer* (1955).

In a suspension of phytoplankton cells in water, a large fraction of the incident light power is scattered within a cone with a half-angle on the order of several tens of degrees. Much of that scattered light is not collected by the detector in a typical spectrophotometer designed for measuring absorption of light by samples that scatter light negligibly. Thus, such a spectrophotometer does not measure the absorption of light by a sample containing particulate material, but a property of the sample that is intermediate between the absorption and attenuation of light and depends on the geometry of the spectrophotometer as well as on the characteristics of the sample. *Shibata* (1958) (see also *Shibata* et al. 1954) proposed a method of measuring the absorption of light in particulate media with a standard spectrophotometer by interposing a diffuser plate between the sample cuvette and the detector. Albeit rarely used today (if any), the Shibata method illustrates the key problems in the measurement of absorption of light by suspensions, hence the discussion that follows here. The Shibata method is based on an assumption that the diffuser plate completely randomizes the angular distribution of light power, i.e., the power measured by a detector behind the plate is proportional to the power of light incident on the plate and is independent of the angular power distribution of the incident light. By the virtue of the diffuser being located close to the sample cuvette and the dominance of the forward scattering in scattering by phytoplankton cells, the light power incident on the diffuser includes almost all scattered light power. Thus the reduction of the incident light power by the sample is essentially due to the absorption of light by the sample (aside from a significant reduction due to the detector intercepting a small fraction of the light field whose angular structure is "randomized" by the diffuser; this latter reduction is however independent of the absorption by the sample and only affects the signal-to-noise level of the spectrophotometer). We have used the word *essentially* because the measured property of the sample remains an approximation of the absorption of light by the sample. Recall that the light power measured behind the diffuser plate located next to the sample is a fraction of the scattered light power *intercepted by the plate*. This scattered light power depends on the geometry of the system and on the characteristics of the particles, i.e., VSF of the suspension (see section 4.2.1). In addition, the diffuser is not ideal in the sense of being able to completely randomize the photon paths.

At a given geometry of the diffuser plate system, the greater the fraction of the total scattered light that propagates in directions near the direction of the incident light, the closer the measured attenuation of the sample is to the absorption of light by the suspension. For a suspension of spherical particles with a mean size greater than 2 μm, and with a refractive index representative of algal cells (1.02 and 1.05, relative to water), over 99.5% of the scattered light is contained within a cone with a half-angle of 43° about the direction of the incident beam (e.g., *Bricaud* et al. 1983). Such collection angle can be readily obtained with a diffuser plate system. However, for particles as small as bacterioplankton (an average size of 0.6 μm), only about 94% is contained in such a cone (*Morel* and Ahn 1990). Since the absorption of light by the suspension is low, a 6% contribution of the scattered light can make a significant error in the absorption measurements with this method.

Spectrophotometer manufacturers offer a "scattered transmission accessory" which permits the placement of the sample cuvette near the photodetector. This assures that light scattered by the particles in the sample is collected from a cone with a relatively large angle. Another way is to place the sample next to an entrance port of an integrating sphere to collect a significant amount of the scattered light (e.g., *Stramski* and Reynolds 1993). As we mentioned it earlier in this section, *Stramski* and Piskozub (2003) discuss key aspects of the use of an integrating sphere in this respect and of the scattering corrections that must be applied to compensate for incomplete collection of the scattered light by the sphere.

Only relatively concentrated phytoplankton suspensions can be analyzed with the "diffuser" method or with a scattered transmission accessory, because the pathlength through the sample has to be short so that most of the scattered light can be collected with a reasonably sized diffuser. Such pathlengths are typically on the order of 1 cm in experiments conducted in laboratory with phytoplankton cultures. Concentration of particles on filters, and following resuspension of the particles in a small volume, as well as concentration of particles through centrifugation, has been attempted, but the efficiencies of such procedures are questionable (*Bricaud* and Stramski 1990).

A simple solution to the low concentration problem was to measure the absorption spectra of marine particles deposited, by filtration of large volumes of water, on membrane filters cleared with immersion oil (*Yentsch* 1962). The membrane filter (replaced by glass-fiber filter in applications following the original work by Yentsch) took the role of the diffuser plate. This technique is widely used in studies of marine particles (*Lohrenz* 2000, *Tassan* and Ferrari 1995, *Bricaud* and Stramski 1990, *Kiefer* and SooHoo 1982). However, multiple scattering of light inside the filter significantly increases the effective absorption of light by particles, because the effective pathlength of light for a given filter thickness is extended by a factor ranging from 2 to 6 in the various studies. In effect, the absorption coefficient is "amplified" as compared with that in suspension (*Butler* 1962). The amplification factor depends on the type of filter, its particle load (*Mitchell* and Kiefer 1988, 1984), the wavelength of light, the type of particles (*Bricaud* and Stramski 1990), and filter batch (*Roesler* 1998). *Cleveland* and Weidemann (1993) reviewed the prior research in this topic and proposed a wavelength-specific quadratic relationship between the optical density of the suspension and the optical density of a glass-fiber filter with the filtered particles. A similar corrective equation, which additionally accounted for the geometric pathlength (the volume of water that was filtered divided by the area of filter covered by the filtered particles), was proposed by *Arbones* et al. (1996). *Kishino* et al. (1985) modified the glass-fiber filter method by extracting the light-absorbing pigments with methanol. That permitted to differentiate between the absorption of light by phytoplankton and by detritus. The effectiveness of this method is unclear (e.g., *Tassan* and Ferrari 1995, *Bricaud* and Stramski 1990) because some phytoplankton species feature cellulose walls that hinder pigment extraction. Consequently, *Tassan* and Ferrari (1995) proposed to use wet bleaching of the pigments instead. They also developed an extension (transmission-reflectance, T-R) of the glass-fiber filter technique in a manner that allowed to compensate for backscattering of light by the filter with the sample. This obviated the need for compensating for backscattering by the filter (and its load) by subtracting an apparent absorption value at 750 nm, where the phytoplankton is presumed not to absorb light. Although such subtraction may be adequate for clear, open ocean waters, where the phytoplankton is the major source of absorption, it is questionable in coastal areas, where detritus contributes significantly to the absorption. The developments in the glass-fiber filter technique have been recently reviewed by *Lohrenz* (2000) who also proposed a correction equation that accounts for the filter loading effect.

An alternative technique of measuring the absorption of light in samples of seawater was proposed some time ago (*Haardt* and Maske 1987, *Maske* and Haardt 1987). In this technique, superior to the filter and the diffuser plate techniques, a long-pathlength cuvette (20 cm) is placed inside a large integrating sphere. The long pathlength permits one to analyze natural water samples, as opposed to concentrated samples. A collimated beam of light traverses the cuvette at an angle about 5° in relation to the cuvette axis (85° incidence angle on the windows). With this arrangement, most of the forward-scattered light, reflected at the exit window of the cuvette, is collected. With an incidence angle of 90°, that light would have exited the integration sphere through the entry port. The technique of measuring the absorption of light by suspensions that involve placing the sample inside an integrating sphere has been recently used for measurement of small absorption by mineral samples (*Babin* and Stramski 2004, 2002) as well as by detrital particles (*Nelson* and Robertson 1993). As these measurements were done on prepared samples, a small (1 cm), standard cuvette could be used. *Babin* and Stramski (2004) also found that tilting the cuvette (by 9°) greatly minimizes losses of the scattered light through the integrating sphere ports.

Instead of measuring the absorption of light by a suspension, microphotometric techniques can be used (*Lohrenz* et al. 1999, *Iturriaga* and Siegel 1989, 1988) to measure the absorption spectra of *single* marine particles in a wavelength range of

400 to 700 nm. Iturriaga and Siegel concentrated particles onto a 0.4 μm Nuclepore membrane filter and then transferred the particles onto a glass plate coated with gelatin. Some size-dependent particle selection occurs, and the particle transfer is not 100% efficient. The light absorption spectra of individual immobilized particles were determined using a microscope equipped with a spectrophotometric accessory. With such an arrangement, the contribution of light scattering by a particle is reduced because (1) the particle is immersed in gelatin, whose refractive index (1.036 relative to seawater) is similar to that of the phytoplankton and (b) scattered light is collected from within a cone with a half-angle of 50°, which is a typical acceptance angle for a high-magnification microscope objective operated with immersion oil. These factors are expected to further reduce the contribution of light scattering to the attenuation. Such measurements led, for example, to the determination of the following spectrum of the imaginary part of the refractive index of detrital particles (*Stramski* et al. 2004a):

$$n''(\lambda) = 0.010658\ e^{-0.007186\ \lambda} \tag{6.55}$$

with the real part n' assumed to be 1.04 (relative to water), and the wavelength in nm.

In the microphotometry technique, a value of particle transmittance at 750 nm is subtracted from the absorption spectrum to compensate for the remaining small contribution of scattering to the measured spectral attenuation values. The geometrical cross-section of a particle can also be measured, permitting the calculation of the spectral absorption efficiency of the particle. The spectra of the absorption efficiency of cultured phytoplankton (diatom *Thalassiosira pseudonana*) obtained using this technique were comparable to those obtained with the filter technique (*Mitchell* and Kiefer 1984). Alas, the microphotometric technique is very time consuming and cannot be applied to particles smaller than about 2 μm because their angular scattering patterns become too flat for the effective collection of the scattered light required by this technique.

Recently, single-particle measurements have been performed on particles trapped by IR light beams (*Iturriaga* et al. 1997). In principle, this approach permits measurements on live cells in suspension. Developments in the optical trapping technology even permit arbitrary orientation of single particles (*Galajda* and Ormos 2003). However, optical particle traps necessarily utilize high-intensity IR beams to create the trapping force. The high power of the trapping beam may be a problem when studying live cells because cell components may be damaged (e.g., *König* 1997, *König* et al. 1996, 1995).

6.3.3. Refractive index of marine particles from the angular scattering pattern

The basis of the determination of the refractive index of particles from their angular light scattering pattern is the significant dependence of that pattern on the refractive index. However, the scattering pattern is also a strong function of the particle size and depends on the particle shape, structure, and orientation with respect to the incident light beam. Indeed, the scattered light intensity at a given angle depends on the interference of the secondary waves scattered by the volume elements of the particle, which are illuminated by the incident wave and also, generally to a much lesser extent, by secondary waves scattered by other volume elements of the particle. The effect of illumination by the secondary waves is especially low in the case of aquatic particles because of their low relative refractive index. Given that the particle is sufficiently large, i.e., there are many volume elements, the result of that interference may be a highly oscillatory function of the scattering direction. Thus, the dependence of the angular pattern on the scattering direction is a consequence of variations in the direction-dependent phase differences between the secondary waves.

The other optical properties of a particle are directly or indirectly integrals of the angular scattering pattern. The integration smoothes much of these direction-dependent oscillations of the scattering pattern. Such oscillations are also smoothed in the angular pattern for a size, shape, and/or orientation distribution of the particles. A convincing example of this latter smoothing process is the VSF (see section 4.4.1) of a polydisperse suspension, such as those present in natural waters.

The determination of the refractive index from the angular scattering pattern can thus proceed in a variety of ways. For perfect spheres, such as small air bubbles or oil droplets in water and liquid droplets in air, high-accuracy wavelength spectra of single-particle properties such as the differential scattering cross-section, or the wavelength spectra of the scattering cross-section, have been used for a high-accuracy determination of the refractive index and particle size. Methods based on mapping of the particle refractive index and size onto a domain of light scattering intensity at two angles have been employed with a degree of success for moderately non-spherical aquatic particles and extrapolated for irregularly shaped natural particles. Finally, first-order-of magnitude methods based on fitting the VSF of natural suspensions have been developed. We discuss these topics in the following sections. The approximate real refractive indices of various aquatic particles, determined from the measurements of scattering and/or attenuation of light, are listed in Table A.6.

6.3.3.1. Single-particle methods

The refractive index and diameter of spheres and cylinders can be determined very accurately (to 1 part in 10^5) by using high-resolution spectroscopic techniques. Indeed, the optical properties of spheres such as the differential scattering cross-section at a fixed large angle ($>\sim 90°$), or the scattering cross-section, display sharp periodic oscillations as functions of x for a sufficiently large relative particle size, x, and refractive index. These oscillations, termed optical resonances (e.g., *Kerker* 1969) result from interference between the forward diffracted light wave and waves propagating along the sphere (cylinder) surface. Such surface waves were visualized by *Ashkin* and Dziedzic (1981) as well as by *Fahlen* and Bryant (1966). The morphology and positions of these resonances depend strongly on the refractive index (Figure 6.15) of the particle.

This sensitivity of the light scattering cross sections of a homogeneous sphere to the refractive index and diameter of the sphere has been used, for example, by *Chýlek* et al. (1983), to simultaneously determine the refractive index and size of ~11 μm oil droplets held in optical traps with an accuracy of better than 5×10^{-5}. Figure 6.15 reproduces the seminal results of their calculations. Similar results have been obtained for infinite cylinders (as represented by optical fibers; e.g., *Owen* et al. 1981, *Ashkin* et al. 1981).

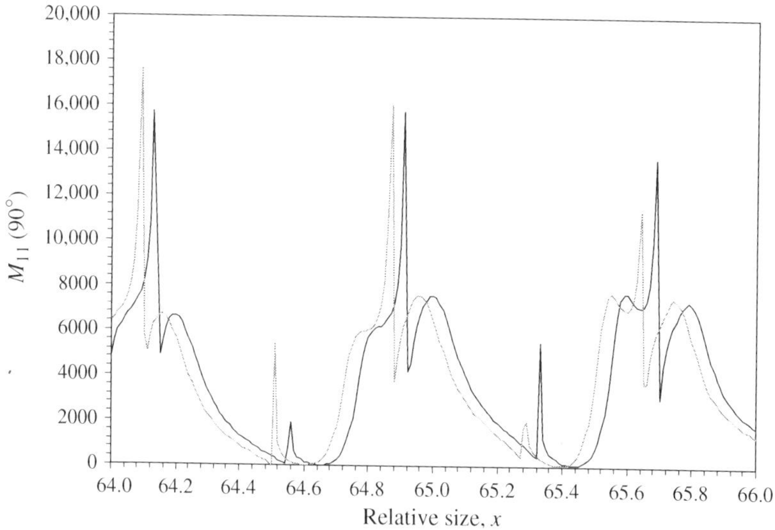

Figure 6.15. The morphology of oscillations (optical resonances, *Kerker* 1969) in the top-left element $M_{11}(90°)$ of the scattering matrix for a homogeneous sphere strongly depends on the sphere diameter and refractive index as shown here for $n = 1.4 - i0$ (*black curve*) and $n = 1.401 - i0$ (*gray curve*) as a function the relative size of the sphere, according to the Mie theory, as calculated at an increment of $\Delta x = 0.01$ with a light scattering calculator for homogeneous spheres (MJC Optical Technology). The black curve reproduces (at a different *y*-axis scale) a section of the top panel of Fig. 3 of *Chýlek* et al. (1983), who used the unique morphology of each group of the optical resonances to determine the refractive index and diameter of ~11 μm oil droplets in air to within 5×10^{-5} by analyzing wavelength spectra of $M_{11}(90°)$ for an optically trapped spherical droplet. In that method, the refractive index and diameter of the sphere must be known to a reasonable accuracy to limit the search range in *x*. The *x* increment corresponds to a wavelength resolution of 0.1 nm. Only resonances of the third and higher order can be displayed at this resolution. Resonances of the lower orders are too sharp to be shown.

The differential scattering cross-section (angular scattering pattern) of single spheres displays similar high sensitivity to sphere diameter and refractive index. An example of this latter sensitivity is shown in Figure 6.16. This phenomenon was used as early as 1975 to accurately size dielectric spheres in flight (*Marshall* et al. 1975). The technique of matching a theoretical to an experimental high-resolution angular scattering pattern has been used to examine the evaporation of liquid droplets (e.g., *Taflin* et al. 1988) and measurement of absorption of light by a single-droplet sample (e.g., *Arnold* et al. 1984). *Ray* et al. (1991) performed a systematic study of the precision of the determinations of the sphere size and refractive index and specified the angular and spectral resolutions required for the determination of the sphere parameters through either the angular pattern or the optical resonance matching. *Steiner* et al. (1999) extended the use of the Fourier transform, applied by *Taflin* et al. (1988), to filter out the experimental pattern noise, and in doing so improved the accuracy estimates of the particle size and refractive index of the sphere. These techniques, working very well for spheres, could find application in the studies of the gas exchange across the gas–water interface for gas bubbles in water. The nanometer-scale sensitivity of the sphere

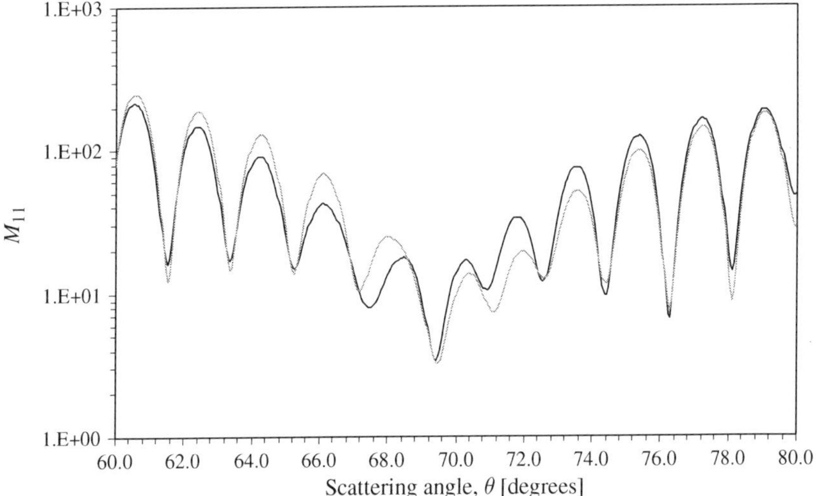

Figure 6.16. The morphology and positions of oscillations in the top-left element $M_{11}(\theta)$ of the scattering matrix for a homogeneous sphere strongly depend on the refractive index as shown in a sample sphere with relative size of $x = 100$ and $n = 1.100 - i0$ (*black curve*) or $n = 1.101 - i0$ (*gray curve*), according to the Mie theory, as calculated at an increment of $\Delta\theta = 0.1°$ with a light scattering calculator for homogeneous spheres (MJC Optical Technology). Note the magnitude order reversal and position shift of the oscillations in these sample patterns at about $\theta = 70°$.

size determination of these techniques can also potentially be used for studies of organic matter deposition on gas bubbles in water.

Note that the ranges of parameters representative of the techniques just discussed are applicable to aerosols with relative refractive indices ~1.5. In porting the optical resonance results to aquatic particles, one must note that the smallest relative particle size at which the resonances appear increases with the decreasing refractive index; hence although resonances can be clearly seen at $x \sim 65$ for a sphere with a relative refractive index ~1.4, by lowering the index to ~1.1, which is more representative for aquatic particles, we move the size range at which the resonances develop fully up to $x \sim 200$.

Unfortunately, the high accuracy of the particle size and refractive index determination for single spheres has not been matched for polydisperse and/or irregularly shaped particles. In the first case, optical resonances are rapidly smoothed with the increasing size distribution width. Interestingly, this "inconvenience" has been turned into a sensing tool (*Lettieri* and Marx 1986) because the degree of smoothing of the resonance pattern is a function of the PSD width (assuming that such a width is reasonably narrow in order not to obliterate the resonances altogether).

In the second case, the optical resonances quickly disappear because the surface waves do require a spherical or cylindrical surface to propagate along. *Mishchenko* and Lacis (2003) demonstrated that optical resonances are essentially leveled off by particle shape deviations from spherical that are as small as 0.003 (the difference between the ratio of the two radii of the spheroid and unity). Generally, the effects of the particle shape and structure, which are difficult to generalize (e.g., *Wiscombe* and Mugnai 1988), are most significant in the angular range in which the effects of the refractive index are most pronounced.

The approaches involving the identification and matching of the high-resolution spectral and angular scattering have two major disadvantages: (1) they apply only to perfect spheres (or cylinders; the "perfection" is achieved if the deviations of the particle surface from spherical or cylindrical are much smaller than the wavelength of light, e.g., *Chýlek* et al. 1978) and (2) they require time consuming identification of curve patterns. In passing, we note that some attempts at automating the process have been made (e.g., *Hill* et al. 1985—who extended the applicability of an algorithm by *Conwell* et al. 1984 developed for sizing spheres whose refractive index is known).

Another approach to the refractive index and sphere size determination from light scattering, applicable also to "imperfect" spheres, utilizes the fact that for certain two-dimensional domains of light scattering at two well-chosen angular ranges, a one-to-one mapping can be established between such domains and refractive index, n', and sphere diameter, D or x. Hence, by measuring these intensities, one can simply identify the refractive index and sphere diameter from such a map. For short, we will refer to this method as flow-cytometric mapping technique (FCMT) because it has been developed and used with flow cytometers.

The FCMT, originally developed for red blood cell characterization (*Tycko* et al. 1985), has been applied to marine particles shortly after (*Ackleson* and Spinrad 1988) as an extension of a simplified technique developed by *Spinrad* and Brown (1986). An "enhanced" version of this method has been recently used to determine the characteristics of phytoplankton cells, as well as detrital, and mineral particles in seawater (*Green* et al. 2003b). In the case of non-spherical cells, the derived values are claimed to represent an "effective" diameter and an "effective" refractive index. The refractive index values obtained this way are shown in Table A.6.

A sample map, similar to that used by *Ackleson* and Spinrad (1988) is shown in Figure 6.17. Note residual oscillations on the constant-n curves despite integrating of the light scattered intensity over ranges of the scattering angle. In the sample shown, these oscillations are caused by oscillations of the scattered light intensity at a large angle. *Tycko* et al. (1985) used two small-angle ranges which produce mapping curves that are much smoother and necessarily different from those shown in Figure 6.17. Thus, the map depends significantly both on the selection of the scattering angle ranges and on the region of the $(n,\ x)$ space to be mapped. As noted by *Tycko* et al. (1985), the mapping $[M^I{}_{11}(\theta_1, \Delta\theta_1; x, n), M^I{}_{11}(\theta_2, \Delta\theta_2; x, n)] \to [x, n]$, where superscript "I" indicates that M_{11} is integrated over a range $\Delta\theta$ of the scattering angle, may be unsolvable for certain selections of the scattering angle ranges. We should stress that the integration of M_{11} must be weighed by $r(\theta)$, the angular response of a detector in a flow cytometer to light scattered within the $\Delta\theta$ range. A rigorous condition of solvability of the mapping is expressed by requiring that the Jacobian:

$$J = \left| \frac{\partial M^{\mathrm{I}}{}_{11}(\theta_1)}{\partial x} \frac{\partial M^{\mathrm{I}}{}_{11}(\theta_2)}{\partial n} - \frac{\partial M^{\mathrm{I}}{}_{11}(\theta_1)}{\partial n} \frac{\partial M^{\mathrm{I}}{}_{11}(\theta_2)}{\partial x} \right| \tag{6.56}$$

is non-zero.

There are important differences between the original application of the FCMT for the red blood cell characterization (*Tycko* et al. 1985) and its subsequent applications to aquatic particles. First, the measurements of Tycko et al. were performed on isovolumetrically sphered red blood cells; hence the $[M^I{}_{11}(\theta_1), M^I{}_{11}(\theta_2)] \to [x, n]$ mapping, where θ is the scattering angle, could be justifiably generated with Mie theory. Secondly, the nephelometer used by Tycko et al. was designed with a view toward optimization of the mapping sensitivity to the cell parameters; hence the scattering angles were carefully chosen and the optics of the nephelometer was accordingly defined.

In contrast, aquatic applications of the mapping technique, although originally used for nearly spherical phytoplankton cells (*Ackleson* and Spinrad 1988, *Spinrad* and Brown 1986), were later adopted for the phytoplankton "as is" (*Green* et al. 2003b), which required several correction factors to be applied to the Mie-based estimates, not counting the guess estimates of the angular responses of

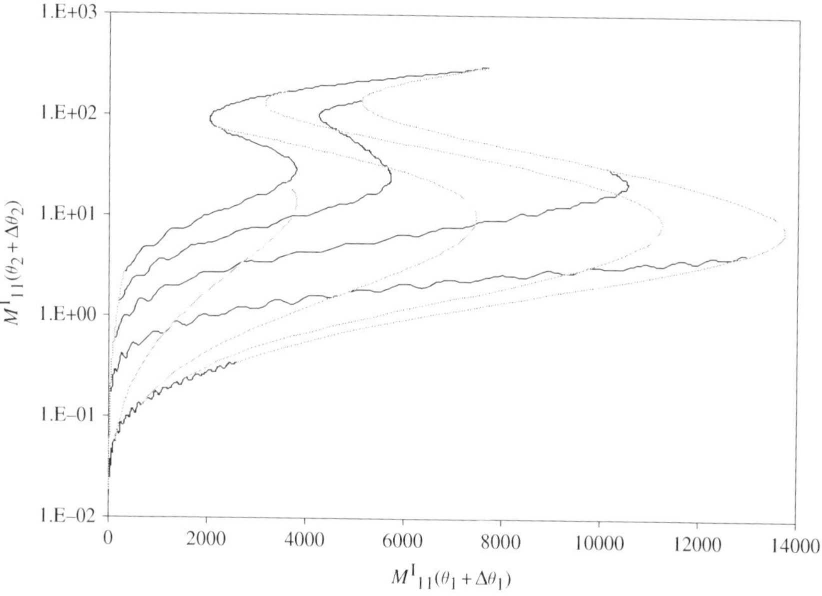

Figure 6.17. Mapping of the refractive index, n, of non-absorbing spheres and the sphere's relative size, x, to measures of light scattering (here the M_{11} element of the scattering matrix) into two annular solid angles (M^{I}_{11} is the integral of M_{11} over a solid angle $2\pi \sin\theta\Delta\theta$). Such mapping has been used to simultaneously determine n and x of single non-absorbing particles with flow cytometers. Each *black* curve represents a constant x (12, 24, 36, 48, and 60, bottom to top). At a wavelength of 500 nm, these values correspond roughly to $D = 2, 4, 6, 8$, and $10\,\mu$m. Each *gray* curve represents a constant (real) n (1.01, 1.03, 1.05, 1.07, and 1.09, left to right). The solid angles are defined as follows: $\theta_1 = 2°$, $\Delta\theta_1 = 5°$ and $\theta_2 = 75°$, and $\Delta\theta_2 = 35°$. Selection of the solid angles is critical for the mapping to be unique. The mapping shown here, close to that used by *Ackleson* and Spinrad (1988), has been calculated with a light scattering calculator for homogeneous spheres (MJC Optical Technology).

the detectors. The aquatic applications were also implemented by using standard flow-cytometer optics, optimized for their main target market: clinical analysis. Optical design details (such as the angular response of the scattered light detector) were unavailable to the researchers. This required reverse engineering of the key aspects of the optics design of these instruments. Thus, estimates of the refractive index obtained with the aquatic applications of the technique, especially those used for natural assemblages of phytoplankton and other particles, should be viewed as gross approximations.

The FCMT was applied in a field study of particles occurring in the North Sea (*Ackleson* et al. 1988b) and in the coastal waters of the western North Atlantic (*Green* et al. 2003a, 2003b). In the study of *Ackleson* et al., the various types of particles were differentiated using fluorescence signatures. The distribution of the refractive index of particles containing both phycoerythrin and chlorophyll peaked at about 1.085 (relative to seawater) and had a half-width of about 0.01. The size distribution of these particles had a maximum at about 7.5 μm and a half-width of about 2 μm. From a plot of the average refractive index and of the average size of these cells against the depth, shown by Ackleson et al., it is apparent that the refractive index covaries with the inverse of the particle size. A similar relationship has been proposed by *Stramski* and Kiefer (1990) [equation (6.65), further on this chapter]. The depth profile of the refractive index had a minimum at about 10 m.

6.3.3.2. Many particle methods

The determination, using any method, of the refractive index and size distributions of many marine particles simultaneously presents a significant challenge because of the diversity of the particles and the fact that all particles contribute simultaneously to the scattering pattern. Methods based on measurements of the angular pattern of light scattering of suspensions of many particles have historically relied on a trial-and-error approach. This approach implies usually a significant computational effort: over 1000 VSFs have been calculated for a study of that type (*Brown* and Gordon 1974), which incidentally pales in comparison with the contemporary computational efforts discussed in section 6.4.1. To limit the number of degrees of freedom, the PSD was measured, typically with a Coulter counter, and used to calculate the scattering patterns for various refractive indices of the particles. Given the present understanding of the role of zone counters, such as the Coulter counter, in disrupting aggregates of aquatic particles, the results of these studies should be regarded as gross estimates.

Routine use of such methods to determine the refractive index of marine particles remains elusive. In passing, we note that a similar method of inverting the VSF data is widely used in laser diffractometry (see section 5.7.9), with a major simplification: the refractive index is assumed constant for the entire population of the particles. Thus, the inversion is aimed solely at the determination of the PSD. As much as this can be a plausible approach in the case of a homogeneous particle population, results of such an inversion applied to particle populations in natural waters may be questionable.

Various approaches were tried in order to account for the diversity of the refractive indices of naturally occurring marine particles. In one of the early attempts (*Gordon* and Brown 1972), the same refractive index was assigned to all particles (Table A.7). The errors in the fitting the scattering pattern were less than 20%. Gordon and Brown assigned different refractive indices to consecutive particle size ranges and concluded that all particles greater than 2.5 μm were inorganic. A subsequent study, in which different size distributions were assigned to particles of different refractive indices (*Zaneveld* et al. 1974) confirmed that conclusion. Interestingly, those studies were devoted to explaining a single VSF measured in the Sargasso Sea waters (*Kullenberg* 1968) which was designated as representing unusual oceanic conditions (*Zaneveld* et al. 1974). No simultaneously determined PSD was available for those scattering function data.

This latter shortcoming was solved, at least partially (in the sense of an incomplete particle size range in which the measured PSD was available), in the following research (*Jonasz* and Prandke 1986, *Reuter* 1980a, 1980b, *Brown* and Gordon 1974) where PSDs were measured for the same samples or in the same area for which the VSF was measured. Modeling of the VSF performed in those studies led to a number of important conclusions. First, the slope of the power-law size distribution of particles smaller than 0.65 μm (the lower limit of the Coulter counter data) must be less than 4, otherwise the size distribution of these particles would have to be terminated in the vicinity of $D = 0.1$ μm (*Brown* and Gordon 1974). Second, the refractive index of the large particles ($D > 4$ mm) was found to be difficult to estimate, because the contribution of these particles to the VSF is small in the angular range ($>5°$) typically covered by the VSF measurements (*Jonasz* and Prandke 1986, *Brown* and Gordon 1974). Third, the size range of particles that make a significant contribution to the VSF depends on the size distribution and the scattering angle. Only particles with diameters in a range on the order of 0.1 to 10 μm were found to contribute significantly to the scattering function in the case of surface Baltic waters (*Jonasz* and Prandke 1986, *Jonasz* 1980). Particles smaller than about 2 μm dominated light scattering at angles $>10°$. Recent theoretical estimates (*Stramski* and Kiefer 1991, *Morel* and Ahn 1991) indicate that particles with diameters in the range of 0.1 to 10 μm contribute most to the scattering coefficient.

The refractive index obtained from angular light scattering by using the Coulter counter-derived PSDs is most likely overestimated. Such overestimation of the imaginary part is a consequence of the particle non-sphericity and occurs because the projected areas of the particles are underestimated when a size distribution is measured with the Coulter counter (*Jonasz* and Prandke 1986). The values of the VSF of particles significantly increase with increasing refractive index, up to some value of the index. A further increase in the index does not cause a corresponding increase in the calculated light scattering. In the study by *Jonasz* and Prandke (1986) of the particulate matter in the Baltic waters, the calculated VSF which was closest to the experimental data at scattering angles smaller than about 15° in the winter was about half of that measured. In contrast, the summer VSF could be closely reproduced with the calculated function for summer particles. In the case of summer particles, an underestimation of the projected areas of the particles could be compensated for by assuming a refractive index higher than the actual refractive index of the particles. Such compensation was not possible for winter particles, whose actual refractive index was higher than that of summer particles.

The reverse also holds. Given the refractive index of the particles, the failure to account for an increased projected area of particles, as compared with that of

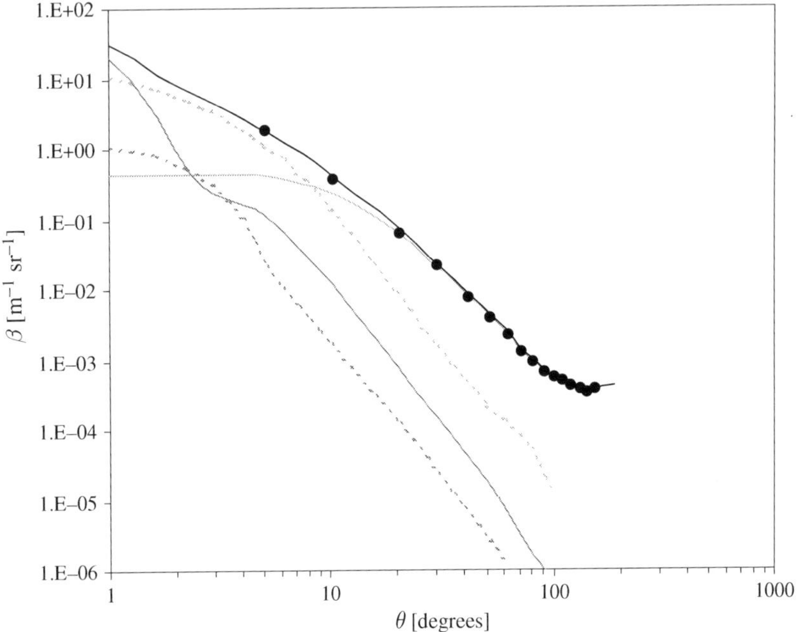

Figure 6.18. The refractive index of natural suspensions of marine particles has been determined by fitting (via trial-and-error) of the measured volume scattering function of the suspension with a combination of scattering functions calculated with Mie theory for measured particle size distributions. This figure shows a sample fit for the surface waters of the Baltic Sea in summer (redrawn from *Jonasz* and Prandke 1986, *Jonasz* 1980). Solid circles—experimental data (average of several volume scattering functions measured *in vitro*), *solid black curve*—the sum of the calculated volume scattering functions for the small size fraction of the particles (*light gray curve* corresponds to a power-law particle size distribution in a diameter D range of 0.1 to 2 µm, $n = 1.1 - i0$), medium size fraction (*light gray dashed curve*, power-law PSD, $D = 2$ to 10 µm, $n = 1.05 - i0.005$), large size fraction (*gray solid curve*, power-law PSD, $D > 10$ µm, $n = 1.03 - i0.01$), and phytoplankton (*gray dashed curve*, Gaussian PSD, $D_{\text{peak}} = 6.2$ µm, $n = 1.01 - \text{i}0.05$).

equal volume spheres, may result in an underestimation of up to ~300% of light attenuation by natural marine suspensions, when such attenuation is calculated by using Coulter counter PSD data (*Jonasz* 1987c).

A note of caution is in order here. Despite "reasonable" fits to the experimental VSFs (as evidenced by the sample results shown in Figure 6.18) one should keep in mind that such fits are based on key far-reaching assumptions:

(1) *particles are assumed to be homogeneous spheres*. Significant differences have been noted to date between the VSFs calculated for homogeneous

spheres and for other particle shapes (see section 6.4.1). This implies that the results of sphere-based fits should be strictly understood as indicative of an "equivalent" sphere in a broad sense.

(2) *partition of the total size distribution into "fractions" is arbitrary*, albeit it is based on reasonable assumptions regarding the particle populations.

In fitting an experimental VSF, relevant experimental PSD data provide a constraint, which can prevent a reasonable fit, as in the case of winter VSF in the surface Baltic waters (*Jonasz* and Prandke 1986). Jonasz and Prandke advanced a hypothesis that, given a plausible underestimation of the geometrical cross-sections of the natural (and nonspherical) particles by the Coulter counter measurement of the PSD, the refractive index of the winter particles was too high to compensate for that underestimation, while in the summer data case the refractive index was sufficiently low, so that such a compensation was possible.

This hypothesis follows from considering the effect of the real part of the refractive index, n', on the scattering efficiency of the particles (Figure 6.10) that reaches a "saturation" level at a n' value that decreases with increasing particle size. When the PSD-induced constraint of the fit to an experimental VSF was removed, it was possible to obtain a much better fit to the winter VSF data for the Baltic surface waters (*Jonasz* 1980).

6.3.4. *Immersion refractometry*

6.3.4.1. *Biological particles*

Barer and Joseph (1954) pioneered the immersion refractometry of biological cells, a technique which permits the determination of the refractive index of living cells. In this technique, cells suspended in a solution of proteins (usually: bovine serum albumin) are observed using a phase-contrast microscope. The cell "vanishes" when its refractive index matches that of the solvent. The refractive index, n', of the solution can be expressed using the following equation, resulting from the Gladstone–Dale formula (6.25):

$$n' - n'_{\mathrm{w}} = n_c C \tag{6.57}$$

where n'_{w} is the refractive index of the solvent (water), C [g cm^{-3}] is the solute concentration, and $n_c = dn'/dC$ [cm^3 g^{-1}] is the refractive index increment.

The immersion refractometry of living cells relies upon the refractive index increment of all proteins being nearly the same (0.185 cm^3 g^{-1}, see also Table 6.8), almost independent of the temperature (between 5 and 25°C), and of the protein concentration (up to a nearly saturated solution at 0.55 g cm^{-3}). The dispersion of the refractive index of protein solutions is relatively low (*Aas* 1981).

Table 6.8. Refractive index increments of substances relevant to immersion refractometry of living cells.

Compound	**Wavelength, nm**	**Temperature, °C**	**Refractive index increment, n_c, cm^3g^{-1}**
Amino acids[a]	Visible	–	0.17
BSA[a]	436	25	0.1924
BSA[b]	488	18	0.192 ± 0.006
BSA[a]	546	25	0.1854
BSA[a]	578	20	0.187
BSA[b]	1060	18	0.181 ± 0.005
β-lactoglobulin[a]	436		0.189
β-lactoglobulin[b]	488	18	0.184 ± 0.013
β-lactoglobulin[a]	546	–	0.1822
β-lactoglobulin[b]	1060	18	0.171 ± 0.012
carbohydrates[a]	Visible	–	0.11
DNA[b]	488	18	0.183 ± 0.018
Lipids[a]	Visible	–	0.15
Protein[a]	Visible	–	0.185
Sucrose[b]	488	18	0.144 ± 0.001

[a] *Barer* and Joseph (1954).
[b] *Coles* et al. (1975).
BSA = bovine serum albumin.

Having determined the refractive index of the cells, (6.57) can be used to determine the solid content (and thus water content) of the cells. The refractive index increment of the cell material is about 0.18 $cm^3\ g^{-1}$, slightly less than that of the proteins because some cell components have lower refractive index increments than that of the proteins. The refractive index increment, n_c, varies from about 0.17 (amino acids), through 0.15 (lipids) to 0.11 (carbohydrates and inorganic constituents such as salts). *Coles* et al. (1975) measured n_c for several compounds important in immersion refractometry. These and other values are listed in Table 6.8

The refractive indices of several species of bacteria and spores determined using this technique at a wavelength of 534 nm are in a range of 1.386 to 1.40 (vegetative cells) and 1.512 to 1.540 (spores) (*Ross* and Billing 1957). The higher refractive indices of spores are due to their water content being lower than that of vegetative cells. A similar tendency is shown by the refractive index of phytoplankton as a function of their water content (*Aas* 1981).

The immersion refractometry technique has been somewhat automated in recent years by measuring the changes in the transmission of a suspension of the cells as a function of the refractive index of the solvent (*Robertson* et al. 1998, *Gerhardt* et al. 1982, and references therein). In the method of Gerhardt et al., the optical density ($= cx \log e$, $c =$ attenuation coefficient, $x =$sample thickness) of a suspension of cells is measured as a function of the increasing refractive index of the solvent. A trend for the linear portion of the graph of the optical density of the suspension vs. the refractive index of the solution is extrapolated to the optical density of 0. The extrapolated value of the refractive index is assumed to represent the particles. With this technique, there is no need for the solution to attain a refractive index equal to that of the particles. However, the refractive index determined in this manner is under estimated because the optical density of a suspension is a non-linear function of the refractive index of the liquid, as follows from (6.34) or (6.40). Indeed, the refractive index of polystyrene latex was underestimated by Gerhardt et al. when using this procedure ($n = 1.522$ vs. an actual value of $n = 1.5905$).

A better approach is to examine the transmission of light by the suspension as a function of the refractive index of the solvent in the vicinity of the null point (*Jonasz* et al. 1997, *Waltham* et al. 1994, *Bateman* et al. 1966), where the refractive index of the liquid is nearly equal that of the particles (Figure 6.19). The optical density of the suspension, OD, can then be closely approximated

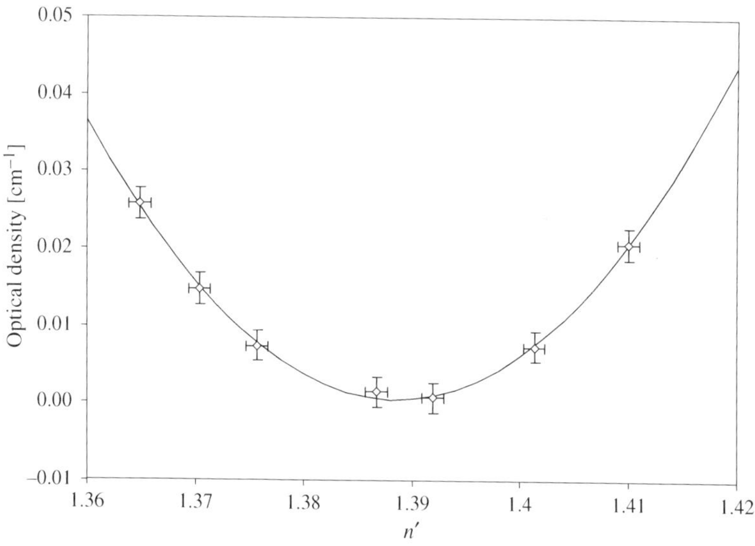

Figure 6.19. Immersion refractometry of heterotrophic marine bacteria. Each data point (*open circle* with error bars) represents a value of the optical density of a suspension of bacteria in an aqueous solution of albumin of specific concentration and thus of a specific refractive index. The curve represents the best fit that is used to determine the real part of the refractive index ($n' = 1.3886$) of the bacteria (redrawn after *Jonasz* et al. 1997).

(*Jonasz* et al. 1997) by the following quadratic equation in the real part of the refractive index, n_s', of the solvent:

$$\mathrm{OD} = C_s AN \left\{ \langle D^4 \rangle \left[n_s'^2 - 2\langle n' \rangle n_s' + \left(\sigma_{n'}{}^2 + \langle n' \rangle \right) \right] + \frac{4\lambda}{3\pi} \langle D^3 \rangle \langle n'' \rangle \right\} \quad (6.58)$$

where $\langle x \rangle$ denotes the average of x, n' and n'' are the real and imaginary parts of the refractive index of the particles respectively, $\sigma_{n'}{}^2$ is the variance of n', C_s is a constant depending on the particle shape ($C_s = {}^1\!/_2$ for uniform spheres and 4/3 for randomly oriented uniform cylinders), $A = \pi^3 \log(eL\lambda^{-2})$, N is the particle number concentration, and λ is the wavelength of light in air. The optical density is defined as follows:

$$\mathrm{OD} = cL \log e \quad (6.59)$$

where c is the attenuation coefficient of the suspension and L is the pathlength of light in the suspension. Note that the use of these formulas implies that the particle concentration is sufficiently small for the multiple scattering of light by the suspension to be negligible.

The immersion refractometry technique has been applied to determine the refractive indices of free-living marine bacteria (*Jonasz* et al. 1997) and bacteria *Escherichia coli* (*Robertson* et al. 1998, *Waltham* et al. 1994) (Table A.6). A sample application of this technique (after *Jonasz* et al. 1997) is shown in Figure 6.19.

6.3.4.2. Atmospheric particles

The immersion refractometry technique is routinely used in mineralogy to determine the refractive index of inorganic material. With mineral particles, the effect of the solvent on the properties of the particles is much less critical than it is with biological cells. That allows for the use of solvents (some of which are toxic) with refractive indices from a wider range than that available for biological particles.

The real part of the refractive index of soil aerosol particles was determined using a microscope-based immersion refractometry technique to be about 1.525 (*Grams* et al. 1974). *Patterson* et al. (1977) determined the real part of the refractive index of Saharan aerosols, using immersion refractometry, to be between 1.558 and 1.562 at a wavelength of 550 nm and 1.544 at 589 nm.

6.3.5. Refractive index of phytoplankton calculated from the composition of the cells

A different approach to the determination of the refractive index of particles was taken by *Aas* (1996, 1981), who estimated the refractive indices of several species of phytoplankton using the Lorentz–Lorenz formula for the refractive index of a mixture (6.26). The phytoplankton cell was assumed to be a "mixture" of the various components, without regard for the fact that some components may be assembled into granules whose size is several orders of magnitude greater than that of the largest molecule. Complex protein molecules have sizes on the order of 10 nm. Additional information on the components (organelles) of eukaryotic cells is shown in Table 6.15

Differences between the real refractive indices of a mixture characteristic of the phytoplankton cell, calculated by using various refractive index mixing rules, are on the order of 0.003 (relative to water). As noted by *Aas* (1981), uncertainty in the cellular water content has a much greater effect on the value of the refractive index than that of the mixing rule selection.

The refractive indices of the major components of phytoplankton (*Aas* 1981) are listed in Table 6.9 The dispersion of the refractive index of phytoplankton components in the visible is generally low. It is most pronounced for protein, for which the difference between the refractive index (relative to water) at 300 nm and at 700 nm amounts to 0.025. The corresponding differences for the other components of phytoplankton are less than 0.008. That difference for seawater at a salinity of 35 ppt and temperature of 25°C is about 0.03.

By using the composition of phytoplankton (Table 6.10) and the Lorentz–Lorenz mixing rule (6.26), the absolute refractive indices of several phytoplankton species were found by *Aas* (1996) to be within a range of 1.34 to 1.56, for the water content of the cells ranging from 1 to 0. The refractive index of phytoplankton decreases approximately linearly with increasing water content, when the latter is in a realistic range of 0.7 to 0.9. These values of the refractive index are within the range reported by other researchers. The effect of absorption of light in the pigments on the real part of the refractive index of phytoplankton and on light

Table 6.9. Ranges of the refractive indices of the major components of phytoplankton in the visible (after *Aas* 1981).

Component	**Refractive index relative to air**
Water[a]	1.34
Calcite	1.59–1.61
Carbohydrate	1.53–1.57
Fat	1.46–1.48
Pigments	1.52
Protein	1.51–1.55
Silica (opal)	1.40–1.46

[a] More exact estimates are discussed in section 6.2.

Kitchen and Zaneveld (1992) quote other estimates of the refractive indices of the phytoplankton cell components and note their wide range.

Table 6.10. Compositions of various phytoplankton groups (*Aas* 1981).

Component	Concentration [% of dry mass]		
	Soft surface plankton	**Hard surface plankton**	**Coccolithophorids**
Proteins	15 to 84	21 to 64	54
Carbohydrates	8 to 65	1 to 33	17
Fat	3 to 34	2 to 13	5
Pigments	<3	<14	1
Silica (opal)	–	1 to 67	–
Calcite	–	–	23

scattering efficiency by phytoplankton was deemed to be small (see also *Bricaud* and Morel 1986, and Figure 6.14).

The refractive index, n', of marine cyanobacteria at 440 nm substantially covaries with the intracellular carbon concentration, C_C (Table 6.11). The basis of such a correlation is given in equation (6.57). A similarly strong correlation was found between the imaginary part of the refractive index and the

Table 6.11. Real part, n', of the refractive index of phytoplankton as a function of the mass concentration of the intracellular carbon, C_C[kg m^{-3}].

Species	Reference	r^2	$n'_\lambda = n'_0 + n'_1 C_C$		
			n'_0	n'_1	λ [nm]
Natural coccoid bacteria	*Morel* and Ahn (1991, 1990)	–	1.0	2.33×10^{-4}	415
Nannochloris sp.	*DuRand* and Olson 1998	0.75	1.02	1.17×10^{-4}	665
Synechocystis sp.	*Stramski* and Morel (1990)	0.88	1.005	2.43×10^{-4}	440
Synechococcus	*Stramski* et al. (1995)	0.42	1.019	1.74×10^{-4}	660
Varions phytoplankton cultures	*Green* et al. (2003b)	0.61, 14 data points	0.974	3.46×10^{-4}	488
Thalassiosira pseudonana	*Stramski* et al. (2002)	–	1.007	1.928×10^{-4}	660
Thalassiosira pseudonana	*Reynolds* et al. (1997)	0.95	1.01	1.3×10^{-4}	660
Thalassiosira pseudonana	*Stramski* and Reynolds (1993)	0.80	1.01	1.75×10^{-4}	660

Table 6.12. Imaginary part, n'', of the refractive index of phytoplankton as a function of the intracellular chlorophyll, C_{Chla}, concentration [kg m^{-3}].

Species	**Reference**	r^2	$n''_\lambda = n''_0 + n''_1 C_{Chla}$		
			n''_0	n''_1	λ[nm]
Synechococcus	*Stramski* et al. (1995)	0.973	-4.766×10^{-4}	8.614×10^{-4}	660
Varions phytoplankton cultures	*Green* et al. (2003b)	0.47, 11 data points	-9.4×10^{-4}	1.529×10^{-3}	488
Thalassiosira pseudonana	*Stramski* et al. (2002)		7.331×10^{-4}	7.154×10^{-4}	674
Thalassiosira pseudonana	*Reynolds* et al. (1997)	0.99	2.4×10^{-4}	6.5×10^{-4}	673
Thalassiosira pseudonana	*Stramski* and Reynolds (1993)	0.37	2.5×10^{-4}	8.45×10^{-4}	673

chlorophyll *a* concentration, C_{Chla} (Table 6.12). The basis of this latter correlation is in equation (6.3).

Given the similarities between these correlations established for two quite different phytoplankton species, *Synechocystis* sp., and *Thalassiosira pseudonana*, *Stramski* (1999) proposed to use these significant relationships for the determination of the intracellular carbon and chlorophyll concentrations of the phytoplankton cells (making appropriate exceptions for certain species of phytoplankton, such as coccolitophorids). We recapitulate his findings in the $n = f(C)$ format:

$$n' = 0.9995 + 2.43 \times 10^{-4} C_C \tag{6.60}$$

and

$$n'' = 0.00117 + 0.001 C_{Chla} \tag{6.61}$$

where, in both cases, the units of concentration of carbon and chlorophyll are [kg m^{-3}]. Recently, *DuRand* et al. (2002) extended these correlations by including results for *Micromonas pusilla*. They obtained very similar relationships:

$$n' = 0.9939 + 2.53 \times 10^{-4} C_C \tag{6.62}$$

with $r^2 = 0.63$, and

$$n'' = 0.000257 + 0.000804 C_{Chla} \tag{6.63}$$

with $r^2 = 0.93$.

Such findings have practical importance not only for the determination of the carbon and chlorophyll contents of phytoplankton cells but also for an approximate evaluation of the real part of the refractive index from abundant flow-cytometric measurement data for a wide range of phytoplankton species. Indeed, as noted by *DuRand* et al. (2002), the carbon content, m_C, and forward light scattering (FLS, also referred to as FALS) measurement are significantly correlated [$\log\text{FLS} = f(\log C_i)$, $r^2 = 0.98$ for several species of phytoplankton]. In using such relationships one should note exceptions, such as armoured cells, or cells with vacuoles or fat globules.

The specific absorption coefficient of chlorophyll a at 673 nm, estimated from the relationship $n''(C_{\text{Chl}a})$ as obtained by *Stramski* and Reynolds (1993), Table 6.12, yields a value of $0.021\,\text{m}^2(\text{mg}\,Chl\,a)^{-1}$. That value is reasonably close to the universally accepted value of $0.0202\,\text{m}^2(\text{mg}\,Chl\,a)^{-1}$ (*Jeffrey* and Humphrey 1975). Note that the real part of the refractive index was also found to covary with the cellular concentration of chlorophyll (*Bricaud* et al. 1988), although the pigment constitutes only about 0.03 to 0.6% of the cell volume. Such a correlation has its basis in a statistical relationship between the particulate organic carbon (c_{POC}) and chlorophyll a concentration, $C_{\text{Chl}a}$ (*Morel* and Ahn 1990):

$$c_{\text{POC}} = 90 C_{\text{Chl}a}{}^{0.57} \tag{6.64}$$

The constant component $1 + \varepsilon$ of n' in (6.51) correlates well with the water content of the cells (*Bricaud* et al. 1988, *Aas* 1981). The effect of the water content on the refractive index of a cell leads to the following relationship between the refractive index of bacteria (relative to water) and their cell diameter, D (*Stramski* and Kiefer 1990):

$$n' - 1 = 0.025 D^{-1.2} \tag{6.65}$$

when the cell diameter, D, is in a range of 0.3 to 1 μm. This equation was derived by using the literature data on the water content and cell volume (*Simon* and Azam 1989), and a relationship between the refractive index and cellular water content of bacterial cells. Such a relationship stems from the hypothesis that the water content is a key factor in the growth and metabolic processes of bacteria: fast growing cells tend to be larger and contain less water than cells which are starved. However, (6.65) has not yet been confirmed using optical measurements.

A similar relationship between the imaginary part of the refractive index, n'', at 675 nm and the cell size was derived by *Kronfeld* (1988). He used results obtained by *Hitchcock* (1982) on the size dependency of the cellular content of chlorophyll, m_{Chl} [mg] for two phytoplankton genera: diatoms and dinoflagellates:

$$\log m_{\text{Chl}} = A + B \log V \tag{6.66}$$

where V [mm^3] is the cell volume, with $A = -2.09$ and $B = 0.79$ for diatoms ($r^2 = 0.99$, $n = 11$), and $A = -1.61$ and $B = 0.81$ for dinoflagellates ($r^2 = 0.88$, $n = 8$). See Table 6.18 for additional regressions of this type. As pointed out by Kronfeld, these results are consistent with those reported by *Morel* and Bricaud (1981) and *Haardt* and Maske (1987). Since,

$$a_c(\lambda) = a_c^{\;0}(\lambda)\frac{m_{Chl}}{V} \tag{6.67}$$

where $a_c^{\;0}$ is the chlorophyll concentration-specific absorption coefficient, this leads to:

$$n''(\lambda) = \frac{\lambda}{4\pi} a_c^{\;0}(\lambda) 10^A V^{B-1} \tag{6.68}$$

6.3.6. Mineral particles

Soil-based mineral particles are an important component of aerosols and form a significant fraction of natural hydrosols (see also section 6.4.3.4), especially in land-locked water bodies or in the coastal ocean areas. Data on the refractive indices of mineral aerosol particles are listed in Table A.6.

The refractive indices of aerosols have been determined by using methods similar to those applicable to aquatic particles. The real part of the refractive index is determined by immersion refractometry (section 6.3.4.2). The complex refractive index is obtained by fitting Mie angular scattering patterns to experimental data for the aerosol (section 6.3.6). The imaginary part, n'', of the refractive index of aerosols has been determined by measuring the absorption of an aerosol sample diluted in a base of a weakly absorbing and strongly scattering powder, or the combination of immersion refractometry and fitting of the VSF (e.g., *Grams* et al. 1974). *Grams* et al. obtained this way $n'' \sim 0.005$, with an uncertainty factor of 2 for soil particles at wavelengths of 488 and 514 nm.

This spectral reflectance method (*Lindberg* 1975, *Lindberg* and Laude 1974) is conceptually equivalent to the filter method of *Yentsch* (1962) for phytoplankton (see section 6.3.2.4). Incidentally, the filter-based method of measuring the absorption of light by aerosols is also widely used in atmospheric research (e.g., *Arnott* et al. 2003). In the powder method, powdered $BaSO_4$ is typically used. The dilution of the aerosol sample depends on the absorption of light by the aerosol and ranges from 1:10^2 to 1:10^5 by volume. A simple function of the measured diffuse reflectance of the diluted sample is proportional to a ratio of the scattering coefficient of the diluent to the Kubelka–Munk (KM, for example, *Kortüm* 1969) absorption coefficient of the sample. The KM absorption coefficient is roughly twice that used in the Beer–Lambert law (6.2). The residual absorption of light in the diluent is also accounted for.

The spectral imaginary refractive index of typical soil aerosols is between 0.11 and 0.02 (at a wavelength of 500 nm), depending on the particle size (*Lindberg* and Gillespie 1977). The data of Lindberg and Gillespie suggest that the PSD is composed of a number of narrow size distributions representing particles of different composition. A similar situation occurs with respect to aquatic particles.

An imaginary part of the refractive index with a similar magnitude was obtained for Saharan aerosols by *Patterson* et al. (1977). It was found to decrease exponentially from about 0.02 at 300 nm to about 0.0035 at 600 nm. This exponential decline is consistent with the absorption spectra obtained for samples of Saharan dust by *Babin* and Stramski (2004, 2002). The real part of the refractive index resulting from the measurements of Patterson et al. was between 1.52 and 1.57 in the visible. The values of the complex refractive index of these aerosols are of the same order of magnitude as those obtained by *Grams* et al. (1974), who used a light scattering-based method to determine the imaginary refractive index of soil particles *in situ*, near the ground.

Among materials which can be present in particulate form in the atmosphere and be deposited into natural water bodies are carbon, iron oxides, MNO_2 (group 1), lead compounds (group 2), mercury, vanadium, and copper oxides (group 3) (*Gillespie* and Lindberg 1992). *Babin* and Stramski (2003), who analyzed absorption of light by mineral particles sampled from various soils, found iron to be a major, albeit not the only, chromophore in samples of soil and dust particles of diverse origins. The imaginary refractive indices, n'', of the group 1 materials vary between 0.01 and 1.5 for the various materials, with little wavelength dependency in the spectral range of 250 to 560 nm. The n'' varies from about 0.25 to less than 0.001 for groups 2 and 3, whose members absorb light mostly in the short wavelength section of the examined range. In comparison, clay minerals have n'' of between 10^{-4} and 10^{-5} (*Gillespie* et al. 1974, *Lindberg* and Smith 1974). The absolute n'' of quartz is expected to be less than 10^{-7} in the examined spectral range.

Particles that are products of combustion, such as soot or fly ash, are an important component of aerosols and thus the aquatic particle populations. Refractive indices of single nearly spherically symmetrical sub-micron fly-ash particles was determined to be in a range $n' = 1.48$ to 1.57 and $n'' = 0$ to 0.01 at 633 nm (HeNe laser emission wavelength) (*Wyatt* 1980). The accuracy of these values is ± 0.01 for the real part and ± 0.002 for the imaginary part of the refractive index.

6.3.7. Refractive index of aquatic particles: key issues

6.3.7.1. The magnitude of the imaginary part of the refractive index of phytoplankton

The magnitude of the imaginary part, n'', of the refractive index of phytoplankton deserves a few comments. *Bohren* and Huffman (1983) point out that the n'' of inorganic compounds may assume either very small values ($< 10^{-4}$) or values on the order of 1. However, the n'' inferred for phytoplankton cells by various methods is notoriously within a range of 0.001 and 0.05. A 10-µm thick phytoplankton cell made of material with $n'' = 0.005$ would transmit about 27% of the incident light flux, neglecting reflection at the cell–medium interface, as follows from (6.2) and (6.3). Measurements of light absorption by individual phytoplankton cells (*Iturriaga* and Siegel 1988) do indeed show transmission of such a magnitude (down to about 40%) at the absorption peak of chlorophyll at about 440 nm. Thus, experimental evidence supports typically obtained values of the imaginary part of the refractive index of marine phytoplankton.

Light is absorbed in phytoplankton cells virtually only by pigments (dyes, most notably chlorophylls), whose molecular structure permits wide absorption bands in the visible. The n'' of phytoplankton has also been estimated by using an approach independent of the procedure described in the previous section. Given the literature data on the cellular chlorophyll *a* concentration ($3\,\mathrm{kg\,m^{-3}}$) and on the specific absorption coefficient of chlorophyll ($0.0207\,\mathrm{m^2\,mg^{-1}}$ at 663 nm), the absorption coefficient of the cell material at $\lambda = 633$ nm was estimated to be on the order of $6 \times 10^4\,\mathrm{m^{-1}}$ (*Morel* and Bricaud 1986). Such a value corresponds to an imaginary part of the refractive index on the order of 0.002. This value is of the same order of magnitude as the values obtained using the BMS procedure. *Quinby*-Hunt et al. (1989) estimated the n'' of nearly spherical *Chlorella* cells (diameter of 4–5 μm), by using the spectral absorption data of *Privoznik* et al. (1978), to be an order of magnitude higher (>0.03) than that.

6.3.7.2. Effect of the physiological state of phytoplankton on their refractive index

Physiological changes in phytoplankton may have a significant effect on the refractive index and size of the cells. *Ackleson* et al. (1988a) used the flow-cytometric technique of *Ackleson* and Spinrad (1988). Several thousand cultured phytoplankton cells of *Dunaliella tertiolecta* (spheroidal shape, see Figure 6.29) in the exponential growth phase were analyzed using an EPICS V flow cytometer (Coulter Electronics, Hialeah, Florida). The population of cells was divided into two sub-populations, one of which was diluted in a ratio of 1:10. The refractive index of cells in the diluted population was about 1.05. That of the stock population was about 1.065 (relative to water). The cell volumes of the diluted population were between 1.2 and 1.4 times greater than those of the stock population.

The dependency of the refractive index of marine photosynthetic cyanobacteria *Synechocystis* on their physiological state, as affected by growth irradiance, has also been established by using the BMS method (*Stramski* and Morel 1990, *Stramski* et al. 1988; see also section 6.3.2.4). The refractive index spectra of cells grown at a two irradiance levels, a low irradiance of $20\,\mathrm{\mu E\,m^{-2}\,s^{-1}}$ and a high irradiance of $700\,\mathrm{\mu E\,m^{-2}\,s^{-1}}$, were significantly different. Most notably, peaks between 600 and 700 nm in the high-irradiance n'' spectrum were reduced to almost 1/10 of those observed in the low-irradiance case. The refractive index, n', of *Synechocystis* was between 1.045 and 1.055 for the culture grown at $20\,\mathrm{\mu E\,m^{-2}\,s^{-1}}$

and between 1.055 and 1.060 at 700 μE m^{-2} s^{-1}. *Ackleson* et al. (1988a, see also references in *Fuhrmann* et al. 2004) note that chloroplasts in phytoplankton exposed to high irradiance contract and aggregate on a time scale of the order of 1 h. Thus, there is a biological basis for the observed changes in optical properties of phytoplankton.

Experiments with other phytoplankton cultures (a marine diatom *Thalassiosira pseudonana*, *Stramski* an Reynolds 1993) established a clear link between the diel cycle of the cells and their refractive index as well as their size. The real part of the refractive index, n'(660 nm), of the diatom varied repetitively within a range of $\sim$1.035 to $\sim$1.055, with the minima corresponding to the dark periods. The imaginary part of the refractive index at the absorption wavelength of the pigments (440 and 675 nm) varied similarly, although less evidently, between 0.008 and 0.013 and 0.006 and 0.001 respectively. Variations of the imaginary part of the refractive index at other wavelengths were much less pronounced. Similar results were obtained from other studies (*DuRand* and Olson 1998, diel variations in the refractive index of *Nannochloris* sp. of n' from $\sim$1.04 to $\sim$1.06, n'' from $\sim$0.002 to $\sim$0.0035). However, *DuRand* et al. (2002) found no correlation between the real or imaginary parts of the refractive index and the cell diel cycle for *Micromonas pusilla*.

A clear link has also been established between the temperature and nitrogen concentration and the refractive index of *T. pseudonana* grown under a saturating irradiance (*Stramski* et al. 2002, *Reynolds* et al. 1997).

6.3.7.3. Variability of the estimates of the refractive index

It is important to note wide variations between the various estimates of the refractive index of phytoplankton as shown in Figure 6.20. Although a part of this variability is likely to result from the various physiological states of the phytoplankton cultures at the measurement times (see section 6.3.7.2), it is unclear whether such a factor can explain the variability of the estimates. We note that the cell composition of a small soft-bodied flagellate *Isochrysis galbana*, for which large variations in the refractive index estimates exists as shown in Figure 6.20, can indeed vary significantly depending on the cell growth phase (for example, *Liu* and Lin 2001). In the stationary growth phase, round granules of oil (lipid) appear and can constitute on the order of 20% of the cell volume.

6.4. Morphologies of aquatic particles

Large archives of microscopic images of aquatic particles have been established by various researchers. Sample images can, for example, be viewed in the papers of *Lenz* (1972), *Wellershaus* et al. (1973) (both using optical microscopy), *McCrone* and Delly (1973), the latter being an extensive atlas of photographs of small particles, including phytoplankton cells, obtained using an optical microscope

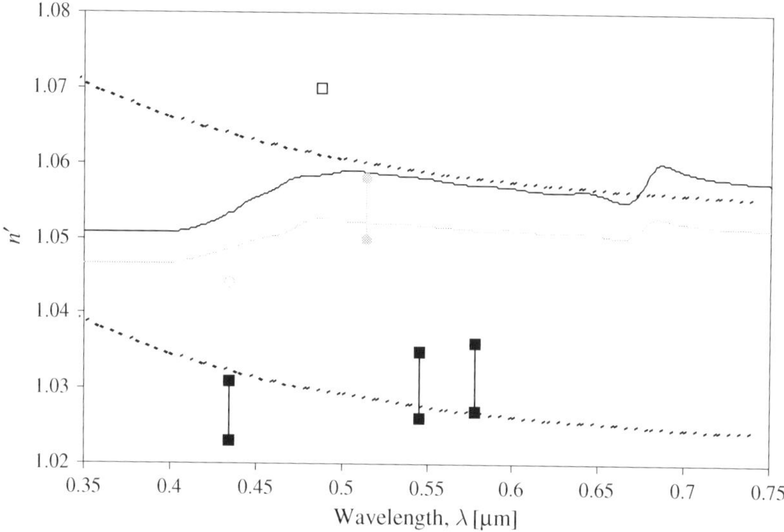

Figure 6.20. Comparison between various estimates of the real part, n', of the refractive index, relative to that of seawater, of selected phytoplankton: *Isochrysis galbana: black* curve—*Stramski* et al. (2001, based on data from *Ahn* et al. 1992), *black* open square—*Green* et al. 2003b), *black* solid squares—*Carder* et al. (1972); *Emiliania huxleyi: light gray* curve—*Stramski* et al. (2001, based on data from *Ahn* et al. 1992), *light gray* open circle—*Bricaud* and Morel (1986), *light gray* solid circle—*Ackleson* and Spinrad (1988). Dashed curves represent the real part of the relative refractive index of flagellates as calculated from the phytoplankton composition by *Aas* (1996) for 60% water content (upper curve) and 80% water content (lower curve). The dispersion of the refractive index of seawater (after *Quan* and Fry 1995) and protein (after *Aas* 1996), the main component of most phytoplankton, are both taken into account.

and a scanning electron microscope. A concise atlas of scanning electron microscope microphotographs by *Cheng* (1980) of typical atmospheric particles contains images of typical particles of fly ash, sand dust, cement, smoke particles from automobile exhaust, and pollen. *Simpson* (1982) and *Richardson* (1987) show microphotographs, obtained with a scanning electron microscope, of mostly inorganic marine particles concentrated on membrane filters. Images of aggregates can be viewed in *Heissenberger* et al. (1996a), *Grout* et al. (2001), *Droppo* and Ongley (1994). With the advent of the Internet, the number of image sources increased considerably. We refer to representative sources of phytoplankton cells' images in section 6.4.3.3.

A casual examination of microscopic images of aquatic particles (Figure 6.21) reveals that many of these particles have complex morphologies. Small particles (~1 μm) in that image appear roughly spherical or spheroidal. Larger particles,

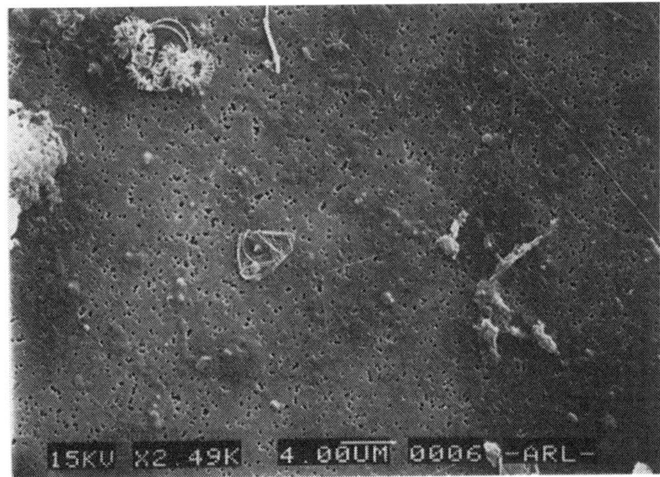

Figure 6.21. Particles from coastal Atlantic waters off Nova Scotia, Canada (a SEM image: M. Jonasz). A number of small (~1 μm) particles appear roughly spherical, but larger particles are highly non-spherical (e.g., coccoliths, a small pile of which is located at the top left, and several spines, the longest of which can be seen across the top right part of the image).

such as coccoliths or long thin spines, are highly non-spherical. Indeed, the non-sphericity of marine particles appears to increase with particle size in range of 0 to 10 μm (*Jonasz* 1987a). The non-sphericity parameter was defined as a ratio of the projected area, P, averaged over all orientations of a particle, to the projected area, P_v, of a sphere with volume equal to that of the particle. The non-sphericity parameter equals unity for spheres and is identical with the shape factor, s, of *Aas* (1984) [equation (6.77)] up to a constant factor of $(\pi/4)/(\pi/6)^{-2/3} = 1.21$. The non-sphericity parameter of marine particles was found to increase according to a power law (*Jonasz* 1987a) from a value of about 1 ($s = 1.21$) for particles with a spherical equivalent particle diameter of 0.1 μm to about 2 ($s = 2.42$) for a 10 μm particle. Most particles were found to be moderately non-spherical.

Note that conclusions regarding the size dependency of non-sphericity of marine particles were reached based on examinations of relatively large particles. Many smaller particles, such as marine colloids, possess highly non-spherical shapes (e.g., *Grout* et al. 2001). Thus, the situation may be more complex than such simplified approach might suggest. *Harris* (1977), who used a transmission electron microscope to analyze sub-micron particles in samples from the deep waters (600 to 3600 m) of the Gulf of Mexico, notes that "*. . . a suspended particle is best characterized as an oblong flake that becomes progressively thinner as its size decreases. The only particles that approximated spheres were intact coccolithophorids.*"

Recent data of *Syvitski* et al. (1995), relevant to the other end of the particle size range, indicate that non-sphericity of marine particles may decrease with increasing particle size for particles >1000 μm, i.e., in the marine snow (aggregates) range. That work suggests that the ratio of the minimum to maximum dimensions of such particles ranges between 0.2 and 1. The decrease in the non-sphericity of the large particles may be related to the state of turbulence and thus the magnitude of the shear forces controlling the maximum dimension of very large, non-animate particles. Interestingly, *Syvitski* et al. (1995) state that there are sampling occasions for which there is no relationship between the particle non-sphericity and the particle size.

6.4.1. Light scattering and the particle shape

Simple geometrical shapes, such as that of a sphere, are merely approximations of actual and generally complex particle shapes and structures which are routinely encountered in aquatic environments. We should stress that the particle shape and structure are in this context understood as a three-dimensional map of the refractive index. For simplicity, we should refer to this index map as the particle shape. In the simplest case one would assume a homogeneous particle, i.e., a volumetric distribution of spatially constant refractive index.

From the point of view of modeling light scattering, the variability and complexity of the particle shapes presents significant challenges. The first of these challenges is theoretical: models of the interaction of light with small particles are limited to a relatively few well-defined particle shapes such as a sphere (Mie theory, e.g., *Bohren* and Huffman 1983), a layered sphere (*Bhandari* 1985, *Toon* and Ackerman 1981, *Aden* and Kerker 1951), aggregates of spheres (*Xu* 1998, 1995, *Fuller* 1991, *Fuller* and Kattawar 1988), a spheroid (*Asano* and Sato 1980, *Asano* 1979, *Asano* and Yamamoto 1975), an infinite cylinder (*Bohren* and Huffman 1983), and axially symmetrical shapes modeled with Chebyshev polynomials (*Mugnai* and *Wiscombe* 1989, 1986, *Wiscombe* and Mugnai 1988, 1986). Approximate models for particles of any shape have been developed for particles which are relatively small compared with the wavelength of light:

- Rayleigh–Gans–Debye "RGD" theory, for example, *Bohren* and Huffman (1983)
- Coupled dipole model *Purcell* and Pennypacker 1973—more recently referred to as the discrete dipole approximation, (e.g., *Draine* 2000, *Draine* and *Flatau* 1994)
- T-matrix or extended boundary conditions method "EBCM": *Mishchenko* et al. (2004, an annotated list of references on the subject), *Schulz* et al. (1998, randomly oriented spheroids), *Mishchenko* et al. (1994, randomly oriented

spheroids), *Johnson* (1988), *Barber* and Yeh (1975), *Peterson* and Ström (1973), and *Waterman* (1971)
- Finite difference time domain "FDTD": *Yang* et al. (2001), *Drezek* et al. (1999), *Dunn* and Richards-Kortum (1996), and *Taflove* and Umashankar (1989).

Recent reviews of numerical methods for light scattering calculations (*Kahnert* 2003) and of light scattering theories (*Wriedt* 1998) have many more references for an interested reader to ponder. *Wriedt* (2000) lists relevant numerical programs.

Two models of light scattering (RGD and DDA) are especially interesting for our discussion here and also from the point of view of physical insight into the process of the interaction of light with particles whose sizes are comparable to the wavelength of light. Thus, we should consider these models at greater length. The RGD and DDA models are frequently used to calculate light scattering by particles with shapes radically different from spherical and for a surprisingly large relative refractive index (e.g., *Hirst* et al. 1994).

Both models adopt an "atomic" view of a particle, i.e., one in which the particle is thought to be composed of parcels of matter small enough that the electric field of the incident wave does not vary appreciably over the spatial extent of each parcel. This allows the use of the "electrostatic" approximation (*Bohren* and Huffman 1983) for each such parcel. Note that the "atomic" approach allows each parcel to have a different refractive index, i.e., it allows for arbitrary structures and shapes of the particle "envelope".

When a particle is illuminated, each parcel ("atom") of the particle can be thought of as the source of a scattered wave. The interaction of light with a particle in the framework of the RGD and DDA models of light scattering can thus be understood to be the result of coherent summation of scattered waves generated by all parcels of the particle. The important distinction between the two models is in the definition of the electric field of the effective wave at the location of each parcel.

If this field is assumed to be the same as that of the incident light wave, then the RGD model can be used. This assumption is valid when the particle size, x, and refractive index, n, are such that

$$|n-1| << 1 \tag{6.69}$$

$$x\,|n-1| << 1 \tag{6.70}$$

where n is the refractive index of the particle relative to that of the surrounding medium. This latter condition states that the electromagnetic wave does not undergo appreciable phase change inside the particle, i.e., it should behave as if there is no particle there. The first condition can be interpreted (*Bohren* and Huffman 1983) as a requirement that there is no appreciable reflection at the medium–particle interface.

If the effective wave field at the location of a parcel is the actual field, a coherent sum of the incident wave field and fields of the waves scattered by all other parcels, we have the "exact" DDA model. We put the emphasis around "exact" because the exactness of the DDA model ends right there, and its results depend on the parcel graininess, d (i.e., spacing between the centers of the closest parcels of the particle) in relation to the wavelength and the refractive index of the particle. Hence, the DDA model requires that

$$|n|\,kd \leq 1 \tag{6.71}$$

i.e., the phase change of the electromagnetic wave at a distance corresponding to the parcel size is small. In practice, $|n|kd \sim 0.5$ reproduces the differential cross-section satisfactorily in the small-angle range. However, to model light scattering in specific angular ranges, such as backward scattering, much finer parcels, i.e., $|n|kd << 1$ may be needed (B. T. Draine, personal communication).

The scattered waves are summed according to their phases to obtain the scattering amplitudes S_1 and S_2, which—in the RGD model—can be expressed as follows:

$$S_1 = -\frac{ik^3}{2\pi}\sum_j (n_j - 1)v_j \exp(i\delta_j) \tag{6.72}$$

$$S_2 = -\frac{ik^3}{2\pi}\sum_j (n_j - 1)v_j \exp(i\delta_j)\cos\theta \tag{6.73}$$

where we simplified to $\exp(i\delta_j)$ form factor $f_j = (1/v_j)\int e^{i\delta}dv_j$, with the integration carried over v_j, in the expressions for S_1 and S_2 (*Bohren* and Huffman 1983). The phase delay δ_j for the jth parcel of the particle depends on both the parcel location and the scattering angle, θ. Parameter k is the wave number corresponding to the medium surrounding the particle and n_j is the complex refractive index of the jth parcel.

In the RGD approach, as represented by equations (6.72) and (6.73), the particle "is not there," i.e., phase delays δ_j are merely due to the geometry of the particle in relation to the incident and scattered wave directions. When referred to a common origin, say at the 0th parcel, $\exp(i\delta_j) = \exp(i\delta_0)\exp[i(\delta_j - \delta_0)]$, these phases translate to phase differences, $\delta_j - \delta_0 = k(R_j - R_0)$, where R is the variable part of the pathlength of a scattered wave (Figure 6.22). The pathlength difference terms,

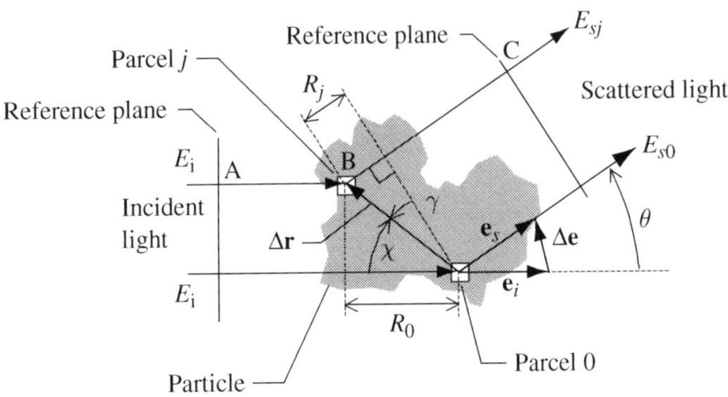

Figure 6.22. The "atomic" view of light scattering by a particle, the basis of both the Rayleigh–Gans–Debye (RGD) and discrete dipole approximation (DDA) models. The waves scattered by the two parcels of the particle are summed according to their respective phase delays. In the RGD model, the interference between two such waves, E_{s1} and E_{s2}, shown here, depends on the difference of the relevant phase delays: $\delta = k(R_j - R_0) = k\Delta\mathbf{r} \bullet \Delta\mathbf{e} = k|\Delta\mathbf{r}||\Delta\mathbf{e}|\cos(\Delta\mathbf{r}, \Delta\mathbf{e}) = k|\Delta\mathbf{r}|2\sin(\theta/2)\cos(\Delta\mathbf{r}, \Delta\mathbf{e})$, where k is the wave number of the incident light in the medium *surrounding* the particle, $R_j - R_0$ is the geometric distance factor of the phase delay, δ, that is different for each pathlength (such pathlength is denoted by points A, B, and C in the case of parcel 1) from the incident wave reference plane to the scattered wave reference plane, $\Delta\mathbf{r} = \mathbf{r}_j - \mathbf{r}_0$, where $\mathbf{r}$ is the parcel position vector in a reference system, $\Delta\mathbf{e} = \mathbf{e}_i - \mathbf{e}_s$, with e_i and e_s being unit vectors. Thus, the phase difference in the RGD model is due solely to geometry, because the incident plane is assumed to incur no phase delay within the space region occupied by the particle due to the particle presence. In the DDA model, the actual electric field at the location of a parcel is used, leading to a much more complex algorithm.

$R_j - R_0$, increase with the distance between the jth parcel and the 0th parcel, and depend on the scattering angle θ, as follows:

$$\begin{aligned} R_j - R_0 &= \Delta\mathbf{r} \bullet \Delta\mathbf{e} \\ &= |\Delta\mathbf{r}|\, 2\sin\frac{\theta}{2}\cos(\Delta\mathbf{r}, \Delta\mathbf{e}) \end{aligned} \tag{6.74}$$

where $\Delta\mathbf{r} = \mathbf{r}_j - \mathbf{r}_0$, with $\mathbf{r}$ being a position vector of the parcel, $\Delta\mathbf{e} = \mathbf{e}_s - \mathbf{e}_i$, with $\mathbf{e}$ being the unit vector indicating the direction of wave propagation.

For small scattering angles, the pathlength difference terms, $R_j - R_0$, are small, because the term $|\Delta\mathbf{e}| = 2\sin(\theta/2)$, independent of parcel location, i.e., of the length and orientation of vector $\Delta\mathbf{r}$, is very small. This means that the effect of phase errors caused by neglecting the influence of the particle on the wave velocity is small in this angular range. However, at the opposite end of the scattering angle range, $\theta \sim \pi$, we have $|\Delta\mathbf{e}| \sim 2$, and such errors may have quite a significant effect.

In the DDA case, which involves "exact" numerical solutions of the Maxwell equations, errors arise due to the finite graininess of the parcels, as well as the purely numerical errors in solving these differential equations. In the small-angle range, these errors, expressed at a relative scale, have little influence because the magnitude of the differential cross-section is large. However, they may become quite significant in any other range where the scattering cross-section is small (e.g., *Gordon* and Du 2001, *Draine* and Flatau 1994).

It thus follows that results of simulations with either RGD or DDA models of light scattering relevant to the small-angle range are expected to be more accurate than results of calculations performed for the large scattering angle. For particle sizes comparable to the wavelength of light, almost all of the scattering cross-section of the particle comes from that small-angle range. With these remarks in mind, we can now continue the discussion of the main topic of this section.

The availability of a proper mathematical model of light interaction with small particles is only a half-way step toward successful modeling of light scattering by non-spherical particles. One is still faced with the task of specifying the shapes and orientations for all particles which are to significantly interact with a beam of light, i.e., provide a shape-orientation-size distribution of the relevant particles. No particle size analysis technique with such capabilities is yet available. In addition, the orientations of particles change in time as do shapes of many fragile marine particles, which may also be severely altered by sampling and handling as discussed in Chapter 5 (section 5.6.1). *Kranck* and Milligan (1991) suggest that "*the very irregular shapes of many of the open-water particles complicates the task* [of particle size analysis]. *Loosely aggregated organic matter frequently forms stringers, veils, and irregular bodies for which the standard geological concept of equivalent diameter is difficult to determine and probably irrelevant.*" This pessimistic outlook applies also to modeling of light scattering that, in addition, is complicated by the need to specify the particle orientation.

In many cases, particles are likely to assume random orientations providing some relief for the modeler, at least in light scattering models which utilize analytical integration over particle orientation (e.g., randomly oriented spheroids by T-matrix, *Mishchenko* and Travis 1994). Note that the assumption of random orientation can be challenged. First, certain results of measurements of the scattering matrix for marine particles were interpreted as being an indication of partial orientation of the particles, perhaps by the gravity field (*Kadyshevich* et al. 1976). This interpretation was challenged by later measurements (*Fry* and Voss 1984) which did not support the preferred orientation hypothesis. However, *in situ* observations of large particles, such as marine snow, confirm the prevalent vertical orientation of certain particles (e.g., "stringers," *Syvitski* et al. 1983). Certain aquatic microbes (algae and bacteria) have both mechanical means and sensors to orient themselves in relation to the magnetic field of the Earth and other fields of importance for these organisms (e.g., *Geobacter* bacterium, *Bazylinski* and *Frankel* 2000). On the other hand, in certain applications, such as single-particle measurements (flow cytometry), particles may be oriented by the flow field in the sensing zone. In such cases, particle orientation is of importance and needs to be included as a parameter in modeling light scattering by the particles.

Having established that particle morphology can affect light scattering by small particles, we should now ask how important is it in the modeling of light scattering. Perhaps this question should be rephrased as follows. Given that the homogeneous sphere is the simplest shape, how much must the particle morphology differ from this "standard" in order to merit trying to account for the effect of this shape on light scattering by the particle? A simple answer to this question invokes the size scale of the deviations of the particle from a homogeneous sphere relative to the wavelength of light. Indeed, if the relative size scale of these deviations is much smaller than the wavelength of light, the particle shape enters the picture merely through its effect on the polarizability of the material. For example, the strength of the response of a small homogeneous needle to the electric field of a long electromagnetic wave will depend on the needle axis orientation in relation to that field. However, the scattering matrix of the particle will still have the same form as far as the angular dependency of the elements is concerned.

If the particle size is comparable to or much larger than wavelength, the answer to the question stated above cannot be given by such simple argumentation and frequently must be arrived at by comparing the measured and calculated light scattering properties of the particles. One should note that the rate of change of relationships between the light scattering and other properties of particles as "functions" of the deviation of the particle shape from homogeneous sphere may vary, allowing one to neglect particle shape effects for certain light scattering properties, or for certain ranges of these properties. The case in point is the prevalence of diffraction as a component process of light scattering in the small-angle range. Consider a prolate spheroid ("egg" shape) with the long axis oriented along the direction of the incident light. The small-angle scattering pattern of this spheroid will be "nearly" identical to that of a sphere with the diameter equal to the small diameter of the spheroid. The same can be said about a disk so oriented that the disk plane is perpendicular to the incident light direction. Yet, the "large"-angle scattering is quite different in both cases.

An interesting study, in this respect, was performed by *Yang* et al. (2004a) regarding the convergence of the light scattering properties of a series of Platonic solids, i.e., polyhedrons tetrahedron (four facets), hexahedron (six facets), octahedron (eight facets), dodecahedron (twelve facets), and icosahedron (twenty facets), to those of sphere. By using the FDTD technique referenced earlier in this section, they found that the convergence depends on the definition of the equivalent sphere (size, volume, projected area, and volume-to-surface ratio equivalent). The phase function of the icosahedron (20 facets) was a good approximation of that of sphere only when the sphere had the same volume as the icosahedron. The attenuation efficiency of the sphere was best approximated by that of the dodecahedron (12 facets) for the volume-to-surface ratio equivalent sphere. Even if this study was carried out for merely three refractive indices, representing low and high absorption, and two relative particle sizes, it is interesting to note that the differences between the attenuation efficiency of the best equivalent sphere and those of the Platonic solids ranged from ~0.3 to ~30%.

Another extreme particle shape is a particle with a solid body (a "sphere") and projections (spines) protruding from it. Some plankton cells have such shapes. The effect of such projections (having the same refractive index as that of the core) on the scattering cross-section of the whole particle was investigated by *Latimer* (1984) by using the RGD model (for the relative particle size, $x < 9$) and the ADA ($x > 3$), this latter valid for x and the refractive index, n, which fulfill the following condition: $x|n - 1| << 1$. Latimer concluded that for very small particles ($x < 1$), the effect of projections on the integral scattering cross-section is negligible. When the particle size is such that the sphere enclosing the whole particle (with projections) has $x \sim 1$, the light scattering is at a minimum. With the particle size increasing further, the ratio of the scattering cross-section of smooth sphere with volume equal to that of the spiny particle follows a series of damped oscillations and is expected to eventually settle at a limit equal to the ratio of the average projected area of the spiny and the smooth spheres. The major effect of the spines was to spread out the refractive index distribution away from the sphere center; the actual details of the spines were suggested to be less important. This conclusion is a fitting vindication for the Physicist (mentioned by *Kerker* et al. 1979) who approximated a horse by a sphere.

However, if the scattering matrix is concerned, projections—such as random surface roughness—can have perceptible effects (*Li* et al. 2004) even if the characteristic "depth," $\eta = 2\pi|n|\mathrm{SD}(r)/\lambda$ of the projections (where n is the refractive index of the particle relative to that of the surrounding medium, and $\mathrm{SD}(r)$ is standard deviation of the rough sphere surface radius, r) is less than the wavelength of light, λ. Specifically, Li et al. found that for spheres with a relative size parameter $x = 2\pi/\lambda > 15$ and a random roughness of the surface with the characteristic size $\eta > 1$, even the differential scattering cross-section $\sigma(\theta)$, where θ is the scattering angle, may "significantly" deviate from the smooth sphere results, although from the results they show, it seems that such deviations are less than a factor of ~2.

The effect of forming projections can be understood as adding a coating to a spherical core, with the refractive index of the coating being "diluted" according to a mixing rule (see section 6.3.1.1). Indeed, calculations performed by *Latimer* (1984) by using the coated sphere theory of *Aden* and Kerker (1951) in a limited particle size range confirm the general behavior of the scattering cross-section ratio derived by using the "exact" RGD approach. Note that given this interpretation, the coated sphere approach also models the optical properties of aggregates (*Latimer* and Wamble 1982). In passing, we note that another approach to the modeling of light scattering by aggregates has been the use of Mie theory with the refractive index of the sphere accounting for the presence of the interstitial water in the aggregate (e.g., *Dobbins* and Megaridis 1991).

Gordon and Du (2001) considered another case important for marine optics: the scattering by detached coccolith disks of a coccolitophore (*E. huxleyi*), which have the shape of a disk with a complex structure. In fact, it is the populations of detached disks that contribute to light scattering by blooms of coccolitophorids rather than the whole cells encased in the calcite armor (*Ackleson* et al. 1994). Indeed, *Volten* et al. (1998) did not find much difference between the VSFs of *E. huxleyi* with and without coccoliths. With a heroic effort (after 58 days of the CPU time), Gordon and Du concluded that an equal volume sphere approach would overestimate the orientation-averaged scattering cross-section of a single coccolith by a factor of about ~1.75 when compared with experiment, while the various disk-like models they used would provide a range of ~0.88 to ~1.1. They also note that the equal-projected-area sphere model would overestimate the cross-section by a factor of ~3.8. An interesting result has been obtained for the backscattering cross-section, where both their disk-like models and the equal-volume sphere underestimated the experimental value: the disk-like models by a factor of ~3 to ~1.4, equal-volume sphere ~ 3, while the equal-projected-area sphere model overestimated it by a factor of 2.4. A notable finding is that various equivalent spheres, as in other studies, produce results that vary in accuracy depending on the light scattering property modeled.

The decreased sensitivity to the particle shape with the particle shape deviations from spherical decreasing in relation to the wavelength of light is a recurring theme of other studies too. *Kahnert* et al. (2002a) point out that characteristic features of angular light scattering by large polygonal particles (such as hexagonal prisms: a halo peak at about 46°) vanish when the particle size decreases to become comparable with the wavelength of light (see also *Mishchenko* and Macke 1999). Yet, certain shape-dependent features remain, for example, the dependence of optical properties of polydisperse spheroids on their aspect ratio (*Schulz* et al. 1999, *Mishchenko* et al. 1997) if all particles are of the same shape. Hence, a so-called "shape hypothesis" has evolved, whereas the shape effects in the case of irregular particles are simulated by using *mixtures* of particle shapes. In support of this hypothesis, *West* et al. (1997) note that the phase functions of cubes and aggregates with similar real refractive index are surprisingly similar. *Mishchenko* et al. (1997) successfully simulated the phase function of soil particles by using a size-shape distribution of randomly oriented spheroids (see also *Mishchenko* 1993). Yet, Mishchenko et al. found that integral scattering can be approximated to within about 10% by that of equivalent spheres in the case of non-spherical aerosols. In comparison, the ratio of the non-spherical phase function to that of equivalent spheres varied between ~2 and ~ 0.5.

Experimental data on the light scattering patterns of aquatic particles (section 4.4.1) suggest that the non-polarized scattering pattern (VSF) for a large number of (randomly oriented) irregular particles which scatter light incoherently is not very sensitive to the details such as particle shapes and their orientations. That insensitivity is obviously created by the incoherent averaging of the effects of particle shape, structure, and composition. However, such averaging does not lead to the light scattering pattern of spheres, as in the example of randomly oriented spheroids, noted in the previous section. For example, despite the averaging of the optical properties of particles, particle shape, and structure, in particular of phytoplankton, may have a measurable effect on the polarized light scattering (*Quinby*-Hunt et al. 1989) and on light scattering and absorption near the absorption peak of chlorophyll (*Zaneveld* and Kitchen 1995).

The statistical interaction of suspended particles with a beam of light calls for a statistical light scattering model (*Bohren* and Singham 1991, *Drossart* 1990). Although such a model of sufficient generality is not yet available, some results are encouraging. For example, *Huffman* and Bohren (1980) developed a model of light attenuation by a suspension of randomly oriented ellipsoids continuously distributed in shape and size. That model yields correct predictions for the attenuation cross-section spectrum of crystalline quartz particles in the infrared. It is unfortunate that this approach cannot be used for visible wavelengths and for the optically important size range of the aquatic particle, because the conditions for the Rayleigh scattering model used by *Huffman* and Bohren (1980) are not met there.

With no simple and expedient alternative for evaluating light scattering by natural aquatic particles, the venerable Mie theory of light scattering by homogeneous spheres (section 3.2.1) still reigns largely unchallenged when it comes to modeling of light scattering by natural assemblies of aquatic particles. However, one should keep firmly in mind that there is plentiful evidence that it badly misrepresents certain key features of light scattering by natural particles, for example, it may produce angular scattering patterns that are too oscillatory and may overestimate backscattering. Even in the small-angle range, where it is frequently hailed as a reasonable approximation, it can misrepresent the scattering intensity and polarization by as much as one and even nearly two orders of magnitude as suggested by a recent systematic study of light scattering by phytoplankton and silt (*Volten* et al. 1998). In fact, *Mugnai* and Wiscombe (1989) commented that the sphere may be the "most unrepresentative shape possible" when it comes to light scattering simulations.

However, when one considers the effort that goes in simulating light scattering "correctly" the question reduces to whether one is to obtain right now relatively inaccurate results for a large class of interesting cases or "accurate" results for one or two cases. Indeed, relevant samples of the CPU time required on "fast" workstations range from up to 14 h for T-matrix calculations analytically averaging orientations for randomly oriented and size-distributed spheroids (*Mishchenko* and Travis 1994) to 58 days(!) for DDA calculations numerically averaging orientations of calcite disks (*Gordon* and Du 2001). One notable exception is the T-matrix algorithm for particle shapes with point-group symmetries (*Kahnert* et al.

2001), i.e., symmetries which allow symmetry operations (such as reflection and rotation) that leave at least one point of the particle fixed. An example of such a shape is a cylinder with a polygonal cross-section. At its best, this method allows calculation of an orientation-averaged T-matrix for such a cylinder in a few seconds of CPU time. Yet, a study of *Kahnert* et al. (2002a), who used this method to calculate scattering matrices of 315 orientation-averaged and size-distributed polygonal cylinders, took 32 h of CPU time. Incidentally, this is a substantial improvement over the DDA and FDTD, which would have taken about 6 months and 1 year of CPU time, respectively!

Although the CPU speed improves almost by the day, these estimates do not seem to promote the routine use of such calculations in modeling of light scattering by non-spherical aquatic particles, as of yet. Certainly, such comments may prompt a justified question: *What good is it to get wrong results fast?* Let us rephrase it into the following pragmatic question: "If it can be fast, then *how wrong* can it be?" If one can live with results that are off by a factor of ~10 in angular scattering and ~0.5 to ~2 in the integral scattering cross-section, then the expediency may be an attractive substitute for accuracy until a better solution is found.

A study by *Kahnert* et al. (2002a), although limited by the speed of the numerical method they used (see comments earlier in this section, regarding the CPU time usage), confirms that optimistic outlook for the Mie theory. They examined the orientation-averaged integral light scattering properties of size-distributed mixtures of five polygonal cylinders of different shapes with moderate aspect ratios and found that Mie theory can predict these properties within several percent. Kahnert et al. also point out, in reference to other studies, that for subwavelength-sized particles, the volume-equivalent sphere is the best substitute, while for particles larger than the wavelength, the projected-area sphere is the best. This corresponds to the well-known first-order behavior of light scattering, with the light scattering parameters increasing roughly with the particle volume for subwavelength particle sizes, and with the projected area, for particles larger than the wavelength of light. In a subsequent study by these authors (*Kahnert* et al. 2002b), they examined the same problem for selected elements of the scattering matrix of few types of polygonal cylinders. Although homogeneous spheres provided a passable fit to the M_{11} element of that matrix, which corresponds to the VSF, other elements of the matrix, describing polarization effects such as M_{22}, were approximated much less accurately (this element is unity for homogeneous spheres, when expressed as M_{22}/M_{11}). Kahnert et al. found, however, that shape-distributed spheroids provided a surprisingly good fit to the scattering matrix.

Given such findings, the relative simplicity of Mie theory, and the blazing speed with which various Mie programs (relative to the speed of programs implementing other models) can produce quantitative results, it is not a coincidence that various attempts were made to adapt Mie theory for the treatment of non-spherical particles. The major difference in light scattering by sphere and a non-spherical particle is in a highly oscillatory structure of the angular and spectral light scattering properties of spheres for a relative size of the sphere comparable to the wavelength of light. Such oscillations are referred to as optical resonances (see section 6.3.3.1). These resonances, due to interference of the diffracted and surface waves (e.g., *van de Hulst* 1957), are highest when the imaginary part of the refractive index, $n'' = 0$. The magnitudes of the resonances decrease with increasing imaginary part of the refractive index because less light power is available for interference. The optical resonances are observed for single spheres or monodisperse populations of spheres. In the case of polydisperse spheres, the convolution with the PSD acts as a smoothing process, reducing the magnitude of the resonances, or even extinguishing them completely for sufficiently wide distributions. Thus, the particle non-sphericity, manifesting itself as the absence of such resonances, can be simulated either by making n'' greater than it really is or—in the case of particle populations—by making the size distribution wider than it is. Needless to say, such "workarounds" introduce errors when properties of natural particle populations are inferred from light scattering and attenuation by these populations.

Surface waves are quickly damped by the irregularities of the particle shape. Thus, optical resonances vanish for a collection of randomly oriented non-spherical particles. An *ad hoc* solution to extend the use of Mie theory to non-spherical particles would be to remove the surface waves as proposed by *Chýlek* et al. (1976). Accordingly, these authors recommended clipping the Mie expansion coefficients a_p and b_p in (3.13) and (3.14) at 1/2 for $p > \sim x$ for the relative particle size x in excess of ~ 3. With this modification, Chýlek et al. obtained "reasonable" representations of the angular scattering patterns of non-spherical aerosol particles.

However, as found by *Welch* and Cox (1978), this modification of Mie theory brings absorption effects "through the back door" this time as "unexpected" apparent absorption (in contrast to the "expected" apparent absorption introduced by deliberately setting the imaginary part of the refractive index, n'', to a non-zero value). Suppression of resonances affects the absorption efficiency calculations more significantly the smaller is the real part of the refractive index, n'. In the regions of parameters of interest to us, $n' < 2.0$ and $n'' << 1$, it causes the absorption efficiency to increase spuriously by several orders of magnitude. This phenomenon was shown by *Acquista* (1978) to be a direct consequence of clipping the resonant peaks of coefficients a_p and b_p.

Drossart (1990) proposed another, little-known statistical extension of Mie theory for non-spherical particles. He postulated that each term in the sums (3.13) and (3.14) be multiplied by a phase modification factor $\exp(i\sigma_p)$, where σ_p are random phase delays whose probability distribution is related to the surface roughness of the particle. Recently, a similar statistical approach was adopted in extending the ADA (see sections 3.3.1 and 6.3.2.1) to account for the random irregularities in the particle shape (*Yang* et al. 2004).

Finally, we should mention another possibility in calculating the scattering properties of aquatic particles: a summation of experimentally determined, concentration-specific optical properties of major species of particles, weighed by the concentration of the species. This approach is slowly becoming feasible with a steady accumulation of experimental data on the light scattering properties of aquatic particles. Recent work of *Volten* et al. (1998) and *Schreurs* (1996) regarding the scattering matrices of phytoplankton and silt naturally points in this direction. Sadly, such data are still generally limited to a single wavelength, which for phytoplankton, is a setback given their complex absorption spectra. Another drawback of many experimental data is the lack of measurements in the small-angle range ($< 5°$) that is very important in many radiative transfer problems concerning natural waters. We should note that this "experimental" approach is not free from fundamental problems, such as the dependence of the optical properties and PSD of living cells on their physiological state.

6.4.2. Particle shape descriptors

The proliferation of inexpensive CCD imagers and of image analysis created conditions for more rigorous approaches to particle shape characterization, such as the use of the Fourier series (*Orford* and Whalley 1991, *Ehrlich* and Weinberg 1970) to approximate and analyze the outline of the particle image. In this case, the outline of a particle is expressed as a function of angle (0 to 360°) by using a radius projected from an arbitrary "center" of the particle. Complications arise, if the outline is not single-valued, that is, the radius projected from the center point intersects the outline at multiple points, as can be envisioned in the case of heavily indented outlines.

Fractals have also been used to analyze the particle shape and texture (*Orford* and Whalley 1991, *Kaye* et al 1994, *Hunt* and Johnson 1996) and to model the interaction of complex aggregates of particles with light (e.g., *West* 1991). *Kennedy* and Lin (1992) compared the Fourier and fractal descriptions of particle shapes and found that both methods yield similar information. However, the fractal description is applicable to a wider range of particle shapes than is the Fourier description. Chebyshev polynomials have been used to create three-dimensional, axially symmetrical particle shapes in a theoretical study of light scattering by moderately non-spherical particles (*Wiscombe* and Mugnai 1986).

Thomas et al. (1995) demonstrated the usability of the Fourier descriptors method to study and classify complex particle shapes. In this latter method, which allows multi-valued particle shape outlines, the particle outline is represented with a contour $C(z)$ in the complex ($z = x + iy$) plane. This function of a complex variable is then decomposed into Fourier harmonics by using the fast fourier transform (FFT) algorithm. With this method, there is no need to locate the center of the particle before the application of the Fourier analysis to the particle outline as in the conventional Fourier technique, nor is there any need to limit the range of harmonic frequencies a *priori*. An example of the usage of the Fourier analysis for cell shape classification is the work of *Blackburn* et al. (1998) in the quantification of bacterial cell morphologies in the Baltic Sea.

By analyzing the characteristics of powders, *Hentschel* and Page (2003) identified two key particle shape descriptors: aspect ratio, r_A:

$$r_A = \frac{\max D_F}{\min D_F} \tag{6.75}$$

where D_F is the Feret diameter, and form factor, F:

$$F = \frac{4\pi A}{P^2} \tag{6.76}$$

where A is the area of the particle image and P is the perimeter of that image. The aspect ratio is sensitive to the particle elongation, while the form factor is sensitive to the particle outline roughness.

Aas (1984) defined a shape factor, s, as follows:

$$s = \frac{P}{V^{2/3}} \tag{6.77}$$

In Eq. 6.77, the symbol P means the projected area of a particle, where, V is the particle volume. For a sphere, $s = (9\pi/16)^{1/2} \sim 1.21$. The shape factor, s, is identical to that proposed by *Jonasz* (1987a) to within a constant factor. Given the Cauchy theorem (see section 3.3.1) for convex three-dimensional shapes, $\langle P \rangle = (1/4)S$, where $\langle P \rangle$ is the orientation-averaged projected area of a shape and S is the shape surface, it follows from (6.77):

$$S = 4\,\langle s \rangle\, V^{2/3} \tag{6.78}$$

where $\langle s \rangle$ is the orientation-averaged shape factor. The smallest value of $\langle s \rangle$ is obtained for a sphere. Thus, when using that value to calculate the projected area, or total surface of non-spherical particles from the particle volume evaluated with a Coulter counter (a volume-sensitive particle counter), the result is bound to be underestimated by a factor equal to the ratio of $\langle s \rangle$ to that of the sphere. Aas evaluates this underestimation at 20 to 50% for marine particles.

The shapes of phytoplankton cells can be derived from the cell growth models as shown by *Pelce* and Sun (1993). A tip-growth model considered by those authors treats the cell wall as a deformable, freely moving boundary, whose normal velocity is a function of the local curvature of the boundary and its derivatives. Recently, *Pappas* (2005) discussed the geometry of diatom cells and proposed a number of parametric equations for modeling of the diatom cell shape. For example, the cell shape of *Gyrosigma* (see also Figure 6.30) can be modeled by the following equations:

$$x = \tan(t/2) \tag{6.79}$$

$$y = \mathrm{sech}(0.8t)\tanh(20u) + \cos(t/4) \tag{6.80}$$

$$z = \sin t + 3t \tag{6.81}$$

where $t = (-\pi, \ \pi)$ and $u = (-2\pi, \ 2\pi)$. *Pappas* (2005) shows the plot of this equation at $dt = \pi/128$ and $du = \pi/300$.

It is important to realize that the shape descriptors discussed above, aside from equations (6.79) through (6.81), refer strictly to a two-dimensional projection of the particle onto a surface. This is a consequence of observing three-dimensional particles with microscopy methods that provide (mostly) two-dimensional images of particles (except for interference microscopy which requires knowledge of the refractive index of the particle, relative to the surrounding medium for the determination of the particle "depth"). Thus, the third dimension is typically guessed at, for example as the smaller of the Feret diameters in two perpendicular directions. Scanning electron microscopy gives a feel of the third dimension ("depth") but not means of measuring it. Scanning confocal microscopy provides means for measuring the "depth," but practical considerations (time consumed) and accessibility of this microscopy tool make it a rare analysis, as one can judge from a virtual absence of this technique in analyses of aquatic and anthropogenic particles.

Incidentally, an estimate of the particle size in the third dimension can be obtained with conventional optical microscopy by mechanical shifting of the focal plane of the objective lens for each particle, i.e., focusing at the "top" of the particle and then at the "bottom" of it—not an attractive proposition in routine particle size analyses. This method of measuring the particle "depth" may be difficult to apply for particles that are not immobilized, such as live cells. Image analysis-aided microscopy is one such example.

An attractive alternative for particle shape/structure description has been recently proposed as moments of the three-dimensional distribution of mass within the particle (*Taylor* 2002). A similar approach has been used in the modeling of light scattering by irregular particles (e.g., *Muinonen* et al. 1996) where the particle is described by a three-dimensional log-normal distribution of the particle radius, r. Here the particle shape is (statistically) described completely by specifying the mean $\log r$, variance of $\log r$, and the covariance matrix for the ensemble of the discrete set of the radii at specified directions from the particle "center." These types of shape descriptors are especially attractive not only for the DDA model of light scattering by non-spherical particles, but also for ray-optics models (*Muinonen* et al. 1996) because the local slopes of the particle surface can be related to the particle radius distribution.

6.4.3. Observed particle morphologies and compositions

6.4.3.1. Viruses

The viruses in aquatic environments have relatively simple shapes (Figure 6.23). Tens of morphological types of viruses have been identified (*Wommack* and Colwell 2000). The typical morphology can be described as an icosahedral or spiral "head" with a diameter on the order of 20 to 200 nm attached to a cylindrical

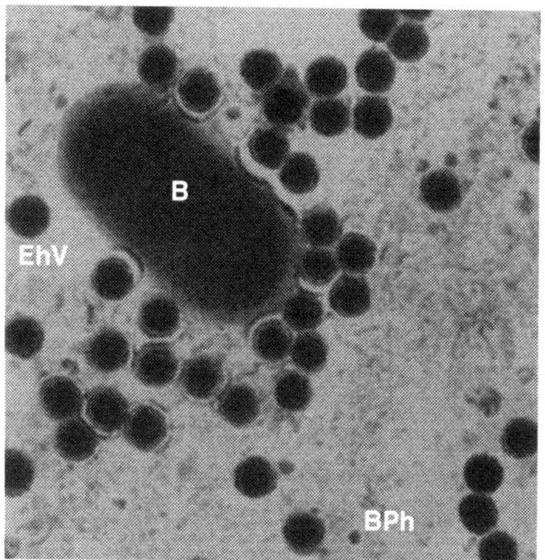

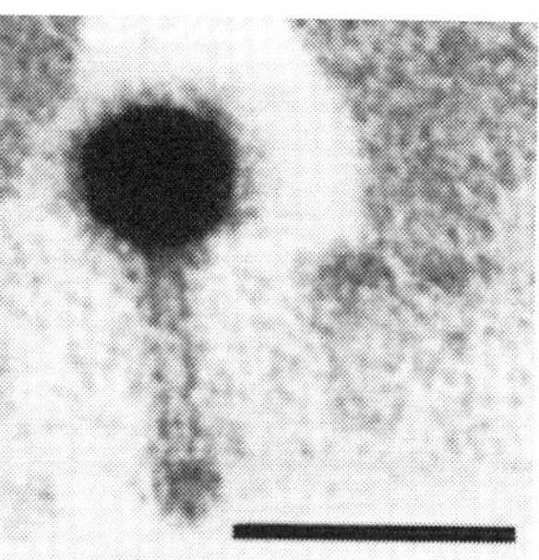

Figure 6.23. Marine viruses. A TEM image of *E. huxleyi* (EhV) viruses, head diameter 180 to 200 nm, bacteriophages (BPh), head diameter ~50 nm, and an unidentified bacteria (B). Scale bar in the bottom left corner—200 nm (photo: courtesy of M. Heldal and G. Bratbak, Bergen Univ., Norway). Many marine viruses have this form, characteristic of the T4 virus: an icosahedral 'head'(seen here) and a narrow cylindrical 'tail' (e.g., *Alonso* et al. 2002). Narrow (~10 nm) cylindrical virus tails are not visible at this resolution.

"tail" of a much smaller diameter (~10 nm) and a length on the order of 100 nm. Very thin fibers, ~1 nm in diameter, may be attached to the "free" end of the tail. Extremely large virus-like particles (340 to 400 nm diameter. heads with 2.2 to 2.8 μm long tails) have been observed in Norwegian and Danish coastal waters (*Bratbak* et al. 1992). Most marine viruses (*Wommack* and Colwell 2000, *Balch* et al. 2000) possess the head-with-tail or lone-head morphologies. Wommack and Colwell note that the absence of the tail can be a consequence of inadvertent separation of tails from heads during sample preparation.

A virus head contains DNA (~25%) and water (~75%), enclosed in a rigid protein shell, based on the typical T4 virus (*Myoviridae* family) composition (*Earnshaw* et al. 1978). The ratio of the DNA to protein content is on the order of 50:50 (*Ackermann* and Dubow 1978). These data enabled *Stramski* and Kiefer (1991) to estimate the refractive index of viruses at 1.05, relative to water, by using equation (6.8).

6.4.3.2. Bacteria

The majority of aquatic bacteria assume simple shapes:

- spherical (ovoid) (*coccus*, e.g., *Micrococcus* sp., diameter of ~0.2 μm, *Prochlorococcus*, diameter of ~0.5 μm, *Synechococcus* sp., spheroid with the largest dimension on the order of 1 μm),

- curved rods (*vibrio*, e.g., *Vibrio alginolyticus*),
- rods, typically with rounded ends (*baccilus*, *Pseudomonas* sp., diameter 0.3 μm, length of 1.1 μm: *Kopelevich* et al. 1987, heterotrophic marine bacteria: *Stramski* and Kiefer 1990)
- spirals (e.g., *Rhodospirillum*) (Figure 6.24, see also Table 6.14).

Natural populations of aquatic bacteria typically contain mixtures of these shapes (e.g., *Jochem* 2001, *Guyard* et al. 1999, *Sieburth* 1979). *Jochem* (2001) distinguished three dominant particle shapes in the Gulf of Mexico by epifluorescence

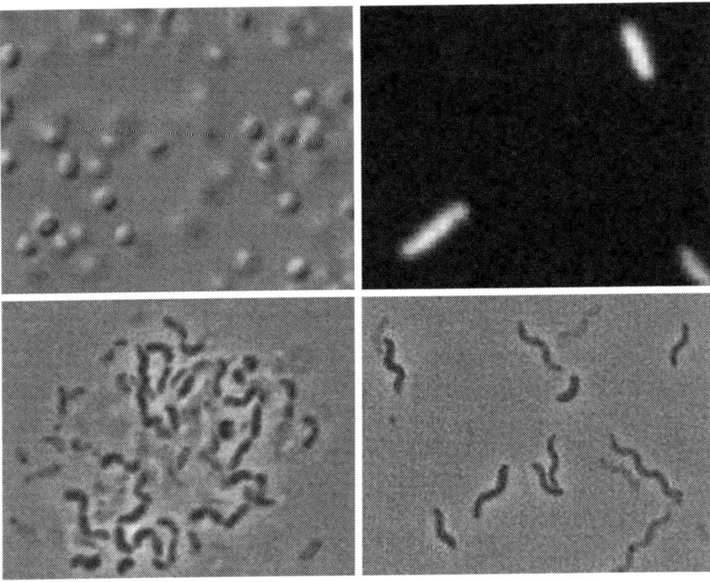

Figure 6.24. Bacterial cell shapes. *Top left*: spherical marine cyanobacterium *Prochlorococcus* (photo: courtesy of D. Patterson and R. Andersen, Provasoli-Guillard National Center for Culture of Marine Phytoplankton, USA). *Top right*: rod-shaped heterotrophic marine bacteria isolated from seawater off Bermuda by J. Fuhrman (an epifluorecence microscopy photo: courtesy of D. Stramski, SIO, USA; see also *Jonasz* et al. 1997 for images of aggregates of these bacteria). A well-researched bacterium *Escherichia coli* has a similar rod-shape. *Bottom left: Vibrio alginolyticus*, an agar-degrading species found in marine environments. The group of cells shown is encased in an extracellular polysaccharide envelope (slime) that they synthesized. Many cells are dividing so they are seen in pairs (photo: courtesy of J. Deacon, Univ. Edinburgh, UK). *Bottom right*: a photoheterotroph *Rhodospirillum* a purple non-sulphur bacterium that grows in shallow anaerobic organic-rich pools, obtaining energy from photoreactions but using organic substances, such as acetate, for cellular synthesis. *Rhodospirillum* swims in a corkscrew manner, using its polar flagella (photo: courtesy of J. Deacon, Univ. Edinburgh, UK).

microscopy: cocci (spheroids, 40 to 60% of the population), rod-shaped bacteria, and curved bacteria. Rods and curved bacteria had similar shares (18 to 25%). Sub-populations of similar shapes consisted of multi-species mixtures as indicated by the variability in size and shape of cells within each shape group.

Jochem (2001) found that regional differences in the shares of bacterial shape groups were less pronounced than changes with depth. Cocci (spheroids) were less dominant toward the coast and within the chlorophyll subsurface maximum, where the supply of dissolved organic substrates is presumed to be higher. Jochem hypothesized that this distribution might be explained by the higher efficiency of the compact, spheroidal shape for intake of nutrients due to the higher surface-to-volume ratio as compared with rod-shaped and curved bacteria. This enables compact-shaped cells to better cope with low nutrient concentrations.

Marine bacteria found in the western and north Atlantic waters are thought to be generally spheroidal, with the ratio of the major-to-minor dimension of about 2 (*Johnson* and Sieburth 1982). However, a sizeable population of rods exists there too (*Rheinheimer* and Schmaljohann 1983) as well as in other waters (Baltic Sea: ~24%, *Schmaljohann* 1984; southeastern Mediterranean: ~13%, *Robarts* et al. 1996, western and central Pacific: ~10%, *Jiao* and Ni 1997). On the other hand, *Blackburn* et al. (1998) found mostly curved and rod-shaped bacteria in the Baltic Sea, although they seem to have included coccoid bacteria into the rod class. The length-to-width ratio of rods ranged from 1 to 12.

Shapes of heterotrophic bacteria in natural populations from the Pacific waters off California, grown in filtered un-enriched seawater, were found to range from spheroidal to cylindrical with rounded ends, with most organisms being nearly spherical (*Stramski* and Kiefer 1990). Stramski and Kiefer found that cell non-sphericity increases with particle size (as also found by *Jonasz* 1987a for other marine particles): a length-to-width ratio on the order of 2 was determined for the largest bacteria with a cell width of about 0.5 μm.

The cell shape is affected by the growth history: the length-to-width ratio distribution peaked in a range of 2.5 to 2.0 for starved bacteria, while for the fast-growing bacteria, that distribution peaked in a range of 1.25–1.1 (*Stramski* et al. 1992c). The cell size parameters, such as cell diameter and length, for rod-shaped bacteria may be linked by a definite relationship (e.g., *E. coli*; *Trueba* and Woldringh 1980). Cell size and shape seem to be also affected by the nutrient concentration and grazing pressures (*Jürgens* and Matz 2002). Scarcity of nutrient results in smaller cells that more efficiently assimilate nutrients via molecular diffusion, employed by bacteria for feeding themselves. When nutrients are abundant, more complex morphologies (curved cells, spirals, and filamentous colonies) are "adopted" to reduce grazing. As noted by Jürgens and Matz, such changes in cell morphology may result from bacteria detecting certain molecular signatures of the predators, aside from the natural selection process: inedible shapes survive.

The bacterial cell is a relatively simple structure from the point of view of optical modeling (Figure 6.25 and Figure 6.26) with a caveat that some bacteria are

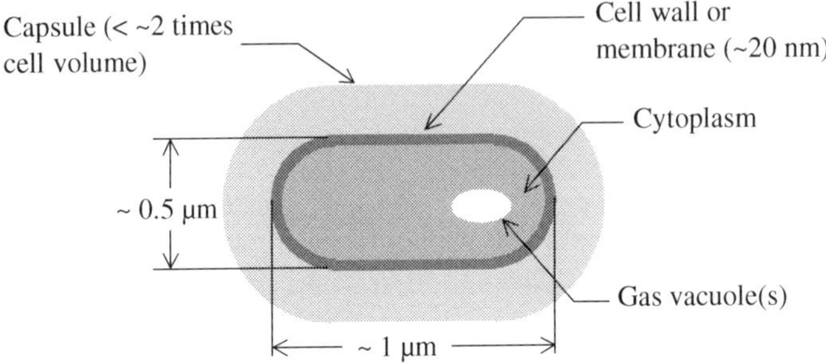

Figure 6.25. Schematic structure of a bacterial cell from the point of view of the optically important components. The cell wall/membrane structures are shown in more detail in Figure 6.26. In addition to the optically contrasting gas vacuoles, the cytoplasm (~70% water, ~30% protein) contains tightly coiled DNA and food storage bodies (e.g., *Madigan* and Martinko 2005).

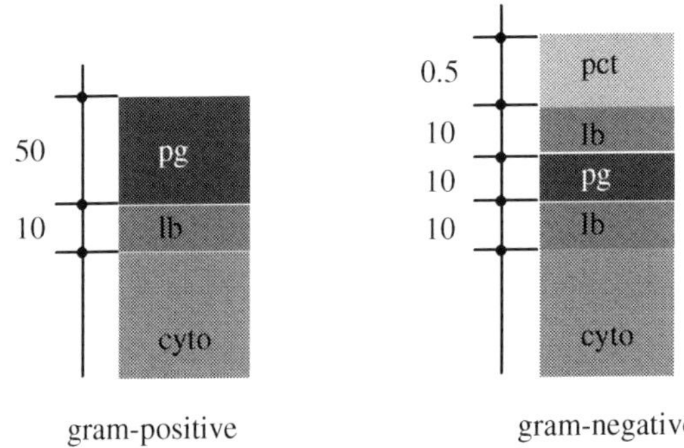

Figure 6.26. Schematic structures of bacterial cell wall/membrane from the point of view of the optically important components (order-of-magnitude thicknesses of the various layers are in nm; the drawing is not to scale): *cyto*—cytoplasm, *lb*—lipid bilayer, *pg*—peptidoglycan, *pct*—polysaccharide coating (chemically bound to the outer lipid bilayer—a lipopolysaccharide complex).

magnetotactic and may contain mineral (magnetic) particles with a refractive index sharply higher than that of the other cell components. The crystalline particles have a characteristic size of about 50 nm and may be aligned into strings ~1 μm long. Magnetic particles used by bacteria typically consists of magnetite (Fe_3O_4), greigite (Fe_3S_4), and pyrrhotite (Fe_7S_8) (e.g., *Bazylinski* and Frankel 2000). These particles enable magnetotactic bacteria to orient themselves in relation to the magnetic field of the Earth. However, the structure of bacterial cell, even without optically dense magnetic particles, can significantly affect the angular pattern of light scattering by the cells (e.g., *Allman* et al. 1993).

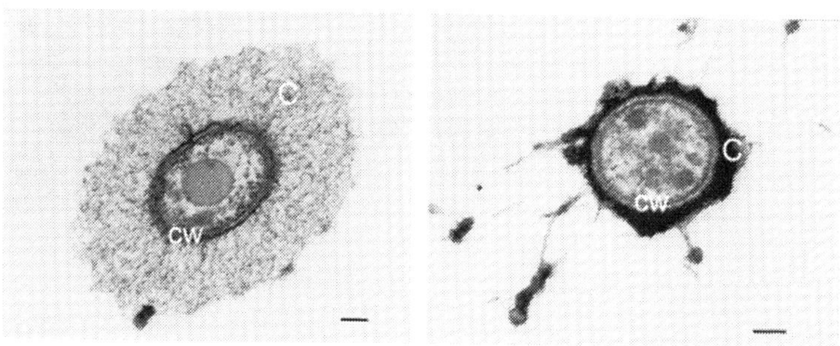

Figure 6.27. Cell structures of free-living marine bacteria (TEM images, scale bars 100 nm): *C*—capsule, *cw*—cell wall (*Heissenberger* et al. 1996b, Fig. 3AB, reproduced by permission).

The cell proper is typically surrounded by a capsule (a watery layer of polysaccharide) of rather variable shape and thickness (Figure 6.27), that can easily account for two times the cell volume (e.g., *Heissenberger* et al. 1996b). The capsule has important physiological functions, such as protecting the cell from infection by viruses. The presence of the capsule is linked to the cell "lifestyle". Heissenberger et al. found that about 35% of intact free-living cells do not have the capsule vs. $<5\%$ of bacteria associated with marine snow. The capsule seems to be lost rapidly upon the cell death.

The cell proper is delineated by a cell wall or membrane. In gram-positive bacteria, this structure consists of a lipid bilayer that encloses the cytoplasm and of a thick layer of peptidoglycan that provides the rigidity. In gram-negative cells, this layer is much thinner and is surrounded by another bilayer, this time of lipopolysaccharide, with the polysaccharide side facing the medium surrounding the cell. The thickness of the cell membrane, on the order of several tens of nm, can vary, even for the same species, with the physiological phase. *Williams* et al. (1999) found the membrane thickness to vary from ~12 nm in the exponential growth phase to 35 nm in the stationary growth phase.

Heissenberger et al. (1996b) found that about 1/4 of bacterial cells in a marine environment are just empty cell walls (remnants of viral lysis, etc.). Empty cell walls are eventually degraded by other bacteria and supply the dissolved organic matter pool or form aggregates. It seems (*Jørgensen* et al. 2003) that the cell walls of gram-positive bacteria (such as *Baccilus* sp.) are recycled faster than those of gram-negative bacteria (e.g., *Erythrobacter* sp.). This, as suggested by Jørgensen et al., might be due to the protective role of the lipopolysaccharide outer membrane in the cell walls of gram-negative bacteria.

Bacteria, when starved, produce survival cells (spores) that can be activated when living conditions return to normal. The spores contain little water (~15 to 25 g/100 ml; e.g., *Ross* and Billing 1957) in contrast to the vegetative cells and are surrounded by a thick coat (70 to 200 nm; e.g., *Driks* 1999) of dense proteinous material. The effective real refractive indices, n', of the spores cluster around 1.5, very close to the refractive index of the proteins and much higher than that of the vegetative cells. For example, the refractive index of *Baccilus subtilis* spores was determined to decrease from 1.6 at 0.2 μm to 1.52 in a wavelength range of 0.5 to 2.5 μm (*Tuminello* et al. 1997, *Katz* et al. 2005 obtained a similar value of 1.515). *Ross* and Billing (1957) report values of n' of 1.49 to 1.540 at 0.542 μm. *Tuminello* et al. (1997) determined the imaginary refractive index, n'', of *B. subtilis* spores to range from ~0.038 at 0.2 μm to ~0.01 at 2.5 μm.

The refractive indices of the cell wall fragments have been estimated by immersion refractometry to range between 1.356 and 1.382 at 700 nm (*Marquis* 1973), resulting in the refractive index increment of 0.0018. The refractive index of the cell wall polymers (mostly peptidoglycan) was estimated at 1.45 to 1.46 by *Marquis* (1973). Please also see other estimates of the refractive indices of cell wall and cytoplasm in Table 6.15

The cells may project into the surrounding medium a number of appendages, the major being flagella and as many as several hundred smaller appendages, pili. The flagellum is used by the cell for quite rapid movement (~20 body lengths/s). The pili are used for adhesion to surfaces.

The cytoplasm consist of mostly water (~ 70%). Proteins, DNA, RNA, and various low-molecular-weight molecules make up the balance (~30%, see also Table 6.13).

The cell DNA content, m_{DNA} [fg], frequently used for the determination of the dry mass of the cell, varies depending roughly on the cell dry mass, m_{dry} [fg], as follows:

$$m_{DNA} = 0.373 m_{dry}^{0.53} \tag{6.82}$$

[$r^2 = 0.55$, $n = 663$ for $\log m_{DNA} = f(\log m_{dry})$; *Posch* et al. (2001)]. Thus, DNA constitutes roughly as much as 87% of the dry mass of a small bacterial cell ($m_{dry} = 5$ fg) but only as little as ~1.5% of the dry mass of a large cell (1000 fg). Note that the correlation between V_{DNA} [%] and the cell volume, V [μm^3], is

Table 6.13. The composition of well-researched bacteria: *Escherichia coli* (including the cell wall material) (after *Moore* 1999).

Species	**Fraction of mass [%]**	**Molecular weight**
Water	70	18
Inorganic ions and small Molecules of other kinds	7	~145
DNA	1	3×10^9
RNA	6	$\sim 1 \times 10^6$
Proteins	16	3×10^4

higher [$r^2 = 0.71$, $n = 1622$ for $\log V_{DNA} = f(\log V)$; lacustrine bacteria, *Trevors* and Psenner (2001)]:

$$V_{DNA} = 0.783V^{-0.568} \tag{6.83}$$

In passing, we note the following relationship between the bacterial cell volume, V [μm^3] and cell dry weight, m_{dry} [fg]:

$$m_{dry} = 435V^{0.86} \tag{6.84}$$

with $r^2 = 0.97$ for $n = 1155$ observations in lacustrine bacteria and *E. coli* cultures. This empirical relationship was obtained by using a relationship between the optical density of TEM images and the dry mass of the cell (*Loferer*-Kröβbacher et al. 1998).

Stramski (1999) proposed a relationship between the carbon content of phytoplankton cells and the real part of their refractive index. Even if such a relationship has not been derived for bacteria in general, limited data are available for some marine cyanobacteria (*Synechococcus* and *Synechocystis*, Table 6.11). With a view toward utilization of such relationships for estimating the refractive index of aquatic bacteria, we will now quote several useful relationships between the cell volume and carbon content.

Posch et al. (2001) summarize a wide range of formulas for converting the bacterial cell volume, V, to carbon content, m_C, ranging from linear ($m_C = AV$) to power-law ($m_C = aV^b$) functions. They note that results obtained with the various formulas may differ significantly, for example, m_C [fg] may vary from 3 to 36 pg at a volume of 0.05 μm^3. Posch et al. thus propose to use the following two staining dye-specific formulas for evaluation of the cell carbon content, m_C [fg], from estimates of the cell volume, V [μm^3], based on epifluorescence microscopy observations on stained bacterial cells:

$$m_C = 218V^{0.86} \tag{6.85}$$

for DAPI (4, 6-diamidino-2-phenylindole)-stained cells, and

$$m_C = 120V^{0.72} \tag{6.86}$$

for AO (acridine orange)-stained cells. This dichotomy results from the different degree to which cells are stained with a specific dye.

By formally combining equations of type (6.85), (6.86), i.e., $m_C = aV^b$ with an equation of type $n'_\lambda = n'_0 + n'_1 C_C$ (Table 6.11) for the real part of the refractive index (relative to water) as a function of the intracellular carbon concentration, C_C, we obtain:

$$\begin{aligned} n' &= n'_0 + n'_1 \frac{aV^b}{V} \\ &= n'_0 + n'_1 a V^{b-1} \end{aligned} \tag{6.87}$$

in terms of the cell volume, and

$$n' = n'_0 + n'_1 a \left(\frac{\pi}{6}\right)^{b-1} D^{3(b-1)} \tag{6.88}$$

in terms of the cell "diameter," D. By taking the average of the coefficients a and b from equations (6.85) and (6.86), and the values of the n'_0 and n'_1 coefficients for a marine cynaobacterium *Synechococcus* from Table 6.11, we obtain the following numerical relationship

$$n' = 1.019 + 0.033 D^{-0.63} \tag{6.89}$$

which is similar to that [(6.65)] obtained by *Stramski* and Kiefer (1990) on different grounds [albeit the exponent here is half of that in (6.65)]. For a 1 μm bacterium, equation (6.89) yields $n' \sim 1.05$.

Based on the density of protein of 1.22 g cm^{-3} (after *Aas* 1996), the density of the cytoplasm is thus expected to be about 1.066 g cm^{-3}, compared to a range of 1.03 to 1.10 g cm^{-3} given by *Boney* (1989) and to the density-gradient measurements of whole bacterial cells: 1.06 to 1.10 g cm^{-3} (*Robertson* et al. 1998). Thus, the absolute refractive index of cytoplasm [by using (6.7)] is on the order of 1.41 (~1.05, relative to water). *Ross* (1954) obtained, with an interference microscope, a value of 1.3535 (i.e., ~1.01, relative to water) for the refractive index of a living cell. *Kitchen* and Zaneveld (1992) quote a similar value (1.015, relative to water).

6.4.3.3. Phytoplankton

Before the discovery of bacterioplankton, phytoplankton was thought to be the most numerous contributor to the livestock of aquatic particles and thus was extensively researched. *Boney* (1989) and *Drebes* (1974) discuss the various aspects of phytoplankton, including the morphology and structure of the cells. Many images of phytoplankton cells obtained with optical and scanning electron microscopy are available in *Bold* and Wynne (1985; algae) and *Round* et al. (1992; diatoms) and in online image libraries, for example:

- http://www-cyanosite.bio.purdue.edu
- http://www.sb-roscoff.fr
- http://www.bgsu.edu/departments/biology/facilities/algae_link.html

- http://www.phycology.net
- http://rbg-web2.rbge.org.uk/ADIAC/db/adiacdb.htm

Photographs of phytoplankton cells, obtained with optical microscopy, transmission electron microscopy, or scanning electron microscopy, have been published in scientific periodicals (examples are in *Hoepffner* and Haas 1990, *Hood* et al. 1991, *Johnson* and Sieburth 1982, and *Richardson* 1987). As it stands, information about the shapes and structures of phytoplankton cells is scattered throughout the vast biological literature. Table A.8, while not pretending to be exhaustive, contains an annotated list of references containing information on the shapes and structures of some phytoplankton species, including some of the species listed in Table A.3 (scattering functions) and Table A.5 (particle size distributions).

Virtually everyone who has seen images of phytoplankton cells has marveled at the *natural* beauty of the intricate patterns of these cells (Figure 6.28), visible especially well in scanning electron micrographs that have a definite three-dimensional appearance. The diatoms' silica shells, despite their apparent fragility, are surprisingly strong, deterring grazing by all but largest zooplankton. Indeed, the shells were found to have easily survived pressures on the order of 700 tonnes m^{-2} (*Hamm* et al. 2003). Hamm et al. who analyzed the fascinating architecture of diatom shells by using a finite element method, widespread in mechanical engineering, note that intricate patterns of these shells must have evolved via successful optimization of the use of silica as the building material for achieving the maximum strength of construction, i.e., a principle that governs macroscopic engineering design as well and results in creations of similarly *functional* beauty. Albeit the form of diatom shell structures span so wide a range, each finds its successful niche in the ecosystem, undoubtedly fitting the function of the species survival. Indeed, as architect Frank Llloyd Wright stated: "*Form is function*."

The shapes of phytoplankton cells range from moderately thin disks, through a variety of shapes, including spherical and spheroidal, to very long macroscopic scale filaments. The filaments are usually colonies of many cells. e.g., *Skeletonema costatum*, whose silica-shelled cells are of cylindrical shape (diameter 6 μm, length 12 μm) form long colonies of up to 15 cells (e.g., *Reuter* 1980b). Phytoplankton cells contain green-colored inclusions which absorb light (chloroplasts) and may contain gaseous inclusions (serving as flotation devices) which refract it (gas vacuoles, fat globules). Some phytoplankton cells may carry hard silica shells (diatoms), cellulose armor plates (dinoflagellates), or calcite armor plates (coccolithophores). The cell shape is to a significant extent a property specific for a genus and undergoes relatively minor variations from a species to another within that genus. Until recently (i.e., before genetic makeup of an organism could be easily deciphered), this fact had been the basis of taxonomy of the phytoplankton and other organism. We should note that in diatoms, the division of cells generate

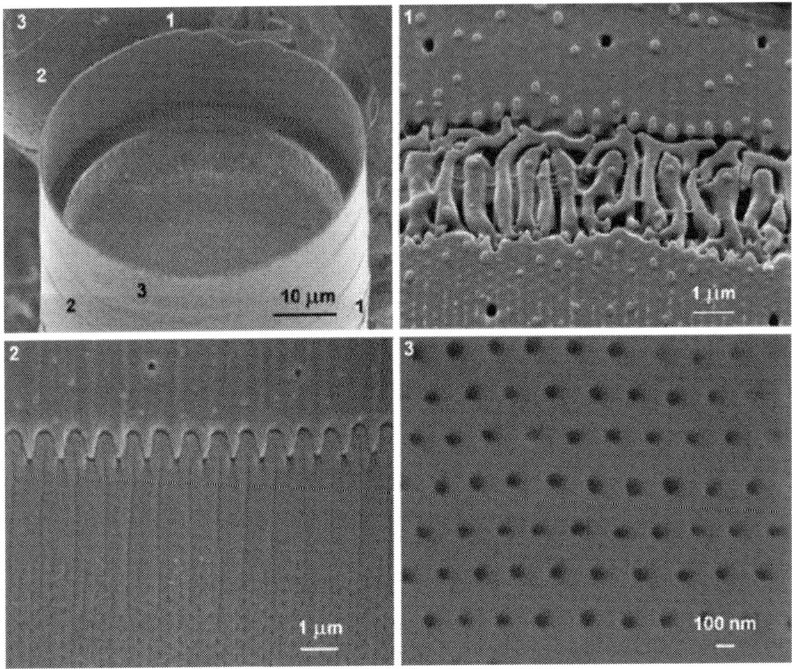

Figure 6.28. SEM images of diatoms, such as this freshwater diatom *Ellerbeckia arenaria* (top left; all images reproduced from *Gebeshuber* et al. 2003, Fig. 1, by permission), invariably invoke impressions of natural beauty. While watching such intricate structures, it is difficult to remember that diatom shells are not objects of art but results of "endless" blind experimentation of nature with results that are judged according to improvements in the chances of survival of the species. The diatom shell is composed of biomineralized silica (amorphous glass). We quote the original "nomenclature" of Gebeshuber et al.: the "tower of life" (top left), the "holy syllable" (top right), "microman" complete with eyes, teeth and beard (bottom left), and nearly perfect hexagonal pattern (bottom right). The images on the top right and in the bottom row are magnified regions denoted by the ordinals in the top left image.

a descending sequence of cell sizes, until a limit is reached and the original cell size is recreated, by sexual reproduction (e.g., *Zurzulo* and Bowler 2001).

The phytoplankton cell/colony shapes have various degrees of symmetry. There is experimental evidence (e.g., *Padisák* et al. 2003a, 2003b) that the degree of symmetry affects the hydrodynamic resistance of the cells and colonies. This should not be a surprise: an arrow-head *is* asymmetric! In fact, for some phytoplankton, such as a colonial diatom *Asterionella* (Figure 6.30), the symmetry of cell colonies may decrease in response to environmental pressures and aging. In modeling of light scattering, the consideration of symmetry is one of the crucial topics. First, highly symmetrical cell shapes, such as the sphere, allow one to use "exact" and relatively simple light scattering models, such as Mie theory. Second, as recently shown by *Kahnert* et al. (2001), the CPU time needed for modeling of light scattering by particles with complex shapes via computationally intensive methods, such as the T-matrix method, can be drastically reduced if the cell has certain symmetries, specifically, a point-group symmetry (section 6.4.1).

Spherical cells have the highest degree of symmetry of finite-extent geometrical shapes, i.e., the greatest number of operations that leave the object unchanged, for example, a rotation by an angle. In fact, an infinite set of rotations about an infinite set of angles can be defined for sphere. Centric diatoms (pillbox) have radial symmetry in the gridle plane, as far as the general cell shape (circular pillbox) is concerned. However, when small elements of the shell (frustule) ornamentation are taken into account, the number of symmetry operations is considerably reduced. In addition, diatoms have generally bilateral symmetry (reflection operation) about the gridle plane (if one neglect the fact that the upper frustule (epitheca—pillbox cover) has a greater radius than the lower frustule (hypotheca—pillbox bottom). Note that we used the "upper" and "lower" designations quite arbitrarily as the cell can potentially assume any orientation in space. This may not be true for certain cell/colony shapes whose symmetry is limited (*Padisák* et al. 2003a). Such situations may exist, for example, in mass settling of certain diatom colonies promoted by a reduction in the symmetry of the colonies, which simultaneously favors certain orientations while settling. Pennate (pen-like) diatoms have bilateral symmetry about their gridle planes and some pennate species may also have bilateral symmetry about a plane perpendicular to the gridle plane. Note that symmetry of a colony of cells may differ from that of the individual cells of the colony (e.g., a spiral-shaped colony of a diatom *Chaetoceros debilis*).

The cell shape is thought to have several functions, including protection from grazing, control of settling (that we discussed above), maximization of the nutrient transfer rate (e.g., *Pahlow* et al. 1997), as well as optimization of the collection of sunlight for photosynthesis (e.g., *Reynolds* 1989). Thus, for a given phytoplankton species, the cell shape is a result of optimization in a multi-dimensional space of key factors affecting the species success in the environment. It is also a function of the availability of nutrients and supply of "building materials" (e.g., *Sommer* 1998). The nutrient transfer rate, at the length scale of phytoplankton cells, is controlled mainly by diffusion; hence, the maximization of the surface-to-volume area is advantageous. This seems to work well for solitary cells, but in the case of chain-shaped cell colonies, only those with ample spaces between cells, such as *Skeletonema costatum*, may enjoy a higher nutrient transfer rate than solitary cells (*Pahlow* et al. 1997).

Cells may exist as unicellular bodies or colonies of many thousands of individual cells (e.g., spherical colonies of *Phaeocystis* can reach a 10 mm diameter, *Hamm* et al. 2000). In some cases, the various cells of the colony have different function and structure. Colonies can even procreate as colonies not as individual cells that aggregate at a later stage. Interestingly, the unicellular vs. colonial life style may be switched on and of by the presence of predators, sensed by the phytoplankton as the presence of certain compounds (kairomones) exuded by the predators, for example, cyanobacterium *Microcystis aeruginosa* aggregates in the presence of a grazer (*Ochromonas* sp., *Burkert* et al. 2001). In the presence of moderate concentrations of the grazer, the population of *M. aeruginosa* initially declines and then levels off as the colonies are formed, deterring grazing. The response has a limited speed, as populations of *M. aeruginosa* exposed to high initial concentration of *Ochromonas* sp. are simply grazed down. Another interesting example is the induction of colonies and of cell shape changes (enlargement of spines) in *Scenedesmus* sp. by the presence of a predator (*Daphnia*, *Wiltshire* and Lampert 1999, see also *Hamm* 2000). This brings an additional level of complication to the task of specifying the cell morphology because the morphology may not in every case be thought of as a species-dependent property.

Aas (1984) used data of *Eppley* et al. (1967) and *Paasche* (1960) to calculate the cell shape factor, s [equation (6.77)], of several phytoplankton species. The shape factor ranged between 1.21 (spherical coccolithophorid *Gonyaulax polyhedra*) and about 3.6 (filamentous *Rhizosolenia hebeta f. semispina*). *Sommer* (1998) gives data on the surface-to-volume ratios for 23 species of phytoplankton whose cell size spans almost three decades.

Kronfeld (1988, pp. 147–150) gives concise phytoplankton cell shape descriptions for species typical of the Baltic Sea. In a much more detailed approach, *Hillebrand* et al. (1999) systematically classified 20 cell shapes for >850 phytoplankton genera. This shape collection has been recently expanded to 31 by *Sun* and Liu (2003). Although such classifications are intended to provide a means of simplifying calculations of the cell volume, they can be used as a first-order guide to the diversity of phytoplankton cell shapes. It turns out that the sphere is not as uncommon a cell shape as one might expect (Table 6.14). Indeed, a cursory review of the vast pictographic material on phytoplankton suggests that it relatively frequent, either as a "homogeneous" sphere (e.g., *Cyanidium* sp.) or as a spherical shell (colonies of *Volvox* sp.). From Table 6.14 it is also clear that the number of cell morphologies is radically smaller for cyanobacteria than for other phytoplankton. It is striking that the oblate spheroid is missing altogether from the possible shape collection listed by *Hillebrand* et al. (1999).

One has to acknowledge that a "shape-squaring" approach, i.e., squeezing the actual cell shape into a simple geometric model, is far from ideal. Indeed, *Sun* and Liu (2003) note: "... *Often there is the dilemma of whether to assign a phytoplankton cell shape to a complex but similar geometric model or to a simple, conveniently measurable, but inadequate model or shape....*" This sounds quite similar to concerns specific in modeling of light scattering by phytoplankton cells.

Examples of the diversity of the phytoplankton cell shapes are shown in Figure 6.29 and Figure 6.30). These figures (especially Figure 6.30) hint at the difficulty of approximating a phytoplankton cell shape by a simple geometric

Table 6.14. Frequency [%] of occurrence of various "basic" cell shape approximations based on compilation for over 850 phytoplankton (including cyanobacteria) species examined by *Hillebrand* et al. (1999).

Shape	**Cyanobacteria (155 species)**	**Phytoplankton (~700 species)**
Sphere	12.9	14.4
Prolate spheroid	32.9	12.5
Oblate spheroid	0.0	0.0
Ellipsoid	0.0	10.2
Cylinder	54.2	8.5[a]
Elliptic section cylinder	0.0	13.4
Box	0.0	4.7
Twelve other shapes	0.0	36.4

[a]Rounded-ends cylinder makes up 0.7% of the cylindrical shapes.
Note that the oblate spheroid shape is a rather unlikely shape for a phytoplankton cell as none was so approximated by *Hillebrand* et al. Other shapes are mostly combinations of the "basic shapes".

model. Even for simple shapes, there seem to be problems: we were hard pressed to find convincing examples of ellipsoids conforming to the classification of cell shapes by *Hillebrand* et al. (1999).

Colonial forms of plankton add additional complexity to the problem of particle shape specification. Indeed, the colonial forms range from short, tens of micrometers (e.g., *Chaetoceros* sp.) to long, millimeter-sized cylindrical chains, through (sometimes) incomplete two-dimensional star-shaped formations like *Asterionella*, to ribbons, spheres (e.g., *Phaeocystis globosa*), spherical shells, like *Volvox*, and irregular clumps of single cells.

The structure of the diatom shell surface exhibits periodicity comparable to the wavelength of light (Figure 6.28). This structure exists on surfaces of geometrical extent much larger than the wavelength of light. Such structures are likely to contribute, through interference of light, to the optical properties of the cells, and to the modification of the light spectrum inside the cells. Periodic structures such as those are known to significantly affect the spectrum of the reflected light in biological objects such as butterfly wings (e.g., *Vukusic* et al. 2001). Indeed, *Fuhrmann* et al. (2004) found that the ornamentation of the shell of a large diatom *Coscinodiscus granii* behaves as a photonic crystal and may affect the spatial and spectral distribution of light inside and outside the cell. This seems consistent with reports cited by Fuhrmann et al. that the position of light-collecting organelles (chloroplasts) inside the diatom may be modified by the organism in response to such a spatial distribution of light power.

The diatom shell is not "naked" as SEM images would lead us to believe. The outer surface of the shell is covered and interspersed (e.g., *Reimann* et al.

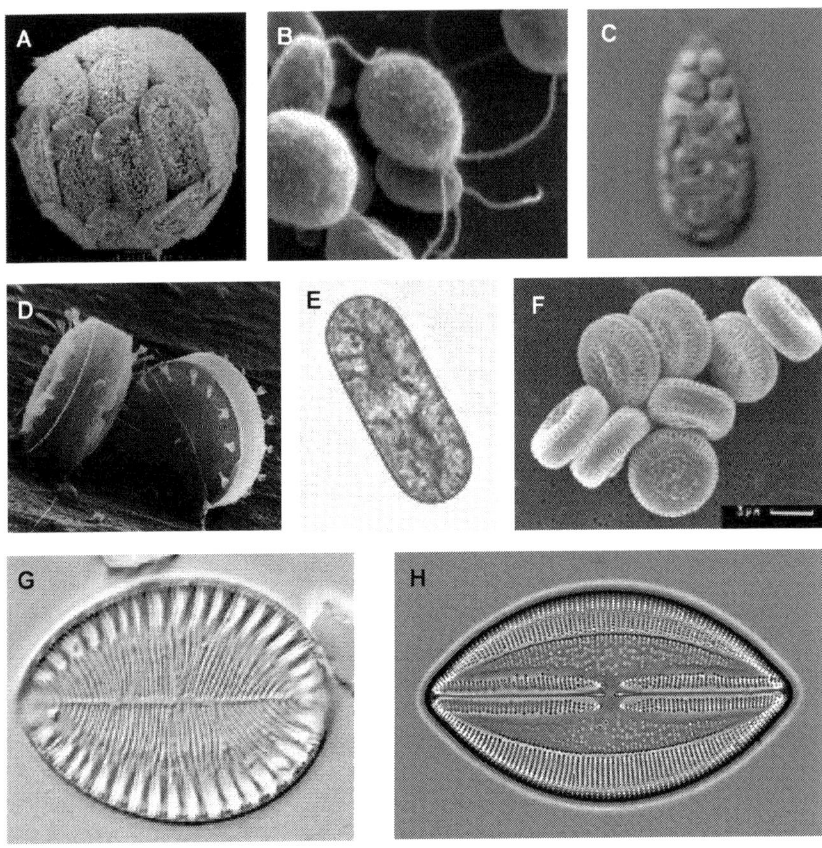

Figure 6.29. Simple phytoplankton cell shapes. ***Sphere***: **A**—a coccolitophorid with calcite scales, there may be several layers of such scales (see also Figure 6.30B) attached to the cell, diameter ~16 μm (photo: M. Jonasz). Both cells themselves (armoured—such as this one, and naked—e.g., *Chlorella* and *Cyanidium caldarium*) and cell colonies (e.g., *Volvox* that forms thin-shell colonies with a diameter ~500 μm) can be spherical. ***Prolate spheroids***: **B**—*Chlamydomonas reinhardii.*, diameter ~4 μm, (photo: courtesy of Dartmouth E. M. Facility, Dartmouth College, USA), **C**—*Dunaliella tertiolecta*, length ~10 μm (photo: courtesy of D. Patterson and R. Andersen, Provasoli-Guillard National Center for Culture of Marine Phytoplankton, USA). ***Cylinders***: **D**—*Thalassiosira* sp., diameter ~26 μm (photo: M. Jonasz), **E**—*Cylindrocystis* sp. (photo: courtesy of J. Kinross, Napier Univ., UK), **F**—centric diatom *Cyclotella choctawhatcheeana*, diameter ~6 μm (photo: courtesy of R. Hansen and S. Busch, IOW, Germany). ***Elliptic-base pillboxes***: **G**—diatom *Surirella crumena*, length ~33 μm (photo: courtesy of M. McQuoid, Göteborg Univ., Sweden), **H**—diatom *Lyrella*, (photo source: ADIAC, CEC contract MAS3-CT97-0122).

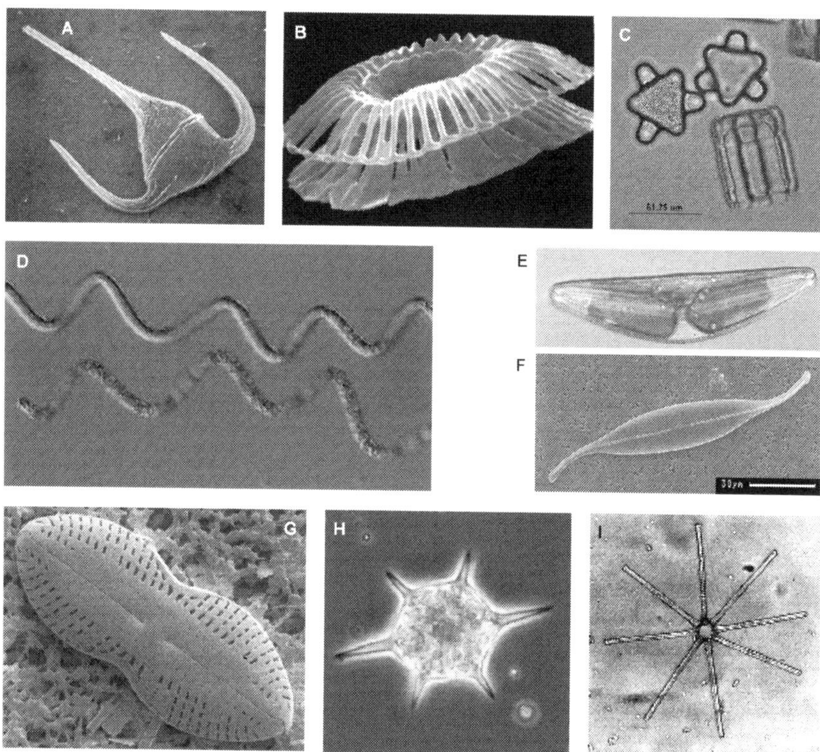

Figure 6.30. Phytoplankton cell shapes—a tiny glimpse of the nature's free-hand "industrial design" portfolio. **A**—a dinoflagellate *Ceratium* sp., image diag. ~300 μm (SEM photo: courtesy of I. Zimmerlin, Rouen Univ., France). **B**—a side-view of a detached coccolith of *E. huxleyi*, diameter ~2.5 μm (SEM photo: courtesy of J. Young, Nat. Hist. Museum, UK). There are between ~20 and ~200 coccoliths per cell (see also Figure 6.29A). Detached coccoliths are believed to be the major cause of backscattering by seawater during coccolithophorids' blooms (e.g., *Tyrrell* et al. 1999). **C**—*Hydrosera whampoensis*, side and end views, 'diam.' ~60 μm (photo: courtesy of M. Vis, Ohio Univ., USA). **D**—*Spirulina* sp. (photo: courtesy of I. Inouye, Tsukuba Univ., Japan). **E**—diatom *Cymbella lanceolata*, the dark areas contain chlorophyll (photo: courtesy of Y. Tsukii, Hose Univ. Japan). **F**—diatom *Gyrosigma* sp., bar 30 μm (SEM photo: courtesy of R. Hansen and S. Busch, IOW, Germany). **G**—diatom *Diploneis didyma*, length ~40 μm (SEM photo: courtesy of M. McQuoid, Göteborg Univ., Sweden). **H**—*Dictyocha speculum*, a skeleton bearing stage, cell core diameter ~25 μm (photo: courtesy of R. Hansen and S. Busch, IOW, Germany). **I**—a plane star-shaped colony of diatom *Asterionella* (photo: courtesy of M. Vis, Ohio Univ. USA). Incomplete stars and stars with a different number of the cells (including three-dimensional, as opposed to plane configurations) are also possible.

1965) with an organic (polysaccharide) coating (Figure 6.32) that maintains the shell cohesivity (the shell is made of nanoparticles of silica) and protects the shell from dissolution or invasion by pathogens. A similar organic coating is present on the equally intricately patterned coccoliths (e.g., *Henriksen* et al. 2004b). An interested reader may find detailed accounts of the shape and structure of coccoliths for three marine coccolitophores in *Henriksen* et al. (2003, 2004a).

The cell surface of armoured dinoflagellates has rounded or polygonal indentations (outline diameter ~1 μm) and may have a network of fine pores (<0.5 to 1 μm, e.g., *Faust* 1993) similar to those featured in diatom frustules. The structure of a phytoplankton cell (Figure 6.31 and Figure 6.33) is much more complicated than that of the bacterial cell. Complications are brought about by the many organelles of the cell: nucleus, mitochondria, endoplasmic reticulum (ER), lysosomes, etc. that evolved as a consequence of the large size of the eukaryotic cell. Surprisingly, very little is known of the optical properties of these organelles in the case of phytoplankton, let alone about their effect on light scattering by phytoplankton. To our knowledge, only *Witkowski* et al. (1993) investigated experimentally the effect of intracellular bodies (spores) on the dynamic light scattering by a culture of *C. vulgaris*, a unicellular spheroidal algae with an average "diameter" on the order of 2 to 3 μm. They found that such bodies (average "diameter" on

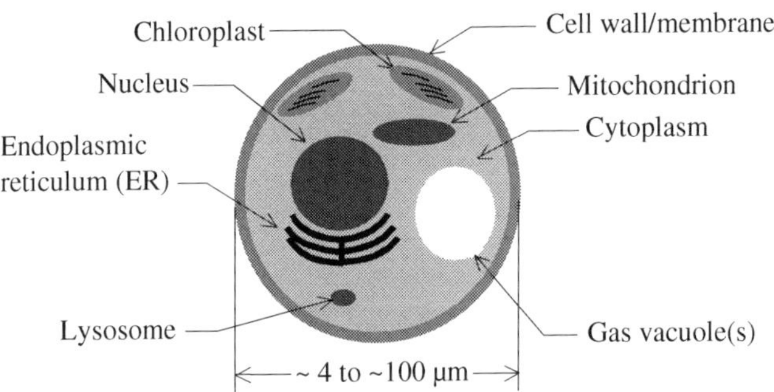

Figure 6.31. Schematic structure of a phytoplankton cell from the point of view of the optically important components. There is a vast variability of the cell shapes. Hence, and also because there are several phytoplankton species with spherical cells (e.g., *Cyanidium caldarium*), the cell is pictured as a sphere. The cell wall/membrane consists of a lipid bilayer cytoplasmic membrane enclosing the cytoplasm, and may also include, or be protected from the outside by a layer of mineral/organic scales (see Figure 6.32 for the representative cell wall configurations). The endoplasmic reticulum (a highly folded organelle, enclosed in an extensive membrane) is located near the nucleus and receives from it molecular "instructions" regarding the synthesis of the various cell materials. The chloroplast contains the chlorophyll dye incorporated into lamellar structures. Characteristic sizes and real refractive indices of the phytoplankton components are listed in Table 6.15

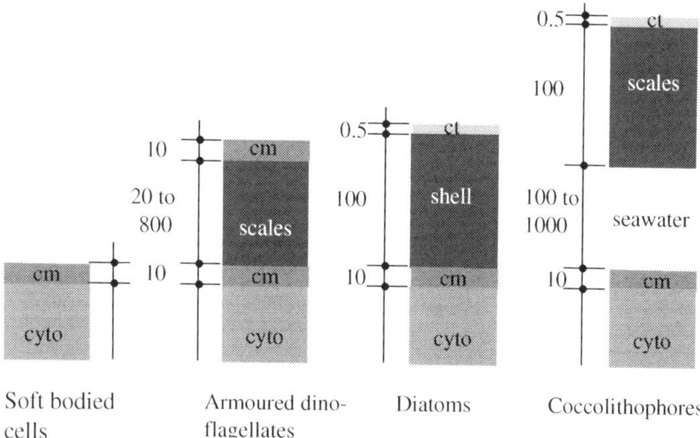

Figure 6.32. Schematic structures of phytoplankton cell wall/membrane from the point of view of the optically important components (order-of-magnitude thicknesses of the various layers are in nm; the drawing is not to scale but relations between the thicknesses of the layers for a cell wall type are maintained): *cyto*— cytoplasm, *cm*—cellular membrane, a lipid bilayer (~10 nm thick), *ct*—polysaccharide coating (~0.5 nm thick—coccolithophores, *Henriksen* et al. 2004b; <~10 nm thick—diatoms; *Gebeshuber* et al. 2003). The thin coating is found only in diatoms and coccolithophores. The scale plates (theca) of dinoflagellates: 20 nm (*Kwok* and Wong 2003) to 800 nm (*Klut* et al. 1989) thick, made typically of cellulose, are located between the cellular membranes. The outer membrane is the original cell cytoplasm membrane of the daughter cell. The shell (frustule) of a diatom (~0.1 μm thick) is made of silica. It consists of two fitting halves, much like a pillbox. In the coccolithophore cell wall case, the scale layer is made of individual interlocking circular coccolith twin-disks (calcite), ~2 μm diameter, with complex structures and the overall thickness of ~0.1 μm. The layer of coccoliths, which may contain several sub-layers and be commensurately thicker than 100 nm, is separated from the cell well by a layer of seawater whose thickness range from approximately the thickness of the shell (*E. huxleyi*) to the radius of the cytoplasm-filled cell membrane (*Oolithus fragilis*, e.g., *Henriksen* et al. 2003).

the order of 0.1 to 0.5 μm) contributed measurably to the autocorrelation function of dynamic light scattering during a period of intense growth of the cells (the exponential phase of the culture). Studies on mammal cells (e.g., *Mourant* et al. 2000) indicate that scattering of light by the internal structures of the cell (nucleus, mitochondria, etc.) accounts for over 55% of light scattering by the cell at the large angles (>40°). The nucleus itself contributes ~40%. Mourant et al. postulate that the remaining portion of the large-angle scattering is accounted for by other organelles of the cell, such as mitochondria.

Given the scarcity of information on the optical properties of the phytoplankton cells, it is quite useful to note that structures of eukaryotic cells (such as those of phytoplankton) across a wide range of biological genera and their chemical compositions change little, as if nature has settled on the mass production of the first cell design that worked (e.g., *Ruppert* et al. 2004). The similarities between the various cell designs can probably be explained by the "engineering" design limitations on the eukaryotic cell structure. First, the minimum cell size is limited by the size and number of key molecules required to perform the cell functions, and by their properties (e.g., *Moore* 1999, Boal 1999). An example of such a limitation is the safe folding radius of the DNA.

Second, once the cell size is to increase above a certain limit below which the transfer of nutrients to the inner components of the cells can be assured by diffusion alone, specialized structures have to be used in order to overcome the diffusion-related limitation. These structures are essentially cells inside the cell (most contain cell-like membranes); hence similar minimum size limits apply to each of them and the cell size that contains such structures must be substantially larger than that of prokaryotic (bacterial) cells. As usual, there are exceptions: giant prokaryotes (bacteria) do exist (e.g., sediment-dwelling cylindrical *Thiomargarita namibiensis*, diameter of $\sim$750 μm, *Schultz* et al. 1999).

Similarity between the eukaryotic cell design, most notably the gross composition ($\sim$70% water and $\sim$30% other components, mostly proteins) thus justifies an extrapolation to phytoplankton of the properties measured for cells typical of multicellular genera, such as mammals. Key parameters of cell components are listed in Table 6.15

Optical properties of phytoplankton cells as whole entities are significantly affected by the phytoplankton composition as we discussed it in section 6.3.5. As far as optics is concerned, the cell composition can be described in the first approximation by stating intracellular concentrations of two major components of the cells: carbon and chlorophyll. In fact, *Stramski* (1999) proposed to derive the carbon content of a phytoplankton cell from the real part of its refractive index (see section 6.3.5).

It has been established quite some time ago that the cellular carbon content, m_C, is closely related to the cell volume, V (*Mullin* et al. 1966):

$$m_C = aV^b \tag{6.90}$$

The values of the coefficients are listed in Table 6.16

It can be seen that the rate of the cellular carbon content increase with cell volume ranges from being approximately proportional to the cell cross-section (or surface area) which changes as D^2 (*Mullin* et al. 1966) to being nearly proportional to the cell volume (*Montagnes* et al. 1994).

The differences between these two relationship are summarized in Table 6.17 As noted by *Montagnes* et al. (1994), among others, differences between various $m_C = f(V)$ approximations reported in the literature result from many factors. Even if those factors related to differences in experimental procedures (for example, shrinking of cells fixed with some preservatives) are ruled out, there still remain the effects of the physiological state and life history of the cells. In addition to the

Table 6.15. Key optical parameters of eukaryotic cell components.

Cell component	Characteristic size [μm]	Refractive index, n'
Cell wall/membrane[m]	~0.01[a] to ~0.06[a]	1.37[d], 1.45[l], 1.46[k], 1.66[h]
Cellulose	–	1.55[i] to 1.57[i]
Calcite	–	1.601[i]
Silica	–	1.40[i] to 1.43[i]
Cytoplasm	~3 to ~100[n]	1.36[b, d, g], 1.37[j], 1.38[b]
Nucleus	~0.5 to ~10	1.38[f], 1.39[g, j], 1.41[f]
Mitochondria	~0.5 to ~1.5	1.38[f] to 1.42[b]
Endoplasmic reticulum (ER)	~0.2 to ~1	1.66[h]
Lysosmes	~0.2 to ~0.5	1.378[e]
Energy storage granule (starch)	~0.5[p]	1.51[i] to 1.54[i]
Chloroplasts	~0.5 to ~5	1.42[c, d]

[a]Thickness.
[b]Rat liver cells (*Dunn* 1997).
[c]Spinach leaf chloroplasts (*Bryant* et al. 1969).
[d]*Chlorella* (*Charney* and Brackett 1961), relative to water values: cell wall $n' = 1.022$, cytoplasm $n' = 1.015$, chloroplasts $n' = 1.056$ to 1.060.
[e]Rat liver cells (*Pryor* et al. 2000).
[f]*Drezek* et al. (1999).
[g]*Drezek* et al. (2003) from immersion refractometry for cervical cells.
[h]*Hoelzl* et al. (1966); the high refractive index of ER is due to the fact that ER contains an extensive folded lipid bilayer membrane.
[i]*Aas* (1996).
[j]Chinese hamster ovary cells (*Brunsting* and Mullaney 1974), immersion refractometry.
[k]*Maier* et al. (1994), quoted from references therein.
[l]*Braun* and Fromherz (1997), quoted from a reference therein.
[m]The cell membrane is referred to exclusive of the skeletal material: cellulose (cell wall component in dinoflagellates), calcite (coccolithophores), and silica (opal; in diatoms).
[n]Order of magnitude of the characteristic size of single cells. Colonial species, some encased in cytoplasm-like gel (like the *Volvox*, colony sphere diameter ~1 mm) can reach sizes of several millimeters.
[p]For example, *Ral* et al. 2004.
Parameter n' is the real part of the refractive index.

cursory evidence discussed in section 5.8.4.4, the effect of the cell physiology has been recently quantified by *Davidson* et al. (2002). These latter authors conclude that although the published $m_C = f(V)$ relationships predict the cell carbon content reasonably well (with the biovolume calculation in mind) when plentiful nutrients were available, these relationships are not good predictors of the intracellular carbon content under nutrient depletion conditions.

Such a wide range of estimates of m_C make the evaluation of m_C from cell volume quite inaccurate. In fact, this inaccuracy was one of the rationales for

Table 6.16. Coefficients of the power-law relationship $m_C = aV^b$ between the cell carbon content, m_C [pg] and cell volume, V [μm^3].

Reference	***a* [pg]**	***b***
Mullin et al. 1966[a]	0.513	0.76
Verity et al. 1992	0.433	0.863
Montagnes et al. 1994[b]	0.109	0.991

[a]Fourteen species of marine phytoplankton (38 organisms) with the diameters in a range of 3.4 to 216 μm. The error of the exponent, b, is 0.15 at 95% confidence level.

[b]Thirty small (2 to 60 μm) marine phytoplankton species. $r^2 = 0.937$ for $\log m_C = f(\log V)$.

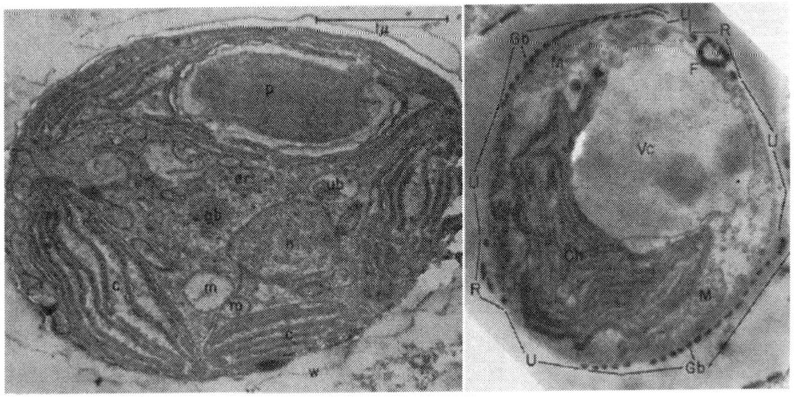

Figure 6.33. Structures of phytoplankton cells in TEM images of cell cross sections. *Left*—soft-bodied *Chlorella pyrenoidosa*: *c*—chloroplast, *er*—endoplasmic reticulum, *gb*—Golgi bodies, *m*—mitochondrium, *p*— pyrenoid, *ub*—unidentified cytoplasmic bodies, *w*— low-density cell wall (reproduced from *Mercer* et al. 1962, Fig. 6, by permission). *Right*—pennate diatom *Cylindrotheca fusiformis*, cell diameter ~7 μm, length (perpendicular to the paper plane) ~30 μm: *Ch*—chloroplast, *Fb*—fibulae, *Gb*—girdle bands (where the two halves of the diatom shell are connected), *M*—mitochondrium, *R*—raphe, *U*—unsilicified part of the cell wall, *Vc*—vacuole (reproduced from *Reimann* et al. 1965, Fig. 2, by permission).

proposing to determine the intracellular carbon concentration from the refractive index of the cell (*Stramski* 1999). However, combining formally (6.90), in the *Montagnes* et al., (1994) version, with (6.62), i.e., by obtaining an equation of type (6.87) and taking account of the units used in these equations, the real part of the refractive index (relative to water) of phytoplankton evaluates to ~1.055, which is not an unrealistic value.

Table 6.17. Differences between the cell carbon contents predicted for phytoplankton cells by two representative approximations.

Cell diameter [μm]	Cell volume [μm^3]	Cell carbon content [pg]		
		Mullin et al. (1966)	*Verity* et al. (1992)[a]	*Montagnes* et al. (1994)
3	14.14	3.84	12.5	1.50
60	113 097.	3 554.	7455.	11 102.

[a]These values have been obtained accounting for the fact that the results of *Verity* et al. (1992) have been obtained for preserved cells, with volume shrunk by some 30%.

The second important component of the phytoplankton cell is chlorophyll. Concentration of intracellular chlorophyll decreases sharply with increasing cell size, approximately as a power function of the cell diameter. Based on data compiled by *Malone* (1980), the intracellular chlorophyll concentration C_{Chl} [$\mathrm{kg\,m^{-3}}$] can be approximately expressed ($r^2 = 0.64$, $n = 28$) as:

$$C_{\mathrm{Chl}} = aD^b \tag{6.91}$$

where the coefficients are listed in Table 6.18.

Table 6.18. Coefficients of the power-law relationship $C_{\mathrm{Chl}} = aD^b$ between the intracellular chlorophyll concentration, C_{Chl} [$\mathrm{kg\ m^{-3}}$], and spherical equivalent cell diameter, D[μm].

Reference	a [$\mathrm{kg\ m^{-3}}$]	b	r^2	n_{obs}
Malone (1980)[a]	79.95	−1.12	0.64	28
Hitchcock (1982); diatoms[b]	9.311	−0.834	0.99	11
Hitchcock (1982); dinoflagellates[b]	27.75	−0.249	0.88	8
Moal et al. (1987)[b]	10.67	−0.630	–	–
Montagnes et al. (1994)[b,c]	4.527	−0.570	0.91	30

[a]Based on data of *Paasche* (1960), *Eppley* and Sloan (1966), *Eppley* et al. (1969), and *Taguchi* (1976). When compared to other results, both coefficients are driven up by the high c_{Chl} data of Paasche.

[b]For 30 small (2 to 60 μm) marine phytoplankton species. As quoted by *Montages* et al. (1994).

[c]Converted to Chl*a* concentration from the Chl*a* mass per cell.

The r^2 refers to a linearized relationship: $\log C_{\mathrm{Chl}} = \log a + b \log D$ (or $b' \log V$ for the data of *Montagnes* et al. 1994).

More recently, *Finkel* et al. (2004) showed that the size scaling of the intracellular carbon concentration that maximizes the rate of photosynthesis at moderate irradiances must decrease as an inverse of the cell size:

$$C_{Chl} \propto D^{-1} \tag{6.92}$$

Such a dependence is indeed consistent with the experimental results listed in Table 6.18 as well as those analyzed by *Finkel* et al. (2004) *Taguchi* (1976) and *Fujiki* and Taguchi (2002). These latter results yield a logarithmic slope of 1.01 ± 0.15.

We should finally acknowledge in a little more detail, the above-mentioned effect of irradiance. This effect has been shown by *Finkel* et al. as well as by other authors (e.g., *Fujiki* and Taguchi 2002) to result in an exponential-like decrease of the intracellular chlorophyll concentration with the increasing irradiance, from a maximum at a low irradiances. Overall, the intracellular pigment concentration can vary as much as five to nine fold at a time scale of hours to days (*Finkel* et al. 2004).

The differences between the intracellular chlorophyll concentrations predicted by regressions referred to in Table 6.18 are listed in Table 6.19.

Equation (6.91) can be formally combined with equation (6.67) to yield an equation of type (6.68) expressing the dependence of the imaginary part of the refractive index of the cell on the cell size.

6.4.3.4. Minerals

Mineral particles can be found in all natural waters albeit understandably their relative concentrations are the greater the closer a water body is to a source of the particles, for example a sea shore. In open waters, the minerals are entrained into a water body by currents and by aerosols (e.g., *Prospero* 1999). *Richardson*

Table 6.19. Differences between the intracellular chlorophyll concentration predicted for phytoplankton cells by the regressions referred to in Table 6.18.

Cell diameter [μm]	**Cell volume [μm³]**	**Intracellular chlorophyll concentration [kg m⁻³]**				
		***Malone* et al. (1966)**	***Hitchcock* (1982)**		***Moal* et al. (1987)**	***Montagnes* et al. (1994)**
			Diatoms	**Dino-flagellates**		
3	14.14	23.4	3.44	4.66	4.27	14.8
60	113097.	0.815	1.63	0.706	0.351	2.69

The cell volume should be regarded as an independent variable, as the cell diameter is an equivalent spherical diameter.

(1987), who examined the types of suspended particles in the North Atlantic, based on SEM analyses, found that mineral grains constitute ~16% of the particles, although together with coccoliths (~48%) they would dominate the number concentration of the particulate matter in that study. Coccoliths (calcite), albeit products of biomineralization, can certainly be classified as minerals (please refer to section 6.4.3.3 for detail). Most of the calcite load in the open ocean waters is likely provided by detached coccoliths (*Bishop* et al. 1977).

In inland waters, minerals may constitute as much as 40% of the particulate load (e.g., *Whiles* and Dodds 2002) to 70% (e.g., *Breitenbach* et al. 1999). Rivers are a major source of minerals in seawater: the rate of fluvial transport of minerals to the oceans is at least an order of magnitude greater than that of the aeolian transport (e.g., *Ratmeyer* et al. 1999). However, mineral particles transported by rivers are efficiently deposited on the continental shelf and slope. Thus, the aeolian supply of minerals prevails in the open sea.

A long-term study by *Ratmeyer* et al. (1999) of the deep waters of the Eastern Atlantic off North Africa indicates that the median particle size of mineral grains is systematically in the region of 10 to 20 μm, however particles with sizes as large as 60 μm have been routinely found. It is interesting to note that size; and possibly also chemical composition of mineral particles in suspension may be poorly reflected in the makeup of the top layer of the sediment (*Ratmeyer* et al. 1999). These authors point out that such a divergence is due to the fact that the sediment makeup in the minimum sampleable thickness of a sediment layer corresponds to a much larger time period than that characteristic of the suspension residence time.

The mineralogical composition of the mineral fraction of suspended particulate matter is generally complex and may be indicative of the source of the particles, as pointed—for example—by *Blanco* et al. (2003). *Breitenbach* et al. (1999) reported the mineralogical composition of minerals in an estuary (Odra River, Baltic Sea) to vary from 30% illite, 38% kaolinite, 14% chlorite, and 17% smectite at the river mouth to 19% illite, 30% kaolinite, 48% chlorite, and 2% smectite close to the estuary outlet.

Asian deserts and arid regions generate dust storms, documented for over 2000 years (*Gao* and Anderson 2001) which are a significant source of dust for the Pacific. *Stramski* et al. (2004b) cite a general source (*Claquin* et al. 1999) indicating that erodible Asian soils have complex compositions dependent on the particle size. *Depetris* (1996), who reviewed the fluvial input to the open ocean, points out at the dominance of illite, followed by smectite, and kaolinite in the suspended mineral matter. We should note here the confusing mineralogical terminology: classification of a mineral type depends on a set of characteristics that can vary. For example, smectite is referred to as a mineral, but also as a mineral group, containing, for example, montmorillonite.

Analysis of the elemental composition of Saharan dust (e.g., *Blanco* et al. 2003), a major source of minerals in the Atlantic and Mediterranean Sea, indicates that roughly 30% of the dust particles contain aluminum with approximately the same percentage containing silicon, clearly indicating the prevalence of aluminosilicates (feldspar, clays), i.e., generally the products of soil erosion (e.g., *Jambers* et al. 1995). A similar prevalence of alumino silicates is also typical of Asian lithogenic aerosol (e.g., *Clarke* et al. 2004). *Lambert* et al. (1981) report aluminosilicates as one of the major fraction of suspended particulate matter in the open ocean.

Particles created by naturally occurring (forest fires) and anthropogenic biomass burning and other combustion sources also contribute to the total load of mineral particles in natural waters.

Single-mineral grains can usually be identified by their sharp edges, identifying cleavage planes of the crystals of which such grains are generally composed (e.g., *Okada* et al. 2001, *Volten* et al. 2001, see also Figure 6.34). Most single mineral grains are coated with organic substance (e.g., *Jambers* and van Grieken 1996, *Hunter* 1991, *Pierce* and Siegel 1979), which may smooth the grain shape somewhat. Given that the refractive index of the coating may be quite similar to that of many minerals, the coating may potentially affect scattering properties, especially for the large particles (e.g., *Kahnert* et al. 2002a). *Pierce* and Siegel (1979), who examined estuarine and oceanic waters by using SEM, found mineral grains in the small-size end of the size spectrum (0.4 to 10 μm, occasionally to 20 μm).

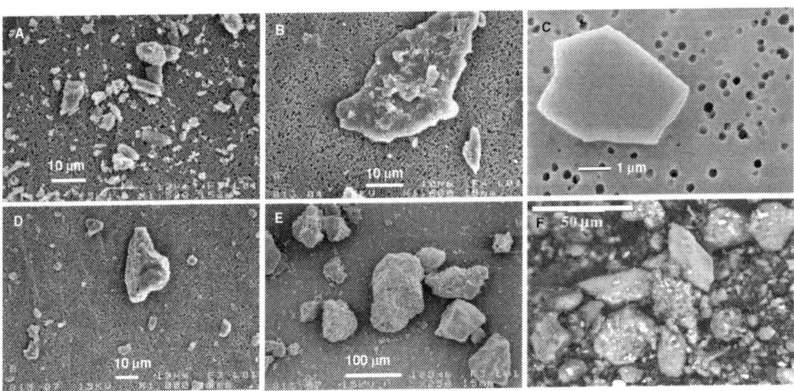

Figure 6.34. Typical shapes and structures of mineral particles (SEM photos): **A**—feldspar (aerosol), **B**—quartz (aerosol), **C**—unidentified mineral grain from a water sample off Nova Scotia (photo: M. Jonasz), **D**—red clay, **E**—Saharan dust particles (aerosol), **F**—Saharan dust collected in rainfall over Italy (*Blanco* et al. 2003, Fig. 5a, reprinted by permission). The shapes of fluvially transported lithogenic particles are similar to those in the aerosol, a major source of mineral particles in the open ocean. Images: A, B, D, and E are reprinted by permission from www.astro.uva.nl/scatter/.

Wells and Goldberg (1992) also found many particles in the 0.1 to 0.5 μm size range to be mineral (composed almost exclusively of iron). Some of these particles had crystalline structure. Aluminosilicates were found in the 0.2 to 1 μm size range. The significance of finding minerals in the sub-micron size range can be properly perceived, when one realizes that the concentrations of particles smaller than 0.1 μm reach $10^8\,cm^{-3}$ in seawater.

Mineral grains of kaolinite and montmorillonite have disk-like shapes (e.g., *Jonasz* 1987b—kaolinite, hexagonal plates). These disks can range in diameter from a fraction of 1 μm (e.g., *Lyubovtseva* and Plakhina 1976) to several micrometers (e.g., *Jonasz* 1987b). Kaolinite is a common clay, frequently present in the riverine suspensions (e.g., *Contado* et al. 1997, Po river suspensions, Italy). The kaolinite plates usually form stacks, whose thickness can range from less than 0.002 μm to several micrometers. Clay particles swell considerably on extended exposure to water. For example, montmorillonite platelets can increase their thickness ~10-fold (*Lyubovtseva* and Plakhina 1976). Other minerals may form needle-shaped particles. For example, palygorskite (*Lyubovtseva* and Plakhina 1976) may exist as needle-shaped particles with diameters as small as 0.03 to 0.01 μm, for a 30–100 length-to-diameter ratio.

Lambert et al. (1981), who examined air-dried particles in GEOSEC samples obtained in deep and bottom waters of various parts of the world ocean by using a scanning electron microscope equipped with an X-ray analysis accessory, identified several types of mineral particles:

- *aluminosilicate particles*: finer than about 10 μm, have spherical to highly irregular forms and considerable thicknesses
- *quartz grains*: rare, fine particles
- *opaline fragments*
- *aggregates*: virtually all particles greater than 10 μm, highly irregular shapes, diverse compositions ranging from organic to mineral.

This last category, aggregates (see section 6.4.3.5), is blurring the distinction between mineral and other particles. Aggregates can contain essentially all types of particles and form microenvironments in natural waters.

Many mineral particles that are products of combustion are either nearly perfect homogeneous spheres (coal burning products) or porous spheres, spherical shells—generally with textured surfaces (oil burning). Black carbon soot particles are fractal aggregates of very small (nanometer-sized) carbon spheres. Mineral particles that are products of combustion are frequently spherical or are aggregates of spheres, such as soot. Shapes of typical mineral particles, both naturally occurring and anthropogenic, are shown in Figure 6.34 and Figure 6.35.

The shape parameter, s [equation (6.77)] of some mineral particles was found to vary widely: 0.05 (hornblende) to 19.1 (mica) (*Aas* 1984). However, the range of s for natural samples is likely to be limited. *Blanco* et al. (2003), who examined

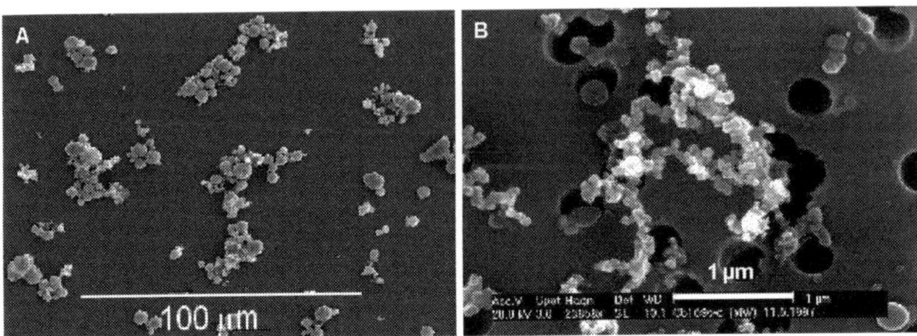

Figure 6.35. Typical shapes of mineral particles resulting from combustion processes: **A**—fly ash aggregates (http://www.astro.uva.nl/scatter/, reprinted by permission), **B**—black carbon soot aggregates (photo: courtesy of S. Weinbruch., Technical Univ. Darmstadt, Germany, note a large difference in the photo scale as compared to that of panel A). Single particles produced by combustion processes frequently assume spherical shapes (e.g., *Sciare* et al. 2003, Fig. 6a) or the shape of a thin porous spherical shell (*Umbria* et al. 2004, Fig. 11).

Saharan desert dust collected in rainfall samples at the south of Italy, found that the shape factor, defined as $1/F$, where F is defined by equation (6.76), has a Gaussian-like probability distribution with the mean factor value of 1.25 and a full-width-at-half-maximum of about 0.5. The maximum recorded shape factor $1/F$ was 2.5. The shape factor $1/F = 1$ for a sphere and is close to unity for rounded particle shapes. *Okada* et al. (2001) who studied the shape of mineral aerosol particles with the help of a scanning electron microscope found that the aspect ratio (the ratio b/a of the longest section to a perpendicular one) is distributed (for $b/a < 5$) according to a probability density function of:

$$p(b/a) = \frac{2}{\dfrac{1}{f_1(b/a)} + \dfrac{1}{f_2(b/a)}} \tag{6.93}$$

where

$$f_1(b/a) = 0.0061e^{5.76\frac{b}{a}} \tag{6.94}$$

$$f_2(b/a) = 755e^{-3.713\frac{b}{a}} \tag{6.95}$$

Minerals have high refractive indices as compared with other aquatic particles. This applies certainly to the real and may apply to the imaginary parts. Given the unspecified orientation of particles with respect to the incident light in most cases, the average refractive index of anisotropic minerals (crystals) is of interest here. According to *Aas* (1981) for uniaxial crystals (e.g., calcite), characterized by two values of the refractive index, the orientation-averaged refractive index is expressed as follows:

$$n' = \frac{2n'_o + n'_e}{3} \tag{6.96}$$

where n'_o is the "ordinary" refractive index (frequently referred to as ω), measured for the polarization of light perpendicular to a principal plane of the crystal and n'_e is the "extraordinary" refractive index (ε), measured for the polarization of light parallel to the principal plane (e.g., *Hecht* 1987).

For biaxial crystals (aragonite), characterized by three values of the refractive index, the orientation-averaged refractive index can be expressed (*Aas* 1981) as a simple arithmetic average of these three values:

$$n' = \frac{n'_1 + n'_2 + n'_3}{3} \tag{6.97}$$

The refractive indices of common minerals are listed in Table 6.20. Refractive index of combustion products are listed in Table A.7. Please also see section 6.3.6 for a brief overview. As pointed out by *Volten* et al. (2001), the real part of the refractive indices of major minerals vary negligibly over the visible spectral range, perhaps with the exception of magnetite. As is evident from Table 6.20, the real part of the refractive index for most minerals is between 1.4 and 1.8 in the visible.

The imaginary part of the refractive index of minerals is generally small ($n'' \sim 10^{-2}$ to 10^{-5}) in the visible as pointed out by *Volten* et al. (2001). However, iron-containing minerals can have n'' as high as 0.4 throughout the visible spectral range (e.g., *Gillespie* and Lindberg 1992). Thus, the spectral dependence of n'' can be highly influenced by the presence of iron and also chromium (*Volten* et al. 2001). See also section 6.3.6 for additional information about the n'' of mineral particles. One should bear in mind that variations of Im(n) within a single mineral species can be as high as one order of magnitude, according to the evidence cited by *Volten* et al. (2001).

6.4.3.5. Aggregates

It has been known for a long time (e.g., see historic references in *Alldredge* and Silver 1988, as well as *Pierce* and Siegel 1979) that many aquatic particles are aggregates (also referred to as flocs and marine or lacustrine snow). Indeed, analyses of carefully sampled particles in high-turbidity waters (e.g., *Droppo* et al. 1996) and measurements of the size distribution of particles *in situ* and of the corresponding disaggregated samples *in vitro* (e.g., *Syvitski* et al. 1995, *Kranck* and Milligan 1992) provide significant evidence that most aquatic particles are aggregates.

Bishop et al. (1977) found in an open ocean study that approximately 93% of the particulate carbonate in the diameter range of >53 μm, mostly coccoliths,

Table 6.20. Refractive indices ($n = n' - in''$, relative to air) in the visible and densities of common minerals.

Mineral	n'	n''	**Density [kg m^{-3}]**	**Mineral**	n'	n''	**Density [kg m^{-3}]**
Albite	1.53		2.69	Hornblende	1.65	3.2	
Anhydrite	1.59		2.96	Illite	1.57		2.85
Apatite	1.64		3.15	Kaolinite	1.56		2.6
Aragonite	1.63		2.93	Limonite	2.05		3.8
Augite	1.71		3.4	Magnesite	1.65		3
Calcite	1.60		2.71	Microcline	1.52		2.56
Calcite[a]	1.61			Mont morillonite	1.51		2.04
Chromite	2.11		5.09	Muscovite	1.59		2.85
Corundum	1.75		4.05	Oligoclase	1.54		2.69
Dolomite	1.64		2.94	Opal	1.43		2.1
Feldspar[c]				Orthoclase	1.52		2.56
Fluorite	1.43		3.18	Prochlorite (Ripidolite)	1.63		2.8
Gibbsite	1.57		2.4	Quartz	1.55		2.63
Gypsum	1.52		2.32	Siderite	1.77		3.96
Halite	1.54		2.16	Talc	1.57		2.82
Hematite[b]	2.32		0.36	Zircon	1.94		4.65

[a] After *Gordon* and Du (2001).

[b] After *Quirantes*-Sierra and Mora (1995).

[c] Feldspars are compounds of aluminosiliates and potassium, or sodium, or calcium. Feldspars are common components of the Earth crust and significant components of granite. Weathering of feldspars generates clays, i.e., minerals such as kaolinite, illite, and montmorillonite.

Unless otherwise indicated, the data are reported as shown in *Woźniak* and Stramski (2004). For uniaxial and biaxial crystals, the refractive index is calculated according to (6.96) and (6.97), respectively. Note that a simple average is used for uniaxial crystals in *Woźniak* and Stramski (2004). Minerals representative of marine suspensions include quartz, calcite, illite, kaolinite, and montmorillonite (*Babin* and Stramski 2004).

was in the form of aggregate particles. Spherical foraminifera, with 1 μm thick calcite walls contributed less than 7% of the total calcite: the whole organisms contributed less than 6% and the cell fragments contributed less than 1%. *Lambert* et al. (1981) also identifies aggregates as a significant type of suspended marine particles.

Aquatic aggregates range in size from <0.1 μm (e.g., *Kim* et al. 1995, *Wells* and Goldberg 1992) to several millimeters (e.g., *Heffler* et al. 1991, *Alldredge* and Silver 1988). Aggregates of aquatic particles have frequently been classified as fractal objects (e.g., *Li* et al. 1998—marine particles, *de Boer* and Stone 1999—freshwater aggregates) and assigned fractal dimensions (see section 5.5.2.2).

The shapes of aggregates vary widely and can range from delicate networks of "primary" particles, through "spheres" and comet-like shapes, the latter also referred to as "stringers" (e.g., *Syvitski* et al. 1983), to strands and plates (*Alldredge* and Silver 1988). The average shape factor, F [eq. (6.76)], for freshwater aggregates assumes moderate values ($\sim$0.6, *Droppo* et al. 1996; $F = 1$ for a sphere, $F = 0$ for a line). The shape factor has been found to decrease as a power of the aggregate size (e.g., *Droppo* et al. 2002):

$$F = aD^{-b} \tag{6.98}$$

where b is on the order of 0.27.

Aquatic aggregates blur the distinction between organic and inorganic matter as well as between live and dead matter. They may contain both mineral grains and aggregates, inanimate particulate organic matter, as well as both live and dead bacteria, phyto- and zooplankton. Even though purely mineral aggregates do occur, especially in the sub-micron size range, large aggregates collect "everything" as they settle (or raise; *Azetsu*-Scott and Passow 2004). There is a notable exception here. The makeup of aggregates formed by phytoplankton (mostly diatoms) at the end of their blooms is much more uniform, at least in the early stages of the life histories of these aggregates.

It is important to note that the shape and composition of aggregates are not constant in time as they pick up or lose components during their "life" history (*Droppo* 2001). In retrospect, aggregates are not exceptions in this respect. Bacteria also undergo shape/size changes during their life cycle, and so do phytoplankton cells. In the case of diatoms, this even leads to predictable shape/size changes, because a daughter cell uses one-half of the shell (theca). However, in contrast, the shape/size changes of the aggregates can only be quantified on an average basis.

First, the shape and size of the aggregate is limited by the turbulence-induced shear stress (e.g., *Fugate* and Friedrichs 2003). This is frequently quantified by invoking the Kolmogorov turbulence length scale, L_{T}, which places an upper limit on the size of most aggregates. In the open ocean, $L_{\mathrm{T}} \sim 1$ to 6 mm (e.g., *Gabaldón* Casasayas 2001), while in coastal environments L_{T} may be as low as 0.1 to 1 mm (*Fugate* and Friedrichs 2003). The Kolmogorov scale is the distance limit for the energy dissipation by turbulence. At distances smaller than L_T, fluid motion energy is dissipated as heat, i.e., molecular diffusion dominates. At distances greater than L_T, fluid motion energy is dissipated by turbulence, hence turbulent diffusion dominates. The effect of turbulence on the aggregate size has been observed in laboratory experiment (e.g., *Manning* and Dyer 1999).

Second, aggregates undergo differential settling shear because the random arrangement of masses within the aggregate typically leads to some parts of the aggregate being pulled by other parts. Indeed, there is evidence that marine aggregates change size and shape with the depth in the ocean from the large, loose aggregates at the surface to smaller, more compact aggregates deep in the water column (e.g., *Honjo* 1996).

Third, aggregates are food for some aquatic organisms (*Dilling* and Alldredge 2000), whose feeding activity may cut off some components or just sever a connection between the various aggregate parts.

Most aggregates, in terms of the number concentration (i.e., sub-micron particles), are likely organic. *Chin* et al. (1998) showed that dissolved organic matter in the ocean, which is estimated to contain most of the ocean's organic carbon, can spontaneously form colloidal aggregates. In a study of *Grout* et al. (2001) in the western Mediterranean, $\sim$20% of the total organic carbon was found to be in a particle size range >1000 Da ($\sim$100 nm). Grout et al. identified three morphotypes for these particles:

1. *globular* (10 to 200 nm)
2. *open-structure aggregates* (OSA, >200 nm) of more than 50 "round entities," i.e., more densely packed networks similar to aggregates in this size range identified by *Kim* et al. (1995) in riverine water and by *Leppard* et al. (1997) in the eastern Mediterranean Sea. Fractal dimensions of the OSA aggregates were about 1.44, i.e., lower than that resulting from a diffusion-limited aggregation ($\sim$1.8, for example, *Logan* and Wilkinson 1990).
3. *crystalline spherulites* ($\sim$500 nm diameter) composed of compactly packed needles with diameters on the order of 30 nm and length of 150 nm (Figure 6.36).

Leppard et al. (1997) also document the typical morphologies of sub-micron colloids isolated from the western Mediterranean Sea waters. These morphologies

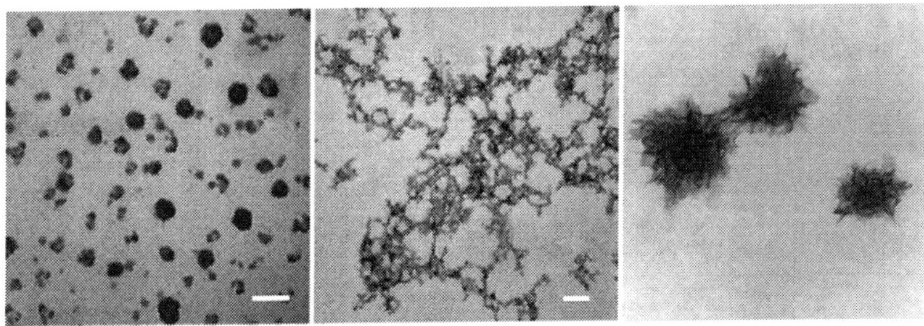

Figure 6.36. Typical shapes of sub-micron marine aggregates examined by *Grout* et al. (2001, Fig. 3b, 4b, and 5b) in the western Mediterranean waters (TEM images reproduced from *Grout* et al. 2001, by permission). *Left*—globules and small aggregates of several globules, *middle*—aggregates of more than 50 "rounded entities," *right*—crystalline spherules (individual needles are about 30 nm in diameter and 150 nm long), "diameter" of about 500 nm. Scale bars—200 nm each. The aggregates are made of organic matter.

range from colloidal aggregates of thin nano-sized strings to dense aggregates of small "rounded entities."

At the other end of the size scale (several millimeters), aggregates (marine snow) can be frequently observed in coastal waters. The primary particles in such aggregates are themselves size distributed. Typical primary particles with sizes of typically less than 5 μm were found in a coastal area study (*Syvitski* et al. 1995), although particles as large as 60 μm were occasionally found. A flake of marine snow can contain hundreds of primary particles of different origin and composition. A sample selection of shapes and structures of these amorphous aggregates is shown in Figure 6.37.

The glue holding together the large aggregates is widely believed to be mostly polysaccharide-containing exudates (gels) of phytoplankton cells. This has been convincingly shown by *Hamm* (2002), who experimentally investigated diatom-induced intense aggregation of a suspension of kaolinite. The polysaccharide gels form particles (TEP, transparent exopolymer particles, e.g., *Passow*

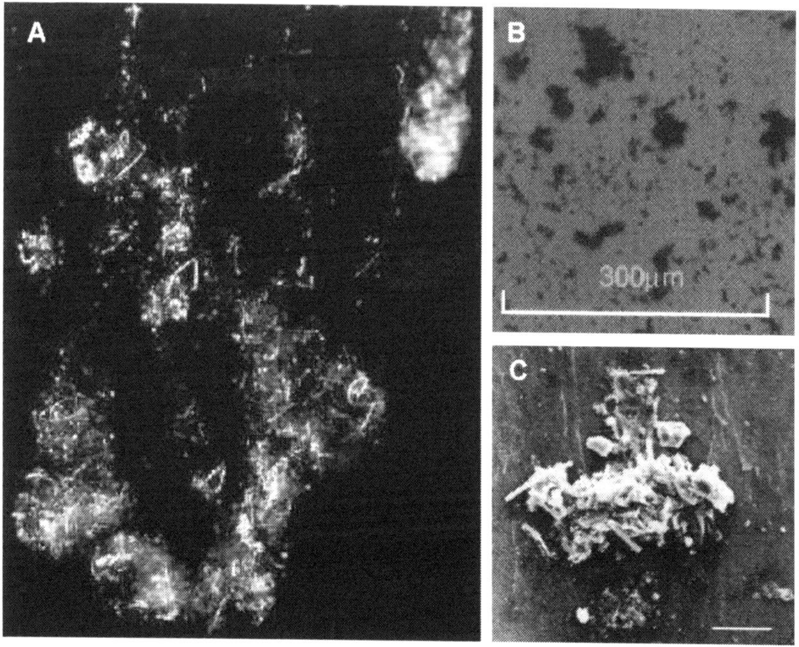

Figure 6.37. Typical shapes the large aquatic aggregates. **A**—marine snow aggregate, image width ~1 cm (photo: courtesy of A. Alldredge, Univ. of California at Santa Barbara, USA), **B**—aggregates from Sixteen Mile Creek, Ontario, Canada (*Droppo* et al. 1996, Fig. 2b, reprinted by permission), **C**—a marine aggregate from the western Atlantic waters, bar length 20 μm (SEM photo: M. Jonasz).

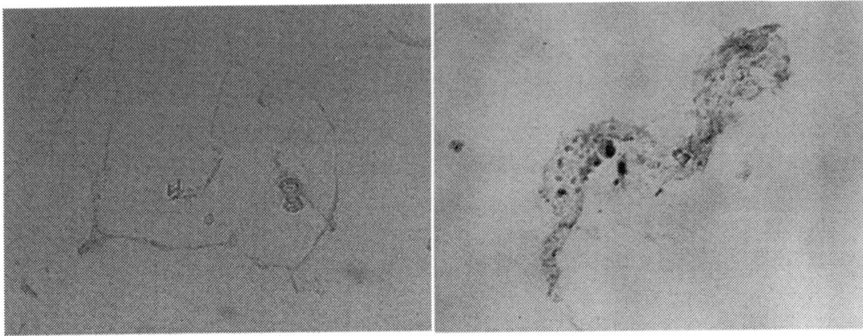

Figure 6.38. Optical microscopy images of transparent exopolymer particles (TEP) in seawater, made visible by staining with alcian blue. *Left*—TEP strands, the very small dark points outlining the strands are mostly bacteria attached to TEP (image width = ~0.35 mm, *Alldredge* et al. 1993, Fig. 1h, reprinted by permission). *Right*—sa sheet-like TEP with attached solid particles (scaled image width = ~0.52 mm, Passow 2002, Fig. 1b, reprinted by permission).

2002, *Passow* et al. 2001) that also exist as independent (precursors?) of the large aggregates (Figure 6.38). *Faganeli* et al. (1995) found that glucose (~ 60% by weight), followed by mannose (~ 14%) and fructose (~ 12%), are the major monosaccharide components of these organic gels in the large, northern Adriatic aggregates. Stability of the organic gels is probably enhanced by the presence of calcite, quartz, and clay particles (*Kovac* et. al. 2002).

Aside from their solid content, the large aggregates are mostly water as is clearly demonstrated by their low "excess" density (the difference between the effective aggregate density and that of water), or effective density (the aggregate mass divided by its envelope volume). In fact, the effective density of these aggregates decreases as the power of the aggregate size (e.g., *Dyer* and Manning 1999, *Droppo* et al. 1997, *Ten Brinke* 1994). The exponent of the power law is about unity (*Fennessy* et al. 1994, tidal estuary aggregates). Both in freshwater and in marine environments, aggregates with "diameters" ~ 1 mm attain an effective density of about 10 to 100 kg m^{-3}.

Finally, a special group of aggregates are fecal pellets of zooplankton. These compact, typically spheroidally shaped, high-density aggregates may constitute a major component of the particle populations in the particle size range of several μm and above in certain areas. For example, a large-volume (> 400 l) *in situ* filtration study conducted in the waters of the equatorial Atlantic (*Bishop* et al. 1977) revealed that the major components of the particle population in the size range > 20 μm are fecal pellets and fecal matter, foraminifera, and radiolarians.

The average length-to-diameter ratio of usually fecal pellets was estimated at about 3. The length-to-width ratio of fecal matter particles was estimated as 1 to 5, with 87% of particles having a ratio of 2. *Bishop* et al. (1978) found that fecal matter particles were usually flake-shaped. The flake thickness, h [μm], was found to be linearly related to the disk diameter, d [μm], as follows:

$$h = 0.052d + 45 \tag{6.99}$$

6.5. Problems

1. *Effect of the dispersion of the refractive index on the attenuation efficiency*
By using the ADA (section 6.3.2.1), examine the effect of the dispersion of refractive index of protein on the attenuation efficiency of a protein sphere in water as a function of the sphere diameter at wavelengths of 400, 550, and 700 nm.

2. *Optical cross-section of a particle vs. optical coefficient of a suspension*
Prove that an optical coefficient, c_y, where y stands for absorption, scattering, or attenuation, of a suspension of N identical particles per unit volume can be expressed as follows:

$$c_y = N\sigma_y \tag{6.100}$$

where y is either an absorption, scattering, or attenuation and σ is the optical cross-section of the particle.

3. *The imaginary part of the refractive index of the particle material*
The imaginary part, n'', of the refractive index of a substance can be determined from the following equation (see section 1.5.1):

$$a = \frac{4\pi}{\lambda} n''$$

where a is the absorption coefficient and λ is the wavelength of light in the material. Thus, once we have determined the absorption coefficient, a, of the suspension (say of identical homogeneous particles), why not use this equation to determine the imaginary part, n'', of the refractive index of the material of particles in the suspension, instead of a more complicated way of calculating n'' from the absorption efficiency, Q_a, of the particles, which we suggested in section 6.3.2.1?

4. *Complex refractive index of phytoplankton from cell volume*
Real part: By combining formally (6.90), in the *Montagnes* et al. (1994) version, with (6.62) obtain an equation of type (6.87). By taking into account the units used in these equations, calculate the real part of the refractive index of phytoplankton relative to that of water in the visible (~1.34).
How significantly does it change with cell volume ("diameter")?
What is the range ("error") of that estimate as a function of cell volume?

Imaginary part: Equation (6.91) can be formally combined with equation (6.67) to yield an equation of type (6.68) expressing the dependence of the imaginary part of the refractive index of the cell on the cell size. Derive such an equation and express quantitatively the imaginary part of the refractive index of phytoplankton as a function of cell volume.

How significantly does it change with cell volume ("diameter")?

What is the range ("error") of that estimate as a function of cell volume?

Appendix

Table A.1. Sources of nephelometer design information.

Reference	Nephelometer design summary	Angular range [°]	Light source	Detector
Agrawal and Trowbridge (2002)	Small-angle (laser diffractometer design: LISST-100), many particle, *in situ*	0.097–19.5	10 mW diode laser 670 nm	Custom PD array
Altendorf et al. (1996)	Single particle, micromachined, for flow-cytometric applications	5–10, 50–60	1.2 mW diode laser 640 nm	PD
Atkins and Poole (1952)	Polar, many particles	20–145	36 W tungsten lamp	PMT
Aughey and Baum (1954)	High-resolution (0.02°) polar, many particles	0.05–140	Hg lamp	PMT
Bantle et al. (1982)	Polar, rotating detector; simultaneous angular and dynamic light scattering, many particles	10–150	Kr or HeNe laser	PMT
Barthel et al. (1998)	Optimized for discrimination of fibers, single-particle, aerosol	3 scattering and 8 azimuthal angles	200 mW Nd:YAG laser	PMT
Bartholdi et al. (1980)	Polar, angular distribution of scattered light mapped onto a circular photodiode array by using an ellipsoidal mirror, single particle	30 angle ranges from 4.4–7.3 to 172.7–175.6	200 mW Ar laser	Custom Si PD array

(*Continued*)

Table A.1. Continued

Reference	Nephelometer design summary	Angular range [°]	Light source	Detector
Bauer and Ivanoff (1965), *Bauer* and Morel (1967)	Small-angle, Fourier transform via a plano-convex lens, many particles, *in situ*	1.5–14		Photo-graphic film
Baum and Billmeyer (1961)	Polar, rotating detector, many particles	20–165	Hg lamp	PMT
Beardsley (1968)	Polar, rotating detector, polarized light (scattering matrix) many particles	20–130		
Brice et al. (1950)	Polar, rotating detector, polarized light, many particles	0, 45, 90, 135	Hg lamp	PMT
Brogioli et al. (2002)	Small-angle, based on the measurement of the power spectrum of a speckle field generated by the sample, many particles	<~24	10 mW HeNe	CCD
Brunsting and Mullaney (1972)	Polar, photographic film positioned along a semi-circle at the center of which is a small cuvette illuminated with a laser beam, many particles	3–80 and 100–177	5 mW HeNe laser	Photo-graphic film
Burns et al. (1976)	Polar, rotating detector, many particles	1–140	5-ns, 1 kW peak, pulsed dye laser	PMT
Beyer (1987)	Polar, fixed-angle, single particle	3 angles: 10, 20, and 40	8 mW He–Cd 442 nm	PMT
Chernyshev et al. (1995), *Maltsev* (2000)	Polar, light scattered by a moving particle into a conical solid angle annulus is measured as a function of time, i,e., the scattering angle, single particle	10–120	12 mW HeNe laser	PMT
Conklin et al. (1998)	Small-angle, many particle, *in vitro*			CCD

Table A.1. Continued

Reference	Nephelometer design summary	Angular range [°]	Light source	Detector
Davidson et al. (1971)	Polar, fixed-angle, polarized light, single particle	90	Ar laser	PMT
Diehl et al. (1979)	Polar, fixed-detectors located around a sample cuvette, polarized light, single particle	45, 90, 135, 225, 170, 315	Laser	Si PD
Drezek et al. (1999)	Polar, fiber bundle rotating about sample cell collects scattered light, longitudinal adjustment of the fiber end position sets the acceptance angle, many particles	1.5–160	5 mW HeNe	PD
Dueweke et al. (1997)	A small-angle Fourier transform (FT) system and a conventional polar nephelometer with fixed observation angles, many particles	0.1(0.1)5, 45, 90, 135, 175(0.1) 179.9	5mW diode-pumped Nd:YAG laser	CMOS linear array
Duntley (1963)	Small-angle, annular stop at focal plane of a lens, many particles, *in situ*	0.5	Tungsten lamp	Photo-tube
Eisert (1979)	Forward scattering, fiber optics rings coupled to PMTs collect light scattered into concentric angular zones, single particle		Laser	PMT
Estes et al. (1997)	Forward scattering with a lens transforming scattered light directions into positions in its focal plane, many particles	< 10	1 mW HeNe 543 nm	CCD
Fry et al. (1992b)	Scattering function at 0°, stationary incident beam fanned out by a photorefractive crystal, light scattered by particles (in Brownian motion) passes undeviated, many particles	0	10 mW Ar laser 514 nm	PMT
Gayet et al. (1997)	Polar, paraboloidal mirror redirects scattered light onto optical fibers connected to photodiodes, many particles, *in situ*, aerosol	33 angles from 4.3 to 169	1.2 W diode laser, 785 nm	PD

(*Continued*)

Table A.1. Continued

Reference	Nephelometer design summary	Angular range [°]	Light source	Detector
Grams et al. (1974)	Polar, rotating detector (rotation plane offset from the scattering plane by the use of folding mirrors), many particles, *in situ*, aerosol	10–170	1 W Ar laser	PMT
Grams et al. (1975)	Polar, rotating detector, many particles, *in situ*, aerosol	15–165	5 mW HeNe laser	PMT
Grasso et al. (1997, 1995)	Polar, scattered light is reflected by an ellipsoidal mirror onto a linear photodiode array via a system of a biconvex and a plano-convex cylindrical lens, many particles	10–170	10 mW HeNe	Si PD array
Gucker et al. (1973)	Polar, angle selected with a system of an ellipsoidal mirror and two flat mirrors (fixed), and two annular apertures (fixed and rotating), logarithmic amplifier limits the range of the scattering signal, many particles, *in situ*, aerosol	7–173 and 187–352	50 mW HeNe laser	PMT
Haller et al. (1983)	Polar, fixed-angle, thermostated-sample, static and dynamic light scattering, detectors receive scattered light through optical fibers installed at fixed scattering angles, many particles	16 angles: 2.6 to 163	HeNe laser	PMT
Hansen and Evans (1980)	Polar, rotating detector, polarized light scattering, many particles, aerosol	2–178)	Pulsed Ar laser	PMT
Hespel et al. (2001)	Mechanical scanning of scattered light reflected by a set of mirrors tangential to an ellipsoid, linearly polarized light, many particles	20 angles: −40 to 45	HeNe laser	PMT

Table A.1. Continued

Reference	Nephelometer design summary	Angular range [°]	Light source	Detector
Hirst et al. (1994), see also *Kaye* and Hirst (1995)	Polar, mapping two-dimensional light scattering at an intensified CCD detector via an ellipsoidal mirror, single particle, aerosol	30–140, 82% of the two-dimensional angular pattern	9 mW diode laser	Intensified CCD camera, PMT
Hodara (1973)	Small-angle, MTF-based measurement system, many particles	0.1–100		
Hodkinson (1963)	Polar, rotating detector, rectangular cell also rotates to halve the scattering angle complement, many particles	0–70	Incandescent lamp	PMT
Holland and Gagne (1970), *Holland* and Draper (1967)	Polar, rotating detector, full scattering matrix measurements, many particles, aerosol	20–165	Halogen lamp	PMT
Holoubek et al. (1999), *Konák* et al. (2001)	Forward scattering, a conical lens transforms scattered light directions into positions at the incident beam axis, many particles	0.5–45	HeNe laser	CCD
Holve and Self (1979a, 1979b), *Holve* and Davis (1985)	Small-angle scattering for particle counting, algorithmic compensation for off-axis particle trajectory, single particle, aerosol, *in situ*	10–20	2 mW HeNe laser	PMT
Huang et al. (1994)	Small-angle, MTF-based system, many particles, *in situ*	0.1–5	White light	CCD array
Hull et al. (2004)	Fixed-angle design with a polarization modulator, many particles	13 unspecified angles	532 nm laser	PMTs

(*Continued*)

Table A.1. Continued

Reference	Nephelometer design summary	Angular range [°]	Light source	Detector
Hunt and Huffman (1973)	Polar, rotating photodetector, a set of four photoelastic polarization modulators allow simultaneous measurement of several components of the scattering matrix, many particles	5–168	Hg lamp	PMT
Jerlov (1961)	Polar, rotating detector, many particles, *in situ*	10–165	Tungsten lamp	PMT
Jerrard and Sellen (1962)	Polar, polarized light, null modulation method, many particles	unspecified	250 W Hg lamp	PMT
Jonasz (1991b)	Polar, rotating periscope, stationary detector, polarized light, many particles	5–175	5 mW HeNe laser	Si PD
Jones et al. (1994), see also *Leong* et al. (1995)	Polar, fixed angles, single particle, aerosol	15 angles (23–129)	Laser diode 840 nm	CCD
Jones and Savaloni (1989)	Forward scattering, polarized light for fiber discrimination, single particle, aerosol		Laser	PMT
Katz et al. (1984)	Polar, rotating detector, circular polarization, many particles	3–27 and 155–170	2 W argon laser	PMT
Kaplan et al. (2000)	Polar, rotating detector, scattering matrix, phase-modulated encoding, many particles	20–140	Ar laser 514 nm	PMT
Kaye and Havlik (1973)	Forward scattering, annular apertures select light scattered at a particular angle, many particles	<1	5 mW HeNe laser	PMT

Table A.1. Continued

Reference	Nephelometer design summary	Angular range [°]	Light source	Detector
Kirmaci and Ward (1979)	Polar, fixed detector, angle selected with a rotating flat mirror and fixed conically arranged mirrors; single particle, aerosol	9–176°, 96 fixed mirrors, each with a FOV of 1° × 1°	4 ns 20 kW peak, pulsed N laser	PMT
Klotz (1978)	Polar, rotating detector, polarization modulator to simultaneously measure various polarized components of the scattering function, many particles	5–170	Kr laser	PMT
Kozlyianinov (1957)	Polar, rotating lamp, fixed detector, many particles	0.5–144.5	Incan-descent lamp	Visual photo-meter
Kuik et al. (1991)	Polar, four rotating detectors allow reduction of the measurement time, polarization modulation measurements of the scattering matrix, many particles, aerosol jet	5–175	HeNe laser	PMT
Kullenberg (1968)	A combination of a small angle and a polar nephelometer, many particles, *in situ*. *Small-angle*: a set of conical mirrors coaxial with the beam axis defines scattering angle range *Polar*: rotating detector	1, 2.5, 3.5, 25–135	0.4 mW HeNe laser	PMT
Kullenberg (1984)	Polar, rotating detector, many particles, *in situ*	8–160	Incan-descent lamp	

(*Continued*)

Table A.1. Continued

Reference	Nephelometer design summary	Angular range [°]	Light source	Detector
Lant et al. (1997)	Polar, parabolic mirror is used for both beam delivery and collection of scattered light into GRIN terminated fiber rotating about the mirror axis, many particles	10–170	Diode-pumped Nd:YAG laser 532 nm	Photon counting APD
Lee and Lewis (2003)	Polar, rotating periscope, fixed detector (the scattered light collection area is offset in respect of the rotation axis, a movable aperture masks the incident beam in the offset direction as the scattering angle decreases—this allows measurements at angles smaller than 1°), many particles, *in situ*	0.6–177	Incan-descent lamp	PMT
Loken et al. (1976)	Polar, particle motion in front of a detector aperture scans the scattering pattern, single particle	~ 1°–67	200 mW Ar laser	PMT
Ludlow and Kaye (1979), *Ulanowski* et al. (2002)	Rotating disk scans fixed light guides within	174 fixed angles	Ar laser 514 nm	PMT
Maffione and Dana (1997)	Hydroscat-6, single fixed-angle, 6 wavelength backscattering, many particles, *in situ*	140 ± 4	LED	PD
Maffione and Honey (1992)	Back-scattering angle, many particles, *in situ*	179–180	Nd:YAG laser 532 nm	CCD
Mankovsky (1971)	Polar, rotating periscope, fixed detector, many particles	1.5–152.5		PMT

Table A.1. Continued

Reference	Nephelometer design summary	Angular range [°]	Light source	Detector
Mankovsky and Haltrin (2002a, 2002b)	Polar, many particle, *in situ*	2–162.5	520 nm, Incandescent lamp	PMT
McCluney (1974)	Small-angle, Fourier transform via a lens, angle selected by a system of one scanning mask and one fixed mask at the focal plane of the FT lens, masks permit the separation of signals from the resulting angular channels, by imparting a different frequency on each signal, many particles	3 angular channels	HeNe laser	PMT
McCluney (1975)	Polar, two fixed angles, single particle	2, 90	Laser	
McIntyre and Doderer (1959)	Polar, rotating detector, many particles	10–150	Hg lamp	PMT
Mertens and Phillips (1972)	Polar, rotating detector also used to measure the beam spread function which was inverted to yield a small-angle scattering function, many particles, *in situ*	1.7–170	Ar laser	Radiometer
Misconi et al. (1990)	Polar, rotating detector, single particle levitated with a laser beam, aerosol	16–170	7.5 W Ar laser	PMT
Moser (1974)	Polar, small cuvette is illuminated at an angle selected by a system of two fixed annular conical mirrors and a rapidly rotating flat mirror, many particles, aerosol	20–160 and 200–340	HeNe laser	PMT

(*Continued*)

Table A.1. Continued

Reference	Nephelometer design summary	Angular range [°]	Light source	Detector
Mujat and Dogariu (2001)	Polar, full scattering matrix simultaneously via polarization modulation with liquid crystal modulators and analysis with a photoelastic modulator, many particles			
Mullaney et al. (1969)	Fixed-angle flow-cytometric system, hydrodynamic focusing of particles with a 250 μm aperture, single particle	0.5–2 and 90	HeNe laser	Si PD
Mullaney et al. (1976)	Small-angle, Fourier transform, single particle	0.28–21.1	5 mW HeNe	Photo-conductive ring array
Pan et al. (2003)	Polar, mapping two-dimensional light scattering at an intensified CCD detector via an ellipsoidal mirror, single particle, aerosol	48–164, 63% of the 2D angular pattern	Nd: YAG diode pumped laser	Inten-sified CCD camera, PMT
Petzold (1972)	Polar, rotating periscope, fixed detector, many particles, *in situ*	10–170	100 W halogen lamp	PMT
Petzold (1972)	Small-angle, an annular stop at the focal plane of a lens, many particles, *in situ*	0.085, 0.17, 0.34	Halogen lamp	PMT
Pinnick et al. (1976)	Polar, rotating detector, polarized light, single particle, aerosol	10–170	2 W Ar laser	PMT
Prandke (1980)	Polar, rotating periscope, fixed detector, many particles	5–175	0.4 mW HeNe laser	PMT
Pritchard and Elliot (1960)	Polar, rotating detector, many particles, *in situ*, aerosol	2–160	Incandescent lamp	PMT

Table A.1. Continued

Reference	Nephelometer design summary	Angular range [°]	Light source	Detector
Privoznik et al. (1978)	Polar, rotating periscope (fiber-optics), fixed detector, sample in constant temperature bath, many particles	7.2–172.8	Xe lamp	PMT
Rozenberg et al. (1970)	Polar, rotating detector, measurement of the entire scattering matrix, full measurement cycle of about 4 h	25–145		
Salzman et al. (1975)	Small-angle, scattered light is detected with a photodiode array of concentric rings, single particle	0–21.1	5 mW HeNe laser	Si PD array
Sasaki et al. (1960, 1968)	Polar, rotating detector, conical sample cell aids in minimizing the stray light reflected by the cell walls, many particles	30–150	Hg lamp	PMT
Schreurs (1996)	Polar, rotating detector, sample cell in a bath of index matching fluid, many particles	20–160	HeNe laser	PMT
Sherman et al. (1968)	Polar, rotating detector, many particles	15–60	30 mW HeNe laser, 75 W Xe lamp	PMT
Sonntag and Russel (1986)	Small-angle, angle selected by two annular apertures, *in vitro*, many particles	2, 3, 4, 5, 6, 8, 10	5 mW HeNe laser	PMT
Spinrad et al. (1978)	Small-angle, Fourier transform, angle selected by scanning a pinhole at a focal plane of a lens, many particles	0.2–0.7	HeNe laser	PMT

(*Continued*)

Table A.1. Continued

Reference	Nephelometer design summary	Angular range [°]	Light source	Detector
Steen and Lindmo (1979), *Steen* (1986)	A microscope-based fixed-angle range is selected by annular apertures, single particle, for flow-cytometric applications	>2, >15	100 W Hg lamp	PMT
Tanis (1992)	Small-angle MTF-based nephelometer uses sine bar pattern generated with a hologram, many particles	0.005–0.5	100 mW HeNe laser	PD array
Thompson et al. (1980)	Polar, rotating detector, a set of four photoelastic polarization modulators for the simultaneous measurement of all components of the scattering matrix, many particles	10–155	Ar laser	PMT
Tontrup and Gruy (2000)	Multiple backscattering by a suspension, delivery and collection of light by a silica fiber bundle with numerical aperture of 0.22.	0–12.7	HeNe laser	Si PD
Tycko et al. (1985)	Two fixed angles, optimized for flow cytometry of red blood cells, single particle	3–5.5 and 5.5–9	Laser diode 840 nm	PD
Tyler and Austin (1964)	Polar, deep-sea design, rotating detector, many particles, *in situ*	20–160	150 W halogen lamp	PMT
Tyler and Richardson (1958)	Polar, rotating detector, Waldram stop makes the scattering volume independent of the scattering angle, many particles, *in situ*	20–170		PMT

Table A.1. Continued

Reference	Nephelometer design summary	Angular range [°]	Light source	Detector
Wang and Hencken (1986)	Small-angle, Fourier transform light-scattering particle sizing photometer, two-color coaxial laser beams discriminate against off-axis particles by triggering a measurement of light scattering from a wide outer beam when a particle passes through a narrow, inner beam, single particle	1.2-	Ar (measurement) and HeNe (particle position sensing) lasers	PMT
Watson et al. (2004)	Polar, mechanically scanned, variable density neutral filter equalizes the scattered light power received by the detector, single particle in an optical trap	0.5–160	40 mW 658 nm laser	PMT
Weiss (1981)	Polar, rotating detector, linear polarization elements of the scattering matrix, electrostatically levitated single particle, aerosol	20–160	15 mW HeNe laser	PMT
Witkowski (1986)	Polar, rotating detector, simultaneous elastic and dynamic light scattering, many particles	20–160	HeNe laser	PMT
Witkowski et al. (1998)	Polar, rotating detector, simultaneous full scattering matrix measurements via frequency modulation of polarizer–analyzer system, many particles	10–170	HeNe	PMT

(*Continued*)

Table A.1. Continued

Reference	Nephelometer design summary	Angular range [°]	Light source	Detector
Wyatt et al. (1988)	Polar, fixed-angles, two-dimensional angular space, each probe connected by an optical fiber to a PMT, measurements triggered when sufficient signal is received from selected probe, single-particle, aerosol jet	16 angles in two-dimensional angular space	10 mW HeCd or 10 mW HeNe laser	PMT
Wyatt and Jackson (1989)	Polar, fixed-angle multi-probe (GRIN lens, fiber optics) system, two-dimensional angular pattern space, each probe connected to a PMT, measurements triggered when sufficient signal is received, single particle	15 angles in two-dimensional angular space, 23–128	5 mW HeNe laser	PMT

See also *Kerker* (1997) for a historical review of light scattering instrumentation for aerosol studies, *Knollenberg* and Veal (1992) for nephelometer designs optimized for particle counting, and *Hodkinson* and Greenfield (1965) for typical designs of lamp-based small-angle nephelometers for aerosol particle counting applications. Abbreviations: APD = avalanche photodiode, PD = photodiode, PMT = photomultiplier, FOV = field of view. The term *rotating* refers to the rotation about an axis perpendicular to the scattering plane. This axis is simultaneously the axis of the sample cell or a sample jet. All references not specifically indicated to refer to aerosol measurements are designed for use in water. All references not specifically marked as *in situ* refer to *in vitro* instruments. Biofouling and its prevention in *in situ* instruments is discussed by *Dolphin* et al. (2001), *Barth* et al. (1997), *McLean* et al. (1997), and *Ridd* and Larcombe (1994).

Table A.2. Sources of experimental data on the volume scattering functions of sea water.

Reference	Study area, number, and mode of measurements	Angular range [°]	Wavelength [nm]	Data format
Atkins and Poole (1952)[d]	English Channel (13)	20–145	460[e]	Graph[a]
Austin (1973)[d]	Pacific off Colombia and Panama (22), Atlantic: Gulf of Mexico, Sargasso Sea (5), *in situ*	10–170	520	Table[a]
Beardsley (1968)[d]	Boston Harbor (1), Charles River, Massachusetts (1), Atlantic off Massachusetts (2), and Cape Cod (1), USA	20–130	546	Graph[a]
Duntley (1963)[d]	Lake Winnipesauke, USA (1), *in situ*	0.5, 20–160	522	Graph[a]
Estes et al. (1997)	Buzzards Bay, MA, USA (1)	0.5–10	534.5	Graph
Gohs et al. (1978)[d]	199 functions by Prandke, Baltic	Varies within 5–175	633[HeNe]	Table[a]
Lee et al. (2003)	Atlantic off New Jersey, USA (60)	0.6–177.3	550 ± 5	Table
Lee and Lewis (2003)	Halifax harbor, Halifax, NS, Canada (1) Atlantic off New Jersey, USA (12)	0.6–177.3	550 ± 5	Graph
Haltrin (1997)	Eight functions by V. Mankovsky: Southern Ocean (2), Indian Ocean (1), Atlantic (2), Lake Baikal (2), Black Sea (1) and 12 functions by various authors	2–162.5	Unknown	Graph,[a] regression coeffi-cients
Huang et al. (1994)	East China Sea (4) *in situ*	0.1–5	White light	Graph[a]

(*Continued*)

Table A.2. Continued

Reference	Study area, number, and mode of measurements	Angular range [°]	Wavelength [nm]	Data format
Jerlov (1961)	North Atlantic off Madeira (1) *in situ*	10–165	633^{HeNe}	Graph[a]
Jonasz (1991b)	North Atlantic off Nova Scotia, Canada (19)	10–165	633^{HeNe}	Table, Graph[a]
Jonasz and Prandke (1986)[d]	Baltic Sea, surface average (one summer, one winter)	5–175	633^{HeNe}	Table, Graph[a]
Kadyshevich et al. (1971)	Black Sea (5 + 1 scattering matrix)	25–145	Within 436–578	Graph[r]
Kadyshevich et al. (1976)[d]	Atlantic, Pacific averages for 0, 10, 100–200, and 300–2000 m + 1 scattering matrix	25–145	546	Graph[r]
Kadyshevich (1977)	Baltic Sea (2 + 1 scattering matrix)	30–140	546	Graph[r]
Kozlyianinov (1957)[d]	East China Sea (1) polarized	0.5–144.5	Unknown	Graph[r]
Kullenberg (1968)[d]	Sargasso Sea (3) *in situ*	0.5–165	633^{HeNe}	Table, graph[a]
Kullenberg (1969)[d]	Baltic Sea (15) *in situ*	Varies within 1–165	440, 525, 655, 488^{Ar} 633^{HeNe}	Table, graph[a]
Kullenberg (1984)[d]	Drake Passage (9), Pacific off Peru (8) upwelling, all *in situ*	10–160	550, 655	Graph[a]
Kullenberg and Olsen (1972)[d]	Mediterranean Sea (16) *in situ*	2.5–150	633^{HeNe}	Table, graph[a]
Mankovsky (1971)	Black Sea (3), North Atlantic (6)	1.5–162.5	520	Graph[r]
Mertens and Phillips (1972)[d]	Atlantic off Bahamas (5) *in situ*, polarized	2.5–172	488^{Ar}	Graph[a]
Morel (1973)	English Channel (2), Mediterranean Sea (13), Indian Ocean (11)	30–150	366, 436, 546	Graph[r]
Morrison (1970)[d]	Long Island Sound (1), Atlantic off New York (Argus Island) (1), USA, all *in situ*	0.2–100	546	Graph[a]

Table A.2. Continued

Reference	Study area, number, and mode of measurements	Angular range [°]	Wavelength [nm]	Data format
Petzold (1972)[d]	Atlantic off Bahamas (3), Pacific off California, USA (2), San Diego harbor, USA (3), all *in situ*	0.08–170	510	Table, graph[a]
Pickard and Giovando (1960)	British Columbia fjord, Canada (two *in vitro*)	20–140	546	Graph[r]
Reuter (1980a)[d]	Baltic Sea, Eckenförde Bay, off Kiel, Germany (one polarized)	7–170	488[Ar]	Graph[a]
Reese and Tucker (1970)[d]	San Diego Bay, USA (8)	10–155	Green light	Graph[a] (in *Morel* 1973)
Sasaki et al. (1960)	Pacific (Japan Trench), four *in vitro*	30–150	531, 652	Graph[r]
Sasaki et al. (1968)	Pacific off Japan (nine *in vitro*)	30–150	463	Graph[r]
Spilhaus (1968)[d]	Atlantic (two *in vitro*)	30–150	546	Graph[a]
Sugihara et al. (1982a)[d]	Pacific off Japan (11 *in vitro*, including 1 polarized), results much different than those of any other researcher	20–150	366, 436, 546, 578	Graph[a]
Tucker (1973)	Pacific off San Diego, USA (66)	10–160	534	
Tyler (1961)[d]	Pacific off California, USA (4), Lake Pend Oreille, USA (2), all *in situ*	20–180	534	Table, Graph[a]
Voss and Fry (1984)	Atlantic (two *in vitro*)	10–160	488[Ar]	Graph[r]
Voss and Fry (1984)	Pacific (two *in vitro* + scat. matrices)	10–160	488[Ar]	Table, graph[r]
Whitlock et al. (1981)[d]	Back River, Virginia, USA (eight *in situ*)	0.37–155	450 (50) 800	Table[a]

[a] Absolute.

[d] Some or all data included in a computer-readable database (*Jonasz* 1996, 1992).

[e] Estimated from data in the reference.

[r] Relative.

Unless specifically indicated, the measurements are preformed *in vitro*. If the light source is a laser, its type is indicated in the superscript of the wavelength.

Table A.3. Sources of experimental data on the volume scattering functions of particle species in sea water and other natural waters.

Reference	Species, number, and mode of measurements	Angular range [°]	Wavelength [nm]	Data format
	Viruses			
Balch et al. (2000)	*E. coli* viruses: MS2 (25–30 nm) and T4 (100 nm) (1 β) Y1 (50–80 nm) and C2 (110 nm) marine bacteriophages (1 β each)	30–150	546	Graph[a]
	Bacteria and spores			
Bickel and Strattford (1981)	*Bacillus aureus*, spores and cells (M_{34})	10–170	Unknown	Graph[r]
Bickel and Strattford (1981)	*Bacillus intersporum*, spores (M_{34})	10–170	Unknown	Graph[r]
Bickel and Strattford (1981)	*Bacillus megaterium*, spores (M_{34})	10–170	Unknown	Graph[r]
Kopelevich et al. (1987)	*Bacillus mycoides* (1 β)	0.4–6.5	550	Graph[a]
Bickel and Strattford (1981)	*Bacillus stearothermophillus* (M_{34})	10–170	Unknown	Graph[r]
Bickel and Strattford (1981)	*Bacillus subtilis*, spores (M_{34})	10–170	Unknown	Graph[r]
Bickel et al. (1976)	*Bacillus subtilis* [2 β, m_{12}, m_{34}, $(m_{13}+m_{33})/(m_{11}+m_{31})$]	2–140	442[HeCd]	Graph[r]
Kirmaci and Ward (1979)	*Cladosporium* spores ($1\beta_{\parallel}$ $\beta_{\perp}$)	9–176	627[dye]	Graph[r]
Morel and Bricaud (1986)	Cyanobacteria (1 β)	25–155	546	Graph[r]
Kopelevich et al. (1987)	*Deleya vinustus* (1 β)	0.4–6.5	550	Graph[a]

Table A.3. Continued

Reference	Species, number, and mode of measurements	Angular range [°]	Wavelength [nm]	Data format
	Bacteria and spores			
Van De Merwe et al. (1989)	*Escherichia coli* ($3\,m_{34}$)	20–140	442^{HeCd}, 633^{HeNe}	Graph[r]
Cross and Latimer (1972)	*Escherichia coli* ($1\ \beta$)	5–80	404, 542	Graph[r]
Lyubovtseva and Plakhina (1976)	*Escherichia coli*	25–145	540	Graph[r]
Kopelevich et al. (1987)	*Flavobacterium* sp. ($1\ \beta$)	0.4–6.5	550	Graph[a]
Kopelevich et al. (1987)	*Micrococcus* sp. ($1\ \beta$)	0.4–6.5	550	Graph[a]
Kopelevich et al. (1987)	*Moraxella* sp. ($1\ \beta$)	0.4–6.5	550	Graph[a]
Balch et al. (2002)	*Photobacterium* sp., also infected by a virus	10–170	514^{Ar}	Graph[a]
Balch et al. (2002)	*Pseudomonas parfectomarinas*, also infected by a virus	10–170	514^{Ar}	Graph[a]
Kopelevich et al. (1987)	*Pseudomonas* sp. ($1\ \beta$)	0.4–6.5	550	Graph[a]
Bickel and Strattford (1981)	*Salmonella typhi* (M_{34})	10–170	Unknown	Graph[r]
Balch et al. (2002)	*Synachococcus 1331*, also infected by a virus	10–170	514^{Ar}	Graph[a]
Fry and Voss (1985)	*Synecococcus* sp. (one full scattering matrix)	2–160	488^{Ar}	Graph[r]
Balch et al. (2002)	*Vibrio harveyi*, also infected by a virus	10–170	514^{Ar}	Graph[a]
Kopelevich et al. (1987)	*Vibrio* sp ($1\ \beta$)	0.4–6.5	550	Graph[a]

(*Continued*)

Table A.3. Continued

Reference	Species, number, and mode of measurements	Angular range [°]	Wavelength [nm]	Data format
	Phytoplankton			
Vaillancourt et al. (2004)	*Alexandrium tamarense* (1 β, $m^{-1}sr^{-1}cell^{-1}$),	34–144	$514^{Ar,lin.pol.}$	Graph[a]
Schreurs (1996)	*Anabaena flos aquae* (1 M_{11}, m_{12})	20–160	633^{HeNe}	Graph
Volten et al. (1998)	*Anabaena flos aquae* (1 M_{11}, M_{12}/M_{11})	20–160	633^{HeNe}	Graph, table
Król (1998)	*Anacystis nidulans* (1 β)	0.25–160	633^{HeNe}	Graph[a]
Price et al. (1978)	*Anacystis nidulans* (single cells) 100β	0.14–19.5	633^{HeNe}	Graph[r] cluster bands
Król (1998)	*Anacystis variabilis* (1 β)	0.25–160	633^{HeNe}	Graph[a]
Sugihara et al. (1982a)	*Ankistrodesmus falcatus* (1 β)	20–150	376,436, 546, 578	Graph
Schreurs (1996)	*Asterionella formosa* (1 M_{11}, M_{12}/M_{11})	20–160	633^{HeNe}	Graph
Volten et al. (1998)	*Asterionella formosa* (1 M_{11}, M_{12}/M_{11})	20–160	633^{HeNe}	Graph, table
Sugihara et al. (1982b)	*Chaetoceros socialis* (1 $\beta_{\|\|}$, $\beta_{\perp}$)	20–150	376, 436, 546, 578	Graph
Price et al. (1978)	*Chlamydomonas reinhardii* single cells (100 β)		633^{HeNe}	Graph,[r] cluster bands
Quinby-Hunt et al. (1989)	*Chlorella* (1 $M_{11} \times \sin\theta$, m_{12}, m_{22}, m_{34})	30–165	442^{HeCd}	Graph
Witkowski et al. (1993)	*Chlorella* (1 M_{11}, M_{12}, M_{34})	10–155	633^{HeNe}	Graph
Sugihara et al. (1982a)	*Chlorella* (1 β), the scattering function is much steeper in both forward and especially backward directions than those measured by any other researcher	20–150	376, 436, 546, 578	Graph

Table A.3. Continued

Reference	Species, number, and mode of measurements	Angular range [°]	Wavelength [nm]	Data format
	Phytoplankton			
Witkowski et al. (1998)	*Chlorella kesleri* (one full matrix M as scattering cross-sections)	10–170	488, 514, 633	Graph[a]
Privoznik et al. (1978)	*Chlorella pyrenoidosa* (one phase function)	7.2–172.8	633^{HeNe}	Graph
Price et al. (1978)	*Chlorella pyrenoidosa*, single cells (400 β)	0.1–19.5	633^{HeNe}	Graph, cluster bands
Witkowski et al. (1998)	*Chlorella vulgaris* (one full matrix M scaled as scattering cross-sections)	10–170	488^{Ar}, 514^{Ar}, 633^{HeNe}	Graph[a]
Witkowski et al. (1998)	*Chrococcuss minor* (one full matrix M scaled as scattering cross-sections)	10–170	488^{Ar}, 514^{Ar}, 633^{HeNe}	Graph[a]
Król (1998)	*Chrococcuss minor* (1 β)	0.25–160	633^{HeNe}	Graph[a]
Król (1998)	*Coscinodiscus granii* (1 β)	6–160	633^{HeNe}	Graph[a]
Shapiro et al. (1991)	*Crypthecodinium cohnii* (1 M_{11}, m_{14})	40–120	457, 488, 514	Graph, averages
Burns et al. (1976)	*Cyanidium caldarium* (1 β)	~1–135	560^{dye}	Graph[r]
Król (1998)	*Cyclotella menengihianina* (1 β)	6–160	633^{HeNe}	Graph[a]
Balch et al. (1999)	*Emiliania huxleyi* (1 β)	30–135	546	Graph[r]
Volten et al. (1998)	*Emiliania huxleyi* with and without coccolithes (1 M_{11})	20–160	633^{HeNe}	Graph, table
Schreurs (1996)	*Emiliania huxleyi* with and without coccolithes (1 M_{11}, m_{12})	20–160	633^{HeNe}	Graph
Voss et al. (1998)	*Emiliania huxleyi* and coccoliths (1 β)	10–170	440	Graph[r]
Shapiro et al. (1991)	*Gonyaulax polyedra* (1 M_{11}, m_{14})	40–120	457, 488, 514	Graph, averages

(*Continued*)

Table A.3. Continued

Reference	Species, number, and mode of measurements	Angular range [°]	Wavelength [nm]	Data format
	Phytoplankton			
Vaillancourt et al. (2004)	*Gymnodinium simplex* (1 β, $m^{-1}sr^{-1}cell^{-1}$)	34–144	$514^{Ar,\ lin.pol.}$	Graph[a]
Vaillancourt et al. (2004)	*Hemiselmis virescens* (1 β, $m^{-1}sr^{-1}cell^{-1}$)	34–144	$514^{Ar,\ lin.pol.}$	Graph[a]
Vaillancourt et al. (2004)	*Katodinium rotundatum* (1 β, $m^{-1}sr^{-1}cell^{-1}$)	34–144	$514^{Ar,\ lin.pol}$	Graph[a]
Schreurs (1996)	*Melosira granulata* (1 M_{11}, m_{12})	20–160	633^{HeNe}	Graph
Volten et al. (1998)	*Melosira granulata* (1 M_{11}, m_{12})	20–160	633^{HeNe}	Graph, table
Volten et al. (1998)	*Microcystis aeruginosa* with and without gas vacuoles (1 M_{11} Graph, table, m_{12} graph)	20–160	633^{HeNe}	Graph, table
Schreurs (1996)	*Microcystis ridii* with and without gas vacuoles (1 M_{11}, m_{12})	20–160	633^{HeNe}	Graph
Volten et al. (1998)	*Microcystis* sp. (1 M_{11}, m_{12})	20–160	633^{HeNe}	Graph, table
Schreurs (1996)	*Oscillatoria agardhii* (1 M_{11}, m_{12})	20–160	633^{HeNe}	Graph
Volten et al. (1998)	*Oscillatoria agardhii* (1 M_{11}, m_{12})	20–160	633^{HeNe}	Graph, table
Schreurs (1996)	*Oscillatoria amoena* (1 M_{11}, m_{12})	20–160	633^{HeNe}	Graph
Volten et al. (1998)	*Oscillatoria amoena* (1 M_{11}, m_{12})	20–160	633^{HeNe}	Graph, table
Vaillancourt et al. (2004)	*Pelagomonas calceolate* (1 β, $m^{-1}sr^{-1}cell^{-1}$)	34–144	$514^{Ar,\ lin.pol}$	Graph[a]
Schreurs (1996)	*Phaeocystis* (1 M_{11}, m_{12})	20–160	633^{HeNe}	Graph
Volten et al. (1998)	*Phaeocystis* (1 M_{11}, m_{12})	20–160	633^{HeNe}	Graph, table

Table A.3. Continued

Reference	Species, number, and mode of measurements	Angular range [°]	Wavelength [nm]	Data format
	Phytoplankton			
Schreurs (1996)	*Phaeodactylum* (1 M_{11}, m_{12})	20–160	633[HeNe]	Graph
Volten et al. (1998)	*Phaeodactylum* (1 M_{11}, m_{12})	20–160	633[HeNe]	Graph, table
Volten et al. (1998)	*Phrochorothrix hollandica* (1 M_{11})	20–160	633[HeNe]	Table
Schreurs (1996)	*Phrochorothrix hollandica* (1 M_{11}, m_{12})	20–160	633[HeNe]	Graph
Balch et al. (1999)	*Pleurochrysis* sp. (1 β)	30–135	546	Graph[r]
Fry and Voss (1985)	*Porphyridium cruentum* (one full matrix M)	2–160	488[Ar]	Graph[r]
Vaillancourt et al. (2004)	*Prasinococcus capsulatus* (1 β, $m^{-1}sr^{-1}cell^{-1}$)	34–144	514[Ar, lin.pol]	Graph[a]
Shapiro et al. (1991)	*Prorocentrum micans* (1 M_{11}, m_{14})	40–120	457, 488, 514	Graph, averages
Shapiro et al. (1990)	*Prorocentrum micans* (1 M_{11} sin θ, m_{14})	>> 20–165	514[Ar]	Graph
Vaillancourt et al. (2004)	*Pycnococcus provasolii* (1 β, $m^{-1}sr^{-1}cell^{-1}$)	34–144	514[Ar, lin.pol]	Graph[a]
Król (1998)	*Scenedesmus microspina* (1 β)	0.25–160	633[HeNe]	Graph[a]
Schreurs (1996)	*Selenastrum capricornutum* (1 M_{11}, m_{12})	20–160	633[HeNe]	Graph
Volten et al. (1998)	*Selenastrum capricornutum* (1 M_{11}, m_{12})	20–160	633[HeNe]	Graph, table
Sugihara et al. (1982b)	*Skeletonema costatum* (1 $\beta_{\parallel}$, $\beta_{\perp}$	20–150	376, 436, 546, 578	Graph

(*Continued*)

Table A.3. Continued

Reference	Species, number, and mode of measurements	Angular range [°]	Wavelength [nm]	Data format
	Phytoplankton			
Balch et al. (1999)	*Syracosphaera elongata* (1 β)	30–135	546	Graph[r]
Balch et al. (1999)	*Thoracosphaera* sp. (1 β)	30–165	546	Graph[r]
Schreurs (1996)	*Volvox aureus* (1 M_{11}, m_{12})	20–160	633^{HeNe}	Graph
Volten et al. (1998)	*Volvox aureus* (1 M_{11}, m_{12})	20–160	633^{HeNe}	Graph, table
	Minerals and other types of non-living particles			
West et al. (1997)	Aluminum oxide in air (M_{11}, degree of linear polarization, i.e, m_{11})	15–170	470, 652, 937	Graph[r]
Shapiro et al. (1990)	Alumina particles (4 $M_{11}\sin\theta$, m_{14})	>> 20–165	514^{Ar}	Graph
West et al. (1997)	Aluminum silicate in air (M_{11}, degree of linear polarization, i.e. m_{11})	15–170	470, 652, 937	Graph[r]
Hodkinson (1963)	Anthracite (3 β)	0–90	White light, 365, 436, 546	Graph[r]
Hodkinson (1963)	Bituminous coal (3 β)	0–90	White light, 365, 436, 546	Graph[r]
West et al. (1997)	Calcium carbonate in air (M_{11}, degree of linear polarization, i.e, m_{11})	15–170	470, 652, 937	Graph[r]
Hodkinson (1963)	Diamond (3 β)	0–90	White light, 365, 436, 546	Graph[r]
Muñoz et al. (1999)	Feldspar, average diameter 3 μm (M_{11}, m_{12}, m_{22}, m_{33}, m_{34}, m_{44})	5–175	633^{HeNe}	Graph[a]
Volten et al. (1996)	Feldspar aerosol, median diameter 3.8 μ*m* (M_{11}, m_{12})	~10–~170	633^{HeNe}	Graph

Table A.3. Continued

Reference	Species, number, and mode of measurements	Angular range [°]	Wavelength [nm]	Data format
	Minerals and other types of non-living particles			
Volten et al. (2001)	Feldspar aerosol, effective diameter[l] ~1 μm (M_{11}, m_{12}, m_{22}, m_{33}, m_{34})	5–173	441.6[HeCd], 633[HeNe]	Graph
Gibbs (1978)	Glass spheres (1 β)	0, 25–140	White light	Graph[r]
Gibbs (1978)	Mica flakes (1 β)	0, 25–140	White light	Graph[r]
Lyubovtseva and Plakhina (1976)	Montmorillonite (2 β, m_{12}, m_{22}, m_{33})	25–145	540	Graph[r]
Lyubovtseva and Plakhina (1976)	Palygorskite (1 β, m_{33}, m_{12}, m_{22})	25–145	540	Graph[r]
Sugihara et al. (1982a)	Clay, potter's (1 β)	20–150	376, 436, 546, 578	Graph
Muñoz et al. (1999)	Clay, red, average diameter 5.1 μm (M_{11}, m_{12}, m_{22}, m_{33}, m_{34}, m_{44})	5–175	633[HeNe]	Graph[a]
Muñoz et al. (1999)	Quartz, average diameter 9.7 μm (M_{11}, m_{12}, m_{22}, m_{33}, m_{34}, m_{44})	5–175	633[HeNe]	Graph[a]
Gibbs (1978)	Quartz, crushed (1 β)	0, 25–140	White light	Graph[r]
Kuik et al. (1991)	Quartz aerosol, randomly oriented, radius 15 + 0.8 SD μm (M_{11}, m_{12}, m_{22}, m_{33}, m_{44})	5–175	633[HeNe]	Graph[a]
Volten et al. (1996)	Quartz aerosol, median diameter 13.6 μm (M_{11}, m_{12})	~10 – ~170	633[HeNe]	Graph
Volten et al. (2001)	Quartz aerosol, effective diameter[l] ~2.3 μm (M_{11}, m_{12}, m_{22}, m_{33}, m_{34})	5–173	441.6[HeCd], 633[HeNe]	Graph

(*Continued*)

Table A.3. Continued

Reference	Species, number, and mode of measurements	Angular range [°]	Wavelength [nm]	Data format
	Minerals and other types of non-living particles			
Volten et al. (1999)	Rutile (TiO_2), birefringent ellipsoidal particles, average equivalent diameter 221 nm, M_{11}, m_{12}, m_{22}, m_{33}, m_{34}, m_{44}	15–165	633^{HeNe}	Graph
Volten et al. (2001)	Sahara sand aerosol, effective diameter[t] ~8.2 μm (M_{11}, m_{12}, m_{22}, m_{33}, m_{34})	5–173	441.6^{HeCd}, 633^{HeNe}	Graph
Sugihara et al. (1982a)	Sericite (1 β)	20–150	376, 436, 546, 578	Graph
Schreurs (1996)	Westershelde silt 3–5 μm and 5–12 μm (M_{11}, m_{12})	20–160	633^{HeNe}	Graph
Volten et al. (1998)	Westershelde silt 3–5 μm and 5–12 μm (M_{11}, m_{12})	20–160	633^{HeNe}	Graph, table
	Large particles, aggregates, marine snow			
Hou (1997) *Hou* et al. (1997)	Particles larger than 280 μm	30–140	Laser diode 685	Graphs, fits

[a] absolute.
[e] estimated from data in the reference.
[r] relative.
[t] average radius weighed by the circularized cross-section area (πr^2), as defined by *Hansen* and Travis (1974).

See also a WWW database of the scattering matrices of many mineral species (*Volten* et al. 2005).

Some particle species other than those suspended in natural waters are included for comparison. The table is ordered alphabetically according to the species name, not the reference's first author name. The lowercase m denotes the scattering matrix elements divided by M_{11}. The dimension of the scattering function is $m^{-1}sr^{-1}$ unless specifically indicated. If the light source is a laser, its type is indicated in the superscript of the wavelength.

Table A.4. Sources of experimental data on the particle size distributions in natural waters.

Reference	Study area, number, and mode of measurements	Diameter range [μm]	Measurement method	Data format
Atteia and Kozel (1997) and *Atteia* et al. (1998)	Karstic aquifier, Switzerland (six *in vitro*), see also references therein for other data sources on freshwater colloids	0.5–10	GS	Graph
Bader (1970)[d]	Atlantic off the Bahamas (six *in vitro*)	1–20	CC	Graph
Bale and Morris (1991)	Tamar Estuary (seven *in vitro*, seven discrete samples, seven primary particles)	1–100	*In situ* LD	Graph
Bishop et al. (1978)	Southeast and equatorial Atlantic (11 *in vitro*). Particles sampled by an *in situ* large-volume filtration system.	>53 μm	OM	Graph
Bradtke (2004)	Gdansk Bay, Baltic Sea (2000) *in vitro*)	2.5–33	CC	Database
Brun-Cottan (1976)	Western Mediterranean Sea (three *in vitro*)	1.5–15	CC	Graph
Carder et al. (1971)[d]	Eastern equatorial Pacific, Pacific off Galapagos, 0–2000 (59 *in vitro*)	2–10	CC 100 μm	Graph
Cavender-Bares et al. (2001)	Western North Atlantic (42 *in vitro*)	0.2–4.5	OFC (Epics V)	Graph
Chen et al. (1994)	Elbe estuary (six *in situ*)	12–500	*In situ* camera	Graph[r]
Chen et al. (1994)	Elbe estuary (nine *in vitro*)	2–32	CC	Graph
Chung (1982)[d]	Indian Ocean (central basin) (six *in vitro*, stored samples)	0.7–35	Scanning counter, Spectrex	Graph

(Continued)

Table A.4. Continued

Reference	Study area, number, and mode of measurements	Diameter range [μm]	Measurement method	Data format
Courp et al. (1993)	Gironde estuary (three *in situ*)	2.5–512	*In situ* camera	Graph
Courp et al. (1993)	Mediterranean Sea (Pyrenean margin) (one *in situ*)	2.5–512	*In situ* camera	Graph
Eisma et al. (1990)	Schelde River, Elbe estuary (two *in situ*)	3.6–512	*In situ* camera	Graph
Gordon et al. (1972)[d]	Atlantic off the Bahamas, 0–300 m (three *in vitro*)	0.7–10	CC 20, 30 μm	Table
Harris (1977)[d]	Mexican Gulf, 600–3600 m; six *in vitro*	0.02–5	TEM	Table
Hood (1986)[d,u]	Pacific off California, 0–50 m (22 *in vitro*)	2–160	CC 100 400 μm	Table
Hood et al. (1991)[d]	Pacific off California, 5–90 m (13 *in vitro*)	4–80	CC 280 μm	Graph
Jonasz (1978)[d,u]	Atlantic (a transect from the Antarctic waters to the English Channel) (50 *in vitro*)	2–32	CC 100 μm	Table
Gohs et al. (1978)[d]	Baltic Sea, 0–40 m (90 *in vitro*)	2–32	CC 100 μm	Table
Jantschik et al. (1992)	Bay of Biscay, 1–100 m (five *in vitro*)	2–50	GS	Graph
Kahru et al. (1991)[d]	Atlantic (Azores front) (six *in vitro*)	1–105	HIAC 320 μm	Graph
Kranck (1987)	Indus River (four *in vitro*, single grain only	0.5–50	CC	Graph
Kranck and Milligan (1992)	San Francisco Bay (over 50 *in situ*)	100–1000	*In situ* camera	Graph, fit para-meters
Kranck and Milligan (1988)	Bedford Basin (12 *in situ*)	0.6–110	CC 30, 200, 1000 μm	Graph
Kranck and Milligan (1986)[d,u]	North Atlantic, 0–2000 m (46 *in vitro*)	0.6–180	CC 30, 200, 1000 μm	Table
Kranck et al. (1992)	Nith River (Ontario, Canada) (23 *in situ*)	50–800	*In situ* camera	Graph

Table A.4. Continued

Reference	Study area, number, and mode of measurements	Diameter range [μm]	Measurement method	Data format
Kranck et al. (1992)	Coastal North Pacific (Skagitt Bay) (nine *in situ*)	50–1000	*In situ* camera	Graph
Kranck et al. (1992)	Coastal South Atlantic (Amazon shelf) (seven *in situ*)	50–1000	*In situ* camera	Graph
Lerman et al. (1977)	Equatorial North Atlantic (15 *in vitro*)	2–12	CC	Graph, fit parameters
Longhurst et al. (1992)	North Atlantic off Halifax, Nova Scotia (four *in vitro*)	0.4–1	CC + EZ	Graph
McCave (1983)	North Atlantic, Nova Scotian Rise (41 *in vitro*)	1.26–32	CC	Graph[a]
McCave (1985)	North Atlantic, Nova Scotian Rise (10 *in vitro*)	1.59–256	CC	Graph[a]
Richardson (1987)	North Atlantic (Iceland Rise) and Northwestern Atlantic (13 *in vitro*)	1–16	CC 50 μm aperture	Graph[r]
Risović (1993)	Pacific off Rarotonga (data of *Shifrin* et al. 1974)	0.4–25		Table
Sheldon (1970)[d,u]	South Atlantic, 0 m (25 *in vitro*)	0.8–101	CC	Table
Sheldon (1972, 1975)[d,u]	North Atlantic, 0–5 m (23 *in vitro*)	0.9–90	CC	Table
Sheldon (1970)[d,u]	South Pacific, 0 m (53 *in vitro*)	0.6–90	CC	Table
Stramski and Sedlák (1994)	Pacific off California (one *in vitro* on processed sample)	0.5–5	CC	Graph
Spinrad et al. (1989a)	Western North Atlantic (5000 m) (nine *in situ*)	3–200	Settling tube	one table + eight graphs
Sugihara and Tsuda (1979)[d]	Pacific off Hawaii; surface (22 *in vitro*)	2.4–34	CC 100 μm	Table
Syvitski et al. (1995)	Bedford Basin (23 *in situ*)	50–3000	*In situ* camera	Graph

(*Continued*)

Table A.4. Continued

Reference	Study area, number, and mode of measurements	Diameter range [μm]	Measurement method	Data format
Wellershaus et al. (1973)[d]	Atlantic (Gulf of Cadiz); average of one station between 10 and 5000 m (two *in vitro*)	2.8–32	M	Graph
Wellershaus et al. (1973)[d]	Indian Ocean (one *in vitro*)	2.8–32	M	Graph
Wells and Goldberg (1992)	Pacific off San Diego (three *in vitro*)	0.005–0.1	TEM	Graph

[a] Absolute.
[d] Some or all data included into the computer-readable database (*Jonasz* 1992).
[r] Relative.
[u] Unpublished data.

Aperture size (if applicable) is given where available following the method abbreviation. Abbreviations: CC = Coulter counter, EZ = Elzone counter, OFC = optical flow cytometer, GS = Galai scanning counter, LD = laser diffractometer, OM = optical microscope, TEM = transmission electron microscope. Extended comments, identified by the reference name, are appended to the table. Many plytoplankton-related works refer to the "size distribution" of plankton. However, the size resolution of the accompanying data is rather low, typically reporting cell concentrations in some of the major size classes of the phytoplankton (pico-, nano-, microplankton).

Additional comments follow, identified by the reference name:

Bishop et al. (1978): The particle size is defined as the largest dimension of the particle. The particle size distributions of fecal pellets, fecal matter, and foraminifera fragments, which were the three major components of the suspended matter in this size range (see also *Bishop* et al. 1977), were approximated using a multi-segment power law.

Bradtke (2004) analyzed particle size distributions by using the power law, a sum of log-normal functions (with an algorithm developed by *Jonasz* and Fournier 1996), and principal components method. She found that the two latter methods each provide best approximation to the size distributions measured. She also identified a number of typical components and associated some with phytoplankton species.

Chung (1982): Particle size distributions were measured with a laser particle counter (Protron, Spectrex Corporation) in samples of seawater of the Indian Ocean (GEOSECS station 453 in the central basin). The samples were held in storage for several months before the particle size analysis was performed. Sub-samples were taken without mixing the samples. Thus, only *permanently suspended* particles were analyzed, and the results do not represent the whole population of particles. The author used the power law approximation to the data.

Gordon and Brown (1972) found the power law to be a reasonable approximation to the particle size distributions measured using a Coulter counter in the Sargasso Sea waters.

Gordon et al. (1972): Measured particle size distributions in a particle diameter range of 0.65 to 10 μm in the Atlantic waters off the Andros Island (the Bahamas). A Coulter counter with a 20 and a 50 μm apertures was used. The samples have consistently contained very large number of particles with diameters smaller than 1 μm (see also *Koike* et al. 1990—the first name, Isao, of Koike is mistakenly used as the second name in that publication, and in *Longhurst* et al. 1992). Gordon et al. assert that these *fine particles are not products of breakup of the large particles* in the Coulter counter. The particle size distributions were well approximated using a three-segment power law. The slope, m, of the size distribution of particles with diameters between 0.65 and 1 μm was on the order of 7. This slope was the highest of the three slopes for each size distribution analyzed.

Jonasz (1983a) analyzed over 160 particle size distributions measured in the Baltic waters in different seasons during the years 1975 to 1978. The measurements were performed using a Coulter counter model ZBI equipped with a 100 μm aperture. The particle size distributions were found to be well approximated by using a sum of a two-segment power function and of a sum of Gaussian functions. The "break-point" between the two segments was determined by finding the minimum of the approximation error as a function of the distribution of the data points between the two segments.

Jonasz (1983b) found that a two-segment power law approximates well particle size distributions of Antarctic fjord (Ezcurra Inlet, King George Island, South Shetland archipelago). A Coulter counter model ZBI with a 100 μm aperture was used. In that fjord, freshwater from melting ice fields washed mineral particles from the shores into the ocean. The particles were carried out of the inlet mostly in the surface layer, few tens of meters deep.

Kitchen and Zaneveld (1990) measured 194 particle size distributions in the Northeastern Pacific Gyre at depths ranging from 1 to 130 m. They fitted a two-segment power-law function to their data divided into two size classes: <6 μm and >6 μm. The slope of the first segment increased with depth from about 3 in the surface layer to about 4 at 130 m, the slope of the second segment was about 5 in the examined layer.

Lerman et al. (1977) measured over 50 size distributions in the equatorial Atlantic waters at depths ranging from about 30 to over 5000 m. A Coulter counter was used to measure the size distribution in a particle diameter range of 2.26 to about 14 mm. The particle size distribution were well approximated using the power law. Interestingly, these authors found that the slope, m, is essentially constant with depth, contrary to the observations of *Kitchen* and *Zaneveld* (1990), while the concentration of the particles decreased significantly with depth.

McCave (1983) measured numerous particle size distributions in the nepheloid layer in a section across the Nova Scotian Rise area in the north Atlantic. The samples were taken from a layer several hundred meters thick above the sea bottom. The measurements were performed using a Coulter counter model TAII equipped with a 70 μm aperture, permitting particle size analysis in a range of 1.26 to 32 μm. For each sample, two or more duplicates were run, each using a sample volume of 5 cm^3. Two types of particle size distributions were distinguished: (1) particle size distribution which are well approximated using a one-segment power law, with an average slope of 3.93. This type of the size distribution is characteristic of old suspension, and (2) particle size distributions which are well approximated using a two-segment power law, with a knee at about 3 to 4 mm. This type of the size distribution is characteristic of freshly re-suspended ("new suspension") material in the nepheloid layer. Many particles in these suspensions were aggregates.

Reuter (1980b) measured size distributions of suspended particles in the coastal Baltic waters during a phytoplankton bloom. He used a Coulter counter to determine the distributions in a particle diameter range of about 2 to 20 μm. The particle size distributions in a diameter range <5 μm can be approximated using the power law. Large numbers of flagellates were detected in this diameter range. The particle size distributions of particles >5 μm have a form which could be approximated using a combination of power law and a Gaussian function (*Jonasz* 1980, 1983a) or by the log-normal function (discussed further in this text). The particles in this diameter range were mainly dinoflagellates, large flagellates, and diatoms.

Richardson (1987) measured particle size distributions using a Coulter counter model TAII, with a 50 μm aperture, in the north Atlantic waters: east of New York, and south of Iceland, along sections across the two continental rise areas. Particles with diameters of 1 to 20 μm were examined. Volume size distributions below the mixed layer column, expressed as particle volumes in logarithmically equal diameter intervals, are nearly flat (see *also Sheldon* el al. 1972).

Sugihara and Tsuda (1979) analyzed, using a Coulter counter model ZB with a 100 μm aperture, 22 samples obtained in the Pacific surface waters north of Hawaii. The authors found that their particle size distributions were well approximated in a diameter range of 2.42 to 15.3 μm using a power law, but also with an exponential function.

Spinrad et al. (1989a) used an *in situ* optical settling tube to measure size distributions of marine particles near the sea bottom in the Nova Scotian Rise area. The particle size distributions, $FD(D)$, derived from the optical settling tube measurements, can be approximated using a power law with a slope, m, on the order of 3.5. This slope is somewhat smaller than that of particle size distributions measured in this area with a Coulter counter. The authors concluded that this discrepancy was due to two factors: (1) the settling tube measured undisturbed particles, and (2) the settling tube response is proportional to the projected area of the particles, which is greater than that of the equal volume spheres. It was possible to explain a significant part of this discrepancy by using an estimate of the ratio of the projected area of marine particles to that of equal volume spheres (*Jonasz* 1987a). All particle size distributions determined with the settling tube show minor peaks or shoulders as previously determined using a Coulter counter for samples taken in the studied area (*McCave* 1983).

Spinrad et al. (1989b) measured cumulative particle size distributions using a Coulter counter (1.5 to 22 μm) in the waters of Peru upwelling. The particle size distributions were well approximated using a power law. The clearest, deepest waters were characterized by particle size distributions with the highest slopes, as were the most turbid waters. The particles outside the Coulter range influenced the transmission of light. In the waters, where the Coulter data suggested a constant particle concentration, the transmission data implied a concentration maximum. A high correlation between the bacterial abundance (<1.5 μm) and the transmission data was established, highlighting the role of small particles in attenuation of light by sea water.

Wellershaus et al. (1973) measured particle size distributions in the Atlantic waters, off Portugal (Gulf of Cadiz), and in the Indian Ocean. Samples were collected from depths ranging from 10 to 200 m (Indian Ocean) and to 5000 m (Atlantic Ocean). The particles were either collected on cellulose membrane filters and air dried, or allowed to settle to the bottom of a plankton settling tube. Optical microscopes were used to obtain the microphotographs of the particles. The magnified prints were analyzed with a Zeiss particle size analyzer (image analyzer) TGZ-3. With this instrument, the particle images are classified into 48 consecutive size classes, depending on the image diameter.

The air drying does not affect solid particles, which are believed to be mostly inorganic. Some organic particles, however, shrink and become nearly invisible when dried. Of the aggregates of solid particles glued with an organic matter, mostly the solid particles remain visible and contribute to the size distribution. Thus, it was suggested that a comparison of the size distribution of the air-dried particles with that of wet particles in seawater may provide information about the composition of the particulate material.

The average distributions of 65 samples from five stations in the Indian Ocean and of seven samples from one station in the Atlantic Ocean can be well approximated using the power law in a diameter range of 2 to 40 μm. The concentration of wet particles was found to be about an order of magnitude greater than that of the air-dried particles.

The particle size distributions, presented as volume histograms, have slopes similar to those obtained by *Sheldon* and Parsons (1967b) using a Coulter counter. This similarity was achieved, when the particle volume was calculated according to the formula: particle cross-section times the particle thickness. The particle thickness was estimated using the focusing mechanism of the microscope.

Table A.5. Sources of data on the size distribution of individual species of aquatic particles.

Reference	**Species, number, and mode of measurements**	**Diameter range [μm]**	**Measurement method**	**Arguments, data formats**
	Viruses			
Bratbak et al. (1990) (their Fig. 5)	Viruses (11 *in vitro*)	0.05–0.12		w, l, graph
Alonso et al. (2002)	Marine viruses	Head diameter 0.03–0.08, Tail length 0.03–0.1	TEM	Head diameter, Tail length histo.
Balch et al. (2000)	Marine viruses	0.02–0.15	FFF	
	Bacteria			
Fuhrman (1981)	Marine bacteria, free living (five *in vitro*)			D_s, graphs
Sieracki and Viles (1992)	Marine bacteria, free living (one *in vitro*)			D_v; graph^r
Stramski and Kiefer (1990)	Marine bacteria, free living (one *in vitro*)			w, l graphs
Stramski et al. (1992c)	Marine bacteria, free living (six *in vitro*)			Axial ratio, D_v; graphs
Sieracki et al. (1985)	Marine bacteria, free living (three natural, eight growth experiment *in vitro*)			D_v; graph^r
Morel and Ahn (1990)	Bacterioplankton, marine, free living (nine *in vitro*)			D_v; graph^r
Bloem et al. (1995)	Bacteria, soil (four *in vitro*)			w, l, graph
Morel and Bricaud (1986)	Cyanobacteria, marine (one *in vitro*)			D_v; graph

Table A.5. Continued

Reference	Species, number, and mode of measurements	Diameter range [μm]	Measurement method	Arguments, data formats
	Bacteria			
Harvey et. al. (1967)	*Escherichia coli* (one *in vitro*)			V, graph
Morel et al. (1993)	*Prochlorococcus* (three *in vitro*)		CC	D_v; graph
Morel et al. (1993)	*Synechococcus* (four *in vitro*)			D_v, graph
Stramski et al. (1995)	*Synechococcus* (six *in vitro*)			D_v; graph
Stramski et al. (1992a)	*Synechococcus* (two *in vitro*)			D_v; graph
Ahn et al. (1992)	*Synechococcus*, $\langle D_v \rangle = 1.05$	0.5–1.5	CC	D_v, graph^r
Stramski and Morel (1990)	*Synechocystis* (two *in vitro*)			D_v graph
Ahn et al. (1992)	*Synechocystis*, $\langle D_v \rangle = 1.39$	0.7–2	CC	D_v, graph^r
	Phytoplankton			
Ahn et al. (1992)	*Anacystis marina*, $\langle D_v \rangle = 1.43$	0.7–2	CC	D_v, graph^r
Morel and Ahn (1991)	ciliates (4 *in vitro*)		CC	D_v, graph^r
Ahn et al. (1992)	*Chroomonas fragraioides*, $\langle D_v \rangle = 5.57$	3–10	CC	D_v, graph^r
Lal and Lerman (1975) (data of *Lisitzyn* (1972)	diatoms (three *in vitro*)		M	Undefined size (0.2 to 40 μm)
Ahn et al. (1992)	*Dunaliela bioculata*, $\langle D_v \rangle = 6.71$	4–10	CC	D_v, graph^r
Stramski et al. (1993)	*Dunaliella tertiolecta* (two *in vitro*)		CC	D_v,
Bricaud and Morel (1986)	*Emiliania huxleyi* (one *in vitro*)		CC	D_v; graph
Ahn et al. (1992)	*Emiliania huxleyi*, $\langle D_v \rangle = 4.93$	2–8	CC	D_v, graph^r

(*Continued*)

Table A.5. Continued

Reference	Species, number, and mode of measurements	Diameter range [μm]	Measurement method	Arguments, data formats
	Phytoplankton			
Lal and Lerman (1975)	foraminiferas (one *in vitro*) (data of *Lisitzyn* (1972), data and a power-law approximation)	0.2–40	M	Undefined size
Ahn et al. (1992)	*Hymenomonas elongata*, $\langle D_v \rangle = 11.77$	6–20	CC	D_v, graph[r]
Ahn et al. (1992)	*Isochrysis galbana*, $\langle D_v \rangle = 4.45$	2–8	CC	D_v, graph[r]
Gaedke (1992)	Lacustrine phytoplanton and zooplankton (one *in vitro*)	~0.6 to ~40	CC, OM	Graph[a] of biomass
Tittel et al. (1998)	Lacustrine phytoplanton and zooplankton (seven *in vitro*) + normalized biomas spectra slopes for 28 lakes in the northern Germany	~0.31 to ~635	OM	graph[a] of biomass
Campbell et al. (1989)	Marine phytoplankton homogeneous (14 *in vitro*)			V, graph
Fururya and Marumo (1983)	Marine phytoplankton, western Pacific (33 *in vitro*)	2–128	OM	D_s, graph
Rodríguez et al. (1998)	Western Mediterranean (Alboran Sea) *in vitro*	~0.5 to ~60	FC, OM	V, graph[a]
Rodríguez et al. (2002)	Antarctic (Gerlache Strait)	~0.4 to ~120	FC, OM	V, graph[a]
Quinones et al. (2003)	Western Atlantic (four *in vitro*)	~0.2 to ~8000	OM	V, graph[a]

Table A.5. Continued

Reference	Species, number, and mode of measurements	Diameter range [μm]	Measurement method	Arguments, data formats
	Phytoplankton			
Dubelaar and van der Reijden (1995)	*Microcystis aeruginosa* (one *in vitro*)	4–500	FC, OM	V, graph[a]
Morel and Ahn (1991)	Nanoflagellates (nine *in vitro*)		CC	D_v, graph[r]
Bricaud and Morel (1986)	*Platymonas suecica* (one *in vitro*)		CC	D_v; graph
Ahn et al. (1992)	*Prorocentrum micans*, $\langle D_v \rangle = 27.64$	10–40	CC	D_v, graph[r]
Bricaud and Morel (1986)	*Skeletonema costatum* (one *in vitro*)		CC	D_v; graph
Stramski and Reynolds (1993)	*Thallasiosira pseudonana* (two *in vitro*)		CC	D_v; graph
Lambert et al. (1981)	Aggregates (four *in vitro*)		SEM	D_a; fit parameter
Hou (1997)	Aggregates (combined *in vitro*—small size, *in situ*—large size)		CC, MP	D_a; graph[r] and fit parameter
	Aggregates and organic matter			
Lambert et al. (1981)	organic matter (five *in vitro*)		SEM	D_a; graph[r] and fit parameter
Mari and Burd (1998)	Transparent exopolymer particles (18 *in vitro*)	1–100	M	graph[a]
Passow and Alldredge (1994)	Transparent exopolymer particles (13 *in vitro*)		M	Maximum l, graph
	Bubbles			
Terrill et al. (2001)	Bubbles in rough sea (Pacific off California, wind speed $U_{10} = 15\,\mathrm{m\,s^{-1}}$), four *in situ*	25–400	Acoustic transmission	D, graph[a]

(*Continued*)

Table A.5. Continued

Reference	Species, number, and mode of measurements	Diameter range [μm]	Measurement method	Arguments, data formats
	Minerals			
Vagle and Farmer (1992)	$U_{10} = 11\,\mathrm{m\,s^{-1}}$, *in situ*	< 260	Acoustic backscattering	
Lambert et al. (1981)	Aluminosilicates (eight *in vitro*, Atlantic)		SEM	D_a; fit parameter
Lambert et al. (1981)	Aluminosilicates (eight *in vitro*, Pacific)		SEM	D_a; fit parameter
Volten et al. (2001)	Feldspar dust (one *in vitro*)		LD	Graph[r,b]
Lambert et al. (1981)	Geothite (FeO_2H) (three *in vitro*)		SEM	D_a; fit parameter
Volten et al. (2001)	Red clay (biotite, illite, quartz) dust (one *in vitro*)		LD	Graph[r,b]
Volten et al. (2001)	Sahara sand dust (one *in vitro*)		LD	Graph[r,b]
Lambert et al. (1981)	SiO_2 particles (three *in vitro*)		SEM	D_a; fit parameter
Taguas et al. (1999)	Soil (3)		Unknown	*Cumulative mass* (*D*)
Hill et al. (1984b)	Soil particles (1)			width/length, D_v, graph

[a] Absolute data.
[r] Relative data.
[b] As a function of the circular equivalent diameter (Table 5.1).

The table is ordered alphabetically according to the species name, not the reference's first author name. Abbreviations: D_a = projected area equivalent diameter; D_v volume-equivalent diameter, l = length, V = volume, w = width, CC = Coulter counter, EZ = Elzone counter, FFF = field flow fractionation, GS = Galai scanning counter, LD = laser diffractometer, M = optical microscope, SEM = scanning electron microscope, TEM = transmission electron microscope. Aperture size (if applicable) is given where available following the method abbreviation. See also a table of absolute abundances and cell sizes of 42 species of marine phytoplankton in a sample of seawater from the Pacific Ocean off Hawaii in *Takahashi* and Bienfang (1976). A composite graph of 13 size distributions of the size distributions of representative marine phytoplankton species is shown by *Stramski* et al. (2001) along with the average diameters and abundance ranges of these species.

Table A.6. Refractive indices, $n = n' - in''$, of various species of aquatic particles.

Reference	Species	n'	n''	Wavelength [nm]	Method
	Viruses				
Stramski and Kiefer (1990)	*Viral head*	1.05		Visible	From composition
Balch et al. (2000)	*Viruses in cultures*	1.165 to 1.025		514	VSF fitting
Balch et al. (2000)	*Viruses in cultures*	1.257 to 1.026		633	VSF fitting
	Bacteria[w]				
Kopelevich et al. (1987)	*Deleya vinustus*	1.03		330 to 690	Fitting $Q_c(n')$
Waltham et al. (1994)	*Escherichia coli*	1.064 ± 0.015		460	IMRE
Bateman et al. (1966)[r]	*Escherichia coli*	1.045		546	IMRE
Bateman et al. (1966)[r]	*Escherichia coli*	1.049 ± 0.002		589	IMRE
Kopelevich et al. (1987)	*Micrococcus* sp.	1.02		330 to 690	Fitting $Q_c(n')$
Jonasz et al. (1997)	*marine bacteria*	1.035[n]		Visible	IMRE
Stramski and Kiefer (1990)	*marine bacteria*	1.042 to 1.068	0.0001		BMS
Kopelevich et al. (1987)	*Pseudomonas* sp.	1.05		330 to 690	Fitting $Q_c(n')$
Ross and Billing (1957)	Spores	1.512 to 1.540		534	IMRE
Green et al. (2003b)	*Synechococcus* sp. culture	1.063	0.0033	488	FCMT
Ross and Billing (1957)	Vegetative cells	1.386[a] to 1.40[a]		534	IMRE

(*Continued*)

Table A.6. Continued

Reference	Species	n'	n''	Wavelength [nm]	Method
	Phytoplankton[w]				
Bricaud et al. (1983)	*Coccolithus huxleyi*		0.00300	435	BMS
Bricaud et al. (1983)	*Coccolithus huxleyi*		0.0004	600	BMS
Spinrad and Brown (1986)	*Chlorella*	1.047 to 1.086		514	FCMT[c]
Quinby-Hunt et al. (1989)	*Chlorella*	Core 1.08, shell 1.13	Core 0.05, shell 0.04	441.8	Fitting the scattering matrix
Bricaud et al. (1988)	*Chaetoceros curvisetum*	1.021			BMS
Bricaud et al. (1988)	*Chaetoceros lauderi*	1.0045			BMS
Stramski et al. (2001)[r]	*Dunaliela bioculata*	1.038		550	BMS
Stramski et al. (2001)[r]	*Dunaliela bioculata*		0.0105 0.0078	440 675	BMS
Bricaud et al. (1988)	*Dunaliela salina*	1.092			BMS
Ackleson and Spinrad (1988)	*Dunaliela salina*	1.062 to 1.065		514	FCMT
Ackleson et al. (1988a)	*Dunaliela tertiolecta*	1.05 to 1.065		514	FCMT
Green et al. (2003b)	*Dunaliela tertiolecta*	1.037	0.0042	488	FCMT
Ackleson and Spinrad (1988)	*Emiliania huxleyi*	1.05 to 1.058		514	FCMT
Bricaud and Morel (1986)	*Emiliania huxleyi*	1.044	0.0047	435	BMS
Bricaud et al. (1983)	*Hymenomonas elongata*		0.00528	435	BMS
Bricaud et al. (1983)	*Hymenomonas elongata*		0.0009	600	BMS
Carder et al. (1972)	*Isochrysis galbana*	1.023 to 1.031		436	Fitting $Q_c(n')$

Table A.6. Continued

Reference	Species	n'	n''	Wavelength [nm]	Method
	Phytoplankton[w]				
Carder et al. (1972)	*Isochrysis galbana*	1.026 to 1.035		546	Fitting $Q_c(n')$
Carder et al. (1972)	*Isochrysis galbana*	1.027 to 1.036		578	Fitting $Q_c(n')$
Green et al. (2003b)	*Isochrysis galbana*	1.070	0.0041	488	FCMT
Green et al. (2003b)	*Monochrysis lutheri*	1.032	0.0022	488	FCMT
DuRand and Olson (1998)	*Nannochloris* sp.	~ 1.04 to ~ 1.06	~ 0.002 to ~ 0.0035	665	BMS
Green et al. (2003b)	*Nannochloris* sp.	1.063	0.0039	488	FCMT
Bricaud et al. (1988)	*Pavlova pinguis*	1.05			BMS
Bricaud et al. (1988)	*Pavlova lutheri*	1.045			BMS
Bricaud et al. (1988)	*Pavlova pavrum*	1.048			BMS
Bricaud et al. (1983)	*Platymonas sp.*		0.00392	435	BMS
Bricaud et al. (1983)	*Platymonas sp.*		0.0015	600	BMS
Bricaud and Morel (1986)	*Platymonas suecica*	1.071	0.0116	435	BMS
Bricaud et al. (1988)	*Porphyridium cruentum*	1.057		–	BMS
Jonasz (1986)	*Rhodomonas sp.*	1.08		white light	CPSD
Bricaud and *Morel* (1986)	*Skeletonema costatum*	1.028	0.00216	435	BMS
Ackleson and Spinrad (1988)	*Thalassiosira pseudonana*	1.06 to 1.065		514	FCMT
Green et al. (2003b)	*Thalassiosira pseudonana*	1.032	0.0022	488	FCMT

(*Continued*)

Table A.6. Continued

Reference	Species	n'	n''	Wavelength [nm]	Method
	Minerals[a]				
Bricaud et al. (1983)	*Traselmis maculata*		0.00324	435	BMS
Bricaud et al. (1983)	*Traselmis maculata*		0.0011	600	BMS
Lindberg and Smith (1974)[e]	Kaolinite		$1.2 \cdot 10^{-5}$	500	DR
Lindberg and Smith (1974)[e]	Kaolinite		$0.3 \cdot 10^{-5}$ to 10^{-5}	1000	DR
Gillespie et al. (1974)[e]	Montmorillonite sample #20		$2 \cdot 10^{-5}$	500	DR
Gillespie et al. (1974)[e]	Montmorillonite sample #20		$0.4 \cdot 10^{-5}$	1000	DR
Gillespie et al. (1974)[e]	Montmorillonite sample #22		$9 \cdot 10^{-5}$	1000	DR
Gillespie and Lindberg (1992)	Quartz		$< 10^{-7}$	Visible	DR
	Aerosols[a]				
Gillespie and Lindberg (1992)	Amorphous carbon		1.5	250	DR
Gillespie and Lindberg (1992)	Amorphous carbon		1	700	DR
Dubovik et al. (2002)	Desert dust	1.36 to 1.58			Retrieval from lidar data
Dubovik et al. (2002)	Desert dust		0.0025–0.0029 0.0007–0.0012 0.0006–0.001 0.0006–0.001	440 670 870 1020	Retrieval from sun photometer data

Table A.6. Continued

Reference	Species	n'	n''	Wavelength [nm]	Method
Gillespie and Lindberg (1992)	Fe_3O_4		0.4	250	DR
Gillespie and Lindberg (1992)	Fe_3O_4		0.4	700	DR
Gillespie and Lindberg (1992)	FeS_2		0.1	250	DR
	Aerosols[a]				
Gillespie and Lindberg (1992)			0.1	700	DR
Wyatt (1980)	Fly ash	1.48 to 1.57	0 to 0.01	633	Fitting VSF
Gillespie and Lindberg (1992)	MnO_2		0.09	250	DR
Gillespie and Lindberg (1992)	MnO_2		0.11	700	DR
Patterson et al. (1977)	Saharan dust		0.02	300	DR
Patterson et al. (1977)	Saharan dust	1.56		550	IMRE
Patterson et al. (1977)	Saharan dust	1.54		589	IMRE
Patterson et al. (1977)	Saharan dust		0.0035	600	DR
Lindberg and Gillespie, (1977)	Soil aerosols		0.02 to 0.11	500	DR
Grams et al. (1974)	Soil aerosols	1.525		488	IMRE
Grams et al. (1974)	Soil aerosols		0.005 ± 0.010		Fitting VSF

(*Continued*)

Table A.6. Continued

Reference	Species	n'	n''	Wavelength [nm]	Method
	Marine particles[w]				
Zaneveld and Pak (1971)	All particles	1.01 to 1.05		436 to 546	See section 6.3.2.2
Twardowski et al. (2001)	All particles (PSD $\sim D^{-3.3}$ to $\sim D^{-4.1}$)	1.05 to 1.15		Visible	See section 6.3.2.2
Ackleson et al. (1988b)	Containing chlorophyll and phycoerythrin (North Sea)	1.079 to 1.092			FCMT

[a]Relative to air.
[c]An early, limited version of the FCMT technique, based on the relationship between the refractive index and the shape of a relationship between a small angle and 90° scattering.
[d]Too high to estimate.
[e]Estimated from the data in the reference.
[n]Negligible.
[r]Based on data from *Ahn* et al. (1992).
[w]Relative to water.

See the sources of the light scattering properties in Table A.3, size distributions for some of these species in Table A.5, and shape/size information in Table A.8. Aerosols are also included because they constitute an important particle source for natural waters. Methods of the refractive index determination: BMS—a method of Bricaud, Morel, and Stramski (*Stramski* et al., 1988, *Bricaud* and Morel, 1986) that inverts the attenuation and absorption efficiencies into the n' and n'' (see section 6.3.2.4), DR—diffuse reflectance, FCMT—flow-cytometry mapping technique: particle size and a refractive index are determined from a map of scattering intensities at two angles (*Spinrad* and Brown 1986, *Tycko* et al. 1985), IMRE—immersion refractometry, CPSD—refractive index is obtained by comparing PSDs obtained with an optical and a resistive particle counters. Other abbreviations: VSF—volume scattering function.

Table A.7. The refractive index, n, of natural populations of marine particles from the angular light scattering pattern (volume scattering function).

Reference	Study area(s), season(s), layer(s)	Wavelength [nm]	n	Size range [μm]	Size distribution
Brown and Gordon (1974)	Off the Bahama Islands	488	1.01	0.01 to 0.65	$1.2 \times 10^5 D^{-4}$
Brown and Gordon (1974)	Off the Bahama Islands	488	1.01	0.65 to 1.25	$\sim D^{-7.5}$
Brown and Gordon (1974)	Off the Bahama Islands	488	1.01[a]	1.25 to 3.75	$\sim D^{-2.7}$ to $\sim D^{-3.5}$
Brown and Gordon (1974)	Off the Bahama Islands	488	1.15[b]	1.25 to 3.75	$\sim D^{-2.7}$ to $\sim D^{-3.5}$
Brown and Gordon (1974)	Off the Bahama Islands	488	$1.03 - i0.01$	3.75 to 17	$\sim D^{-3}$ to $\sim D^{-4}$
Jonasz and Prandke (1986), *Jonasz* (1980)	Baltic: summer, surface layer	633	1.1	0.1 to 2	$5.6 \times 10^4 D^{-4.1}$
Jonasz and Prandke (1986), *Jonasz* (1980)	Baltic: summer, surface layer	633	$1.05 - i0.005$[e]	2 to 10	$3.7 \times 10^4 D^{-2.7}$
Jonasz and Prandke (1986), *Jonasz* (1980)	Baltic: summer, surface layer	633	$1.03 - i0.01$	10 to 32	$4.4 \times 10^6 D^{-4.9}$
Jonasz and Prandke (1986), *Jonasz* (1980)	Baltic: winter, surface layer	633	1.1	0.1 to 2	$9.6 \times 10^4 D^{-4.1}$
Jonasz and Prandke (1986), *Jonasz* (1980)	Baltic: winter, surface layer	633	1.1	2 to 10	$3.7 \times 10^4 D^{-2.7}$
Jonasz and Prandke (1986), *Jonasz* (1980)	Baltic: winter, surface layer	633	1.1	10 to 32	$4.4 \times 10^5 D^{-4.9}$
Kullenberg (1974)	Baltic Sea	488, 525, 633	1.03 to 1.10	1 to 38	$\sim D^{-1.25}$ to $\sim D^{-1.5}$

(*Continued*)

Table A.7. Continued

Reference	**Study area(s), season(s), layer(s)**	**Wavelength [nm]**	*n*	**Size range [μm]**	**Size distribution**
Reuter (1980a, 1980b)	Baltic, coastal waters	488	$1.15 - i0.001$[c]	0.02 to 1.5	$\sim D^{-3.8}$
Reuter (1980a, 1980b)	Baltic, coastal waters	488	1.05[d]	0.02 to 1.5	$\sim D^{-3.8}$
Reuter (1980a, 1980b)	Baltic, coastal waters	488	1.20[c]	1 to 2.5	Gaussian
Reuter (1980a, 1980b)	Baltic, coastal waters	488	1.05[d]	1 to 2.5	Gaussian
Reuter (1980a, 1980b)	Baltic, coastal waters	488	1.05	>2.5	Unspecified
Zaneveld et al. (1974)	Sargasso Sea	633	1.15	0.08 to 10	$\sim D^{-3.5}$
Reuter (1980a, 1980b)	Baltic, coastal waters	488	1.075	0.08 to 10	$\sim D^{-3.9}$
Reuter (1980a, 1980b)	Baltic, coastal waters	488	1.05	0.08 to 10	$\sim D^{-3.7}$
Kullenberg (1974)	Mediterranean Sea	633	1.2	1 to 38	$\sim D^{-1.6}$ to $\sim D^{-2.2}$
Brown and Gordon (1973)	Sargasso Sea	633	$1.01 - i0.01$	0.1 to 2.5	$4.8 \times 10^4 D^{-4}$
Brown and Gordon (1973)	Sargasso Sea	633	$1.01 - i\,0.01$	0.1 to 2.5	$4.8 \times 10^4 D^{-4}$
Brown and Gordon (1973)	Sargasso Sea	633	1.15	2.5 to 10.0	$4.8 \times 10^4 D^{-4}$
Gordon and Brown (1972)	Sargasso Sea	633	$1.05 - i\,0.01$	0.08 to 10	$3.3 \times 10^4 D^{-4}$

[a] 2/3 of the experimental particle size distribution in this size range.
[b] 1/3 of the experimental particle size distribution in this size range.
[c] 20% of the particles.
[d] 80% of the particles.
[e] The Gaussian component in this size range did not contribute substantially to the volume scattering function.

Table A.8. Sources of the data on the shapes and structures of some species of particles occurring in sea water.

Reference	Species	Characteristic dimensions [μm]	Shape, structure comments	Shape data	Structure data
	Bacteria				
Venkataraman et al. (1974)	*Anacystis nidulans*	*D* 0.5 to 1, *L* 1 to 4	Rounded-cups cylinder	TEM photo	TEM photo
Kopelevich et al. (1987)	*Baccilus mycoides*	*D* 1.8, *L* 5	Spheroid	–	–
Kopelevich et al. (1987)	*Deleya vinustus*	*D* 0.8, *L* 2.3	Spheroid with several flagellae	–	–
Lyubovtseva and Plakhina (1976)	*Escherichia coli*	*D* 0.5, *L* 2 to 4	Cylinder	–	–
Kopelevich et al. (1987)	*Flavo-bacterium*	*D* 0.8, *L* 2	Rod	–	–
Kopelevich et al. (1987)	*Micrococcus* sp.	*D* 0.2	Spheroid	–	–
Kopelevich et al. (1987)	*Moraxella* sp.	*D* 1.7, *L* 2.2	Spheroid	–	–

(*Continued*)

Table A.8. Continued

Reference	Species	Characteristic dimensions [μm]	Shape, structure comments	Shape data	Structure data
	Bacteria				
Kopelevich et al. (1987)	*Pseudomonas* sp.	*D* 0.3, *L* 1	Spheroid with single flagellum	–	–
Chisholm et al. (1988)	*Synechococcus* sp.	*D* 1	Spheroid	TEM photo	Layered sphere; TEM photo
Kopelevich et al. (1987)	*Vibrio* sp.	*D* 0.7, *L* 1	Curved cylinder	–	–
	Phytoplankton				
Johnson and Sieburth (1982)	Algae (scaled: probably organic scales)	*D* 1 (cell) *D* 0.15 (scales)	Spheroid	TEM photo	TEM photo
?, (1975)	*Ankistrodemus falcatus*	1: 10 diameter to length ratio	Needle	OM photo	OM photo
?, (1975)	*Chamydomonas reinhardii*		Spheroid with two sets of thin flagellas	TEM photo	TEM photo
Priddle and Fryxell (1985)	*Chaetoceros socialis* (diatom)	*D* 5, *L* 3 forms chains of several cells; can form filaments	Disk with three curved spines	Drawing	

Bricaud et al. (1988)	*Chaetoceros curvisetum*	*D* 7.5, *L* 15	Cylinder; wall covered with silica plates; vacuoles; many small chloroplasts	–	–
Bricaud et al. (1988)	*Chaetoceros lauderi*	*D* 25, *L* 50	cylinder (see *C. curvisetum*)	–	–
Quinby-Hunt et al. (1989)	*Chlorella* (green algae)		Spheroid	OM, SEM photo	TEM photo
Johnson and Sieburth (1982)	*Chlorella-like picoplankton*	*D* 1	Spheroid	TEM photo	TEM photo
Hoepffner and Haas (1990)	*Chrysomulina vexilifera*	*D* 3 (whole cell); *D* 0.5 (disk organic scales)	Spheroid, two thin threads, and scales	TEM photo	–
Bricaud et al. (1983)	*Coccolithus huxleyi*	*D* 3 to 4	Sphere	–	–
Hoepffner and Haas (1990)	*Cosmoeca ventricosa* (chaono-flagellate)	*D* 3, *L* 6 (cell); *D* 12, *L* 12 (basket)	Spheroidal cell with thin flagellum, in a basket	TEM photo	–
Bricaud et al. (1988)	*Dunaliela salina*	*D* 10	ovoid	–	–

(*Continued*)

Table A.8. Continued

Reference	Species	Characteristic dimensions [μm]	Shape, structure comments	Shape data	Structure data
	Phytoplankton				
Hoepffner and Haas (1990), *Bricaud* and Morel (1986)	*Emiliania huxleyi* (coccolitho-phorid)	*D* 4 to 6 (whole cell) *D* 2 (calcite scale)	Spheroid coated with scales	TEM photo	
Bricaud et al. (1983)	*Hymenomonas elongata*	*D* 12 to 15	Spheroid	–	–
Carder et al. (1972)	*Isochrysis galbana*	*D* 4.2 to 4.6	Spheroid	–	–
Liu and Lin (2001)	*Isochrysis galbana*	*D1* 5 to 6x *D2* 2 to 4x *D3* 2.5 to 3	Ellipsoidal, no cell wall, lipid bodies 0.3 to 5 μm appearing in stationary growth	Phase microscope photo	TEM photo
Hoepffner and Haas (1990)	*Meringo-sphaera mediterranea*	*D* 4 spine length 20	Spheroid	SEM photo	
Johnson and Sieburth (1982)	*Micromonas pusilla* (micro-flagellate)	*D* 1.5, *L* 3	Spheroid	TEM photo	TEM photo
Hoepffner and Haas (1990)	*Minidiscus triaculatus* (diatom)	*D* 7	Disk	TEM photo	
Subba Rao et al. (1991)	*Nitzschia pungens* (diatom)	*D* 5, *L* 100	Needle	OM photo	OM photo
Hoepffner and Haas (1990)	*Nitzschia bifurcata* (diatom)	*W* 2, *L* 4, *T* 1	Shuttle	TEM photo	

Hoepffner and Haas (1990)	*Oxytoxum* sp. (diatom)	*D* 6, *L* 12	Shuttle	TEM photo	
Bricaud et al. (1988)	*Pavlova pinguis*	*D* 3.6	Spheroid	–	–
Bricaud et al. (1988)	*Pavlova lutheri*	*D* 4.5	Spheroid	–	–
Bricaud et al. (1988)	*Pavlova pavrum*	*D* 5.7	Spheroid	–	–
Hoepffner and Haas (1990)	*Phaeocystis poucheti*	*D* 4 to 6	Spheroid	TEM photo	TEM photo
Bricaud et al. (1983)	*Platymonas* sp.	*D* 6 to 7.5	Spheroid	–	–
Bricaud and Morel (1986)	*Platymonas suecica*	*D* 3.4	Spheroid	–	–
Dodge (1973)	*Porphyridium cruentum* (red algae)	*D* 5	Sphere, one very large chloroplast	TEM photo	TEM photo
Jonasz (1986)	*Rhodomonas* sp.	*D* 4.5, *L* 7.3	Spheroid	SEM photo	–
Lebour (1930), *Bricaud* and Morel (1986)	*Skeletonema costatum*	*D* 5.5 to 16, forms long colonies	Cylinder	drawing	
Round et al. (1992)	*Skeletonema* sp.	*D* 6	Round cylinder	SEM photo	SEM photo

(Continued)

Table A.8. Continued

Reference	Species	Characteristic dimensions [μm]	Shape, structure comments	Shape data	Structure data
	Phytoplankton				
Faust (1993)	*Synophysis microcephalu* (dino-flagellate)	*W* 33 to 35 *L* 42 to 44 *T* 20 to 30	Ellipsoid	SEM photo	–
Stramski and Reynolds (1993)	*Thalassiosira pseudonana* (diatom)	*D* 4	Spheroid	–	–
Bricaud et al. (1983)	*Traselmis maculata*	*D* 8 to 10	Spheroid		
	Minerals				
Jonasz (1986, 1987b)	Kaolinite	*D* 10, *T* 0.1	Hexagonal disks	SEM photo	–
Lyubovtseva and Plakhina (1976)	Montmorillonite	*D* 0.2. *T* 0.001	Hexagonal disks	–	–
Lyubovtseva and Plakhina (1976)	Palygorskite	*D* 0.01 to 0.03 *L* 1	Cylinder	–	–

Abbreviations: *L* = length, *D* = diameter, *T* = thickness, *W* = width, OM – optical microscopy, SEM = scanning electron microscopy.

Bibliography

A+

Aas E. 1981. *The refractive index of phytoplankton*. Ref. 46, University of Oslo, Institute of Geophysics, Oslo, Norway, 61 pp.

Aas E. 1984. *Influence of shape and structure on light scattering by marine particles*. Ref. 53, University of Oslo, Institute of Geophysics, Oslo, Norway, 112 pp.

Aas E. 1987. Two-stream irradiance model for deep waters. *Appl. Opt.* **26**: 2095–2101.

Aas E. 1996. Refractive index of phytoplankton derived from its metabolite composition. *J. Plankton Res.* **18**: 2223–2249.

Aas E. 2000. Spectral slope of yellow substance: problems caused by small particles. In: *Ocean Optics XV*, CD-ROM, Office of Naval Research, Washington DC.

Abramowitz M., Stegun I. A. 1964. *Handbook of mathematical functions with formulas, graphs and mathematical tables*. National Bureau of Standards, Washington DC.

Ackermann H. W., Dubow M. S. 1987. *Viruses of prokaryotes*. Vol. II: Natural groups of bacteriophages, CRC Press, Boca Raton, 242 pp.

Ackleson S. G., Balch W. M., Holligan P. M. 1994. Response of water-leaving radiance to particulate calcite and chlorophyll *a* concentrations: A model for Gulf of Maine coccolithophore blooms. *J. Geophys. Res. C* **99**: 7483–7499.

Ackleson S. G., Robins D. B., Stephens J. A. 1988b. Distributions in phytoplankton refractive index and size within the North Sea. *Ocean Optics IX. Proc. SPIE* **925**: 317–325.

Ackleson S. G., Spinrad R. W. 1988. Size and refractive index of individual marine particulates: a flow cytometric approach. *Appl. Opt.* **27**: 1270–1277.

Ackleson S. G., Spinrad R. W., Yentsch C. M., Brown J., Korjeff-Bellows W. 1988a. Phytoplankton optical properties: flow-cytometric examinations of dilution-induced effects. *Appl. Opt.* **27**: 1262–1269.

Acquista C. 1978. Validity of modifying Mie theory to describe scattering by nonspherical particles. *Appl. Opt.* **17**: 3851–3852.

Aden A. L., Kerker M. 1951. Scattering of electromagnetic waves from two concentric spheres. *J. Appl. Phys.* **22**: 1242–1246.

Agrawal Y. C., McCave I. N., Riley J. B. 1991. Laser diffraction size analysis. In: *Principles, methods, and application of particle size analysis*. Syvitski J. P. M. (ed.) Cambridge University Press, New York, pp. 119–128.

Agrawal Y. C., Pottsmith H. C. 2000. Instruments for particle size and settling velocity observations in sediment transport. *Mar. Geol.* **168**: 89–114.

Agrawal Y. C., Trowbridge J. C. 2002. The optical volume scattering function: temporal and vertical structure in the water column at LEO-15. In: *Proceedings of the XVI Ocean Optics Conference*, Poster 65.

Agusti S., Duarte C. M, Kalff J. 1987. Algal cell size and the maximum density and biomass of phytoplankton. *Limnol. Oceanogr.* **32**: 983–986.

Agusti S., Kalff J. 1989. The influence of growth on the size dependence of maximal algal density and biomass. *Limnol. Oceanogr.* **34**: 1104–1108.

Aharonson E. F., Karasikov N., Roitberg M., Shamir J. 1986. GALAI-CIS-1 – A novel approach to aerosol particle size analysis. *J. Aerosol Sci.* **17**: 530–536.

Ahn Y.-H., Bricaud A., Morel A. 1992. Light backscattering efficiency and related properties of some phytoplankters. *Deep-Sea Res.* **39**: 1835–1855.

Aitchinson J., Brown J. A. C. 1957. *The lognormal distribution.* Cambridge University Press, Cambridge, Massachusetts, 176 pp.

Aizaki M., Otsuki A., Fukushima T., Hosomi M., Muroaka K. 1981. Application of Carlson's trophic state index to Japanese lakes and relationships between the index and other parameters. *Verh. Int. Verein. Limnol.* **21**: 675–681.

Akers R. J., Sinclair I., Stenhouse J. I. T. 1992. The particle size analysis of flocs using the light obscuration principle. In: *Particle size analysis.* Stanley-Wood N. G., Lines R. W. (eds) The Royal Society of Chemistry, Cambridge, pp. 246–255.

Al Ani S., Dyer K. R., Huntley D. A.1991. Measurement of the influence of salinity on floc density and strength. *Geo-Mar. Lett.* **11**: 154–158.

Albertano P., Di Somma D., Capucci E. 1997. Cyanobacterial picoplankton from the Central Baltic Sea: cell size classification by image-analyzed fluorescence microscopy. *J. Plankton Res.* **19**: 1405–1416.

Alford M. H., Gerdt D. W., Adkins C. M. 2006. An ocean refractometer: Resolving millimeter-scale turbulent density fluctuations via the refractive index. *J. Atmos. Ocean. Tech.* **23**: 121–137.

Alger T. W. 1979. Polydisperse particle-size-distribution function determined from intensity profile of angularly scattered light. *Appl. Opt.* **18**: 3404–3501.

Alldredge A. L., Passow U., Logan B. E. 1993. The abundance and significance of a class of large transparent organic particles in the ocean. *Deep-Sea Res.* **40**: 1131–1140.

Alldredge A., Silver M. W. 1988. Characteristics, dynamics, and significance of marine snow. *Progr. Oceanogr.* **20**: 41–82.

Allen T. 1990a. *Particle size measurements.* Chapman and Hall, London.

Allen T. 1990b. An industrial perspective on optical particle sizing. In: *Proceedings of the 2nd International Congress on Optical Particle Sizing*, Tempe, Arizona, USA, 5–9 March 1990, Arizona State University, Tempe, Arizona, pp. 359–371.

Allman R., Manchee R., Lloyd D. 1993. Flow-cytometric analysis of heterogeneous bacterial populations. In: *Flow cytometry in microbiology.* Lloyd D. (ed.) Springer-Verlag, Berlin, pp. 27–47.

Alonso M. C., Rodríguez J., Borrego J. J. 2002. Characterization of marine bacteriophages isolated from the Alboran Sea (Western Mediterranean). *J. Plankton Res.* **24**: 1079–1087.

Altendorf E., Iverson E., Schutte D., Weigl B., Osborn T. D., Sabet, R., Yager P. 1996. Optical flow cytometry utilizing microfabricated silicon flow channels. *Proc. SPIE* **2678**: 267–273.

Anderson T. W. 1958. *An introduction to multivariate statistical analysis.* J. Wiley, New York.

Anonymous. 1985. *The international system of units (SI) in oceanography.* UNESCO Technical Papers in Marine Science 45, 124 pp.

Arbones B., Figueiras F. G., Zapata M. 1996. Determination of phytoplankton absorption coefficient in natural seawater samples: evidence of a unique equation to correct for the pathlength amplification on glass-fiber filters. *Mar. Ecol. Prog. Ser.* **137**: 293–304.

Archie G. E. 1942. The electrical resistivity log as an aid in determining some reservoir characteristics. *Trans. Am. Inst. Min. Metallurg. Petrol. Eng.* **146**: 54–62.

Arnold S., Neuman M., Pluchino A. B. 1984. Molecular spectroscopy of a single aerosol particle. *Opt. Lett.***9**: 4–6.

Arnott W. P., Moosmuller H., Sheridan P. J., Ogren J. A., Raspert R., Slaton W. V., Hand J. L., Kreidenweis S. M., Collett J. L., Jr. 2003. Photoacoustic and filter-based ambient aerosol light absorption measurements: Instrument comparisons and the role of relative humidity. *J. Geophys. Res. D* **108**: 4034, 10.1029/2002JD002165.

Asano S. 1979. Light scattering properties of spheroidal particles. *Appl. Opt.* **18**: 712–723.

Asano S., Sato M. 1980. Light scattering by randomly oriented spheroidal particles. *Appl. Opt.* **14**: 29–49.

Asano S., Yamamoto G. 1975. Light scattering by a spheroidal particle. *Appl. Opt.* **14**: 29–49.

Ashkin A., Dziedzic J. M. 1981. Observation of optical resonances of dielectric spheres by light scattering. *Appl. Opt.* **20**: 1803–1814.

Ashkin A., Dziedzic J. M., Stolen R. H. 1981. Outer diameter measurement of low birefringence optical fibers by a new resonant backscattering technique. *Appl. Opt.* **20**: 2299–2303.

Ashkin A., Dziedzic J. M., Yamane T. M. 1987. Optical trapping and manipulation of single cells using infrared laser beams. *Nature* **330**: 769–771.

Atkins W. R. G., Poole H. H. 1952. An experimental study of the scattering of light by natural waters. *Proc. Royal Soc. London B* **140**: 321–338.

Atteia O., Kozel R. 1997. Particle size distributions in waters from a karstic aquifer: from particles to colloids. *J. Hydrol.* **201**: 102–119.

Atteia O., Perret D., Adatte T., Kozel R., Rossi P. 1998. Characterization of natural colloids from a river and spring in a karstic basin. *Environ. Geol.* **34**: 257–269.

Aughey W. H., Baum F. J. 1954. Angular-dependence light scattering – a high-resolution recording instrument for the angular range 0.05–140°. *J. Opt. Soc. Am.* **44**: 833–837.

Austin R. W. 1973. Transmittance, scattering, and ocean color. SCOR Discoverer Expedition, May 1970. In: *Measurements of photosynthesis, available radiant flux, and supporting oceanographical data.* Tyler, J. E. (ed.) Data Report, SCOR Expedition, May 1970: Scripps Institution of Oceanography, San Diego, California, USA, pp. J1–J132.

Austin L. G. 1998. Conversion factors to convert particle size distributions measured by one method to those measured by another method. *Part. Part. Syst. Charact.* **15**: 108–112.

Austin R. W., Halikas G. 1976. *The index of refraction of sea water.* SIO Report 76-1, January 1976, Scripps Institution of Oceanography, San Diego, California, USA, 171 pp.

Azam F., Hodson R. E. 1977. Size distribution and activity of marine microheterotrophs. *Limnol. Oceanogr.* **22**: 492–501.

Azam F., Worden A. Z. 2004. Microbes, molecules, and marine ecosystems. *Science* **303**: 1622–1624.

Azetsu-Scott K., Passow U. 2004. Ascending marine particles: Significance of transparent exopolymer particles (TEP) in the upper ocean.*Limnol. Oceanogr.* **49**: 741–748.

Azzam R. M. A. 1978. Photopolarimetric measurement of the Mueller matrix by Fourier analysis of a single detected signal.*Opt. Lett.* **2**: 148–150.

Azzam R. M. A. 1997. Mueller-matrix ellipsometry: a review. In: *Polarization: Measurement, Analysis, and Remote Sensing*. Goldstein D. H. and Chipman R. A. (eds) Proc. SPIE **3121**: 396–399.

B+

Babin M., Morel A., Fournier-Sicre V., Fell F., Stramski D. 2003. Light scattering properties of marine particles in coastal and open ocean waters as related to the particle mass concentration. *Limnol. Oceanogr.* **48**: 843–859.

Babin M., Stramski D. 2002. Light absorption by aquatic particles in the near-infrared spectral region.*Limnol. Oceanogr.* **47**: 911–915.

Babin M., Stramski D. 2003. Relationship between light absorption and chemical composition of mineral particles. In: *Ocean Optics XVI* 18–22 November 2002, Santa Fe, NM, USA. Poster 199.

Babin M., Stramski D. 2004. Variations in the mass-specific absorption coefficient of mineral particles suspended in water. *Limnol. Oceanogr.* **49**: 756–767.

Bagnold R. A., Barndorff-Nielsen O. 1980. The pattern of natural size distributions. *Sedimentology* **27**: 199–207.

Bader H. 1970. The hyperbolic distribution of particle sizes. *J. Geophys. Res.* **75**: 2822–2830.

Baier V. 1996. *Discrimination and description of the form of suspension particles with a videomicroscope and digital image processing*. (In German: *Bestimmung und Beschreibung der Form von Schwebstoffpartikeln mittels Unterwasservideomikroskop und digitalen Bildverarbeitung*). Universitat der Bundeswehr Munchen, Institut fur Wasserwessen Mittelungen, Heft 57/1996, 182 pp.

Baier V., Bechteler W., Sattel H. 1996. An underwater-videomicroscope to determine the size and shape of suspended particles by means of digital image processing. Extended abstract for the *International Offshore and Polar Engineering Conference*. Los Angeles, CA, USA, May 26–31, 7 pp.

Baird M. E., Emsley S. M. 1999. Towards a mechanistic model of plankton population dynamics. *J. Plankton Res.* **21**: 85–126.

Baker E. T., Cannon G. A., Curl H. C., Jr. 1983. Particle transport processes in a small marine bay. *J. Geophys. Res.* **88**: 9661–9669.

Baker E. T., Lavelle J. W. 1984. The effect of particle size on the light attenuation coefficient of natural suspensions.*J. Geophys. Res.* **89**: 8197–8203.

Balch W. M., Drapeau D. T., Cucci T. L., Vaillancourt R. D., Kilpatrick K. A., Fritz J. J. 1999. Optical backscattering by calcifying algae: Separating the contribution of particulate inorganic and organic carbon fractions. *J. Geophys. Res. C* **104**: 1541–1558.

Balch W. M., Vaughn J. M., Novotny J. F., Drapeau D. T., Goes J. I., Booth E., Lapierre J. M., Vining C. L., Ashe A., Vaughn J. M., Jr. 2002. Fundamental changes in light scattering associated with infection of marine bacteria by bacteriophage. *Limnol. Oceanogr.* **47**: 1554–1561.

Balch W. M., Vaughn J., Novotny J., Drapeau D. T., Vaillancourt R., Lapierre J., Ashe A. 2000. Light scattering by viral suspensions. *Limnol. Oceanogr.* **45**: 492–498.

Baldwin R. B. 1965. Mars: an estimate of the age of its surface. *Science* **149**: 1498–1950.

Baldy S. 1988. Bubbles in the close vicinity of breaking waves: Statistical characteristics of the generation and dispersion mechanism. *J. Geophys. Res. C* **93**: 8239–8248.

Bale A. J., Morris A. W. 1991. In situ size measurements of suspended particles in estuarine and coastal waters using laser diffraction. In: *Principles, methods, and application of particle size analysis*. Syvitski J. P. M. (ed.) Cambridge University Press, New York, pp. 197–208.

Bale A. J., Uncles R. J., Widdows M., Brinsley D., Barrett C. D. 2002. Direct observation of the formation and break-up of aggregates in an annular flume using laser reflectance particle sizing. In: *Fine sediment dynamics in the marine environment*. Winterwerp J. C., Kranenburg C. (eds.) Elsevier Science, Amsterdam, pp. 189–201.

Ballenegger V. C., Weber T. A. 1999. The Ewald-Oseen extinction theorem and extinction lengths. *Am. J. Phys.* **67**: 599–605.

Bantle S., Schmidt M., Burchard W. 1982. Simultaneous static and dynamic light scattering. *Macromolecules* **15**: 1604–1609.

Barer R., Joseph S. 1954. Refractometry of living cells. Part I: Basic principles. *Quart. J. Microsc. Sci.* **95**: 399–423.

Barber P. W., Hill S. C. 1990. *Light scattering by particles: computational methods*. World Scientific, Singapore.

Barber P., Yeh C. 1975. Scattering of electromagnetic waves by arbitrarily shaped dielectric bodies. *Appl. Opt.* **14**: 2864–2872.

Barer R. 1955. Spectrophotometry of clarified cell suspensions.*Science* **121**: 709–715.

Barer R., Joseph S. 1954. Refractometry of living cells. Part I: Basic principles. *Quart. J. Microsc. Sci.* **95**: 399–423.

Barnard A. H., Pegau W. S., Zaneveld J. R. V. 1998. Global relationships on the inherent optical properties of the oceans. *J. Geophys. Res. C* **103**: 24,955–24,968.

Barnard J. C., Harrison L. C. 1988. Monotonic responses from monochromatic optical particle counters. *Appl. Opt.* **27**: 584–592.

Barndorff-Nielsen O. 1977. Exponentially decreasing distributions for the logarithm of particle size. *Proc. Royal Soc. London* **353**: 401–419.

Barndorff-Nielsen O. 1978. Hyperbolic distributions and distributions on hyperbolae. *Scand. J. Statistics* **5**: 151–157.

Barrett P., Glennon B. 1999. In-line FBRM monitoring of particle size in dilute agitated suspensions.*Part. Part. Syst. Charact.* **16**: 207–211.

Barth H., Grisard K., Holtsch K., Reuter R., Stute U. 1997. Polychromatic transmissometer for in situ measurements of suspended particles and gelbstoff in water. *Appl. Opt.* **36**: 7919–7928.

Barth H. G., Sun S.-T. 1991. Particle size analysis. *Anal. Chem.* **63**: 1R–10R.

Barthel H., Sachweh B., Ebert F. 1998. Measurement of airborne mineral fibres using a new differential light scattering device. *Meas. Sci. Technol.* **9**: 210–220.

Bartholdi M. G., Salzman G. C., Hiebert R. D., Kerker M. 1980. Differential light scattering photometer for rapid analysis of single particles in flow. *Appl. Opt.* **19**: 1573–1581.

Barton J. P. 2002. Electromagnetic field calculations for an irregularly shaped, near-spheroidal particle with arbitrary illumination. *J. Opt. Soc. Am. A* **19**: 2429–2435.

Batchelder J. S., Taubenblatt M. A. 1989. Interferometric detection of forward scattered light from small particles. *Appl. Phys. Lett.* **55**: 215–217.

Bateman J. B., Wagman J., Carstensen E. L. 1966. Refraction and absorption of light in bacterial suspensions. *Kolloid-Zeitschrift und Zeitschrift fur Polymer* **208**: 44–58.

Bates A. P., Hopcraft K. I., Jakeman E. 1997. Particle shape determination from polarization fluctuations of scattered radiation. *J. Opt. Soc. Am. A* **14**: 3372–3378.

Batz-Sohn C. 2003. Particle sizes of fumed oxides: A new approach using PCS signals. *Part. Part. Syst. Charact.* **20**: 370–378.

Bauer D., Ivanoff A. 1965. On the measurements of the light scattering coefficient of seawater at angles between 14 and 1° 30′. (In French: *Au sujet de la mesure du coefficient de diffusion de la lumiere par les eaux de mer pour des angles compris entre 14 et 1 degres 30′*) *Estr. C. R. Acad. Sc. Paris* **260**: 631–634.

Bauer D., Morel A. 1967. A study of the scattering function of seawater at small angles. (In French: *Etude aux petit angles de l'indicatrice de diffusion de la lumiere par les eaux de mer.*) *Ann. Geophys.* **23**: 109–123.

Baum F. J., Billmeyer F. W. 1961. Automatic photometer for measuring the angular dissymmetry of light scattering. *J. Opt. Soc. Am.* **51**: 452–456.

Bazylinski D. A., Frankel R. B. 2000. Biologically controlled mineralization of magnetic iron minerals by magnetotactic bacteria. In: *Environmental microbe–metal interactions.* Lovley D. R. (ed.) ASM Press, Washington DC, pp. 109–143.

Beardsley G. F., Jr. 1966. *The polarization of the near asymptotic light field in sea water.* Ph.D. Thesis, MIT, Boston, Massachusetts, USA.

Beardsley G. F., Jr. 1968. Mueller scattering matrix of sea water. *J. Opt. Soc. Am.* **58**: 52–57.

Beardsley G. F., Jr., Zaneveld J. R. V. 1969. Theoretical dependence of the near-asymptotic apparent optical properties on the inherent optical properties of sea water. *J. Opt. Soc. Am.* **49**: 373–377.

Beaulieu S., Mullin M., Tang V., Pyne S., King A., Twining B. 1999. Using an optical plankton counter to determine the size distributions of preserved zooplankton samples. *J. Plankton Res.* **21**: 1939–1956.

Beckett R., Murphy D., Tadjiki S., Chittleborough D., Giddings J. C. 1997. Determination of thickness, aspect ratio and size distributions for platey particles using sedimentation field-flow fractionation and electron microscopy. *Coll. Surf. A* **120**: 17–26.

Beckett R., Nicholson G., Hart B. T., Hansen M. E., Giddings J. C. 1988. Separation and size characterization of colloidal particles in river water by sedimentation field-flow fractionation. *Water Res.* **22**: 1535–1545.

Beckett R., Nicholson G., Hotchin D. M., Hart B. T. 1992. The use of sedimentation field-flow fractionation to study suspended particulate matter. *Hydrobiologia* **235/236**: 697–710.

Belgrano A., Allen A. P., Enquist B. J., Gillooly J. F. 2002. Allometric scaling of maximum population density: a common rule for marine phytoplankton and terrestrial plants. *Ecol. Lett.* **5**: 611–613.

Bell G. J., Anderson E. C. 1967. Cell growth and division: I. A mathematical model with applications to cell distributions in mammalian suspension cultures. *Biophys. J.* **7**: 329–351.

Berge L. I., Feder J., Jossang T. 1989. A novel method to study single-particle dynamics by resistive pulse technique. *Rev. Sci. Instrum.* **60**: 2756–2763.

Berge L. I., Jossang T., Feder J. 1990. Off-axis response for particles passing through long apertures in Coulter-type counters. *Meas. Sci. Technol.* **1**: 471–474.

Bergh O., Borsheim G., Bratbak G., Heldal M. 1989. High abundances of viruses found in aquatic environments. *Nature* **340**: 467–468.

Berne B. J., Pecora R. 1976. *Dynamic light scattering*. J. Wiley, New York.

Berry H. G., Gabrielsa G., Livingston A. E. 1977. Measurement of the Stokes parameters of light. *Appl. Opt.***16**: 3200–3205.

Beuttell R. G., Brewer A. W. 1949. Instrument for the measurement of the visual range. *J. Sci. Instrum.* **26**: 357–359.

Beyer G. L. 1987. Particle counter for rapid determination of size distributions. *J. Colloid Interface Sci.* **118**: 137–147.

Bhandari R. 1985. Scattering coefficients for a multi-layered sphere. *Appl. Opt.* **24**: 1960–1964.

Bickel W. S., Davidson J. F., Huffman D. R., Kilkson R. 1976. Application of polarization effects in light scattering: A new biophysical tool. *Proc. Nat. Acad. Sci. USA* **73**: 486–490.

Bickel W. S., Stafford M. E. 1981. Polarized light scattering from biological systems: a technique for cell differentiation. *J. Biol. Phys.* **9**: 53–66.

Bigg P. H. 1967. Density of water in SI units over the range 0–40°C. *Br. J. Appl. Phys.* **18**: 521–525.

Binder B. J., Chisholm S. W., Olson R. J., Frankel S. L., Worden A. Z. 1996. Dynamics of picophytoplankton, ultraphytoplankton and bacteria in the central equatorial Pacific. *Deep-Sea Res. II* **43**: 907–931.

Bird D. F., Kalff J. 1984. Empirical relationships between bacterial abundance and chlorophyll concentration in fresh and marine waters. *Can. J. Fish. Aquat. Sci.* **41**: 1015–1023.

Bishop J. K. B., Edmond J. M., Ketten D. R., Bacon M. P., Silker W. B. 1977. The chemistry, biology, and vertical flux of particulate matter from the upper 400 m of the equatorial Atlantic Ocean. *Deep-Sea Res.* **25**: 511–548.

Bishop, J. K. B., Ketten D. R., Edmond J. M. 1978. The chemistry, biology, and vertical flux of particulate matter from the upper 400 m of the Cape Basin in the southeast Atlantic Ocean. *Deep-Sea Res.* **24**: 1121–1161.

Black D. L., McQuay M. Q., Bonin M. P. 1996. Laser-based techniques for particle size measurement: A review of sizing methods and their industrial applications. *Prog. Energy Combust. Sci.* **22**: 267–306.

Blackburn N., Fenchel T., Mitchell J. 1998. Microscale nutrient patches in planktonic habitats shown by chemotactic bacteria. *Science* **282**: 2254–2256.

Blackburn N., Hagström Å., Wikner J., Cuadros-Hansson R., Bjørnsen P. K. 1998. Rapid determination of bacterial abundance, biovolume, morphology, and growth by neural network-based image analysis. *Appl. Environ. Microbiol.* **64**: 3246–3255.

Blair D. P., Sydenham P. H. 1975. Phase sensitive detection as a means to recover signals buried in noise. *J. Phys. E.: Sci. Instrum.* **8**: 621–627.

Blanco A., De Tomasi F., Filippo E., Manno D., Perrone M. R., Serra A., Tafuro A. M., Tepore A. 2003. Characterization of African dust over southern Italy.*Atmos. Chem. Phys. Discuss.* **3**: 4633–4670.

Blanco J. M., Echevarria F., Garcia C. M. 1994. Dealing with size-spectra: Some conceptual and mathematical problems. *Sci. Mar.* **58**: 17–29.

Bloem J., Veninga M., Shephard J. 1995. Fully automatic determination of soil bacterium numbers, cell volumes, and frequencies of dividing cells by confocal laser scanning microscopy and image analysis. *Appl. Environ. Microbiol.* **61**: 926–936.

Boal D. 1999. Mechanical characteristics of very small cells. In: *Size limits of very small microorganisms. Proceedings of a Workshop.* Washington, DC, October 22–23, 1998, National Academy Press, Washington, DC, pp. 26–31.

Bogucki D. J., Domaradzki J. A., Stramski D., Zaneveld J. R. 1998. Comparison of near-forward light scattering on oceanic turbulence and particles. *Appl. Opt.* **37**: 4669–4677.

Bogucki D., Domaradzki A., Zaneveld R., Dickey T. 1994. Light scattering by turbulent flow. *Ocean Optics XII. Proc. SPIE* **2258**: 247–255.

Bohling B. 2005. Variations in grain size analysis with a time-of-transition laser sizer (Galai CIS-50) using a gravitational flow system. *Part. Part. Syst. Charact.* **21**: 455–462.

Bohren C. F. 1987. Multiple scattering of light and some of its observable consequences. *Am. J. Phys.* **55**: 524–533.

Bohren C. F., Huffman D. 1983. *Absorption and scattering of light by small particles.* Wiley, New York.

Bohren C. F., Singham S. A. B. 1991. Backscattering by nonspherical particles: A review of methods and suggested new approaches. *J. Geophys. Res. C* **96**: 5269–5277.

Boido J. 1947. A study of the settling velocity of lamellar and aspherical particles. (In French: *Etude de la vitesse de chute des particules lamellaires et aciculaires*) *Peintures et fragments*, T5, N° 2–5, Avril 1947.

Boivin L. P., Davidson W. F., Storey R. S., Sinclair D., Earle E. D. 1986. Determination of the attenuation coefficient of visible and ultraviolet radiation in heavy water. *Appl. Opt.* **25**: 877–882.

Bold H. C., Wynne M. L. 1985. *Introduction to the algae: structure and reproduction.* Prentice Hall, Englewood Cliffs, New Jersey, 189 pp.

Bondoc L. L., Jr., Fitzpatrick S. 1998. Size distribution analysis of recombinant adenovirus using disc centrifugation. *J. Ind. Microbiol. Biotechnol.* **20**: 317–322.

Boney A. D. 1989. *Phytoplankton.* Edward Arnold, London, 118 pp.

Boon J. P., Nossal R., Chen S.-H. 1974. Light scattering spectrum due to wiggling motions of bacteria. *Biophys. J.* **14**: 847–864.

Borgmann U. 1987. Model on the slope of, and biomass flow up, the biomass size spectrum. *Can. J. Fish. Aquat. Sci.* **44**: 136–140.

Born M., Wolf E. 1980. *Principles of optics* (6th ed.) Pergamon Press, New York, 808 pp.

Boss E., Pegau W. S. 2001. Relationship of light scattering at an angle in the backward direction to the backscattering coefficient. *Appl. Opt.***40**: 5503–5507.

Boss E., Pegau W. S., Lee M., Twardowski M., Shybanov E., Korotaev G., Baratange F. 2004. Particulate backscattering ratio at LEO 15 and its use to study particle composition and distribution. *J. Geophys. Res. C* **109**: C01014.

Boss E., Twardowski M. S., Herring S. 2001. Shape of the particulate beam attenuation spectrum and its inversion to obtain the shape of the particulate size distribution. *Appl. Opt.* **40**: 4885–4893.

Botet R., Rannou P., Cabane M. 1997. Mean-field approximation of Mie scattering by fractal aggregates of identical spheres. *Appl. Opt.* **36**: 8791–8797.

Bottlinger M., Umhauer H. 1989. Single particle light scattering size analysis: quantification and elimination of the effect of particle shape and structure.*Part. Part. Syst. Charact.* **6**: 100–109.

Boudreau P. R., Dickie L. M., Kerr S. R. 1991. Body-size spectra of production and biomass as system-level indicators of ecological dynamics. *J. Theor. Biol.* **152**: 329–339.

Bouvier T., Troussellier M., Anzil A., Courties C., Servais P. 2001. Using light scatter signal to estimate bacterial biovolume by flow cytometry. *Cytometry* **44**: 188–194.

Bowen J. D., Stolzenbach K. D., Chisholm S. W. 1993. Simulating bacterial clustering around phytoplankton cells in a turbulent ocean. *Limnol. Oceanogr.* **38**: 36–51.

Boxman A., Merkus H. G., Verheijen P. J. T., Scarlett B. 1991. Deconvolution of light-scattering patterns by observing intensity fluctuations. *Appl. Opt.* **30**: 4818–4823.

Boyd C. M., Johnson G. W. 1995. Precision of size determination of resistive particle counters. *J. Plankton Res.* **17**: 41–58.

Bradtke K. 2004. The suspension field and its influence on optical properties of coastal waters as exemplified by the waters of the Gdańsk Bay, Baltic Sea. (In Polish: *Pole zawiesiny i jego wpływ na właściwości optyczne wód przybrzeżnych na przykładzie Zatoki Gdańskiej*). Ph.D. Thesis, Department of Physical Oceanography, Institute of Oceanography, Gdansk University, Gdynia, Poland, 173 pp.

Brasil A. M., Farias T. L., Carvalho M. G. 1999. A recipe for image characterization of fractal-like aggregates. *J. Aerosol Sci.* **30**: 1379–1389.

Bratbak G., Haslund O. H., Heldal M., Næss A, Roeggen T. 1992. Giant marine viruses? *Mar. Ecol. Prog. Ser.* **85**: 201–202.

Bratbak G., Heldal M., Norland S., Thingstad T. F. 1990. Viruses as partners in spring bloom microbial trophodynamics. *Appl. Environ. Microbiol.* **56**: 1400–1405.

Braun D., Fromherz P. 1997. Fluorescence interference-contrast microscopy of cell adhesion on oxidized silicon. *Appl. Phys. A* **65**: 341–348.

Breitenbach E., Lampe R., Leipe T.1999. Investigations on mineralogical and chemical composition of suspended particulate matter (SPM) in the Odra estuary. *Acta Hydrochim. Hydrobiol.* **27**: 298–302.

Brett M. T., Lubnow F. S., Villar-Argaiz M., Müller-Solger A., Goldman C. R. 1999. Nutrient control of bacterioplankton and phytoplankton dynamics. *Aquatic Ecol.* **33**: 135–145.

Bricaud A., Babin M., Morel A., Claustre H. 1995. Variability in the chlorophyll-specific absorption coefficients of natural phytoplankton: analysis and parametrization. *J. Geophys. Res.* **100**: 13,321–13,332.

Bricaud A., Bedhomme A. L., Morel A. 1988. Optical properties of diverse phytoplanktonic species: experimental results and theoretical interpretation. *J. Plankton Res.* **10**: 851-873.

Bricaud A., Morel A. 1986. Light attenuation and scattering by phytoplanktonic cells: a theoretical modeling. *Appl. Opt.* **25**: 571–580.

Bricaud A., Morel A., Prieur L. 1983. Optical efficiency factors of some phytoplankters. *Limnol. Oceanogr.* **28**: 816–832.

Bricaud A., Stramski D. 1990. Spectral absorption coefficients of living phytoplankton and non-algal biogenous matter: A comparison between the Peru upwelling area and the Sargasso Sea. *Limnol. Oceanogr.* **35**: 562–582.

Brice B. A., Halwer M., Speiser R. 1950. Photoelectric light scattering photometer for determining high molecular weights. *J. Opt. Soc. Am.* **40**: 768–778.

Brogioli D., Vailati A., Giglio M. 2002. Heterodyne near-field scattering. *Appl. Phys. Lett.* **81**: 4109–4111.

Brown O. B., Gordon H. R. 1973. Two component Mie scattering models of Sargasso Sea particles.*Appl. Opt.* **12**: 2461–2465.

Brown O. B., Gordon H. R. 1974. Size-refractive index distribution of clear coastal water particulates from light scattering. *Appl. Opt.* **13**: 2874–2881.

Brown D. J., Vickers G. T. 1998. The use of projected area distribution functions in particle shape measurement. *Powder Technol.* **98**: 250–257.

Brown W. K., Wohletz K. H. 1995. Derivation of the Weibull distribution based on physical principles and its connection to the Rosin-Rammler and the lognormal distributions. *J. App. Phys.* **78**: 2758–2763.

Brun-Cottan J. 1971. Study of the particle size distribution as measured with a Coulter counter. (In French: *Etude de la granulometrie des particules, mesures effectuées avec un compteur Coulter*). *Cah. Oceanogr.* **23**: 193–205.

Brun-Cottan J. C. 1976. Stokes settling and dissolution rate model for marine particles as a function of size distribution. *J. Geophys. Res.* **81**: 1601–1606.

Brun-Cottan J. C. 1986. Vertical transport of particles within the ocean. In: *The role of air-sea exchange in geochemical cycling*. Buat-Mdnard P. (ed.) D. Reidel Publishing Co., pp. 83–111.

Brunsting A., Mullaney P. F. 1972. A light scattering photometer using photographic film. *Rev. Sci. Instrum.* **43**: 1514–1519.

Brunsting A., Mullaney P. F. 1974. Differential light scattering from spherical mammalian cells.*Biophys. J.* **14**: 439–453.

Bryant F. D., Seiber B. A., Latimer P. 1969. Absolute optical cross sections of cells and chloroplasts.*Arch. Biochem. Biophys.* **135**: 97–108.

Buck K. R., Chavez F. P. 1994. Diatom aggregates in the open ocean. *J. Plankton Res.* **16**: 1449–1457.

Buiteveld H., Hakvoort J. H., Donze M. 1994. The optical properties of pure water. In: *Ocean Optics XII*. Jaffe J. J. (ed.) *Proc. SPIE* **2258**: 174–183.

Bukata R. P., Bruton J. E., Jerome J. H. 1980. Conceptual approach to the simultaneous determination of the backscatter and absorption coefficients of natural waters. *Appl. Opt.***19**: 1550–1559.

Bunville L. G. 1984. Commercial instrumentation for particle size analysis. In: *Modern methods of particle size analysis*. Barth, H. G. (ed.) *Chemical Analysis* **73**, Wiley, New York, pp. 1–42.

Burkert U., Hyenstrand P., Drakare S., Blomqvist P. 2001. Effects of the mixotrophic flagellate Ochromonas sp. on colony formation in *Microcystis aeruginosa*. *Aquat. Ecol.* **35**: 11–17.

Burkholz A., Polke R. 1984. Laser diffraction spectrometers/experience in particle size analysis. *Part. Charact.* **1**: 153–161.

Burns M., Francke R., Austin M., Ryan L. W., Jr., Feld M. S. 1976. Laser spectrophotometer for measuring differential cross section of biological particles. *Appl. Opt.* **15**: 672–676.

Bush J. 1951. Derivation of a size-frequency curve from the cumulative curve. *J. Sediment. Petrol.* **21**: 178–182.

Bushell G. C., Amal R. 1998. Fractal aggregates of polydisperse particles. *J. Colloid Interface Sci.* **205**: 459–469.

Bushell G. C., Yan Y. D., Woodfield D., Raper J., Amal R. 2002. On techniques for the measurement of the mass fractal dimension of aggregates. *Adv. Colloid Interface Sci.* **95**: 1–50.

Butler W. 1962. Absorption of light by turbid materials. *J. Opt. Soc. Am.* **52**: 292–299.

Butler A. A., Butler P. B., Luerkens D. W. 1989. Single-particle statistics derived from light-blockage instruments. *Powder Technol.* **57**: 143–149.

C+

Camacho J. 2001. Scaling in steady-state aggregation with injection. *Phys. Rev. E* **63**, paper 046112, 4 pp.

Camacho J., Solé R. V. 2001. Scaling laws in ecological size spectra. *Europhys. Lett.* **55**: 774–780.

Campbell J. W. 1995. The lognormal distribution as a model for bio-optical variability in the sea. *J. Geophys. Res.* **100C**: 13,237–13,254.

Campbell J. W., Yentsch C. M. 1989. Variance within homogeneous phytoplankton populations, I: Theoretical framework for interpreting histograms. *Cytometry* **10**: 587–595.

Campbell J. W., Yentsch C. M., Cucci T. L. 1989. Variance within homogeneous phytoplankton populations, III: Analysis of natural populations. *Cytometry* **10**: 605–611.

Carder K. L. 1978. A holographic micro-velocimeter for use in studying ocean particle dynamics. *Ocean Optics V. Proc. SPIE* **160**: 63–66.

Carder K. L., Beardsley G. F., Jr., Pak H. 1971. Particle size distributions in the eastern equatorial Pacific. *J. Geophys. Res.* **76**: 5070–5077.

Carder K. L., Betzer P. R., Eggiman D. W. 1974. Physical, chemical and optical measures of suspended particle concentrations: their intercomparison and application to the West African shelf. In: *Suspended solids in water.* Gibbs J. R. (ed.) Plenum Press, New York.

Carder K. L., Costello D. K. 1994. Optical effects of large particles. In*: Ocean optics* Spinrad R. W., Carder L. K., Perry M. J. (eds.) Oxford University Press, New York, pp. 243–257.

Carder K. L., Meyers D. J. 1979. New optical techniques for particle studies in the bottom boundary layer. *Ocean Optics VI. Proc. SPIE* **208**: 151–158.

Carder K. L., Steward R. G., Betzer P. R. 1982. In situ holographic measurements of the sizes and settling rates of oceanic particulates.*J. Geophys. Res. C* **87**: 5681–5685.

Carder K. L., Steward R. G., Betzer P. R., Johnson D. L., Prospero J. M. 1986. Dynamics and composition of particles from an aeolian input event into the Sargasso Sea. *J. Geophys. Res. D* **91**: 1055–1066.

Carder K. L., Stewart R. G., Harvey G. R., Ortner P. B. 1989. Marine humic and fulvic acids: their effect on remote sensing of ocean chlorophyll. *Limnol. Oceanogr.* **34**: 68–81.

Carder K. L., Tomlinson R. D., Beardsley G. F., Jr. 1972. A technique for the estimation of indices of refraction of marine phytoplankters. *Limnol. Oceanogr.* **17**: 833–839.

Carrias J.-F., Serre J.-P., Sime-Ngando T., Amblard C. 2002. Distribution, size, and bacterial colonization of pico- and nano-detrital organic particles (DOP) in two lakes of different trophic status. *Limnol. Oceanogr.* **47**: 1202–1209.

Casperson L. W. 1977. Light extinction in polydisperse particulate systems. *App. Opt.* **16**: 3183–3189.

Cavender-Bares K. K., Frankel S. L., Chisholm S. W. 1998. A dual sheath flow cytometer for shipboard analyses of phytoplankton communities from the oligotrophic oceans. *Limnol. Oceanogr.* **43**: 1383–1388.

Cavender-Bares K., Rinaldo A., Chisholm S. W. 2001. Microbial size spectra from natural and nutrient enriched ecosystems. *Limnol. Oceanogr.* **46**: 2001, 778–789.

Cecchi G. C. 1991. Error analysis of the parameters of a least-squares determined curve when both variables have uncertainties.*Meas. Sci. Technol.* **2**: 1127–1128.

Ceronio A. D., Haarhoff J. 2005. An improvement on the power law for the description of particle size distributions in potable water treatment. *Water Res.* **39**: 305–313.

Chalmers J. J., Haam S., Zhao Y., McCloskey K., Moore L., Zborowski M., Williams P. S. 1999. Quantification of cellular properties from external fields and resulting induced velocity: cellular hydrodynamic diameter. *Biotechnol. Bioeng.* **64**: 509–518.

Chamberlin T. C. 1899. An attempt to frame a working hypothesis of the cause of glacial periods on an atmospheric basis. *J. Geology* **7**: 545–584.

Chandrasekhar S. 1960. *Radiative transfer.* Dover Publications, New York, 393 pp.

Charlson R. J. 1993. *An annotated bibliography for the integrating nephelometer.* TSI Application Paper #2, 20 pp.
Charney E., Brackett F. S. 1961. The spectral dependence of scattering from a spherical alga and its implications for the state of organization of the light-absorbing pigments. *Arch. Biochem. Biophys.* **91**: 1–12.
Chen T. W. 1984. Generalized eikonal approximation. *Phys. Rev.* **30**: 585–592.
Chen T. W. 1989. High energy light scattering in the generalized eikonal approximation. *Appl. Opt.* **28**: 4096–4102.
Chen T. W. 1993. Simple formula for light scattering by a large spherical dielectric. *Appl. Opt.* **32**: 7568–7571.
Chen S., Eisma D. 1995. Fractal geometry of in situ flocs in the estuarine and coastal environments. *Neth. J. Sea Res.* **32**: 173–182.
Chen S., Eisma D., Kalf J. 1994. In situ distribution of suspended matter during the tidal cycle in the Elbe estuary. *Neth. J. Sea Res.* **32**: 37–48.
Chen T. W., Smith W. S. 1992. Large-angle light scattering at large size parameters. *Appl. Opt.* **31**: 6558–6560.
Cheng R. J. 1980. Physical properties of atmospheric particulates. In: *Light scattering by irregularly shaped particles.* Schuerman D. W. (ed.) Plenum Press, New York, pp. 69–78.
Chernyshev A. V., Prots V. I., Doroshkin A. A., Maltsev V. P. 1995. Measurement of scattering properties of individual particles with a scanning flow cytometer. *Appl. Opt.* **34**: 6301–6305.
Chevaillier J.-P., Fabre J., Hamelin P. 1986. Forward scattered light intensities by a sphere located anywhere in a Gaussian beam. *Appl. Opt.* **25**: 1222–1225.
Chilton F., Jones D. D., Talley W. K. 1969. Imaging properties of light scattered by the sea. *J. Opt. Soc. Am.* **59**: 891–898.
Chin A. D., Butler P. B., Luerkens D. W. 1988. Influence of particle shape on size measured by the light-blockage technique. *Powder Technol.* **54**: 99–105.
Chin W. C., Orrelana M. V., Verdugo V. 1998. Spontaneous assembly of marine dissolved organic matter into polymer gels. *Nature* **391**: 568–572.
Chisholm S. W. 1992. Phytoplankton size. In: *Primary production and biogeochemical cycles in the sea.* Falkowski P., Woodhead A. D. (eds.) Plenum Press, New York.
Chisholm S. W., Olson R. J., Zettler E. R., Goericke R. Waterbury J. B., Wechselmeyer N. A. 1988. A novel free-living prochlorophyte abundant in the oceanic euphotic zone. *Nature* **334**: 340–343.
Cho B. C., Azam F. 1990. Biogeochemical significance of bacterial biomass in the ocean's euphotic zone. *Mar. Ecol. Prog. Ser.* **63**: 253–259.
Christiansen C., Hartmann D. 1991. The hyperbolic distribution. In: *Principles, methods, and application of particle size analysis.* Syvitski J. P. M. (ed.) Cambridge University Press, New York, pp. 237–248.
Chýlek P, Dobbie J. S., Geldart D. J. W, Tso H. C. W., Videen G. 2000. Effective medium approximation for heterogeneous particles. In: *Light scattering by nonspherical particles: theory, measurements, and applications.* Mishchenko M. (ed.) Academic Press, San Diego, pp. 273–308.
Chýlek P., Grams G., Pinnick R. G. 1976. Light scattering by irregular randomly oriented particles. *Science* **193**: 480–482.
Chýlek P., Kiehl J. T., Ko M. K. W. 1978. Optical levitation and partial-wave resonances. *Phys. Rev. A* **18**: 2229–2233.

Chýlek P., Ramaswamy V., Ashkin A., Dziedzic J. M. 1983. Simultaneous determination of refractive index and size of spherical dielectric particles from light scattering data. *Appl. Opt.* **22**: 2302–2307.

Chýlek P., Srivastava V., Pinnick R. G., Wang R. T. 1988. Scattering of electromagnetic waves by composite spherical particles: experiment and effective medium approximations. *Appl. Opt.* **27**: 2396–2404.

Chung Y. 1982. Suspended particulates in ocean waters: abundance and size distribution as determined by a laser particle counter. *Proc. Geol. Soc. China* **25**: 102–110.

Claquin T. 1999. *Modeling of the mineralogy and the radiative forcing of desert dust.* Ph.D. Thesis, Paris University, Paris, France.

Claquin T., Schulz M., Balkanski Y. J. 1999. Modeling the mineralogy of atmospheric dust sources. *J. Geophys. Res.* **104**: 22, 243–22, 256.

Clarke A. D., Shinozuka Y., Kapustin V. N., Howell S., Huebert B., Doherty S., Anderson T., Covert D., Anderson J., Hua Z., Moore K. G. II, McNaughton C., Carmichael G. 2004. Size-distributions and mixtures of dust and black carbon aerosol in Asian outflow: Physio-chemistry and optical properties. *J. Geophys. Res. D* **109**, 15S09, doi:10.1029/2003JD004378.

Claustre H., Fell F., Oubelkheir K., Prieur L. 2000. Continuous monitoring of surface optical properties across a geostrophic front: Biogeochemical inferences.*Limnol. Oceanogr.* **45**: 309–321.

Cleveland J. S., Weidemann A. D. 1993. Quantifying absorption by aquatic particles: A multiple scattering correction for glass-fiber filters. *Limnol. Oceanogr.* **38**: 1321–1327.

Clift R., Grace J. R., Weber M. E. 1978. *Bubbles, drops, and particles.* Academic Press, New York.

Cole J. J. 2000. Microbial carbon cycling in pelagic ecosystems: microbial methods for ecosystem scientist. In: *Methods in ecosystem science.* Sala O. E., Jackson R. B., Mooney H. A., Howarth R. W. (eds.) Springer-Verlag, New York, NY, pp. 138–150.

Cole J. J., Findlay S., Pace M. L. 1988. Bacterial production in fresh and saltwater ecosystems: a cross-system overview. *Mar. Ecol. Progr. Ser.* **43**: 1–10.

Cole J. J., Pace M. L., Caraco N. F., Steinhart G. S. 1993. Bacterial biomass and cell size distributions in lakes: More and larger cells in anoxic waters. *Limnol. Oceanogr.* **38**: 1627–l632.

Coles H. J., Jennings B. R., Morris V. J. 1975. Refractive index increment measurements for bacterial suspensions. *Phys. Med. Biol.* **20**: 310–313.

Collins J. F, Richmond M. H. 1962. Rate of growth of *Bacillus* cereus between divisions. *J. Gen. Microbiol.* **28**: 15–33.

Conklin W. B., Oliver J. P., Strickland M. L. 1998. Capturing static light scattering data using a high-resolution charge-coupled device detector. In: *Particle size distribution III*, Provder T. (ed.) ACS Symposium Series 693.

Contado C., Blo G., Fagioli F., Dondi F., Beckett R. 1997. Characterisation of river Po particles by sedimentation field-flow fractionation coupled to GFAAS and ICP-MS. *Coll. Surf. A* **120**: 47–59.

Conwell P. R., Rushforth C. K., Benner R. E., Hill S. C. 1984. Efficient automated algorithm for the sizing of dielectric microspheres using the resonance spectrum. *J. Opt. Soc. Am. A* **1**: 1181–1186.

Cornette W. M., Shanks J. G. 1992. Physically reasonable analytic expression for the single-scattering phase function. *Appl. Opt.* **31**: 3152–3160.

Cornillault J. 1972. Particle size analyzer. *Appl. Opt.* **11**: 265–268.

Costello D. K., Carder K. L., Betzer P. R., Young R. W. 1989. In-situ holographic imaging of settling particles: applications for individual particle dynamics and oceanic flux measurements. *Deep-Sea Res.* **36**: 1595–1605.

Costello D. K., Carder K. L., Steward R. G. 1991. Development of the marine-aggregated-particle profiling and enumerating rover. In: *Underwater imaging, photography, and visibility*. Spinrad R. W. (ed.) *Proc. SPIE* **1537**: 161–172.

Costello D. K., Hou W., Carder K. L. 1994. Some effects of the sensitivity threshold and spatial resolution of a particle imaging system on the shape of the measured particle size distribution. In: *Ocean Optics XII*. Jaffe J. J. (ed.) *Proc. SPIE* **2258**: 768–783.

Costello D. K., Carder K. L., Hou W. 1995. Aggregation of diatom bloom in a mesocosm: Bulk and individual particle optical measurements. *Deep-Sea Res. II* **42**: 29–45.

Courp T., Eisma D., Kalf J. 1993. In situ observation of flocs in natural waters. *Ann. Inst. Oceanogr.* **69**: 184–188.

Cowan M. P., Harfield J. G. 1990. The linearity and response of focused apertures. *Part. Part. Syst. Charact.* **7**: 1–5.

Cowen J. P., Holloway C. 1996. Sequential imaging of marine aggregates by in situ macrophotography and laser confocal and electron microscopy. *Mar. Biol.* **126**: 163–174.

Cox R. A. 1965. The physical properties of sea water. In: *Chemical oceanography*. Riley J. P. and Skirrow G. (eds.) Academic Press, London, pp. 73–117.

Craig C., Alexander S., Hendry D. C., Hobson P. R., Lampitt R. S., Lucas-Leclin B., Nareid H., Nebrensky J. J., Player M. A., Saw K., Tipping K., Watson J. 2000. HoloCam: A subsea holographic camera for recording marine organisms and particles. In: *Optical diagnostics for industrial applications*. Halliwell Neil A. (ed.) *Proc. SPIE* **4076**: 111–119.

Crawford F. S. 1968. *Waves*. McGraw Hill, New York, 600 pp.

Crittenden E. C., Cooper A. W., Milne E. A., Rodeback G. W., Armstead R. L., Kalmback S. H., Land D., Katz B. 1978. Optical resolution in the turbulent atmosphere of the marine boundary layer. Naval Postgraduate School, Monterey, CA, USA, Report No. NPS-61-78-003, February 1978.

Cross D. A., Latimer P. 1972. Angular dependence of scattering from *Escherichia coli* cells. *Appl. Opt.* **11**: 1225–1228.

Crouse R. F., Latimer P. 1990. Particle size determination by Fourier transform of scattered light. *Proceedings of the 2nd International Congress on Optical Particle Sizing*, March 2–8, 1990, Tempe, Arizona, USA, pp. 9–11.

Crow E. L. 1988. Applications [of lognormal distribution] in atmospheric sciences. In: *Lognormal distributions: theory and applications*. Crow E. L. and Shimizu K. (eds.) Marcell Dekker, New York, pp. 331–356.

Cunningham A. 1990. A low-cost, portable flow cytometer specifically designed for phytoplankton analysis. *J. Plankton Res.* **12**: 149–160.

Curcio J. A., Petty C. C. 1951. The near infrared absorption spectrum of liquid water. *J. Opt. Soc. Am.* **41**: 302–304.

Cyr H. 2000. Individual energy use and the allometry of population density. In:*Scaling in biology*. Brown J. H., West G. B. (eds.) Oxford University Press, pp. 267–295.

Cyr H., Downing J. A., Peters R. H. 1997a. Density-body size relationships in local aquatic communities. *Oikos* **79**: 333–346.

Cyr H., Peters. R. H., Downing J. A. 1997b. Population density and community size structure: Comparison of aquatic and terrestrial systems. *Oikos* **80**: 139–149.

D+

Dana D., Mafione R., Coenen P. 1998. A new in-situ instrument for measuring the backward scattering and absorption coefficients simultaneously. In: *Ocean Optics XIV*, Kailua-Kona, Hawaii, November 10–13, 1998.

D'Arrigo J. S. 1984. Surface properties of microbubble-surfactant monolayers at the air-water interface. *J. Colloid Interface Sci.* **100**: 106–111.

D'Arrigo J. S., Saiz-Jimenez C., Reimer N. S. 1984. Surface properties of microbubble-surfactant monolayers at the air-water interface. *J. Colloid Interface Sci.* **100**: 96–105.

Davey H. M., Davey C. L., Kell D. B. 1993. On the determination of the size of microbial cells using flow cytometry. In: *Flow cytometry in microbiology*. D. Lloyd (ed.) Springer-Verlag London, pp.49–65.

Davey H. M., Kell D. B. 1996. Flow cytometry and cell sorting of heterogeneous microbial populations: the importance of single-cell analyses. *Microbiol. Rev.* **60**: 641–696.

Davidson J. A., Collins E. A., Haller H. S. 1971. Latex particle size analysis. Part III: Particle size distribution by flow ultramicroscopy. *J. Polym. Sci.* **35C**: 235–255.

Davidson K., Roberts E. C., Gilpin L. C. 2002. The relationship between carbon and biovolume in marine microbial mesocosms under different nutrient regimes. *Eur. J. Phycol.* **37**: 501–507.

de Boer D. H., Stone M. 1999. Fractal dimensions of suspended solids in streams: comparison of sampling and analysis techniques. *Hydrol. Processes* **13**: 249–254.

Dean P. N. 1990. Commercial [flow cytometric] instruments. In: *Flow cytometry and sorting*. Melamed M. R., Lindmo T., Mendelsohn M. L. (eds.) Wiley-Liss, New York, pp. 171–186.

Deirmendijan D. 1969. *Electromagnetic scattering on spherical polydispersions*. Elsevier, Amsterdam.

del Giorgio P. A., Cole J. J. 1999. Bacterial growth efficiency in natural aquatic ecosystems. *Ann. Rev. Ecol. Syst.* **29**: 503–541.

Deng X., Gan X., Gu M. 2004. Effective Mie scattering of a spherical fractal aggregate and its application in turbid media. *Appl. Opt.* **43**: 2925–2929.

Depetris P. J. 1996. Riverine transfer of particulate matter to ocean systems. In: *Particle flux in the ocean*. Ittekkot V., Schäfer P., Honjo S., Depetris P. J. (eds.) Wiley, New York, pp. 53–69.

Dera J. 1992. *Marine physics*. Elsevier Oceanography Series, Elsevier, Amsterdam, 520 pp.

Deuser W. G., Emeis K., Ittekkot V., and Degens E. T. 1983. Fly-ash particles intercepted in the deep Sargasso Sea. *Nature* **305**: 216–218.

Deweert M. J., Moran S. E., Ulich B. L., Keeler R. N. 1999. Numerical simulations of the relative performance of streak-tube, range-gated, and PMT-based airborne imaging lidar systems with realistic sea surfaces. In: *Airborne and in-water underwater imaging*. Gilbert G. D. (ed.) *Proc. SPIE* **3761**: 115–129.

Diehl S. R., Smith D. T., Sydor M. 1979. Analysis of suspended solids by single-particle scattering. *Appl. Opt.* **18**: 1653-1658.

Dilling L., Alldredge A. L. 2000. Fragmentation of marine snow by swimming macro-zooplankton: A new process impacting carbon cycling in the sea. *Deep-Sea Res. I* **47**: 1227–1245.

Dobbins R. A., Megaridis C. M. 1991. Absorption and scattering of light by polydisperse aggregates. *Appl. Opt.* **30**: 4747–4754.

Dodge J. D. 1973. *The fine structure of algal cells*. Academic Press, New York, 261 pp.

Dodge L. G. 1984. Calibration of the Malvern particle sizer. *Appl. Opt.* **23**: 2415–2419.

Dolphin T. J., Green M. O., Radford J. D. J., Black K. P. 2001. Biofouling sensors: Prevention and analytical correction of data. *J. Coastal Res.* (Special Issue: ICS 2000 Proceedings, New Zealand) **34**: 334–341.

Doss W., Wells W. 1992. Undersea compound radiometer. *Appl. Opt.* **31**: 4268–4274.

Doyle W. T. 1985. Scattering approach to Fresnel's equations and Brewster's law. *Am. J. Phys.* **53**: 463–468.

Draine B. T. 2000. The discrete dipole approximation for light scattering by irregular targets. In: *Light scattering by nonspherical particles: theory, measurements, and geophysical applications.* Mishchenko M. I., Hovenier J. W., Travis L. D. (eds.) Academic Press, NY, pp. 131–145.

Draine B. T., Flatau P. J. 1994. Discrete dipole approximation for scattering calculations.*J. Opt. Soc. Am. A* **11**: 1491–1499.

Drebes G. 1974. *Marine phytoplankton.* Georg Threm, Stuttgart, 186 pp.

Drezek R., Dunn A, Richards-Kortum R. 1999. Light scattering from cells: finite-difference time-domain simulations and goniometric measurements. *Appl. Opt.* **38**: 3651–3661.

Drezek R., Guillaud M., Collier T., Boiko I., Malpica A., Macaulay C., Follen M., Richards-Kortum R. 2003. Light scattering from cervical cells throughout neoplastic progression: influence of nuclear morphology, DNA content, and chromatin texture. *J. Biomed. Opt.* **8**: 7–16.

Driks A. 1999. *Bacillus subtilis* spore coat. *Microbiol. Mol. Biol. Rev.* **63**: 1–20.

Droppo I. G. 2001. Rethinking what constitutes suspended sediment. *Hydrol. Processes* **15**: 1551–1564.

Droppo I. G., Flannigan D. T., Leppard G. G., Jaskot C., Liss S. N. 1996. Floc stabilization for multiple microscopic techniques. *Appl. Envrion. Microbiol.***62**: 3508–3515.

Droppo I. G., Leppard G. G., Flannigan D. T., Liss S. N. 1997. The freshwater floc: a functional relationship of water and organic and inorganic floc constituents affecting suspended sediment properties. *Water Air Soil Pollut.* **99**: 43–54.

Droppo I. G., Ongley E. D. 1994. Flocculation of suspended sediment in rivers of south-eastern Canada. *Water Res.* **28**: 1799–1809.

Droppo I. G., Walling D. E., Ongley E. D. 2002. Suspended sediment structure: Implications for sediment transport/yield modelling. In: *Modelling erosion, sediment transport and sediment yield.* Summer W., Walling D. E. (eds.) *International hydrological programme – technical documents in hydrology, No 60,* UNESCO, Paris, pp. 205–228.

Drossart P. 1990. A statistical model for scattering by irregular particles. *Astrophys. J.* **361**: L29–L32.

Duarte C. M., Agusti S., Peters H. 1987. An upper limit to the abundance of aquatic organisms. *Oecologia* **74**: 272–276.

Dubelaar G. B. J., Gerritzen P. L. 2000. CytoBuoy: a step forward towards using flow cytometry in operational oceanography. *Sci. Mar.* **64**: 255–265.

Dubelaar G. B. J., Gerritzen P. L., Beeker A. E. R., Jonker R. R., Tangen K. 1999. Design and first results of cytobuoy: a wireless flow cytometer for in situ analysis of marine and fresh waters. *Cytometry* **37**: 247–254.

Dubelaar G. B. J., Groenewegen A. C., Stokdijk W., van den Engh G. J., Visser J. W. M. 1989. Optical plankton analyser: a flow cytometer for plankton analysis, II: Specifications. *Cytometry* **10**: 529–539.

Dubelaar G. B. J., Jonker R. R. 2000. Flow cytometry as a tool for the study of phytoplankton. *Sci. Mar.* **64**: 135–156.

Dubelaar G. B. J., van der Reijden C. S. 1995. Size distributions of *Microcystis aeruginosa* colonies: a flow cytometric approach. *Water Sci. Technol.* **32**: 171–176.

Dubovik O., Holben B., Eck T., Smirnov A., Kaufmann Y., King M., Tanre D., Slutsker I. 2002. Variability of absorption and optical properties of key aerosol types observed in worldwide locations. *J. Atmos. Sci.* **59**: 590–608.

Dunn A., Richards-Kortum R. 1996. Three-dimensional computation of light scattering from cells. *IEEE J. Sel. Top. Quantum Electron.* **2**: 898–905.

Dueweke P. W., Bolstad J., Leonard D. A., Sweeney H. E., Boyer P. A., Winkler E. M. 1997. Instrument for underwater high-angular resolution volume scattering function measurements. In: *Ocean Optics XIII*. Steven G. Ackleson (ed.) *Proc. SPIE* **2963**: 658–663.

Dunn A. K. 1997. *Light scattering properties of cells*. Ph.D. Thesis, Biomedical Engineering, University of Texas at Austin, Austin TX, 141 pp.

Duntley S. Q. 1963. Light in the sea. *J. Opt. Soc. Am.* **53**: 214–233.

Dur J. C., Elsass F., Chaplain V., Tessier D. 2004. The relationship between particle-size distribution by laser granulometry and image analysis by transmission electron microscopy in a soil clay fraction. *Eur. J. Soil Sci.* **55**: 265–270.

DuRand M. D., Green R. E., Sosik H. M., Olson R. J. 2002. Diel variations in optical properties of *Micromonas pusilla* (*Prasinophyceae*). *J. Phycol.* **38**: 1132–1142.

DuRand M. D., Olson R. J. 1998. Diel patterns in optical properties of the chlorophyte *Nannochloris* sp.: Relating individual-cell to bulk measurements. *Limnol. Oceanogr.* **43**: 1107–1118.

Durst F., Macagno M. 1986. Experimental particle size distributions and their representation by log-hyperbolic functions. *Powder Technol.* **45**: 223–244.

Duysens L. N. M. 1956. The flattening of the absorption spectrum of suspensions, as compared to that of solutions. *Biochim. Biophys. Acta* **19**: 1–12.

Dyer K. R., Manning A. J. 1999. Observation of the size, settling velocity and effective density of flocs, and their fractal dimensions. *J. Sea Res.* **41**: 87–95.

E+

Earnshaw W. C., King J., Eiserling F. A. 1978. The size of the bacteriophage T4 head in solution with comments about the dimension of virus particles as visualized by electron microscopy. *J. Mol. Biol.* **122**: 247–253.

Eckhoff R. K. 1969. A static investigation of the Coulter principle of particle sizing. *J. Sci. Instrum.* **2**: 973–977.

Edgerton H., Ortner P., McElroy W. 1981. *In situ* plankton camera. *Oceans*, September 1981, 558–560.

Eggersdorfer B., Häder D.-P. 1991. Phototaxis, gravitaxis and vertical migrations in the marine dinoflagellates, *Peridinium faeroenseand Amphidinium caterea. Acta Protozool.* **30**: 63–71.

Ehrlich R. Weinberg B. 1970. An exact method for characterization of grain shape. *J. Sediment. Petrol.* **40**: 205–212.

Einstein A. 1910. Theory of opalescence in homogeneous liquids and liquid mixtures in the vicinity of the critical point (In German: *Theorie der Opaleszenz von homogenen Flüssigkeiten und Flüssigkeitsgemischen in der Nähe der kritischen Zustandes*). *Ann.*

Physik. **33**: 1275-1298. (In English: *Colloid chemistry.* Alexander J. (ed.) Vol. 1, pp. 323–339, Reinhold, New York, 1926).

Eisenberg H. 1965. Equation for the refractive index of water. *J. Chem. Phys.* **43**: 3887–3892.

Eisert W. G. 1979. Cell differentiation based on absorption and scattering. *J. Histochem. Cytochem.* **27**: 404–409.

Eisert W. G., Ostertag R., Niemann E. G. 1975. Simple flow microphotometer for rapid cell population analysis. *Rev. Sci. Instrum.* **46**: 1021–1024.

Eisma D., Bale A. J., Dearnley M. P., Fennessy M. J., Van Leussen W., Maldiney M.-A., Pfeiffer A., Wells J. T. 1996. Intercomparison of *in situ* suspended matter (floc) size measurements. *J. Sea Res.* **36**: 3–14.

Eisma D., Bernard D., Cadeé G. C., Ittekott V., Kalf J., Laane R., Martin J. M., Mook W. G., Van Put A., Schuhmacher T. 1991a. Suspended-matter particle size in some West-European estuaries; Part I: Particle-size distribution. *Neth. J. Sea Res.* **28**: 193–214.

Eisma D., Bernard, D., Cadeé G. C., Ittekott V., Kalf J., Laane R., Martin J. M., Mook W. G., Van Put A., Schuhmacher T. 1991b. Suspended-matter particle size in some West-European estuaries; Part II: A review on floc formation and break-up. *Neth. J. Sea Res.* **28**: 215–220.

Eisma D., Boon J., Gronenwegen R., Ittekot V., Kalf J., Mook W. G. 1983. Observations on macro-aggregates, particle size and organic composition of suspended matter in the Ems Estuary. Mitt. Geol. Palont. Inst. Univ. Hamburg. SCOPE/ENEP Sonderb. **55**: 295–314.

Eisma D., Dyer K. R., Van Leussen W. 1997. The in-situ determination of the settling velocities of suspended fine-grained sediment. In: *Cohesive sediments.* Burt N., Parker R., Watts J. (eds.) Wiley, Chichester, pp. 17–44.

Eisma D., Li A. 1993. Changes in the suspended-matter floc size during the tidal cycle in the Dollard Estuary. *Neth. J. Sea Res.* **31**: 107–117.

Eisma D., Schuhmacher T., Boekel H., van Heerwaarden J., Franken H., Laan M., Vaars A., Eijgenraam F., Kalf, J. 1990. A camera and image-analysis system for in situ observation of flocs in natural waters. *Neth. J. Sea Res.* **27**: 43–56.

Endoh S., Kuga Y., Ohya H., Ikeda C., Iwata H. 1998. Shape estimation of anisometric particles using size measurement techniques.*Part. Part. Syst. Charact.* **15**: 145–149.

Eppeldauer G. P. 2000. Noise-optimized silicon radiometers. *J. Res. Natl. Inst. Stand. Technol.* **105**: 209–219.

Eppeldauer G., Hardis J. E. 1991. Fourteen-decade photocurrent measurements with large-area silicon photodiodes at room temperature. *Appl. Opt.* **30**: 3091–3099.

Eppley R. W., Holmes R. W., Strickland J. D. H. 1967. Sinking rates of marine phytoplankton measured with a fluorometer. *J. Exp. Mar. Biol. Ecol.* **1**: 191–208.

Eppley R. W., Rogers J. N., McCarthy J. J. 1969. Half-saturation constants for uptake of nitrogen and ammonium by marine phytoplankton. *Limnol. Oceanogr.* **1**: 912–920.

Eppley R. W., Sloan P. R. 1966. Growth rates of marine phytoplankton: Correlation with light absorption by cell chlorophyll *a*. *Physiologia Pl* **19**: 47–59.

Errington J. R., Panagiotopoulos A. Z. 1998. A fixed point charge model for water optimized to the vapor-liquid coexistence properties. *J. Phys. Chem. B* **102**: 7470–7475.

Estes L. E., Fain G., Harris J. D. II 1997. A new instrument for optical forward scattering phase function surveys. *Proc. Oceans 97 Conf.*, Oct. 6–9, 1997, MTS/IEEE, 3 pp.

Evans B. T. N., Fournier G. R. 1994. Analytic approximation to randomly oriented spheroid extinction. *Appl. Opt.* **33**: 5796–5804.

Evtyushenkov A. M., Kiyachenko Yu F. 1982. Determination of the dependence of liquid refractive index on pressure and temperature. *Opt. Spectrosc.* **52**: 56–58.

F+

Fabelinskii I. L. 1968. *Molecular scattering of light.* Plenum Press, New York, 622 pp.

Faganeli J., Kovac N., Leskovsek H., Pezdic J. 1995. Sources and fluxes of particulate organic matter in shallow coastal waters characterized by summer macroaggregate formation. *Biogeochem.* **29**: 71–88.

Fahlen T. S., Bryant H. C. 1966. Direct observation of surface waves on water droplets. *J. Opt. Soc. Am.* **56**: 1635–1636.

Falk W. R. 1992. Transformation of histograms. *Rev. Sci. Instrum.* **63**: 4222–4224.

Falkowski P. G., Laws E. A., Barber R. T., Murray J. W. 2003. Phytoplankton and their role in primary, new, and export production. Chapter 4 in *Ocean Biogeochemistry*. Fasham M. J. R. (ed.) Springer, Berlin, pp. 99–121.

Farinato R. S., Roswell R. L. 1975. New values of the light scattering depolarization and anisotropy of water. *J. Chem. Phys.* **65**: 593–595.

Farrow J. B., Warren L. J. 1993. Measurement of the size of aggregates. In: *Coagulation and flocculation.* Dobias B. (ed.) Marcel Dekker, New York, pp. 391–426.

Faust M. A. 1993. Surface morphology of the marine dinoflagellate *Synophysis microcephalus*. *J. Phycol.* **29**: 355–363.

Fearn H., James, D. F. V., Milonni P. W. 1996. Microscopic approach to reflection, transmission, and the Ewald-Oseen extinction theorem. *Am. J. Phys.* **64**: 986–995.

Fedosseev R., Belayev Y., Frohn J., Stemmer A. 2005. Structured light illumination for extended resolution in fluorescence microscopy. *Opt. Lasers Eng.* **43**: 403–414.

Fennessy M. J., Dyer K. R., Huntley D. A. 1994. INSSEV: An instrument to measure the size and settling velocity of flocs *in situ.Mar. Geol.* **117**: 107–117.

Feynman R. P. 1949. Space-time approach to quantum electrodynamics. *Phys. Rev.* **76**: 769–789.

Feynman R. P. 1962. *Quantum electrodynamics.* W. A. Benjamin, Reading, Massachusetts, 198 pp.

Finder C., Wohlgemuth M., Mayer C. 2005. Analysis of particle size distribution by particle tracking. *Part. Part. Syst. Charact.* **21**: 372–378.

Finkel Z. V., Irwin A. J., Schofield O. 2004. Resource limitation alters the 3/4 size scaling of metabolic rates in phytoplankton. *Mar. Ecol. Prog. Ser.* **273**: 269–279.

Finsy R. 1994. Particle sizing by quasi-elastic light scattering. *Adv. Colloid Interface Sci.* **52**: 79–143.

Finsy R., De Jaeger N., Gelade E. 1992. Particle sizing by photon correlation spectroscopy. Part III: Mono and bimodal distributions and data analysis. *Part. Part. Syst. Charact.* **9**: 125–137.

Fofonoff N. P., Millard R. C. 1983. *Algorithms for computation of fundamental properties of seawater.* UNESCO Technical Papers in Marine Science, No. 44, 53 pp.

Forand L., Fournier G. R. 1999. Particle distributions and index of refraction estimation for Canadian waters. In: *Airborne and in-water underwater imaging*. Gary D. Gilbert (ed.). *Proc. SPIE* **3761**: 34–44.

Forand J. L., Fournier G. R., Bonnier D., Pelletier G., Pace P. 1993. NEARSCAT: A full spectrum narrow forward angle transmissometer-nephelometer. In: *IEEE Oceans '93 Proceedings III*, pp. 165–170.

Fournier G. R. 2000. Information theory optimization of phase function measurements. In: *Ocean Optics XV* (CD-ROM), Office of Naval Research, Washington DC.

Fournier G. R., Evans B. T. N. 1991. Approximation to extinction efficiency for randomly oriented spheroids. *Appl. Opt.* **30**: 2041–2048.

Fournier G. R., Evans B. T. N. 1996. Approximations to extinction from randomly oriented circular and elliptical cylinders. *Appl. Opt.* **35**: 4271–4282.

Fournier G. R., Forand J. L. 1994. Analytic phase function for ocean water. In: *Ocean Optics XII*. Jules S. Jaffe (ed.). *Proc. SPIE* **2258**: 194–201.

Frankel R. B., Bazylinski D. A., Johnson M., Taylor B. L. 1997. Magneto-aerotaxis in marine, coccoid bacteria. *Biophys. J.* **73**: 994–1000.

Fry E. S. 1974. Absolute calibration of a scatterance meter. In: *Suspended solids in water*. R. J. Gibbs (ed.). Plenum Press, New York, pp. 101–109.

Fry E. S., Kattawar G. W. 1981. Relationships between elements of the Stokes matrix. *Appl. Opt.* **20**: 2811–2814.

Fry E. S., Kattawar G. W., Pope R. M. 1992a. Integrating cavity absorption meter. *Appl. Opt.* **31**: 2055–2065.

Fry E. S., Padmabandu G. G., Og C. 1992b. Coherent effects in forward scattering. *Ocean Optics XI. Proc. SPIE* **1750**: 170–177.

Fry E. S., Voss K. J. 1985. Measurement of the Mueller matrix for phytoplankton. *Limnol. Oceanogr.* **30**: 1322–1326.

Fugate D. C., Friedrichs C. T. 2002. Determining concentration and fall velocity of estuarine particle populations using ADV, OBS and LISST. *Cont. Shelf Res.* **22**: 1867–1886.

Fugate D. C., Friedrichs C. T. 2003. Controls on suspended aggregate size in partially mixed estuaries. *Estuar. Coast. Shelf Sci.* **58**: 389–404.

Fuhrman J. A. 1981. Influence of method on the apparent size distribution of bacterioplankton cells: Epifluorescence microscopy compared to scanning electron microscopy.*Mar. Ecol. Prog. Ser.* **5**: 103–106.

Fuhrmann T., Landwehr S., El Rharbi-Kucki M., Sumper M. 2004. Diatoms as living photonic crystals. *Appl. Phys. B* **78**: 257–260.

Fujiki T., Taguchi S. 2002. Variability in chlorophyll a specific absorption coefficient in marine phytoplankton as a function of cell size and irradiance. *J. Plankton Res.* **24**: 859–874.

Full W. E., Ehrlich R., Kennedy S. K. 1984. Optimal configuration and information content of sets of frequency distributions. *J. Sediment. Petrol.* **54**: 117–126.

Fuller K. A. 1991. Optical resonances and two-sphere systems. *Appl. Opt. LP* **30**: 4716–4731.

Fuller K., Kattawar G. W. 1988. Consummate solution of the problem of classical electromagnetic scattering by an ensemble of spheres. II: Clusters of arbitrary configuration. *Opt. Lett.* **13**: 1063–1065.

Furuya K., Marumo R. 1983. Size distribution of phytoplankton in the western Pacific ocean and adjacent waters in summer. *Bull. Plankton Soc. Japan* **30**: 21–32.

G+

Gabaix X., Ioannides Y. 2004. Evolution of city size distributions. Chapter 53 in: *Handbook of regional and urban economics*. Henderson V., Thisse J. F. (eds.) Vol. 4, pp. 2342–2375.

Gabas N., Hiquily N., Laguérie C. 1994. Responses of laser diffraction patterns to anisometric particles. *Part. Part. Syst. Charact.* **11**: 121–126.

Gabaldón Casasayas J. E. 2001. *Analysis of the effect of the small-scale turbulence on the phytoplankton dynamics in the open ocean. Modelling and numerical simulation in the vertical dimension.* Ph.D. Thesis, Polytechnic University of Catalunya, Barcelona, Spain, 178 pp., B.37611-2001/84-699-5653-1.

Gaedke U. 1992. Identifying ecosystem properties: a case study using plankton biomass size distributions. *Ecol. Model.* **63**: 277–298.

Galajda P., Ormos P. 2003. Orientation of flat particles in optical tweezers by linearly polarized light. *Opt. Express* **11**: 446–451.

Gaudoin O., Yang B., Xie M. 2003. A simple goodness-of-fit test for the power-law process based on the Duane plot. *IEEE Trans. Reliab.* **52**: 69–74.

Gao Y., Anderson J. R. 2001. Characteristics of Chinese aerosols determined by individual-particle analysis. *J. Geophys. Res. D* **106**: 18,037–18,045.

Garbow N., Müller J., Schätzel K., Palberg T. 1997. High-resolution particle sizing by optical tracking of single colloidal particles. *Physica A* **235**: 291–305.

Gardner W. D. 1977. Incomplete extraction of rapidly settling particles from water samplers. *Limnol. Oceanogr.* **22**: 764–768.

Gartner J. W., Carder K. L. 1979. A method to determine specific gravity of suspended particles using an electronic particle counter. *J. Sediment. Petrol.* **49**: 631–633.

Gartner J. W., Cheng R. T., Wangh P.-E., Richter K. 2001. Laboratory and field evaluations of the LISST-100 instrument for suspended particle size determinations.*Mar. Geol.* **175**: 199–219.

Gayet J. F., Crépell O., Fournol J. F., Oshchepkov S. 1997. A new airborne polar nephelometer for the measurements of optical and microphysical cloud properties. Part I: Theoretical design.*Ann. Geophysicae* **15**: 451–459.

Gebeshuber I. C., Kindt J. H., Thompson J. B., Del Amo Y., Stachelberger H., Brzezinski M. A., Stucky G. D., Morse D. E., Hansma P. K. 2003. Atomic force microscopy study of living diatoms in ambient conditions.*J. Microsc.* **212**: 292–299.

Gerhardt P., Beaman T. C., Corner T. R., Greenamyre J. T., Tisa L. S. 1982. Photometric immersion refractometry of bacterial spores. *J. Bacteriol.* **150**: 643–648.

Ghormley J. A., Hochanadel C. J. 1971. Production of H, OH, and H_2O_2 in the flash photolysis of ice. *J. Chem. Phys.* **75**: 40–44.

Gianinoni I., Golinelli E., Melzi G., Musazzi S., Perini U., Trespidi F. 2003. Optical particle sizers for on-line applications in industrial plants. *Opt. Lasers Eng.* **39**: 141–154.

Gibbs R. J. 1972. The accuracy of particle-size analysis utilizing settling tubes. *J. Sediment. Petrol.* **42**: 141–145.

Gibbs R. J. 1978. Light scattering from particles of different shapes. *J. Geophys. Res.* **8**: 501–502.

Gibbs R. J. 1981. Floc breakage by pumps.*J. Sediment. Petrol.* **51**: 670–672.

Gibbs R. J. 1982a. Floc stability during Coulter counter measurements. *J. Sediment. Petrol.* **52**: 657–660.

Gibbs R. J. 1982b. Floc breakage during HIAC light-blocking analysis.*Environ. Sci. Technol.* **16**: 298–299.

Gibbs R. J., Matthews M. D., Link D. A. 1971. The relationship between sphere size and settling velocity. *J. Sediment. Petrol.* **41**: 7–18.

Gibbs R. J., Konwar L. 1983. Sampling of mineral flocs using Niskin bottles. *Environ. Sci. Technol.* **17**: 374–375.

Gibbs R. J., Tshudy D. M., Konwar L., Martin J.-M. 1989. Coagulation and transport of sediments in the Gironde Estuary. *Sedimentology* **36**: 987–999.

Giddings J. C., Yang F. J. F., Myers M. N. 1974. Sedimentation field fractionation. *Analyt. Chem.* **46**: 1917–1924.

Gillespie J. B., Jennings S. G., Lindberg J. D. 1978. Use of an average complex refractive index in atmospheric propagation calculations. *Appl. Opt.* **17**: 989–991.

Gillespie J. B., Lindberg J. D. 1992. Ultraviolet and visible imaginary refractive index of strongly absorbing atmospheric particulate matter. *Appl. Opt.* **31**: 2112–2115.

Gillespie J. B., Lindberg J. D., Smith M. S. 1974. Visible and near-infrared absorption coefficients of montmorillonite and related clays. *Am. Mineral.* **59**: 1113–1116.

Gillies G. T., Allison S. W. 1986. Precision limits of lock-in amplifiers below unity signal-to-noise ratio. *Rev. Sci. Instrum.* **57**: 268–270.

Gimsa J. 1999. New light-scattering and field-trapping methods access the internal structure of submicron particles, like Influenza viruses. In: *Electrical bio-impedance methods. Applications to medicine and biotechnology.* Riu P. J., Rosell J., Bragós R., Casas Ó. (eds.) *Ann. New York Acad. Sci.* **873**: pp. 287–298.

Gimsa J., Prüger B., Eppmann P., Donath E. 1995. Electrorotation of particles measured by dynamic light scattering - a new dielectric spectroscopy technique. *Colloids Surf. A* **98**: 243–249.

Gin K. Y. H., Guo J., Cheong H.-F. 1998. A size-based ecosystem model for pelagic waters. *Ecol. Model* **112**: 53–72.

Glatter O., Spurej E., Pfeiler G. 1990. Activation studies of human platelets using electrophoretic and quasi-elastic light scattering. In: *Proceedings of the 2nd International Congress on Optical Particle Sizing*, March 5–8, 1991, Tempe, AZ, USA, 395–403.

Gohs L., Dera J., Gedziorowska D., Hapter R., Jonasz M., Prandke H., Siegel H., Schenkel G., Olszewski J., Woźniak B., Zalewski M. S. 1978. Investigation of relationships between optical, physical, biological, and chemical environmental factors in the Baltic in years 1974, 1975, and 1976 (In German: *Untersuchen zur Wechselwirkung zwischen der optischen, physikalischen, biologischen, und chemischen Umweltfaktoren in der Ostsee aus den Jahren 1974, 1975 und 1976*). *Geod. Geophys. Veroff.* Serie 4, Heft 25, 176 pp.

Golibersuch D. C. 1973. Observation of aspherical particle rotation in Poisseuille flow via the resistance pulse technique (I). Application to human erythrocytes. *Biophys. J.* **13**: 265–280.

Goransson B. 1990. Improved accuracy in the measurements of particle size distribution with a Coulter counter equipped with a hydrodynamic focussed aperture. *Part. Part. Syst. Charact.* **7**: 6–10.

Gordon D. C., Jr. 1970. A microscopic study of organic particles in the North Atlantic Ocean. *Deep-Sea Res.* **17**: 175–185.

Gordon H. R., Bader H., Brown O. B. 1972. An experimental and theoretical study of suspended particulate matter in the Tongue of the Ocean and its influence on underwater visibility. In: *Measurements of the volume scattering function of sea water. Technical*

Report 334. Mertens, L. E. and Phillips, D. L. (eds). Range Measurements Laboratory, Patrick Air Force Base Florida, USA, 62 pp.

Gordon H. R., Brown O. B. 1972. A theoretical model of light scattering by Sargasso Sea particulates. *Limnol. Oceanogr.* **17**: 826–832.

Gordon H. R., Clark D. K., Brown J. W., Brown O. B., Evans R. H., Broenkow W. W. 1983. Phytoplankton pigment concentrations in the Middle Atlantic Bight: comparison of ship determinations and CZCS estimates. *Appl. Opt.* **22**: 20–36.

Gordon H. R., Morel A. 1983. Remote assessment of ocean color for interpretation of satellite visible imagery. In: *Lecture notes on coastal and estuarine studies*, Vol. 4, Springer-Verlag, New York, 114 pp.

Gordon H. R., Du T. 2001. Light scattering by nonspherical particles: Application to coccoliths detached from *Emiliania huxleyi. Limnol. Oceanogr.* **46**: 1438–1454.

Gordon J. I., Johnson R. W. 1985. Integrating nephelometer: theory and implications. *Appl. Opt.* **24**: 2721–2730.

Gouesbet G., Gréhan G. 2000. Generalized Lorenz–Mie theories, from past to future. *Atomization and Sprays* **10**: 277–333.

Gouesbet G., Maheu B. 1988. Light scattering from a sphere arbitrarily located in a Gaussian beam, using a Bromwich formulation. *J. Opt. Soc. Am. A* **5**: 1427–1443.

Gouesbet G., Meunier-Guttin-Cluzel S., Gréhan G. 2001. Generalized Lorenz-Mie theory for a sphere with an eccentrically located inclusion, and optical chaos. *Part. Part. Syst. Charact.* **18**: 190–195.

Gould R. W., Arnone R. A., Martinolich P. M. 1999. Spectral dependence of the scattering coefficient in Case 1 and Case 2 waters. *Appl. Opt.* **38**: 2377–2383.

Graeme J. 1995. *Photodiode amplifiers OP AMP solutions.* McGraw-Hill, New York, 252 pp.

Graeme J. G., Tobey G. E., Huelsman L. P. 1971. *Operational amplifiers. Design and applications.* McGraw-Hill, New York, 473 pp.

Grams G. W., Blifford Jr., I. H., Gillette D. A., Russel P. B. 1974. Complex index of refraction of airborne soil particles. *J. Appl. Meteorol.* **13**: 459–471.

Grams G. W., Dascher A. J., Wyman C. M. 1975. Laser polar nephelometer for airborne measurements of aerosol optical properties. *Opt. Eng.* **14**: 85–90.

Grasso V., Neri F., Fucile E. 1995. Simple angle resolved light scattering photometer using a photodiode array. In: *Air Pollution and Visibility Measurements. Proc. SPIE* **2506**: 763–772.

Grasso V., Neri F., Fucile E. 1997. Particle sizing with a simple differential light scattering photometer: homogeneous spherical particles. *Appl. Opt.* **36**: 2452–2458.

Green R. E., Sosik H. M., Olson R. J. 2003a. Contributions of phytoplankton and other particles to inherent optical properties in New England continental shelf waters. *Limnol. Oceanogr.* **48**: 2377–2391.

Green R. E., Sosik H. M., Olson R. J., DuRand M. D. 2003b. Flow cytometric determination of size and complex refractive index for marine particles: comparison with independent and bulk estimates. *Appl. Opt.* **42**: 526–541.

Grout H., Sempere R., Thill A., Calafat A., Prieur L., Canals M. 2001. Morphological and chemical variability of colloids in the Almeria–Oran Front in the eastern Alboran Sea (SW Mediterranean Sea).*Limnol. Oceanogr.* **46**: 1347–1357.

Grundinkina N. P. 1956. Absorption of ultraviolet radiation by water. *Opt. Spektrosk.* **1**: 658–662.

Gualtieri D. M. 1987. Precision of lock-in amplifiers as a function of signal-to-noise ratio. *Rev. Sci. Instrum.* **58**: 299–300.

Gucker F. T., O'Kunski C., Pickard H. B., Pitts J. N. 1947. A photoelectric counter for colloidal particles. *Am. J. Chem.* **69**: 2422–2431.

Gucker F. T., Tuma J., Lin H.-M., Huang Ch.-M., Ems S. C., and Marshall T. R. 1973. Rapid measurement of light scattering diagrams from single particle in an aerosol stream and determination of latex particle size. *Aerosol Sci.* **4**: 389–404.

Gurgul H. 1993. Description of quantity and dispersion distribution changes of mineral suspensions occurring in the Ezcurra Inlet waters, King George Island within a year cycle.*Korean J. Polar Res.* **4**: 3–14.

Gurwich I., Kleiman M., Shiloah N., Cohen A. 2000. Scattering of electromagnetic radiation by multilayered spheroidal particles: Recursive procedure. *Appl. Opt.* **39**: 470–477.

Guyard S., Mary P., Defives C., Hornez J. P. 1999. Enumeration and characterization of bacteria in mineral water by improved direct viable count method. *J. Appl. Microbiol.* **86**: 841–850.

H+

Haardt H., Maske H. 1987. Specific in vivo absorption coefficient of chlorophyll a at 675 nm. *Limnol. Oceanogr.* **32**: 608–619.

Hale G. M., Querry M. R. 1973. Optical constants of water in the 200-nm to 200-μm wavelength region. *Appl. Opt.* **12**: 555–563.

Haller H. R., Destor C., Cannell D. S. 1983. Photometer for quasielastic and classical light scattering.*Rev. Sci. Instrum.* **54**: 973–983.

Halley J., Inchausti P. 2002. Lognormality in ecological time series. *Oikos* **99**: 518–530.

Haltrin V. I. 1997. Theoretical and empirical phase functions for Monte Carlo calculations of light scattering in seawater. In: *Proceedings of the Fourth International Conference Remote Sensing for Marine and Coastal Environments: Technology and Applications*, Vol. I, ISSN 1066-3711, pp. 509–518.

Haltrin V. I. 1998. Self-consistent approach to the solution of the light transfer problem for irradiances in marine waters with arbitrary turbidity, depth and surface illumination: I. Case of absorption and elastic scattering. *Appl. Opt.* **37**: 3773–3784.

Haltrin V. I. 1999. Chlorophyll-based model of seawater optical properties. *Appl. Opt.* **38**: 6826–6832.

Haltrin V. I., Lee M. E., Mankovsky V. I., Shybanov E. B., Weidemann A. D. 2003. Integral properties of angular light scattering coefficient measured in various natural waters. In: *Proceedings of the II International Conference "Current Problems in Optics of Natural Waters"* ONW'2003, St. Petersburg, Russia, 2003. Levin I., Gilbert G. (eds.), pp. 252–257.

Hamm C. E. 2000. Architecture, ecology and biogeochemistry of *Phaeocystis* colonies. *J. Sea Res.* **43**: 307–315.

Hamm C. E. 2002. Interactive aggregation and sedimentation of diatoms and clay-sized lithogenic material.*Limnol. Oceanogr.* **47**: 1790–1795.

Hamm C. E., Merkel R., Springer O., Jurkojc P., Maier C., Prechtel K., Smetacek V. 2003. Architecture and material properties of diatom shells provide effective mechanical protection.*Nature* **421**: 841–843.

Hansen M. Z., Evans W. H. 1980. Polar nephelometer for atmospheric particulate studies. *Appl. Opt.* **19**: 3389–3395.

Hansen J. E., Travis L. D. 1974. Light scattering in planetary atmospheres. *Space Sci. Rev.* **16**: 527–610.

Haracz R. D., Cohen L. D., Cohen A. 1985. Scattering of linearly polarized light from randomly oriented cylinders and spheroids. *J. Appl. Phys.* **58**: 3322–3327.

Harris J. E. 1977. Characterization of suspended matter in the Gulf of Mexico. II: Particle size distribution of suspended matter from deep water. *Deep-Sea Res.* **24**: 1055–1061.

Harvey A. H., Gallagher J. S., Sengers J. M. H. L. 1998. Revised formulation for the refractive index of water and steam as a function of wavelength, temperature and density. *J. Phys. Chem. Ref. Data.* **27**: 761–774.

Harvey R. J., Marr A. G., Painter P. R. 1967. Kinetics of growth of individual cells of *Escherichia coli* and *Azotobacter agilis*. *J. Bacteriol.* **93**: 605–617.

Hassen M. A., Davis R. H. 1989. Effects of particle interactions on the determination of size distributions by sedimentation. *Powder Technol.* **58**: 285–289.

Heath A. R., Fawell P. D., Bahri P. A., Swift J. D. 2002. Estimating average particle size by focused beam reflectance measurement (FBRM).*Part. Part. Syst. Charact.* **19**: 84–95.

Hecht E. 1987. *Optics*. Addison-Wesley Reading, Massachusetts, 676 pp.

Heffels C. M. G. 1995. *On-line particle size and shape characterization by narrow angle light scattering*. Ph.D. Thesis, Delft University of Technology, 169 pp.

Heffler D. E., Syvitski J. P. M., Asprey K. W. 1991. The floc camera assembly. In: *The principles, methods, and applications of particle size analysis*. Syvitski J. P. M. (ed.) Cambridge University Press, New York, pp. 209–222.

Heidt L. J., Johnson A. M. 1957. Optical study of the hydrates of molecular oxygen in water. *J. Am. Chem. Soc.* **79**: 5587–5593.

Heintzenberg J. 1994. Properties of the log-normal particle size distribution. *Aerosol Sci. Technol.* **21**: 46–48.

Heintzenberg J., Charlson R. J. 1996. Design and application of the integrating nephelometer: a review. *J. Atmos. Ocean. Technol.* **13**: 987–1000.

Heissenberger A., Leppard G. G., Herndl G. J. 1996a. Ultrastructure of marine snow: II. Microbiological considerations. *Mar. Ecol. Prog. Ser.* **135**: 299–308.

Heissenberger A., Leppard G. G., Herndl G. J. 1996b. Relationship between the intracellular integrity and the morphology of the capsular envelope in attached and free-living marine bacteria.*Appl. Environ. Microbiol.* **62**: 4521–4528.

Henriksen K., Stipp S. L. S., Young J. R., Bown P. R. 2003. Tailoring calcite: Nanoscale AFM of coccolith biocrystals. *Am. Mineral.* **88**: 2040–2044.

Henriksen K., Stipp S. L. S., Young J. R., Marsh M. E. 2004a. Biological control on calcite crystallization: AFM investigation of coccolith polysaccharide function. *Am. Mineral.* **89**: 1709–1716.

Henriksen K., Young J. R., Bown P. R., Stipp S. L. S. 2004b. Coccolith biomineralisation studied with atomic force microscopy. *Palaeontol*ogy **47**: 725–743.

Hentschel M. L., Page N. W. 2003. Selection of descriptors for particle shape characterization. *Part. Part. Syst. Charact.* **20**: 25–38.

Henyey L., Greenstein J. 1941. Diffuse radiation in the galaxy. *Astrophys. J.* **93**: 70–83.

Herman A. W. 1992. Design and calibration of a new optical plankton counter capable of sizing small zooplankton. *Deep-Sea Res.* **39**: 395–415.

Herman A. W., Beanlands B., Phillips E. F. 2004. The next generation of optical plankton counter: the Laser-OPC. *J. Plankton Res.* **26**: 1135–1145.

Herman A. W., Cochrane N. A., Sameoto D. D. 1993. Detection and abundance estimation of euphausiids using an optical plankton counter. *Mar. Ecol. Prog. Ser.* **94**: 165–173.

Herzberg G. 1950. *Spectra of diatomic molecules*. D. Van Nostrand Company, Inc, Princeton, 658 pp.

Hespel L., Delfour A., Guillame B. 2001. Mie light-scattering granulometer with adaptive numerical filtering. II. Experiments. *Appl. Opt.* **40**: 974–985.

Hielscher A. H., Eick A. A., Mourant J. R., Shen D., Freyer J. P., Bigio I. J. 1997. Diffuse backscattering Mueller matrices of highly scattering media. *Opt. Express* **1**: 441–453.

Hill S. C., Brenner R. E., Rushforth C. K., Convell P. R. 1984a. Structural resonances observed in the fluorescence emission from small spheres on substrates. *Appl. Opt.* **23**: 1680–1682.

Hill S. C., Hill A. C., Barber P. W. 1984b. Light scattering by size/shape distributions of soil particles and spheroids. *Appl. Opt.* **23**: 1025–1031.

Hill S. C., Rushforth C. K., Benner R. E., Conwell P. R. 1985. Sizing dielectric spheres and cylinders by aligning measured and computed resonance locations: algorithm for multiple orders. *Appl. Opt.* **24**: 2380–2390.

Hill P. S., Syvitski J. P. M., Cowan E. A., Powell R. D.1998. In situ observations of floc settling velocities in Glacier Bay, Alaska. *Mar. Geol.* **145**: 85–94.

Hillebrand H., Dürselen C. D., Krischtel D., Pollingher U, Zohary T. 1999. Biovolume calculation for pelagic and benthic microalgae.*J. Phycol.* **35**: 403–424.

Hirleman D. E., Oechsle V., Chigier N. A. 1984. Response characteristics of laser diffraction particle size analyzers: optical sample volume extent and lens effects. *Opt. Eng.* **23**: 610–619.

Hirst E., Kaye P. H. 1996. Experimental and theoretical light scattering profiles from spherical and non-spherical particles. *J. Geophys. Res. D* **101**: 19,231–19,235.

Hirst E., Kaye P. H., Guppy J. R. 1994. Light scattering from nonspherical airborne particles: experimental and theoretical comparisons. *Appl. Opt.* **33**: 7180–7186.

Hitchcock G. L. 1982. A comparative study of the size-dependent organic composition of the marine diatoms and dinoflagellates. *J. Plankton Res.* **4**: 363–377.

Hobson P. R., Lampitt R. S., Rogerson A., Watson J., Fang X., Krantz E. P. 2000. Three-dimensional spatial coordinates of individual plankton determined using underwater hologrammetry. *Limnol. Oceanogr.* **4–5**: 1167–1174.

Hodara H. 1973. Experimental results of small angle scattering. In: *Optics of the sea.* AGARD Lecture Series No.61, pp. 3.4.1–3.4.17.

Hodkinson J. R. 1963. Light scattering and extinction by irregular particles larger than the wavelength. In: *Proceedings of the International Conference on Electromagnetic Scattering*, August 1962, Potsdam, NY. Kerker, M. (ed.) Pergamon Press, Oxford, pp. 87–100.

Hodkinson J. R., Greenfield J. R. 1965. Response calculation for light scattering aerosol counters and photometers. *App. Opt.* **4**: 1463–1474.

Hoelzl Wallach D. F., Kamat V. B., Gail M. H. 1966. Physicochemical differences between fragments of plasma membrane and endoplasmic reticulum. *J. Cell Biol.* **30**: 601–621.

Hoepffner N., Haas L. W. 1990. Electron microscopy of nanoplankton from the North Pacific central gyre.*J. Phycol.* **26**: 421–439.

Hoepffner N., Sathyendranath S. 1991. Effect of pigment composition on absorption properties of phytoplankton. *Mar. Ecol. Prog. Ser.* **73**: 11–23.

Hofmann A., Dominik J. 1995. Turbidity and mass concentration of suspended matter in lake water: A comparison of two calibration methods. *Aquat. Sci.* **57**: 54–69.

Hofstraat J. W., de Vreeze M. E. J., van Zeijl W. J. M., Peperzak L., Peeters J. C. H., Balfoort H. W. 1991. Flow cytometric discrimination of phytoplankton classes by fluorescence and excitation properties *J. Fluoresc.* **1**: 249–265.

Hofstraat J. W., Zeijl van W. J. M., Vreeze de M. E. J., Peeters J. C. H., Peperzak L., Colijn F., Rademaker T. W. M. 1994. Phytoplankton monitoring by flow cytometry. *J. Plankton Res.* **16**: 1197–1224.

Højerslev N., Aas E. 1998. Spectral light absorption by Gelbstoff in coastal waters displaying highly different concentrations. In: *Ocean Optics XIV* CD-ROM, Office of Naval Research, Washington DC, 8 pp.

Højerslev N. K., Aas E. 2001. Spectral light absorption by yellow substance in the Kattegat-Skagerrak area. *Oceanologia* **43**: 39–60.

Holl L. J., McCormick N. J. 1994. Ocean optical-property estimation with the Zaneveld–Wells algorithm. *App. Opt.* **34**: 5433–5441.

Holland A. C. 1980. Problems in calibrating a polar nephelometer. In: *Light scattering by irregularly shaped particles,* D. W. Schuerman (ed.) Plenum Press, New York, pp. 247–254.

Holland A. C., Draper J. S. 1967. Analytical and experimental investigation of light scattering from polydispersions of Mie particles. *Appl. Opt.* **6**: 511–518.

Holland A. C., Gagne G. 1970. The scattering of polarized light by polydisperse systems of irregular particles. *Appl. Opt.* **9**: 1113–1121.

Holler S., Pan Y. L., Chang R. K., Bottiger J. R., Hill S. C., Hillis D. B 1998. Two-dimensional angular optical scattering for the characterization of airborne microparticles. *Opt. Lett.* **23**: 1489–1491.

Holm-Hansen O., Mitchell B. G. 1991. Spatial and temporal distribution of phytoplankton and primary production in the Western Bransfield Strait region. *Deep-Sea Res.* **38**: 49–57.

Holoubek J., Konak C., Stepanek P. 1999. Time-resolved small-angle light scattering apparatus.*Part. Part. Syst. Charact.***16**: 102-105.

Holve D. J., Davis G. W. 1985. Sample volume and alignment analysis for an optical particle counter sizer and other applications.*Appl. Opt.* **24**: 998–1005.

Holve D. J., Self S. A. 1979a. Optical particle sizing for in situ measurements. Part 1. *Appl. Opt.* **18**: 1632–1645.

Holve D. J., Self S. A. 1979b. Optical particle sizing for in situ measurements. Part 2. *Appl. Opt.* **18**: 1646–1652.

Honey R. C., Sorenson G. P. 1970. Optical absorption and turbulence induced narrow angle forward scatter in the sea. In: *Electromagnetics of the sea.* AGARD Conf. Proc. **77**, pp. 39.1–39.3.

Honjo S. 1996. Fluxes of particles to the interior of the open oceans. In: *Particle flux in the ocean.* Ittekkot V., Schäfer P., Honjo S., Depetris P. J. (eds.) Wiley, New York, pp. 91–113.

Honjo S., Doherty K. W., Agraval Y. C., Asper V. L. 1984. Direct optical assessment of large amorphous aggregates (marine snow) in the deep ocean. *Deep-Sea Res.* **31**: 67–76.

Hood R. R. 1986. Personal communication.

Hood R. R., Abbot M. R., Huyer A. 1991. Phytoplankton and photosynthetic response in the costal transition zone off northern California in June 1987. *J. Geophys. Res. C* **96**: 14,769–14,780.

Horák D., Peška J., Švec F., Štamberg J. 1982. Effect of the porosity of discrete particles upon their apparent dimensions as measured by the Coulter principle. *Powder Technol.* **31**: 263–268.

Horne R. A. 1969. *Marine chemistry*. Wiley, New York.

Hou W. 1997. Characteristics of large particles and their effects on the submarine light field. Ph.D. Thesis, Department of Marine Science, University of South Florida, Florida, USA, 165 pp.

Hou W., Carder K. L., Costello D. K. 1997. Scattering phase function of very large particles in the ocean. In: *Ocean Optics XIII*. Steven G. Ackleson (ed.) *Proc. SPIE* **2963**: 579–584.

Hovenier J. W. 1999. Basic relationships for matrices describing scattering by small particles. In: *Light scattering by nonspherical particles: Theory, measurements, and applications.* Michael I. Mishchenko, Joop W. Hovenier, Larry D. Travis (eds.) Academic Press, San Diego, pp. 1–3.

Hovenier J. W., van de Hulst H. C., van der Mee C. V. M. 1986. Conditions for the elements of the scattering matrix. *Astron. Astrophys.* **157**: 301–310.

Hovenier J. W., van der Mee C. V. M. 1996. Testing scattering matrices: a compendium of recipes. *J. Quant. Spectrosc. Radiat. Transfer* **55**: 649–661.

Huang X., Zhang J., Chen W., Zhang T., He M., Liu Z. 1994. A new type of scatterometer for measuring the small angle volume scattering function of seawater and the experiments in the East China Sea. In: *Ocean Optics XI. Proc. SPIE* **2258**: 556–559.

Huffman D. R., Bohren C. F. 1980. Infrared absorption spectra of non-spherical particles treated in the Rayleigh-ellipsoid approximation. In: *Irregularly shaped particles.* Schuerman, D. (ed.) Plenum Press, New York, pp. 103–111.

Huibers P. D. T. 1997. Models for the wavelength dependence of the index of refraction of water. *Appl. Opt.* **36**: 3785–3787.

Hulburt E. O. 1945. Optics of distilled and natural water. *J. Opt. Soc. Am.* **35**: 698–705.

Hull P., Shepherd I., Hunt A. 2004. Modeling light scattering from diesel soot particles. *Appl. Opt.* **43**: 3433–3441.

Hunt J. R. 1980. Prediction of oceanic particle size distribution from coagulation and sedimentation mechanisms. In:*Advances in chemistry.* Kavanaugh M. C., Leckie J. O. (eds.) American Chemical Society Washington, DC, **198**: 243–257.

Hunt J. R. 1982. Self-similar particle size distribution during coagulation: theory and experimental verification. *J. Fluid Mech.* **122**: 169–185.

Hunt A. J., Huffman D. J. 1973. A new polarization-modulated light scattering instrument. *Rev. Sci. Instrum.* **44**: 1753–1762.

Hunt A., Johnson D. L. 1996. Characterizing the outlines of degraded fine-particles by fractal dimension. *Scanning Electron Microsc.* **10**: 69–83.

Hunter K. A. 1991. Surface charge and size spectra of marine particles. In: *The analysis and characterization of marine particles.* Hurd D. C., Spencer D. W. AGU, Washington DC, pp. 259–262.

Hurley A. C. 1976. *Introduction to the electron theory of small molecules.* Academic Press, New York, 329 pp.

Hurley J. 1970. Sizing particles with a Coulter counter.*Biophys. J.* **10**: 74–79.

Hüller R., Gloßner E., Schaub S., Weingärter J., Kachel V. 1994. A macro flow planktometer – a new device for volume and fluorescence analysis of macro plankton including triggered video imaging in flow. *Cytometry* **17**: 109–118.

I+

IAPWS. 1997. Release on the refractive index of ordinary water substance as a function of wavelength, temperature and pressure. IAPWS, Erlangen, Germany, September 1997, 7 pp.

Inaba K., Matsumoto K. 1995. Effect of particle shape on particle size analysis using the electrical sensing zone method and laser diffraction method. *J. Soc. Powder Technol. Jpn.* **32**: 722–730.

Inoue S. 1989. Imaging of unresolved objects, superresolution, and precision of distance measurements with video microscopy. In: *Fluorescence microscopy of living cells in culture. Part B: Quantitative fluorescence microscopy – imaging and spectroscopy. Methods in cell biology.* Taylor D. L., Wang, Y. (eds.) Academic Press, New York, pp. 85–112.

Irvin J. A., Quickenden T. I. 1983. Linear least squares treatment when there are errors in both x and y. *J. Chem. Edu.* **60**: 711–712.

Iturriaga R. I., Berwald J., Sonek G. J. 1997. New technique for the determination of spectral reflectance of individual and bulk particulate suspended matter in natural water samples. In: *Ocean Optics XIII.* Steven G. Ackleson (ed.) *Proc. SPIE* **2963**: 455–460.

Iturriaga R., Siegel D. A. 1988. Discrimination of the absorption properties of marine particulates using a microphotometric technique. In: *Ocean Optics IX. Proc. SPIE* **925**: 277–287.

Iturriaga R., Siegel D. A. 1989. Microphotometric characterization of phytoplankton and detrital absorption properties in the Sargasso Sea. *Limnol. Oceanogr.* **34**: 1706–1726.

J+

Jackson G. A. 1990. A model of the formation of marine algal flocs by physical coagulation processes. *Deep-Sea Res.* **37**: 1197–1211.

Jackson G. A. 1995. Coagulation of marine algae. In: *Aquatic chemistry: principles and applications of interfacial and inter-species interactions in aquatic system.* Huang C. P., O'Melia C. R., Morgan J. J. (eds.), *Am. Chem. Soc.*, Washington, pp. 203–217.

Jackson G. A., Burd A. B. 1998. Aggregation in the marine environment. *Environ. Sci. Technol.* **32**: 2805–2814.

Jackson G. A., Lochmann S. E. 1993. Modeling coagulation in marine ecosystems. In: *Environmental particles.* Buffle J., van Leeuwen H. P. (eds.), Lewis Publ., Boca Raton, Florida, pp. 387–414.

Jackson G. A., Logan B. E., Alldredge A. L., Dam H. G. 1995. Combining particle size spectra from a mesocosm experiment measured using photographic and aperture impedance (Coulter and Elzone) techniques. *Deep-Sea Res. II* **42**: 139–157.

Jackson G. A., Maffione R. D., Costello K., Alldrdge A. L., Logan B. E., Dam H. G. 1997. Particle size spectra between 1 um and 1 cm at Monterey Bay determined using multiple instruments. *Deep-Sea Res.* **44**: 1739–1767.

Jacques S. L., Alter C. A., Prahl S. A. 1987. Angular dependence of HeNe laser light scattering by human tissue. *Lasers Life Sci.* **1**: 309–333.

Jambers W., Van Grieken R. E. 1996. Single particle characterization of inorganic North Sea suspensions. In: *Proceedings of the 1996 Workshop on Progress in Belgian Oceanographic Research*, Editions Derouaux Ordina, Liège, p. 115–117.

Jambers W., De Bock L., Van Grieken R. E. 1995. Recent advances in the analysis of individual environmental particles: a review. *Analyst* **120**: 681–692.

Jambers W., De Bock L., Van Grieken R. E. 1996. Applications of micro-analysis to individual environmental particles. *Fresenius' J. Anal. Chem.* **355**: 521–527.

James M. B., Griffiths D. J. 1992. Why the speed of light is reduced in a transparent medium. *Am. J. Phys.* **60**: 309–313.

Jantschik R., Nyffeler F., Donard O. F. X. 1992. Marine particle size measurement with a stream-scanning laser system.*Mar. Geol.* **106**: 239–250.

Jeffrey S. W., Humphrey G. F. 1975. New spectrophotometric equations for chlorophylls a, b, c1, c2 in higher plants, algae and natural phytoplankton. *Biochim. Physiol. Pflanz.* **167**: 374–384.

Jenkins F. A., White H. E. 2001. *Fundamentals of optics.* McGraw-Hill, New York, 768 pp.

Jerlov N. G. 1961. Optical measurements in the eastern North Atlantic. *Medd. Oceanogr. Institut Göteborg,* **30**: 1-40.

Jerlov N. G. 1963. Optical oceanography. *Oceanogr. Mar. Biol. Ann. Rev.* **1**: 89–114.

Jerlov N. G. 1968. Optical oceanography. Elsevier, Amsterdam, 194 pp.

Jerlov N. G. 1976. *Marine optics.* Elsevier, Amsterdam, 231 pp.

Jerrard H. G., Sellen D. B. 1962. A new apparatus for light scattering studies. *Appl. Opt.* **1**: 243–247.

Jiang Q., Logan B. E. 1996. Fractal dimensions of aggregates from shear devices. *J. Am. Water Works Assoc.* **88**: 100–113.

Jiao N. Z., Ni I. H. 1997. Spatial variation of size-fractionated chlorophyll, cyanobacteria and heterotrophic bacteria in the central and western Pacific. *Hydrobiol*ogy **352**: 219–230.

Jillavenkatesa A., Lum L.-S. H., Dapkunas S. 2001. *Particle size characterization.* Special Publication 960-1, NIST, Washington, USA, 165 pp.

Jochem F. J. 2001. Morphology and DNA content of bacterioplankton in the northern Gulf of Mexico: analysis by epifluorescence microscopy and flow cytometry. *Aquat. Microb. Ecol.* **25**: 179–194.

Johnson B. D. 1986. Bubble populations: Background and breaking waves. In: *Oceanic whitecaps and their role in air-sea gas exchange.* Monahan E. C., MacNiocalli G. (eds.) D. Reidel Publishing Co., Hingham, Massachusetts, pp. 69–73.

Johnson B. R. 1988. Invariant imbedding: T matrix approach to electromagnetic scattering. *Appl. Opt.* **27**: 4861–4873.

Johnson B. D., Cooke R. C. 1980. Organic particles and aggregate formation resulting from the dissolution of bubbles in sea water. *Limnol. Oceanogr.* **25**: 653–661.

Johnson B. D., Wangersky P. J. 1987. Microbubbles: Stabilization by monolayers of adsorbed particles. *J. Geophys. Res. C* **92**: 14, 641–14, 647.

Johnson C. P., Li X., Logan B. E. 1996. Settling velocities of fractal aggregates. *Environ. Sci. Technol.* **30**: 1911–1918.

Johnson P. W., Sieburth J. McN. 1982. In situ morphology and occurrence of eucaryotic phototrophs in bacterial size in the picoplankton of estuarine and oceanic waters. *J. Phycol.* **18**: 318–327.

Jonasz M. 1978. Particle size distributions measured with a Coulter counter along a transect of the Atlantic Ocean, from Antarctica to the English Channel, 2nd Polish Antarctic Expedition 1977–78, unpublished data.

Jonasz M. 1980. *Characterization of physical properties of marine particles using light scattering.* (In Polish: *Wykorzystanie zjawiska rozpraszania światła do wyznaczenia własności fizycznych zawiesin morskich*) Ph.D. Thesis, Institute of Oceanology, Polish Academy of Sciences, Sopot, Poland, 214 pp.

Jonasz M. 1983a. Particle size distribution in the Baltic. *Tellus B* **35**: 346–358.

Jonasz M. 1983b. Particulate matter in the Ezcurra Inlet: Concentration and size distributions. *Oceanologia* (Poland) **15**: 65–74.

Jonasz M. 1986. Role of nonsphericity of marine particles in light scattering and a comparison of results of its determination using SEM and two types of particle counters. *Ocean Optics VIII Proc. SPIE* **637**: 148–154.

Jonasz M. 1987a. Nonsphericity of suspended marine particles and its influence on light scattering. *Limnol. Oceanogr.* **32**: 1059–1065.

Jonasz M. 1987b. Nonspherical sediment particles: comparison of size and volume distributions obtained with an optical and a resistive particle counters. *Mar. Geol.* **78**: 137–142.

Jonasz M. 1987c. The effect of nonsphericity of marine particles on light attenuation. *J. Geophys. Res. C* **13**: 14,637–14,640.

Jonasz M. 1990. Volume scattering function measurement error: effect of angular resolution of the nephelometer. *Appl. Opt.* **29**: 64–70.

Jonasz M. 1991a. Light scattering: a basis of effective particle characterization techniques. *Am. Lab.* **24**: 30–36.

Jonasz M. 1991b. *Processed and analyzed results obtained with a wide-angle, polarization sensitive scattering meter in the Atlantic waters south of Nova Scotia, 3 to 6 May 1986.* Report for the Defense Research Establishment Valcartier under contract W7701-1-3213/01-XLA. M. Jonasz Consultants, Beaconsfield, QC, Canada, 42 pp.

Jonasz M. 1992. *Scattering functions of sea water, and size distributions, refractive indices, shapes, and compositions of marine particles.* Report for the Defense Research Establishment Valcartier under contract W7701-1-4528/01-XSK. M. Jonasz Consultants, Beaconsfield, QC, Canada, 600 pp.

Jonasz M. 1996. *A database of scattering functions.* M. Jonasz Consultants, Beaconsfield, QC, Canada.

Jonasz M., Fournier G. 1996. Approximation of the size distribution of marine particles by a sum of log-normal functions. *Limnol. Oceanogr.* **41**: 744–754.

Jonasz M., Fournier G. F. 1999. Approximation of the size distribution of marine particles by a sum of log-normal functions (Errata: Corrections and additional results). *Limnol. Oceanogr.* **44**: 1358–1358.

Jonasz M., Fournier G., Stramski, D. 1997. Photometric immersion refractometry: a method of determining the refractive index of marine microbial particles from beam attenuation. *Appl. Opt.* **36**: 4214–4225.

Jonasz M., Prandke H. 1986. Comparison of measured and computed light scattering in the Baltic. *Tellus* B**38**: 144–157.

Jonasz M., Zalewski M. S. 1978. Stability of the shape of particle size distribution in the Baltic. *Tellus* **30**: 569–572.

Jones A. R. 1999. Light scattering for particle characterisation. *Prog. Energy Comb. Sci.* **25**: 1–53.

Jones M. R., Leong K. H., Brewster M. Q., Curry B. P. 1994. Inversion of light-scattering measurements for particle size and optical constants: experimental study. *Appl. Opt.* **33**: 4035–4041.

Jones A. R., Savaloni H. 1989. A light scattering instrument to discriminate and size fibres. Part 2: Experimental system.*Part. Part. Syst. Charact.* **6**: 144–150.

Jonkman J. E. N., Swoger J., Kress H., Rohrbach A., Stelzer E. H. K. 2003. Resolution in optical microscopy. In: *Biophotonics, methods in enzymology*. Marriott G., Parker I. (eds) Vol. 360, pp. 416–446.

Jørgensen N. O. G., Stepanauskas R., A. Pedersen U., Hansen M., Nybroe O. 2003. Occurrence and degradation of peptidoglycan in aquatic environments. *FEMS Microbiol. Ecol.* **46**: 269–280.

Jullien R., Botet R. 1987. *Aggregation and fractal aggregates*. World Scientific Publishing Co., Singapore, 120 pp.

Jumars P. A., Deming J. W., Hill P. S., Karp-Boss L., Yager P. L., Dade W. B. 1993. Physical constraints on marine osmotrophy in an optimal foraging context. *Mar. Microb. Food Webs* **7**: 121–159.

Junge C. E. 1963. *Air chemistry and radioactivity*. Academic Press, New York, 382 pp.

Jürgens K., Matz C. 2002. Predation as a shaping force for the phenotypic and genotypic composition of planktonic bacteria. *Antonie van Leeuwenhoek* **81**: 413–434.

K+

Kachel V. 1982. Sizing of cells by the electrical resistance pulse technique. Methodology and application in cytometric systems. In: *Cell analysis*, Vol. 1. Catsimpoolas N. (ed.) Plenum Press, New York, pp. 195–331.

Kachel V. 1990. Electrical resistance pulse sizing: Coulter sizing. In:*Flow cytometry and sorting*. Melamed M. R., Lindmo T., Mendelsohn M. L. (eds.) Wiley-Liss, New York, pp. 45–80.

Kachel V. 2001. Personal communication.

Kachel V., Benker G., Lichtnau K., Valet G., Glossner E. 1979. Fast imaging in flow: a means of combining flow-cytometry and image analysis. *J. Histochem. Cytochem.* **27**: 335–341.

Kachel V., Benker G., Weiss W., Glossner E., Valet G., Ahrens O. 1980. Problems of fast imaging in flow. *Flow Cytometry* **4**: 49–55.

Kachel V., Glossner E., Kordwig E., Ruhenstroth-Bauer G. 1977. Fluvo-Metricell, a combined cell volume and cell fluorescence analyzer. *J. Histochem. Cytochem.* **25**: 804–812.

Kachel V., Kiemle C., Rembold H., Eichmeier J., Messerschmied T., Flüchter J. 1992. A computerized analyzer for size measurements of fish larvae and young fishes. In: *Trends in ichtyology: an international perspective*. Schröder J. H., Bauer J., Schartl M. (eds.) GSF Forschungzentrum für Umwelt und Gesundheit GmbH, Nuerbergh, and Blackwell Sci. Publications, Oxford, pp. 359–368.

Kachel V., Menke E. 1979. Hydrodynamic properties of flow-cytometry instruments. In: *Flow cytometry and sorting*. Melamed M. R., Lindmo T., Mendelsohn M. L. (eds.) Wiley-Liss, New York, pp. 41–59.

Kadyshevich Y. A. 1977. Light scattering matrices of inshore waters of the Baltic Sea. *Izv. Acad. Sci. USSR: Atmosph. Ocean. Phys.* **13**: 77–78.

Kadyshevich Y. A., Lyubovtseva Yu S., Plakhina I. N. 1971. Measurements of matrices for light scattered by sea water. *Izv. Acad. Sci. USSR: Atmos. Oceanic Phys.* **7**: 557–561.

Kadyshevich Y. A., Lyubovtseva Yu S., Rozenberg G. V. 1976. Light-scattering matrices of Pacific and Atlantic Ocean waters. *Izv. Acad. Sci. USSR: Atmosph. Ocean. Phys.* **12**: 106–111.

Kahnert F. M. 2003. Numerical methods in electromagnetic scattering theory. *J. Quant. Spectrosc. Rad. Transfer* **79-80**: 775–824.

Kahnert F. M., Stamnes J. J., Stamnes K. 2001. Application of the extended boundary condition method to homogeneous particles with point-group symmetries. *Appl. Opt.* **40**: 3110–3123.

Kahnert F. M., Stamnes J. J., Stamnes K. 2002a. Can simple particle shapes be used to model scalar optical properties of an ensemble of wavelength-sized particles with complex shapes? *J.Opt.Soc.Am.A* **19**: 521–531.

Kahnert M., Stamnes J. J., Stamnes K. 2002b. Using simple particle shapes to model the Stokes scattering matrix of ensembles of wavelength-sized particles with complex shapes: possibilities and limitations. *J. Quart. Spectrosc. Radiat. Transfer.* **74**: 167–182.

Kahru M., Nommann S., Zeitzschel B. 1991. Particle (plankton) size structure across the Azores front (Joint Global Ocean Flux Study North Atlantic Bloom Experiment). *J. Geophys. Res. C* **96**: 7083–7088.

Kamiuto K. 1987. Study of the Henyey–Greenstein approximation to scattering phase functions. *J. Quant. Spectrosc. Radiat. Transfer* **37**: 411–413.

Kaplan B., Compain E., Drevillon B. 2000. Phase-modulated Mueller ellipsometry characterization of scattering by latex sphere suspensions. *Appl. Opt.* **39**: 629–636.

Karuhn R. F., Berg R. H. 1984. Practical aspects of ELECTROZONE size analysis. In: *Particle characterization in technology*. Beddow J. K. (ed.). CRC Press, Boca Raton, FL, USA, pp. 157–181.

Kattawar G. W. 1975. A three-parameter analytic phase function for multiple scattering calculations.*J. Quant. Spectrosc. Radiat. Transfer* **15**: 839–849.

Kattawar G. W., Plass, G. N. 1967. Electromagnetic scattering from absorbing spheres. *Appl. Opt.* **6**: 1377–1382.

Katz A., Alimova A., Xu M., Gottlieb P., Rudolph E., Steiner J. C., Alfano R. R. 2005. In situ determination of refractive index and size of *Bacillus* spores by light transmission. *Opt. Lett.* **30**: 589–591.

Katz J. E., Wells S., Ussery D., Bustamante C., Maestre M. F. 1984. Design and construction of a circular intensity differential scattering instrument. *Rev. Sci. Instrum.* **55**: 1574-1579.

Katz J., Donaghay P. L., Zhang J., King S., Russell K. 1999. Submersible holocamera for detection of particle characteristics and motions in the ocean. *Deep-Sea Res. I* **46**: 1455–1481.

Kay D. B., Cambier J. L., Wheeless L. 1979. Imaging in flow. *J. Histochem. Cytochem.* **27**: 329–334.

Kaye B. H. 1999. *Characterization of powders and aerosols*. Wiley-VCH GmbH, Weinheim, Germany, 323 pp.

Kaye B. H., Alliet D., Switzer L., Turbitt-Daoust C. 1997. The effect of shape on intermethod correlation of techniques for characterizing the size distribution of powder. Part 1: Correlating the size distribution measured by sieving, image analysis, and diffractometer methods.*Part. Part. Syst. Charact.* **14**: 219–224.

Kaye B. H., Alliet D., Switzer L., Turbitt-Daoust C. 1999. The effect of shape on intermethod correlation of techniques for characterizing the size distribution of powder. Part 2: Correlating the size distribution measured by diffractometer methods, TSI-Amherst spectrometer and Coulter counter.*Part. Part. Syst. Charact.* **16**: 266–272.

Kaye B. H., Clark G. G., Kydar Y. 1994. Strategies for evaluating boundary fractal dimensions by computer aided image analysis.*Part. Part. Syst. Charact.* **11**: 411–417.

Kaye W., Havlik A. J. 1973. Low angle laser light scattering - absolute calibration. *Appl. Opt.* **12**: 541–550.

Kaye P. H., Hirst E. 1995. *Apparatus and method for the analysis of particle characteristics using monotonically scattered light.* United States Patent 5,471,299, 14 pp.

Kaye B. H., Trottier R. 1995. The many measures of fine particles.*Chem. Eng.* April: 78–86.

Kennedy S. K., Lin W.-H. 1992. A comparison of Fourier and fractal techniques in the analysis of closed forms. *J. Sediment. Petrol.* **62**: 842–848.

Kennedy S. K., Mazzullo J. 1991. Image analysis method of grain size measurement. In: *Principles, methods, and application of particle size analysis.* Syvitski J. P. M. (ed.) Cambridge University Press, New York, pp. 76–87.

Kenney B. C. 1982. Beware of spurious self-correlations! *Water Resour. Res.* **18**: 1041–1048.

Kepkay P. E. 1994. Particle aggregation and the biological reactivity of colloids. *Mar. Ecol. Prog. Ser.* **109**: 293–304.

Kerker M. 1969. *The scattering of light and other electromagnetic radiation.* Academic Press, New York, 666 pp.

Kerker M. 1997. Light scattering instrumentation for aerosol studies: An historical overview.*Aerosol Sci. Technol.* **27**: 522–540.

Kerker M., Chew H., Mcnulty P. J., Kratohvil J. P., Cooke D. D., Sculley M., Lee M.-P. 1979. Light scattering and fluorescence by small particles having internal structure. *J. Histochem. Cytochem.* **27**: 250–263.

Kerner M., Hohenberg H., Ertl S., Reckermann M., Spitzy A. 2003. Self-organization of dissolved organic matter to micelle-like microparticles in river water. *Nature* **422**: 150–154.

Kiefer D. A., Berwald J. 1992. A random encounter model for the microbial planktonic community. *Limnol. Oceanogr.* **37**: 457–467.

Kiefer D. A., SooHoo J. B. 1982. Spectral absorption by marine particles of coastal waters of Baja California. *Limnol. Oceanogr.* **27**: 492–499.

Kilps J. R., Logan B. E., Alldredge A. L. 1994. Fractal dimensions of marine snow determined from image analysis of in situ photographs. *Deep-Sea Res. I* **41**: 1159–1169.

Kim J. P., Lemmon J., Hunter K. A. 1995. Size-distribution analysis of sub-micron colloidal, particles in river water. *Environ. Technol.* **16**: 861–868.

Kino G. S., Corle T. R. 1989. Confocal scanning optical microscopy. *Phys. Today* September: 55–62.

Kiørboe T. 2001. Formation and fate of marine snow: small-scale processes with large-scale implications. *Sci. Mar.* **65** (Suppl. 2): 57–71.

Kirk J. T. O. 1983a. *Light and photosynthesis in marine ecosystems.* Cambridge Press, New York, 525 pp.

Kirk J. T. O. 1983b. Monte Carlo modeling of the performance of a reflective tube absorption meter.*Appl. Opt.* **18**: 3328–3331.

Kirmaci I., Ward G. 1979. Scattering of 0.627-mm light from spheroidal 2-μm cladosporium and cubical 4-μm NaCl particles. *Appl. Opt.* **18**: 3328–3331.

Kishino M. 1980. Studies of the optical properties of sea water: Application of Mie theory to suspended particles in sea water. *Sci. Papers Inst. Phys. Chem Res.* **74**: 31–45.

Kishino M., Takahashi M., Okami N., Ichimura S. 1985. Estimation of the spectral absorption coefficients of phytoplankton in the sea.*Bull. Mar. Sci.* **37**: 634–642.

Kitchen J. C., Menzies D., Pak H., Zaneveld J. R. V. 1975. Particle size distributions in a region of coastal upwelling analyzed by characteristic vectors. *Limnol. Oceanogr.* **20**: 775–783.

Kitchen J. C., Zaneveld J. R. V. 1990. On the noncorrelation of the vertical structure of light scattering and chlorophyll a in case I waters. *J. Geophys. Res. C* **95**: 20,237–20,246.

Kitchen J. C., Zaneveld J. R. V. 1992. A three-layered sphere model of the optical properties of phytoplankton. *Limnol. Oceanogr.* **37**: 1680–1690.

Kitchen J. C., Zaneveld J. R. V., Pak H. 1982. Effect of particle size distribution and chlorophyll content on beam attenuation spectra. *Appl. Opt.* **21**: 3913–3918.

Kittleman L. R., Jr. 1964. Application of Rosin's distribution in size-frequency analysis of clastic rocks. *J. Sediment. Petrol.* **34**: 483–502.

Kjaergaard H. G., Henry B. R., Wei H., Lefebvre S., Carrington T., Jr., Mortensen O. S., Sage M. L. 1994. Calculation of vibrational fundamental and overtone band intensities of H_2O. *J. Chem. Phys.* **100**: 6228–6239.

Klett J. D., Sutherland R. A. 1992. Approximate methods for modeling the scattering properties of nonspherical particles: evaluation of the Wentzel-Kramers-Brillouin method. *Appl. Opt.* **31**: 373–386.

Klotz R. 1978. Investigation on the scattering function of artificial and natural marine particles in the optical wavelength range with applications for the Kieler Bucht. (In German: *Untersuchen uber Streufunktionen im optischen Spectralbereich an kunstlichen und naturlichen marinen Partikelsuspensionen mit Beispielen aus der Kieler Bucht*). Thesis, Kiel University (identical with Reports Sonderforschungsbereich 95, No.46, Kiel University, November 1978), 157 pp.

Klut M. E., Bisalputra T., Antia N. J. 1989. Some details of the cell surface of two marine dinoflagellates *Bot. Mar.* **32**: 89–95.

Knollenberg R. G., Veal D. L. 1992. Optical particle monitors, counters, and spectrometers: performance characterization, comparison, and use. *J. IES* **35**: 64–81.

Knowles S. C., Wells J. T. 1998. In situ aggregate analysis camera (ISAAC): A quantitative tool for analyzing fine-grained suspended material. *Limnol. Oceanogr.* **43**: 1954–1962.

Knox C. 1966. Holographic microscopy as a technique for recording dynamic microscopic subjects. *Science* **153**: 989–990.

Koh P. T. L., Lwin T., Albrecht D. 1989. Analysis of weight frequency particle size distributions: With special application to bimodal floc size distributions in shear-flocculation. *Powder Technol.* **59**: 87–95.

Koike I. Hara S., Terauchi K., Kogure K. 1990. Role of sub-micrometre particles in the ocean. *Nature* **345**: 242–244.

Konák C., Holoubek J., Stepánek P. 2001. A time-resolved low-angle light scattering apparatus. Application to phase separation problems in polymer systems. *Collect. Czech. Chem. Cornmum.* **66**: 973–982.

Kondolf G. M., Adhikari A. 2000. Weibull vs. lognormal distributions for fluvial gravels *J. Sediment. Res.* **70**: 456–460.

Kopelevich O. V. 1976. Optical properties of water in the 250-600 nm range. *Opt. Spectrosc.* **41**: 391–392.

Kopelevich O. V., Burenkov V. I. 1971. On the nephelometric method of determination of the total scattering coefficient of light by sea water. (In Russian: *O nephelometricheskoj metode opredelenia obshchego pokazatela rassejania sveta morskoj vodoi*) *Izv. AN SSSR. Atmos. Oceanic Phys.* **7**: 1280–1286.

Kopelevich O. V., Burenkov W. I. 1972. Statistical characteristics of light scattering by sea water. (In Russian: *Statisticzeskije characteristiki rassiejania sveta morskoy vodoy*). In: Shifrin, K. S. (ed.) *Optics of the ocean and the atmosphere*, Nauka, Leningrad, pp. 126–136.

Kopelevich O. V., Mezhericher, E M. 1983. Calculation of spectral characteristics of light scattering by sea water. (In Russian: *Rasschiot spectralnovo pokazatiela rassiejania sveta morskoj vodoi*) *Izv. AN SSSR Fiz. Atmos. Okeana.* **19**: 144–148.

Kopelevich O. V., Rodionov V. V., Stupakova T. P. 1987. Effect of bacteria on optical characteristics of ocean water. *Oceanology* **27**: 696–700 (English translation of *Okeanologiya*).

Koppes L. J., Woldringh C. L., Grover N. B. 1987. Predicted steady-state cell size distributions for various growth models. *J. Theor. Biol.* **129**: 325–335.

Korn G. A., Korn T. M. 1968. *Mathematical handbook for scientists and engineers.* McGraw-Hill, New York, 830 pp.

Korson L., Drost-Hansen W., Millero F. J. 1969. Viscosity of water at various temperatures. *J. Phys. Chem.* **73**: 34–39.

Kortüm G. 1969. *Reflectance spectroscopy: Principles, methods, applications.* Springer Verlag, New York, 366 pp.

Kou L., Labrie D., Chýlek P. 1993. Refractive indices of water and ice the 0.65- to 2.5-μm spectral range. *Appl. Opt.* **32**: 3531–3540.

Kovac N., Baita O., Faganelia J., Sketb B., Orel B. 2002. Study of macroaggregate composition using FT-IR and ^{1}H-NMR spectroscopy.*Mar. Chem.* **78**: 205–215.

König K. 1997. Two-photon near infrared excitation in living cells. *J. Near Infrared Spectrosc.* **5**: 27–34.

König K., Liang H., Berns M. W., Tromberg B. 1995. Cell damage by near-IR microbeams. *Nature* **377**: 20–21.

König K., Liang H., Berns M. W., Tromberg B. J. 1996. Cell damage in near infrared multimode optical traps due to multi-photon absorption. *Opt. Lett.* **21**: 1090–1092.

Köylü U. O., Faeth G. M., Farias T. L., Carvalho M. G. 1995. Fractal and projected structure properties of soot aggregates. *Combust. Flame* **100**: 621–633.

Kozlyianinov M. V. 1957. New instrument for measurements of optical properties of sea water. (In Russian: *Novyi pribor dla izmerenyia opticheskikh svoistv morskoyi vody*).*Trudy Inst. Okeanologyi Akad. Nauk SSR* **25**: 134–142.

Krambeck C., Krambeck H-J., Overbeck J. 1981. Microcomputer-assisted determination of plankton bacteria on scanning electron micrographs. *Appl. Environ. Microbiol.* **42**: 142–149.

Kranck K. 1986. Generation of grain size distribution of fine grained sediments. In: *River sedimentation.* Wang, S. Y., Shen, H. W., and Ding, L. Z. (eds.) School of Engineering, University of Mississippi, Mississippi, pp. 1776–1784.

Kranck K. 1987. Granulometric changes in fluvial sediments from source to deposition. *Mitt. Geol.-Paleont. Inst. Univ. Hamburg* **64**: 45–56.

Kranck K. 1993. Flocculation and sediment particle size. *Archiven für Hydrobiologie* (supplement) **75**: 299–309.

Kranck K., Milligan T. G. 1985. Origin of grain size spectra of suspension deposited sediment. *Geo-Mar. Lett.* **5**: 61–66.

Kranck K., Milligan T. G. 1986. Personal communication.

Kranck K., Milligan T. G. 1979. The use of the Coulter counter in studies of marine particle-size distributions in aquatic environments. In: *Bedford Institute Report Series BI-R 79*, Bedford Inst. Oceanogr., Dartmouth, Nova Scotia, Canada, 48 pp.

Kranck K., Milligan T. G. 1988. Macroflocs from diatoms: in situ photography of particles in Bedford Basin, Nova Scotia. *Mar. Ecol. Prog. Ser.* **44**: 183–189.

Kranck K., Milligan T. G. 1991. Grain size in oceanography. In: *Principles, methods, and application of particle size analysis.* Syvitski, J. P. M. (ed.) Cambridge University Press, New York, pp. 332–345.

Kranck K., Milligan T. G. 1992. Characteristics of suspended particles at an 11-hour anchor station in San Francisco Bay, California. *J. Geophys. Res.* **97**: 11,373–11,382.

Kranck K., Petticrew E., Milligan T. G., Droppo I. G. 1992. *In situ* particle size distributions resulting from flocculation of suspended sediment. In: *Nearshore and estuarine cohesive sediment transport.* Mehta, A. J. (ed.) Coastal Estuarine Studies, Series 42, American Geophysical Union, Washington, pp. 60–75.

Kratohvil J. P. Dezelic G., Kerker M. Matijevic E. 1962. Calibration of light-scattering instruments: a critical survey.*J. Polym. Sci.* **57**: 59–78.

Kratohvil J. P., Kerker M., Oppenheimer L. E. 1965. Light scattering by pure water. *J. Chem. Phys.* **43**: 914–921.

Krey J. 1961. Detritus in the sea. (In German: *Der Detritus im Meere*). *J. Cons. perm. int. l'Explor. Mer* **26**: 262–280.

Kreyszig E. 1972. *Advanced engineering mathematics.* Wiley, New York, 866 pp.

Kronfeld U. 1988. Optical properties of oceanic suspended matter and their implications for the remote sensing of phytoplankton. (In German *Die optischen Eigenschaften der ozeanischen Schebstoffe und ihre Bedeutung für die Fernenkundungen von Phytoplankton*). Thesis, Earth Sciences Department, Hamburg University, 156 pp.

Król T. 1998. *Light scattering by phytoplankton.* (In Polish: *Rozpraszanie światła przez fitoplankton*). Polish Academy of Sciences, Institute of Oceanology, Sopot, 147 pp.

Krumbein W. C. 1934. Size-frequency distributions of sediments. *J. Sediment. Petrol.* **4**: 65–77.

Krumbein W. C. 1936. The application of logarithmic moments to size-frequency distributions of sediments. *J. Sedimen. Petrol.* **4**: 35–47.

Ku J. C., Felske J. D. 1986. Determination of refractive indices of Mie scatterers from Kramers-Kronig analysis of spectral extinction data. *J. Opt. Soc. Am. A* **3**: 617–623.

Kubitschek H. E. 1962. Loss of resolution in Coulter counter. *Rev. Sci. Instrum.* **33**: 576–577.

Kuik F., Stammes P., Hovenier J. W. 1991. Experimental determination of scattering matrices of water droplets and quartz particles. *Appl. Opt.* **30**: 4872–4881.

Kullenberg G. 1968. Scattering of light by Sargasso Sea water. *Deep-Sea Res.* **15**: 423–432.

Kullenberg G. 1969. *Light scattering in the central Baltic.* Rep. No. 5, Inst. Phys. Oceanogr., Copenhagen Univ., Copenhagen, 16 pp.

Kullenberg G. 1970. *A comparison between observed and computed light scattering functions.* Rep. No. 13, Inst. Phys. Oceanogr., Copengahen Univ., Copenhagen, 22 pp.

Kullenberg G. 1974. Observed and computed scattering functions. In: *Optical aspects of oceanography.* Jerlov N. G., Steeman-Nielsen E. (eds.) Academic Press, London, pp. 25–49.

Kullenberg G. 1978. Light scattering observations in frontal zones. *J. Geophys. Res.* **83**: 4683–4691.

Kullenberg G. 1984. Observations of light scattering functions in two oceanic areas. *Deep-Sea Res.* **31**: 295–316.

Kullenberg G., Olsen N. B. 1972. *A comparison between observed and computed light scattering functions - II.* Rep. No. 19, Inst. Phys. Oceanogr., Copengahen Univ., Copenhagen, 27 pp.

Kusmartseva O., Smith P. R. 2001. Robust method for non-sphere detection by photon correlation between scattered polarization states. *Meas. Sci. Technol.* **12**: 1–5.

Kusters K. A., Wijers J. G., Thoenes D. 1991. Particle sizing by laser diffraction spectrometry in the anomalous regime. *Appl. Opt.* **30**: 4839–4847.

Kwok A. C. M., Wong J. T. Y. 2003. Cellulose synthesis is coupled to cell cycle progression at G1 in the dinoflagellate *Crypthecodinium cohnii. Plant Physiol.* **131**: 1681–1691.

L+

Lal D., Lerman A. 1975. Size spectra of biogenic particles in ocean water and sediments. *J. Geophys. Res.* **80**: 423–430.

Lalor E., Wolf E. 1972. Exact solution of the equations of molecular optics for refraction, and reflection of an electromagnetic wave on a semi-infinite dielectric. *J. Opt. Soc. Am.* **62**: 1165–1174.

Lamarre E., Melville W. K. 1991. Air entrainment and dissipation in brealing waves. *Nature* **351**: 469–472.

Lambert C. E., Jehanno C., Silverberg N., Brun-Cottan J. C., Chesselet R. 1981. Log-normal distributions of suspended particles in the open ocean.*J. Mar. Res.* **39**: 77–98.

Langston P. A., Burbidge A. S., Jones T. F., Simmons M. J. H. 2001. Particle and droplet size analysis from chord measurements using Bayes' theorem. *Powder Technol.* **116**: 33–42.

Lant C. T., Smart A. E., Cannell D. S., Meyer W. V., Doherty M. P. 1997. Physics of hard spheres experiment: a general-purpose light-scattering instrument. *App. Opt.* **36**: 7501–7507.

Latimer P. 1984. Light scattering by a homogeneous sphere with radial projections. *Appl. Opt.* **23**: 442–447.

Latimer P., Wamble F. 1982. Light scattering by aggregates of large colloidal particles. *Appl. Opt.* **21**: 2447–2455.

Law D. J., Bale A. J. 1998. In situ characterization of suspended particles using focused-beam, laser reflectance particle sizing. In: *Sedimentary processes in the intertidal zone.* Black K. S., Patterson D. M., Cramp A. (eds.) *Geol. Soc. London, Special Publ.* **139**: 57–68.

Law D. J., Bale A. J., Jones S. E. 1997. Adaptation of focused beam reflectance measurement to in-situ particle sizing in estuaries and coastal waters. *Mar. Geol.* **140**: 47–59.

Lawler D. F. 1997. Particle size distributions in treatment processes: theory and practice. In: *Proceedings of the 4th International Conference: The Role of Particle Characteristics in Separation Processes*, IAWQ–IWSA Joint Specialist Group on Particle Separation, Jerusalem, 28–30 October 1996.

Lawless P. A. 2001. High-resolution reconstruction method for presenting and manipulating particle histogram data.*Aerosol Sci. Technol.* **34**: 528–534.

Le J., Wehr J. D., Campbell L., Calder L. 1994. Uncoupling of bacterioplankton and phytoplankton production in fresh waters is affected by inorganic nutrient limitation. *Appl. Environ. Microbiol.* **60**: 2086–2093.

Leathers R. A., McCormick N. J. 1997. Ocean inherent optical property estimation from irradiances. *Appl. Opt.* **36**: 8685–8698.

Lebour M. V. 1930. *The planktonic diatoms of northern seas*. Dulau & Co., London, UK.

Lee L. G., Berry G. M., Chen Ch-H. 1989. Vita blue: a new 633-nm excitable fluorescent dye for cell analysis.*Cytometry* **10**: 151–164.

Lee M. E., Haltrin V. I., Shybanov E. B., Weidemann A. D. 2003. Light scattering phase functions of turbid coastal waters. In: *OCEANS 2003* MTS-IEEE Conference Proceedings on CD-ROM, pp. 2835–2841, San Diego, California, September 22–26, 2003; ISBN: 0-933957-31-9; Holland Enterprises, Escondido, CA, USA; file P2835-2841.pdf.

Lee M. E., Lewis M. R. 2003. A new method for the measurement of the optical volume scattering function in the upper ocean. *J. Atmos. Ocean. Technol.* **20**: 563–671.

Lee S., Fuhrman J. A. 1987. Relationship between biovolume and biomass of naturally derived marine bacterioplankton. *Appl. Environ. Microbiol.* **53**, 1298–1303.

Lee S., Kang Y.-C., Fuhrman J. A. 1995. Imperfect retention of natural bacterioplankton cells by glass fiber filters. *Mar. Ecol. Prog. Ser.* **119**: 285–290.

Legendre L., Courties C., Troussellier M. 2001. Flow cytometry in oceanography 1989–1999: Environmental challenges and research trends. *Cytometry* **44**:164–172.

Leif R. C., Thomas R. A. 1973. Electronic cell-volume analysis by use of AMAC I transducer. *Clin. Chem.* **19**: 853–870.

Leifer I., de Leeuw G., Cohen L. 2000. Secondary bubble production from breaking waves: The bubble burst mechanism. *Geophys. Res. Lett.* **27**: 4077–4080.

Lenz J. 1972. The size distribution of particles in marine detritus. *Memorie Dell'Instituto Italiano di Idrobiologia Dott*, **29** (Supplement), 17–35.

Leong K. H., Jones M. R., Holdridge D. J., Ivey M. 1995. Design and test of a polar nephelometer. *Aerosol Sci. Tech.* **23**: 341–356.

Leppard G. G., West M. M., Flannigan D. T., Carson J., Lott J. N. A. 1997. A classification scheme for marine organic colloids in the Adriatic Sea: colloid speciation by transmission electron microscopy. *Can. J. Fish. Aquat. Sci.* **54**: 2334–2349.

Lepple F. K., Millero F. J. 1971. The isothermal compressibility of seawater near one atmosphere. *Deep-Sea Res.* **18**: 1233–1254.

Lerman A., Carder K. L., Betzer P. R. 1977. Elimination of fine suspensoids in the oceanic water column.*Earth Planet. Sci. Lett.* **37**: 61–70.

Lerner R. M., Summers J. D. 1982. Monte Carlo description of time- and space-resolved multiple forward scatter in natural waters.*Appl. Opt.* **21**: 861–869.

Lettieri T. R., Marx E. 1986. Resonance light scattering from a liquid suspension of microspheres. *Appl. Opt.* **25**: 4325–4331.

Levoni C., Cervino M., Guzzi R., Torricella F. 1997. Atmospheric aerosol optical properties: a database of radiative characteristics for different components and classes. *Appl. Opt.* **30**: 8031–8041.

Lewis D. W., McConchie D. 1994. *Analytical sedimentation.* Chapman & Hall, New York, 223 pp.

Li C., Kattawar G. W., Yang P. 2004. Effects of surface roughness on light scattering by small particles. *J. Quant. Spectrosc. Radiat. Transfer* **89**: 123–131.

Li W. K. W. 1990. Bivariate and trivariate analysis in flow cytometry: Phytoplankton size and fluorescence. *Limnol. Oceanogr.* **35**: 1356–1367.

Li W. K. W. 1994. Phytoplankton biomass and chlorophyll concentrations across the North Atlantic. *Sci. Mar.* **58**: 67–79.

Li W. K. W., E. Head J. H., Harrison W. G. 2004. Macroecological limits of heterotrophic bacterial abundance in the ocean. *Deep-Sea Res. I* **51**: 1529–1540.

Li X., Passow U., Logan B. E. 1998. Fractal dimensions of small (15–200 μm) particles in Eastern Pacific coastal waters. *Deep-Sea Res. I* **45**: 115–131.

Li X., Logan B. E. 1995. Size distributions and fractal properties of particles during a simulated phytoplankton bloom in a mesocosm *Deep-Sea Res. II* **42**: 125–137.

Lim M., Amal R., Pinson D., Cathers B. 2002. Discrete particle method modelling of flocs behaviour. In: *Proceedings of the 9th APCChE Congress and CHEMECA 2002*, 29 September–3 October 2002, Christchurch, NZ, Gostomski P. A., Morison K. R. (eds.) Univ. Canterbury, Christchurch, NZ, ISBN 0-473-09252-2, Paper # 581, 22 pp.

Limpert E, Stahl W. A., Abbt M. 2001. Log-normal distributions across the sciences: keys and clues. *Bioscience* **51**: 341–352.

Lin C.-H., Lee G.-B. 2003. Micromachined flow cytometers with embedded etched optic fibers for optical detection. *J. Micromech. Microeng.* **13**: 447–453.

Lindberg J. D. 1975. Absorption coefficient of atmospheric dust and other strongly absorbing powders: an improvement on the method of measurement.*Appl. Opt.* **14**: 2813–2815.

Lindberg J. D., Gillespie J. B. 1977. Relationship between particle size and imaginary refractive index in atmospheric dust.*Appl. Opt.* **16**: 2628–2630.

Lindberg J. D., Laude L. S. 1974. Measurement of the absorption coefficient of atmospheric dust.*Appl. Opt.* **13**: 1923–1927.

Lindberg J. D., Smith M. S. 1974. Visible and near infrared absorption coefficients of kaolinite and related clays.*Am. Mineral.* **59**: 274–279.

Lindmo T., Peters D. C., Sweer R. G. 1990. Flow sorters for biological cells. In: *Flow cytometry and sorting*. Melamed M. R., Lindmo T., Mendelsohn M. L. (eds.) Wiley-Liss, New York, pp. 145–170.

Lines R. W. 1992. The electrical sensing zone method (the Coulter principle). In: *Particle size analysis*. Stanley-Wood N. G., Lines R. W. (eds.) Royal Society of Chemistry, Cambridge, UK, pp. 350–373.

Lisitzyn A. P. 1972. Sedimentation in the world ocean. In: *Spec. Publ. 17*, Soc. Econ. Paleont. and Mineral. Tulsa, Oklahoma, USA.

Liss S., Droppo I. G., Flannigan D. T., Leppard G. G. 1996. Floc architecture in wastewater and natural riverine systems. *Environ. Sci. Technol.* **30**: 680–686.

Liu Y, Arnott P. W., Hallett J. 1998. Anomalous diffraction theory for arbitrarily oriented finite circular cylinders and comparison with exact T-matrix results. *Appl. Opt.* **37**: 5019–5030.

Liu C.-P., Lin L.-P. 2001. Ultrastructural study and lipid formation of *Isochrysis* sp. CCMP1324. *Bot. Bull. Acad. Sin.* **42**: 207–214.

Liu C., Liu Z., Bo F., Wang Y., Z J. 2002. Super-resolution digital holographic imaging method. *Appl. Phys. Lett.***81**: 3143–3145.

Lloyd P. J. 1982. Response of electrical sensing zone method to non-spherical particles. In: *Particle size analysis*. Stanley-Wood N. G., Allet T. (eds.) Wiley-Heyden, London, UK, pp. 199–208.

Lock J. A., Hodges J. T. 1996. Far-field scattering of a non-Gaussian off-axis axisymmetric laser beam by a spherical particle.*Appl. Opt.* **35**: 6605–6616.

Loferer-Krößbacher M., Klima J., Psenner R. 1998. Determination of bacterial cell dry mass by transmission electron microscopy and densitometric image analysis. *Appl. Environ. Microbiol.* **64**: 688–694.

Logan B. E. 1993. Theoretical analysis of size distributions determined with screens and filters. *Limnol. Oceanogr.* **38**: 372–381.

Logan B. E., Kilps J. R. 1995. Fractal dimensions of aggregates formed in different fluid mechanical environments. *Water Res.* **29**: 443–453.

Logan B. E., Passow U., Alldredge A. L. 1994. Variable retention of diatoms on screens during size separation. *Limnol. Oceanogr.* **39**: 390–395.

Logan B. E., Wilkinson D. B. 1990. Fractal geometry of marine snow and other biological aggregates. *Limnol. Oceanogr.* **35**: 130–136.

Lohrenz S. E. 2000. A novel theoretical approach to correct for pathlength amplification and variable sampling loading in measurements of particulate spectral absorption by the quantitative filter technique. *J. Plankton Res.* **22**: 639–657.

Lohrenz S. E., Fahnenstiel G. L., Kirkpatrick G. J., Carroll C. L., Kelly K. A. 1999. Microphotometric assessment of spectral absorption and its potential application for characterization of harmful algal blooms. *J. Phycol.* **35**: 1438–1446.

Loizeau J.-L., Arbouville D., Santiago S., Vernet J.-P. 1994. Evaluation of a wide range laser diffraction grain size analyzer for use with sediments. *Sedimentology* **41**: 353–361.

Loken M. R., Sweet R. G., Herzenberg L. A. 1976. Cell discrimination by multiangle light scattering. *J. Histochem. Cytochem.* **24**: 284–291.

Longhurst A. R., Isao K., Li W. K. W., Rodríguez J., Dickie P., Kepay P., Partensky F., Bautista B., Ruiz J., Wells M., Bird D. F. 1992. Sub-micron particles in northwest Atlantic shelf waters.*Deep-Sea Res.* **39**: 1–7.

López-Amorós R., Comas J., Carulla C., Vives-Rego J. 1994. Variations in flow cytometric forward scatter signals and cell size in batch cultures of Escherichia coli. *FEMS Microbiol Lett.* **117**: 225–229.

Ludlam T., Slansky R. 1977. Search and classification mechanism in multiparticle data. *Rhys. Rev. D* **16**: 100–113.

Ludlow I. A., Kaye P. H. 1979. A scanning diffractometer for the rapid analysis of microparticles and biological cells. *J. Colloid Interface. Sci.* **69**: 571–589.

Lunau M., Rink B., Grossart H.-P., Simon M. 2003. How to sample marine microaggregates in shallow and turbid environments? - Problems and solutions. *Extended abstracts of the Workshop of BioGeoChemistry of Tidal Flats.* Carl von Ossietzky University, Oldenburg, Germany, 4 pp.

Lunau M., Sommer A., Lemke A., Grossart H.-P., Simon M. 2004. A new sampling device for microaggregates in turbid aquatic systems. *Limnol. Oceanogr. Methods* **2**: 387–397.

Lunven M., Gentien P., Kononen K., Le Gall E., Daniélou M. M. 2003. In situ video and diffraction analysis of marine particles. *Estuar. Coast. Shelf. Sci.* **57**: 1127–1137.

Lwin T. 2003. Parameterization of particle size distributions by three methods. *Math. Geol.* **35**: 719–736.

Lyubovtseva Y. S., Plakhina I. N. 1976. Measurements of the light scattering matrices of suspensions of nonspherical particles.*Oceanology* **15**: 111–115 (English translation of *Okeanologyia*).

M+

MacDonald M. P., Spalding G. C., Dholakia K. 2003. Microfluidic sorting in an optical lattice.*Nature* **426**: 421–424.

MacIntyre S., Alldredge A. L., Gotschalk C. C. 1995. Accumulation of marine snow at density discontinuities in the water column. *Limnol. Oceanogr.* **40**: 449–468.

Mackowski D. W. 1991. Analysis of radiative scattering of multiple sphere configurations. *Proc. Roy. Soc. London A* **433**: 599–614.

Mackowski D. W. 1995. Electrostatics analysis of radiative absorption by sphere clusters in the Rayleigh limit: application to soot particles. *Appl. Opt.* **34**: 3535–3545.

Madigan M. T., Martinko J. M. 2005. *Brock biology of microorganisms* (11th ed.) Pearson Prentice Hall, Upper Saddle River, NJ, 1074 pp.

Maffione R. A., Dana D. R. 1997. Instruments and methods for measuring the backward-scattering coefficient of ocean waters. *Appl. Opt.***36**: 6057–6067.

Maffione R. A., Dana, D. R., Voss J. M., Frysinger G. S. 1993. Instrumented remotely operated vehicle for measuring inherent and apparent optical properties of the ocean. In: *Underwater Light Measurements*. Eilertsen H. C. (ed.) *Proc. SPIE* **2048**: 124–137.

Maffione R. A., Honey R. C. 1992. Instrument for measuring the volume scattering function in the backward direction. *Ocean Opt. XI* **1750**: 15–26.

Magni V., Cerullo G., De Silvestri S. 1992. High-accuracy fast Hankel transform for optical beam propagation. *J. Opt. Soc. Am. A* **9**: 2031–2033.

Maier J., Walker S., Fantini S., Franceschini M., Gratton E. 1994. Possible correlation between blood glucose concentration and the reduced scattering coefficient of tissues in the near infrared. *Opt. Lett.* **19**: 2062–2064.

Malkiel E., Alquaddoomi O., Katz. J. 1999. Measurements of plankton distribution using submersible holography. *Meas. Sci. Technol.* **10:** 1153–1161.

Malone T. C. 1980. Algal size. In: *Studies in ecology*. Barth, H. G. (ed.) University of California Press, California, USA, pp. 433–463.

Maltsev V. P. A. 2000. Scanning flow cytometry for individual particle analysis. *Rev. Sci. Instrum.* **71**: 243–255.

Mamaev O. I. 1975. *Temperature-salinity analysis of world ocean waters*. Elsevier, Amsterdam, 374 pp.

Mandelbrot B. B. 1997. Fractals and scaling in finance. Springer Verlag, New York, 561 pp.

Mankovsky V. I. 1971. Relationships between the total scattering coefficient and the angular coefficients of light scattering. (In Russian: *O sootnoshenii meshdu integralnem pokazatelem rasseyianyia sveta morskoy vodoy e pokazatelem resseyanyia v fixirovannom napravlenyi*). *Morskoyie Gidrofiz. Issled., Akad. Nauk Ukrainskoyi SSR* **6**: 145–154.

Mankovsky V. I., Haltrin V. I. 2002a. Phase functions of light scattering measured in waters of World Ocean and Lake Baykal. In: *2002 IEEE International Geoscience and Remote Sensing Symposium and the 24th Canadian Symposium on Remote Sensing*, June 24–28, 2002, Toronto, Canada, Paper I2E09_1759.

Mankovsky V. I., Haltrin V. I. 2002b. Light scattering phase functions measured in waters of Mediterranean Sea. In: *Proceedings of OCEANS 2002 MTS-IEEE*, October 29–31, 2002, Biloxi, Mississippi, USA, Vol. 4, IEEE, pp. 2368–2373.

Mankovsky V. I., Semenikhina V. M., Neuyimin O. G. 1970. A marine nephelometer in Russian. *Morskoyie Gidrofiz. Issled., Akad. Nauk Ukrainskoyi SSR* **2**: 171–181.

Mankovsky V. I., Solovev M. V. 2003. Seawater phase scattering functions in the Black Sea. In: *Proceedings of the II International Conf. "Current Problems in Optics of Natural Waters,"* ONW'2003, Levin I., Gilbert G. (eds.) St. Petersburg, Russia, 2003, pp. 293–296.

Manning A. J., Dyer K. R. 1999. A laboratory examination of floc characteristics with regard to turbulent shearing. *Mar. Geol.* **160**: 147–170.

Maranger R., Bird D. F. 1995. Viral abundance in aquatic systems: a comparison between marine and fresh waters. *Mar. Ecol. Prog. Ser.* **121**: 217–226.

Maranger R., Bird D. F., Juniper S. K. 1994. Viral and bacterial dynamics in Arctic sea ice during the spring algal bloom near Resolute, N. W. T., Canada. *Mar. Ecol. Prog. Ser.* **111**: 121–127.

Mari X., Burd A. 1998. Seasonal size spectra of transparent exopolymeric particles (TEP) in a coastal sea and comparison with those predicted using coagulation theory. *Mar. Ecol. Prog. Ser.* **163**: 63–76.

Mari X, Kiørboe T. 1996. Abundance, size distribution and bacterial colonization of transparent exopolymeric particles (TEP) during spring in the Kattegat. *J. Plankton Res.* **18**: 969–986.

Marquis R. E. 1973. Immersion refractometry of isolated bacterial cell walls.*J. Bacteriol.* **116**: 1273–1279.

Marquet P. A. 2000. Invariants, scaling laws, and ecological complexity. *Science* **289**: 1487–1488.

Marshall T. R., Parmenter Ch. S., Seaver M. 1975. Precision measurement of particulates by light scattering at optical resonance. *Science* **190**: 375–377.

Marston P. L. 1979. Critical angle scattering by a bubble: physical-optics approximation and observations. *J. Opt. Soc. Am.* **69**: 1205–1211.

Marston P. L., Langley D. S., Kingsbury D. L. 1982. Light scattering by bubbles in liquids: Mie theory, physical-optics approximations, and experiments. *Appl. Sci. Res.* **38**: 373–383.

Martin J. S. 1970. *Empirical Bayes methods.* Methuen, London.

Marie D., Brussaard C. P. D., Thyrhaug R., Bratbak G., Vaulot D. 1999. Enumeration of marine viruses in culture and natural samples by flow cytometry. *Appl. Environ. Microbiol.* **65**: 45–52.

Martinis M., Risović D. 1998. Fractal analysis of suspended particles in seawater. *Fizika B* **7**: 65–72.

Marzo G. A., Blanco A., De Carlo F., D'Elia M., Fonti S., Marra A. C., Orofino V., Politi R. 2004. The optical constants of gypsum particles as analog of Martian sulfates. *Adv. Space Res.* **33**: 2246–2251.

Maske H., Haardt H. 1987. Quantitative in vivo absorption spectra of phytoplankton: Detrital absorption and comparison with fluorescence excitation spectra. *Limnol. Oceanogr.* **32**: 620–633.

Masuda H., Gotoh K. 1999. Study on the sample size required for the estimation of mean particle diameter. *Adv. Powder Technol.* **10**: 159–173.

Masuda H., Iinoya K. 1971. Theoretical study of the scatter of experimental data due to particle-size-distribution. *J. Chem. Eng. Jpn.* **40**: 60–66.

Matlack D. E. 1974. Deep ocean optical measurement (DOOM) report on North Atlantic, Caribbean, and Mediterranean cruises. *Tech. Report. Naval Ordnance Laboratory* NOLTR 74-42-62, White Oak, Silver Spring, MD, USA, 103 pp.

Matsuyama T., Yamamoto H., Scarlett B. 2000. Transformation of diffraction pattern due to ellipsoids into equivalent diameter distribution for spheres. *Part. Part. Syst. Charact.* **17**: 41–46.

Matthäus W. 1974. Empirical equations for the refractive index of seawater. (In German: *Empirische Gleigungen für den Brechungsindex des Meerwassers*). *Beitrage Zur Meereskunde* **33**: 73–78.

Matthews B. A., Rhodes C. T. 1970. Some observations on the use of the Coulter counter model B in coagulation studies. *J. Colloid Interface Sci.* **32**: 339–348.

McCave I. N. 1983. Particulate size spectra behavior and origin of nepheloid layers over the Nova Scotian continental Rise. *J. Geophys. Res. C* **88**: 7647–7666.

McCave I. N. 1984. Size spectra and aggregation of suspended particles in the deep ocean. *Deep-Sea Res.* **31**: 329–352.

McCave I. N. 1985. Properties of suspended sediment over the Hebble area on the Nova Scotian Rise. In: *Deep ocean sediment transport – preliminary results of the high energy benthic boundary layer experiment.* Nowell A. R. M., Hollister C. D. (eds). *Mar. Geol.* **66**: 169–188.

McCave I. N., Bryant R. J., Cook H. F., Coughanover C. A. 1986. Evaluation of a laser-diffraction-size analyzer for use with natural sediments. *J. Sediment. Petrol.* **56**: 561–564.

McCave I. N., Syvitski J. P. M. 1991. Principles and methods of geological particle size analysis. In: *Principles, methods, and application of particle size analysis.* Syvitski J. P. M. (ed.). Cambridge University Press, New York, pp. 3–21.

McCluney W. R. 1974. Multichannel forward scattering meter for oceanography. *Appl. Opt.* **13**: 548–555.

McCluney W. R. 1975. Two-channel laboratory scattering meter. *Rev. Sci. Instrum.* **46**: 1231–1236.

McCrone W. C., Delly J. G. 1973. *The particle atlas.* Ann Arbor Science Publishers, Ann Arbor, Michigan, USA.

McCormick N. J. 1987. Inverse radiative transfer with a delta-Eddington phase function. *Astrophys. Space Sci.* **129**: 331–334.

McCormick N. J., Rinaldi G. E. 1989. Seawater optical property estimation from in situ measurements. *Appl. Opt.* **28**: 2605–2613.

McIntyre D., Doderer G. C. 1959. Absolute light-scattering photometer: I. Design and operation. *J. Res. Nat. Bur. Stand.* **62**: 153–159.

McKellar B. J., Box M. A., Bohren C. F. 1982. Sum rules for optical scattering amplitudes. *J. Opt. Soc. Am.* **72**: 535–538.

McLean J. W., Freeman J. D., Walker R. E. 1998. Beam spread function with time dispersion. *Appl. Opt.* **37**: 4701–4711.

McLean J. W., Voss K. J. 1991. Point spread function in ocean water: a comparison between theory and experiment. *Appl. Opt.* **30**: 2027–2030.

McLean S., Schofield B., Zibordi G., Lewis M. R., Hooker S., Weidemann A. D. 1997. Field evaluation of antibiofouling compounds on optical instrumentation. In: *Ocean Optics XIII.* Ackleson S. G. (ed.) *Proc. SPIE* **2963**: 708–713.

McNeil G. T. 1977. Metrical fundamentals of underwater lens system. *Opt. Eng.* **16**: 128–139.

Meade M. L. 1982. Advances in lock-in amplifiers. *J. Phys. E: Sci. Instrum.* **15**: 395–403.

Medwin H. 1977. *In situ* acoustic measurements of microbubbles at sea. *J. Geophys. Res. C* **82**: 971–976.

Mehu A., Johannin-Gilles A. 1968. Variations of the refractive index of the Copenhagen Standard Seawater, and their dilutions as a function of the length of the wavelength, temperature, and chlorinity. (In French: *Variations de l'indice de refraction de l'eau de mer, Etalon de Copenhage, et ses dilutions en fonction de la longeur d'onde, de la temperature, et de la chlorinite*). *Cah. Oceanogr.* **20**: 803–812.

Melamed M. R., Mullaney P. F., Shapiro H. M. 1990. An historical review of the development of flow cytometers and sorters. In: *Flow cytometry and sorting*. Melamed M. R., Lindmo T., Mendelsohn M. L. (eds.) Wiley-Liss, New York, p. 1–10.

Mercer W. B. 1966. Calibration of Coulter counters for particles $\sim 1\mu$ in diameter. *Rev. Sci. Instrum.* **37**: 1515–1520.

Mercer F. V., Bogorad L., Mullens R. 1962. Studies with *Cyanidium caldarium*: I. The fine structure and systematic position of the organism. *J. Cell Biol.* **13**: 393–403.

Merkus H. G., Liu H., Scarlett B. 1990. Improved resolution and accuracy in electrical sensing zone particle counters through hydrodynamic focussing.*Part. Part. Syst. Charact.* **7**: 11–15.

Mertens L. E., Phillips D. L. 1972. *Measurements of the volume scattering function of sea water*. Technical Report 334, Range Measurements Laboratory, Patrick Air Force Base, Florida, USA, 29 pp.

Mertens L. E., Replogle F. S., Jr. 1977. Use of point spread and beam spread functions for analysis of imaging systems in water. *J. Opt. Soc. Am.* **67**: 1105–1117.

Middelberg A. P. J., Bogle I. D. L., Snoswell M. 1990. Sizing biological samples by photosedimentation techniques. *Biotechnol. Prog.* **6**: 255–261.

Middleton G. V. 1970. Generation of the log-normal frequency distribution in sediments. In: *Topics in mathematical geology*. Romanova, M. A. and Sarmand, V. (eds.) Consultants Bureau, New York, pp. 34–42.

Mie G. 1908. On the optics of turbid media, special colloidal metallic solutions. (In German: *Beitrage zur Optik trüber Medien speziell kolloidaler Metallösungen*). *Ann. Phys.* **25**: 377–445.

Mielenz K. D. 1998. Algorithms for Fresnel diffraction at rectangular and circular apertures. *J. Res. Natl. Inst. Stand. Technol.* **103**: 497–509.

Millard R. C., Seaver G. 1990. An index of refraction algorithm for seawater over temperature, pressure, salinity, density, and wavelength. *Deep-Sea Res.* **37**: 1909–1926.

Millero F. J. 2001. *The physical chemistry of natural waters*. Wiley-Interscience, New York., 654 pp.

Millero F. J., and Poisson A. 1981. International one-atmosphere equation of state of seawater. *Deep-Sea Res.* A **28**, 625–629.

Milligan T. G. 1995. An examination of the settling behaviour of a flocculated suspension. *Neth. J. Sea Res.* **33**: 163–171.

Milligan T. G., Kranck K. 1991. Electroresistance particle size analyzers. In: *Principles, methods, and application of particle size analysis*. Syvitski J. P. M. (ed.) Cambridge University Press, New York, pp. 109–118.

Misconi N. Y., Oliver J. P., Ratcliff K. F., Rusk E. T., Wang W.-X. 1990. Light scattering by laser levitated particles.*Appl. Opt.* **29**: 2276-2281.

Mishchenko M. I. 1993. Light scattering by size-shape distributions of randomly oriented axially symmetric particles of a size comparable to a wavelength. *Appl. Opt.* **32**: 4652–4666.

Mishchenko M. I., Lacis A. A. 2003. Morphology-dependent resonances of nearly spherical particles in random orientation. *Appl. Opt. LP* **42**: 5551–5556.

Mishchenko M. I., Macke A. 1999. How big should hexagonal ice crystals be to produce halos? *Appl. Opt.* **38**: 1626–1629.

Mishchenko M. I., Travis L. D. 1994. Light scattering by polydispersions of randomly oriented spheroids with sizes comparable to wavelengths of observation. *Appl. Opt.* **33**: 7206–7225.

Mishchenko M. I., Travis L. D., Kahn R. A., West R. A. 1997. Modeling phase functions for dustlike tropospheric aerosols using a shape mixture of randomly oriented polydisperse spheroids. *J. Geophys. Res. D* **102**: 16,831–16,847.

Mishchenko M. L., Travis L. D., Mackowski D. W. 1994. T-matrix computations of light scattering by large nonspherical particles: recent advances. In: *Passive infrared remote sensing of clouds and the atmosphere II.* David K. Lynch (ed.) *Proc. SPIE* **2309**: 72–83.

Mishchenko M. I., Videen G., Babenko V. A., Khlebtsov N. G., Wriedt Th. 2004. T-matrix theory of electromagnetic scattering by particles and its applications: a comprehensive reference database. *J. Quant. Spectrosc. Radiat. Transfer* **88**: 357–406.

Mita A. 1982. Light absorption properties of inhomogeneous spherical aerosol particles. *J. Meteorol. Soc. Jpn.* **60**: 765–776.

Mita A., Isono K. 1980. Effective complex refractive index of atmospheric aerosols containing absorbing substances. *J. Meteorol. Soc. Jpn.* **58**: 69–80.

Mitchell B. G., Kiefer D. A. 1984. Determination of absorption and fluorescence excitation spectra for phytoplankton. In: *Marine phytoplankton and productivity.* Holm-Hansen O., Bolis L., Gilles R. (eds.). Springer-Verlag, Berlin, p. 157–169.

Mitchell B. G., Kiefer D. A. 1988. Chlorophyll a specific absorption and fluorescence excitation spectra for light-limited phytoplankton. *Deep-Sea Res.* **35**: 639–663.

Mitzenmacher M. 2004. A brief history of generative models for power law and lognormal distributions. *Internet Math.* **1**: 226–251.

Moal J., Martin-Jezequal V., Harris R. P., Samain J.-F., Poulet S. A. 1987. Interspecific and intra-specific variability of the chemical composition of marine phytoplankton. *Oceanologica Acta* **10**: 339–346.

Mobley C. D. 1994. *Light and water: Radiative transfer in natural waters.* Academic Press, San Diego.

Mobley C. D., Sundman L. K., Boss E. 2002. Phase function effects on oceanic light fields. *Appl. Opt.* **41**: 1035–1050.

Montagnes D. J. S., Berges J. A., Harrison P. J., Taylor F. J. R.1994. Estimating carbon, nitrogen, protein, and chlorophyll a from volume in marine phytoplankton. *Limnol. Oceanogr.* **39**: 1044–1060.

Montesinos E., Esteve I., Guerrero R. 1983. Comparison between direct methods for determination of microbial cell volume: electron microscopy and electronic particle sizing. *Appl. Environ. Microbiol.* **45**: 1651–1658.

Moore P. B. 1999. A biophysical chemist's thoughts on cell size. In: *Size limits of very small microorganisms. Proceedings of a Workshop.* Washington, D.C., October 22–23, 1998, National Academy Press, Washington, DC, pp. 16–20.

Morel A. 1965. Interpretation of variations of the form of the volume scattering function of sea water. (In French: *Interpretation des variations de la forme de l'indicatrice de diffusion de la lumiere par les eaux de mer*). *Ann. Geophys.* **21**: 281–284.

Morel A. 1973. The scattering of light by sea water. Experimental results and theoretical approach. (In French: *Diffusion de la lumière par les eaux de mer; résultats expérimentaux et approche théorique*) In: *Optics of the sea, interface and in-water transmission and imaging*. AGARD Lecture Series No.61, Published in August 1973, translated to English by G. Halikas, 161 pp.

Morel A. 1974. Optical properties of pure water and pure sea water. In: *Optical aspects of oceanography*. Jerlov N. G., Steeman-Nielsen E. (eds.) Academic Press, London.

Morel A., Ahn Y.-H. 1990. Optical efficiency factors of free-living marine bacteria: Influence of bacterioplankton upon the optical properties and particulate organic carbon in oceanic waters. *J. Mar. Res.* **48**: 145–175.

Morel A., Ahn Y.-H. 1991. Optics of heterotrophic nanoflagellates and ciliates: A tentative assessment of their scattering role in oceanic waters compared to those of bacterial and algal cells. *J. Mar. Res.* **49**: 177–202.

Morel A., Ahn Y.-H., Partensky F., Vaulot D., Claustre H. 1993. *Prochlorococcus* and *Synechococcus*: A comparative study of their optical properties in relation to their size and pigmentation. *J. Mar. Res.* **51**: 617–649.

Morel A., Bricaud A. 1981. Theoretical results concerning light absorption in a discrete medium, and application to specific absorption of phytoplankton. *Deep-Sea Res.* **28**A: 1375–1393.

Morel A., Bricaud A. 1986. Inherent properties of algal cells including picoplankton: theoretical and experimental results. *Can. Bull. Fish. Aquatic Sci.* **214**: 521–559.

Morel A., Gentili B. 1993. Diffuse reflectance of oceanic waters. II: Bidirectional aspects. *Appl. Opt.* **32**: 6864–6879.

Morel A., Prieur L. 1977. Analysis of variations in ocean color.*Limnol. Oceanogr.* **22**: 709–722.

Morel A., Smith R. C. 1982. Terminology and units in optical oceanography. *Mar. Geod.* **5**: 335–349.

Morrison R. E. 1967. *Studies on the optical properties of sea water, Argus Island in the north Atlantic and in Long Island and Block Island Sounds*. Ph.D. Thesis, New York University, New York University, New York.

Morrison R. E. 1970. Experimental studies on the optical properties of sea water. *J. Geophys. Res.* **75**: 612–628.

Moser H. O. 1974. Instrument for measuring angular intensity distributions of light sources within a few milliseconds.*Appl. Opt.* **13**: 173–176.

Mostajir B., Dolan J. R., Rassoulzadegan F. 1995a. A simple method for the quantification of a class of labile marine pico- and nano-sized detritus: DAPI yellow particles (DYP). *Aquat. Microb. Ecol.* **9**: 259–266.

Mostajir B., Dolan J. R., Rassoulzadegan F. 1995b. Seasonal variations of pico- and nano-detrital particles (DAPI yellow particles, DYP) in the Ligurian Sea (NW Mediterranean). *Aquat. Microb. Ecol.* **9**: 267–277.

Mostajir B., Fagerbakke K. M., Heldal M., Thingstad T. F., Rassoulzadegan F. 1998. Elemental composition of individual pico- and nano-sized marine detrital particles in the northwestern Mediterranean Sea. *Oceanologica Acta* **21**: 589–596.

Mourant J. R., Canpolat M., Brocker C., Esponda-Ramos O., Johnson T. M., Matanock A., Stetter K., Freyer J. P. 2000. Light scattering from cells: the contribution of the nucleus and the effects of proliferative status. *J. Biomed. Opt.* **5**: 131–137.

Mueller J. J. 1974. *The influence of phytoplankton on ocean color spectra.* Ph.D. Thesis, Oregon State Univ., Corvallis, OR, USA, 239 pp.

Mugnai A., Wiscombe W. J. 1980. Scattering of radiation by moderately nonspherical particles. *J. Atmosph. Res.* **37**: 1291–1307.

Mugnai A., Wiscombe W. J. 1986. Scattering from nonspherical Chebyshev particles. 1: Cross sections, single-scattering albedo, asymmetry factor and backscattering fraction. *Appl. Opt.* **25**: 1235–1244.

Mugnai A., Wiscombe W. J. 1989. Scattering from nonspherical Chebyshev particles. 3: Variability in angular scattering patterns. *Appl. Opt.* **28**: 3061–3073.

Mühlenweg H., Hirleman E. D. 1998. Irregularly-shaped particles measured by laser diffraction spectroscopy. In *Proceedings on the 5th International Congress on Optical Particle Sizing*, University of Minnesota, Minneapolis, MN, USA, 1998, pp. 147–150.

Muinonen K., Nousiainen T., Fast P., Lumme K., Peltoniemi J. I. 1996. Light scattering by Gaussian random particles: ray optics approximation. *J. Quant. Spectrosc. Rad. Transfer* **55**: 577–601.

Mujat M., Dogariu A. 2001. Real-time measurement of the polarization transfer function. *Appl. Opt.* **40**: 34–44.

Mullaney P. F., Crowell J. M., Salzman G. C. 1976. Pulse-height light-scatter distributions using flow-system instrumentation. *J. Histochem. Cytochem.* **24**: 298–304.

Mullaney P. F., Dean P. N. 1969. Cell sizing: A small-angle light-scattering method for sizing particles of low refractive index. *Appl. Opt.* **8**: 2361–2362.

Mullaney P. F., Van Dilla M. A., Coulter J. R., Dean P. N. 1969. Cell sizing: a light scattering photometer for rapid volume determination. *Rev. Sci. Instrum.* **40**: 1029–1032.

Mullin M. M., Sloan P. R., Eppley R. W. 1966. Relationships between carbon content, cell volume, and area in phytoplankton.*Limnol. Oceanogr.* **11**: 307–311.

Mulhearn P. J. 1981. Distribution of microbubbles in coastal waters. *J. Geophys. Res. C* **86**: 6429–6434.

Muñoz O., Volten H., Rol E., de Haan J., Vassen W., Hovenier J. W. 1999. Experimental determination of scattering matrices of mineral particles. In: *Light scattering by non-spherical particles: Theory, measurements, and applications.* Michael I. Mishchenko, Joop W. Hovenier, Larry D. Travis (eds.) Academic Press, San Diego, pp. 265–268.

N+

Naganuma T. 1997. Abundance and production of bacterioplankton along a transect of Ise Bay, Japan. *J. Oceanogr.* **53**: 579–583.

Nagel M., Ay P. 2000. 3D morphological investigation of flocs. *Imag. Microsc.* **2**: 28–31.

Naito M., Hayakawa O., Nakahira K., Mori H., Tsubaki J. 1998. Effect of particle shape on the particle size distribution measured with commercial equipment. *Powder Technol.* **100**: 52–60.

Nakroshis P., Amoroso M., Legere J., Smith C. 2003. Measuring Boltzmann's constant using video microscopy of Brownian motion. *Am. J. Phys.* **71**: 568–573.

Nebrensky J. J., Player M. A., Saw K., Tipping K., Watson J. 2000. HoloCam: A subsea holographic camera for recording marine organisms and particles. In: *Optical diagnostics for industrial applications.* Halliwell, Neil A. (ed.) *Proc. SPIE* **4076**: 111–119.

Neihoff R. A., Loeb G. I. 1972. The surface charge of particulate matter in sea water. *Limnol. Oceanogr.* **17**: 7–16.

Neihoff R. A., Loeb G. I. 1974. Dissolved organic matter in sea water and the electric charge of immersed surfaces.*J. Mar. Res.* **32**: 5–12.

Nelder J. A., Mead R. 1965. Downhill simplex method. *Comput. J.* **7**: 308–315.

Nelson J. R., Robertson C. Y. 1993. Detrital spectral absorption: Laboratory studies of visible light effects on phytodetritus absorption, bacterial spectral signal, and comparison to field measurements. *J. Mar. Res.* **51**: 181–207.

Nieto M. M., Simmons L. M., Jr. 1979. Eigenstates, coherent states and uncertainty products for the Morse oscillator. *Phys. Rev. A.* **19**: 438–444.

Nishizawa S., Fukuda M., Inoue N. 1954. Photographic study of suspended particles and plankton in the sea. *Bull. Faculty Fish. Hokkaido Univ.* **5**: 36–40.

Nyffeler F., Ruch P. 1989. Size distribution of the suspended particles in the water masses of the northeast Atlantic. In: *Interim oceanographic description of the North-East Atlantic site for the disposal of low-level radioactive waste.* Nuclear Energy Agency, Paris, pp. 123–127.

O+

O'Connor C. L., Schlupf J. P. 1967. Brillouin scattering in water: the Landau-Placzek ratio. *J. Chem. Phys.* **47**: 31–38.

Oh C., Sorensen C. 1997. The effect of overlap between monomers on the determination of fractal cluster morphology. *J. Colloid Interface Sci.* **193**: 17–25.

O'Hern T. J., d'Agostino L., Acosta A. J. 1988. Comparison of holographic and Coulter counter measurements of cavitation nuclei in the ocean. *J. Fluids Eng.* **110**: 200–207.

Oishi T. 1990. Significant relationship between the backward scattering coefficient of sea water and the scatterance at 120°. *Appl. Opt.* **29**: 4658–4665.

Okada K., Heintzenberg J., Kai K., Qin Y. 2001. Shape of atmospheric mineral particles in three Chinese arid-regions. *Geophys. Res. Lett.* **28**: 3123–3126.

Oldenbourg R., Ruiz T. 1989. Birefringence of macromolecules: Wiener's theory revisited, with applications to DNA and tobacco mosaic virus. *Biophys. J.* **56**: 195–205.

Olson R. J., Chisholm S. W., Frankel S. L., Shapiro H. M. 1983. An inexpensive flow cytometer for the analysis of fluorescence signals in phytoplankton: Chlorophyll and DNA distributions. *J. Exp. Mar. Biol. Ecol.* **68**: 129–144.

Olson R. J., Sosik H. M. 2000. An *in situ* flow cytometer for the optical analysis of individual particles in coastal waters. (Abstract) ASLO, Albuquerque, NM, February 12–16, 2001.

Olson R. J., Sosik H. M., Williams E. J., III. 2001. An *in situ* flow cytometer for the optical analysis of individual particles in seawater (posted at http://www.whoi.edu/science/B/Olsonlab/insitu2001.htm).

Olson R. J., Vaulot D., Chisholm S. W. 1985. Marine phytoplankton distributions measured using shipboard flow cytometry. *Deep-Sea Res.* **32**: 1273–1280.

Olson R. J., Zettler E. R., Anderson O. K. 1989. Discrimination of eukaryotic phytoplankton cell types from light scatter and autofluorescence properties measured by flow cytometry. *Cytometry* **10**: 636–643.

Olson R. J., Zettler E. R., DuRand M. D. 1993. Phytoplankton analysis using flow cytometry. In: *Aquatic microbial ecology.* Kemp P. K., Sherr B. D., Sherr E. B., Cole J. J. (eds.) Lewis Publishers, Boca Raton, pp. 175–186.

Orford J. D., Whalley W. B. 1991. Quantitative grain form analysis. In: *Principles, methods, and application of particle size analysis.* Syvitski, J. P. M. (ed.) Cambridge University Press, New York, pp. 174–193.

Owen J. F., Barber P. W., Messinger B. J., Chang R. K. 1981. Determination of optical-fiber diameter from resonances in the elastic scattering spectrum. *Opt. Lett.* **6**: 272–274.

Ozmidov R. W. 1967. Experimental investigation of the horizontal turbulent diffusion in the sea and in an artificial, shallow water body. (In Russian: *Experimentalnoyie issledovanyie ghorizontalnoyi turbulentnoyi diffuzyi w morie i iskustviennom vodoyiomyie nebolshoyi ghloubiny*). *Izv. AN SSR, Seria Geophys.* **6**: 756–760.

P+

Paasche E. 1960. On the relation between primary production and standing stock of phytoplankton. *J. Conseil Perm. Internat. l'Explor. Mer.* **26**: 33–48.

Padisák J., Soróczki-Pintér É., Rezner Z. 2003a. Sinking properties of some phytoplankton shapes and the relation of form resistance to morphological diversity of plankton – an experimental study. *Hydrobiology* **500**: 243–257.

Padisák J., Soróczki-Pintér É., Rezner Z. 2003b. Sinking properties of some phytoplankton shapes and the relation of form resistance to morphological diversity of plankton – an experimental study (Erratum). *Hydrobiology* **501**: 219–219.

Pahlow M., Riebesell U., Wolf-Gladrow D.-A. 1997. Impact of cell shape and chain formation on nutrient acquisition by marine diatoms. *Limnol. Oceanogr.* **42**: 1660–1672.

Pan Y., Aptowicz K. B., Chang R. K., Hart M., Eversole J. D. 2003. Characterizing and monitoring respiratory aerosols by light scattering. *Opt. Lett.* **28**: 589–591.

Pappas J. 2005. Geometry and topology of diatom shape and surface morphogenesis for use in applications of nanotechnology. *J. Nanosci. Nanotechnol.* **5**: 120–130.

Pareto V. 1986. The course of political economy. (In Italian: *Corso di economia politica*). Unione Tipographico-Editrice Torinese, Torino.

Passow U. 2002. Transparent exopolymer particles (TEP) in aquatic environments. *Prog. Oceanogr.* **55**: 287–333.

Passow U., Alldredge A. L. 1994. Distribution, size and bacterial colonization of transparent exopolymer particles {TEP) in the ocean. *Mar. Ecol. Prog. Ser.* **113**: 185–198.

Passow U., Shipe R. F., Murray A., Pak D. K., Brzezinski M. A., Alldredge A. L. 2001. The origin of transparent exopolymer particles (TEP) and their role in the sedimentation of particulate matter. *Continent. Shelf Res.* **21**: 327–346.

Patel C. K. N., Tam A. C. 1979. Optical absorption coefficients of water. *Nature* **280**: 302–304.

Patterson E. M., Gillette D. A., Stockton B. H. 1977. Complex index of refraction between 300 and 700 nm for Saharan aerosols.*J. Geophys. Res.* **82**: 3153–3160.

Paul J. H., Jeffrey W. H. 1984. Measurement of diameters of estuarine bacteria and particulates in natural water samples by use of a submicron particle analyzer. *Curr. Microbiol.* **10**: 7–12.

Pawlak B., Kopec J. 1998. Size distributions of *Scenedesmus obliquus* cells: experimental results from optical microscopy and their approximations using the ϕ-normal distribution. *Oceanologia* **40**: 345–353.

Pecora R. 2000. Dynamic light scattering measurement of nanometer particles in liquids. *J. Nanopart. Res.* **2**: 123–131.

Peeters J. C. H., Dubelaar G. B. J., Ringelberg J., Visser J. W. M. 1989. Optical plankton analyzer: a flow cytometer for plankton analysis, I: Design considerations. *Cytometry* **10**: 522–528.

Pegau W. S., Gray D., Zaneveld J. R. V. 1997. Absorption and attenuation of visible and near-infrared light in water: dependence on temperature and salinity. *Appl. Opt.* **36**: 6035–6046.

Pelce P., Sun J. 1993. Geometrical models for the growth of unicellular algae.*J. Theor. Biol.* **160**: 375–386.

Pelssers E. G. M., Stuart M. A. C., Fleer G. J. 1990a. Single particle optical sizing: I. Design of an improved SPOS instrument and application to stable dispersions.*J. Colloid Interface Sci.* **137**: 350–361.

Pelssers E. G. M., Stuart M. A. C., Fleer G. J. 1990b. Single particle optical sizing: II. Hydrodynamic forces and application to aggregating dispersions.*J. Colloid Interface Sci.* **137**: 362–372.

Pendleton J. D. 1982. Mie scattering into solid angles. *J. Opt. Soc. Am.* **72**: 1029–1033.

Peterson B, Ström S. 1973. T matrix for electromagnetic scattering from an arbitrary number of scatterers and representation of E(3). *Phys. Rev. D* **8**: 3661–3677.

Pereira F., Gharib M. 2004. A method for three-dimensional particle sizing in two-phase flows. *Meas. Sci. Technol.* **15**: 2029–2038.

Perrin F. 1942. Polarization if light scattered by isotropic opalescent materials. *J. Chem. Phys.* **10**: 415–427.

Petzold, T. J. 1972. *Volume scattering functions for selected ocean waters.* SIO Ref. 72–78, Scripps Institution of Oceanography, San Diego, 79 pp.

Phinney D. A., Cucci T. L. 1989. Flow cytometry and phytoplankton. *Cytometry* **10**: 511–521.

Pickard G. L., Giovando L. F. 1960. Some observations of turbidity in British Columbia inlets. *Limnol. Oceanogr.* **5**: 162–170.

Pickett-Heaps J. D. 1975. *Green algae: structure, reproduction and evolution in selected genera.* Sinauer Associates, Inc., Sunderland, Massachusetts, 606 pp.

Pierce J. W., Siegel F. R. 1979. Particulate material suspended in estuarine and oceanic waters.*Scan. Electr. Microsc.* **1**: 555–562.

Pike E. R., Pomeroy W. R. M., Vaughan, J. M. 1975. Measurement of Rayleigh ratio for several pure liquids using a laser and monitored photon counting. *J. Chem. Phys.* **62**: 3188–3192.

Pinnick R. G., Carroll D. E., Huffman D. J. 1976. Polarized light scattered from monodisperse randomly oriented nonspherical aerosol particles: measurements.*Appl. Opt.* **15**: 384–393.

Pinnick R. G., Auvermann H. J. 1979. Response characteristics of Knollenberg light-scattering aerosol counters. *J. Aerosol Sci.* **10**: 55–74.

Pitter M. C. P., Hopcraft K. I., Jakeman E., Walker J. G. 1999. Structure of polarization fluctuations and their relation to particle shape. *J. Quant. Spectrosc. Rad. Transfer* **63**: 433–444.

Plass G. N., Kattawar G. W., Humphreys T. J. 1985. Influence of the oceanic scattering phase function on the radiance.*J. Geophys. Res. C* **90**: 3347–3351.

Platt T., Denman K. 1977. Organization in the pelagic ecosystem. *Helgolander wiss. Meeresuntersuchen* **30**: 575–581.

Platt T., Denman K. 1978. The structure of pelagic marine ecosystem. *Rapp. P.-v. Réun. Cons. int. Explor. Mer* **173**: 60–65.

Pope R. M., Fry E. S. 1997. Absorption spectrum (380–700 nm) of pure water. II. Integrating cavity measurements. *Appl. Opt.* **36**: 8710–8723.

Posch T., Loferer-Krößbacher M., Gao G., Alfreider A., Pernthaler J., Psenner R. 2001. Precision of bacterioplankton biomass determination: a comparison of two fluorescent dyes, and of allometric and linear volume-to-carbon conversion factors. *Aquat. Microb. Ecol.* **25**: 55–63.

Poulet S. A., Chanut J.-P., Morissette M. 1986. Size spectra of particles in the estuary and Gulf of Saint Lawrence. I. Variations with space. (In French: *Etude des spectres de taille des particules en suspension dans l'estuaire et le golfe du Saint-Laurent. I. Variations spatiales*).*Oceanol. Acta* **9**: 179–189.

Prahl S. A. 1988. Light transport in tissue. Ph.D. Thesis, University of Texas at Austin, 221 pp.

Prandke H. 1980. Development of a laboratory light scattering photometer for applications in marine research. (In German: *Konstruktion eines laborstreulichtphotometer fur den einsatz in der meeresforschung*). *Beitrage zur Meereskunde* **43**: 109–122.

Press W. H., Flannery B. P., Teukolsky S. A., Vetterling W. H. 1989. *Numerical recipes in C.* Cambridge University Press, New York, 994 pp.

Preston-Thomas H. 1990a. The international temperature scale of 1990 (ITS-90). *Metrologia* **27**: 3–10.

Preston-Thomas H. 1990b. Erratum: The international temperature scale of 1990 (ITS-90). *Metrologia* **27**: 107–107.

Price B. J., Kollman V. H., Salzman G. C. 1978. Light-scatter analysis of microalgae: Correlation of scatter patterns from pure and mixed asynchronous cultures. *Biophys. J.* **22**: 29–36.

Priddle J., Fryxell G. 1985. Handbook of common plankton diatoms of the Southern Ocean: Centrales except the genus *Thalassiosira*. British Antarctic Survey, High Cross, Cambridge University.

Princen L. H., Kwolek W. F. 1965. Coincidence corrections for particle size determination with the Coulter counter.*Rev. Sci. Instrum.* **36**: 646–653.

Pritchard B. S., Elliot W. G. 1960. Two instruments for atmospheric optics measurements. *J. Opt. Soc. Am.* **50**: 191–202.

Privoznik K. G., Daniel K. J., Incropera F. P. 1978. Absorption, extinction and phase function measurements for algal suspensions of *Chlorella pyrenoidosa*.*J. Quant. Spectrosc. Rad. Transfer* **20**: 345–352.

Prospero J. M. 1996. The atmospheric transport of particles to the ocean. In: *Particle flux in the ocean.* Ittekkot V., Schäfer P., Honjo S., Depetris P. J. (eds.) SCOPE Report 57, John Wiley & Sons, Chichester, pp. 19–52.

Prospero J. M. 1999. Long-range transport of mineral dust in the global atmosphere: Impact of African dust on the environment of the southeastern United States. *Proc. Natl. Acad. Sci. USA* **96**: 3396–3403.

Prothero J. 1986. Methodological aspects of scaling in biology. *J. Theor. Biol.* **118**: 259–292.

Pryor P. R., Mullock B. M., Bright N. A., Gray S. R., Luzio J. P. 2000. The role of intraorganellar Ca2+ in late endosome–lysosome heterotypic fusion and in the reformation of lysosomes from hybrid organelles. *J. Cell Biol.* **149**: 1053–1062.

Pugh P. R. 1978. The application of particle counting to an understanding of the small-scale distribution of plankton. In: *Spatial patterns in plankton communities.* Steel J. H. (ed.). Plenum Press, New York, pp. 111–129.

Purcell E. M., Pennypacker C. R. 1973. Scattering and absorption of light by non-spherical dielectric grains. *Astrophys. J.* **186**: 705–714.

Q+

Querry M. R., Cary P. G., Waring R. C. 1978. Split-pulse laser method for measuring attenuation coefficient of transparent liquids: application to deionized filtered water in the visible region.*Appl. Opt.* **17**: 3587–3592.

Querry M. R., Wieliczka D. M., Segelstein D. J. 1991. [Refractive index of] Water (H2O). In: *Handbook of optical constants of solids.* Palik, E. D. (ed.) Academic Press, New York, pp. 1059–1077.

Qian H. 2001. Relative entropy: free energy associated with equilibrium fluctuations and nonequilibrium deviations. *Phys. Rev. E* **63**: 042103, 4 pp.

Quan X., Fry E. S. 1995. Empirical equation for the index of refraction of sea water. *Appl. Opt.* **34**: 3477–3480.

Quickenden T. I., Irvin J. A. 1980. The ultraviolet absorption spectrum of liquid water. *J. Chem. Phys.* **72**: 4416–4428.

Quinby-Hunt M. S., Hunt A. J., Lofftus K., Shapiro D. 1989. Polarized-light scattering studies of marine *Chlorella. Limnol. Oceanogr.* **34**: 1587–1600.

Quinones R. A., Platt T., Rodríguez J. 2003. Patterns of biomass-size spectra from oligotrophic waters of the Northwest Atlantic. *Progr. Oceanogr.* **57**: 405–427.

Quirantes A., Bernard S. 2004. Light scattering by marine algae: two-layer spherical and nonspherical models.*J. Quant. Spectrosc. Radiat. Transfer* **89**: 311–321.

Quirantes Sierra A., Delgado Mora A. V. 1995. Size-shape determination of nonspherical particles in suspension by means of full and depolarized static light scattering. *Appl. Opt.* **34**: 6256–6262.

R+

Ral J.-P., Derelle E., Ferraz C., Wattebled F., Farinas B., Corellou F., Buléon A., Slomianny M.-C., Delvalle D., d'Hulst C., Rombauts S., Moreau H., Ball S. 2004. Starch division and partitioning. A mechanism for granule propagation and maintenance in the picophytoplanktonic green alga *Ostreococcus tauri. Plant Physiol.* **136**: 3333–3340.

Raschke E. 1978. *Terminology and units of radiation quantities and measurements.* International Association of Meteorology and Atmospheric Physics (IAMAP), Boulder, Colorado, USA, 17 pp.

Ratmeyer V., Fischer G., Wefer G. 1999. Lithogenic particle fluxes and grain size distributions in the deep ocean off NW Africa: Implications for seasonal changes of aeolian dust input and downward transport. *Deep-Sea. Res. I* **46**: 1289–1337.

Ratmeyer V., Wefer G. 1996. A high resolution camera system (ParCa) for imaging particles in the ocean: System design and results from profiles and a three-month deployment. *J. Mar. Res.* **54**: 589–603.

Ravisankar M., Reghunath A. T., Sathianaden K., Nampori V. P. N. 1988. Effect of dissolved NaCl, $MgCl_2$ and $Na2SO_4$ in sea water on the optical attenuation in the region from 430 to 630 nm.*Appl. Opt.* **27**: 3887–3894.

Ray A. K., Souyri A., Davis E. J., Allen T. M. 1991. Precision of light scattering techniques for measuring optical parameters of microspheres. *Appl. Opt.* **30**: 3974–3983.

Reali G. C. 1982. Exact solution of the equations of molecular optics for refraction and reflection of an electromagnetic wave on a semi-infinite dielectric. *J. Opt. Soc. Am.* **72**: 1421–1424.

Reali G. C. 1992. Reflection from dielectric materials. *Am. J. Phys.* **60**: 532–536.

Reese J. W., Tucker S. P. 1970. Light measurements off the San Diego coast.*NUC Technical Publication* **203**: 1–37.

Reimann B. E. F., Lewin J. C., Volcani B. E. 1965. Studies on the biochemistry and fine structure of silica shell formation in diatoms: I. The structure of the cell wall of *Cylindrotheca fusiformis*. *J. Cell Biol.* **24**: 39–55.

Reuter R. 1980a. Characterization of marine particle suspension by light scattering. I. Numerical predictions from Mie theory.*Oceanol. Acta* **3**: 317–324.

Reuter R. 1980b. Characterization of marine particle suspensions by light scattering. II. Experimental results.*Oceanol. Acta* **3**: 325–332.

Reynolds C. S. 1989. Physical determinants in phytoplankton succession. In: *Plankton ecology – succession in plankton communities*. Sommer U. (ed.) Springer Verlag, Berlin, pp. 9–56.

Reynolds L. O., McCormick N. J. 1980. Approximate two-parameter phase function for light scattering.*J. Opt. Soc. Am.* **70**: 1206–1212.

Reynolds R. A., Stramski D., Kiefer D. A. 1997. The effect of nitrogen limitation on the absorption and scattering properties of the marine diatom *Thalassiosira pseudonana*. *Limnol. Oceanogr.* **42**: 881–892.

Rheinheimer G., Schmaljohann R. 1983. Investigations on the influence of coastal upwelling and polluted rivers on the microflora of the northeastern Atlantic off Portugal—I. Size and composition of the bacterial population. *Bot. Mar.* **26**: 137–152.

Richardson L. F. 1926. Atmospheric diffusion shown on distance-neighbour graph. *Proc. Roy. Soc. London A* **110**: 709–737.

Richardson M. J. 1987. Particle size, light scattering and composition of suspended particulate matter in the North Atlantic.*Deep-Sea Res.* **34**: 1301–1329.

Ridd P. V., Larcombe P. 1994. Biofouling control for optical backscatter turbidity sensors. *Mar. Geol.* **116**: 255–258.

Riebesell U. 1991. Particle aggregation during a diatom bloom. I. Physical aspects. *Mar. Ecol. Prog. Ser.* **69**: 273–280.

Riley J. P. 1975. Analytical chemistry of sea water. In: *Chemical oceanography*. Riley J. P., Skirrow G. (eds.) Academic Press, London, pp. 193–547.

Rinaldo A., Maritan A., Cavender-Bares K. K., Chisholm S. W. 2002. Cross-scale ecological dynamics and microbial size spectra in marine ecosystems.*Roy. Soc. Proc.: Biol. Sci.* **269**: 2051–2059.

Risović D. 1993. Two-component model of sea particle size distribution.*Deep-Sea Res.* **40**: 1459–1473.

Robarts R. D., Zohary T., Waiser M. J., Yacobi Y. Z. 1996. Bacterial abundance, biomass, and production in relation to phytoplankton biomass in the Levantine Basin of the southeastern Mediterranean Sea. *Mar. Ecol. Prog. Ser.* **137**: 273–281.

Robertson B. R., Button D. K., Koch A. L. 1998. Determination of the biomasses of small bacteria at low concentrations in a mixture of species with forward light scatter measurements by flow cytometry. *Appl. Environ. Microbiol.* **64**: 3900–3909.

Rodríguez J., Blanco J. M., Jimenez-Gomez F., Echevarria F., Gil, J., Rodríguez V., Ruiz J., Bautista B., Guerrero F. 1998. Patterns in the size structure of the phytoplankton community in the deep fluorescence maximum of the Alboran Sea (southwestern Mediterranean). *Deep-Sea Res. I* **45**: 1577–1593.

Rodríguez J., Jiménez-Gómez F., Blanco J. M., Figueroa F. L. 2002. Physical gradients and spatial variability of the size structure and composition of phytoplankton in the Gerlache Strait (Antarctica). *Deep-Sea Res. II* **49**: 693–706.

Roesler C. S. 1998. Theoretical and experimental approaches to improve the accuracy of particulate absorption coefficients from the quantitative filter technique. *Limnol. Oceanogr.* **43**: 1649–1660.

Rosen J. M., Pinnick R. G., Garvey D. M. 1997. Nephelometer optical response model for the interpretation of atmospheric aerosol measurements. *Appl. Opt.* **36**: 2642–2649.

Rosin P., Rammler E. 1933. The grading of the grinding stock in the light of the probability teachings (in German: *Die Kornzusammensetzung des Mahlgutes im Lichte der Wahrscheinlichkeitslehre*). *Kolloid. Ztschr.* **67**: 16–26.

Ross K. F. A. 1954. Measurement of the refractive index of cytoplasmic inclusions in living cells by the interference microscope.*Nature* **174**: 836–837.

Ross W. D. 1978. Note: Logarithmic distribution functions for particle size. *J. Colloid Interface Sci.* **67**: 181–182.

Ross K. F. A., Billing E. 1957. The water and solid content of living bacterial spores and vegetative cells as indicated by refractive index measurements.*J. Gen. Microbiol.* **16**: 418–425.

Round F. E., Crawford R. M., Mann D. G. 1992. *The diatoms*. Cambridge University Press, New York, USA, 747 pp.

Rozenberg G. V., Lyubovtseva Y. S., Kadyshevich E. A., Amnuil N. R. 1970. Measurement of light scattering matrices in the laboratory. (in Russian *Opredlenye matric rasseyaniya sveta laboratornemy sredami)Izv AN SSSR, Fizika Atmos. Okeana* **6**: 1255–1261.

Ruan J. Z., Litt M. H., Krieger I. M. 1988. Pore size distributions of foams from chord distributions of random lines: mathematical inversion and computer simulation. *J. Colloid Interface Sci.* **126**: 93–100.

Ruf A., Worlitschek J., Mazzotti M. 2000. Modeling and experimental analysis of PSD measurements through FBRM. *Part. Part. Syst. Charact.* **27**: 167–179.

Ruppert E. E., Fox R. S., Barnes R. D. 2004. Invertebrate zoology: A functional evolutionary approach (7th ed.). Thomson/Brooks/Cole, Florence, KY, USA, 928 pp., ISBN: 0-03-025982-7.

S+

Sabetta L., Fiocca A., Margheriti L., Vignes F., Basset A., Mangoni O., Carrada G. C., Ruggieri N., Ianni C. 2005. Body size–abundance distributions of nano- and micro-phytoplankton guilds in coastal marine ecosystems. *Estuar. Coast. Shelf Sci.* **63**: 645–663.

Salzman G. C., Crowell J. M., Goad C. A., Hansen K. M., Hiebert R. D., LaBauve P. M., Martin J. C., Ingram M., Mullaney P. F. 1975. A flow-system multiangle light-scattering instrument for cell characterization. *Clin. Chem.* **21**: 1297-1304.

Sandler B. M., Selivanovsky D. A., Stunzhas P. A., Krupatkina D. K. 1992. Gas vacuoles in marine phytoplankton: Ultrasonic resonance measurements. *Oceanology* **32**: 60–65.

Santos N. C., Castanho M. A. R. B. 1996. Teaching light scattering spectroscopy: the dimension and shape of tobacco mosaic virus.*Biophys. J.* **71**: 1641–1646.

Sasaki T., Okami N., Matsumura S. 1968. Scattering functions for deep sea water of the Kuroshio. *La Mer* **6**: 165–175.

Sasaki T., Okami N., Oshiba G., Watanabe S. 1960. Angular distribution of scattered light in deep sea water. *Rec. Oceanogr. Works Jpn.* **5**: 1–10.

Saunders P. M. 1981. Practical conversion of pressure to depth. *J. Phys. Oceanogr.* **11**: 573–574.

Schätzel K., Neumann W.-G., Müller J., Materzok B. 1992. Optical tracking of single Brownian particles. *Appl. Opt.* **31**: 770–778.

Scherbaum O., Rasch G. 1957. Cell size distribution and single cell growth in *Tetrahymena pyriformis GL. Acta Pathol. Microbiol. Scand.* **41**(3): 161–182.

Schiebener P., Straub J., Levelt-Sengers J. M. H., Gallagher J. S. 1990. Refractive index of water and steam as function of wavelength, temperature and density. *J. Phys. Chem. Ref. Data* **19**: 677–717.

Schlager K. L., Schneider J. B. 1995. A selective survey of the finite-difference-time-domain literature. *IEEE Ant. Propag. Mag.* **37**: 39–57.

Schmid P. E., Tokeshi M., Schmid-Araya J. M. 2000. Relation between population density and body size in stream communities. *Science* **289**: 1557–1560.

Schmidt R., Bottlinger M. 1996. The use of a confocal-laser-scanning-microscope to determine the 3-dimensional shape of fibre aggregates. *J. Aerosol Sci.* **27**: S331–S332.

Schmidt R., Housen K. 1995. Problem solving with dimensional analysis. *Ind. Physicist* **1**: 21–24.

Schnars U., Jüptner W. P. O. 2002. Digital recording and numerical reconstruction of holograms.*Meas. Sci. Technol.* **13**: R85–R101.

Schoonmaker J. S., Hammond R. R., Heath A. L., Martinez-Bazan C., Lasheras J. C., Rohr J. 1998. An investigation of the effect of bubble size distributions on optical backscattering. In: *Ocean Optics XIV*, Kailua-Kona, Hawaii, November 10–13, 1998.

Schreurs R. 1996. *Light scattering by algae: Fitting experimental data using Lorenz-Mie theory*. Vrije Universiteit, Amsterdam, 80 pp.

Schultz H. N., Brinkhoff T., Ferdelman T. G., Hernández-Mariné M., Teske A., Jørgensen B. B. 1999. Dense populations of giant sulfur bacterium in Namibian Shelf sediment. *Science* **284**: 493–495.

Schulz F. M., Stamnes K., Stamnes J. J. 1998. Scattering of electromagnetic waves by spheroidal particles: a novel approach exploiting the T- matrix computed in spheroidal coordinates. *Appl. Opt.* **37**: 7875–7896.

Schulz F. M., Stamnes K., Stamnes J. J. 1999. Shape dependence of the optical properties in size-shape distributions of randomly oriented prolate spheroids, including highly elongated shapes. *J. Geophys. Res. D* **108**: 9413–9421.

Sciandra A., Lazzara L., Claustre H., Babin M. 2000. Responses of growth rate, pigment composition and optical properties of *Cryptomonas* sp. to light and nitrogen stresses. *Mar. Ecol. Prog. Ser.* **201**: 107–120.

Sciare J., Cachier H., Oikonomou K., Ausset P., Sarda-Estève R., Mihalopoulos N. 2003. Characterization of carbonaceous aerosols during the MINOS campaign in Crete, July–August 2001: a multi-analytical approach. *Atmos. Chem. Phys.* **3**: 1743–1757.

Seaver G. 1987. The optical determination of temperature, pressure, salinity and density in physical oceanography. *MTS J.* **21**: 69–79.

Sengers J. W., Watson J. T. R. 1986. Improved international formulations for the viscosity and thermal conductivity of the water substance. *J. Phys. Chem. Ref. Data* **15**: 1291–1314.

Shannon C. E. 1949. Communication in the presence of noise. *Proc. IRE* **37**: 10–21.
Shannon C. E. 1948. A mathematical theory of communication. *Bell Syst. Tech. J.* **27**: 379–423, 623–656.
Shapiro H. M. 2003. *Practical flow-cytometry*. Wiley, New York, 736 pp.
Shapiro D. B., Hunt A. J., Quinby-Hunt M. S., Hull P. G. 1991. Circular polarization effects in the light scattering from single and suspensions of dinoflagellates. In: *Underwater imaging, photography, and visibility. Proc. SPIE* **1537**: 30–41.
Shapiro D. B., Quinby-Hunt M. S., Hunt A. J. 1990. Origin of the induced circular polarization in the light scattering from a dinofiagellate. In: *Ocean Optics X. Proc. SPIE* **1302**: 281–289.
Sharma S. K. 1992. On the validity of the anomalous diffraction approximation. *J. Mod. Opt.* **39**: 2355–2361.
Sheldon R. W. 1970. Personal communication.
Sheldon R. W. 1972. Size separation of marine seston by membrane and glass-fiber filters. *Limnol. Oceanogr.* **17**: 494–498.
Sheldon R. W. 1975. Personal communication.
Sheldon R. W., Parsons T. R. 1967a. *Practical manual on the use of the Coulter counter in marine sciences*. Coulter Electronics, Toronto, Canada, 66 pp.
Sheldon R. W., Parsons T. R. 1967b. A continuous size spectrum of particulate matter in the sea. *J. Fish. Res. Bd. Can.* **24**: 909–915.
Sheldon R. W., Prakash A., Sutcliffe W. H., Jr. 1972. The size distribution of particles in the ocean. *Limnol. Oceanogr.* **17**: 327–340.
Sheldon R. W., Sutcliffe W. H., Jr. 1969. Retention of marine particles by screens and filters. *Limnol. Oceanogr.* **14**: 441–444.
Sheldon R. W., Sutcliffe W. H., Paranjpe M. A. 1977. Structure of pelagic food chains and relationship between plankton and fish production. *J. Fish. Res. Bd. Can.* **34**: 2344–2353.
Shen J., Riebel U. 2003. The fundamentals of particle size analysis by transmission fluctuation spectrometry. Part 3: A theory on transmission fluctuations in a Gaussian beam and with signal filtering. *Part. Part. Syst. Charact.* **20**: 94–103.
Sherman G. C., Harris F. S., Jr., Morse F. L. 1968. Scattering of coherent and incoherent light by latex hydrosols.*Appl. Opt.* **7**: 421–423.
Shettle E. P., Weinman J. A. 1970. The transfer of solar irradiance through inhomogeneous turbid atmospheres evaluated by Edditngton's approximation. *J. Atmos. Sci.* **27**: 1048–1055.
Shibata K. 1958. Spectrophotometry of intact biological materials: absolute and relative measurements of their transmission, reflection and absorption spectra. *J. Biochim.* **45**: 599–623.
Shibata K., Benson A. A., Calvin M. 1954. The absorption spectra of suspensions of living micro-organisms.*Biochem. Biophys. Acta* **15**: 461–470.
Shifrin K. S., Kopelevich O. V., Burenkov V. I., Mashtakov Y. L. 1974. The light scattering functions and structure of the sea hydrosol (in Russian). *Izv. Akad. Nauk SSSR, Fizika Atmosfery i Okeana* **10**: 25–35.
Shimizu K., Crow E. L. 1988. History, genesis, and properties [of lognormal distribution]. In: *Lognormal distributions: theory and applications*. Crow, E. L. and Shimizu, K. (eds.). Marcell Dekker, New York, pp. 1–25.

Shvalov A. N., Surovtsev I. V., Chernyshev A. V., Soini J. T., Maltsev V. P. 1999. Particle classification from light scattering with the scanning flow cytometer. *Cytometry* **37**: 215–220.

Sieburth J. M. S. 1979. *Sea microbes*. Oxford University Press, New York, USA, 491 pp.

Sieracki C. K., Sieracki M. E. 1997. High-throughput volume particle in-flow imaging system. In: *Ocean Optics XIII*. Ackleson S. G. (ed.) *Proc. SPIE* **2963**: 886–891.

Sieracki C. K., Sieracki M. E., Yentsch C. S. 1998. An imaging-in-flow system for automated analysis of marine microplankton.*Mar. Ecol. Prog. Ser.* **168**: 285–296.

Sieracki M. E., Johnson P. W., Sieburth J. McN. 1985. Detection, enumeration, and sizing of planktonic bacteria by image-analyzed epifluorescence microscopy.*Appl. Environ. Microbiol.* **49**: 799–810.

Sieracki M. E., Viles C. 1992. Distribution and fluorochrome-staining properties of submicrometer particles and bacteria in the North Atlantic. *Deep-Sea Res.* **39**: 1919–1929.

Sieracki M. L., Reichenbach S. E., Webb K. L. 1989. Evaluation of automated threshold selection methods for accurately sizing microscopic fluorescent cells by image analysis.*Appl. Environ. Microbiol.* **55**: 2762–2772.

Simon M., Azam F. 1989. Protein content and protein synthesis rates of planktonic marine bacteria.*Mar. Ecol. Prog. Ser.* **51**: 201–213.

Simon M., Grossart H.-P., Schweitzer B., Ploug H. 2002. Microbial ecology of organic aggregates in aquatic ecosystems.*Aquat. Microb. Ecol.* **28**: 175–211.

Simpson W. R. 1982. Particulate matter in the oceans – sampling methods, concentration, size distribution and particle dynamics. *Oceanogr. Mar. Biol. Ann. Rev.* **20**: 119–172.

Singer J. K. 1986. *Results of an intercalibration experiment: An evaluation of the reproducibility of data generated from instruments used in textural analyses.* Rice University Sedimentology Report, Rice University.

Singer J. K., Anderson J. B., Ledbetter M. T., McCave I. N., Jones K. P., Wright R. 1988. An assessment of analytical techniques for the size analysis of fine grained sediments. *J. Sediment. Petrol.* **58**: 534–543.

Singer J. M., Veekman F. C. A., Lichtenbelt J. W. T., Hesselink F. T., Wiersma P. H. 1973. Kinetics of flocculation of latex particles by human ganna globulin. *J. Colloid Interface Sci.* **45**: 608–613.

Sloot P. M. A., Hoekstra A. G., van der Liet H., Figdor C. G. 1990. Arbitrarily-shaped particles measured in flow-through systems. In: *The 2nd Internaltional Congress on particle sizing*. Hirleman E. D. (ed.) Arizona State University, Tempe, Arizona, March 1990, pp. 605-611.

Schmaljohann R. 1984. Morphological investigations on bacterioplankton of the Baltic Sea, Kattegat and Skagerrak. *Bot. Mar.* **27**: 425–436.

Smart M. M., Rada R. G., Nielsen D. N., Claflin T. O. 1985. The effect of commercial and recreational traffic on the resuspension of sediment in Navigation Pool 9 of the Upper Mississippi River. *Hydrobiologia* **126**: 263–274.

Smayda T. J. 1970. The suspension and sinking of phytoplankton in the sea. *Oceanogr. Mar. Biol. Ann. Rev.* **8**: 353–414.

Smith R. C., Baker K. S. 1981. Optical properties of the clearest natural waters (200–800 nm). *Appl. Opt.* **20**: 177–184.

Smoluchowski M. 1908. Molecular kinectic theory of opalescence in gases at critical point as well as some related features. (In German: *Molekular-kinetische Theorie der Opaleszenz von Gasen im kritischen Zustande, sowie einiger verwandter Erscheinungen) Ann. Phys.* **25**: 205–226.

Smythe W. R. 1964. Flow around a spheroid in a circular tube. *Phys. Fluids* **7**: 633–638.

Smythe W. R. 1972. Off-axis particles in Coulter-type counters. *Rev. Sci. Instrum.* **43**: 817–818.

Sogandares F. M., Fry E. S. 1997. Absorption spectrum (340–640 nm) of pure water. I. Photothermal measurements. *Appl. Opt.* **36**: 8699–8709.

Sokolov R. N., Kudryavltskiy F. A., Petrov G. D. 1971. Submarine laser instrument for measuring the size spectra of particles suspended in sea water. *Izv. AN USSR Atmos. Ocean. Phys.* **7**: 1015–1018 (in Russian).

Sommer U. 1998. Silicate and the functional geometry of marine phytoplankton.*J. Plankton Res.* **20**: 1853–1859.

Sonek G. J., Liu Y., Iturriaga R. H. 1995. In situ microparticle analysis of marine phytoplankton cells with infrared laser-based optical tweezers. *Appl. Opt.* **34**: 7731–7741.

Sonntag R. C., Russel W. B. 1986. Structure and breakup of flocs subjected to fluid stresses. I. Shear experiments. *J. Colloid Interface. Sci.* **113**: 399–413.

Sonntag R. C., Russel W. B. 1987. Structure and breakup of flocs subjected to fluid stresses. III. Converging flow. *J. Colloid Interface Sci.***115**: 390–395.

Sournia A., Chretiennot-Dinet G., Ricard M. 1991. Marine phytoplankton: How many species in the world ocean? *J. Plankton. Res.* **13**: 1093–1099.

Sparks R. G., Dobbs C. L. 1993. The use of laser backscatter instrumentation for the on-line measurement of the particle size distribution of emulsions. *Part. Part. Syst. Charact.* **10**: 279–289.

Spicer P. T., Pratsinis S. E., Wilemse A. W., Merkus H. G., Scarlett B. 1999. Monitoring the dynamics of concentrated suspensions by enhanced backward light scattering. *Part. Part. Syst. Charact.* **16**: 201–206.

Spielman L., Goren S. L. 1968. Improving resolution in Coulter counting by hydrodynamic focusing. *J. Colloid Interface Sci.* **26**: 175–182.

Spilhaus A. F., Jr. 1968. Observations of light scattering in sea water.*Limnol. Oceanogr.* **13**: 418–422.

Spinrad R. W., Bartz R., Kitchen J. C. 1989a. In situ measurements of marine particle settling velocity and size distributions using the remote optical settling tube.*J. Geophys. Res. C* **95**: 931–938.

Spinrad R. W., Brown J. F. 1986. Relative real refractive index of marine microorganisms: a technique for flow cytometric estimation.*Appl. Opt.* **25**: 1930–1934.

Spinrad R. W., Glover H., Ward B. W., Codispoti L. A., Kullenberg G. 1989b. Suspended particle and bacterial maxima in Peruvian coastal waters during a cold water anomaly. *Deep-Sea Res.* **36**: 715–733.

Spinrad R. W., Zaneveld J. R. V., Pak H. 1978. Volume scattering function of suspended particulate matter at near-forward angles: a comparison of experimental and theoretical values.*Appl. Opt.* **17**: 1125–1130.

Squires K., Eaton J. 1991. Preferential concentration of particles by turbulence. *Phys. Fluids A* **3**: 1169–1178.

Stanley E. M. 1970. *The refractive index of pure water and seawater as a function of high pressure and moderate temperature. Part II. Pure water for a wavelength 6328 Å.* Naval Ship. Res. Develop. Center Report 3066.

Stavn R. H. 1993. Effects of Raman scattering across the visible spectrum in clear ocean water: a Monte Carlo study. *Appl. Opt.* **32**: 6853–6863.

Stavn R. H., Keen T. R. 2004. Suspended minerogenic particle distributions in high-energy coastal environments: Optical implications. *J. Geophys. Res.* **109**: C05005.

Stavn R. H., Weidemann A. D. 1992. Raman scattering in ocean optics: quantitative assessment of internal radiant emission. *Appl. Opt.* **31**: 1294–1303.

Steen H. B. 1986. Simultaneous separate detection of low angle and large angle light scattering in an arc lamp-based flow cytometer. *Cytometry* **7**: 445–449.

Steen H. B. 1990. Characteristics of flow cytometers. In: *Flow cytometry and sorting.* Melamed M. R., Lindmo T., Mendelsohn M. L. (eds.) Wiley-Liss, New York, pp. 11–26.

Steen H. B. 2000. Flow cytometry of bacteria: Glimpses from the past with a view to the future. *J. Microbiol. Methods* **42**: 65–74.

Steen H. B., Lindmo T. 1979. Flow-cytometry: A high-resolution instrument for everyone.*Science* **204**: 403–404.

Steiner B., Berge B., Gausmann R., Rohmann J., Rühl E. 1999. Fast *in situ* sizing technique for single levitated liquid aerosols. *Appl. Opt.* **38**: 1523–1529.

Steinkamp J. A. 1984. Flow-cytometry.*Rev. Sci. Instrum.* **55**: 1375–1400.

Steinkamp J. A., Fulwyler M. J., Coulter J. R., Hiebert R. D., Horney J. L., Mullaney P. F. 1973. A new multiparameter separator for microscopic particles and biological cells. *Rev. Sci. Instrum.* **44**: 1301–1310.

Stelzer E. H. K. 2002. Beyond the diffraction limit? *Nature* **417**: 806–807.

Stemmann L., Jackson G. A., Gorsky G. 2004b. A vertical model of particle size distributions and fluxes in the midwater column that includes biological and physical processes Part II: application to a three year survey in the NW Mediterranean Sea. *Deep-Sea Res. I* **51**: 885–908.

Stemmann L., Jackson G. A., Ianson D. 2004a. A vertical model of particle size distributions and fluxes in the midwater column that includes biological and physical processes—Part I: model formulation. *Deep-Sea Res. I* **51**: 865–884.

Sternberg R. W., Baker E. T., McManus D. A., Smith S., Morrison D. R. 1974. An integrating nephelometer for measuring particle concentrations in the deep sea. *Deep-Sea Res.* **21**: 887–892.

Stewart G. L., Beers J. R., Knox C. 1973. Application of holographic techniques to the study of marine plankton in the field and laboratory. In: *Developments in laser technology – II. Proc. SPIE* **41**: 183–194.

Stewart C. C., Stewart S. J., Habberset R. C. 1989. Resolving leukocytes using axial light loss. *Cytometry* **10**: 426–432.

Stigliani D. J., Jr., Mittra R., Semonin R. G. 1970. Particle-size measurement using forward scatter holography. *J. Opt. Soc. Am.* **60**: 1059–1067.

Stockham J. D., Fochtman E. G. (eds) 1979. *Particle size analysis.* Ann Arbor Science Publishers, Inc., Ann Arbor, Michigan, 140 pp.

Stoderegger K. E., Herndl G. J. 1999. Production of exopolymer particles by marine bacterioplankton under contrasting turbulence conditions. *Mar. Ecol. Prog. Ser.* **189**: 9–16.

Stone J., Pochapsky T. E. 1969. Brillouin scattering by seawater. *Limnol. Oceanogr.* **14**: 783–787.

Stramski D. 1994. Gas microbubbles: An assessment of their significance to light scattering in quiescent seas. In: *Ocean Optics XII. Proc. SPIE* **2258**: 704–710.

Stramski D. 1999. Refractive index of planktonic cells as a measure of cellular carbon and chlorophyll a content. *Deep Sea Res. I* **46**: 335–351.

Stramski D., Boss E., Bogucki D., Voss K. J. 2004a. The role of seawater constituents in light backscattering in the ocean. *Prog. Oceanogr.* **61**: 27–56.

Stramski D., Bricaud A., Morel A. 2001. Modeling the inherent optical properties of the ocean based on the detailed composition of the planktonic community. *Appl. Opt.* **40**: 2929–2945.

Stramski D., Kiefer D. A. 1990. Optical properties of marine bacteria. In:*Ocean Optics X, Proc. SPIE* **1302**: 250–268.

Stramski D., Kiefer D. A. 1991. Light scattering by microorganisms in the open ocean.*Prog. Oceanogr.* **28**: 343–383.

Stramski D., Mobley C. D. 1997. Effects of microbial particles on oceanic optics: A database of single-particle optical properties. *Limnol. Oceanogr.* **42**: 538–549.

Stramski D., Morel A. 1990. Optical properties of photosynthetic picoplankton in different physiological states as affected by growth irradiance.*Deep-Sea Res.* **37**: 245–266.

Stramski D., Morel A., Bricaud A. 1988. Modelling the light attenuation by spherical phytoplanktonic cells: a retrieval of the bulk refractive index.*Appl. Opt.* **27**: 3954–3956.

Stramski D., Piskozub J. 2003. Estimation of scattering error in spectrophotometric measurements of light absorption by aquatic particles from three-dimensional radiative transfer simulations. *Appl. Opt.* **42**: 3634–3646.

Stramski D., Rassoulzadegan F., Kiefer D. 1992a. Changes in the optical properties of a particle suspension caused by protist grazing.*J. Plankton Res.* **14**: 961–977.

Stramski D., Reynolds R. A. 1993. Diel variations in the optical properties of a marine diatom. *Limnol. Oceanogr.* **38**: 1347–1364.

Stramski D., Reynolds R. A., Kahru M., Mitchell B. G. 1999. Estimation of particulate organic carbon in the ocean from satellite remote sensing. *Science* **285**: 239–242.

Stramski D., Rosenberg G., Legendre L. 1993. Photosynthetic and optical properties of the marine chlorophyte *Dunaliella tertiolecta* grown under fluctuating light caused by surface-wave focusing. *Mar. Biol.* **115**: 363–372.

Stramski D., Sciandra A., Claustre H. 2002. Effects of temperature, nitrogen, and light limitation on the optical properties of the marine diatom *Thalassiosira pseudonana. Limnol. Oceanogr.* **47**: 392–403.

Stramski D., Sedlák M. 1994. Application of dynamic light scattering to the study of small marine particles. *Appl. Opt.* **33**: 4825–4834.

Stramski D., Sedlák M. 1995. Application of dynamic light scattering to the study of small marine particles: errata. *Appl. Opt.* **34**: 571–571.

Stramski D., Sedlák M., Tsai D., Amis E., Kiefer D. 1992c. Dynamic light scattering by cultures of heterotrophic marine bacteria. In: *Ocean Optics XI. Proc. SPIE* **1750:** 73–85.

Stramski D., Shalapyonok A., Reynolds R. 1995. Optical characterization of the oceanic unicellular cyanobacterium *Synechococcus* grown under a day-night cycle in natural irradiance. *J. Geophys. Res. C* **100**: 13,295–13,307.

Stramski D., Tęgowski J. 2001. Effects of intermittent entrainment of air bubbles by breaking wind waves on ocean reflectance and underwater light field.*J. Geophys. Res. C* **106**: 31,345–31,360.

Stramski D., Woźniak S. B., Flatau P. J. 2004b. Optical properties of Asian mineral dust suspended in seawater.*Limnol. Oceanogr.* **49**: 749–755.

Stratton J. 1941. *Electromagnetic theory*. McGraw-Hill, New York, 615 pp.

Subba Rao D. V., Partensky F., Wohlgeschaffen G., Li W. K. 1991. Flow cytometry and microscopy of gametogenesins in Nitzschia pungens, a toxic, bloom-forming marine diatom. *J. Phycol.* **27**: 21–26.

Sugihara S., Kishino M., Okami N. 1982a. Back-scattering of light by particles suspended in water.*Sci. Pap. Inst. Phys. Chem. Res.* **76**: 1–8.

Sugihara S., Kishino M., Okami N. 1982b. Experimental study of volume scattering function of particles suspended in water.*Sci. Pap. Inst. Phys. Chem. Res.* **76**: 96–99.

Sugihara S., Tsuda R. 1979. Size distribution of suspended particles in the surface water of the central North Pacific Ocean. *Sci. Pap. Inst. Phys. Chem. Res.* **73**: 1–8.

Sun J., Liu D. 2003. Geometric models for calculating cell biovolume and surface area for phytoplankton. *J. Plankton Res.* **25**: 1331–1346.

Sydor M., Arnone R. A. 1997 Effect of suspended particulate and dissolved organic matter on remote sensing of coastal and riverine waters. *Appl. Opt.* **36**: 6905–6912.

Syvitski J. P. M., Asprey K., Clattenburg D. A 1991a. Principles, design, and calibration of settling tubes. In: *Principles, methods, and application of particle size analysis.* Syvitski, J. P. M. (ed.) Cambridge University Press, New York, pp. 45–63.

Syvitski J. P. M., Asprey K. W., LeBlanc K. W. G. 1995. In-situ characteristics of particles settling within a deep-water estuary. *Deep-Sea Res.* **42**: 223–256.

Syvitski J. P. M., Hutton E. W. H. 1996. In situ characteristics of suspended particles as determined by the floc camera assembly FCA. *J. Sea Res.* **36**: 131–142.

Syvitski J. P. M., LeBlanc K. W., Asprey K. 1991b. Interlaboratory, interinstrument calibration experiment. In: *Principles, methods, and application of particle size analysis.* Syvitski, J. P. M. (ed.) Cambridge University Press, New York, pp. 174–193.

Syvitski J. P. M., Silverberg N., Ouellet G., Asprey K. W. 1983. First observations of benthos and seston from a submersible in the lower St. Lawrence Estuary. *Geogr. Phys. Quat.* **37**: 227–240.

Szudy J., Bayliss W. E. 1975. Uniform Franck-Condon treatment of pressure broadening of spectral lines. *J. Quant. Spectrosc. Radiat. Transfer* **15**: 641–668.

T+

Tada K., Monaka K., Morishita M., Hashimoto T. 1998. Standing stocks and production rates of phytoplankton and abundance of bacteria in the Seto Inland Sea, Japan. *J. Oceanogr.* **54**: 285–295.

Taflin D. C., Zhang S. H., Allen T., Davis E. J. 1988. Measurement of droplet interfacial phenomena by light-scattering techniques. *Am. Inst. Chem. Eng. J.* **34**: 1310–1320.

Taflove A., Umashankar K. R. 1989. Review of FD-TD numerical modeling of electromagnetic wave scattering and radar cross section. *Proc. IEEE* **77**: 682–699.

Taguas F. J., Martin M. A., Perfect E. 1999. Simulation and testing of self-similar structures for soil particle-size distributions using iterated function systems.*Geoderma* **88**: 191–203.

Taguchi S. 1976. Relationship between photosynthesis and cell size of marine diatoms. *J. Physiol.* **12**: 185–189.

Takahashi M., Bienfang P. K. 1983. Size structure of phytoplankton biomass and photosynthesis in subtropical Hawaiian waters.*Mar. Biol.* **76**: 203–211.

Tanis F. J. 1992. Holographic instrument to measure small angle scattering. In: *Ocean Optics XI. Proc. SPIE* **1750**: 2–14.

Tanner W. F. 1969. The particle size scale. *J. Sediment. Petrol.* **39**: 809–812.

Tassan S., Ferrari G. M. 1995. An alternative approach to absorption measurements of aquatic particles retained on filters. *Limnol. Oceanogr.* **40**: 1358–1368.

Tatarski V. I. 1961. *Wave propagation in turbulent media.* McGraw-Hill, New York, 520 pp.

Taubenblatt M. A., Batchelder J. S. 1990. Phase shift due to a particle in a gaussian beam: Calculation and measurement. In: *Proceedings of the 2nd International Congress on Optical Particle Sizing*. Hirleman E. D. (ed.) Arizona State University, Tempe, AZ, USA, 1990, pp. 101–107.

Taylor M. A. 2002. Quantitative measures for shape and size of particles. *Powder Technol.* **124**: 94–100.

Ten Brinke W. B. M. 1994. *In situ* aggregate size and settling velocity in the Oosterschelde tidal basin (The Netherlands). *Neth. J. Sea Res.* **32**: 23–35.

Tenchov B. G., Yanev T. K. 1986. Weibull distribution of particle sizes obtained by uniform random fragmentation. *J. Colloid Interface Sci.* **111**: 1–7.

Terrill E. J., Melville W. K. 2000. A broadband acoustic technique for measuring bubble size distributions: Laboratory and shallow water measurements. *J. Atmos. Ocean. Technol.* **17**: 220–239.

Terrill E. J., Melville W. K., Stramski D. 1998. Bubble entrainment by breaking waves and their effects on the inherent optical properties of the upper ocean. In: *Ocean Optics XIV*, Kailua-Kona, Hawaii, November 10–13, 1998.

Terrill E. J., Melville W. K., Stramski D. 2001. Bubble entrainment by breaking waves and their influence on optical properties in the upper ocean. *J. Geophys. Res. C* **106**: 16815–16823.

Thiebaux M. L., Dickie L. M. 1993. Structure of the body-size spectrum of the biomass in aquatic ecosystems: a consequence of allometry in predatory–prey interactions. *Can. J. Fish. Aquat. Sci.* **50**: 1308–1317.

Thill A., Veerapaneni S., Simon B., Wiesner M., Bottero J. Y., Snidaro D. 1998. Determination of structure of aggregates by confocal scanning laser microscopy. *J. Colloid Interface Sci.* **204**: 357–362.

Thomas J. C. 1987. The determination of log-normal particle size distributions by dynamic light scattering. *J. Colloid Interface Sci.* **117**: 187–192.

Thomas R. A., Cameron B. F., Leif R. C. 1974. Computer based electronic cell volume analysis with the AMAC II transducer. *J. Histochem. Cytochem.* **22**: 626–641.

Thomas M. C., Wiltshire R. J., Williams A. T. 1995. The use of Fourier descriptors in the classification of particle shape. *Sedimentology* **42**: 635–645.

Thompson B. J., Ward J. H., Zinky W. R. 1967. Application of hologram techniques for particle size analysis.*Appl. Opt.* **6**: 519–526.

Thompson R. C., Bottiger J. R., Fry E. S. 1980. Measurements of polarized light interactions via the Mueller matrix. *Appl. Opt.* **19**: 1323–1332.

Thormählen I., Straub J., Grigull U. 1985. Refractive index of water and its dependence on wavelength, temperature, and density. *J. Phys. Chem. Ref. Data* **14**: 933–945.

Tilton L. W., Taylor J. K. 1938. Refractive index and dispersion of distilled water for visible radiation at temperatures 0 to 60°C. *J. Res. Natl. Bur. Std.* **20**: 419–477.

Tittel J., Zippel B., Geller W., Seeger J. 1998. Relationships between plankton community structure and plankton size distribution in lakes of northern Germany. *Limnol. Oceanogr.* **43**: 1119–1132.

Toon O. B., Ackerman T. P. 1981. Algorithms for the calculation of scattering by stratified spheres.*Appl. Opt.* **20**: 3657–3660.

Tontrup C., Gruy F. 2000. Light backscattering by fine non-absorbing suspended particles. *Powder Technol.* **107**: 1–12.

Trabjerg I., Højerslev N. K. 1996. Temperature influence on light absorption by fresh water and seawater in the visible and near-infrared spectrum. *Appl. Opt.* **35**: 2653–2658.

Trainer M. N. 2001. *The effects of particle shape on particle size resolution using angular scattering measurements.* Pittcon 2001, March 4, 2001.

Trevors J. T., Psenner R. 2001. From self-assembly of life to present-day bacteria: a possible role for nanocells. *FEMS Microbiol. Rev.* **25**: 573–582.

Trueba F. J., Woldringh C. 1980. Changes in cell diameter during the division cycle of *Escherichia coli. J. Bacteriol.* **142**: 869–878.

Tsai C.-H. 1996. An assessment of a time-of-transition laser sizer in measuring suspended particles in the ocean. *Mar. Geol.* **134**: 95–112.

Tsai C.-H., Rau S.-R. 1992. Evaluation of Galai CIS-1 for measuring size distribution of suspended primary particles in the ocean. *TAO* **3**: 147–163.

Tschudi T., Herziger G., Engel A. 1974. Particle size analysis using computer synthetized holograms.*Appl. Opt.* **13**: 245–248.

Tucker S. P. 1973. Measurements of the absolute volume scattering function for green light in southern California coastal waters. Ph.D. Thesis, Oregon State University, Oregon State University, Corvallis, Oregon, USA, 211 pp.

Tuminello P. S., Arakawa E. T., Khare B. N., Wrobel J. M., M. R. Querry, Milham M. E. 1997. Optical properties of *Bacillus subtilis* spores from 0.2 to 2.5 μm. *Appl. Opt.* **36**: 2818–2824.

Twardowski M. S., Boss E., MacDonald J. B., Pegau W. S., Barnard A. H., Zaneveld J. R. V. 2001. A model for estimating bulk refractive index from the optical backscattering ratio and the implications for understanding particle compositions in Case I and Case II waters. *J. Geophys. Res. C* **106**: 14,129–14,142.

Tycko D. H., Metz M. H., Epstein E. A., Grinbaum A. 1985. Flow-cytometric light scattering measurements of red blood cell volume and hemoglobin concentration. *Appl. Opt.* **24**: 1355–1365.

Tyler J. E. 1961. Scattering properties of distilled and natural waters.*Limnol. Oceanogr.* **6**: 451–456.

Tyler J. E. 1963. Design theory for a submersible scattering meter.*Appl. Opt.* **2**: 245–248.

Tyler J. E., Austin R. W. 1964. A scattering meter for deep water.*Appl. Opt.* **3**: 613–620.

Tyler J. E., Richardson W. H. 1958. Nephelometer for the measurement of volume scattering function *in situ*.*J. Opt. Soc. Am.* **48**: 354–357.

Tyrell T., Holligan P. M., Mobley C. D. 1999. Optical impacts of oceanic coccolithophore blooms. *J. Geophys. Res. C* **104**: 3223–3241.

Tyson J. J., Hannsgen K. B. 1985. The distribution of cell size and generation time in a model of the cell cycle incorporating size control and random transitions. *J. Theor. Biol.* **113**: 29–62.

U+

Ulanowski Z., Greenaway R. S., Kaye P. H., Ludlow I. K. 2002. Laser diffractometer for single-particle scattering measurements. *Meas. Sci. Technol.* **13**: 292–296.

Ulloa O., Sathyendranath S., Platt T., Quinones R. A. 1992. Light scattering by marine heterotrophic bacteria.*J. Geophys. Res. C* **97**: 9619–9629.

Umhauer H., Bottlinger M. 1990. The effect of particle shape and structure on the results of single particle light scattering size analysis. *Proceedings of the 2nd International Congress on Optical Particle Sizing, March 5-8, 1990, Tempe, Arizona*, Arizona State University, Tempe, Arizona, pp. 425–434.

Umhauer H., Bottlinger M. 1991. The effect of particle shape and structure on the results of single particle light scattering analysis. *Appl. Opt.* **30**: 4980–4986.

Umhauer H., Gutsch A. 1997. Particle characterization by projected area determination. *Part. Part. Syst. Charact.* **14**: 105–115.

Utterback C. L., Thompson T. G., Thomas B. A. 1934. Refractivity-chlorinity-temperature relationship of ocean waters. *J. Cons. Perm. Int. Explor. Mer* **9**: 35–38.

V+

Vagle S., Farmer D. M., 1992. The measurement of bubble-size distributions by acoustic backscatter. *J. Atmos. Ocean. Technol.* **9**: 630–644.

Vaillancourt R. D., Balch W. M. 2000. Size distribution of marine submicron particles determined by flow field-flow fractionation. *Limnol. Oceanogr.* **45**: 485–492.

Vaillancourt R. D., Brown C. W., Guillard R. R. L., Balch W. M. 2004. Light backscattering properties of marine phytoplankton: relationships to cell size, chemical composition and taxonomy. *J. Plankton Res.* **26**: 191–212.

van Andel T. H. 1973. Texture and dispersal of sediments in the Panama Basin. *J. Geol.* **81**: 434–457.

van de Hulst H. C. 1957. *Light scattering by small particles.* Dover, New York, 470 pp.

Van De Merwe W. P., Huffman D. R., Bronk B. V. 1989. Reproducibility and sensitivity of polarized light scattering for identifying bacterial suspensions. *Appl. Opt.* **28**: 5052–5057.

van der Plaats G., Herps H. 1983. A study of the sizing process of an instrument based on the electrical sensing zone principle. Part 1: The influence of particle material. *Powder Technol.* **36**: 131-136.

van der Plaats G., Herps H. 1984. A study of the sizing process of an instrument based on the electrical sensing zone principle. Part 2: The influence of particle porosity. *Powder Technol.* **38**: 73–76.

van Hout R., Katz J. 2004. A method for measuring the density of irregularly shaped biological aerosols such as pollen. *J. Aerosol Sci.* **35**: 1369–1384.

van Leussen W. 1999. The variability of settling velocities of suspended fine-grained sediment in the Ems estuary. *J. Sea Res.* **41**: 109–118.

Venkataraman G. S., Goyal S. K., Kaushik B. D., Roychoudry P. 1974. *Algae: form and function.* Today & Tomorrow's Printer's and Publishers, New Delhi, India.

Verity P. G., Robertson C. K., Tronzo C. R., Andrews M. G., Nelson J. R., Sieracki M. E. 1992. Relationships between cell volume and the carbon and nitrogen content of marine photosynthetic nanoplankton. *Limnol. Oceanogr.* **37**: 1434–1446.

Vesely P., Boyde A. 1996. Video rate confocal laser scanning reflection microscopy in the investigation of normal and neoplastic living cell dynamics. *Scan. Microsc. Suppl.* **10**: 201–211.

Vickers G. T. 1996. The projected area of ellipsoids and cylinders. *Powder Technol.* **86**: 195–200.

Vickers G. T., Brown D. J. 2001. The distribution of projected area and perimeter of convex, solid particles. *Proc. Roy. Soc. A* **457**: 283–306.

Vidondo B., Prairie Y. T., Blanco J. M., Duarte C. M. 1997. Some aspects of the analysis of biomass size-spectra in aquatic ecology. *Limnol. Oceanogr.* **42**: 184–192.

Vignati D., Pardos M., Diserens J., Ugazio G., Thomas R., Dominik J. 2003. Characterisation of bed sediments and suspension of the river Po (Italy) during normal and high flow conditions. *Water Res.* **37**: 2847–2864.

Vigneau E., Loisel C., Devaux M. F., Cantoni P. 2000. Number of particles for the determination of size distribution from microscopic images. *Powder Technol.* **107**: 243–250.

Viles C. L., Sieracki M. E. 1992. Measurement of marine picoplankton cell size by using a cooled, charge-coupled device camera with image analyzed fluorescence microscopy. *Appl. Environ. Microbiol.* **58**: 584–592.

Volten, H., de Haan, J. F., Hovenier, J. W. , Schreurs, R., Vassen, W., Dekker, A. G., Hoogenboom, H. J., Charlton. F., Wouts, R. 1998. Laboratory measurements of angular distributions of light scattered by phytoplankton and silt. *Limnol. Oceanogr.* **43**: 1180–1197.

Volten, H., de Haan, Vassen, W., Lumme, K., Hovenier, J. W. 1996. Experimental determination of polarized light scattering by irregular particles. *J. Aerosol Sci. Suppl.* **27**: S527–S528.

Volten H., Jalava J.-P., Lumme K., de Haan J. F., Vassen W., Hovenier J. W. 1999. Laboratory measurements and T-matrix calculations of the scattering matrix of rutile particles in water. *Appl. Opt.* **38**: 5232–5240.

Volten H., Muñoz E., de Haan J. F., Vassen W., Hovenier J.W., Muinonen K., Nousiainen T. 2001. Scattering matrices of mineral aerosol particles at 441.4 nm and 632.8 nm. *J. Geophys. Res. D* **106**: 17,375–17,401.

Volten H., Muñoz O., Hovenier J. W., de Haan J. F., Vassene W., van der Zande W. J., Waters L. B. F. M. 2005. WWW scattering matrix database for small mineral particles at 441.6 and 632.8 nm. *J. Quant. Spectrosc. Radiat. Transfer* **90**: 191–206.

Voshchinnikov N. V., Farafonov V. G. 1985. Light scattering by dielectric spheroids. Part 1. *Opt. Spectrosc.* **58**: 81–85.

Voss K. J., Balch W. M., Kilpatrick K. A. 1998. Scattering and attenuation properties of *Emiliania huxleyi* cells and their detached coccoliths. *Limnol. Oceanogr.* **43**: 870–876.

Voss K. J., Fry E. S. 1984. Measurement of the Mueller matrix for ocean water. *Appl. Opt.* **23**: 4427–4439.

Vrieling E., Vriezekolk G., Gieskes W., Veenhuis M., Harder W. 1996. Immuno-flow cytometric identification and enumeration of the ichthyotoxic dinoflagellate *Gyrodinium aureolum Hulburt* in artificially mixed algal populations. *J. Plankton Res.* **18**: 1503–1512.

Vukusic P., Sambles J. R., Lawrence C. R., Wootton R. J. 2001. Limited-view iridescence in the butterfly *Ancyluris meliboeus*. *Proc. R. Soc. London B* **269**: 7–14.

W+

Waldram J. M. 1945. Measurement of the photometric properties of the upper atmosphere. *Trans. Illum. Eng. Soc. (London)* **10**: 147–188.

Wales M., Wilson J. N. 1961. Theory of coincidence in Coulter counter. *Rev. Sci. Instrum.* **32**: 1132–1136.

Wales M., Wilson J. N. 1962. Coincidence in Coulter counter.*Rev. Sci. Instrum.* **33**: 575–576.

Walker J. G., Chang P. C. Y., Hopcraft K. I., Mozaffari E. 2004. Independent particle size and shape estimation from polarization fluctuation spectroscopy. *Meas. Sci. Technol.* **15**: 771–776.

Wallqvist A., Berne B. J. 1993. Effective potentials for liquid water using polarizable and nonpolarizable models. *J. Phys. Chem.* **97**: 13841–13851.

Walstra P., Oortwign H. 1969. Estimating globule-size distribution of oil-in-water emulsions by coulter counter. *J. Colloid Interface Sci.* **29**: 424–431.

Waltham Ch., Boyle J., Ramey B., Smit J. 1994. Light scattering and absorption caused by bacterial activity in water. *Appl. Opt.* **33**: 7536–7541.

Wang J. C. F., Hencken K. R. 1986. In situ size measurements using a two-color laser scattering technique. *Appl. Opt.* **25**: 653–675.

Waterbury J. B., Watson S. W., Guillard R. R., Brand L. E. 1979. Widespread occurrence of a unicellular marine planktonic cyanobacteria. *Nature* **277**: 293–294.

Waterman P. C. 1971. Symmetry, unitarity and geometry in electromagnetic scattering. *Phys. Rev. D* **3**: 825–839.

Watson G. N. 1952. *A treatise on the theory of Bessel functions.* Cambridge University Press, Cambridge, 804 pp.

Watson D., Hagen N., Diver J., Marchand P., Chachisvilis M. 2004. Elastic light scattering from single cells: Orientational dynamics in optical trap. *Biophys. J.* **87**: 1298–1306.

Waxler R. M., Weir C. E., Schamp H. W., Jr. 1964. Effect of pressure and temperature upon the optical dispersion of benzene, carbon tetrachloride and water. *J. Res. Natl. Bur. Std. A* **68**: 489–498.

Webb R. H. 1996. Confocal optical microscopy. *Rep. Prog. Phys.* **59**: 427–471.

Wedd M. W. 2001. Procedure for predicting a minimum volume or mass of sample to provide a given size parameter precision. *Part. Part. Syst. Charact.* **18**: 109–113.

Weibull W. 1939. *A statistical theory of the strength of materials.* The Royal Swedish Institute of Engineering Research Proc. 151, Stockholm.

Weiner B. B., Fairhurst D., Tscharnuter W. W. 1991. Particle size analysis with a disc centrifuge: importance of the extinction efficiency. Chapter 12 in *Particle Size Distribution II.* Prowder T. (ed.) *ACS Symposium Series* **472**: 184–195.

Weiner B. B., Tscharnuter W. W., Karasikov N. 1998. Improvements in accuracy and speed using the time-of-transition method and dynamic image analysis for particle sizing: some real-world examples. In: *Particle Size Distribution III.* Provder T. (ed.) *ACS Symposium Series* **693**: 88–103.

Weiss K. 1981. *Laser scattering experiments with single particles for the interpretation of the optical properties of the interplanetary dust.* (In German: *Laserstreuexperimente an Einzelteilchen zur Interpretation des optischen Verhaltens interplanetare Staubes*). Bereich Extraterresrische Physik, Ruhr-Universität Bochum, 98 pp.

Weiss D. G., Maile W., Wick R. A. 1989. Video microscopy. In: *Light microscopy in biology. A practical approach.* Lacey A. J. (ed.). IRL Press, Oxford, UK, pp. 221–278.

Welch R. M., Cox S. K. 1978. Nonspherical extinction and absorption efficiencies. *Appl. Opt.* **17**: 3159–3168.

Wellershaus S., Goke L., Frank P. 1973. Size distribution of suspended particles in sea water. *"Meteor"Forschungsergebnisse D* **16**: 1–16.

Wells M. L. 1998. A neglected dimension. *Nature* **391**: 530–531.

Wells M. L., Goldberg E. D. 1991. Occurrence of small colloids in sea water. *Nature* **353**: 342–344.

Wells M. L., Goldberg E. D. 1992. Marine submicron particles.*Mar. Chem.* **40**: 5–18.

Wells M. L., Goldberg E. D. 1993. Colloid aggregation in seawater. *Mar. Chem.* **41**: 353–358.

Wells W. H. 1969. Loss of resolution in water as a result of multiple small-angle scattering. *J. Opt. Soc. Am.* **59**: 686–691.

Wells W. H. 1973. Theory of small angle scattering. In: *AGARD Lecture Series No. 61: Optics of the Sea*, pp. 3.3-1–3.3-19.

Wells W. H. 1983. Techniques for measuring radiance in sea and air. *Appl. Opt.* **22**: 2313–2321.

West R. 1991. Optical properties of aggregate particles whose outer diameter is comparable to the wavelength. *Appl. Opt.* **30**: 5316–5324.

West R. A., Doose L. R., Eibl A. M., Tomasko M. G., Mishchenko M. I. 1997. Laboratory measurements of mineral dust scattering phase function and linear polarization. *J. Geophys. Res. D* **102**: 16,871–16,881.

Wheeless L. L., Jr. 1990. Slit-scanning. In: *Flow cytometry and sorting.* Melamed M. R., Lindmo T., Mendelsohn M. L. (eds.) Wiley-Liss, New York, pp. 109–125.

Whiles M. R., Dodds W. K. 2002. Relationships between stream size, suspended particles, and filter-feeding macroinvertebrates in a great plains drainage network. *J. Environ. Qual.* **31**:1589–1600.

Whitby K. T., Vomela R. A. 1967. Response of single particle optical counters to nonideal particles.*Environ. Sci. Technol.* **1**: 801–814.

Whitlock C. H., Poole L. R., Usry J. W., Houghton W. M., Witte W. G., Morris W. D., Gurganus E. A. 1981. Comparison of reflectance with backscatter and absorption parameters for turbid waters.*Appl. Opt.* **20**: 517–522.

Wiebe W. J., Pomeroy L. R. 1972. Microorganisms and their association with aggregates and detritus in the sea: A microscopic study. In *Detritus and its role in the aquatic ecosystems.* Melchiorri-Santolini U., Hopton J. W. (eds.) *Mem. Ist. Ital. Idrobiol.* (suppl.) **29**: 325–352.

Wietzorrek J., Plesnila N., Baethmann A., Kachel V. 1999. A new multiparameter flow cytometer: Optical and electrical cell analysis in combination with video microscopy in flow. *Cytometry* **35**: 291–301.

Wietzorrek J., Stadler M., Kachel V. 1994. Flow cytometric imaging, implemented on the EurOPA flow cytometer – a novel tool for identification of marine organisms. *Proceedings OCEANS '94 OSATES*, Vol. I, pp. 688–693.

Williams I., Paul F., Lloyd D., Jepras R., Critchley I., Newman M., Warrack J., Giokarini T., Hayes A. J., Randerson P. F., Venables W. A. 1999. Flow cytometry and other techniques show that *Staphylococcus aureus* undergoes significant physiological changes in the early stages of surface-attached culture. *Microbiology* **145**: 1325–1333.

Wiltshire K. H., Lampert W. 1999. Urea excretion by Daphnia: A colony-inducing factor in Scenedesmus?*Limnol. Oceanogr.* **44**: 1894–1903.

Windom H. L. 1969. Atmospheric dust records in permanent snowfields: Implications to marine sedimentation. *Geol. Soc. Am. Bull.* **80**: 761–782.

Winterwerp J. C., Bale A. J., Christie M. C., Dyer K. R., Jones S., D. Lintern G., Manning A. J., Roberts W. 2002. Flocculation and settling velocity of fine sediment. In: *Fine sediment dynamics in the marine environment.* Winterwerp J.C. Kranenburg C. (eds.) Elsevier, pp. 25–40.

Wiscombe W. J. 1980. Improved Mie scattering algorithms.*Appl. Opt.* **19**: 1505–1509.

Wiscombe W. J., Chýlek P. 1977. Mie scattering between any two angles. *J. Opt. Soc. Am.* **67**: 572–573.

Wiscombe W. J., Mugnai A. 1986. *Single scattering from nonspherical Chebyshev particles: A compendium of calculations*. NASA Reference Publication 1157, Greenbelt, MD, USA, 278 pp.

Wiscombe W. J., Mugnai A. 1988. Scattering from nonspherical Chebyshev particles. 2: Means of angular scattering patterns.*Appl. Opt.* **27**: 2405–2421.

Witek Z., Krajewska-Soltys A. 1989. Some examples of the epipelagic plankton size structure in high latitude oceans. *J. Plankton Res.* **11**: 1143–1155.

Witkowski K. 1986. Simultaneous integral and quasielastic light scattering measurements for macromolecular solutions. Part 2. Experimental results for latex suspensions and polystyrene dilute solutions. *Optik* **73**: 133–137.

Witkowski K., Król T., Zielinski A., Kuten E. 1998. A light-scattering matrix for unicellular marine phytoplankton. *Limnol. Oceanogr.* **43**: 859–869.

Witkowski K., Wolinski L., Turzynski Z., Gedziorowska D., Zielinski A. 1993. The investigation of kinetic growth of *Chlorella vulgaris* by the method of integral and dynamic light scattering. *Limnol. Oceanogr.* **38**: 1365–1372.

Wommack K. E., Colwell R. R. 2000. Virioplankton: Viruses in aquatic ecosystems. *Microbiol. Mol. Biol. Rev.* **64**: 69–114.

Woodward J. C., Walling D. E. 1992. A field sampling method for obtaining representative samples of composite fluvial suspended sediment particles for SEM analysis. *J. Sediment. Petrol.* **62**: 742–743.

Worm J., Søndergaard M. 1998. Alcian Blue-stained particles in a eutrophic lake. *J. Plankton Res.* **20**: 179–186.

Woźniak B. 1977. The influence of components of seawater on the underwater light field. (In Polish, *Wpływ składnikow wody morskiej na podwodne pole światła*). Ph.D. Thesis, Institute of Oceanology PAS, Sopot), 175 pp.

Woźniak S. B., Stramski D. 2004. Modeling the optical properties of mineral particles suspended in seawater and their influence on ocean reflectance and chlorophyll estimation from remote sensing algorithms. *Appl. Opt.* **43**: 3489–3503.

Wriedt T. 1998. A review of elastic light scattering theories. *Part. Part. Syst. Charact.* **15**: 67–74.

Wriedt T. 2000. A list of electromagnetic scattering programmes: http://diogenes.iwt.uni-bremen.de/~wriedt/.

Wu J., Partovi F., Field M. S., Rava R. P. 1993. Diffuse reflectance from turbid media: an analytical model of photon migration. *Appl. Opt.* **32**: 1115–1121.

Wyatt P. J. 1968. Differential light scattering: a physical method for identifying bacterial cells. *Appl. Opt.* **7**: 1879–1896.

Wyatt P. J. 1980. Some chemical, physical, and optical properties of fly ash particles. *Appl. Opt.* **19**: 975–983.

Wyatt P. J. 1998. Submicrometer particle sizing by multiangle light scattering following fractionation. *J. Colloid Interface Sci.* **197**: 9–20.

Wyatt P. J., Jackson C. 1989. Discrimination of phytoplankton via light-scattering properties. *Limnol. Oceanogr.* **34**: 96–112.

Wyatt P. J., Schehrer K. L., Phillips S. D., Jackson C., Chang Y.-J., Parker R. G., Phillips D. T., Bottiger J. R. 1988. Aerosol particle analyzer. *Appl. Opt.* **27**: 217–221.

Wynn E. J. W. 2003. Relationship between particle-size and chord-length distributions in focused beam reflectance measurement: stability of direct inversion and weighting. *Powder Technol.* **133**: 125–133.

Wynn E. J. W., Hounslow M. J. 1997. Coincidence correction for electrical-zone (Coulter-counter) particle-size analysers. *Powder Technol.* **93**: 163–175.

X+

Xu Y.-L. 1995. Electromagnetic scattering by an aggregate of spheres. *Appl. Opt.* **34**: 4573–4588.

Xu Y.-L. 1998. Electromagnetic scattering by an aggregate of spheres: asymmetry parameter. *Phys. Lett. A* **249**: 30–36.

Xu W., Jericho M. H., Meinertzhagen I. A., Kreuzer H. J. 2002. Digital in-line holography for biological applications. *Proc. Natl. Acad. Sci.* **98**: 11301–11305.

Xu W., Jericho M. H., Meinertzhagen I. A., Kreuzer H. J. 2003. Tracking particles in 4-D with in-line holographic microscopy. *Opt. Lett.* **28**: 164–166.

Y+

Yamamoto H., Matsuyama T., Wada M. 2002. Shape distinction of particulate materials by laser diffraction pattern analysis. *Powder Technol.* **122**: 205–211.

Yamasaki A., Fukuda H., Fukuda R., Mijajima T., Nagata T., Ogawa H., Koike I. 1998. Submicrometer particles in northwest Pacific coastal environments: Abundance, size distribution, and biological origins. *Limnol. Oceanogr.* **43**: 536–542.

Yang P., Kattawar G. W., Wiscombe W. J. 2004a. Effect of particle asphericity on single-scattering parameters: comparison between platonic solids and spheres. *Appl. Opt.* **43**: 4427–4435.

Yang P., Liou K. N., Mishchenko M. I., Gao B.-C. 2001. Efficient finite-difference time-domain scheme for light scattering by dielectric particles: Application to aerosols. *Appl. Opt.* **39**: 3727–3737.

Yang P., Zhang Z., Baum B. A., Huang H.-L., Hu Y. 2004b. A new look at anomalous diffraction theory (ADT): Algorithm in cumulative projected-area distribution domain and modified ADT. *J. Quant. Spectrosc. Radiat. Transfer* **89**: 421–442.

Yee K. S. 1966. Numerical solution of initial boundary value problems involving Maxwell's equations in isotropic media. *IEEE Trans. Ant. Propag.* AP-**14**: 302–307.

Yentsch C. S. 1962. Measurement of visible light absorption by particulate matter in the ocean. *Limnol. Oceanogr.* **7**: 207–217.

Yentsch C. M., Horan P. K., Muirhead K., Dortch Q., Haugen E., Legendre L., Murphy L. S., Perry M. J., Phinney D. A., Pomponi S. A., Spinrad R. W., Wood A. M., Yentsch C. S., Zahuranec B. J. 1983. Flow cytometry and cell sorting: a technique for analysis and sorting of aquatic particles. *Limnol. Oceanogr.* **28**: 1275–1280.

Yoshida H., Masuda H., Fukui K., Tokunaga Y., Takarada K., Sakurai T., Matsumoto H. 2003. Particle size measurement of standard reference particle candidates with improved size measurement devices. *Adv. Powder Technol.* **14**: 17–31.

Young A. T. 1981. Rayleigh scattering.*Appl. Opt.***20**: 533–535.

Z+

Zalewski M. S. 1977. *Analysis of the particle size distribution of particles suspended in seawater*. (In Polish: *Analiza widm rozmiarów cząstek unoszonych w wodzie morskiej*). Ph.D. Thesis, Department of Oceanology, Institute of Geophysics, Polish Acad. Sci., Warsaw, 152 pp.

Zaneveld J. R. V. 1974. New developments in the theory of radiative transfer in the oceans. In: *Optical aspects of oceanography*. Jerlov N. G., Nielsen K. S. (eds.) Academic Press, New York, pp. 121–134.

Zaneveld J. R. V., Kitchen J. C. 1995. The variation in the inherent optical properties of phytoplankton near an absorption peak as determined by various models of cell structure. *J. Geophys. Res. C* **100**: 13,309–13,320.

Zaneveld J. R. V., Kitchen J. C., Moore C. 1994. The scattering error correction of reflecting-tube absorption meters. In: *Ocean Optics XII*. Jaffe J. J. (ed.) *Proc. SPIE* **2258**: 44–55.

Zaneveld J. R. V., Pak H. 1971. Method for the determination of the index of refraction of particles suspended in the ocean.*J. Opt. Soc. Am.* **63**: 321–324.

Zaneveld J. R. V., Roach D. M., Pak H. 1974. The determination of the index of refraction distribution of oceanic particulates.*J. Geophys. Res.* **79**: 4091–4095.

Zaneveld J. R. V., Spinrad R. W., Bartz R. 1982. An optical settling tube for the determination of particle size distributions.*Mar. Geol.* **49**: 357–376.

Zerull R. H., Giese R. H., Weiss K. 1977. Scattering measurements of irregular particles vs. Mie theory. In: *Optical polarimetry. Proc. SPIE* **112**: 191–199.

Zhang X., Lewis M., Johnson B. 1998. The effect of bubbles on light scattering in the ocean. In: *Ocean Optics XIV*, Kailua-Kona, Hawaii, November 10–13, 1998.

Zhang X., Lewis M., Lee M., Johnson B., Korotaev G. 2002. The volume scattering function of natural bubble populations. *Limnol. Oceanogr.* **47**: 1273–1282.

Zhang Y., West R., Williams R. A., Spink M. 2003. A multi-electrode approach for on-line characterisation of size and shape. *Part. Part. Syst. Charact.* **20**: 3–11.

Zipf G. K. 1949. Human behavior and the principle of least effort. Addison-Wesley, Cambridge, MA.

Zurzolo C., Bowler C. 2001. Exploring bioinorganic pattern formation in diatoms. A story of polarized trafficking. *Plant Physiol.* **127**: 1339–1345.

Zuur E. A. H., Nyffeler F. 1992. Theoretical distributions of suspended particles in the ocean and a comparison with observations. *J. Mar. Syst.* **3**: 529–538.

List of major symbols and abbreviations

Values of the universal constants are given here following the 2002 edition of a report of the Committee on Data for Science and Technology (CODATA) as published by the National Institute of Standards and Technology (NIST) of USA in their webpage http://physics.nist.gov/cuu/Constants.

Greek symbols

β	scattering function, see equation (4.2)
$\beta_{\text{iso}}(\pi/2)$	isotropic portion of the 90° scattering, see equation (2.46)
β_{T}	the isothermal compressibility
Γ	gamma function, see equation (5.192)
Γ	autocorrelation function of the fluctuations of the temporal light scattering signal, see equation (5.117)
Δ	translational diffusion coefficient, see equation (5.120)
ε	usually measurement error
η	dynamic viscosity
θ	scattering angle
θ_0	angle between the axis of symmetry of the cylinder and the direction of the incident radiation
θ_{i}	incidence angle
δ	anisotropy factor, see equation (1.61)
δ	parameter of the Fournier–Forand approximation to the scattering function, see equation (3.156)
δ	phase difference
θ_{B}	Brewster angle, see equation (1.33)
θ_{C}	critical angle, see equation (1.31)
Θ	scattering angle parameter defined in (3.53)
λ	wavelength of light
ν	frequency
ν	kinematic viscosity

ν_{nm}	is the frequency corresponding to the energy difference between state n and state m
ν	parameter of the Fournier–Forand approximation to the scattering function, see equation (3.156)
μ	$n' - 1$
μ	reduced mass of a molecule, see equation (2.11)
ξ	direction vector
ρ	density of a medium
ρ	complex phase shift parameter, defined in (3.38)
ρ	reflectance or bidirectional reflectance
ρ	resistivity of a medium
ρ'	real phase shift parameter, defined in (3.39)
σ	variance of the logarithm of the particle diameter, a parameter of a log-normal component of a particle size distribution, see equation (5.177)
σ_{abs}	absorption cross-section
σ_{scat}	scattering differential cross-section, usually used as $\sigma_{scat}(\theta)$, where θ is the scattering angle
ϕ	azimuth angle
ϕ	phase
ϕ	the *phi* scale of the particle size, see equation (5.18)
χ_l^2	the chi-square test statistics for l degrees of freedom
χ_p	Ricatti–Bessel function
ψ_p	Ricatti–Bessel function
ψ_i	energy levels
ω	angular frequency, $\omega = 2\pi\nu$
ω	photon state density, see equation (1.4)
ω	see equation (3.42)
$\omega_{\|\|p}, \omega_{\perp p}$	integral reflection factor as the surface of a particle for the polarization of light parallel and perpendicular, respectively, to the scattering plane; p indicates the surface and reflection order as follows: $p = 0$, light reflected by the frontal part of the particle surface, $p = 2$, light reflected by the back part of the particle surface
Ω	solid angle

Latin symbols

a	absorption coefficient, see equation (1.8) and also equation (1.23)
a	particle radius
a'	average volume polarizability
a_p	amplitude of the vector spherical harmonics with even symmetry, see equation (3.9)

A	amplification
A_m	molar refractivity
A_n	the *n*th order function in the Deirmendijan algorithm for the calculation of the Mie scattering coefficients a_n and b_n
A_{shadow}	area of the geometric shadow of a particle
ADA	anomalous diffraction approximation
$\mathbf{A}$	vector wave amplitude
$\mathbf{A}$	vector wave potential, see equation (1.14)
b	scattering coefficient, see equation (1.10)
b_b	backward scattering coefficient, see equation (4.7)
b_f	forward scattering coefficient, see equation (4.7)
b_N	normalized biomass spectrum
b_p	amplitude of the vector spherical harmonics with odd symmetry, see equation (3.10)
B	backscattering probability
B	biomass
B_i	for $i = 0, 1, 2$; parameters of a log-normal component of a particle size distribution, see equation (5.178)
B	rotational energy constant, see equation (2.31)
c	attenuation coefficient, see equations (1.11) and (1.12)
c	velocity of light in vacuum ($299792458\,\mathrm{m\,s^{-1}} \sim 3 \times 10^8\,\mathrm{m\,s^{-1}}$)
$c(D)$	distribution of the particle cross-section
C	solute concentration
C_C	carbon concentration
C_{Chla}	chlorophyll *a* concentration
$C(u)$	Fresnel integral
C_p	mass concentration of proteins
CLD	chord length distribution
CPU	central processing unit
$\mathrm{cov}(x, y)$	covariance of x and y
C_{abs}	absorption cross-section of a particle
C_{scat}	scattering cross-section of a particle
d	three-dimensional fractal dimension of an aggregate
d_2	two-dimensional fractal dimension of an aggregate
d_{KL}	Kullback–Liebler distance, see equation (5.152)
d_{KS}	Kolmogorov–Smirnov distance, see equation (5.149)
D	particle diameter
D_C	circular equivalent particle diameter
D_F	Feret diameter
D_g	gyration diameter of an aggregate
D_{peak}	particle diameter corresponding to the peak of a log-normal component of a particle size distribution, see equation (5.177)
D_S	spherical equivalent particle diameter

DDA	discrete dipole approximation
e	electron charge ($1.60217653 \times 10^{-19}$C)
E	electric field
E	irradiance (see section 1.4)
E_d	downwelling irradiance
E_u	upwelling irradiance
$E_{\parallel}$	component of the scattered light wave field that is parallel to the scattering plane, see equation (1.47)
$E_{\perp}$	component of the scattered light wave field that is perpendicular to the scattering plane, see equation (1.46)
$E_3(z)$	the third-order exponential integral
$\mathbf{E}$	electric wave vector
E_H	Helmholtz free energy
E_p	photon energy
f	electrical current frequency
$f_r(D)$	the rth moment of the particle size distribution
F	light flux, i.e., the rate of light power flow through a surface
F	shape factor, see equation (6.76)
FDTD	finite difference time domain
FFF	flow-field fractionation
g	asymmetry factor of the scattering function (mean cosine), see equation (4.8)
g	acceleration of gravity ($9.80665\,\text{ms}^{-2}$)
g	cell growth rate
$G(v)$	vibrational energy level distribution, see equation (2.21)
h	Planck's constant ($6.6260693 \times 10^{-34}$ Js)
$h(D)$	histogram-type size distribution
$\mathbf{H}_1$	Struve function
i	imaginary unity, $i = \sqrt{(-1)}$
i_1	normalized scattering intensity, see equation (3.15)
i_2	normalized scattering intensity, see equation (3.16)
$\mathbf{i}$	unit vector along the x-axis
I	moment of inertia of a molecule, see equation (2.31)
I	the first element of the Stokes vector, see equation (4.59)
I	electric current
I	intensity of an electromagnetic wave (in the physics sense, see section 1.4)
I_0	incident wave intensity (in the physical sense, see section 1.4)
j_p	spherical Bessel function
$\mathbf{j}$	unit vector along the y-axis
J	rotational energy state index
J_i	Bessel function of the i-th order
k	wave number, see equation

k	scale factor of the power-law particle size distribution
k_f	fractal prefactor, see equation (5.30)
K	Boltzmann constant ($1.3806505 \times 10^{-23}\, JK^{-1}$)
K	scant bulk modulus of water, see equation (5.129)
$\mathbf{k}$	wave vector
$\mathbf{k}$	unit vector along the z-axis
L	radiance
L_n	n-th moment of a radiance field, see equation (4.48)
m	slope of the power-law particle size distribution
m_C	cell carbon content
m_{Chl}	cell chlorophyll content
m_{DNA}	cell DNA content
m_{dry}	dry cell mass
M_{ij}	ij-th element of the scattering matrix
$\mathbf{M}$	dipole moment, see equation (2.2)
$\mathbf{M}$	vector harmonic, see equation (3.1)
$\mathbf{M}$	scattering matrix, see equation (4.73)
n	refractive index, usually relative to that of water, see equation(1.20)
n'_e	extraordinary (real) refractive index
n'_o	ordinary (real) refractive index
$n(\nu)$	number density of photon states (ν is the photon frequency)
$n(D)$	frequency particle size distribution (D is the particle diameter); sometimes also used as n if the context prevents confusion
$\mathbf{N}$	vector harmonic, see equation (3.2)
$N(D)$	cumulative particle size distribution (D is the particle diameter)
NEP	noise-equivalent power
N_m	number density of atoms in state m
n'	real part of the refractive index, see equation (1.20)
n''	imaginary part of the refractive index, see equation (1.20) and also equation (1.23)
n_{max}	scale factor of a log-normal component of a particle size distribution, see equation (5.177)
p	pressure
OD	optical density, see equation (6.59)
$p(N)$	probability of finding N particles within a volume of suspension
$p(\theta)$	phase function, see equation (4.5)
P	power
P	aggregate porosity
P	projected area of a particle
P_B	bacterial production, see equation (5.143)
P_L	degree of linear polarization, see equation (4.121)
P_P	phytoplankton production, see equation (5.143)
P_p	Legendre polynomial

P_p^q	associated Legendre polynomial
PMT	photomultiplier
PSD	particle size distribution
q	scattering vector amplitude, see equation (5.119)
Q	the second element of the Stokes vector, see equation (4.59)
Q_{abs}	absorption efficiency of a particle
Q_{attn}	attenuation efficiency of a particle
Q_{scat}	scattering efficiency of a particle
r	distance in a medium
r	particle radius, can also be denoted by a
r^2	determination coefficient, see equation (5.148)
r	reflection coefficient for the wave amplitude, see equation (1.26)
$r_{\|\|}$	reflection coefficient for the wave amplitude at a polarization parallel to the wave propagation plane, see equation (1.28)
$r_{\perp}$	reflection coefficient for the wave amplitude at a polarization perpendicular to the wave propagation plane, see equation (1.29)
$\mathbf{r}$	position vector
R	reflection coefficient for light intensity, see equation (1.27)
R	electrical resistance
R_f	feedback resistor resistance
$R_{\|\|}$	reflection coefficient for light intensity at a polarization parallel to the wave propagation plane, see equation
$R_{\perp}$	reflection coefficient for light intensity at a polarization perpendicular to the wave propagation plane, see equation (1.30)
R	Rayleigh ratio, i.e., the isotropic portion of the 90° scattering, see equation (2.46)
$\mathbf{R}^{nm}$	transition matrix, see equation (2.3)
Re	Reynolds number, see equation (5.130)
RGD	Rayleigh–Gans–Debye
s	shape factor, see equation (6.77)
$\hat{s}$	mean surface area per particle, $\hat{s}$, of an ensemble of randomly oriented convex particles
S	scattering amplitude of a scalar wave, see equation (1.35)
S	salinity
S	particle surface
$\mathbf{S}$	Stokes vector, see equation (4.73)
$S(u)$	Fresnel integral
SD(x)	standard deviation of x
SEM	scanning electron microscope
SNR	signal-to-noise ratio, see equation (4.15)
S_{o2}	solubility of oxygen in water, see equation (2.88)
S_1	component of the scattered light wave amplitude factor for the polarization parallel to the scattering plane, see equation (1.47)

S_2	component of the scattered light wave amplitude factor for the polarization perpendicular to the scattering plane, see equation (1.46)
t	time
T	absolute temperature in degrees Kelvin
T_c	temperature in degrees Celsius
TEM	transmission electron microscope
u	scalar wave amplitude, see equation (1.34)
U	the third element of the Stokes vector, see equation (4.59)
$v(D)$	distribution of particle volume
$\mathrm{var}(x)$	variance of variable x
V	the fourth element of the Stokes vector, see equation (4.59)
V	particle volume or simply volume
V	potential
V	voltage
V_d	Lennard–Jones potential, see equation (2.1)
VSF	volume scattering function
W	molecular weight
W_{a-b}	probability of spontaneous emission per unit time and per unit frequency interval in a transition from state a to state b
x	particle size relative to the wavelength of light, see equation (1.56)
x	spatial coordinate
y	see equation (3.49)
y	spatial coordinate
y_p	spherical Bessel function
z	see equation (3.40)
z	distance
z	spatial coordinate
z_p	denotes spherical Bessel function j_p or y_p

Index

Zeitfracht Medien GmbH
Ferdinand-Jühlke-Straße 7
99095 Erfurt, Deutschland
produktsicherheit@kolibri360.de